Methods in Enzymology

Volume 171
BIOMEMBRANES
Part R
Transport Theory:
Cells and Model Membranes

METHODS IN ENZYMOLOGY

EDITORS-IN-CHIEF

John N. Abelson   Melvin I. Simon

DIVISION OF BIOLOGY
CALIFORNIA INSTITUTE OF TECHNOLOGY
PASADENA, CALIFORNIA

FOUNDING EDITORS

Sidney P. Colowick and Nathan O. Kaplan

Methods in Enzymology

Volume 171

Biomembranes

Part R
Transport Theory:
Cells and Model Membranes

EDITED BY

Sidney Fleischer

Becca Fleischer

DEPARTMENT OF MOLECULAR BIOLOGY
VANDERBILT UNIVERSITY
NASHVILLE, TENNESSEE

Editorial Advisory Board

ACADEMIC PRESS, INC.
Harcourt Brace Jovanovich, Publishers
San Diego New York Berkeley Boston
London Sydney Tokyo Toronto

ACADEMIC PRESS, INC.
San Diego, California 92101

United Kingdom Edition published by
ACADEMIC PRESS LIMITED
24-28 Oval Road, London NW1 7DX

LIBRARY OF CONGRESS CATALOG CARD NUMBER: 54-9110

ISBN 0-12-182072-6 (alk. paper)

PRINTED IN THE UNITED STATES OF AMERICA
89 90 91 92   9 8 7 6 5 4 3 2 1

Table of Contents

Section I. Transport Theory

Section II. Model Membranes and Their Characteristics

Section III. Isolation of Cells

Section IV. Polar Cell Systems

Section V. Modification of Cells

Contributors to Volume 171

Article numbers are in parentheses following the names of contributors.
Affiliations listed are current.

PER-ÅKE ALBERTSSON (26), *Department of Biochemistry, Chemical Center, University of Lund, S-221 00 Lund, Sweden*

OLAF SPARRE ANDERSEN (4), *Department of Physiology and Biophysics, Cornell University Medical College, New York, New York 10021*

PETER F. BAKER[1] (40), *Medical Research Council (MRC) Secretory Mechanisms Group, Department of Physiology, Division of Biomedical Sciences, King's College London, Strand, London WC2R 2LS, England*

PETER H. BARRY (35), *School of Physiology and Pharmacology, The University of New South Wales, Sydney, New South Wales, Australia 2033*

R. BENZ (14), *Lehrstuhl für Biotechnologie, Universität Würzburg, D-8700 Würzburg, Federal Republic of Germany*

DAVID CAFISO (16), *Department of Chemistry, University of Virginia, Charlottesville, Virginia 22901*

S. ROY CAPLAN (20), *Department of Membrane Research, The Weizmann Institute of Science, Rehovot 76100, Israel*

G. CARRASQUER (31), *Department of Medicine, University of Louisville, Louisville, Kentucky 40292*

N. CHEJANOVSKY (41), *Laboratory of Molecular and Cellular Biology, National Institute of Diabetes and Digestive and Kidney Diseases, National Institutes of Health, Bethesda, Maryland 20892*

A. M. CHERET (21), *Chargé de Recherche, Institut National de la Santé et de la Recherche Médicale (INSERM) U10, Hôpital Bichat, F-75877 Paris Cedex 18, France*

V. CITOVSKY (41), *Division of Molecular Plant Biology, University of California, Berkeley, Berkeley, California 94720*

CHRIS CLAUSEN (32), *Department of Physiology and Biophysics, School of Medicine, Health Sciences Center, State University of New York (SUNY) at Stony Brook, Stony Brook, New York 11794*

NORBERT A. DENCHER (13), *Biophysics Group, Department of Physics, Freie Universität Berlin, D-1000 Berlin 33, Federal Republic of Germany*

ROSA DEVÉS (5), *Department of Physiology and Biophysics, Faculty of Medicine, University of Chile, Casilla 70055, Santiago, Chile*

SIDNEY FLEISCHER (1), *Department of Molecular Biology, Vanderbilt University, Nashville, Tennessee 37235*

J. KEVIN FOSKETT (39), *Division of Cell Biology, Research Institute, The Hospital for Sick Children, Toronto, Ontario, Canada M5G 1X8*

ILAN FRIEDBERG (43), *Department of Microbiology, The George S. Wise Faculty of Life Sciences, Tel-Aviv University, Tel-Aviv 69978, Israel*

PETER FROMHERZ (18), *Abteilung für Biophysik, Universität Ulm, D-7900 Ulm, Donau 7, Federal Republic of Germany*

E. FRÖMTER (33), *Zentrum der Physiologie, Johann Wolfgang Goethe-Universität, D-6000 Frankfurt am Main 70, Federal Republic of Germany*

ARLYN GARCIA-PEREZ (28), *Laboratory of Kidney and Electrolyte Metabolism, National Heart, Lung and Blood Institute, National Institutes of Health, Bethesda, Maryland 20892*

JERRY D. GARDNER (29), *Digestive Diseases Branch, National Institute of Diabetes and*

[1] Deceased.

Digestive and Kidney Diseases, National Institutes of Health, Bethesda, Maryland 20892

L. G. M. GORDON (33), Department of Physiology, University of Otago, Medical School, Dunedin, New Zealand

NORDICA GREEN (37), Laboratory of Kidney and Electrolyte Metabolism, National Heart, Lung and Blood Institute, National Institutes of Health, Bethesda, Maryland 20892

ROBERT B. GUNN (8), Department of Physiology, Emory University School of Medicine, Atlanta, Georgia 30322

JOSEPH S. HANDLER (37), Laboratory of Kidney and Electrolyte Metabolism, National Heart, Lung and Blood Institute, National Institutes of Health, Bethesda, Maryland 20892

KURT HANNIG (25), Max-Planck-Institut für Biochemie, D-8033 Martinsried bei München, Federal Republic of Germany

HANS-G. HEIDRICH (25), Max-Planck-Institut für Biochemie, D-8033 Martinsried bei München, Federal Republic of Germany

LEON A. HEPPEL (43), Section of Biochemistry, Molecular and Cell Biology, Cornell University, Ithaca, New York 14853

PAUL KARL HORAN (27), Zynaxis Cell Science, Inc., Malvern, Pennsylvania 19355

WILLIAM P. JENCKS (7), Graduate Department of Biochemistry, Brandeis University, Waltham, Massachusetts 02254

BRUCE D. JENSEN (27), Zynaxis Cell Science, Inc., Malvern, Pennsylvania 19355

ROBERT T. JENSEN (29), Digestive Diseases Branch, National Institute of Diabetes and Digestive and Kidney Diseases, National Institutes of Health, Bethesda, Maryland 20892

ROSE M. JOHNSTONE (30), Department of Biochemistry, McGill University, Montreal, Quebec, Canada H3G 1Y6

MARTIN KLINGENBERG (2), Institut für Physikalische Biochemie, Universität München, D-8000 München 2, Federal Republic of Germany

DEREK E. KNIGHT (40), Medical Research Council (MRC) Secretory Mechanisms Group, Department of Physiology, Division of Biomedical Sciences, King's College London, Strand, London WC2R 2LS, England

H.-A. KOLB (14), Department of Biology, University of Konstanz, D-7750 Konstanz 1, Federal Republic of Germany

TAKAHITO KONDO (10), First Department of Medicine, Hokkaido University School of Medicine, Sapporo 060, Japan

G. KOTTRA (33), Zentrum der Physiologie, Johann Wolfgang Goethe-Universität, D-6000 Frankfurt am Main 70, Federal Republic of Germany

REINHARD KRÄMER (19), Institut für Biotechnologie 1 der Kernforschungsanlage Jülich GmbH, D-5170 Jülich, Federal Republic of Germany

R. M. KRUPKA (5), Research Center, Agriculture Canada, London, Ontario, Canada N6A 5B7

PHILIP C. LARIS (30), Department of Biological Sciences, University of California, Santa Barbara, Santa Barbara, California 93106

P. LÄUGER (14), Department of Biology, University of Konstanz, D-7750 Konstanz 1, Federal Republic of Germany

JOHN P. LEADER (38), Department of Physiology, University of Otago, Medical School, Dunedin, New Zealand

MIGUEL J. M. LEWIN (21), Unité de Recherches de Gastro-Entérologie, Institut National de la Santé et de la Recherche Médicale (INSERM) U10, Hôpital Bichat, F-75877 Paris Cedex 18, France

SIMON A. LEWIS (36), Department of Physiology and Biophysics, University of Texas Medical Branch, Galveston, Texas 77550

ABRAHAM LOYTER (41), Department of Biological Chemistry, Institute of Life Sciences, The Hebrew University of Jerusalem, Jerusalem 91904, Israel

KEVIN D. LUSTIG (43), *Departments of Biochemistry and Food Science and Nutrition, University of Missouri-Columbia, Columbia, Missouri 65211*

ANTHONY D. C. MACKNIGHT (38), *Department of Physiology, University of Otago, Medical School, Dunedin, New Zealand*

ALAN MCLAUGHLIN (16), *Laboratory of Metabolism, National Institute of Alcohol Abuse and Alcoholism, Rockville, Maryland 20852*

STUART MCLAUGHLIN (16), *Department of Physiology and Biophysics, School of Medicine, Health Sciences Center, State University of New York (SUNY) at Stony Brook, Stony Brook, New York 11794*

R. B. MCMICHENS (15), *Laboratory of Molecular Biophysics, The University of Alabama at Birmingham, School of Medicine, Birmingham, Alabama 35294*

RICHARD E. PAGANO (42), *Department of Embryology, Carnegie Institution of Washington, Baltimore, Maryland 21210*

DEMETRIOS PAPAHADJOPOULOS (9), *Cancer Research Institute, University of California, San Francisco, San Francisco, California 94143*

OLE H. PETERSEN (34), *Department of Physiology, University of Liverpool, Liverpool L69 3BX, England*

DANIELA PIETROBON (20), *CNR Unit for the Study of the Physiology of Mitochondria, Institute of General Pathology, University of Padua, Padua, Italy*

THERESA P. PRETLOW (22), *Institute of Pathology, Case Western Reserve University, School of Medicine, Cleveland, Ohio 44106*

THOMAS G. PRETLOW (22), *Institute of Pathology, Case Western Reserve University, School of Medicine, Cleveland, Ohio 44106*

WARREN S. REHM (31), *Department of Medicine, University of Louisville, Louisville, Kentucky 40292*

HAGAI ROTTENBERG (17), *Pathology Department, Hahnemann University, Philadelphia, Pennsylvania 19102*

KEITH R. RUNYAN (8), *Department of Medicine, Emory University School of Medicine, Atlanta, Georgia 30322*

GEORGE SACHS (1), *UCLA School of Medicine, University of California, Los Angeles, Los Angeles, California 90024, and Center for Ulcer Research and Education (CURE), Veterans Administration, VA Wadsworth Hospital Center, Los Angeles, California 90073*

MARTIN J. SANDERS (23), *Department of Metabolic Diseases and General Pharmacology, Pfizer Central Research, Groton, Connecticut 06340*

CARL SCHEFFEY (39), *Department of Biological Sciences, Columbia University, New York, New York 10027*

HANSGEORG SCHINDLER (11), *Institute of Biophysics, University of Linz, A-4040 Linz/Auhof, Austria*

M. SCHWARTZ (31), *Department of Medicine, University of Louisville, Louisville, Kentucky 40292*

ANTHONY W. SCOTTO (12), *Division of Digestive Diseases, Department of Medicine, Cornell University Medical College, New York, New York 10021*

WILLIAM L. SMITH (28), *Department of Biochemistry, Michigan State University, East Lansing, Michigan 48824*

ANDREW H. SOLL (23), *UCLA School of Medicine, University of California, Los Angeles, Los Angeles, California 90024, and Center for Ulcer Research and Education (CURE), Veterans Administration, VA Wadsworth Hospital Center, Los Angeles, California 90073*

WILLIAM K. SONNENBURG (28), *Department of Biochemistry, Michigan State University, East Lansing, Michigan 48824*

WILLIAM S. SPIELMAN (28), *Department of Physiology, Michigan State University, East Lansing, Michigan 48824*

G. STARK (14), *Department of Biology, University of Konstanz, D-7750 Konstanz 1, Federal Republic of Germany*

RODERICK E. STEELE (37), *Laboratory of Technical Development, National Heart,*

Lung and Blood Institute, National Institutes of Health, Bethesda, Maryland 20892

WILFRED D. STEIN (3), Department of Biological Chemistry, Institute of Life Sciences, The Hebrew University of Jerusalem, Jerusalem 91904, Israel

T. L. TRAPANE (15), Laboratory of Molecular Biophysics, The University of Alabama at Birmingham, School of Medicine, Birmingham, Alabama 35294

DAN W. URRY (15), Laboratory of Molecular Biophysics, The University of Alabama at Birmingham, School of Medicine, Birmingham, Alabama 35294

PAUL S. USTER (42), Liposome Technology, Inc., Menlo Park, California 94025

JAYDUTT V. VADGAMA (6), Division of Nephrology and Hypertension, Department of Medicine, LAC Harbor-UCLA Medical Center, UCLA School of Medicine, University of California, Los Angeles, Torrance, California 90509

C. M. VENKATACHALAM (15), Polygen Corporation, Waltham, Massachusetts 02154

GARY A. WEISMAN (43), Departments of Biochemistry and Food Science and Nutrition, University of Missouri-Columbia, Columbia, Missouri 65211

JOHN WELLS (24), Department of Internal Medicine, Pulmonary Division, University of Texas Medical Branch, Galveston, Texas 77550

NANCY K. WILLS (36), Department of Physiology and Biophysics, University of Texas Medical Branch, Galveston, Texas 77550

ANTHONY WINISKI (16), Department of Biochemistry, State University of New York (SUNY) at Stony Brook, Stony Brook, New York 11794

MARTIN C. WOODLE (9), Liposome Technology, Inc., Menlo Park, California 94025

DAVID ZAKIM (12), Division of Digestive Diseases, Department of Medicine, Cornell University Medical College, New York, New York 10021

Preface

Biological transport is part of the Biomembranes series of *Methods in Enzymology*. It is a continuation of methodology concerned with membrane function. This is a particularly good time to cover the topic of biological membrane transport because there is now a strong conceptual basis for its understanding. The field of transport has been subdivided into five topics.

1. Transport in Bacteria, Mitochondria, and Chloroplasts
2. ATP-Driven Pumps and Related Transport
3. General Methodology of Cellular and Subcellular Transport
4. Cellular and Subcellular Transport: Eukaryotic (Nonepithelial) Cells
5. Cellular and Subcellular Transport: Epithelial Cells

Topic 1 covered in Volumes 125 and 126 initiated the series. Topic 2 is covered in Volumes 156 and 157. Topic 3 is covered in Volumes 171 and 172. The remaining two topics will appear in subsequent volumes of the Biomembranes series.

Topic 3 provides general methodology and concepts. Its coverage includes transport theory and general methodology for the study of biomembranes. It is divided into two parts: this volume (Part R) which covers Transport Theory: Cells and Model Membranes and Volume 172 (Part S) which covers Transport: Membrane Isolation and Characterization.

We are fortunate to have the good counsel of our Advisory Board. Their input ensures the quality of these volumes. The same Advisory Board has served for the complete transport series. Valuable input on the outlines of the five topics was also provided by Qais Al-Awqati, Ernesto Carafoli, Halvor Christensen, Isadore Edelman, Joseph Hoffman, Phil Knauf, and Hermann Passow.

The names of our advisory board members were inadvertently omitted in Volumes 125 and 126. When we noted the omission, it was too late to rectify the problem. For Volumes 125 and 126, we are pleased to acknowledge the advice of Angelo Azzi, Youssef Hatefi, Dieter Oesterhelt, and Peter Pedersen.

The enthusiasm and cooperation of the participants have enriched and made these volumes possible. The friendly cooperation of the staff of Academic Press is gratefully acknowledged.

These volumes are dedicated to Professor Sidney Colowick, a dear friend and colleague, who died in 1985. We shall miss his wise counsel, encouragement, and friendship.

SIDNEY FLEISCHER
BECCA FLEISCHER

METHODS IN ENZYMOLOGY

VOLUME 69. Photosynthesis and Nitrogen Fixation (Part C)
Edited by ANTHONY SAN PIETRO

VOLUME 70. Immunochemical Techniques (Part A)
Edited by HELEN VAN VUNAKIS AND JOHN J. LANGONE

VOLUME 71. Lipids (Part C)
Edited by JOHN M. LOWENSTEIN

VOLUME 72. Lipids (Part D)
Edited by JOHN M. LOWENSTEIN

VOLUME 73. Immunochemical Techniques (Part B)
Edited by JOHN J. LANGONE AND HELEN VAN VUNAKIS

VOLUME 74. Immunochemical Techniques (Part C)
Edited by JOHN J. LANGONE AND HELEN VAN VUNAKIS

VOLUME 75. Cumulative Subject Index Volumes XXXI, XXXII,
XXXIV–LX
Edited by EDWARD A. DENNIS AND MARTHA G. DENNIS

VOLUME 76. Hemoglobins
Edited by ERALDO ANTONINI, LUIGI ROSSI-BERNARDI, AND EMILIA
CHIANCONE

VOLUME 77. Detoxication and Drug Metabolism
Edited by WILLIAM B. JAKOBY

VOLUME 78. Interferons (Part A)
Edited by SIDNEY PESTKA

VOLUME 79. Interferons (Part B)
Edited by SIDNEY PESTKA

VOLUME 80. Proteolytic Enzymes (Part C)
Edited by LASZLO LORAND

VOLUME 81. Biomembranes (Part H: Visual Pigments and Purple Membranes, I)
Edited by LESTER PACKER

Section I

Transport Theory

[1] Transport Machinery: An Overview

By George Sachs and Sidney Fleischer

Much of enzymology deals with scalar chemical reactions catalyzed by proteins. It is central to the understanding of membrane transport to recognize that there is a vectorial component whereby ions or solutes are translocated through a protein (complex) embedded in a lipid bilayer membrane. The latter separates two compartments and is largely impermeable to the ions or solutes.[1] Thus, transport proteins are vectorial catalysts and specialized theoretical and practical approaches have been developed to study their mechanisms. As in other areas of biology, the advances in membrane science in the last decade or so have been striking. In the 1930s, the concept of the phospholipid bilayer was already established, resulting in the idea of a barrier surrounding living cells.[2] By the early 1970s, it was recognized that the phospholipid bilayer was asymmetric with respect to its constituent phospholipids, and also that the proteins were anisotropically oriented in the bilayer in a variety of ways that could be intrinsic to the bilayer or membrane associated.[3] The fluid mosaic model explicitly stated in 1972[4] represents the current view of membrane structure. Intrinsic proteins can be partially inserted or are oriented transmembrane[5] as required for either transmembrane signaling, in the case of cellular hormonal responses, or for selective transport of nonlipophilic molecules into and out of cells. Since then, there has been a rapidly expanding list of transport processes. The advent of molecular biology has provided a new set of techniques for identification, classification, and investigation of the function of transporters, greatly extending biochemical approaches in this field. The ability to sequence, mutate, and crystalize transporter proteins has also provided hope for future description of transmembrane transport in molecular and submolecular detail, thereby setting new standards and goals in modern transport research. We can now think realistically in terms

[1] P. Mitchell, *Biol. Rev.* **44**, 445 (1966).

[2] E. Gorter and F. Grendel, *J. Exp. Med.* **41**, 439 (1925).

[3] S. Fleischer, W. L. Zahler, and H. Ozawa, *in* "Biomembranes" (L. Manson, ed.), Vol. 2, pp. 105–119. Plenum, New York, 1971.

[4] S. J. Singer and G. L. Nicolson, *Science* **175**, 720 (1972).

[5] J. O. McIntyre, H.-G. Bock, and S. Fleischer, *Biochim. Biophys. Acta* **513**, 255 (1978).

of relating structure to function. Transport function is characterized by kinetics, electrogenicity, specificity, and regulation.

A major difference between eukaryotes and prokaryotes is that with evolution the number of membrane-bounded compartments has increased from just the plasma membrane in prokaryotes to a diversity of membrane-bounded organelles including nuclei, mitochondria, endoplasmic reticulum, Golgi complex, clathrin-coated vesicles, secretory granules, peroxisomes, and lysosomes, as well as specialized membrane systems such as sarcoplasmic reticulum. Beyond this specialization of membrane classes there is also an increase in the diversity of transporters. For example, a variety of regulated channels appear, as well as other signaling systems. Additionally, control of various ionic concentrations such as Ca^{2+} or pH affects cellular differentiation in eukaryotes, resulting in a process intrinsic to development in multicellular organisms. It has been estimated that there are at least 200 membrane polypeptides in the *Escherichia coli* plasma membrane. Given the multiplicity of membrane types in eukaryotes, there are, perhaps, tenfold more membrane polypeptides as compared with prokaryotes. In a number of instances, membranes in higher animals and plants are specialized with regard to membrane functions and are thereby simplified with respect to protein composition. For example, the calcium pump membrane of sarcoplasmic reticulum consists mainly of the calcium pump protein (~90% of the protein).[6] In multicellular organs, it is also common that an organ is composed of more than one cell type. Further diversity is introduced in the allocation of a specific transporter to a type of membrane and/or tissue.

Models of membrane systems involve as a starting point the generation of an experimental model of the lipid bilayer, that is, either planar bilayers or unilammellar bilayer vesicles referred to as liposomes. Although natural membranes are composed of a diversity of lipids, often arranged asymmetrically, it has been the exception rather than the rule to demonstrate a functional correlate of the specificity of the lipid composition in terms of enzymatic function or transport.[7] The usefulness of these lipid membrane model systems has been profound, both for the physical understanding of transport and in the identification of transport proteins. Nevertheless, artifacts can arise due to contaminants present in the lipids used to form bilayers or liposomes. For example, it has been known for many years that

[6] G. Meissner, G. E. Conner, and S. Fleischer, *Biochim. Biophys. Acta* **298**, 246 (1973).
[7] S. Fleischer, J. O. McIntyre, P. Churchill, E. Fleer, and A. Maurer, *in* "Structure and Function of Membrane Proteins" (E. Quagliariello and F. Palmieri, eds.), pp. 283–300. Elsevier, Amsterdam, 1983.

the H^+ permeability of liposomes and bilayers was relatively high, and from this it was argued that natural membranes had an equivalently high permeability, in spite of the fact that several membrane types are able to maintain relatively high proton gradients, in particular the canalicular membrane of the parietal cell. It now appears that the high proton permeability of artificial membranes was due to contaminants that had proton transport capability. Thus, when it is experimentally not feasible to reproduce the characteristics of the natural bilayer, anomalous data must be viewed with caution.[8]

Model peptide ionophores may simulate the vectorial properties of natural protein transporters. These can be classified into carrier or channel types based on whether the ion complex or the ion only moves. It seems unlikely that membrane transporters simulate the carrier model, and channellike structures would appear to be the rule. Gramicidin, which forms dimeric pores in natural and artificial membranes, is the best studied example of a model channel.[9] It is a cation-selective single helical channel $[H^+ > K > Na^+]$ that, during transport, replaces the hydration shell of the ion with the carbonyl oxygens of the peptide bonds that line the channel. Biological transporters are considerably more complex, usually consisting of a number of membrane-spanning helices ranging from six or seven, as in bacterial rhodopsin,[10] to twelve, as in the anion exchanger.[11] Rather than a single membrane-spanning helix forming a porelike structure, as in gramicidin, the folding of the membrane sectors of transport proteins apparently forms the ion or solute transport pathway being composed of several transmembrane-spanning regions. In spite of considerable sequence information on transporters and limited structure information, there is not as yet a well-accepted model for transmembrane transport of a solute by any transporter. Gramicidin illustrates several principles by which membrane channels must operate. There is interaction of ions with charged groups or dipoles (in the case of gramicidin) in the mouth and walls or the channel, resulting in several binding sites. The ionic motion is through a structure fixed with respect to the plane of the membrane and it is very rapid, ranging from 10^6 to as high as 5×10^8 ions/sec in natural channels, corresponding to a conductance of 1 to 500 pS at 1 mV. The cyclic peptide valinomycin is an example of a carrier mechanism wherein the ion and the cyclic peptide move together to provide transmembrane electrogenic transport. Here, the rate of transport is slower (10^5/sec) and is below the noise

[8] J. Gutknecht and A. Walter, *Biochim. Biophys. Acta* **641**, 183 (1981).
[9] D. W. Urry, *Proc. Natl. Acad. Sci. U.S.A.* **68**, 672 (1971).
[10] R. Henderson and P. N. Unwin, *Nature (London)* **257**, 28 (1975).
[11] R. R. Kopito and H. F. Lodish, *Nature (London)* **316**, 234 (1985).

level of electrical measurement, so that single events cannot be recorded by bilayer or patch clamp methods. Biological transporters usually operate at rates much slower than channels. Some are electrogenic and others are electroneutral, whereas all channels by definition are electrogenic. Examples of such slower transporters are the Na^+ glucose symporter, the Na^+/Ca^{2+} and Na^+/H^+ antiporters, and all electrogenic ion pumps. Although carrier in the rate sense, it is likely that the majority of the polypeptide is fixed in the membrane, but that there is a rate-limiting step involving a conformational barrier, thereby slowing the rate of transport.

Identification of transport systems often has as its starting point an intact organ (epithelial sheet or tubule) where it is possible to manipulate ion gradients and electrical potentials. From such data and knowledge of the driving forces, it is possible to derive important conclusions relevant to transport mechanism. For example, in the case of glucose absorption by the small intestine, it was possible to show Na^+ dependence and that transport was electrogenic.[12] Investigation of a similar system in the kidney revealed that two transporters were present.[13] One was responsible for low-affinity, high-capacity transport, but could not account for the final glucose gradient achieved in the distal proximal tubule. The higher concentration gradient was achieved by a high-affinity, low-capacity transporter, with a higher Na^+/glucose stoichiometry necessary for the larger concentration gradient. The conclusion was derived from kinetic measurements and thermodynamics, knowing the Na^+ gradient, the potential across the brush border, and the final tubular glucose concentration gradient. This study illustrates but one application of irreversible thermodynamics. This conceptual framework has also been applied to the multiion transporters and transporters coupled to electrochemical driving forces. For example, catecholamine uptake by chromaffin granules was accounted for by a H^+/catecholamine electrogenic antiporter, driven by the H^+ gradient and the intragranular positive potential, because the H^+ gradient alone was insufficient for the gradient achieved.[14] Thermodynamic analysis is an important adjunct to kinetic analysis of transport.

Na^+/H^+ exchange is involved in pH regulation by most eukaryotic cells. A negative intracellular potential of usually about 60 mV, given a proton conductance, would set cell pH about 1 unit less than ambient. To maintain cell pH at about 7.1, electroneutral Na^+/H^+ exchange occurs across the plasma membrane. However, given a tenfold Na^+ gradient, it would be anticipated that cell pH would be about 1 unit alkaline to

[12] R. Crane, *Fed. Proc., Fed. Am. Soc. Exp. Biol.* **21**, 891 (1962).
[13] R. J. Turner, *Ann. N.Y. Acad. Sci.* **456**, 10 (1985).
[14] R. G. Johnson, *Physiol. Rev.* **68**, 232 (1988).

ambient. The fact that it is usually about 0.3 pH units acid indicates that the Na^+/H^+ exchanger does not equilibrate cell pH. Accordingly, it can be anticipated that the exchanger is regulated so as to be inactive at about pH 7.1 and this was found to be due to an internal regulatory H^+ site, insensitive to Na^+.[15] Similar thermodynamic considerations led to the major discoveries in epithelial Cl^- transport, which could then be applied to Cl^- transport in other cells and organs. Most regions of the intestine are capable of active Cl^- secretion and certain regions of the kidney are capable of active Cl^- absorption. For many years, this was modeled as a primary active pump. However, it was realized and later demonstrated that intracellular Cl^- was above electrochemical equilibrium, so that if a transporter and a conductance were placed asymmetrically on two cell surfaces, the former to augment $[Cl^-]_i$, then the high $[Cl^-]_i$ would be driven out of the cell by the potential difference across the Cl^--conducting membrane.[16] The transporter was found to be usually $NaKCl_2$ cotransport, but occasionally Cl^-/HCO_3 exchange, where the driving force for Cl^- entry via these necessarily electroneutral pathways was the Na^+ and/or Cl^- gradients. Thus, the transepithelial Cl^- transport was due to an asymmetric arrangement of inward electroneutral cotransport and conductive efflux, a secondary active process. Coupling of transporters on opposite sides of the cell membrane is almost universal in epithelia. The $NaKCl_2$ pathway was found to be involved not only in Cl^- secretion, but also in cell volume regulation.[17] Cells shrunken in hypertonic medium often regain volume by a process known as volume regulatory increase (VRI). Analysis of cells such as ascites cells showed that cotransport of Na^+, K^+ and $2Cl^-$ was involved in an electroneutral fashion. Furthermore, the transporter did not equilibrate these gradients in spite of the abolition of Na^+ pump activity. Thus, not only was the transporter regulated by a volume change in terms of being switched on, but, as for the Na^+/H^+ exchanger, it had to possess an internal regulatory site sensitive to ionic concentrations to inhibit its function at the set point of the cell.

The influence of an electric field across the membrane on the transporter is of interest both in a physiological and structural context. Whether a transporter is electrogenic or not determines the driving forces that can operate on the transported species. Thus, with electroneutral Na^+ or H^+-coupled symporter, the maximum gradient that can be achieved is usually about one order of magnitude, i.e., the Na^+ or H^+ gradient. However, when the porter is electrogenic, the gradient can be amplified as a function of both the potential and the stoichiometry of coupling. As an example, the

[15] P. S. Aronson, M. A. Suhm, and J. Nee, *J. Biol. Chem.* **258,** 6767 (1983).
[16] R. Frizzell, M. Field, and S. G. Schultz, *Am. J. Physiol.* **236,** F1 (1976).
[17] K. R. Spring and H. H. Ussing, *J. Membr. Biol.* **92,** 21 (1986).

Na^+/Ca^{2+} antiporter of cardiac sarcolemma operates with a $3:1$ Na^+/Ca^{2+} stoichiometry. It is thus electrogenic and allows a greater than 10,000-fold gradient of Ca^{2+} to be maintained across the plasma membrane.[18] A lesser stoichiometry or electroneutrality would make the Na^+/Ca^{2+} exchanger an inward-leak pathway given a $[Ca^{2+}]_i$ of 100 nM. The above example illustrates the case of an electrogenic transporter, where the equilibrium (but not the activity) is determined by the electrochemical gradient across the membrane. This is different from a voltage-activated pathway, where the transporter changes either orientation, with respect to the membrane, or conformation, as a function of applied voltage, so that its functional state is changed, i.e., the channel opens or closes. In this class of transporter is found a variety of ion channels that are voltage activated as well as electrogenic.[19] From this, the distinction between channels and carriers is, at our present level of structural information, based on rate of transport of the ion that is related to the height of the energy barrier to transport. However, in both there must be a region of the transporter that responds to an applied electric field, a gate in the case of channels, a transport site or barrier that changes cyclically in the case of electrogenic transporter. In the pump cycle, a conformational change limits the turnover rate.

Ion pumps form a distinct class of transporters that interconvert energy from light or chemical reactions into ion gradients, and can consist of one or more different subunits. In general, their turnover number is much slower than that of carriers, at about 100/sec. In an epithelium such as the collecting duct, it has been calculated that the multiplicity of Na^+ channels, $NaKCl_2$ cotransporters, the Na^+ pump corresponds inversely to their relative turnover number.[20] As for other transporters, pumps may be electrogenic or electroneutral, voltage sensitive or insensitive. Since pumps catalyze enzymatic reactions as well as ion translocation, kinetic analysis has allowed identification of the chemical steps in the pump cycle involved in transmembrane transport.[21,22]

Transporters are universal sites of regulation and therefore targets of useful drugs. Regulation may be by voltage, binding of regulating ions, such as H^+ or Ca^{2+}; by chemical modification, such as phosphorylation; or by change of location, as occurs in exocytosis or membrane fusion. Ligands

[18] J. P. Reeves, C. A. Bailey, and C. C. Hale, *J. Biol. Chem.* **261**, 4948 (1986).
[19] B. Hille, "Ionic Channels of Excitable Membranes." Sinauer, Sunderland, Massachusetts, 1984.
[20] D. A. Klaerke and P. L. Jorgensen, *Comp. Biochem. Physiol.* **90**, 757 (1988).
[21] R. L. Post, K. Taniguchi, and G. Toda, *in* "Molecular Aspects of Membrane Phenomena" (H. R. Kabak, ed.), pp. 92–103. Springer Publ., New York, 1975.
[22] J. M. Wood, J. G. Wise, A. E. Senior, M. Futai, and P. D. Boyer, *J. Biol. Chem.* **262**, 2180 (1987).

interacting with transporters such as channels, carriers, or pumps have proved useful clinically and have facilitated the isolation and identification of transporters. Examples of such use of ligands are the recent purification of the NaKCl$_2$ transporter and the Na$^+$ channel of epithelia.[23,24]

There are usually several transporters affecting the distribution of the same ions or solutes across cell membranes, as there are many more than one cell type regulating transepithelial movements. To identify the cell type, as well as to enrich the sources of a transport system, cell isolation and separation techniques have been developed that vary considerably. General methods taking advantage of mass, charge, and density are the most frequently applied, although other approaches, such as using fluorescent cell sorting or ligand chromatography, are finding increasing use. The low cell yield from highly selective separation techniques is supplemented by subsequent cell amplification by culture methods. Perhaps the most selective methods have been applied in culture involving isolation of mutants, single-cell cloning, and computer-controlled laser selection methods. The need to separate cells is useful for physiological correlation. Recent advances, such as affinity chromography, use of monoclonal antibodies, cloning, and expression methodology, have made highly selective methods commonplace.

As organs, epithelia were the model of choice for transport studies, given their polar characteristics and ease of manipulation, taking advantage of the sheetlike characteristics of most epithelia to separate two bathing solutions. In special instances, such as in the kidney, the relevant structures are tubular, and specialized tubule perfusion methods have been developed to study the transport characteristics of separate regions. In some epithelia, such as liver or pancreas, even these methods are technically too demanding and methods have been developed for studying transport in cell clusters or even cell pairs. The methods that have been retained reflect the variety of signals that can be assayed. The most frequently employed method involves electrical measurements ranging from simple voltage and current methods, to determination of individual channel events by noise analysis or patch clamp methods, the latter applied in either the epithelium or in single cells. The patch clamp method has allowed characterization of single-channel conductance in bilayers and has been an important tool in defining regions of interest in cloned and mutated channels, e.g., the acetylcholine receptor.[25]

[23] T. Zeuthen, O. Christensen, and B. Cherksey, *Pfluegers Arch.* **408,** 275 (1987).
[24] D. J. Benos, G. Saccomani, and S. Sariban-Sohraby, *J. Biol. Chem.* **262,** 10613 (1987).
[25] E. Neher, B. Sakman, and J. H. Steinbach, *Pfluegers Arch.* **375,** 219 (1978).

Ion flux measurements form the next most frequent method, with either tracer or ionspecific electrode methods. Since water is an inevitable accompaniment of ion or solute transport, knowledge of water flow and its regulation is an important area, and methods to determine the osmotic and diffusive flow of water are necessary in understanding function of specialized epithelia and cells that are involved in volume and osmotic regulation. Recently, computer processing of optical data has become possible. Hence, microscopic measurements of transport by single cells, even within sheets of epithelia, have been developed using optical probes of pH, $[Ca^{2+}]_i$, $[Cl^-]_i$ potential, and cell volume.[26,27] Measurements of transport in single cells have become routine.

Regulation of transport sometimes involves fusion of separate membrane vesicles by exocytosis and endocytosis. This involves the interplay of several intracellular pathways, mediation by second messengers, as well as energy supply. Analysis of the process may be approached by isolation of the membrane involved, but is often sufficiently complicated so that experiments on permeabilized cells provide crucial information. Various techniques have been developed to permeabilize cells or to remove ionic barriers with ionophores, allowing regulation more or less of the intracellular environment while other chemicals, such as ATP or IP_3, can be added.[28,29] This provides a means of dissecting the regulation of transport and of studying membrane–membrane interactions.

Separation of the transporter from the cell and from its natural membrane environment is essential for further insights. Classically, therefore, a large sector of membrane science has been devoted to isolation of plasma and intracellular membranes in highly enriched form, and preferably in a state where the transport function remains operative. Membrane sidedness is an important factor in this context.

Given that the critical component of any transporter exists embedded in the lipid bilayer, isolation techniques have been developed that take advantage of the membrane location of the molecule and the special characteristics of the particular membrane. Study of the transport process requires that the membrane continues to separate two aqueous compartments, and this is often the case since disrupted membranes spontaneously reseal to form vesicles during the homogenization procedure. Alternatively, reconstitution into liposomes or planar bilayers provides a useful model for study. The use of vesicles to study transport required the devel-

[26] R. Y. Tsien, *Nature (London)* **290,** 527 (1981).

[27] G. Grynkiewicz, M. Penie, and R. Y. Tsien, *J. Biol. Chem.* **260,** 3440 (1985).

[28] P. F. Baker and D. E. Knight, *Nature (London)* **276,** 620 (1978).

[29] H. Streb, R. F. Irvine, M. J. Merridge, and I. Schulz, *Nature (London)* **306,** 67 (1983).

opment of methods that were sufficiently sensitive to monitor uptake or release from the small fraction of volume occupied by vesicles in their bathing solution. A large variety of these methods exist, and combined application of dye and flux methods are often required to define the nature of transport at the vesicle level. The application of dye probe methods to isolated cells required that the dye probes be prepared in membrane-permeable form insensitive to the ion of interest. Intracellular hydrolysis leads to the production of a charged, membrane-impermeant form, now responsive to the concentration of the ion of interest.

It is unusual to find a system where only a single active protein is present in a membrane so that ion or solute movement can be confidently ascribed to the given protein. Different strategies have emerged to identify and isolate the functional polypeptides. A powerful approach is solubilization, purification, and reconstitution, often aided by binding studies with high-affinity inhibitors. It appears that no general recipe exists to ensure survival of the protein in its functional form and each situation requires optimization of conditions, especially the selection of detergent and how it is used, as well as sometimes the addition of lipid during purification. With perseverance, it is now conceivable that, following identification of a transport process, a protein or a group of proteins can be identified and a sufficient quantity identified on single- or two-dimensional gels to allow partial sequencing. From there, it is now routine to be able to sequence the protein by screening an appropriate cDNA or genomic library and to provide a full-length clone for biosynthesis and expression in various cells, such as bacteria, yeast, *Xenopus* oocytes, and mammalian cells in culture. This type of technology promises to provide sufficient material to allow routine application of physical methods to the study of transport proteins.

Studies of membrane protein biosynthesis have shown that membrane insertion into ER is cotranslational. Sometimes the signal sequence is internal and is not processed by signal cleavage enzymes. Yet, we do not know how to reconstruct the folding process that selects for the correct membrane-spanning sequences and excludes other hydrophobic areas of the protein. This restricts the ability to reconstitute transport from engineered and even natural peptides at high efficiency. If such technology were developed, it is apparent that with the techniques discussed is this volume, understanding of transport in molecular terms could rapidly be achieved using engineered proteins.

For naturally abundant proteins, and for study of complex assemblies, it is quite feasible to develop molecular information without the use of bioengineering. A number of proteins have been crystallized in two dimensions so that at least their general shape and arrangement of membrane-spanning components have been defined. In the last few years, three-dimensional crystals have been obtained for a few transport proteins, which should provide detailed structural insights into the transporters. The correlation of defined structure – function relationships in membrane polypeptides generally and transporters and channels specifically looms on the horizon.

[2] Survey of Carrier Methodology: Strategy for Identification, Isolation, and Characterization of Transport Systems

By MARTIN KLINGENBERG

Introduction

The ability of biomembranes to permit selectively the passage of solutes is generally facilitated by the catalytic activity of a particular class of transmembrane proteins. These may form either continuous channels or carrier-type devices. A carrier allows the translocation of only one molecule, or rarely a few molecules, in a catalytic cycle. To this category belong transporters, which are designated not only as carriers, but also as translocators, permeases, and pumps. In this chapter pores and channels are excluded and only the vast group of carrier-linked transport systems will be addressed, with less emphasis on ion pumps.

The characterization of carrier-linked transport goes through various stages, spanning a wide range from identifying the transport phenomena to analyzing the molecular structure and function (see Scheme 1). Historically, development in this field has followed a similar path although today the recombinant DNA approach intervenes at an early stage and strongly facilitates the molecular understanding. In some prokaryotic systems, genetics has played an early major role in singling out transport systems. These results then formed the basis for molecular characterization using modern methods at a considerably later date. Research on transport in eukaryotic cells has profited from genetics to a lesser degree, but, like all other membrane phenomena, has been stimulated enormously by the recombinant DNA approach. In this chapter various methods will be examined in detail using three different transport systems: glucose transport in erythrocytes, ADP/ATP exchange in mitochondria, and lactose transport in *Escherichia coli*.

METHODS IN ENZYMOLOGY, VOL. 171

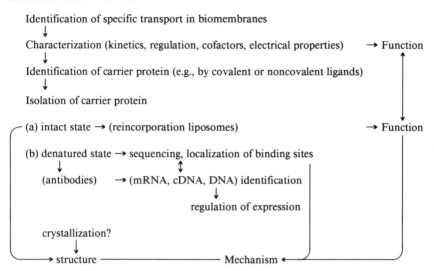

SCHEME 1. Flow scheme of research on transport systems.

Erythrocytes represent a particularly well-investigated example of eukaryotic cell membranes. They are preferred for studying membrane permeability because of the availability of isolated cells, the definition of the membrane versus the inner volume, and the relative simplicity of the transport systems in this membrane. Erythrocytes are relatively stable and easily yield preparations in which the inner volume is depleted of proteins, as in erythrocyte ghosts. Here glucose transport is the prime example for the elucidation of carrier-catalyzed transport.[1] The kinetics of this system has led to a broad discussion of carrier concepts in the literature.[1,2,3] Because of the ease and simplicity in studying transport, many kinetic data have been obtained.

Another eukaryotic transport system, in which progress can be followed from the elucidation of a specific transport to the characterization of the carrier molecules and its mechanism, is the ADP/ATP carrier of the inner mitochondrial membrane.[4,5] This is an obligatory exchange system, in contrast to glucose uptake, and therefore from the beginning has been a particularly strong case for carrier-catalyzed transport. It was here that for

[1] W. D. Stein, "Transport and Diffusion across Cell Membranes," p. 232. Academic Press, Orlando, Florida, 1986.
[2] T. Rosenberg and W. Wilbrandt, *J. Gen. Physiol.* **41**, 289 (1957).
[3] P. G. LeFevre and G. F. McGinniss, *J. Gen. Physiol.* **44**, 87 (1960).
[4] M. Klingenberg, *in* "The Enzymes of Biological Membranes" (A. Martonosi, ed.), Vol. 3, p. 383. Plenum, New York, 1976.
[5] M. Klingenberg, *in* "Membranes and Transport" (A. Martonosi, ed.), Vol. 1, p. 203. Plenum, New York, 1982.

the first time a basic carrier mechanism could be elucidated on a molecular level. The system is also representative of intracellular transport.

A well-characterized example of transport through a prokaryotic membrane is the galactoside system in *E. coli*.[6,7] In this case the genetic approach has been useful from the very beginning in characterizing the gene and therefore the protein-dependent transport system. Here the first carrier protein was detected by covalent labeling.[8] However, it took many years before the carrier was isolated and its molecular characteristics were defined. Site-directed mutagenesis has recently been applied to identify the structure–function relationship in this carrier.[9]

Kinetic Criteria for Carrier-Linked Transport

At the first stage in identifying a specific transport, kinetic criteria are generally applied. One of the simplest approaches is the saturation of transport activity with increasing solute concentration. Unspecific permeation should not be saturated, unless at much higher solute concentrations. A related criterion is transport specificity. High selectivity between chemically similar substrates is indicative of carrier-linked transport, whereas unspecific transport activity varies over a broad range. The specificity criterion is correlated to concentration dependence since generally higher specificity is associated with a low K_m.

A more sophisticated kinetic probe for carrier-catalyzed transport is the acceleration of substrate uptake by the presence of a countersubstrate on the inner side of the membrane.[1,2,10] This internal substrate may be either identical to or slightly different from the external one. The uptake requires a counterexchange of the substrate through the same carrier. The exchange rate can be measured with isotope labeling of either the external or internal substrate. Acceleration of transport by the internal substrate is interpreted in terms of a counterexchange of the external against the internal substrate. Counterexchange is one of the most stringent criteria for a carrier-type transport.[2] Acceleration of uptake by the internal substrate corresponds to a facultative counterexchange. An obligatory counterexchange exists if there is uptake only in the presence of internal substrate. In the majority of cases there is a stoichiometric one-by-one molecule exchange. At very high

[6] J. K. Wright, R. Secker, and P. Overath, *Annu. Rev. Biochem.* **55**, 225 (1986).

[7] H. R. Kaback, *Annu. Rev. Biophys. Chem.* **15**, 279 (1986).

[8] C. F. Fox, J. R. Carter, and E. P. Kennedy, *Proc. Natl. Acad. Sci. U.S.A.* **57**, 698 (1967).

[9] P. V. Viitanen, D. R. Meninck, H. K. Sarkar, W. R. Trumble, and H. R. Kaback, *Biochemistry* **24**, 7628 (1985).

[10] W. R. Lieb, *in* "Red Cell Membranes: A Methodological Approach" (J. C. Ellory and J. D. Young, eds.), p. 135. Academic Press, London, 1982.

concentrations an unspecifically permeant solute may cause an unstoichiometric or nonobligatory counterexchange. Counterexchange or acceleration of uptake by a countersubstrate can also be used to discriminate a carrier transport from a pore-type transport. In a continuous pore the countersubstrate should always inhibit the uptake.

The counterexchange or acceleration of transport by the opposite substrate was investigated in detail for glucose transport in erythrocytes. On this basis the first reaction schemes for the circulating or mobile carrier were derived.[2,3] Although this simple carrier model well explains the basic data, a more detailed analysis of glucose transport and its concentration dependence revealed deviations from the straightforward kinetics and required some modifications from this model in a yet-to-be-defined manner. An obligatory counterexchange was derived for the ADP/ATP exchange in mitochondria from kinetic analysis and also using double-isotope labeling.[11] Here the counterexchange is at least 100 times faster than the unidirectional transport.[4] In this system the K_m for ADP or ATP is also rather low (5 to 15 μM).

Inhibitors as Tools for Identifying Transport Systems

From the beginning inhibitors have been extremely useful in identifying transport systems. In fact, inhibitors played a much more important role in characterizing specific transport than in characterizing enzyme reactions. One reason is the necessity to single out a particular transport system in a membrane from other transport systems. Another reason is that inhibitors have often been the last resort when segregating protein catalyzed from unspecific diffusion. Essentially, two classes of inhibitors have been used, those which bind noncovalently and in competition to the substrate site, and those which modify essential amino acid residues operating either in the binding center or in the translocation process. To the first group belong the more specific and ideal inhibitors, which often originate from biological sources. These inhibitors can be of extreme selectivity and, due to their high affinity, are very important instruments in identifying specific transport processes and their underlying carriers, and thus facilitate the elucidation of the mechanism. Applied in the inhibitor–stop method, inhibitors are used for analyzing the kinetics of transport.[12,13] The transport is arrested by a rapid addition of excess inhibitors before the separation of the vesicles from the medium. A peculiarity of these inhibitors is the membrane sidedness of interaction. A specific inhibitor is effec-

[11] E. Pfaff and M. Klingenberg, *Eur. J. Biochem.* **6**, 66 (1968).
[12] E. Pfaff, H. W. Heldt, and M. Klingenberg, *Eur. J. Biochem.* **10**, 484 (1969).
[13] F. Palmieri and M. Klingenberg, this series, Vol. 56, p. 279.

tive only from one side of the membrane in inhibiting transport and binding to the carrier. This side specificity is the result of the conformational difference of the binding center when facing the outer or inner aspect of the membrane.

Examples of this type of xenobiotic inhibitor can be found for glucose transport in the form of phloretin and cytochalasin B, and the highly specific digitalis glucosides as inhibitors of the Na^+/K^+ pump. These only interact from the outside, but probably not at the binding center for sodium or potassium. Particularly revealing in their applications are the inhibitors for the ADP/ATP carrier in mitochondria. Two groups of inhibitors are found, the atractyloside group and the bongkrekate group, which both interact with high selectivity only with the ADP/ATP carrier.[4,5,14,15] The atractylosides interact only from the outside, whereas the bongkrekates bind only from the inside. Both types of inhibitors are assumed to bind to the substrate-binding center for ADP or ATP, although their structures are different. The only clear common denominator with the substrates is three or four negative charges. Because bongkrekate can penetrate the membranes unspecifically, it is particularly useful in inhibiting transport and is thus ideal for studying the compartmentalization of ADP and ATP within eukaryotic cells. Phloretin and cytochalasin B were also found to bind to the glucose carrier from opposite sides of the membrane, phloretin from the outside and cytochalasin B from the inside.[16,17] Both remove the substrate and displace each other so that here also ligand binding to only one reorienting substrate site is assumed, and the "single-site reorientation model" also is extended to the glucose transporter.[17a]

The differences in structure between the inhibitors and the substrate, for example, in the case of the ADP/ATP carrier, and the great differences between the two types of inhibitors seem to contradict the notion that they bind to the substrate-binding site. However, these differences serve to demonstrate the flexibility of the binding center during the translocation process. Thus "the single-site gated-pore mechanism" was derived for the ADP/ATP carrier,[4] a molecular mechanism which has now been accepted for most other solute carriers. According to this model, the asymmetric substrate located within the asymmetric carrier has different interfaces in the external or internal state, and these are at least partially reflected in the structure of the two inhibitors. In addition, according to the internal

[14] P. V. Vignais, P. M. Vignais, G. Lauquin, and F. Morel, *Biochimie* **55**, 763 (1973).

[15] M. Klingenberg, B. Scherer, L. Stengel-Rutkowski, M. Buchholz, and K. Grebe, *in* "Mechanisms in Bioenergetics" (G. F. Azzone, L. Ernster, S. Papa, E. Quagliariello, and N. Siliprandi, eds.), p. 257. Academic Press, New York, 1973.

[16] F. R. Gorga and G. E. Lienhard, *Biochemistry* **20**, 5108 (1981).

[17] R. M. Krupka and R. Deves, *J. Biol. Chem.* **256**, 5410 (1981).

[17a] S. A. Baldwin and G. E. Lienhard, *Trends Biochem. Sci. (Pers. Ed.)* **6**, 208 (1981).

catalytic energy-transfer theory, the binding center is substratelike only in the transition state but not in the externally or internally localized state.[18] The conformation of these states is more faithfully traced by the inhibitors than by the substrate and therefore can express their full internal binding energy.

Covalent inhibitors have been widely and successfully used in identifying carrier-linked transport. They are mostly amino acid reagents which may interact unspecifically with the membrane proteins or may be targeted specifically to the carrier by differential protection or activation. The best known and nearly universally used inhibitors for transport systems are SH reagents. They have been applied not only for inhibiting transport, but also for identifying the underlying carrier protein. A famous example is the identification of the lactose permease in *E. coli* by difference labeling with *N*-ethylmaleimide (NEM).[8] The ATP/ADP carrier protein was first identified by difference labeling with NEM in extracts from mitochondria.[19] In lactose permease the substrate inhibits the labeling of the specific SH group, and in the ATP/ADP carrier the SH group becomes unmasked by the addition of the substrate. SH reagents have also been applied in conjunction with probing the membrane sidedness. Mercurials interact mainly from the outside, whereas NEM also reaches inner SH groups. Covalent modifiers of lysine or arginine reagents are often used for inhibiting transport quite specifically. Distilbene derivatives, which interact quite specifically with lysine groups, have been used to identify the anion-binding groups in the erythrocyte anion transporter.[20] The membrane-impermeant lysine reagent pyridoxal phosphate can be used to probe only the lysine groups localized toward the outside.[21] In general, by the judicious differential application of more or less membrane-permeant reagents, the specificity of the covalent labeling can be enhanced.

Strategy for Isolation of Carrier Proteins

For determining the transport function of a carrier and the dependence on environmental factors it is useful to isolate the protein and to study its function after reincorporation into artificial phospholipid vesicles. Another reason for isolating membrane protein is to characterize its molecular

[18] M. Klingenberg, *in* "Recent Advances in Biological Membrane Studies: Structure and Biogenesis, Oxidation and Energetics" (L. Packer, ed.), p. 479. Plenum, New York, 1985.
[19] M. Klingenberg, P. Riccio, H. Aquila, B. Schmiedt, K. Grebe, and P. Topitsch, *in* "Membrane Proteins in Transport and Phosphorylation" (G. F. Azzone, M. Klingenberg, E. Quaglariello, and N. Siliprandi, eds.), p. 229. North-Holland Publ., Amsterdam, 1974.
[20] M. L. Jennings and H. Passow, *Biochim. Biophys. Acta* **554,** 498 (1979).
[21] W. Bogner, H. Aquila, and M. Klingenberg, *Eur. J. Biochem.* **161,** 611 (1986).

properties and eventually its molecular mechanism. The isolation of a specific membrane protein requires first its identification among the other membrane proteins. In some membranes the carrier protein is the major one and its identification is no problem. Normally, covalent or noncovalent labels to the carrier are used for tracing the carrier. Also the functional assay in reconstituted vesicles is exploited for this purpose, when there is no other marker available. Isolation of the phosphate carrier from mitochondria, for example, partially follows this route.[22]

The successful isolation of carrier proteins from biomembranes has so far been limited mostly to cases where the carrier is present in sufficiently high amounts. Therefore the best strategy is to choose biological material with a high content of carrier protein. For example, the isolation of the ATP/ADP carrier from mitochrondria profited from the high content of this protein in beef heart mitochondria, where it amounts to about 12% of the total protein. In prokaryotes the method of choice is to use mutants which overexpress the respective carrier protein. In the best known example, a mutant was constructed with a severalfold overexpression of lactose permease.[23] Only under these conditions, after more than a decade of frustrating attempts, could this carrier be isolated. An outline of frequently used steps for the isolation of intact carriers and other membrane proteins is given in Scheme 2.

When choosing membranes in which the protein is present in high content, it is a prudent strategy to purify the membranes as much as possible, separating them from other membranes and proteins. Differential sedimentation centrifugation, sucrose gradient centrifugation, and large-molecular-weight gel filtration are used to segregate the membranes, once a specific assay has been established (e.g., Ref. 23). Real efforts should be invested in this early step in order not to impair the subsequent isolation and purification. In fact, there are cases in the literature wherein this has not been accomplished and the claimed purification was in fact a "dirtification," compared to the content of the protein in the pure membrane.

The most critical step in the isolation is a suitable solubilization of the membrane protein, which presently is generally performed by the use of detergents. The role of the detergent is first to extract the phospholipids from the membrane and then to envelop the membrane proteins in detergent–protein micelles, typically with some phospholipids still attached. For the purpose of reconstitution and for molecular functional studies, the protein should be isolated in an intact state. Therefore, criteria

[22] H. V. J. Kolbe, D. Costello, A. Wong, R. C. Lu, and H. Wohlrab, *J. Biol. Chem.* **259,** 9115 (1984).

[23] R. M. Teather, O. Hamelin, H. Schwarz, and P. Overath, *Biochim. Biophys. Acta* **467,** 386 (1977).

Organelles
↓
Purification of membranes (sucrose gradient, sonication, etc.)
↓
Enrichment of target protein with extracting, nonsolubilizing detergent
↓
Solubilization with detergent and salt
↓
Centrifugation, monodisperse micelle solution
↓
Nonadsorptive chromatography in detergent
↓
Gel filtration, hydroxylapatite chromatography, DEAE–cellulose, ECTEOLA
↓
Sucrose gradient centrifugation for special applications
↓
Detergent exchange (e.g., lauroylmaltoside against Triton X-100)

SCHEME 2. Outline for the isolation of intact carrier proteins.

for intactness must be applied, such as the capability for noncovalent and specific binding of ligands (inhibitors), and transport activity after reincorporation into liposomes. It is also important to establish that "true" solubilzation has been achieved, i.e., that the target protein is in a monodisperse (monomeric or oligomeric) state. Usually, ultracentrifugation at 100,000 g should be able to discriminate the solubilized proteins.

With this in mind, the selection of the right detergent is absolutely crucial. Here the literature comes to quite divergent conclusions about the usefulness of the various detergent types.[24–28] Often a claim is made for the use of only one particular detergent for the isolation of a membrane protein. Therefore, it would seem that no general rules can be applied to the solubilization of membrane proteins. However, this impression is based mainly on an insufficient understanding of the interaction of detergents with the membrane protein and also of the physical chemistry of this process. Also, in some reviews the tolerance of functional membrane proteins for various detergents has not been critically evaluated. When more sensitive criteria are taken into account, the choice of suitable detergents is narrowed, and often very expensive materials are not necessary.

At any rate, the vast experience accumulated regarding the use of

[24] G. Meissner, G. E. Conner, and S. Fleischer, *Biochim. Biophys. Acta* **298**, 246 (1973).
[25] A. Helenius, D. R. McCaslin, E. Fries, and C. Tanford, this series, Vol. 56, p. 734.
[26] L. M. Hjelmeland and A. Chrambach, this series, Vol. 104, p. 305.
[27] J. V. Moller, M. Le Maire, and J. P. Andersen, *in* "Progress in Protein–Lipid Interactions 2" (A. Watts and J. J. De Pont, eds.). Elsevier, Amsterdam, 1986.
[28] Y. Yanagita and Y. Kagawa, *in* "Techniques for the Analysis of Membrane Proteins" (C. I. Ragan and R. J. Cherry, eds.), p. 61. Chapman & Hall, London, 1986.

detergents allows the derivation of some general rules about their choice. The criteria for the retention of carrier function depend on the type of protein and function assayed. With a given protein often one function is tested, whereas other functions may already have been inactivated. As a rule, the more sensitive the assay toward detergents, the better the resulting discrimination. The ADP/ATP carrier offers a hierarchy of functional criteria, such as stability of binding in different conformational states, and the speed of inactivation that permits a grading of the detergents, as shown below.[5] Two requirements have to be compromised, the ability of the detergent to solubilize the protein effectively in monodisperse form and at the same time to maintain its intactness. In general, the best solubilizing detergents are also those which are strongly denaturing. Ionic detergents are very strong solubilizers of biological membranes and consistently denature those proteins. Also, the widely used "ionic" cholate and deoxycholate, as well as the cholate analog 3-[(3-cholamidopropyl) dimethylammonio]-1-propane sulfonate (CHAPS), generally inactivate the carriers. Partially ionic detergents, such as those with cationic head groups, for example, lysolecithin and laurylbetaine, are also denaturing. Still less polar are "dipole" head groups, such as in the aminoxide groups, for instance lauryldimethyl aminoxide (LDAO), which is tolerated by several membrane proteins. Less denaturing is the lauroylamidodimethylpropylaminoxide (LAPAO), although it is a good solubilizer.

In general, the best and least harmful detergents are found among the nonionic detergents. However, here too care must be taken to select the right type since some confusion exists in the use of low-molecular-weight detergents. Best known are the poly(oxyethylene) (POE) derivatives. Stronger solubilizers in this category are those with a nonlinear hydrophobic group, such as isooctylphenyl (e.g., Triton X series) or isotridecyl (e.g., Emulphogen). Detergents with linear alkyl groups are weaker solubilizers (e.g., Brij and Lubrol WX). For mitochondrial and bacterial membranes stronger solubilizers are required than for plasma or nonmitochondrial intercellular membranes, for which the linear alkyl detergents are very useful. The presence of a relatively large salt concentration is important because it increases the solubilizing power of the nonionic detergents. In some cases an extremely high salt concentration may be instrumental for the solubilization.[28a] These detergents in general have the most beneficial effect in retaining protein intactness. Small alkyl detergents (C_8 to C_{10}) are stronger solubilizers but are also slightly more harmful. Still less tolerated by the carrier are small nonionic detergents with a high critical micelle concentration (cmc), such as the popular octylglucoside. This detergent may solubilize but tends to inactivate membrane proteins quite easily.

[28a] R. Krämer, G. Kürzinger, and C. Heberger, *Arch. Biochem. Biophys.* **251,** 166 (1986).

This varying tolerance of intactness among the nonionic detergents illustrates an important factor governing detergent–protein interaction. High-cmc detergents are in general more damaging because of their high concentration of monomers, which can penetrate and unfold the protein structure. This is true for nonionic as well as for ionic high-cmc detergents. The popularity of high-cmc detergents, such as octylglucoside or cholate, is based on the ease with which they can be removed by dialysis from the protein for reincorporation. In many cases, however, reconstitution can be achieved by low-cmc detergents, based on their removal with polystyrene beds. With more stringent conditions for tight vesicles, medium-cmc detergents of the linear alkylpoly(oxyethylene) type are suitable, such as C_8 to E_4 to C_{10} to E_5. The following sequence of intactness is given for various detergents: linear alkyl POE (C_{16} to C_{12}–E_8 to E_{10}) > branched alkyl, acryl POE (isooctylphenyl, isotridecyl–E_8 to E_{10}) > C_{12} maltoside > short linear alkyl POE (C_8 to C_{10}–E_4 to E_5) > dipole detergent (C_{12} aminoxides) > high-cmc nonionic (C_8 to C_{10}–glucopyranosides; C_8 to C_{10} N-methylglucamide) > zwitterionic (C_{12} betaine, CHAPS, lysolecithin, etc.) > weakly ionic (cholate, deoxycholate) > strongly ionic (C_{10} to C_{12} sulfate, sulfonate).

The purification of the protein–detergent micelle is best accomplished by conditions wherein the micelle remains intact. Thus, methods in which the protein micelle is adsorbed to column material are often less suitable than those involving a column passthrough or gel filtration. For this reason, affinity chromatography has not been very useful in the purification of carrier proteins. Any damage to the micelle shell will often result in irreversible alterations in the protein structure. This can be avoided by gel filtration and in some cases by hydroxylapatite passage. The latter is suitable for segregating proteins with a relatively large free surface from more intrinsic proteins that are extensively covered by the detergent. These are concentrated in the passthrough whereas the more hydrophilic proteins are retained. This procedure has been extremely useful for several membrane proteins. Pioneered for the isolation of the ATP/ADP carrier from mitochondria, it has since been applied for the successful isolation of carriers and other proteins from mitochondria and bacterial membranes.[29]

In some cases, purification, at least to a partial degree, is achieved by removing the unwanted proteins by selective solubilization. The procedure can only be applied to membranes which already have a high content of the target protein. It can be combined with subsequent solubilization and chromatography stages as, for example, used for the purification of the lactose permease from *E. coli*.[6] Often insurmountable problems are encountered in the purification of the intact protein and, depending on the

[29] M. Klingenberg, H. Aquila, and P. Riccio, this series, Vol. 56, p. 407.

purpose, compromise procedures must be applied. In these cases, for the reincorporation in liposomes, enriched but not yet pure carrier preparations are used for kinetic and functional studies.

The isolation and purification of denatured protein is used for several different purposes:

1. For chemical studies of the protein, such as amino acid sequencing, and, in conjunction with amino acid sequencing, for the localization in the sequence of covalently incorporated ligands or reagents that serve to elucidate the specific binding sites within the structure. More often, only a part of the protein will be sequenced in order to generate oligonucleotides for the hybridization search in genomic or cDNA libraries. The purpose then is to obtain the sequence of the genome and to derive the primary structure of the carrier. In addition, leader sequences for the protein import are obtained and possibly a further quest into problems such as its genomic organization is undertaken.

2. For the generation of antibodies, which are used for screening carrier-specific mRNA and for the subsequent cloning of cDNA. This second use is widely applied, or rather widely attempted, but will succeed only with sufficient specificity of the antibody mRNA screening.

3. Eventually the denatured carrier protein might be reconverted to structural and functional intactness.

4. An entirely different approach — mentioned briefly here — involves immunoprecipitation. Antibodies are first generated against synthetic peptides containing partial sequences of protein, obtained from cDNA-based structure information. These antibodies are then employed for purification by precipitating the target carrier protein from the solubilizate.

The isolation of denatured protein is relatively straightforward once a specific assay, in particular a covalent marker, is available. Generally the same primary steps as for the isolation of the intact protein, i.e., purification or enrichment of the membranes, should precede the solubilization. Often it is easier first to enrich the protein still in the intact form by using a corresponding detergent and to take advantage of the purification steps possible at this stage, and then to proceed with the isolation of the denatured protein using chromatography, e.g., in formic acid or SDS. If this fails, one should resort to a purification involving highly resolving gel electrophoresis, mostly on SDS, and the extraction of the corresponding protein band from the gel.

Conclusions

The identification of biomembrane carriers serves to characterize the transport processes, their biological implications, and their mechanism.

Together they form the groundwork for kinetic studies, either in the original intact membrane or in the reconstituted system. These include functional studies on the role of activators, inhibitors, and regulators, the influence of membrane potential, the coordination with other transport processes, and the mechanism of transport as described in kinetic terms. The isolation of carriers serves two purposes. First, to generate with the isolated carrier a reconstituted system which allows an in-depth study of its function. Here more leeway for manipulating the conditions is given and a fuller description of the carrier function is possible. Second, to characterize the carrier on a molecular level, which includes first of all structural studies, and on this basis to further the functional studies in order to ultimately understand the carrier mechanism. Interwoven with this approach is the identification of the genome, the elucidation of its structure, and subsequently an understanding of the genomic regulation and organization. This, together with the functional studies, serves to clarify the functional role of the carrier within the cellular and organismic network.

[3] Kinetics of Transport: Analyzing, Testing, and Characterizing Models Using Kinetic Approaches

By WILFRED D. STEIN

Introduction

The aim of this chapter is to provide the reader with a systematic guide to analyzing, testing, and characterizing models for transport systems, using the kinetic approach. First I shall list and discuss the various experimental paradigms that are used to study the kinetics of membrane transport. The systematic use of such procedures gives a surprisingly good qualitative indication as to what type of transport system is present in the experimental situation considered. Various models that have been suggested as being relevant to the study of transport are then discussed, proceeding from the simplest model to quite complex ones. The kinetic predictions for each model are listed. Using the experimental paradigms we show how to reject model after model, until one reaches the simplest model which is *not* rejectable by the available data. The system can then be characterized in terms of the rate constants of the model chosen, using the experimental data. Throughout this chapter the emphasis is on the data and their meaning. Detailed mathematical analysis of the models is available in the references cited.

METHODS IN ENZYMOLOGY, VOL. 171

Experimental Paradigms of Membrane Transport

Kinetics is the study of the rates of movement — in the present context, the rate of movement of a transported substance (the "substrate") across a biological membrane. One needs to measure the amount of substrate that has moved into, or out of, a cell or membrane vesicle in a defined length of time. Amount is generally measured chemically, or by an isotope labeling technique, or by following a change in pH inside or outside the cell, or the change in cell volume as water follows the movement of an osmotically active substance, or by following the movement of a fluorescent substrate. The measurement of time implies that the transported substance and the membrane must be brought together at a defined instant of time and separated from each other at a subsequent, defined instant. Experimental procedures that have been used are detailed in references cited. All of the paradigms discussed below require some such rate measurement to provide the basic experimental data. The various experimental procedures that use these data will now be discussed.

Equilibrium Exchange Procedure

Conceptually, this is the simplest procedure. The cell or vesicle is loaded with a certain concentration of the substrate, placed into the same concentration of substrate (so that an equilibrium exists for this substrate across the cell membrane), and the exchange of isotopic label is then followed by radioactively labeling the inner or outer compartments of the cell or vesicle.[1] Since the system is everywhere in equilibrium, there is no flow of matter or change in cell volume (cell water) to complicate the kinetics. Flow of label into, or out of, the cell is proportional to the concentration of label present in the labeled phase at any time and is, therefore, strictly exponential across each cell membrane. It must, therefore, be characterized by a *single rate constant* whatever the detailed nature of the transport system that is carrying the substrate across the membrane.[2] A plot of the logarithm of the fraction of label remaining (for an efflux experiment) or of the logarithm of 1 *minus* the fraction that remains to enter (for an influx experiment) should be a straight line, the slope of which gives exactly the rate constant, k, for the equilibrium exchange.[2] The equilibrium exchange experiment is an excellent test of two fundamental properties of the system under study.

[1] J. Brahm, *J. Gen. Physiol.* **82,** 1 (1983).
[2] Y. Eilam and W. D. Stein, *in* "Methods in Membrane Biology" (E. D. Korn, ed.), p. 283. Plenum, New York, 1974.

1. The rate constant, k, depends only on the area of the cell or vesicle, its volume, and the capacity of the system to exchange label. If a straight line is not obtained for the experimental plot of the fraction of label remaining, then the conclusion is inescapable that the system is not homogeneous. Either there is a mixture of cells or vesicles present of different surface area-to-volume ratio (that is, a mixture of smaller and larger cells or vesicles), or else the cells themselves differ in the amount of transport system(s) associated with each cell. (The cells may be from a population undergoing differentiation, so that some cells have an effective, while others have a poorer, transport system for the substrate in question.)

2. The system is in equilibrium during the exchange, so that its properties are invariant with time and with direction. The same rate constant for exchange must be obtained for an efflux experiment as for an influx experiment. This *must* be tested for each system and, if the test fails, a search must be initiated for the procedural artifact that has brought about this asymmetry. The cause for the asymmetry could lie in the mixing or stopping procedures, which may be inadequate, or in some trivial dilution or calculation error. A very likely source of asymmetry is that the transport system in question is one of the coupled systems, in which the test substrate and some other substrate, asymmetrically disposed across the membrane, are involved in transmembrane transport. These important coupled transports are discussed later.

The paradigms that will be discussed in the immediately following sections do not yield simple exponential plots nor are they of necessity symmetric, and, therefore, the tests for homogeneity or for the presence of an asymmetric artifact must be made by the equilibrium exchange procedure.

With these tests for linearity and symmetry successfully accomplished, k, the rate constant for equilibrium exchange, should now be determined over a range of concentrations of the substrate. The rate constant may well vary with substrate concentration, as I describe, but at each concentration there must be but a single value for k, the same for influx as for efflux if the system is homogeneous and free from an asymmetric artifact.

Figure 1 depicts in schematic fashion some typical forms, often found in practice, for the variation of rate constant k with substrate concentration S. (Figure 1 also depicts corresponding plots of $1/k$ versus S and of kS, the amount of substrate crossing the cell in unit time, versus S.) I proceed to discuss these, in turn.

k Invariant with S. (See Fig. 1, top row.) This case is characteristic of simple diffusion (or of some transport system far from saturation). The rate

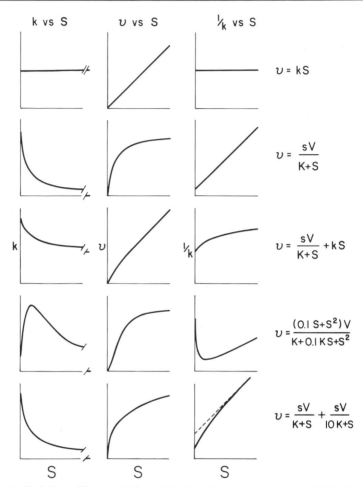

k vs S v vs S $\frac{1}{k}$ vs S

$v = kS$

$v = \dfrac{sV}{K+S}$

$v = \dfrac{sV}{K+S} + kS$

$v = \dfrac{(0.1\,S+S^2)V}{K+0.1\,KS+S^2}$

$v = \dfrac{sV}{K+S} + \dfrac{sV}{10\,K+S}$

S   S   S

FIG. 1. Variation of k, v, and $1/k$ ($=S/v$) with substrate concentration S. The five cases depicted are appropriate to the equations listed on the right of each row. For each plot, K is taken as unity, as is V, so concentrations are in multiples of K and velocities are in multiples of V.

constant k is then proportional to the permeability constant π, for the substrate in question. The product kS increases linearly with S according to the equation: velocity v $(=kS) = \pi S$. A study of how the rates of simple diffusion vary with the nature of the substrate gives much information as to the nature of the membrane's barrier to diffusion (see the chapter by Lieb and Stein[3]). From a knowledge of the simple diffusion characteristics

[3] W. R. Lieb and W. D. Stein, "Transport and Diffusion across Cell Membranes," Chap. 2. Academic Press, Orlando, Florida, 1986.

of the membrane in question, obtained by a systematic study of permeabilities for a broad range of substrate types, one can check whether the rate constant k (that has just been characterized as invariant with substrate concentration) is that predicted for simple diffusion. It can be that k here is far higher than would be expected from a knowledge of the membrane in question. It is then probable that the substrate under study is moving across the membrane by some route other than simple diffusion, that is, by a specific system operating well below its level of saturation. The system should then be further tested by extending the concentration ranger under study, or by the use of a range of transport inhibitors.[4]

k *Falling with S to Zero, $1/k$ Increasing Linearly with S.* (See Fig. 1, row 2.) In this case it is clear that as S increases, each substrate molecule or ion finds itself in an environment in which it becomes increasingly difficult to cross the membrane. Substrate molecules compete with one another for the limited capacity of the transport system, until at limitingly high concentrations, competition is complete and the chances of any particular substrate molecule's being able to find a vacant transporter are zero. The plot of kS, the amount of S transported, versus S is the familiar Michaelis plot $[v = kS = SV^{ee}/(K^{ee} + S)$, where rate constant $k = V_{ee}/(K_{ee} + S)]$, showing a maximum velocity of transport $V_{max} = V_{ee}$ and $K_m = K_{ee}$ for the concentration at which half this maximal velocity is reached. (The subscript ee denotes equilibrium exchange.) The limiting value of k at low substrate concentrations is defined as the limiting permeability π_{ee} ($= V_{ee}/K_{ee}$). The reciprocal plot of $1/k$ versus S is the familiar straight-line plot,[5] with slope of $1/V_{ee}$ and intercept on the abscissa of $-K_{ee}$ and on the ordinate of $1/\pi_{ee}$. Since the equilibrium exchange experiment must give a side-invariant value of k at *any* concentration S, it must do so at *all* concentrations, and the derived parameters V_{ee} and K_{ee} are side invariant, that is, symmetric. If they are not, there is some experimental artifact present.

k *Falling with S to a Nonzero Asymptote.* (See Fig. 1, row 3.) Clearly, in this case, two parallel systems are present. One system saturates with increasing S until it contributes nothing significant to the overall transport. The other system contributes a constant permeability at all substrate concentrations. The overall transport rate v is given, then, as the sum of a saturable and a nonsaturable component. This can be written as $v = kS = SV_{ee}/(K_{ee} + S) + S$. (The permeability π is the value of the asymptote to which k tends as S increases, or is the slope of the plot of v against S, at high concentrations of S. At the lowest substrate concentration, $k = \pi_{ee} + \pi$.)

[4] R. Devés and R. M. Krupka, this volume [5].
[5] I. H. Segel, "Enzyme Kinetics." Wiley, New York, 1975.

Curve-fitting methods enable the three parameters V_{ee}, K_{ee}, and π to be extracted from the experimental data. Just as in the first case, with k invariant with S, one can test whether the nonsaturable component is simple diffusion across the lipid bilayer or occurs via some saturable system operating well below its level of saturation.

k Rising with S and Then Falling as S Increases Further. (See Fig. 1, row 4.) In this case, two (or more) substrate molecules cooperate, rather than compete, to achieve transport. At high levels of S, competition for the limited amount of transport system will in general be found. The curve of v ($=kS$) versus S is sigmoidal. The equation followed is

$$v = kS = (SV_{ee}^1 + S^2 V_{ee}^2)/(K_{ee}^1 + K_{ee}^2 S + S^2)$$

The value of the rate constant k at limitingly low values of S is $\pi_{ee} = V_{ee}^1/K_{ee}^1$, and may be zero if two substrate molecules are mandatory for transport. The maximum velocity of transport at limitingly high values of S is V_{ee}^2. One can obtain these parameters by curve-fitting procedures.

1/k Rising with S along Two Straight Lines. (See Fig. 1, row 5.) It often happens that the plot of $1/k$ versus S is not the simple, rising straight line of the second case given above, but is, rather, best fitted by a pair of straight lines. This finding suggests that at least two saturable transport systems are here carrying substrate across the membrane. The relevant transport equation is

$$v(=kS) = \frac{SV_{ee}^1}{K_{ee}^1 + S} + \frac{SV_{ee}^2}{K_{ee}^2 + S}$$

where the superscripts 1 and 2 refer to the two transport systems concerned. [The rate constant $k = v/S = (A + BS)/(C + DS + S^2)$ where A, B, C, and D are sums and products of the terms in V_{ee} and K_{ee}.] The permeability at the lowest substrate concentrations is given by $\pi_{ee}^1 + \pi_{ee}^2$. The transport parameters for the respective systems can be found by curve-fitting procedures, best performed over a low and high range of substrate concentrations.

These cases comprise most of the situations encountered experimentally. As noted, from the different cases, the experimental parameters π, K_{ee}, V_{ee}, and π_{ee} ($= V^{ee}/K^{ee}$, for the simple Michaelis behavior) can be found from the data and are then available for the further characterization of the system as described below.

Zero Trans Procedures

This valuable procedure is conceptually simple but is fraught with technical difficulties.[2] Transport is measured into, or out of, the cell or vesicle under conditions in which one face of the membrane (the trans face) is free of the substrate in question. The substrate concentration at the other (cis) face is varied and the rate of zero trans transport determined. The technical difficulties arise from the fact that the substrate concentrations at both faces of the membrane are necessarily changing during the course of the experiment. It is impossible to maintain the substrate concentration at zero within the cell or vesicle during uptake (but see Ref. 6 for the use of an enzymatic substrate-trapping procedure). Transport determinations must be made extremely rapidly when only insignificant amounts of substrate have entered the cell (yet in sufficient amounts to measure uptake!), or else integrated rate equation treatments must be undertaken,[2,7,8] treatments which themselves must make assumptions as to the nature of the transport system being investigated. It is easier to maintain the trans concentration at zero if efflux is studied in a large volume of trans medium, but this dilutes the substrate that has left the cell and makes its measurement difficult. The alternative is to measure the (necessarily) small differences in the amount of substrate that remains within the cell.[2] A second problem is that there is no simple test for the homogeneity of the cell preparation being used and artifacts may arise from a lack of homogeneity (which the equilibrium exchange procedure would show up). Third, zero trans measurements may very well be asymmetric, the transport systems themselves being in general asymmetric. It is not clear from the data whether any such asymmetry is real or is an artifact arising out of the technical situation in which quite different experimental procedures are often used for zero trans efflux and influx studies (a large volume of external medium in the former case, an integrated rate equation method in the latter). Finally, there is necessarily net movement of substrate in the zero trans procedure and this can be associated with the (osmotic) movement of water or with pH changes that can themselves affect the measuring system. These effects must be assessed and, if necessary, guarded against.

Systematic zero trans studies have, however, been performed.[2,9,10] The velocity of zero trans movement is determined as a function of substrate concentration. The zero trans movement may or may not saturate with substrate concentration. Plots such as those in Fig. 1 are, in general, found. We can define a permeability π^{zt} for the zero trans (zt) procedure, the ratio of velocity to substrate concentration at limitingly low concentrations of substrate, a maximum velocity of transport in each direction (V^{zt}_{12} and V^{zt}_{21}), and a corresponding half-saturation concentration K^{zt}_{12} and K^{zt}_{21}, and

[6] R. D. Taverna and R. G. Landon, *Biochim. Biophys. Acta* **298,** 412 (1973).

[7] W. R. Lieb and W. D. Stein, *J. Theor. Biol.* **69,** 311 (1977).

[8] B. A. Orsi and K. F. Tipton, this series, Vol. 63, p. 159.

[9] S. J. D. Karlish, W. R. Lieb, D. Ram, and W. D. Stein, *Biochim. Biophys. Acta* **255,** 126 (1972).

[10] L. Lacko, B. Wittke, and H. Kromphardt, *Eur. J. Biochem.* **25,** 447 (1972).

obtain these parameters by appropriate curve-fitting procedures.[9] There may also be a nonsaturable permeability π, dominant at high substrate concentrations.

A transport system may display parameters for zero trans movement in the 1-to-2 direction quite different from those in the 2-to-1 direction. Thus, in general, $V_{12}^{zt} \neq V_{21}^{zt}$ and $K_{12}^{zt} \neq K_{21}^{zt}$. But it will always be the case, for a system that transports only a single substrate, that the permeability π^{zt}, measured as the limiting slope (at low substrate concentration S) of the saturable component, is symmetric. The proof of this assertion is as follows. We have already seen that π^{ee}, for the equilibrium exchange procedure, is symmetric. But π^{ee} is given as the slope of the curve of the amount of substrate flowing versus substrate concentration, at limitingly low substrate concentration. It is measured when, effectively, there is substrate at neither face of the membrane. π^{zt} is, however, measured under precisely the same conditions—where S is set at zero at the trans face and is extrapolated to zero at the cis face. Thus π^{ee} and π^{zt} are measurements of the same flux measured under identical conditions, and must be identical. This is so for a zero trans measurement in the 1-to-2 direction, and also for one in the 2-to-1 direction. We have, therefore, $\pi^{ee} = \pi_{12}^{zt} = \pi_{21}^{zt}$, as a test of the adequacy of our experimental system. This is a model-independent test and will fail only if the movement of the substrate considered is coupled to the movement of some other substrate. Indeed, the test will indicate that coupling is occurring, if true artifacts can be ruled out. If the data pass the test of the identity of π values, the measured zero trans parameters for V_{max} and K_m can be used together with those from the equilibrium exchange experiment in further tests and in characterization of the transport system, as discussed later.

Infinite Trans Procedure

This procedure is often used. Cells (or vesicles) are loaded with a high concentration of unlabeled substrate and are then placed in a medium containing labeled substrate.[11] The latter crosses the membrane, enters the cell, and its specific activity is there greatly reduced by the unlabeled substrate present, which competes with label for efflux. Thus, for the saturable component of entry, the exchange of label for unlabeled substrate brings about a virtually one-way flux of substrate. Uptake of label is generally linear with time for long periods and is easily measured.

Uptake can be studied as a function of the substrate concentration at the outer face of the membrane and characterized in terms of a maximum velocity of uptake V_{12}^{it} and a K_{12}^{it} (where the superscript stands for infinite

[11] H. Ginsburg and D. Ram, *Biochim. Biophys. Acta* **382**, 369 (1975).

trans), the concentration at which half this maximum velocity is reached. For the subsequent interpretation of these parameters in terms of the models for transport, it is essential that the concentration of unlabeled substrate at the trans face be *effectively* infinite. It must be shown that a further increase in the concentration of unlabeled substrate at the trans face makes no discernible change in the transport parameters V_{12}^{it} and K_{12}^{it}. If such a change is found on increasing the trans concentration, the determinations must be repeated, iteratively, until no further change is found. A further test is the obvious one that the maximum velocity of the infinite trans experiment must equal that of the equilibrium exchange procedure, since, in both, substrate is at limitingly high levels at both membrane faces, and a unidirectional flux is being measured. Thus $V^{ee} = V_{12}^{it} = V_{21}^{it}$, and the maximum velocity of the infinite trans experiment is necessarily symmetric, if a single substrate is carried by the transporter and no artifact is present. (Compare this with the findings and logic for the permeabilities in π, measured at *low* substrate concentrations, for the zero trans and equilibrium exchange experiments.) The half-saturation concentrations are not, in general, symmetric and there is no reason that they should be. The infinite trans experiment for efflux is particularly difficult to perform for technical reasons. It *must* be performed into very large concentrations of the unlabeled substrate. This flows into the small volume comprising the cell or vesicle interior, diluting the specific activity of the labeled substrate within the cell, and rendering suspect any attempts to measure efflux accurately from low concentrations of internal substrate. (Note the contrast to the zero trans procedure. There efflux was relatively easy to measure, uptake difficult. Here the reverse is true.) It is often found that the infinite trans uptake is faster than the zero trans flow. The significance of this fact for modeling of transport is discussed below. Note that the saturable component of the overall uptake is emphasized in the infinite trans procedure and, therefore, it seldom happens that a nonsaturable component is seen in parallel with the saturable. Thus curve-fitting is seldom needed to extract the V^{it} and K^{it} parameters.

Infinite Cis Procedure

This procedure uses net movements of substrate rather than the exchange of label with nonlabel, but, as we shall see below, yields transport parameters which can be identical with those for the infinite trans case. Once again, cells or vesicles are loaded with (limitingly) high concentrations of unlabeled substrate and placed in media containing various concentrations of unlabeled substrates, ranging from zero to quite high values.

The net loss of substrate from the cell (or vesicle) is measured.[12,13] Cis is now interior, trans is exterior. In the absence of substrate at the trans face, one is measuring merely $V_{21}^{\pi t}$, the maximum velocity in a zero trans experiment. At a certain concentration of substrate at the trans face, net efflux will be so countered by influx that the net efflux will be reduced to one-half of $V_{12}^{\pi t}$. This concentration defines K_{21}^{ic}, the half-saturation concentration for the infinite cis (ic) experiment, generally found by interpolation in a plot of net efflux against trans substrate concentration. The experiment is easy to perform in the efflux direction and forms the basis of many fine studies of, especially, red cell transport.[12,14] It is difficult to perform for influx but full studies of red cell sugar uptake have been made by integrated rate equation methods[7] and near-initial rate studies.[13] For the infinite cis paradigm, both maximum velocities $[V_{12}^{ic}\ (=V_{12}^{\pi t})$ and V_{21}^{ic} $(=V_{21}^{\pi t})]$ and half-saturation concentrations $(K_{12}^{ic}$ and $K_{21}^{ic})$ can be, and often are, asymmetric.

Counterflow

This is a variant of the infinite trans experiment and, similarly, is an excellent criterion for the operation of the "carrier" model that we shall discuss below. Cells are loaded with high levels of unlabeled substrate and are placed into medium containing a low concentration of the labeled substrate, or a low concentration of a second substrate that may be using the same transport system as the first. The entry of label (or second substrate) is followed over a period of time.[15] Labeled substrate enters the cell in exchange for unlabeled substrate. The specific activity of the label is greatly reduced as it enters the cell and mixes with the unlabeled substrate present. Label (or the second substrate) is thus, in effect, trapped within the cell and eventually reaches high concentrations within the cell, higher than are present outside. It is concentrated within the cell, driven against its concentration gradient by the outflow of unlabeled substrate. As we shall see, no simple diffusion or even channel system (in the absence of a transmembrane potential) can perform such a feat. With the passage of time, the specific activity of labeled substrate within the cell rises and eventually reaches that of the outside. Label begins to flow out of the cell and the transient accumulation of label within the cell decays to zero. An important parameter of the membrane's transport system, its side-

[12] A. K. Sen and W. F. Widdas, *J. Physiol. (London)* **160,** 392 (1962).
[13] H. Ginsburg and W. D. Stein, *Biochim. Biophys. Acta* **382,** 353 (1975).
[14] R. Devés and R. M. Krupka, *Biochim. Biophys. Acta* **510,** 339 (1978).
[15] G. F. Baker and W. F. Widdas, *J. Physiol. (London)* **231,** 143 (1973).

weighted average affinity for substrate, can be computed from the value of the accumulation ratio at the instant that it reaches its peak.[16]

This completes the catalogue of the paradigms of membrane transport. We have seen how to collect the variety of transport parameters that characterize these paradigms: V^{ee}, K^{ee}, V^{zt}_{12}, V^{zt}_{21}, K^{zt}_{12}, K^{zt}_{21}, K^{it}_{12}, K^{it}_{21}, K^{ic}_{12}, and K^{ic}_{21} and the value of the nonsaturable component. In the next section, we analyze the various transport models and show how the paradigms just described can be used as touchstones to distinguish among the various models and how the parameters can then be used to characterize the properties of the models.

Models of the Membrane Transport Process

Simple Diffusion

This is the simplest case. For a system in which the rate constant k is invariant with substrate concentration, that is, in which the rate at which substrate is transported is always directly proportional to substrate concentration, for all the experimental paradigms, and for which no specific inhibitor of transport is found, one may suspect that simple diffusion is occurring. Here transport is by solubilization in, and diffusion across, the lipid bilayer of the cell or vesicle membrane. If simple diffusion is indeed the mechanism of transport, the solubility–diffusion model must be shown to apply. The rate of transport, k, must be predictable from the properties of the membrane in question. We have shown[3] how it is possible to characterize a cell membrane in terms of its solvent properties and in terms of the diffusion barrier that it presents to intramembrane diffusion of permeating substances. For the human red blood cell, for instance, we have shown that the barrier to simple diffusion has solvent properties similar to those of the hydrocarbon hexadecane, and that the barrier discriminates between diffusants according to a high power of the molecular volume, behaving much as does a soft polymeric sheet.[3] Membranes that have not been so well characterized can be, on measuring the permeability properties of a well-chosen range of permeants, and on analyzing the data as in the chapter by Lieb and Stein.[3] If the rate of transport of a particular substance is far higher than that predicted by the simple diffusion properties of the cell membrane in question, it is probably not being transported by that process, and one must go on to test the next most complex model, the simple channel.

[16] W. D. Stein, "Transport and Diffusion across Cell Membranes." Academic Press, Orlando, Florida, 1986.

The Simple Channel

In this model, substrate interacts with some membrane component which provides a route for transport. The membrane component is strictly *inert* as far as the substrate is concerned. The channel is either empty or else is occupied by the substrate but these are the only states available to the channel. (In a subsequent section we discuss the carrier, which, in contrast, undergoes a conformation change between two orientations as the substrate is transported.) The channel, may, however, be more or less selective for the substrate S, or for a range of substrates, discriminating between would-be permeants by virtue of their size of charge.

Analyzing and Testing the Simple Channel. Figure 2 is a schematic model of the channel and of the formal kinetic scheme which describes its action. E is the channel, S the substrate. The rate constants in b and f describe the breakdown or formation of the ES complex at the two faces of the membrane. Table I gives the steady-state kinetic solution of this model[16] as v (the rate of transport), in terms of the rate constants of Fig. 2, which appear in combination in experimentally accessible parameters in Q and R. What do we mean by "experimentally accessible?" Consider the equation at the top of Table I. If we want the zero trans flow in the 1-to-2 direction, we substitute $S_2 = 0$ in this equation. Then, dividing by R_{12}, we obtain $v_{1 \to 2} = (S_1 \times 1/R_{12})/[(Q/R_{12}) + S_1]$. This is the Michaelis form of Fig. 1, row 2 $[v = SV_{\max}/(K_m + S)]$, with V_{12}^{zt} of $1/R_{12}$ and K_{12}^{zt} of Q/R_{12}, and π^{zt} of $1/Q$. We derive Q, R_{12}, and, from the zero *trans* experiment in the 2-to-1 direction, Q again and R_{21}. Thus Q, R_{12}, and R_{21} are experimentally accessible parameters. They define the equation at the head of Table I totally and hence characterize all transport experiments on the simple channel of Fig. 2. It is easy to deal with the experimental paradigms

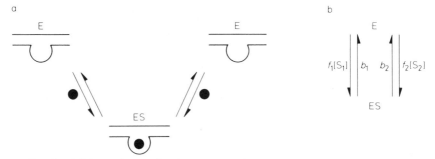

FIG. 2. (a) Scheme for the simple channel, E, interacting with its substrate, S, and (b) formal representation of the scheme in terms of the relevant rate constants as depicted. Subscripts 1 and 2 refer to sides 1 and 2 of the membrane, respectively.

and their interpretation in terms of the Q and R parameters. Take, for instance, the equilibrium exchange experiment where the substrate concentrations are the same as both membrane faces, so that $S_1 = S_2 = S$. Substituting in Table I, we obtain $v_{12} = S/1/(R_{12} + R_{21})/[Q/R_{12} + R_{21}) + S]$. This is a Michaelis form with $V^{ee} = 1/(R_{12} + R_{21})$, $K^{ee} = Q/(R_{12} + R_{21})$, and $\pi^{ee} = 1/Q$. With these solutions for V^{ee}, V_{12}^{zt}, and V_{21}^{zt}, it follows that for transport on the simple channel of Fig. 2, $1/V^{ee} = 1/V_{12}^{zt} + 1/V_{21}^{zt}$ and $1/K^{ee} = 1/K_{12}^{zt} + 1/K_{21}^{zt}$. The maximum velocity for equilibrium exchange is, therefore, always less than, or at most just equal to, the maximal velocity of zero trans movement. A channel is blockable from either end and hence shows a maximum rate of transport when there is *no* substrate at the trans face. This is the distinguishing characteristic of the simple channel. It follows, therefore, that there is no meaning to the infinite trans paradigm for a simple channel. If a channel is saturated at the trans face (and this is the defining situation for the infinite trans paradigm), there is, at finite concentration of substrate at the cis face, no possibility of labeled substrate being transported through the system from cis to trans, against the flow of substrate from the trans face. There is, likewise, no possibility of seeing the counterflow phenomenon (except for certain situations where the transmembrane potential causes an apparent coupling of ion flows[17]).

[17] H. Garty, J. Rudi, and S. J. D. Karlish, *J. Biol. Chem.* **258**, 13049 (1983).

TABLE I

STEADY-STATE VALUES FOR TWO FORMS OF THE SIMPLE CHANNEL[a]

$$v_{1\to 2} = \frac{[S_1]}{Q + R_{12}[S_1] + R_{21}[S_2]}$$

	One state for ES	Two states for ES
$nR_{12} =$	$1/b_2$	$1/b_2 + 1/g_1 + g_2/b_2 g_1$
$nR_{21} =$	$1/b_1$	$1/b_1 + 1/g_2 + g_1/b_1 g_2$
$nQ =$	$1/f_1 + 1/f_2$	$1/f_1 + 1/f_2 + b_1/f_1 g_1$
Constraint	$b_1 f_2 = b_2 f_1$	$b_1 f_2 g_2 = b_2 f_1 g_1$

[a] These results are for the special case of an uncharged (neutral) permeant of Fig. 2; $v_{1\to 2}$ is the unidirectional flux of permeant from solution 1 to solution 2; the reverse flux $v_{2\to 1}$ is obtained by simply interchanging subscripts 1 and 2 in the equation. $[S_1]$ and $[S_2]$ are the concentrations of permeant in solutions 1 and 2, while n is the total concentration of pores. Notice that the steady-state equations for $v_{1\to 2}$ and $v_{2\to 1}$ are independent of the form used when expressed in terms of the observable transport parameters Q, R_{12}, and R_{21}. From Lieb.[22]

The simple channel shows no trans stimulation of transport. (In the equation of Table I, the unidirectional flux v_{12} is steadily *reduced* as the trans concentration S_2 increases.) The ratio of unidirectional fluxes v_{12}/v_{21} is given exactly by the ratio of concentrations S_1/S_2, for all concentrations.

A useful criterion for the applicability of the simple channel proceeds as follows. Measure the velocity of transport from side 1 to side 2 (v_{12}) as a function of the concentration at side 1 (S_1), and at various concentrations of substrate at side 2 (S_2), Plot $1/v$ against $1/S_1$ for each value of S_2. It will be seen, by inverting the equation at the head of Table I, that one obtains the equation $1/v_1 = (Q + R_{21}S_2)/S_1 + R_{12}$. This defines a pencil of lines, all intersecting at a single point on the $1/S_1$ axis to give a single value of the maximum velocity, $1/R_{12}$, at all concentrations of substrate S_2 at the trans face. For the simple channel, S_2 increases the half-saturation concentration, K_m, but not the maximum velocity of transport. It acts as a "competitive inhibitor," in the terminology of the enzymologist.

Characterizing the Simple Channel. If a transport system shows saturation, so that transport is not occurring by simple diffusion, and yet does not show transstimulation of transport, it may be a simple channel. If, on measuring the transport parameters for equilibrium exchange and for the zero trans paradigms, it can be shown that $1/V^{\text{ee}} = 1/V_{12}^{\text{zt}} + 1/V_{21}^{\text{zt}}$, and $1/K^{\text{ee}} = 1/K_{12}^{\text{zt}} + 1/K_{21}^{\text{zt}}$, the system is certainly behaving as a simple channel. It can then be characterized as such. The ratio of V_{12}^{zt} to V_{21}^{zt} gives the ratio of the channel–substrate dissociation constants, the b terms in Fig. 2, and the absolute values of these terms can be determined if the channels can be counted (fixing n in Table I). From the value of Q and the constants listed in Table I, the association constants f_1 and f_2 can then be found. These latter numbers, if known absolutely, can be compared with the values predicted for diffusion-limited chemical reactions,[16] and a deep analysis of the behavior of the channel can be attempted. (Hille[18] should be consulted for an enlightening approach to this problem for ion channels.) Clearly a channel can be asymmetric—there is no reason why b_1 should equal b_2 or f_1 should equal f_2. However, the dissociation constant b_1/f_1 at side 1 must equal that at side 2, b_2/f_2. This is because there is only one form of the empty channel E, and only one form of the occupied channel ES. The free energy for formation of the complex (given by the difference between the standard-state free energies of ES and E) is, therefore, the same at each face of the membrane.

[18] B. Hille, "Ionic Channels of Excitable Membranes." Sinauer, Sunderland, Massachusetts, 1984.

The Complex Channel

Single-Occupancy Channel. There can be two or more forms of the (channel + substrate) complex, ES, namely, ES_1, ES_n, $\cdots$, ES_2, where ES_1 and ES_2 are the forms that are exposed to sides 1 and 2 of the membrane respectively, while ES_n represents one of a set of intermediate forms or states during the movement of substrate S through the channel from one side to the other. This defines the complex single-occupancy channel. It can readily be shown[16,18] that the same equation is derived for all the various complex single-occupancy channel models, but the interpretations of the experimentally accessible transport parameters in Q and R differ for the various cases (Table I gives the interpretation where ES_1 and ES_2 are the intermediate forms). Thus a mere determination of saturation behavior or of the applicability or otherwise of the test that $1/V^{ce} = 1/V_{12}^{ct} + 1/V_{21}^{ct}$ does not distinguish between the simple and complex forms of the channel. In particular, for both of these forms of the channel, the ratio of unidirectional fluxes v_{12}/v_{21} is given exactly by the ratio of substrate concentrations S_1/S_2. Distinction can be made, however, by using more subtle criteria based on the use of the voltage dependence of the channel's kinetic parameters.[16,18]

Multiple-Occupancy Channel. However, a channel may bind, or be occupied by, more than one substrate at any time, with the formation of double-occupancy forms such as S_1ES_2, or even triply or multiply occupied forms. These ions or molecules will interfere with one another's movement across the membrane. The kinetic relations for such multiple-occupancy models are quite complex.[16,18,19] Their distinguishing characteristic is that the ratio of the unidirectional fluxes v_{12}/v_{21} in such a model is not that given by the equation $v_{12}/v_{21} = S_1/S_2$, but some higher power of the ratio of substrate concentrations [namely, $v_{12}/v_{21} = (S_1/S_2)^n$, where n is related to the number of substrate particles than can be present within the channel]. Additional substates of the multiple-occupancy channels can be shown.

Hess and Tsien[20] studied a calcium channel from guinea pig heart ventricle cells. These cells showed a calcium flux (measured as a flow of current) which exhibited the typical Michaelis behavior of Fig. 1 and half-saturated at ~ 10 mM calcium. There is nothing surprising about this. But when Hess and Tsien[20] studied the movement of sodium ions through these same channels, they found that calcium blocked the (slow) sodium flux at a concentration of some 1 to 2 μM. Why should calcium ions compete with one another 1000 times less effectively than they compete

[19] O. S. Andersen, this volume [4].
[20] P. Hess and R. Y. Tsien, *Nature (London)* **309**, 453 (1984).

with sodium ions? Hess and Tsien[20] suggested that this channel may be of multiple occupancy for calcium ions, with two calcium ions binding to the channel with different affinities, the first tightly, the second poorly. Binding of a second calcium ion is possible at low affinity (10 mM) to a channel that has already bound a calcium ion tightly. The binding of this second calcium ion brings about the ejection at the trans face of the first ion bound, and the transmembrane movement of the first calcium ion is recorded as a conductance event. The relative affinities of the first and second ions bound fit with a model in which the first (doubly charged) calcium ion binds to, and charges, the channel, reducing the affinity of the second ion to be bound. The rate constants of transmembrane movement can be extracted from the data. A similar approach has been used to study the gramicidin channel, also of double occupancy for certain ions.[19]

The Simple Carrier

Figure 3 depicts the model of the simple carrier and Table II records the kinetic solution of this model in terms of the rate constants in Fig. 3.[16] The model is the iso Uni Uni reaction scheme of the enzymologist.[5] The essential features of this model is that the *free* carrier exists in two interconvertible conformations E_1 and E_2, with substrate-binding sites accessible at side 1 and at side 2 of the membrane, respectively. The equilibria between substrate and the carrier's binding sites at one face of the membrane are shielded from those at the other face, and a definite, measurable energy barrier exists for the interconversion of the two forms E_1 and E_2. The existence of this barrier distinguishes the carrier from the simple and complex channels discussed earlier, and gives the carrier its unique properties and, in the more complex versions to be discussed in later sections, allows it to bring about the coupling of the flows of two substrates, or of the flow of a substrate and the progress of a chemical reaction. (Certain authors[21] find it useful to emphasize the similarity of channels and carriers, in that the carrier can be thought of as a channel which can undergo a conformation change between two forms E_1 and E_2. I prefer to maintain the distinction between channels and carriers, reserving the former term for those in which there is *no* conformation change that converts the substrate-binding sites from being accessible first at one and then at the other face of the membrane. So many differences in biologically important properties are linked to this new property of the carrier that a distinction between channel and carrier seems to me to be very meaningful.) As Fig. 3

---

[21] P. Läuger, *in* "Ion Channels: Molecular and Physiological Aspects" (W. D. Stein, ed.), p. 309. Academic Press, Orlando, Florida, 1984.

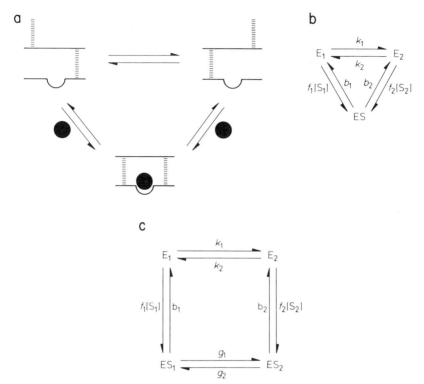

FIG. 3. (a) Scheme for the simple carrier, (b) formal representation of the model with one state of the carrier–substrate complex, and (c) the model with two such states. Definitions of Fig. 2 apply.

TABLE II

STEADY-STATE VALUES FOR TWO FORMS OF THE SIMPLE CARRIER[a]

$$v_{1 \to 2} = \frac{(K + [S_2])[S_1]}{K^2 R_{oo} + K R_{12}[S_1] + K R_{21}[S_2] + R_{ee}[S_1][S_2]}$$

	One state for ES	Two states for ES
$nR_{12} =$	$1/b_2 + 1/k_2$	$1/b_2 + 1/k_2 + 1/g_1 + g_2/b_2 g_1$
$nR_{21} =$	$1/b_1 + 1/k_1$	$1/b_1 + 1/k_1 + 1/g_2 + g_1/b_1 g_2$
$nR_{ee} =$	$1/b_1 + 1/b_2$	$1/b_1 + 1/b_2 + 1/g_1 + 1/g_2 + g_1/b_1 g_2 + g_2/b_2 g_1$
$nR_{oo} =$	$1/k_1 + 1/k_2$	$1/k_1 + 1/k_2$
$K =$	$k_1/f_1 + k_2/f_2$	$k_1/f_1 + k_2/f_2 + b_1 k_1/f_1 g_1$
Constraint	$b_1 f_2 k_1 = b_2 f_1 k_2$	$b_1 f_2 g_2 k_1 = b_2 f_1 g_1 k_2$

[a] These results are for the special case of an uncharged (neutral) permeant. Definitions as in Table I and in Fig. 3. From Lieb.[22]

TABLE III
INTERPRETATION OF EXPERIMENTAL DATA IN TERMS OF BASIC OBSERVABLE
PARAMETERS FOR THE SIMPLE CHANNEL AND CARRIER[a]

Procedure	V_{max}	K_m	Maximum permeability (π)
Simple channel			
Zero trans	$V^{zt}_{1 \to 2} = 1/R_{12}$	$K^{zt}_{1 \to 2} = Q/R_{12}$	$1/Q = \pi^{zt}_{1 \to 2}$
	$V^{zt}_{1 \to 2} = 1/R_{21}$	$K^{zt}_{1 \to 2} = Q/R_{21}$	$1/Q = \pi^{zt}_{2 \to 1}$
Equilibrium exchange	$V^{ee} = 1/R_{ee}$	$K^{ee} = Q/R_{ee}$	$1/Q = \pi^{ee}$
Constraint	$R_{ee} = R_{12} + R_{21}$		
Simple carrier			
Zero trans	$V^{zt}_{1 \to 2} = 1/R_{12}$	$K^{zt}_{1 \to 2} = K(R_{oo}/R_{12})$	$1/KR_{oo} = \pi^{zt}_{1 \to 2}$
	$V^{zt}_{2 \to 1} = 1/R_{21}$	$K^{zt}_{2 \to 1} = K(R_{oo}/R_{21})$	$1/KR_{oo} = \pi^{zt}_{2 \to 1}$
Infinite trans	$V^{it} = v^{it}_{1 \to 2}$	$K^{it}_{1 \to 2} = K(R_{21}/R_{ee})$	$1/KR_{21} = \pi^{it}_{1 \to 2}$
	$= v^{it}_{2 \to 1} = 1/R_{ee}$	$K^{it}_{2 \to 1} = K(R_{12}/R_{ee})$	$1/KR_{12} = \pi^{it}_{2 \to 1}$
Infinite cis	$V^{ic}_{1 \to 2} = 1/R_{12}$	$K^{ic}_{1 \to 2} = K(R_{12}/R_{ee})$	—
	$V^{ic}_{2 \to 1} = 1/R_{12}$	$K^{ic}_{2 \to 1} = K(R_{21}/R_{ee})$	—
Equilibrium exchange	$V^{ee} = 1/R_{ee}$	$K^{ee} = K(R_{oo}/R_{ee})$	$1/KR_{oo} = \pi^{ee}$
Constraint	$R_{ee} + R_{oo} = R_{12} + R_{21}$		

[a] From Lieb.[22]

emphasizes schematically, there is no need for the carrier to move bodily through the membrane (although the substrate must do so). All that is required is for the substrate-binding sites to alter in their transmembrane accessibility. Just as the simple channel occupied with substrate could exist in (at least) two forms, ES_1 and ES_2, so can the simple carrier exist in a multiplicity of transition states ES_1, ES_n, $\cdots$, ES_2. The solution for the two-state ES_1/ES_2 model is also recorded in Table II. The form of the solution is the same for the ES as for the ES_1/ES_2 model, but there is a difference in the interpretation of the experimentally accessible parameters in K and R, in terms of the rate constants of the model in b, f, and k, and g for the ES_1/ES_2 case, where

$$ES_1 \underset{g_2}{\overset{g_1}{\rightleftharpoons}} ES_2$$

Analyzing the Simple Carrier Model. The lower half of Table II records the interpretation of the various experimental paradigms of transport in terms of the experimentally accessible parameters in K and R of the simple carrier. The solutions for the zero trans and equilibrium exchange paradigms are easily derived by, respectively, putting S_1 or $S_2 = 0$, or putting $S_1 = S_2 = S$, in the equation at the top of Table II. The solution for the infinite trans case is achieved as follows: We let S_2 go to infinity (in the equation at the top of Table II) and, after simplifying, obtain the infinite trans unidirectional fluxes $V^{it}_{12} = S_1/(KR_{21} + R_{ee}S_1) = (S_1/R_{ee})/[(KR_{21}/$

$R_{ee}) + S_1$]. This is a typical Michaelis form with maximum velocity of $1/R_{ee}$, and with KR_{21}/R_{ee} as the concentration at which half this maximum velocity is achieved. We note three features of this treatment. First, the simple carrier model of Fig. 3 (and the equation at the top of Table II) enables us, for the first time in our discussion so far, to account for the infinite trans experiment. The fact that the free carrier exists in two states E_1 and E_2 means that the interaction of substrate with its binding site at the trans face of the membrane is shielded from the interaction at the cis face. Thus substrate can find free carrier at the cis face, interact with it, and cross the membrane in an infinite trans flow. Such an explanation eluded us for the simple channel. The existence of an infinite trans flow thus refutes the simple channel and allows the simple carrier model. Second, the maximum velocity of infinite trans exchange, $1/R_{ee}$, is exactly the maximum velocity of the equilibrium exchange experiment—a fact which we derived without reference to any model in discussions of experimental paradigms. Third, the magnitude of this maximum velocity $1/R_{ee}$ can now, on the simple carrier, be bigger (or even far bigger) than the magnitude of the zero trans maximum velocities. (Proof: From Table II, $R_{ee} = R_{12} + R_{21} - R_{oo}$, so that $1/V_{ee} = 1/V_{12}^{zt} + 1/V_{21}^{zt} - R_{oo}$. The term R_{oo} can be very large in comparison with R_{12}, R_{21}, and R_{ee}, so that $1/V_{ee}$ can be far less than $1/V_{12}^{zt}$ and $1/V_{21}^{zt}$.) Consider the equation at the top of Table II, the kinetic solution for the simple carrier. As S_2, the trans concentration, is increased, the unidirectional flux $v_{1\rightarrow2}$ can increase at any cis concentration, S_1, so the simple carrier (unlike the channel) shows as its distinguishing characteristic trans stimulation of a unidirectional flux.

Similarly, the kinetic solution for the infinite cis experiment, listed in Table III, can readily be derived by calculating the net flux in the 1-to-2 direction as resulting from the unidirectional flux in this 1-to-2 direction *minus* the flux in the 2-to-1 direction.[16] Since both infinite trans and infinite cis paradigms are defined for the situation where substrate is at infinite (that is, limitingly high) concentration at one face of the membrane and varied at the other, they measure the same half-saturation concentration and, therefore, $K_{12}^{it} = K_{21}^{ic}$ and $K_{21}^{it} = K_{12}^{ic}$.

A plot of $1/v_{12}$ versus $1/S_1$ at various values of the trans concentration S_2 gives a pencil of lines that, in general, intersect neither on the ordinate nor on the abscissa. (Proof: Find the reciprocal of $1/v_{12}$ in Table II and substitute and simplify to obtain the equation

$$\frac{1}{v_{12}} = \frac{KR_{oo} + R_{21}S_2}{K + S_2} \times \frac{K}{S_1} + \frac{KR_{12} + R_{ee}S_2}{K + S_2}$$

This will have invariant intercepts on the ordinate only if $R_{12} = R_{ee}$, and on the abscissa only if $R_{12} = R_{ee}$ *and* $R_{oo} = R_{21}$.) This corresponds with the case of uncompetitive "inhibition" studied by the enzymologist.[5]

Testing the Simple Carrier Model. If a transport system shows saturation by any of the experimental paradigms discussed earlier, it may well be a simple channel or a simple carrier. If it shows transstimulation, it cannot be a channel (if effects of transmembrane potential[17] are excluded), but may yet be a carrier. How can we test whether it does indeed behave as a simple carrier? We can apply consistency tests, as follows.

There are only four independent experimentally accessible transport parameters in the solutions of Table III and the lower half of Table II, namely, K and three of the four parameters in R (since these are connected by the relationship $R_{ee} + R_{oo} = R_{12} + R_{21}$). Yet there are many more than four ways of determining these. There are, for instance, seven measurable half-saturation concentrations, each of which contains the parameter K. (There is the K_m for the equilibrium exchange experiment and two values each of K_m, for the 1-to-2 and 2-to-1 directions, for each of the zero trans, infinite trans, and infinite cis experiments.) Similarly there are seven ways to determine a maximum velocity, although, in principle, the three ways to determine V^{ee} are equivalent since $V_{12}^{it} = V_{21}^{it} = V^{ee}$, or similarly the zero trans maximum velocities are found also in the infinite cis experiment, since $V_{12}^{zt} = V_{12}^{ic}$ and $V_{21}^{zt} = V_{21}^{ic}$. Lieb[22] showed how it was possible to use the redundancy in this information to test the carrier model. Tables IV and V show the values obtained for the parameter K using all the available information on the uridine transport system of the human red cell. The value K is derived as follows: values of R_{12}, R_{21}, and R_{ee} are first obtained as the reciprocal of the maximum velocities of the relevant transport experiments, and R_{oo} calculated from the relation $R_{oo} = R_{12} + R_{21} - R_{ee}$. Then, from each experimentally determined half-saturation parameter

[22] W. R. Lieb, *in* "Red Cell Membranes: A Methodological Approach" (J. C. Ellory and J. D. Young, eds.), p. 135. Academic Press, London, 1982.

TABLE IV
TESTING THE CARRIER MODEL: URIDINE TRANSPORT ACROSS THE
HUMAN RED BLOOD CELL MEMBRANE[a]

Procedure	Direction	V_{max} (mM/min)	K_m (mM)
		Experimental values ($\pm$ SE)[b]	
Zero trans	In $\rightarrow$ out	$V_{1\rightarrow2}^{zt} = 1.98 \ (\pm0.31)$	$K_{1\rightarrow2}^{zt} = 0.40 \ (\pm0.12)$
	Out $\rightarrow$ in	$V_{2\rightarrow1}^{zt} = 0.53 \ (\pm0.038)$	$K_{2\rightarrow1}^{zt} = 0.073 \ (\pm0.069)$
Infinite cis	In $\rightarrow$ out	$V_{1\rightarrow2}^{ic} \equiv V_{1\rightarrow2}^{zt}$	$K_{1\rightarrow2}^{ic} = 0.252 \ (\pm0.096)$
	Out $\rightarrow$ in	$V_{2\rightarrow1}^{ic} \equiv V_{2\rightarrow1}^{zt}$	$K_{2\rightarrow1}^{ic} = 0.937 \ (\pm0.226)$
Equilibrium exchange	Both	$V^{ee} = 7.54 \ (\pm0.45)$	$K^{ee} = 1.29 \ (\pm0.11)$

TABLE IV   (continued)

Derived resistance parameters ($\pm$ SE) (min/mM)	(msec)
$R_{12} = (V^{zt}_{1\to2})^{-1} = 0.505\ (\pm 0.079)$	$nR_{12} = 10.1\ (\pm 1.6)$
$R_{21} = (V^{zt}_{2\to1})^{-1} = 1.887\ (\pm 0.135)$	$nR_{21} = 37.7\ (\pm 2.7)$
$R_{ee} = (V^{ee})^{-1} = 0.133\ (\pm 0.008)$	$nR_{ee} = 2.7\ (\pm 0.16)$
$R_{oo} = R_{12} - R_{21} - R_{ee} = 2.259\ (\pm 0.157)$	$nR_{oo} = 45.1\ (\pm 3.1)$

Independent estimates of affinity parameter K (μM) ($\pm$ SE)

$$K = K^{zt}_{1\to2}(R_{12}/R_{oo}) = 89\ (\pm 31)$$
$$K = K^{zt}_{2\to1}(R_{21}/R_{oo}) = 61\ (\pm 58)$$
$$K = K^{ic}_{1\to2}(R_{ee}/R_{12}) = 66\ (\pm 28)$$
$$K = K^{ic}_{2\to1}(R_{ee}/R_{21}) = 66\ (\pm 17)$$
$$K = K^{ee}(R_{ee}/R_{oo}) = 76\ (\pm 10)$$

Mean ($\pm$ SD) $= 72 \pm 11\ \mu M$

Bounds for rate constants for uridine transport on the simple carrier[c]

$$27\ (\pm 2) < k_1 < 29\ (\pm 2)\ \text{sec}^{-1}$$
$$99\ (\pm 16) < k_2 < 135\ (\pm 29)\ \text{sec}^{-1}$$
$$b_1, b_2, g_1, g_2 > 370\ (\pm 22)\ \text{sec}^{-1}$$
$$f_1 > 3.7\ (\pm 0.6) \times 10^5\ M^{-1}\ \text{sec}^{-1}$$
$$f_2 > 1.4\ (\pm 0.3) \times 10^6\ M^{-1}\ \text{sec}^{-1}$$

[a] At 25°, interpreted in terms of the simple carrier model. Data of Z. I. Cabantchik and H. Ginsburg, *J. Gen. Physiol.* **69**, 75 (1977), analyzed by Lieb.[22]

[b] Solution 1 is arbitrarily chosen to be the cytoplasm. The unit (mM) in both V_{max} and the resistance parameters is shorthand for millimoles uridine per liter of isotonic cell water. To obtain the values of nR in the table, Lieb used a value of $n = 3.3 \times 10^{-4}$ mmol of uridine transport systems per liter of isotonic cell water. This value was obtained from the average number of nitrobenzylthioinosine high-affinity binding sites (12,500 sites per cell), assuming one site per transport system, and taking the average red cell volume to be 87 μm^3 and the average water content to be 71.7% (v/v).

[c] All inequalities involving b_1, b_2, f_1, f_2, k_1, and k_2 are valid for both the two-complex formulation and the one-complex formulation of the simple carrier. The above values ($\pm$ SE) were calculated using the relationships given in Tables VIII and IX and the experimental values of K, nR_{12}, nR_{21}, nR_{ee}, and nR_{oo} listed above.

TABLE V

URIDINE TRANSPORT IN ERYTHROCYTES FROM FRESH AND OUTDATED HUMAN BLOOD[a]

Term	Fresh blood	Outdated blood
Experimental values ($\pm$SE) (V in mM hr^{-1}, K in mM)[b]		
V_{12}^{zt}	135 ± 16	117 ± 18
V_{21}^{zt}	116 ± 14	35.4 ± 3.9
V^{ee}	462 ± 60	353 ± 41
K_{12}^{zt}	0.14 ± 0.04	0.38 ± 0.10
K_{21}^{zt}	0.17 ± 0.02	0.13 ± 0.03
K^{ee}	0.76 ± 0.07	1.1 ± 0.1
Derived resistance parameters (R in hr mM^{-1})[b]		
R_{12}	0.0074 ± 0.00088 (8.9)	0.0085 ± 0.0013 (10.2)
R_{21}	0.0086 ± 0.00104 (10.3)	0.028 ± 0.003 (33.6)
R_{ee}	0.0022 ± 0.00028 (2.64)	0.0028 ± 0.0003 (3.4)
$\therefore R_{oo}$	0.0138 ± 0.0014 (16.6)	0.034 ± 0.0034 (40.8)
Estimates of affinity parameter (K in mM)		
$K = K_{12}^{zt}(R_{12}/R_{oo})$	0.075 ± 0.024	0.096 ± 0.031
$K = K_{21}^{zt}(R_{21}/R_{oo})$	0.11 ± 0.02	0.11 ± 0.03
$K = K^{ee}(R_{ee}/R_{oo})$	0.12 ± 0.023	0.092 ± 0.016
Mean K	0.101 ± 0.023	0.099 ± 0.008

Bounds for the rate constants of the simple carrier (b, g, and k in sec^{-1}, f in $10^6 M^{-1}$ sec^{-1})[c]

	$97 < k_1 < 131$	$28 < k_1 < 33$
	$112 < k_2 < 160$	$98 < k_2 < 147$
$b_1, b_2, g_1, g_2 >$	378	294
$f_1 >$	0.96	0.30
$f_2 >$	1.1	0.99

[a] At 22°. Data from S. M. Jarvis, J. R. Hammond, A. R. P. Paterson, and A. S. Clanachan, *Biochem. J.* **210**, 457 (1983).

[b] Solution 1 is arbitrarily chosen to be the cytoplasm. The unit (mM) in both V_{max} and the resistance parameters is shorthand for millimoles uridine per liter of isotonic cell water. To obtain the values of nR in the table, Lieb used a value of $n = 3.3 \times 10^{-4}$ mmol of uridine transport systems per liter of isotonic cell water. This value was obtained from the average number of nitrobenzylthioinosine high-affinity binding sites (12,500 sites per cell), assuming one site per transport system, and taking the average red cell volume to be 87 μm^3 and the average water content to be 71.7% (v/v). Values in parentheses correspond to nR terms (in milliseconds).

[c] All inequalities involving b_1, b_2, f_1, f_2, k_1, and k_2 are valid for both the two-complex formulation and the one-complex formulation of the simple carrier. The above values ($\pm$SE) were calculated using the relationships given in Tables VIII and IX and the experimental values of K, nR_{12}, nR_{21}, nR_{ee}, and nR_{oo} listed above.

$(K_{12}^{\pi}, K_{21}^{\pi}$, etc.), a value of K is derived from the relations of Table III, rearranged in the form $K = (K_{12}^{\pi} \times R_{12})/R_{oo} = K^{ee} \times R_{ee})/R_{oo} \cdots$ etc. These values of K for cold-stored erythrocytes are listed in Table IV and in the third column of Table V; for freshly drawn erythrocytes values are listed in the second column of Table V. It can be seen that the experimental data are consistent among themselves and are also consistent with the carrier model. Four independent parameters, K, R_{12}, R_{21}, and R_{ee}, are sufficient to describe all published experiments on the uridine transport system of the human red cell. The system behaves as a simple carrier by all currently available tests. The system is symmetric when present in freshly drawn cells, that is, $R_{12} = R_{21}$, but becomes asymmetric on cold storing, when $R_{12} < R_{21}$. As for both conditions, $R_{ee} < R_{12}$, R_{21}, the system shows trans stimulation in both directions.

A system has been found which fails the test for the simple carrier. The glucose transport systems of the human red cell[7,13] and of a human lymphocyte cell line[23] show saturation behavior and show trans stimulation of unidirectional sugar flux, but they are not simple carriers in the sense of Fig. 3. Tables VI and VII record the results of testing the simple carrier for these two systems, respectively. Again, values of the R parameters are derived from (the reciprocals of) the maximum velocities and values of K are computed from these numbers and the measured half-saturation parameters. The K values so derived do not form a consistent set. The value K for the infinite cis experiment in the 2-to-1 direction (1 being inside the cell) is an order of magnitude lower than it would be were a simple carrier operative here. The data apparently demonstrate that net sugar transport is regulated (depressed) by the binding of sugar at some site at the cytoplasmic surface of the cell, but a full analysis of this phenomenon is not yet available. Lieb[22] has shown also that the choline transport system of the human erythrocyte is more complex than a simple carrier.

Characterizing a Simple Carrier System. For a system such as that which transports uridine across the red cell membrane, which has passed the tests for a simple carrier, the experimentally accessible parameters in K and R can be used to set limits (upper and/or lower bounds) to the values of certain of the individual rate constants of the scheme of Fig. 3. Lieb[22] has used this important approach very successfully. One first needs an estimate of n, the number of transporter sites per cell membrane, so as to obtain absolute values for the R parameters. The R parameters are the *resistances* for the steps involving the conformation changes that interconvert bound and free forms of the carrier and the breakdown of the carrier–substrate complex at the two faces of the membrane. Thus, from Table III,

[23] W. D. Rees and J. Gliemann, *Biochim. Biophys. Acta* **812**, 98 (1985).

TABLE VI
Glucose Transport across the Human Red Blood Cell Membrane[a]

			Experimental values ($\pm$SE)		
Procedure	Direction	State of blood	V_{max} (mM sec^{-1})	K_m (mM)	Reference
Zero trans	In $\rightarrow$ out	Outdated	3.63 ± 0.56	22.9 ± 6.3	b
	In $\rightarrow$ out	Outdated	2.32 ± 0.18	25 ± 3	c
	Out $\rightarrow$ in	Outdated	0.51 ± 0.04	1.60 ± 0.24	b
	Out $\rightarrow$ in	Fresh	0.60 ± 0.002	1.6 ± 0.2	d
Infinite trans	Out $\rightarrow$ in	Fresh	—	1.7 ± 0.3	d
Infinite cis	In $\rightarrow$ out	Outdated	—	1.8 ± 0.3	e
	In $\rightarrow$ out	Fresh	—	1.7	f
	Out $\rightarrow$ in	Outdated	—	2.8 ± 0.6	g
Equilibrium	Both	Outdated	4.33 ± 0.50	38 ± 3	h
exchange	Both	Outdated	5.95 ± 0.17	32 ± 1.1	i
	Both	Outdated	6.0 ± 0.51	34 ± 0.6	j
	Both	Outdated	4.82 ± 0.29	22 ± 1.9	k
	Both	Fresh	4.4 ± 0.7	20 ± 3	d
	Both	Fresh	6.15 ± 0.17	12.7 ± 0.9	k

Derived (sec mM^{-1}) resistance parameters (from means of V_{max} values, from data for outdated blood only)[l]

$$R_{12} = 0.34 \pm 0.10$$
$$R_{21} = 1.96 \pm 0.16$$
$$R_{ee} = 0.19 \pm 0.03$$
$$R_{oo} = 2.11 \pm 0.19$$

Estimates (mM) of affinity parameter K (from means of K and R values, from data for outdated blood only)[l]

$$K = K_{12}^{zt} R_{12} / R_{oo} = 3.87 \pm 1.23$$
$$K = K_{21}^{zt} R_{21} / R_{oo} = 1.49 \pm 0.29$$
$$K = K^{ee} R_{ee} / R_{oo} = 2.84 \pm 0.81$$
$$K = K_{12}^{ic} R_{ee} / R_{12} = 1.01 \pm 0.38$$
$$K = K_{21}^{ic} R_{ee} / R_{21} = 0.27 \pm 0.08$$

[a] At 20°.

[b] From J. R. A. Challis, L. P. Taylor, and G. P. Holman, *Biochim. Biophys. Acta* **602,** 155 (1980).

[c] From S. J. D. Karlish, W. R. Lieb, D. Ram, and W. D. Stein, *Biochim. Biophys. Acta* **255,** 126 (1972).

[d] From L. Lacko, B. Wittke, and H. Kromphardt, *Eur. J. Biochem.* **25,** 447 (1972).

[e] From D. M. Miller, *Biophys. J.* **5,** 407 (1965).

[f] From A. K. Sen and W. F. Widdas, *J. Physiol. (London)* **160,** 392 (1962).

[g] From B. L. Hankin, W. R. Lieb, and W. D. Stein, *Biochim. Biophys. Acta* **288,** 114 (1972).

[h] From D. M. Miller, *Biophys. J.* **8,** 1329 (1968).

[i] From Y. Eilam and W. D. Stein, *Biochim. Biophys. Acta* **266,** 161 (1972).

[j] From Y. Eilam, *Biochim. Biophys. Acta* **401,** 364 (1975).

[k] From M. B. Weiser, M. Razin, and W. D. Stein, *Biochim. Biophys. Acta* **727,** 379 (1983).

[l] Side 1 is taken to be the inner face of the membrane.

TABLE VII
3-O-METHYL-D-GLUCOSE TRANSPORT ACROSS THE CELL
MEMBRANE OF THE CULTURED
HUMAN LYMPHOCYTE[a]

| | | Experimental values | |
		V (mM min^{-1})	K (mM)
Zero trans	In → out	37.9	9.5
Zero trans	Out → in	10.7	2.8
Equilibrium exchange		44.0	9.9
Infinite *cis*	Out → in	—	1.2

Derived resistance parameters (from V_{max} values, min mM^{-1})

$$R_{12} = 0.026$$
$$R_{21} = 0.094$$
$$R_{ee} = 0.023$$
$$R_{oo} = 0.097$$

Independent estimates of affinity parameter K (mM)

$$K = K_{12}^{zt} R_{12}/R_{oo} = 2.58$$
$$K = K_{21}^{zt} R_{21}/R_{oo} = 2.70$$
$$K = K^{ee} R_{ee}/R_{oo} = 2.32$$
$$K = K_{21}^{ic} R_{ee}/R_{21} = 0.29$$

[a] IM-9 line at 37°. Data from Rees and Gliemann.[23]

R_{12} is the resistance for the conformation change of free carrier in the 2-to-1 direction, followed by the movement of substrate on the carrier in the 1-to-2 direction and its release at side 2. The other R parameters have corresponding meanings. The parameter R_{oo} has a particularly simple interpretation. It is the sum of the resistances of conformation changes of the free, unloaded carrier for transformation in the 1-to-2 direction and the 2-to-1 direction. Since the experimentally determined resistance parameters are always the *sum* of the resistances of two or more individual steps, they can be used to set lower limits to the values of the transport steps. Tables VIII and IX record the results obtained by Lieb[22] from the analysis of the simple carrier. Table VIII records the parameters which set lower (and occasionally also upper) limits to the individual rate constants of the simple carrier of Fig. 3. Table IX demonstrates how ratios of the resistance parameters of the model can be used to set bounds (lower bounds and an upper bound) to ratios of these rate constants. Since ratios of R parameters are used in Table IX, the results can be applied even when the absolute number of carriers per cell, n, is *not* known. The application of some of

TABLE VIII
USEFUL BOUNDS FOR MOLECULAR RATE
CONSTANTS OF THE SIMPLE CARRIER MODEL
WHEN THE NUMBER OF CARRIERS IS KNOWN[a,b]

$$b_1, g_2 > 1/nR_{ee}, 1/nR_{21}$$
$$b_2, g_1 > 1/nR_{ee}, 1/nR_{12}$$
$$k_1 > 1/nR_{21}, 1/nR_{oo}$$
$$k_2 > 1/nR_{12}, 1/nR_{oo}$$
$$f_1 > 1/nKR_{21}, 1/nKR_{oo}$$
$$f_2 > 1/nKR_{12}, 1/nKR_{oo}$$
$$k_1 < 1/n(R_{21} - R_{ee})$$
$$k_2 < 1/n(R_{12} - R_{ee})$$

[a] From Lieb.[22]
[b] All inequalities involving b_1, b_2, f_1, f_2, k_1, and k_2 are valid for both the two-complex formulation and the one-complex formulation of the simple carrier of Fig. 3. The lower bounds are always correct, but the two upper bounds are correct only when $R_{21} - R_{ee}$ and $R_{12} - R_{ee}$, respectively, are positive, i.e., when trans acceleration is observed.

TABLE IX
USEFUL BOUNDS FOR CERTAIN RATIOS OF THE
SIMPLE CARRIER MOLECULAR RATE CONSTANTS
WHEN TRANS ACCELERATION OCCURS[a,b]

$$b_1/k_1, g_2/k_1 > (R_{21} - R_{ee})/R_{ee}$$
$$b_2/k_2, g_1/k_2 > (R_{12} - R_{ee})/R_{ee}$$
$$b_1/k_2, g_2/k_2 > R_{12}/R_{ee}$$
$$b_2/k_1, g_1/k_1 > R_{21}/R_{ee}$$
$$f_1/k_1, f_2/k_2 > 1/K$$
$$(R_{21} - R_{ee})/R_{12} < k_2/k_1 < R_{21}/(R_{12} - R_{ee})$$

[a] From Lieb.[22]
[b] All inequalities involving b_1, b_2, f_1, f_2, k_1, and k_2 are valid for both the two-complex formulation and the one-complex formulation of the simple carrier of Fig. 3. The lower bounds are always correct, but the upper bound is correct only when $R_{12} - R_{ee}$ is positive, i.e., when trans acceleration is observed.

these relationships to the data for the uridine transport system of the human red cell are recorded in Tables IV and V. The values of the rate constants k_1 and k_2, which determine the conformation change which interconverts the two forms of the carrier, are well characterized, both for fresh and for cold-stored red cells. They are of order 100/sec, quite a reasonable number for protein conformation changes. The rate of interconversion from "in" to "out" is specifically slowed on cold storing. The *on* rate constants, f_1 and f_2, are at least $10^6 M^{-1} \sec^{-1}$, below the rates of simple diffusion-controlled reactions but quite comparable to the numbers obtained for many enzyme–substrate interaction steps. The rate of interconversion of bound carrier or the rates of release of substrate (the rate constants in b and g, which cannot be distinguished by this analysis, but see later) are of a magnitude expected for conformation changes of proteins. We note that the rate constants in b and g (given lower bounds) are all bigger than those in k (bounded by upper and lower limits). Thus the transmembrane interconversion steps of the forms of the loaded carrier are all faster than the steps for the forms of the unloaded carrier. Binding of the substrate speeds up the rate of the conformation changes of the carrier. This analysis ranks the uridine system as the best characterized by steady-state transport experiments, but the analysis is very general and would repay application to a wider range of transport systems.

The parameter K deserves further attention. Clearly (see Table III) all of the half-saturation constants contain this parameter. Its meaning can be seen if we use Table III to obtain the following relation: $K_{12}^{\pi t}R_{12} = K_{21}^{\pi t}R_{21} = K^{ee}R^{ee} = KR_{oo}$. Here, in the first three terms, the half-saturation constant for each fundamental transport paradigm is multiplied by the resistance parameter found in the corresponding paradigm. In the first term, substrate is present at side 1 only; in the second term, at side 2 only; in the third term, at both sides. It is natural to think of the fourth term as relating to the hypothetical, conceptual experiment where substrate is present at neither face of the membrane. Then R_{oo} is, indeed, the resistance parameter which describes this hypothetical experiment. Thus K is, as it were, the half-saturation concentration applicable when substrate is present at neither face of the membrane. It is the "affinity" of substrate for carrier when no substrate is present to disturb the substrate–carrier interaction. It is an average, taken across the membrane, of the interaction of carrier with substrate at the two membrane faces.

Asymmetry of a Carrier System. The ratio of R_{21}/R_{12} characterizes the asymmetry of those transport parameters that can be asymmetric. From Table III, $V_{12}^{\pi t}/V_{21}^{\pi t} = K_{12}^{\pi t}/K_{21}^{\pi t} = K_{12}^{it}/K_{21}^{it} = K_{21}^{ic}/K_{12}^{ic} = R_{21}/R_{12}$. [Note that, in contrast, the equilibrium exchange parameters V^{ee} and K^{ee} and the permeability π^{ee} ($= KR_{oo}$) are all symmetric.] If, as very often happens, the conformation changes of the loaded carrier are much faster than those of the unloaded carrier, the terms in $1/k$ in R_{12} and R_{21} are large and hence they dominate the ratio R_{21}/R_{12}. (Consider the lowermost inequality in Table IX and let R_{ee} tend to zero with respect to R_{12} and R_{21}.) The asymmetry is, in these circumstances, given by asymmetry $= R_{21}/R_{12} \simeq k_2/k_1$. The ratio k_2/k_1 is nothing other than the equilibrium constant for the conformation change from E_2 to E_1,

$$E_1 \underset{k_2}{\overset{k_1}{\rightleftharpoons}} E_2$$

The equilibrium constant K_f is given exactly by the free energy change for this conformation change $\Delta G = -RT \ln K_f = -RT \ln k_2/k_1$. Thus, the asymmetry determined by any pair of corresponding transport parameters is a measure of (and is determined solely by) the difference in free energy of the unloaded carrier in its two states, one with binding sites facing the cytoplasm or vesicle interior, the other with these sites facing the external medium. If a number of different substrates use the same carrier, and for all of these the value of R_{ee} is considerably lower than that of R_{12} and R_{21} (so that the rates of unloaded carrier conformation change are lower than for the loaded carrier), all must show the same degree of asymmetry. Whatever the absolute value of maximum velocities or half-saturation parameters for these different substrates, the asymmetry in these parameters must be the same for all substrates if the carrier model holds for all of them. The unloaded carrier is the same for all substrates, no mater what the potential load.

The half-saturation concentration for the zero trans or infinite trans experiment will be greater at the side at which the carrier is more stable. (Proof: If side 1 is the more stable, $k_2 > k_1$, hence $R_{21} > R_{12}$ and $K_{12}^{zt} > K_{21}^{zt}$, $K_{12}^{it} > K_{21}^{it}$.) If such half-saturation concentrations are taken to be (reciprocal) measures of the apparent affinity of a carrier for its substrate, the apparent affinity will be lower at the side at which the carrier is more stable, while the maximum velocity of transport will be *greater* from that side. (Proof: If $R_{21} > R_{12}$, $V_{12}^{zt} > V_{21}^{zt}$.)

For all substrates that use the same carrier, if the loaded carrier moves faster than the unloaded carrier, then zero trans maximum velocities, in a particular direction, must be the same. This is because these maximum velocities are determined solely by the rate constants k_1 or k_2, properties only of the unloaded carrier, the same whatever its load. Again, this prediction can be used to test the carrier model.

Antiporter or Exchange-Only Carrier

There are a number of very important transport systems in which only the exchange of labeled substrates, or of one substrate for another species of substrate, is found, that is, there is no (or negligibly small amounts of) net transport. Examples are the ATP/ADP carrier of the mitochondrion,[24] the anion exchanger of the red blood cell,[25] and the Na^+/H^+ antiporter[26] and the Na^+/Ca^{2+} exchanger,[27] which both appear in many cell membranes. For all of these, the simple form

$$E_1 \underset{b_1}{\overset{f_1}{\rightleftharpoons}} ES \underset{f_2}{\overset{b_2}{\rightleftharpoons}} E_2 \qquad \text{(for the one-state complex)}$$

or the form

$$E_1 \underset{b_1}{\overset{f_1}{\rightleftharpoons}} ES_1 \underset{g_2}{\overset{g_1}{\rightleftharpoons}} ES_2 \underset{f_2}{\overset{b_2}{\rightleftharpoons}} E_2 \qquad \text{(for the two-state complex)}$$

define the model completely for one substrate, and Fig. 4 (a and b) depicts these models for two substrates. A test of the applicability of these models is that they show saturation, that they show trans stimulation, and—their defining characteristic—that they show *only* trans stimulation. They are characterized by an effective transport measured as an exchange in the presence of a trans substrate, with no (net) transport in the absence of a

[24] M. Klingenberg, this volume [2].
[25] J. O. Wieth, O. S. Andersen, J. Brahm, P. J. Bjerrum, and C. L. Borders, *Philos. Trans. R. Soc. London, Ser. B.* **299**, 383 (1982).
[26] P. S. Aronson, *Annu. Rev. Physiol.* **47**, 545 (1985).
[27] K. D. Philipson and A. Y. Nishimoto, *J. Biol. Chem.* **257**, 5111 (1982).

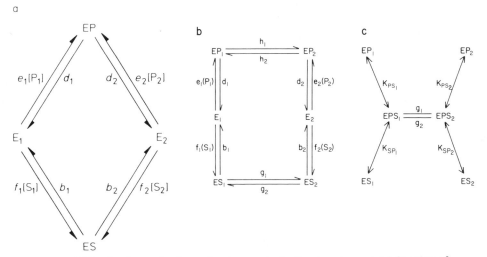

FIG. 4. Kinetic schemes for the carrier–transporter (antiport, exchanger). (a) One state of the carrier–substrate complex; (b) two such states; and (c) the transporting form is the tertiary complex, EPS.

trans substrate. The kinetic solution for the one-substrate case (homoexchange) is simply given as[16]

$$v_{12} = \frac{S_1 S_2 (1/R_{ee})}{K_2 S_1 + K_1 S_2 + S_1 S_2}$$

where $R_{ee} = 1/b_1 + 1/b_2 + 1/g_1 + 1/g_2$, $K_1 = b_1/f_1[g_2/(g_1 + g_2)]$, and $K_2 = b_2/f_2[g_1/(g_1 + g_2)]$. The terms in g drop out if the one-state model is analyzed.) As can readily be seen, the unidirectional flux in the 1-to-2 direction in the presence of an infinite (limitingly high) concentration of substrate at the trans face is obtained by setting $S_2 = \infty$ and is $v_{12}^{it} = (S_1 \times 1/R_{ee})/(K_1 + S_1)$. This is a typical Michaelis form with maximum velocity of $1/R_{ee}$ and half-saturation concentration of K_1. For the infinite trans flux in the opposite direction, the same maximum velocity must be found while the half-saturation concentration is given by K_2. At any other than infinite concentration of S_2, the maximum velocity for the unidirectional flux of S_1 from 1 to 2 is given by $[S_2/(K_2 + S_2)] \times 1/R_{ee}$ and the half-saturation concentration by $K_1 S_2/(K_2 + S_2)$. As S_2 increases from zero to infinity, both V_{max} and K_m for the flux of S_1 increase in strict parallel. Plots of $1/v$ against $1/S_1$ will be sets of parallel straight lines intersecting the $1/v$ and $1/S_1$ axes at smaller and smaller values as S_2 increases. The finding of such plots is a clear indication of the presence of an exchange-only carrier.

The ratio of the apparent affinities of substrate at one side of the membrane coupled with that at the other is, at any value of S_2, given by the ratio of K_1 to K_2. This ratio is itself given by the ratio of rate constants $b_1 g_2 f_2/b_2 g_1 f_1$ in Fig. 4b, or as $b_1 f_2/b_2 f_1$ in Fig. 4a. For both cases this ratio of rate constants is nothing other than the free energy difference between the forms E_1 and E_2. The numerator of this ratio represents the rate for overall reaction which leads to E_1; the denominator is the rate for the formation of E_2. Once again, as for the simple carrier, the asymmetry of the antiporter is determined only by the asymmetric stability of the unloaded carrier. The more stable form demonstrates a lower apparent affinity for substrate, that is, it half-saturates at a higher concentration. Again, this is true for all substrates that use the same antiporter and the aptness of this criterion is a test for the applicability of schemes a and b of Fig. 4. The maximum rate of exchange need not be the same for all substrates. In general these rates are very different for the different substrates that use a common exchanger.

For the exchanger, in contrast to the simple carrier, the permeability at low substrate concentration (given from the equation above as $1/K_1 R_{ee}$ for the 1-to-2 direction) is dependent of the direction in which transport is being measured. It is independent of the concentration of the trans substrate S_2 but is sided, and its asymmetry is given by K_2/K_1, the (reciprocal of the) asymmetry of the half-saturation concentrations.

This is clearly a limiting (perhaps ideal) case. For the anion transporter of the human red cell,[25] for example, all of the above tests that have been applied to the data are found to hold, but there is a small net transport of anions far slower (some 10^4-fold in the case of chlorides) than the exchange reaction. Should the system, then, be tested as an exchanger or as a simple carrier? When tested as a simple carrier the anion transport system fails an elementary test in that the transport rate under zero trans conditions and under equilibrium exchange conditions are not the same at limitingly low substrate concentrations.[28] Indeed, instead of having the same value of $1/KR_{oo}$, they are more than 10^3-fold different (that for exchange being the faster[28]). The simple carrier presumably does not apply to the anion transporter and either net transport or the exchange transport must be proceeding by some route other than the $ES_1 \rightleftharpoons ES_2$ path of the simple carrier of Fig. 3. (See Fröhlich[29] for an explanation of this paradox.)

Distinguishing between Rapid Equilibrium and Rapid Transport Models

There is a troubling loose end that the present discussion has left untied. The impression may have been left that kinetic methods cannot allow a distinction to be made between carrier models in which transport is the rate-limiting step for overall transmembrane movement and those models in which it is the dissociation of substrate at the trans face that is rate limiting. To understand this point, return to Fig. 3 and consider the carrier model in which two forms of the carrier–substrate complex are depicted, ES_1 and ES_2. How can we test whether, for overall transport, it is g_1 and g_2, the steps that interchange the transmembrane conformations of the complex, that are rate limiting, or b_1 and b_2, the breakdown steps? A conventional kinetic analysis provides estimates of K_m and V_{max} only and, as Table III makes clear, these values by themselves cannot enable a distinction to be made between transport-limited or dissociation-limited models—both b and g terms enter directly into the K_m and V_{max} parameters and cannot be separated one from the other. However, a distinction can be made in two ways, as is now described.

Devés and Krupka[30] studied the effects of a series of substrates of a particular transport system (in their case, the choline transporter of the human red cell) on choline efflux from the cell and on the inhibition of the system by an irreversible inhibitor N-ethylmaleimide (NEM). The substrates ranged from those which were very poorly transported and slowed choline efflux by trapping the carrier at the outer face of the cell, to those

[28] J. H. Kaplan, M. Pring, and H. Passow, *FEBS Lett.* **156**, 175 (1983).
[29] O. Fröhlich, *J. Gen. Physiol.* **84**, 877 (1984).
[30] R. Devés and R. M. Krupka, *J. Membr. Biol.* **61**, 21 (1981).

which were rapidly transported and shuttled carrier back to the cytoplasmic face and thus stimulated choline efflux. They showed that the effect of these substrates on stimulating and inhibiting choline efflux was, strictly, linearly proportional to their effect in stimulating or inhibiting the irreversible inhibition by NEM. Devés and Krupka showed that the phenomenon of strict linearity could not be explained on the one-state complex model of Fig. 3, but could be accommodated by the model in which two forms of the carrier–substrate complex were present. This latter model has the necessary flexibility to accommodate the experimental findings of Devés and Krupka, whereas the one-state complex model, with a single parameter b, into which is subsumed both any transmembrane conformation change and a carrier–substrate dissociation, cannot account for the set of data. The full argument is presented in these terms in Stein.[16] Using this analysis, Devés and Krupka[30] were able to show that for the choline transport system, the rate-limiting steps for transport were the interconversion of carrier and carrier–substrate conformations, and not the dissociation of the carrier–substrate complex. Transport in this system is transformation limited and not dissociation limited.

Another approach to the making of this distinction is to adopt the viewpoint that a carrier is nothing other than a type of enzyme, so that the rates of enzyme–substrate association and dissociation should apply in the case of carrier–substrate reactions, too. The argument proceeds along the following lines.[16] For a number of carrier systems, we have sufficient data to derive an estimate of the turnover number of the carrier for a particular substrate. (This turnover number is merely the value of V_{max} for a particular transport paradigm divided by the number of carrier molecules producing that maximum velocity.) For these systems we proceed to set up as the null hypothesis to be tested the statement that the system is here dissociation limited and that hence the parameters in b in Fig. 3 and Table II are far smaller than those in g. The parameter R_{oo} in Table II is determined by terms in rate constants k, while the affinity K will be determined by ratios of k/f, so that $1KR_{oo}$ will be determined by $(1/k)xf/k = f$. We thus measure $1/KR_{oo}$ and compare the derived value of f with the values found for enzyme systems, for the analogous association rate constant between enzyme and substrate. If the estimate of f from measured values of $1/KR_{oo}$ is within the range found for enzyme systems, the null hypothesis cannot be rejected and the transport system may well be dissociation limited. If, on the other hand, the value of f, estimated as $1/KR_{oo}$, is less than that found for enzyme systems, one can suggest that the null hypothesis is wrong — and hence that the transport step may be rate limiting for the transport system under study. Because the test uses the value of $1/KR_{oo}$, which we saw earlier to be given exactly equal to $1/K^{ee}R_{ee}$, $1/K_{12}^{zt}R_{12}$, and $1/K_{21}^{zt}R_{21}$, the test can be applied to any of the experimental paradigms for transport (if the number of transporters is known, so that absolute values of the R parameters can be derived).

Applying this test to a number of transport systems gave the following results[16] for the derived estimates of the association rate constant f, in units of M^{-1} sec^{-1}, on the assumption that transport is dissociation limited: for the human red cell sugar system, at $20°$, $f = 2-5 \times 10^4$ for glucose, 6×10^3 for galactose; for the anion system of that cell, 3×10^3 for Cl$^-$, 1×10^4 for HCO$_3^-$ (both at $0°$), and 2.5×10 for the H$_2$PO$_4^-$ (at $25°$); for the uridine system of that cell, 1×10^6 at $15°$, 6×10^6 at $35°$, and for uridine on the sheep system, 2×10^6 at $37°$; for the mitochondrial ADP/ATP system, 2×10^6 to 1×10^7; for lactose permease, 2×10^2 to 6×10^4; and finally, for Na$^+$,K$^+$-ATPase, (EC 3.6.1.3), 10^4 for Na$^+$ as substrate and 10^6-10^7 for ATP as substrate. These numbers should now be compared with values of f directly measured for enzyme systems[31] which generally lie in the range of 10^6-10^9 M^{-1} sec^{-1}. It would seem that for the sugar system and the anion system of the human red cell and for cation movements on the Na$^+$,K$^+$-ATPase, the dissociation step is unlikely to be rate limiting, whereas transport may well be so. For the nucleotide and nucleoside transport systems (in the human red cell, the mitochondrion and the nucleoside site of the ATPase), it could well be that dissociation of these tightly bound substrates is rate limiting for the overall transport event.

Testing and Characterizing Cotransport Systems

Figure 5 shows various kinetic schemes for the important cotransport systems where two (or more) substrates (the cosubstrates S and A in the diagram) can bind to the carrier, E, to form a tertiary (or higher order) complex, ESA, capable of undergoing conformation changes, $E_1SA_1 \rightleftharpoons ESA_2$, which make the substrate-binding sites alternatively accessible first at one face of the membrane and then at the other.

Figure 5b is a rather general form in which the binary complexes, ES and EA, are also capable of performing the transmembrane conformation changes. Figure 5a is a more restricted case but one often found, in which there is no "slippage," that is, where these ES and EA complexes are not able to "transconform." Figure 5c and d are the still more restricted forms, again often found, where the binding of S and A to E is not *random* as in Fig. 5a and b, but is ordered, with A binding first at both sides in Fig. 5c and S binding first at both sides in Fig. 5d. There are other cases (not

---

[31] C. R. Cantor and P. R. Schimmel, "Biophysical Chemistry." Freeman, San Francisco, California, 1980.

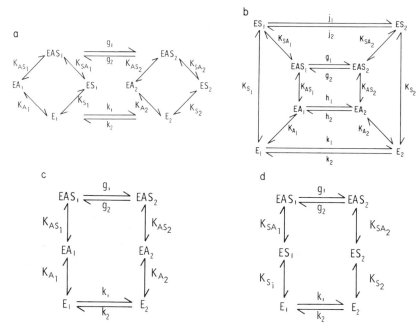

Fig. 5. Kinetic schemes for cotransport in rapid equilibrium. (a) The scheme with no slippage; (b) with slippage; (both a and b) random order; (c) as in a, ordered, S binding first; and (d) as in c, A binding first.

depicted) of ordered bindings, the "glide" cases, in which S binds first at one face and debinds first (binds second) at the other face. The kinetic solutions for the schemes of Fig. 5 are given in Tables X and XI. In addition to the schemes depicted in Fig. 5, in all of which the transport steps are shown as rate limiting, there are all those cases not depicted in which it is the dissociation of the substrates that limit the rate of overall transport, that is, those in which carrier and its cosubstrates are not in rapid equilibrium.

Very naturally, the kinetic problems for the cotransport systems are (1) to test whether the system is indeed one of cotransport, that is, that more than one substrate is involved in the transport event. With this established, one then tests whether (2) the system does or does not demonstrate significant slippage, that is, if Fig. 5b or instead Fig. 5a, c, and d are appropriate. (3) One should then attempt to decide whether or not the system is one of rapid equilibrium. (4) If it is, one can test whether the binding of the cosubstrates is random or ordered and establish the order, if the latter is found to be the case. If it is established that the rapid equilibrium model does not apply or if this point cannot be decided with certainty, the kinetic

TABLE X

KINETIC SOLUTIONS FOR VARIOUS FORMS OF THE SIMPLE COTRANSPORTER WITH NO SLIPPAGE[a,b]

Random order of binding (Fig. 5a)	A binding first (Fig. 5c)	S binding first (Fig. 5d)
$nR_{12} = (1/k_2)(1 + A_2/K_{A_2}) + (1/g_1)(1 + K_{SA_1}/A_1)$	$1/k_2 + (1/g_1)(+K_{SA_1}/A_1)$	$(1/k_2)(1 + A_2/K_{A_2}) + 1/g_1$
$nR_{21} = (1/k_1)(1 + A_1/K_{A_1}) + (1/g_2)(1 + K_{SA_2}/A_2)$	$1/k_1 + (1/g_2)(1 + K_{SA_2}/A_2)$	$(1/k_1)(1 + A_1/K_{A_1}) + 1/g_2$
$nR_{ee} = (1/g_1)(1 + K_{SA_1}/A_1) + (1/g_2)(1 + K_{SA_2}/A_2)$	$(1/g_1)(1 + K_{SA_1}/A_1) + (1/g_2)(1 + K_{SA_2}/A_2)$	$1/g_1 + 1/g_2$
$nR_{oo} = (1/k_1)(1 + A_1/K_{A_1}) + (1/k_2)(1 + A_2/K_{A_2})$	$1/k_1 + 1/k_2$	$(1/k_1)(1 + A_1/K_{A_1}) + (1/k_2)(1 + A_2/K_{A_2})$
$K_1 = (k_1/g_1)K_{A_1}K_{SA_1} = (k_1/g_1)K_{S_1}K_{SA_1}$	$(k_1/g_1)K_{S_1}K_{SA_1}$	$(k_1/g_1)K_{A_1}K_{AS_1}$
$K_2 = (k_2/g_2)K_{A_2}K_{SA_2} = (k_2/g_2)K_{S_2}K_{SA_2}$	$(k_2/g_2)K_{S_2}K_{SA_2}$	$(k_2/g_2)K_{A_2}K_{AS_2}$
Constraint[c]: $K_2 = K_1K_{eq}$, which includes $K_{A_1}K_{AS_1} = K_{S_1}K_{SA_1}$ $K_{A_2}K_{AS_2} = K_{S_2}K_{SA_2}$	$K_2 = K_1K_{eq}$	$K_2 = K_1K_{eq}$

[a] See Fig. 5a, c, and d.
[b] General solution:

$$v_{1\to2}^S = \frac{K_2S_1A_1 + S_1A_1S_2A_2}{K_1K_2R_{oo} + K_2R_{12}S_1A_1 + K_1R_{21}S_2A_2 + R_{ee}S_1A_1S_2A_2}$$

where K_1, K_2, and the terms in R are defined below and $R_{oo} + R_{ee} = R_{12} + R_{21}$; n = total number of cotransporters per unit area of membrane. Terms in k, g, and K are defined in the table or in Fig. 5. For the flux of A (e.g., $v_{1\to2}^A$), interchange A and S everywhere in the table. The solutions given are for the rapid equilibrium model, the situation where the transport steps are rate limiting.

[c] For cotransport, $K_{eq} = 1$; for cochemiport, K_{eq} = dissociation constant for the reaction ATP $\rightleftharpoons$ ADP + P_i.

TABLE XI

KINETIC SOLUTION FOR THE COTRANSPORTER WITH SLIPPAGE[a]

$$v^S_{1 \to 2} = \frac{\bar{S}_1 G_1 (H_2 + \bar{S}_2 G_2)}{H_1 H_2 R_{oo} + H_2 G_1 R_{12} \bar{S}_1 + H_1 G_2 R_{21} \bar{S}_2 + G_1 G_2 R_{ee} \bar{S}_1 \bar{S}_2}$$

$$H_1 = k_1 + h_1 \bar{A}_1 \qquad \bar{A}_1 = A_1/K_{A_1} \qquad \bar{S}_1 = S_1/K_{S_1}$$

$$H_2 = k_2 + h_2 \bar{A}_2 \qquad \bar{A}_2 = A_2/K_{A_2} \qquad \bar{S}_2 = S_2/K_{S_2}$$

$$G_1 = j_1 + g_1 \bar{A}^*_1 \qquad \bar{A}^*_1 = A_1/K_{SA_1}$$

$$G_2 = j_2 + g_2 \bar{A}^*_2 \qquad \bar{A}^*_2 = A_2/K_{SA_2}$$

$$nR_{12} = \frac{1 + \bar{A}^*_1}{G_1} + \frac{1 + \bar{A}_2}{H_2} \quad (= 1/V^{zt}_{12})$$

$$nR_{21} = \frac{1 + \bar{A}^*_2}{G_2} + \frac{1 + \bar{A}_1}{H_1} \quad (= 1/V^{zt}_{21})$$

$$nR_{ee} = \frac{1 + \bar{A}^*_1}{G_1} + \frac{1 + \bar{A}^*_2}{G_2} \quad (= 1/V^{it}_{12} = 1/V^{it}_{21})$$

$$nR_{oo} = \frac{1 + \bar{A}_2}{H_2} + \frac{1 + \bar{A}_1}{H_1}$$

$$K^{zt}_{12} = K_{S_1} \frac{R_{oo}}{R_{12}} \frac{H_1}{G_1}$$

$$K^{it}_{12} = K_{S_1} \frac{R_{21}}{R_{ee}} \frac{H_1}{G_1}$$

[a] From Stein.[16] Definitions and overall assumptions as in Table X. Rate constants in g, h, j, and k refer to Fig. 5b, where we depict also the dissociation constants K_{A_1}, K_{S_1}, K_{SA_1}, etc., transformed into resistance terms from the original solution of R. J. Turner, *Biochim. Biophys. Acta* **689**, 444 (1982).

analysis becomes very difficult in that many possible models give the same kinetic predictions.[32]

A brief survey of the methods of testing and characterizing cotransport systems follows in the spirit in which I have analyzed the simpler channels and carriers above. For a full analysis see Sanders *et al.*[32] and for a valuable application to the sugar plus sodium cotransporters, see Semenza *et al.*[33]

Identification of Cosubstrates. In most cotransport systems that have been studied, one cosubstrate is sodium or a proton while the other can be

[32] D. Sanders, U.-P. Hansen, D. Gradmann, and C. L. Slayman, *J. Membr. Biol.* **77**, 123 (1984).

[33] G. Semenza, M. Kessler, M. Hosang, J. Weber, and U. Schmidt, *Biochim. Biophys. Acta* **779**, 343 (1984).

a sugar, an amino acid, another cation, or a host of other metabolites. (The cochemiporters, in which the transport event, the transport of a cation, is coupled to a chemical event, generally the splitting of formation of ATP, behave kinetically as cotransporters and can be considered together with them.) To test whether the two substrates are cotransported, one can measure the transport of, say, S as a function of S at different fixed values of the cosubstrate A. Data such as those of Fig. 6 are often found. Either K_m or V_{max} (or both) for transport of S will be affected by the concentration of A in a systematic way. (In a subsequent discussion we show how plots such as those of Fig. 6 can enable the establishment of the order of binding of the cosubstrates.) If it can be shown that A affects K_m and/or V_{max} for the transport of S, it can tentatively be held that A and S are cosubstrates, but for *proof* it must be shown that both A and S move in concert and with a defined stoichiometry across the membrane. This can often be very difficult to demonstrate since parallel systems for the transport of S and A can mask or falsely indicate a cotransport of these substrates. An apparent cotransport of proton may be indeed a countertransport of hydroxide ion.[26]

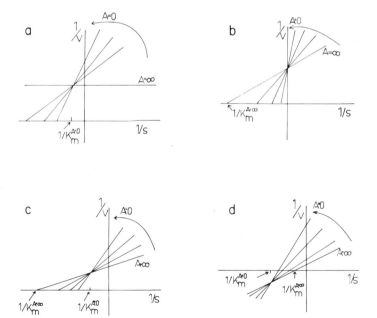

FIG. 6. Plots of $1/v$ versus $1/S$ for cotransport, in rapid equilibrium, at different fixed concentrations of the cosubstrate A (from zero to infinity). (a) S binding first, (b) A binding first, and (c and d) random order. Apparent dissociation constants as given by the intercepts on the $1/S$ axis.

Testing for Slippage. The tests for the presence of slippage are also difficult to perform. It will be very rare to find a system in which cotransport of S and A exists with no slippage and yet there is no *parallel* system for S or A transport. How does one establish that a system which transports S (with no A) is indeed a parallel system and not a cotransporter showing slippage of S, in the absence of A? The tests have to be indirect. To demonstrate that S-alone transport is *not* slippage, one has to demonstrate that this transport and the cotransport of S and A are dissociable, that is, that one and not the other can be inhibited by a particular inhibitor, or be subject independently to genetic mutation, or be physically separable (purified) one from the other. In general, as cotransport systems are subjected to a more intense analysis, minor slippage pathways seem to be revealed (see Ref. 34 for slippage on the cochemiporter of Na, K, and ATP, i.e., Na^+,K^+-ATPase).

Distinguishing Transport-Limited from Dissociation-Limited Reactions. Making the distinction between a rapid equilibrium, transport-limited model for cotransport and a nonequilibrium, dissociation-limited model is crucial for subsequent attempts to describe unambiguously the different orders of binding of the cosubstrates. To make such a distinction proved laborious for the simple carrier systems. It is even more difficult for the cotransporters. An indication that nonequilibrium kinetics are applicable comes from a finding that plots of $1/v$ versus $1/S$ give curved lines in place of the straight lines of Fig. 6. The rapid transport step in the nonequilibrium model samples the populations of EAS forms, and reports on how many at any time are formed by the route $E \rightleftharpoons EA \rightleftharpoons EAS$ and how many by the route $E \rightleftharpoons ES \rightleftharpoons EAS$. The asymptotes of the curved-line plots of $1/v$ versus $1/S$ give the relative contributions of these routes at any concentration of A. In a rapid equilibrium system, the EAS present is an equilibrium-weighted average of that formed by each route. The separate contributions cannot be disentangled and straight-line plots are found. Curved $1/v$ versus $1/S$ plots are not, however, unambiguous indications of a nonequilibrium system since they can arise from the simultaneous presence of two independent cotransport systems (and have been shown so to arise in two cases[35,36]). Another indication of the existence of rapid equilibrium is to proceed along the lines discussed in the preceding section regarding distinguishing between rapid equilibrium and rapid transport models, and to compare computed association constants for substrate–transporter in-

[34] S. J. D. Karlish and W. D. Stein, *J. Physiol. (London)* **328**, 295 (1982).
[35] R. J. Turner and A. Moran, *J. Membr. Biol.* **70**, 37 (1982).
[36] J. D. Kaunitz and E. M. Wright, *J. Membr. Biol.* **79**, 41 (1984).

teractions with those expected for enzyme–substrate associations. As mentioned in the preceding discussion, the lactose + proton cotransporter of *Escherichia coli* is, by this analysis, probably a rapid equilibrium system. For some systems (for example, the Na/K ATP cochemiporter, the ATPase) direct measurements of the rates of the individual transport and dissociation steps have shown that the rapid equilibrium model applies. It is most unlikely to apply to all systems.

Fixing the Order of Binding of Cosubstrates. If it can be established that the system is one of rapid equilibrium and that slippage is negligible, the kinetic solutions listed in Table X will enable a distinction to be made between models in which the binding of S and A is random and those in which either S or A is first. The predictions of these models as plots of $1/v$ versus $1/S$ for the zero trans experiment were given in Fig. 6. For A binding first, the maximum velocity of transport of S is independent of the concentration of A, and the plot is a pencil of straight lines intersecting on the $1/S$ axis. For S binding first, both V_{max} and K_m are affected by A and the lines intersect in the third quadrant of the figure. For the random order of binding, the plots of $1/v$ versus $1/S$ (at different values of A) intersect in the third (or fourth) quadrant, but so do the plots of $1/v$ versus $1/A$ (at different S). In an ordered system, a pencil of lines intersecting on the $1/S$ (or $1/A$) axis must be found for one of the two cosubstrates, whereas in a random system the plot for neither cosubstrate intersects in this manner on the $1/S$ ($1/A$) axis. For the nonequilibrium cases, the kinetic predictions are not as unambiguous. They can be found in Sanders *et al.*[32]

Characterizing a Cotransport System. With the testing of a cotransporter completed, the system can then be characterized. The procedures will follow the lines along which the carrier systems were characterized. For instance, if it has been established that a cotransporter cotransports substrates A and S, that the system is one in which slippage is negligible, that it is in rapid equilibrium with the cosubstrates, and that A binds first at both faces of the membrane, we can apply the kinetic solutions listed in Table X as follows. At each concentration of A, we obtain R_{12}, R_{21}, R_{ee}, and hence R_{oo} and K from plots of the $1/v$ versus $1/S$ of Fig. 6; R_{21}/R_{12} characterizes the asymmetry of the system. On replotting R_{12}, R_{21}, and R_{oo} against A_1 and A_2, we obtain for the slopes of such plots the products $k_1 K_{A_1}$ and $k_2 K_{A_2}$, which are the apparent binding affinities of A to the unloaded transporter on sides 1 and 2. From K and these values of $k_1 K_{A_1}$ and $k_2 K_{A_2}$ we obtain K_{AS_1}/g_1 and K_{AS_2}/g_2, the apparent affinities of binding of S to the A-loaded transporter at sides 1 and 2. In favorable circumstances we can set (narrow) upper and lower bounds to k_1 and k_2, as was discussed in the section on characterizing a simple carrier system, and hence obtain K_{A_2}

and K_{A_1} individually. In general, we are not able by kinetic analysis of $1/v$ versus $1/S$ plots also to determine unambiguously g_1 or g_2 and hence to extricate K_{AS_1} and K_{AS_2}.

The characterization of the system will enable us to account for the origin of any asymmetry that the system manifests in transport rates. [Is this asymmetry determined by the ratio k_2/k_1 and hence by the free energy difference of the unloaded transporter, or in the binding of the cosubstrates (is K_{A_1} unequal to K_{A_2})?] The characterization will enable us to understand how the free energy of binding of the cosubstrates is apportioned as between S and A at the two faces of the membrane. (Is $k_1 K_{A_1} > K_{AS_1}/g_1$ or is $k_2 K_{A_2} < K_{AS_2}/g_2$?) With the system so characterized we can begin to speculate on how it has evolved to its present-day effectiveness as a cotransporter, maximizing the rate of pumping of the physiologically crucial substrate from the whence to the whither side of the membrane.[16]

Conclusions

Kinetic analyses are now available that, when applied in a systematic fashion, enable a transport system to be characterized as a channel, a carrier, or a cotransporter, and enable many of the detailed properties of the model to be found. So far, these procedures have been applied in full analyses only to carriers, but the methods are available in principle for other transport systems as well. These kinetic analyses enable us to characterize the models in considerable detail. It seems well worthwhile to provide such detailed analyses for that interpretation on the molecular level, which is the aim of much transport biology.

[4] Kinetics of Ion Movement Mediated by Carriers and Channels

By Olaf Sparre Andersen

Introduction

Biological membranes are mosaic structures, composed of lipid bilayer leaflets that are spanned by integral membrane proteins.[1] The lipid bilayer provides the barrier property, while the membrane proteins provide the solute translocation functions.

[1] S. J. Singer and G. L. Nicolson, *Science* **175**, 720 (1972).

The lipid bilayer is an efficient barrier against nonselective movement of small ions because the ion concentration in a bilayer is very low.[2,3] The membrane conductance, G, is proportional to the ion concentration, C, and mobility, u, in the hydrocarbon membrane interior. If C is related to the aqueous ion concentration, [I], by the Boltzmann distribution, then

$$G \sim uC = uK[I] = u[I] \exp\{-\Delta E/kT\} \qquad (1)$$

where K is the ion's partition coefficient $(= \exp\{-\Delta E/kT\})$, ΔE denotes the free energy difference for moving an ion from the aqueous phase into the bilayer, k is Boltzmann's constant, and T is the temperature in Kelvin. That C is low, relative to [I], implies that ΔE is large, relative to kT.

The dominant term in ΔE is the electrostatic charging, or Born energy, ΔE_B, for transferring an ion from an (infinite) aqueous medium of dielectric constant ϵ_{H_2O} (~ 80) into an (infinite) medium of dielectric constant ϵ_m (~ 2), e.g.,[4] $\Delta E_B = [(ze)^2/(8\pi r\epsilon_0)](1/\epsilon_m - 1/\epsilon_{H_2O})$, where z is the ion valency, e is the elementary charge, r is the ion radius, and ϵ_0 is the permittivity of free space. ΔE_B is large, $\sim 70\ kT$, for transfer of a monovalent ion of radius 2 Å from dielectric constant 80 to 2. The ion's partition coefficient, K $(= \exp\{-\Delta E_B/kT\})$, is therefore $\sim 10^{-30}$!

Several factors serve to decrease the energy barrier for ion translocation across the bilayer.[3] The two most important factors are as follows: an ion may cross the membrane "wrapped" in a neutral molecule with a high polarizability, such that the effective ion radius is increased[3]; or the ion may cross the membrane through a path of high polarizability, such that the electrostatic energy barrier is reduced and the ion is solvated.[3,5-9]

Integral membrane proteins provide the structural basis for ion translocation by forming (aqueous) paths of high polarizability that connect the two aqueous phases. The proteins catalyze the transmembrane movement of ions by providing an alternate permeation, or reaction, path such that permeating ions do not move through the bilayer per se. Membrane proteins that mediate selective ion movement are therefore enzymes. In some

[2] A. Finkelstein and A. Cass, *J. Gen. Physiol.* **52**, 145 (1968).
[3] A Parsegian, *Nature (London)* **221**, 844 (1969).
[4] J. O. Bockriss and A. K. N. Reddy, "Modern Electrochemistry," Chap. 2. Plenum, New York, 1970.
[5] D. G. Levitt, *Biophys. J.* **22**, 209 (1978).
[6] P. C. Jordan, *Biophys. Chem.* **13**, 203 (1981).
[7] O. S. Andersen, *Annu. Rev. Physiol.* **46**, 531 (1984).
[8] P. Jordan, *J. Membr. Biol.* **78**, 91 (1984).
[9] P. Jordan, *in* "Ion Channel Reconstitution" (C. Miller, ed.), pp. 37–55. Plenum, New York, 1986.

cases, e.g., for ATP-driven ion transport, it is generally recognized that the ion transfer is enzyme-mediated. But all membrane proteins that mediate selective solute movement, e.g., voltage-dependent sodium channels in excitable membranes or anion exchangers in red blood cell membranes, satisfy the criteria for being classified as enzymes. The kinetics of selective ion movement across biological membranes are consequently the kinetics of enzyme-catalyzed substrate turnover.

Enzymatic reactions are conveniently analyzed in terms of a reaction coordinate diagram, where the free energy of the reagents (substrate, enzyme, and possible cofactors) is plotted as a function of the reaction coordinate (e.g., Ref. 10). The free-energy profile for ion movement across a membrane results from the superposition of several contributions. Two of these contributions, a long-range electrostatic energy term E_{el}, and a short-range "chemical" energy term, E_{chem}, are illustrated in Fig. 1 for the case of ion movement mediated by a proteinaceous path. The electrostatic contribution to the free-energy barrier is determined by the protein structure and membrane thickness.[9] The chemical, or solvation, energy is determined by local interactions between the permeating ions and the polar groups that form the boundaries of the permeation path.[5,7,11] The total energy, E, is the sum of E_{el}, E_{chem}, and any other terms that might contribute. A quantitative evaluation of the energy profile depends on structural information that presently is unavailable for any membrane-spanning ion-translocating protein.

Carrier versus Channel

The selective translocation of an ion across a membrane, from one aqueous phase (denoted cis) to the other (denoted trans), can be described by the following steps: recognition (association) ↔ translocation ↔ release (dissociation). Each step involves multiple elementary reactions: the association step involves, for example, a diffusive approach followed by a sequential exchange of H_2O by polar groups, e.g., carbonyl oxygens, as the ion becomes (partially) dehydrated and "solvated" by the ion-translocating molecule.[11,12]

[10] A. Fersht, "Enzyme Structure and Mechanism," 2nd Ed., Chap. 2. Freeman, San Francisco, California, 1985.

[11] O. S. Andersen and J. Procopio, *Acta Physiol. Scand., Suppl.,* **481,** 27 (1980).

[12] H. Diebler, M. Eigen, G. Ilgenfritz, G. Maass, and R. Winkler, *Pure Appl. Chem.* **20,** 93 (1969).

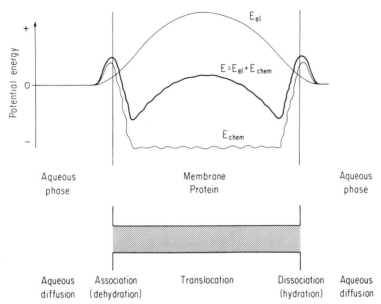

FIG. 1. Schematic representation of the energy profile for ion movement through an integral membrane protein. The lower part of the figure is a highly idealized representation of a membrane-spanning protein. The upper part depicts two contributions to the total energy profile: E_{chem} denotes the qualitative variation in the local (solvation) interactions between a permeating ion and the protein; E_{el} denotes the variation in the electrostatic energy as the ion moves across the membrane (through the protein).[5,6,9] (Modified from Andersen and Procopio.[11])

Discussions of molecular mechanisms underlying selective ion permeation usually focus on two alternatives (e.g., Ref. 13): movement mediated by translatory carriers, where the ion crosses the membrane "wrapped" in the carrier, and movement mediated by fixed channels, where the ion traverses the membrane through a preformed (aqueous) permeation path (see Fig. 2). These terms were long used mostly as formal concepts, but the discovery of molecules that could function as ion carriers or channels in planar lipid bilayers gave molecular reality to them.[15]

[13] E. Heinz, "Mechanics and Energetics of Biological Transport," 159 pp. Springer-Verlag, Berlin and New York, 1978.

[14] O. S. Andersen, in "Renal Function" (G. H. Giebisch and E. Purcell, eds.), pp. 71–99. Independent Publ., Port Washington, New York, 1978.

[15] P. Mueller and D. O. Rudin, in "Current Topics in Bioenergetics" (D. R. Sanadi, ed.), Vol. 3, pp. 157–249. Academic Press, New York, 1969.

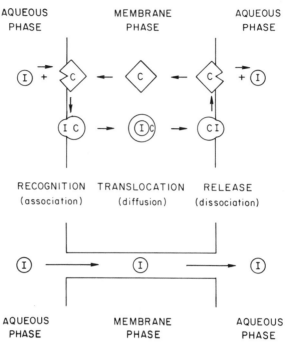

FIG. 2. Schematic representation of two alternative modes for membrane translocation of an ion, I. (Top) Translocation mediated by a mobile carrier, C. (Bottom) Translocation mediated by a channel. In either case, ion translocation involves three similar steps as denoted in the middle of the figure. (Modified from Andersen.[14])

The evidence that some small or medium-sized nonpolar molecules (e.g., nonactin and valinomycin) can function as translatory ion carriers is compelling: the structures of the alkali metal complexes of valinomycin and nonactin are known (e.g., Ref. 16); the molecules can wrap themselves around the ions to provide a polarizable shell, and the hydrophobic exterior permits the ion–carrier complex to partition into hydrocarbon solvents (and the membrane interior); the nactins form stoichiometric complexes with alkali metal cations in bulk solutions, and their ability to increase the membrane's ion permeability parallels closely their bulk phase ion-complexing capabilities[17]; and mobile carriers become ineffective in mediating transmembrane ion movement when the hydrocarbon membrane interior is "frozen."[18]

[16] M. Dobler, "Ionophores and Their Structures," 379 pp. Wiley, New York, 1981.
[17] G. Szabo, G. Eisenman, and S. Ciani, *J. Membr. Biol.* **1**, 346 (1969).
[18] S. Krasne, G. Eisenman, and G. Szabo, *Science* **174**, 412 (1971).

The evidence that other molecules (e.g., alamethicin and gramicidin A) can form transmembrane channels is likewise very strong. The strongest evidence is probably that these compounds increase the membrane conductance in unitary, stepwise increments (Fig. 3[19]). Additional evidence is that the conductance of the unitary events is fairly insensitive to the physical state, thickness, and lipid composition of the bilayer.[18,20,21]

The distinction between ion movement mediated by translatory carriers, such as nonactin, and by membrane-spanning channels, such as those formed by gramicidin A, is unambiguous. The distinction between carrier- and channel-mediated solute movement is less clear when the translocators (transport molecules) are large integral membrane proteins. These proteins span the membrane, and the physical motion of the translatory carrier is replaced by conformational changes in the proteins (see Fig. 4[22-24]). Carrier-mediated solute movement involves, however, a "return" of the substrate-binding site, from the trans to the cis orientation. Therefore, a fairly general criterion to distinguish between carriers and channels is as follows: at any time, an empty carrier is accessible from only one aqueous phase, while an empty channel is accessible from both aqueous phases.[25] When an ion translocator can be occupied by at most one ion, the criterion can be rephrased: a carrier "remembers" the direction of the last translocation event, while a channel has no such "mem-

[19] O. S. Andersen, *Biophys. J.* **41**, 119 (1983).
[20] S. B. Hladky and D. A. Haydon, *Biochim. Biophys. Acta* **274**, 294 (1972).
[21] H.-A. Kolb and E. Bamberg, *Biochim. Biophys. Acta* **464**, 127 (1977).
[22] C. S. Patlak, *Bull. Math. Biophys.* **19**, 209 (1957).
[23] P. R. Steinmetz and O. S. Andersen, *J. Membr. Biol.* **65**, 155 (1982).
[24] J. O. Wieth, P. J. Bjerrum, J. Brahm, and O. S. Andersen, *Tokai J. Exp. Clin. Med. (Suppl.)* **7**, 91 (1982).
[25] P. Läuger, *J. Membr. Biol.* **57**, 163 (1980).

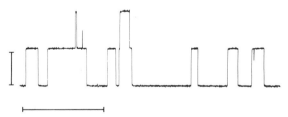

FIG. 3. Current fluctuations (single-channel current transitions) in a planar diphytanoyl-phosphatidylcholine/n-decane bilayer modified by valine gramicidin C (see Andersen[19] for experimental details). $\Delta V = 100$ mV. The calibration bars denote 1.0 pA (vertically) and 5.0 sec (horizontally). 1.0 M NaCl, 25°.

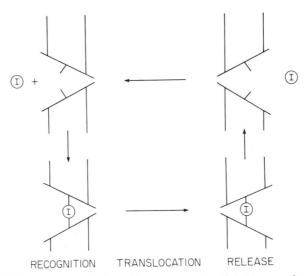

RECOGNITION   TRANSLOCATION   RELEASE

Fig. 4. Schematic representation of ion movement mediated by a conformational carrier.[22] The four states are functionally identical to the four states in the mobile carrier scheme (Fig. 2). The kinetics of the interconversion between adjacent states are likewise identical with the kinetics of the mobile carrier. The conformational changes may involve just the transposition of a few critical groups in a channel (e.g., Refs. 23 and 24).

ory." The distinction between carriers and channels depends on how good the memory is: if a translocator has almost perfect memory, it is very unlikely to cross the membrane in the empty state, and will behave as an exchange carrier; if a translocator has almost total amnesia, it will behave as a channel. There is therefore no absolute separation between carriers and channels. The mobile carriers and simple transmembrane channels, as exemplified by valinomycin and gramicidin A, represent the extremes of a wide variety molecular mechanisms that catalyze selective transmembrane solute movement.[25]

Ion Movement across Energy Barriers

Selective ion movement depends on specific interactions between permeant ions and the ion-translocating molecules. Kinetically, these interactions usually are described in terms of transitions among a relatively small number of states, similar to the kinetic description of other enzyme-catalyzed reactions.[26] The individual kinetic steps, the transitions from one

[26] J. T.-F. Wong, "Kinetics of Enzyme Mechanisms," Chap. 2. Academic Press, New York, 1975.

state to another, are conformational transitions affecting the accessibility of ion-binding sites, translational transitions of ions from one coordinating (binding) site to another, or combinations of such events. The transitions can be represented as passages across energy barriers and analyzed using Eyring's transition state theory[27] or Kramers' electrodiffusive description.[28]

Whatever the model, passive ion movement results from the random (thermal) motion and migrative flux of ions (and ion-translocating proteins). In this section, I first discuss simple electrodiffusive ion movement, then ion movement across energy barriers. Specific models for carrier- and channel-mediated ion movement are presented in the next section.

Simple Electrodiffusion

The Nernst–Planck equations describe ion movement resulting from the thermal motion of ions in a concentration gradient and their migrative flux in an electric field. (Sten-Knudsen[29] should be consulted for a careful exposition of the theory.) For a single permeant species, the flux equation is

$$J = -\{D \, dC/dx + zuCe \, dV/dx\} \tag{2}$$

where J is the net ion flux (from left to right) per unit membrane area, D is the diffusion coefficient, x is the distance (from left to right), and V is the electrostatic potential. (For comparison with models that will be developed later, zeV is the electrostatic contribution to the ion's free energy.) The membrane extends from $x = 0$ to $x = d$. The ion concentrations at the two membrane–solution interfaces, $C(0)$ and $C(d)$, are related to the bulk aqueous concentrations, $[I]_l$ and $[I]_r$, by $C(0) = K[I]_l$ and $C(d) = K[I]_r$, where K is the ion's partition coefficient, and the subscripts "l" and "r" denote the left and the right aqueous phases.

The diffusion coefficient is related to the mobility by the Nernst–Einstein relation $D = ukT$ (e.g., Ref. 29), and Eq. (2) can be integrated across the membrane to give

$$J = ([I]_l - [I]_r \exp\{-z \, \Delta U\})/DN \tag{3}$$

[27] H. Eyring, *J. Chem. Phys.* **3**, 107 (1935).
[28] H. A. Kramers, *Physica* **7**, 284 (1940).
[29] O. Sten-Knudsen, *in* "Membrane Transport in Biology" (G. Giebisch, D. C. Tosteson, and H. H. Ussing, eds.), Vol. 1, pp. 5–112. Springer-Verlag, Berlin and New York, 1978.

where ΔU denotes the dimensionless potential difference $\Delta U = -(U(d) - U(0))$, and where $U(x) = eV(x)/kT$. The denominator, DN, is given by

$$DN = \int_0^d \exp\{zU(x)\}/(DK) \, dx \qquad (4)$$

The numerator in Eq. (3) is a function of the driving forces for solute movement, while DN is determined by the permeability characteristics of the membrane (see below). (Note the sign convention for ΔU! It is chosen such that positive charges flow from left to right when ΔU is positive.)

The net flux, J, is the difference between two unidirectional fluxes (from left to right, $J^\rightarrow$; and from right to left, $J^\leftarrow$):

$$J = J^\rightarrow - J^\leftarrow \qquad (5)$$

Generally, the unidirectional fluxes can be determined from tracer flux measurements.[30] When the Nernst–Planck equations are valid, the unidirectional fluxes are equal to the net fluxes, determined when the trans concentration is zero: $J^\rightarrow = K[I]_l/DN$ and $J^\leftarrow = K[I]_r \exp\{-z \Delta U\}/DN$. The ratio of the unidirectional fluxes is therefore given by Ussing's flux ratio equation[31]:

$$J^\rightarrow/J^\leftarrow = [I]_l/[I]_r \exp\{z \Delta U\} \qquad (6)$$

When $U(x)$ is a linear function of x, Eq. (4) is readily integrated and Eq. (3) becomes the constant-field expression[32]:

$$DN = (1 - \exp\{-z \Delta U\})d/(DKz \Delta U) \qquad (7)$$

The relation between the current density, I ($= zeJ$), and ΔU is then

$$I = Pz^2e \Delta U([I]_l - [I]_r \exp\{-z \Delta U\})/(1 - \exp\{-z \Delta U\}) \qquad (8)$$

where P denotes the ion's permeability coefficient: $P = DK/d$. (Note that $P = 1/DN$ at $\Delta V = 0$.) In symmetrical solutions, when $[I]_l = [I]_r = [I]$, the current expression simplifies to $I = P[I](ze)^2 \Delta V/kT$. That is, the current is a linear function of potential. When $[I]_l \neq [I]_r$, the current–voltage, I–V, relations are nonlinear and the membrane displays rectification. The current is zero at the ion's reversal or Nernst potential, V_{rev}:

$$V_{rev} = -(V(d) - V(0)) = [kT/(ze)] \ln\{[I]_r/[I]_l\} \qquad (9)$$

[30] H. H. Ussing, *Adv. Enzymol.* **13**, 21 (1952).
[31] H. H. Ussing, *Acta Physiol. Scand.* **19**, 43 (1949).
[32] D. E. Goldman, *J. Gen. Physiol.* **27**, 37 (1943).

(Actually, the Nernst potential is determined by the ratio of permeant ion's thermodynamic *activities*. But since the theoretical development is simplified by the use of concentrations rather than activities, I will do so throughout this chapter.) As $\Delta V \rightarrow \pm\infty$, the two branches of the I-V approach the asymptotes: $I = P[I]_l(ze)^2 \Delta V/kT$ as $z\Delta V \rightarrow \infty$; and $I = -P[I]_r(ze)^2 \Delta V/kT$ as $z\Delta V \rightarrow -\infty$.

A membrane's permeability characteristics are quantified by its permeability coefficient or its conductance. When the current–voltage relation is nonlinear, however, the term conductance becomes ambiguous. It is usually defined as either the chord conductance, $G_c(\Delta V)$

$$G_c(\Delta V) = I(\Delta V)/(\Delta V - V_{rev}) \tag{10}$$

or as the slope conductance, $G_s(\Delta V)$

$$G_s(\Delta V) = dI(\Delta V)/dV \tag{11}$$

The physical significance of G_c and G_s is discussed by Finkelstein and Mauro.[33] Here G_c and G_s are equal when the membrane behaves as an ohmic resistor. In this model, $G_c = G_s$ $(= G)$, when $[I]_l = [I]_r = [I]$, in which case

$$G = P[I](ze)^2/kT \tag{12}$$

When $[I]_l \neq [I]_r$, G_c will generally not be equal to G_s, and either conductance will vary as a function of ΔV. Some practical consequences of this voltage dependence are discussed in Ref. 34.

In the vicinity of $\Delta V = V_{rev}$, the small-signal conductance G $(= G_c = G_s)$ can be expressed in terms of the unidirectional ion flux at the reversal potential, J_0 $(= J^\rightarrow = J^\leftarrow)$:

$$G = J_0(ze)^2/kT \tag{13}$$

Equation (13) relates the thermal ion motion, as expressed by J_0, to the membrane's electrical behavior, as expressed by G. The physical content of Eq. (13) is most clearly recognized by rewriting it as $G = zeJ_0/[kT/(ze)]$. Equation (13) is closely related to Eq. (6), which will become important when discussing specific permeation models. The relation between Eqs. (6) and (13) can be seen as follows: $I = ze(J^\rightarrow - J^\leftarrow)$; when $\Delta V \simeq V_{rev}$, then $I \simeq G(\Delta V - V_{rev})$, and Eq. (6) becomes $J^\rightarrow/J^\leftarrow \simeq 1 + (ze/kT)(\Delta V - V_{rev})$

[33] A. Finkelstein and A. Mauro, *in* "Handbook of Physiology: 1. The Nervous System" (E. R. Kandel, ed.), Vol. 1, pp. 161–213. Amer. Physiol. Soc., Washington, D.C., 1977.
[34] O. S. Andersen, W. N. Green, and B. W. Urban, *in* "Reconstitution of Ion Channels" (C. Miller, ed.), pp. 385–404. Plenum, New York, 1986.

or $(J^{\rightarrow} - J^{\leftarrow}) \simeq (ze/kT)J_0(\Delta V - V_{rev})$, which can be rewritten as $I = [(ze)^2/kT]J_0(\Delta V - V_{rev})$.

If several ions are permeant, a membrane's ion selectivity (ability to discriminate among the permeant ions) can be determined from the conductance ratio or from the permeability ratio. The conductance ratio for two ions, A and B, is determined in experiments where either A or B is present at equal concentrations in both aqueous phases. The ratio of the measured conductances, G_A/G_B, is the conductance ratio, $G_A/G_B = K_A D_A/(K_B D_B)$. The permeability ratio, P_A/P_B, is determined in reversal potential experiments where both ions are present simultaneously. When the constant-field approximation is valid, the reversal potential becomes[35]

$$V_{rev} = [kT/(ze)] \ln\{(P_A[A]_r + P_B[B]_r)/(P_A[A]_l + P_B[B]_l)\} \qquad (14a)$$

$$= [kT/(ze)] \ln\{(P_A/P_B[A]_r + [B]_r)/(P_A/P_B[A]_l + [B]_l)\} \qquad (14b)$$

where $P_A = K_A D_A/d$ and $P_B = K_B D_B/d$, such that $P_A/P_B = K_A D_A/(K_B D_B)$. When the ions traverse the membrane independent of each other, that is, when the flux does not saturate as a function of the permeant ion concentration, the permeability ratio is equal to the conductance ratio.

Discrete State Flux Equations

Ion movement through a membrane protein probably occurs as a series of transitions from one coordinating site to another (or to the aqueous phase). This is the physical basis for describing selective ion movement by "discrete state" permeation models, where it is assumed that an ion moving across a membrane will reside most of the time in a few distinct regions, denoted binding sites (Fig. 5), and that the time spent in transition from one site to another is negligible relative to the time spent at the sites. The stationary flux through the membrane is equal to the net flux across any one energy barrier[36]:

$$J = k_{n+1}^n N(n) - k_n^{n+1} N(n + 1) \qquad (15)$$

where N denotes the number of ions per unit membrane area in the nth and $n + 1$ binding site, and k denotes the rate constants for transitions across the barriers separating the energy minima.

There is no fundamental difference between continuum and discrete descriptions of ion movement. Certain features of selective ion movement —e.g., flux saturation as a function of permeant concentration and nonlinear I–V relations in symmetric salt solutions—are, however, most

[35] A. L. Hodgkin and B. Katz, *J. Physiol. (London)* **108**, 37 (1949).

[36] B. J. Zwolinski, H. Eyring, and C. E. Reese, *J. Phys. Colloid Chem.* **53**, 1426 (1949).

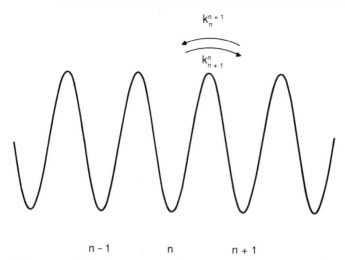

FIG. 5. Energy profile for ion movement along an array of discrete sites.[36] k^n_{n+1} denotes the rate constant for transitions site n to site $n + 1$, while k^{n+1}_n denotes the rate constant for transitions from site $n + 1$ to n.

conveniently introduced into discrete state flux equations *as consequences of approximations and assumptions that are made in the formulation of the kinetic schemes.*

To take a simple example, a membrane-spanning channel is *approximated* as having a single binding site, and it is *assumed* that at most a single ion can reside in the binding site at any time. The channel can thus exist is two states: empty, O, and ion occupied, I, (see Fig. 6). Ion movement through the channel can be described in terms of the transitions between the two states. These transitions can be described using conventional chemical kinetics (e.g., Ref. 26):

$$dW(\text{O})/dt = -(k^{\text{O}}_{\text{Il}}[\text{I}]_l + k^{\text{O}}_{\text{Ir}}[\text{I}]_r)W(\text{O}) + (k^{\text{I}}_{\text{Ol}} + k^{\text{I}}_{\text{Or}})W(\text{I}) \qquad (16)$$

$$dW(\text{I})/dt = (k^{\text{O}}_{\text{Il}}[\text{I}]_l + k^{\text{O}}_{\text{Ir}}[\text{I}]_r)W(\text{O}) - (k^{\text{I}}_{\text{Ol}} + k^{\text{I}}_{\text{Or}})W(\text{I}) \qquad (17)$$

and, by conservation,

$$1 = W(\text{O}) + W(\text{I}) \qquad (18)$$

where W denotes the probability of finding the system in state O and I, respectively, t is time, and k is the association or dissociation rate constant (which is a function of ΔV). The rate constants are subject to the restrictions imposed by the principle of detailed balance (e.g., Ref. 37), which

[37] I. Amdur and G. G. Hammes, "Chemical Kinetics, Principles and Selected Topics," Chap. 1. McGraw-Hill, New York, 1966.

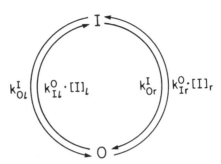

FIG. 6. Schematic representation of the interconnections between the two states in a one-site channel. k_l^O and k_l^O denote the association and dissociation rate constants, at the left and right interfaces, as indicated by the subscripts l and r.

states that the ratio of the counterclockwise and clockwise products of the rate constants must equal $\exp\{z\,\Delta U\}$ to satisfy equilibrium thermodynamic constraints:

$$k_{Il}^O k_{Or}^I/(k_{Ir}^O k_{Ol}^I) = \exp\{z\,\Delta U\} \tag{19}$$

In the stationary state, $dW(\text{I})/dt = 0$ and

$$W(\text{I}) = (k_{Il}^O[\text{I}]_l + k_{Ir}^O[\text{I}]_r)/(k_{Il}^O[\text{I}]_l + k_{Ir}^O[\text{I}]_r + k_{Ol}^I + k_{Or}^I) \tag{20}$$

$$W(\text{O}) = (k_{Ol}^I + k_{Or}^I)/(k_{Il}^O[\text{I}]_l + k_{Ir}^O[\text{I}]_r + k_{Ol}^I + k_{Or}^I) \tag{21}$$

The net flux through the channel, j, is obtained as the net flux across one of the barriers:

$$j = k_{Il}^O[\text{I}]_l W(\text{O}) - k_{Ol}^I W(\text{I}) = k_{Or}^I W(\text{I}) - k_{Ir}^O[\text{I}]_r W(\text{O}) \tag{22}$$

(In the following discussion, the lowercase letters j, i, and g denote the flux, current, and conductance per channel or carrier. The corresponding uppercase letters denote the same quantities per unit membrane area: $J = Nj$, etc, where N denotes the number of channels or carriers per unit membrane area.) The single-channel current, i, is obtained as $i = zej$. When Eqs. (20) and (21) are inserted into Eq. (22), the resulting expression for the current is

$$i = ze(k_{Il}^O k_{Or}^I[\text{I}]_l - k_{Ir}^O k_{Ol}^I[\text{I}]_r)/(k_{Il}^O[\text{I}]_l + k_{Ir}^O[\text{I}]_r + k_{Ol}^I + k_{Or}^I) \tag{23}$$

In symmetrical solutions, the small-signal single-channel conductance, g, is the limiting value of $i/\Delta V$ as $\Delta V \to 0$. Since $k_{Ir}^O k_{Ol}^I \simeq k_{Il}^O k_{Or}^I(1 - ze\,\Delta V/kT)$ [from Eq. (19)]:

$$g = g_{max}[\text{I}]/(K_\text{I} + [\text{I}]) \tag{24}$$

where $g_{max} = [(ze)^2/kT]k_{Ol}^I k_{Or}^I/(k_{Ol}^I + k_{Or}^I)$ and $K_\text{I} = k_{Ol}^I/k_{Il}^O = k_{Or}^I/k_{Ir}^O$ (the rate constants denote the values at $\Delta V = 0$). Equation (24) should be contrasted with Eq. (12). In the present model, the conductance saturates as a function of [I] because there is an upper limit for the number of ions that can reside in the channel — enzyme-catalyzed turnover saturates as a function of substrate concentration because there is an upper limit to the number of substrate molecules that can reside in the active center.

Eyring Transition State Theory

The aim of the kinetic description is to determine the kinetic mechanism, to measure the magnitude of the rate constants, and ultimately to relate their magnitudes to molecular features of the ion translocator. This is where Eyring's transition state theory[27,38] can be useful. The rate constant for ion movement across an elementary barrier k_b^a, from one site at $x = a$ into another at $x = b$, can be expressed using the transition state theory as

$$k_b^a = (kT/h) \exp\{-\Delta E^{\ddagger}/kT\} \qquad (25)$$

where h is Planck's constant and $\Delta E^{\ddagger}$ denotes the height of the free energy barrier (relative to the energy minimum at $x = a$) the ion must traverse to go from a to b. The physical significance of the two terms in Eq. (25) is that $\exp\{-\Delta E^{\ddagger}/kT\}$ denotes the ion's "partition coefficient" between a transition state, at the barrier peak, and the energy minimum at a [cf. Eq. (1)], while kT/h is a frequency factor that denotes how fast the ion traverses the transition state. Equation (25) can be used in two ways: to calculate $\Delta E^{\ddagger}$ from a measured k_b^a, and to calculate either k_b^a from a known $\Delta E^{\ddagger}$, or a *change* in k_b^a from a *change* in $\Delta E^{\ddagger}$. The last possibility is particularly useful in the analysis of current–voltage relations because $\Delta E^{\ddagger}$ varies as a function of ΔU. To a first approximation,

$$\Delta E^{\ddagger}(\Delta U) = \Delta E^{\ddagger}(0) - z\delta_1 kT \, \Delta U \qquad (26)$$

where δ_1 is the fraction of ΔU that falls between the energy minimum at $x = a$ and the energy peak at the barrier separating the site at a from that at b (Fig. 7). [Note that the minus sign in Eq. (26) results from the sign convention for ΔU.] The change in $\Delta E^{\ddagger}$ as a function of ΔU results in a change in k_b^a:

$$k_b^a = \kappa_b^a \exp\{z\delta_1 \, \Delta U\} \qquad (27)$$

[38] J. W. Moore and R. G. Pearson, "Kinetics and Mechanism," 3rd Ed., Chap. 5. Wiley, New York, 1981.

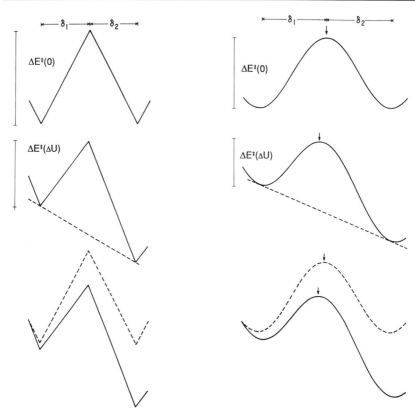

FIG. 7. Energy barriers for ion movement. (Left) Sharp (Eyring-type) barrier. (Right) Broad barrier. (Top) $\Delta U = 0$, $\Delta E^{\ddagger}$ denotes the height of the free energy barrier the ion must traverse. For each profile, δ_1 denotes the fraction of ΔU that falls between the energy minimum toward the left and the peak, while δ_2 denotes the fraction of ΔU that falls between the peak and the minimum toward the right. (Middle) The energy profiles are modified by an applied potential difference. (Bottom) A superposition of the profiles in the presence (——) and absence (---) of an applied potential difference. For the sharp barrier, the position of the peak *does not* vary as a function of ΔU. For the broad barrier, the position of the peak *does* vary as a function of ΔU.

where $\kappa_b^a = (kT/h) \exp\{-\Delta E^{\ddagger}(0)/kT\}$ denotes the magnitude of k_b^a at $\Delta U = 0$.

Equations (26) and (27) should be valid when the ion distribution between the transition state and adjacent energy well is unperturbed by a net flux across the barrier. This situation is usually depicted by a sharp energy barrier (Fig. 7, left-hand side). Generally, if the transition involves (collisional) energy exchange with the environment, the ion distribution at the barrier peak will be perturbed by a net flux (e.g., Ref. 39). This situation is usually depicted by a broad energy barrier. The position of the barrier peak will in this case vary as function of ΔU (Fig. 7, right-hand side), which affects the voltage dependence of the rate constants (vide infra).

Continuum Models

The simple electrodiffusive description of membrane permeation (e.g., the Nernst–Planck equations) does not account for flux saturation or for nonlinear $I-V$ relations in symmetrical solutions. These features can be described using more elaborate electrodiffusive permeation models. An advantage of these descriptions is that they can easily be extended to arbitrary energy profiles.

The failure of the Nernst–Planck equations to describe flux saturation results from a neglect of ion–ion interactions. These interactions can be introduced using the same ad hoc approach as was used in the discrete state description, Eqs. (16)–(18), i.e., by assuming that an ion can enter only an empty channel. Analytically this approximation can be implemented as suggested by Levitt[40]: the ion concentrations at the left and right entrance, $C(0)$ and $C(d)$, are related to the bulk concentrations, $[I]_l$ and $[I]_r$, by $C(0) = W(O)K[I]$ and $C(d) = W(O)K[I]_r$, where $W(O)$ is the probability that the channel is empty. The partition coefficient, K, is defined as $C(0)/[I]_l$—or $C(d)/[I]_r$—in the limit when $W(O) \rightarrow 1$. The (time-averaged) ion concentration in the channel can be obtained by integrating and reorganizing Eq. (1)[40]:

$$C(x) = (C(0) - J \, DN(x)) \exp\{- zU(x)\} \tag{28}$$

where $DN(x) = \int_0^x \exp\{zU(x')\}/D \, dx'$. Introducing Eqs. (3) and (4), the ion's concentration profile in the channel is

$$C(x) = W(O)K\{[I]_l - ([I]_l - [I]_r \exp\{- z \, \Delta U\})DN(x)/DN\}$$
$$\times \exp\{- zU(x)\} \tag{29}$$

and the average ion occupancy, or probability of finding an ion in the channel, $W(I)$, is

$$W(I) = A \int_0^d C(x) \, dx = W(O)[I]_l M(\Delta U, [I]_r/[I]_l) \tag{30}$$

[39] O. S. Andersen, in "Membrane Transport in Biology" (G. Giebisch, D. C. Tosteson, and H. H. Ussing, eds.), Vol. 1, pp. 369–446. Springer-Verlag, Berlin and New York, 1978.
[40] D. G. Levitt, Annu. Rev. Biophys. Chem. 15, 29 (1986).

where A is the channel's cross-sectional area, and

$M(\Delta U, [I]_r/[I]_l)$

$$= KA \int_0^d \{1 - (1 - \exp\{-z\,\Delta U\}[I]_r/[I]_l)DN(x)/DN\}$$

$$\times \exp\{-zU(x)\}\,dx \tag{31}$$

The single-channel flux, j, is equal to AJ. In symmetrical solutions, $W(O) = 1/(1 + [I]M(\Delta U, 1))$, and

$$
\begin{aligned}
j &= j_{max}(\Delta U)M(\Delta U, 1)[I]/(1 + M(\Delta U, 1)[I]) \\
&= j_{max}(\Delta U)[I]/(1/M(\Delta U, 1) + [I])
\end{aligned} \tag{32}
$$

where $j_{max}(\Delta U) = (1 - \exp\{-z\,\Delta U\})A/(DN M(\Delta U, 1))$, and $1/M(\Delta U, 1)$ is the concentration for half-maximal flux. At low permeant ion concentrations, when $M(\Delta U, 1)[I] \ll 1$, Eq. (32) simplifies to become: $j = [I](1 - \exp\{-z\,\Delta U\})A/DN$ [cf. Eq. (3)]. When the constant-field approximation holds, the physical interpretation of the terms in Eq. (32) is simple, $M(\Delta U, 1) = KAd$, and the maximal single-channel conductance, g_{max}, is

$$g_{max} = [(ze)^2/kT]D/d^2 \tag{33}$$

The one-ion assumption implies that the channel's cross-sectional area does *not* enter into the expression for g_{max}! By comparison with Eq. (13), it is apparent that D/d^2 is equal to the maximal unidirectional flux through the channel at $\Delta V = 0$. For a 25-Å-long channel and an effective diffusion coefficient of 2×10^{-6} cm^2 sec^{-1},[41] D/d^2 is 3.2×10^7 sec^{-1} and the maximal conductance is 200 pS. If D in the channel is equal to the bulk diffusion coefficient, $D \approx 2 \times 10^{-5}$ cm^2 sec^{-1}, the maximal conductance is 2 nS!

The expression for the maximal unidirectional flux, D/d^2, is related to the Einstein–Smoluchowski relation for one-dimensional diffusion, $D = \langle x^2 \rangle/(2t)$ (e.g., Ref. 29), where x denotes the distance a particle has diffused during the time interval t and the brackets denote the average value. [The "missing" factor of 2 in Eq. (33) reflects flux saturation; the channel will with equal probability be occupied by an ion from the left *or* the right solution—cf. Eqs. (67) and (68).] Here $d^2/(2D)$ is the average time between passages of an ion from the left to the right aqueous phase, or vice versa, at $\Delta V = 0$. It is related to the first passage time for ion movement through the channel, τ, which in this case is equal to $d^2/[6D]$ (cf. Ref. 42). Here τ denotes the average duration of a successful ion passage through the

---

[41] J. A. Dani and D. G. Levitt, *Biophys. J.* **35**, 501 (1981).
[42] E. Jakobsson and S.-W. Chiu, *Biophys. J.* **52**, 33 (1987).

channel. In other words, the channel will two-thirds of the time be occupied by ions that exit to the same side from which they entered. Analysis of first passage times for arbitrary values of ΔV and energy profiles provides many other insights into electrodiffusive ion movement through channels.[42,43]

The failure of the Nernst–Planck equations to describe nonlinear I–V relations in symmetrical solutions results from a neglect of spatial variations in the ion's energy or mobility (diffusion coefficient). Nonlinear I–V relations result when the ion diffusion coefficient varies as a function of distance, when the ion traverses an energy barrier, and from a combination of these effects. Electrodiffusive ion movement across an energy barrier is described by[28]

$$J = -D(x)\{dC/dx + C(dE/dx + ze\,dV/dx)/kT\} \tag{34}$$

where $E(x)$ denotes the potential energy (potential of mean force[44]) for the ion, apart from the $zeV(x)$ term. Equation (34) can be deceptive in the implication that one can separate the interactions between the permeating ion and its environment into time-averaged interactions that are expressed through the energy term, E, and dynamic interactions that are expressed through the diffusion coefficient, D. The approximations involved in this separation should hold when the average passage time for ion movement is either much shorter or much longer than the characteristic relaxation times for the channel protein and associated H_2O and lipid molecules, etc. When the passage time for ion movement and the channel relaxation times are comparable, the approximations that are involved in the derivation of Eq. (34) break down (cf. Ref. 43). Given this reservation, Eq. (34) can be integrated across the membrane, from $x = 0$ to $x = d$:

$$J = (C(0) - C(d)\exp\{\Delta E/kT - z\,\Delta U\})/\mathrm{DK} \tag{35}$$

where $\Delta E = E(d) - E(0)$, and

$$\mathrm{DK} = \int_0^d \exp\{E(x)/kT + zU(x)\}/D(x)\,dx \tag{36}$$

In Eq. (36), $\exp\{E(x)/kT\}$ is divided by $D(x)$, so in terms of the effect on DK, a local decrease in D is equivalent to an increase in E. Only the energetic term, E, enters into the numerator in Eq. (35), however. One can, therefore, in principle, distinguish between time-averaged energetic interactions and dynamic interactions between a channel and a permeating ion

[43] R. S. Eisenberg, K. E. Cooper, and P. Y. Gates, *Q. Rev. Biophys.* **21**, 331 (1988).
[44] T. L. Hill, "An Introduction to Statistical Thermodynamics," p. 508. Addison-Wesley, Reading, Massachusetts, 1960.

in a saturable channel, when the permeation is studied over a sufficiently large permeant ion concentration range. Flux saturation is introduced as in Eqs. (28)–(33).[40]

This approach is particularly useful when the energy profile is fairly flat, such that the energy minima are shallow and are separated by small energy barriers. In this case, the local ion distribution in the vicinity of the minima cannot be approximated by an equilibrium distribution, and the energy wells cannot be approximated to be spatially defined *sites*.

Discrete Approximations to the Continuum Model

When the free-energy profile has deep and spatially defined minima that are separated by substantial energy barriers, the local ion distribution in the vicinity of the energy minima, at the binding sites, can be approximated by an equilibrium distribution. (As a practical measure, the height of the energy barrier relative to the adjacent wells should be at least $5kT$, since $\exp\{-5\} < 10^{-2}$.) It should in this case be possible to decompose the continuum description into a set of discrete flux equations by integrating Eq. (34) piecewise, from the left aqueous phase to the first energy minimum, and so on, until the right aqueous phase is reached. For each piece, say between $x = a$ and $x = b$, one can use the equivalents of Eqs. (35) and (36) to derive rate constants for ion movement between two binding sites defined by the energy wells.[45,46] For sites at a and b, the rate constants, k_b^a and k_a^b, are determined from

$$J = k_b^a N(a) - k_a^b N(b) \tag{37}$$

where $N(a)$ and $N(b)$ denote the (time-averaged) number of ions in the binding sites [cf. Eq. (16)]. The N values are proportional to the local ion concentrations, C. The problem is to relate each k and N to the energy profile. To relate each N to each C, the energy profile in the binding sites will be *approximated* by a parabola[46]: around $x = a$, $E(x) = E(a) + kT[(x - a)/\rho_a]^2$; around $x = b$, $E(x) = E(b) + kT[(x - b)/\rho_b]^2$, where ρ denotes how E varies with distance from the minima: $E(x)$ has increased by kT when $|x - a| = \rho a$, etc. (Here ρ can also be expressed as $\rho = 1/[2kT]d^2E/dx^2$, evaluated at $x = a$ and $x = b$.) When the ρs are sufficiently small,

$$N(a) \simeq C(a) \int_{-\infty}^{+\infty} \exp\{-[(x - a)/\rho_a]^2\}dx = C(a)\pi^{0.5}\rho_a \tag{38}$$

[45] P. Läuger, *Biophys. Chem.* **15**, 89 (1982).
[46] K. Cooper, E. Jakobsson, and P. Wolynes, *Prog. Biophys. Mol. Biol.* **46**, 51 (1985).

with a similar expression for $N(b)$. To relate each k to each DK, the energy profile in the vicinity of the peak will again be approximated by a parabola, at $x = x_1$: $E^{\ddagger}(x) = E^{\ddagger}(x_1) - kT((x - x_1)/\rho)^2$, where ρ denotes how fast the energy falls off from the maximal value at $x = x_1$. If $D(x)$ is constant $(= D)$ in the vicinity of x_1

$$\text{DK}(0) \simeq [\exp\{(E(x_1) - E(a))/kT\}/D] \int_{-\infty}^{+\infty} \exp\{-[(x - x_1)/\rho]^2\}\, dx$$

$$= \exp\{\Delta E^{\ddagger}/kT\}\pi^{0.5}\rho/D \tag{39}$$

where $\Delta E^{\ddagger} = E(x_1) - E(a)$. Combining Eqs. (35)–(39), the voltage-independent contribution to k_b^a, κ_b^a, is

$$\kappa_b^a = [D/(\pi\rho_a\rho)] \exp\{-\Delta E^{\ddagger}/kT\} \tag{40a}$$

while the voltage-independent contribution to k_a^b is

$$\kappa_a^b \simeq [\kappa_b^a\rho_a/\rho_b] \exp\{\Delta E/kT\} \tag{40b}$$

where $\Delta E = E(b) - E(a)$.

Equations (40a) and (40b) are similar to the Eyring transition state theory expression, Eq. (25). But the preexponential factors, and thus the relation between barrier height and rate constants, are different.[45,46] At 25°, $kT/h = 6.2 \times 10^{12} \text{ sec}^{-1}$, while an upper estimate for $D/(\pi\rho_a\rho)$ is $\sim 6 \times 10^{10} \text{ sec}^{-1}$ ($D \leq 2.10^{-5} \text{ cm}^2 \text{ sec}^{-1}$, and ρ_a and ρ are $\geq 10^{-8}$ cm). The difference is inconsequential given the uncertainties in interpreting (or calculating) $\Delta E^{\ddagger}$. But it serves to emphasize that kinetic data provide information about rate constants, and that deductions about the heights of energy barriers, etc., depend on an underlying physical model.

The voltage variation of k_b^a and k_a^b can be approximated by superimposing a linear electrostatic potential profile on the free-energy profile (cf. Fig. 7, right-hand side). (The electrostatic potential profile is only *approximately* constant,[47] but no loss of generality results from this assumption.) In the vicinity of the peak at x_1, the variation in E as a function of ΔU and x is

$$E(\Delta U, x) = E(0, x_1) - zkT \Delta U[\delta_1 + (x - x_1)/d] - kT[(x - x_1)/\rho]^2 \tag{41}$$

where δ_1 is the fraction of ΔU that falls between a and x_1. The x-dependent part of $E(\Delta U, x)$, $-kT\{z \Delta U(x - x_1)/d + [(x - x_1)/\rho]^2\}$, can be rewritten as $-kT\{[(x - x_1)/\rho + z \Delta U \cdot \rho/(2d)]^2 - [z \Delta U\rho/2d)]^2\}$, and the expression for DK becomes

$$\text{DK}(\Delta U) = \exp\{-z\delta_1 \Delta U - \omega \Delta U^2\}\text{DK}(0) \tag{42}$$

[47] B. Neumcke and P. Läuger, *J. Membr. Biol.* 3, 54 (1970).

where $\omega = (z\rho/(2d))^2.$[48,49] (This approximation is meaningful as long as $z\delta_1 \Delta U - \omega \Delta U^2$ is a monotonic function of ΔU, i.e., as long as the restriction $\omega \leq \delta_1/[2 \Delta U]$ holds.)

An applied electrostatic potential difference will also alter the energy profile at the energy minima that define the binding sites. Analytically, this situation is handled as above. If ρ_a and ρ_b are (much) less than ρ, that is, if the minima are better defined than the peak, one can, to a first approximation, disregard the effect of ΔU on the relation between C and N [Eq. (39)]. In this approximation, the expressions for k_b^a and k_a^b are

$$k_b^a = \kappa_b^a \exp\{z\delta_1 \Delta U - \omega \Delta U^2\} \tag{43a}$$

and

$$k_a^b = \kappa_a^b \exp\{-z\delta_2 \Delta U - \omega \Delta U^2\} \tag{43b}$$

where δ_2 is the fraction of ΔU that falls between x_1 and $b[= (x_1 - b)/d$; see Fig. 7). (For a flat or symmetrical barrier, it is convenient to define $\delta_2 = \delta_1$.) A curvature at the peak alters both k_b^a and k_a^b, such that the ratio k_b^a/k_a^b satisfies the thermodynamic constraint, $k_b^a/k_a^b = (\kappa_b^a/\kappa_a^b)$ $\exp\{z(\delta_1 + \delta_2)\Delta U\}$, independent of barrier shape.

Generally, for arbitrary energy, mobility, and electrostatic potential profiles, the rate constants can be written as

$$k_b^a = \kappa_b^a \exp\{z\delta_1 \Delta U\}S(\Delta U) \tag{44a}$$

and

$$k_a^b = \kappa_a^b \exp\{-z\delta_2 \Delta U\}S(\Delta U) \tag{44b}$$

where the shape function, $S(\Delta U)$ (≤ 1), denotes how the voltage dependence of the k parameters departs from a simple exponential behavior[50]:

$$S(\Delta U) = \exp\{-z\delta_1 \Delta U\}DK(0)/DK(\Delta U) \tag{45}$$

Two limiting cases are of interest: if $\exp\{E(x)/kT\}/D(x)$ has a sharp maximum at x_1, then $DK(\Delta U) \simeq \exp\{z\delta_1 \Delta U\}DK(0)$, and $S(\Delta U) \simeq 1$ [cf. Eq. (27)]; if $E(x)$ and $D(x)$ are constant in the membrane interior, if $\Delta E(x) = E$ and $D(x) = D$ for $a < x < b$, one can approximate the ion movement as a one-step transition if $E \geq 5\ kT$, and $DK(\Delta U) = DK(0)(1 - \exp\{-z \Delta U\})/ (z \Delta U)$ [cf. Eq. (7)], or $S(\Delta U) = z \Delta U/\sinh\{z\delta_1 \Delta U\}$ and the predicted $I - V$ relations are linear!

In principle, there is a one-to-one relation between the barrier shape

[48] D. A. Haydon and S. B. Hladky, *Q. Rev. Biophys.* **5**, 187 (1972).

[49] O. S. Andersen and M. Fuchs, *Biophys. J.* **15**, 795 (1975).

[50] J. E. Hall, C. A. Mead, and G. Szabo, *J. Membr. Biol.* **11**, 75 (1973).

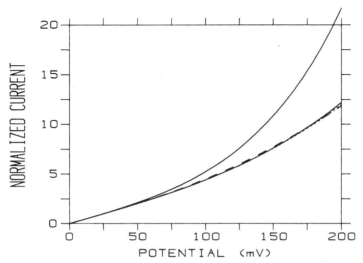

FIG. 8. Current–voltage relations for ion movement across a symmetrical barrier. Upper curve is simulated using Eq. (49), $z = 1$ and $\delta_1 = 0.35$. Lower curves (---) simulated using Eq. (50), $z = 1$, $\delta_1 = 0.35$, $\omega = 0.01$; ($\cdot \cdot \cdot \cdot \cdot$) simulated using Eq. (52), $z = 1$, $\delta_1 = 0.35$, $\delta_2 = 0.154$; (——) simulated using Eq. (47), $z = 1$, $\delta_1' = 0.22$. The three lower curves are effectively indistinguishable.

and the shape function.[51] In practice, fairly different barrier shapes can be used to predict similarly shaped current–voltage relations (e.g., Refs. 49 and 52). Given this uncertainty, it may be expeditious to use the approximations

$$\exp\{z\delta_1' \, \Delta U\} \simeq \exp\{z\delta_1 \, \Delta U\}S(\Delta U) \qquad (46)$$

$$\sinh\{z\delta_1' \, \Delta U\} \simeq \sinh\{z\delta_1 \, \Delta U\}S(\Delta U) \qquad (47)$$

where $\delta_1' < \delta_1$[49,53] (see Fig. 8). For additional information about electrodiffusive ion movement across energy barriers, see Refs. 39, 46, 49, 51, and 54.

A Broad Barrier Can Be Simulated by Incorporating Additional Sites

According to either Eq. (27) or Eq. (43a), k_0^a increases as $z\Delta U$ increases. For a given δ_1, however, if the barrier is broad (if $\omega > 0$) the predicted voltage dependence is less when calculated using Eqs. (43a) and (43b) than using Eq. (27). This has implications for the shape of the current–voltage

[51] D. Attwell and J. Jack, *Prog. Biophys. Mol. Biol.* **34**, 81 (1978).
[52] S. Ginsburg and D. Noble, *J. Membr. Biol.* **29**, 211 (1976).
[53] S. B. Hladky, *Biochim. Biophys. Acta* **375**, 327 (1975).
[54] P. Läuger and B. Neumcke, *in* "Membranes: 2. Lipid Bilayers and Antibiotics" (G. Eisenman, ed.), pp. 1–59. Dekker, New York, 1973.

relations. For ion movement across a symmetrical barrier separating two sites, sites 1 and 3, the current is

$$i = zej = ze\{k_3^1 W_1 - k_1^3 W_3\} \tag{48}$$

where W_1 denotes the probability that site 1 is occupied and site 3 is empty, and vice versa for W_3. We will assume that $W_1 = W_3 = W$, independent of ΔU. To obtain the $i - V$ relations, the explicit voltage dependence of the k parameters must be introduced. If the barrier is sharp and $\delta_2 = \delta_1$, Eq. (48) can be rewritten using Eq. (27),

$$i = 2eW\kappa \sinh\{\delta_1 z \, \Delta U\} \tag{49}$$

where $\kappa = \kappa_3^1 = \kappa_1^3$. If the peak has a finite curvature, which means that $\rho > 0$ (or $\omega > 0$), Eq. (48) can be rewritten using Eqs. (43a) and (43b):

$$i = 2eW\kappa \sinh\{\delta_1 z \, \Delta U\}/\exp\{\omega(\Delta U)^2\} \tag{50}$$

Simulated $i - V$ relations for a monovalent ion, with $\omega = 0$ and 0.01 (or $\rho/d = 0$ and 0.2), are compared in Fig. 8. A broad barrier results in less steep $i - V$ relations.

If the voltage dependence of the rate constants is stipulated to be given by Eq. (27), the less steep voltage dependence can be described by introducing an additional (low-affinity) site, site 2, halfway between sites 1 and 3.[36,55] The expression for the stationary current is obtained by noting that $W_2 = (k_2^1 W_1 + k_2^3 W_3)/(k_1^2 + k_3^2)$, $W_2 \ll W_1$, W_2. The net flux is equal to $k_2^1 W_1 - k_1^2 W_2$, and i is given by

$$i = e(k_2^1 W_1 k_3^2 - k_2^3 W_3 k_1^2)/(k_1^2 + k_3^2) \tag{51}$$

which is equivalent to Eq. (48) when k_3^1 and k_1^3 are defined as $k_3^1 = k_2^1 k_3^2/(k_1^2 + k_3^2)$ and $k_1^3 = k_2^3 k_1^2/(k_1^2 + k_3^2)$. Introducing the explicit voltage dependencies, Eq. (51) becomes

$$i = zeW\kappa_2^1 \sinh\{\delta_1 z \, \Delta U\}/\cosh\{z\delta_2 \, \Delta U\} \tag{52}$$

where δ_2 denotes the fraction of ΔU that falls between site 2 and the adjacent barrier to the left or right. Using Eq. (52) one can simulate $i - V$ relations that are similar to those obtained using Eq. (50) (see Fig. 8). Site 2 was introduced to allow the voltage dependence of the rate constants to be described by Eq. (27); it therefore follows that assumptions about the barrier shape, i.e., about the voltage dependence of the rate constants, have implications for deductions about the number of sites in the permeation path.

[55] S. W. Feldberg and G. Kissel, J. Membr. Biol. **20**, 269 (1975).

Specific Kinetic Models

Eventually, kinetic models for selective transmembrane ion movement will be continuum descriptions incorporating increasingly detailed structural and dynamic information about the ion-translocating protein in question. Irrespective of the power of such models, however, discrete kinetic models will continue to be important as shorthand notations to describe features of selective ion movement. Discrete state models are approximate, however: if the energy minima that serve as ion-binding sites are deep and spatially distinct, the transmembrane ion movement can be approximated as a series of transitions between a small number of states. Despite this simplification, the models quickly become intractable. I will therefore focus on stationary ion movement mediated by two basic permeation models: ion movement mediated by a two-site channel with single-ion occupancy, the simple channel; and ion movement mediated by a one-site carrier, the simple carrier. The two-site channel with double-ion occupancy will be used to illustrate that there is no intuitively obvious relation between ion occupancy and ion movement in more complex schemes.

To facilitate comparison among the schemes, the kinetic equations are expressed for a single membrane-bound ion translocator, using a uniform nomenclature. All equations are written using concentrations rather than activities.

Two-Site–One-Ion Channel

The general features of the energy profile for ion translocation mediated by integral membrane proteins (Fig. 1) suggest that there should be significant energy wells, ion-binding sites, at each end of the permeation path. If a channel is *approximated* as having two binding sites, and if it is *assumed* that at most a single ion can reside in the channel, ion movement through the channel can be described as transitions among three states (e.g., Refs. 19 and 56; see Fig. 9): empty, OO; an ion in the left binding site, IO; and an ion in the right binding site, OI. The kinetic equations associated with this scheme are solved as described for Eqs. (16)–(21). The stationary state probabilities are listed in the Section 1 of the appendix at the end of this chapter. The net flux is equal to the net flux between any two of the three states, e.g., between IO and OI:

$$j = k_{OI}^{IO} W(IO) - k_{IO}^{OI} W(OI)$$
$$= (k_{IO}^{OO} k_{OI}^{IO} k_{OO}^{OI}[I]_l - k_{OI}^{OO} k_{IO}^{OI} k_{OO}^{IO}[I]_r)/DD_1 \qquad (53)$$

[56] S. B. Hladky, *in* "Drugs and Transport Processes" (B. A. Callingham, ed.), pp. 193–210. Macmillan, New York, 1974.

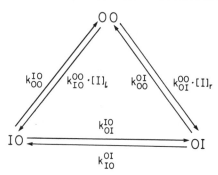

FIG. 9. Schematic representation of the interconnections among the three kinetic states in a two-site channel with single occupancy.

where W denotes the probability of finding the channel in the indicated states [see Eqs. (A1.2) and (A1.3)], and the denominator DD_1 is defined in Eq. (A1.4). Equation (53) can be rewritten in terms of experimental parameters:

$$j = (k_I^l[I]_l - k_I^r[I]_r)/(1 + [I]_l/K_I^l + [I]_r/K_I^r) \tag{54}$$

where k_I^l and k_I^r denote rate coefficients for ion movement through the channel from left to right and right to left, respectively, and K_I^l and K_I^r are the concentrations for half-maximal flux when the trans concentration is zero. The channel's stationary permeability characteristics are described by these parameters, which can be expressed in terms of the underlying rate constants:

$$k_I^l = k_{IO}^{OO} k_{OI}^{IO} k_{OO}^{OI}/(k_{OI}^{IO} k_{OO}^{OI} + k_{IO}^{OI} k_{OO}^{IO} + k_{OO}^{IO} k_{OO}^{OI}) \tag{55a}$$

$$k_I^r = k_{OI}^{OO} k_{IO}^{OI} k_{OO}^{IO}/(k_{OI}^{IO} k_{OO}^{OI} + k_{IO}^{OI} k_{OO}^{IO} + k_{OO}^{IO} k_{OO}^{OI}) \tag{55b}$$

$$K_I^l = (k_{OI}^{IO} k_{OO}^{OI} + k_{IO}^{OI} k_{OO}^{IO} + k_{OO}^{IO} k_{OO}^{OI})/(k_{IO}^{OO}(k_{OI}^{OI} + k_{IO}^{OI} + k_{OO}^{OI})) \tag{56a}$$

$$K_I^r = (k_{OI}^{IO} k_{OO}^{OI} + k_{IO}^{OI} k_{OO}^{IO} + k_{OO}^{IO} k_{OO}^{OI})/(k_{OI}^{OO}(k_{OI}^{IO} + k_{IO}^{OI} + k_{OO}^{IO})) \tag{56b}$$

Importantly, k_I^l and k_I^r are related through the constraint imposed by detailed balance, Eq. (A1.5): $k_I^l/k_I^r = \exp\{z \, \Delta U\}$. For a given ΔV, there are only three independent experimental parameters available to determine five independent rate constants!

At low permeant ion concentrations, when $[I]_l/K_I^l$ and $[I]_r/K_I^r \ll 1$, Eq. (54) reduces to $j = k_I^l[I]_l - k_I^r[I]_r$, which can be rewritten as $j = ([I]_l - [I]_r \exp\{-z \, \Delta U\})k_I^l$ [cf. Eq. (3)]. A further simplification occurs at $\Delta V = 0$, when $k_I^l = k_I^r$, and $j = k_I^l([I]_l - [I]_r)$. In this case k_I^l and k_I^r are the channels' permeability coefficients. Generally, the permeability coefficient, p, is defined as: $p = j/([I]_l - [I]_r)$, when $\Delta V = 0$. For comparison with theoretical expressions, however, it is useful to determine the permeability coefficient in the limit where $([I]_l - [I]_r)/[I]_l \to 0$.[57] In the present model,

$$p_I = p_I^o/(1 + [I]/K_I) \tag{57}$$

where p_I^o denotes the limiting permeability coefficient at low permeant ion concentrations $(= k_I^l = k_I^r)$, and $K_I = K_I^l K_I^r/(K_I^l + K_I^r)$. This latter expression can be rewritten as $K_I = K_{IO}K_{OI}/(K_{IO} + K_{OI})$, where $K_{IO} = \kappa_{OO}^{IO}/\kappa_{IO}^{OO}$ and $K_{OI} = \kappa_{OO}^{OI}/\kappa_{OI}^{OO}$. Additional insight into the significance of p_I^o is obtained by noting that the maximal fluxes when the trans concentrations are zero, $j_{max}^{\rightarrow\prime}$ and $j_{max}^{\leftarrow\prime}$, are equal to $p_I^o K_I^l$ and $p_I^r K_I^r$, respectively, or

$$p_I^o = j_{max}^{\rightarrow\prime}/K_I^l = j_{max}^{\leftarrow\prime}/K_I^r \tag{58}$$

In enzyme kinetic terms, p_I^o is equal to k_{cat}/K_m,[58] and Eq. (58) is the Haldane equation (e.g., Ref. 58).

In symmetrical solutions, the current is

$$i = zej = ze(k_I^l - k_I^r)[I]/(1 + [I]/K_I) \tag{59}$$

where k and K_I vary as functions of ΔV. The small-signal conductance, g, is obtained from Eqs. (59) and (A1.5) [cf. the derivation of Eq. (24)]:

$$g = g_{max}[I]/(K_I + [I]) \tag{60}$$

where g_{max} is the maximal conductance

$$g_{max} = [(ze)^2/kT]p_I^o K_I \tag{61}$$

For the simple channel, the concentration for half-maximal conductance is equal to the dissociation constant for ion binding in the channel, and the single-ion assumption implies that $g_{max}/(p_I^o K_I)$ is a constant $[= (ze)^2/kT]$ for ions of the same valency. This can be used as a diagnostic test (e.g., Ref. 59).

The flux equations simplify considerably when the channel is symmetrical, i.e., when $\kappa_{IO}^{OO} = \kappa_{OI}^{OO}$, etc. In this case,

$$K_I = K_{IO}/2 = \kappa_{OO}^{IO}/(2\kappa_{IO}^{OO}) \tag{62}$$

$$g_{max} = [(ze)^2/kT]j_{max}^{\rightarrow\prime}/2 = [(ze)^2/kT]\kappa_{OI}^{IO}\kappa_{OO}^{OI}/(2(2\kappa_{OI}^{IO} + \kappa_{OO}^{OI})) \tag{63}$$

and

$$p_I^o = \kappa_{IO}^{OO}\kappa_{OI}^{IO}/(2\kappa_{OI}^{IO} + \kappa_{OO}^{OI}) \tag{64}$$

[57] A. Finkelstein and O. S. Andersen, *J. Membr. Biol.* **59**, 155 (1981).

[58] A. Fersht, "Enzyme Structure and Mechanism," 2nd Ed., Chap. 3. Freeman, San Francisco, California, 1985.

[59] R. Coronado, R. L. Rosenberg, and C. Miller, *J. Gen. Physiol.* **76**, 425 (1980).

Any one of these three parameters can be expressed by the other two. When $\kappa_{OO}^{OI}/\kappa_{OI}^{IO} \to 0$, $p_I^o \to \kappa_{IO}^{OO}$, and the channel's ability to discriminate among permeant ions is determined by the ions' association rate constants [see also Bezanilla and Armstrong[60] and the discussion relating to Eqs. (72) and (73)].

Flux saturation implies that different combinations of rate constants determine the rate of ion permeating at low concentrations, when $[I] \ll K_I$, and at high concentrations, when $K_I \ll [I]$ [cf. Eqs. (63) and (64)]. When $[I] \ll K_I$, the rate of ion movement is determined by the entry, translocation, and exit steps, Eq. (64). When $K_T \ll [I]$, the rate of ion movement is determined only by the translocation and exit steps, Eq. (63). Additionally, there is no fixed relation between p_I^o and g_{max} [cf. Eq. (62)]. Depending on the value of K_I, a permeant ion might have a large permeability coefficient and a small or a large single-channel conductance.[61,62]

Another consequence of flux saturation is that the unidirectional ion fluxes differ from the net fluxes determined when the trans concentration is zero. This situation is analyzed by considering two permeant ion species, A and B (see Fig. 16 in Section 2 appendix). The net fluxes of A and B, j_A and j_B, are given by

$$j_A = k_{OA}^{AO} W(AO) - k_{AO}^{OA} W(OA)$$
$$= (k_A^l[A]_l - k_A^r[A]_r)/(1 + [A]_l/K_A^l + [A]_r/K_A^r + [B]_l/K_B^l + [B]_r/K_B^r) \tag{65}$$

and

$$j_B = k_{OB}^{BO} W(BO) - k_{BO}^{OB} W(OB)$$
$$= (k_B^l[B]_l - k_B^r[B]_r)/(1 + [B]_l/K_B^l + [B]_r/K_B^r + [A]_l/K_A^l + [A]_r/K_A^r) \tag{66}$$

where W denotes the probability of finding the channel in the indicated states [see Eqs. (A2.4)–(A2.7)] and k and K correspond to those defined in Eqs. (56) and (57). When A and B are chemically indistinguishable, such that $k_{AO}^{OO} = k_{BO}^{OO} (= k_{IO}^{OO})$, etc., and if $[A]_r = [B]_l = 0$, the denominator in Eqs. (65) and (66) is identical to the denominator in Eq. (56) with $[I]_l = [A]_l$ and $[I]_r = [B]_r$. By definition, the net flux of A is the unidirectional (tracer) flux of I from left to right, $j^\to$,

$$j^\to = k_I^l[I]_l/(1 + [I]_l/K_I^l + [I]_r/K_I^r) \tag{67}$$

[60] F. Bezanilla and C. M. Armstrong, *J. Gen. Physiol.* **60**, 588 (1972).
[61] B. Hille, *J. Gen. Physiol.* **66**, 535 (1975).
[62] S. Cukierman, G. Yellen, and C. Miller, *Biophys. J.* **48**, 477 (1985).

with a corresponding expression for $j_I^{\leftarrow}$. The unidirectional flux is less than the net flux determined when the trans concentration is zero, $j^{\rightarrow\prime}$:

$$j^{\rightarrow\prime} = k_I^l[I]_l/(1 + [I]_l/K_I^l) \qquad (68)$$

The basis for this difference is that the channel may be occupied by ions that entered from the left *or* from the right aqueous phase.

The tracer permeability coefficient, p_I^*, is defined as $j_I^{\rightarrow}/[I]_l$ at $\Delta V = 0$.[30,57] In symmetrical solutions,

$$p_I^* = p_I^o/(1 + [I]/K_I) \qquad (69)$$

and the correlation coefficient, f (e.g., Ref. 63), that relates the p_I^* to p_I

$$f = p_I^*/p_I \qquad (70)$$

is equal to 1! The relation between the unidirectional flux at $\Delta V = 0$ and the single-channel conductance is therefore given by Eq. (12), and the flux ratio is expressed by Ussing's flux ratio equation, Eq. (6).

For two ions of the same valency, A and B, the channel's ion selectivity is quantified by the permeability ratio, p_A/p_B, or the conductance ratio, g_A/g_B. The permeability ratio is determined from the reversal potential measured in the presence of two permeant ions. The zero-current condition is that $j_A = -j_B$ or, after substitution of Eqs. (65) and (66),

$$k_A^l[A]_l + k_B^l[B]_l = k_A^r[A]_r + k_B^r[B]_r \qquad (71)$$

From the detailed balance constraint one finds that $k_A^l/k_A^r = k_B^l/k_B^r = \exp\{z\,\Delta U\}$ and $k_A^l/k_B^l = k_A^r/k_B^r$, and Eq. (71) can be rewritten as

$$U_{rev} = 1/z\,\ln\{((k_A^l/k_B^l)[A]_r + [B]_r)/((k_A^l/k_B^l)[A]_l + [B]_l)\} \qquad (72)$$

Equation (72) is similar to Eq. (14b), and k_A^l/k_B^l can be interpreted as a permeability ratio, p_A/p_B.[64] This interpretation is strengthened by noting that $k_A^l/k_B^l \rightarrow p_A^o/p_B^o$ as $\Delta V \rightarrow 0$ [cf. Eq. (57)]. Generally, p_A/p_B (or k_A^l/k_B^l) varies as function of ΔU,[61,65] because k_A^l will be a different function of ΔU than k_B^l unless the voltage dependence of corresponding rate constants (e.g., k_{OA}^{AO} and k_{OB}^{BO}) as well as the magnitudes of the ratios k_{OO}^{IO}/k_{OI}^{IO} and k_{OO}^{OI}/k_{IO}^{OI} do not vary with ion type. This latter condition is equivalent to the "constant peak offset condition."[61]

If k_A^l/k_B^l is voltage dependent, the magnitude of the permeability ratio depends on the experimental protocol. It is therefore important that the biionic potential (determined when $[A]_l = [B]_r = [I]$ and $[A]_r = [B]_l = 0$)

[63] K. Heckmann, in "Biomembranes" (F. Kreuzer and J. F. G. Slegers, eds.), Vol. 3, pp. 127–153. Plenum, New York, 1972.
[64] C. S. Patlak, *Nature (London)* **188**, 944 (1960).
[65] P. Läuger, *Biochim. Biophys. Acta* **311**, 423 (1973).

according to the present model does not vary as a function of [I]. If U_{rev} is determined in mixed salt conditions, however, it will vary as a function of the aqueous electrolyte composition. This variation in U_{rev} may produce an apparent concentration dependence of p_A/p_B.[66] (Attwell and Jack[51] should be consulted for specific examples of how the potential dependence of k_A^l/k_B^l depends on the barrier structure.)

The ratio of the measured single-channel conductances, g_A/g_B, is given by

$$g_A/g_B = (g_{max,A}/g_{max,B})(K_B + [I])/(K_A + [I]) \tag{73}$$

In contrast to the permeability ratio (determined under biionic conditions), the conductance ratio varies as a function of [I]: as $[I] \rightarrow 0$, $g_A/g_B \rightarrow p_A^0/p_B^0$; as $[I] \rightarrow \infty$, $g_A/g_B \rightarrow g_{max,A}/g_{max,B}$.

The two different quantifiers, the permeability ratio and the conductance ratio, of a channel's ion selectivity reflect different aspects of channel behavior. Importantly, a channel's ability to *discriminate* among two permeant ions in terms of their contributions to the total ion flux [cf. Eqs. (65) and (66)] is quantified by the permeability ratio (disregarding complications resulting from its voltage dependence). Further, if an ion has a high p^0 but a low g_{max}, it will bind strongly in the channel and may act as a (partial) channel blocker.[61,62]

For the simple channel, stationary flux experiments at a given ΔV can be used to determine three independent experimental parameters, while the ion translocation kinetics are determined by five independent rate constants. (For a symmetrical channel, at $\Delta V = 0$, two parameters can be measured while three rate constants determine the kinetics.) It is necessary to have additional experimental information to determine all the rate constants. This information can be obtained from current–voltage, i–V, relations. The principles underlying the analysis can be illustrated by noting how the shape of the i–V relations varies as a function of the rate constants and permeation concentration. For a symmetrical channel with simple exponential voltage dependence of the rate constants [see Eq. (27)] the i–V relation is given by

$$i = \frac{ze\kappa_{OI}^{IO}[I] \sinh\{(\delta_1 + \delta_2 + \delta_3)z \Delta U\}}{K_{IO}(\kappa_{OI}^{IO}/\kappa_{OO}^{OI} \cosh\{(\delta_2 + \delta_3)z \Delta U\} + 0.5) + [I] \cosh\{(\delta_1 + \delta_2)z \Delta U\} + [I](\kappa_{OI}^{IO}/\kappa_{OO}^{OI})(\cosh\{(\delta_1 + \delta_3)z \Delta U\} + \cosh\{(\delta_1 - \delta_3)z \Delta U\}))} \tag{74}$$

where δ_1, δ_2, and δ_3 denote the fractions of ΔU that affect k_{IO}^{OO}, k_{OO}^{IO}, and k_{OI}^{IO}, respectively: $\delta_1 + \delta_2 + \delta_3 = 0.5$. The shape of the i–V relations varies

[66] G. Eisenman and R. Horn, *J. Membr. Biol.* **76**, 196 (1983).

as a function of δ, $\kappa_{OI}^{IO}/\kappa_{OO}^{OI}$, and $[I]/K_{IO}$. (If $\delta_1 = \delta_2 = 0$, the shape of the $i-V$ relations is independent of the magnitude of $[I]/K_{IO}$.) These variations are conveniently displayed in plots of the normalized current, i_n, versus ΔV, where $i_n = ize/(gkT)$ (see Fig. 10).

As $\kappa_{OI}^{IO}/\kappa_{OO}^{OI}$ decreases, the $i_n - V$ relations became steeper, (Fig. 10, top), and the shape becomes less dependent on the magnitude of $[I]/K_{IO}$. When $\kappa_{OI}^{IO}/\kappa_{OO}^{OI}$ is approximately equal to or less than 1, the shape may vary as a function of $[I]/K_{IO}$ (Fig. 10, bottom): when $[I]/K_{IO} < 1$, the $i-V$ relations tend to bend toward the voltage axis and they are said to be *sublinear;* when $[I]/K_{IO} > 1$, the $i-V$ relations tend to bend toward the current axis and they are said to be *superlinear.* The shape changes occur because the rate-determining steps vary as a function of $[I]/K_{IO}$.[19,65,67]

The rate constants can be determined from the absolute magnitudes of the single-channel currents and the concentration-dependent shape changes of the $i-V$ relations. This information is obtained at a cost, however, as it becomes necessary to determine additional parameters, including at least two of the δ parameters in a symmetrical channel. In practice, all δ values must be determined because one rarely, if ever, will have sharp barriers, which, in any case, must be established experimentally. The δ estimates should thus be interpreted as empirical approximations [cf. Eqs. (46) and (47)], and the sum $\delta'_1 + \delta'_2 + \delta'_3$ will generally be less than 0.5 (cf. Ref. 68). Further, the shape changes are usually not very large (cf. Fig. 10, bottom) and it is therefore necessary to obtain data over an extended voltage range.[19,69]

Mobile Carrier

A simple carrier can be *approximated* as having a single binding site that alternately is accessible from one or the other aqueous phase.[70,71] If only a single ion can bind to the carrier, ion movement through the membrane can be described as transitions among four states (see Fig. 11): empty and facing the left aqueous phase, O|; empty and facing the right aqueous phase, |O; occupied and facing the left aqueous phase, I|; and occupied and facing the right aqueous phase, |I. The kinetic equations

[67] S. B. Hladky, B. W. Urban, and D. A. Haydon, in "Membrane Transport Processes" (C. F. Stevens and R. W. Tsien, eds.), Vol. 3, pp. 89–103. Raven, New York, 1979.
[68] B. W. Urban, S. B. Hladky, and D. A. Haydon, *Biochim. Biophys. Acta* **602**, 331 (1980).
[69] O. S. Andersen, *Biophys. J.* **41**, 147 (1983).
[70] W. Wildbrandt and T. Rosenberg, *Pharmacol. Rev.* **13**, 109 (1961).
[71] P. G. LeFevre, *Curr. Top. Membr. Transp.* **3**, 109 (1975).

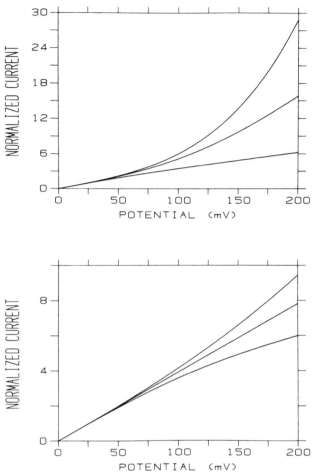

FIG. 10. Current–voltage relations predicted for the simple channel. The curves were calculated using Eq. (72) and plotted as i_n versus ΔV. The following parameters were held constant, $K_{IO} = 0.1\ m$, $\kappa_{OO}^{OI} = 10^7\ \text{sec}^{-1}$, $\delta_1 = 0.05$, $\delta_2 = 0.1$, and $\delta_3 = 0.35$. (Top) [I] = $0.1\ m$, while κ_{OI}^{IO} was $10^5\ \text{sec}^{-1}$ (upper curve), $10^6\ \text{sec}^{-1}$ (middle curve), and $10^7\ \text{sec}^{-1}$ (lower curve). (Bottom) $\kappa_{OI}^{IO} = 5 \times 10^6\ \text{sec}^{-1}$, while [I] was $1\ m$ (upper curve), $0.1\ m$ (middle curve), and $0.01\ m$ (lower curve). $i_n - V$ relations at [I] $> 1\ m$ and at [I] $< 0.01\ m$ are similar to the upper and lower curves, respectively.

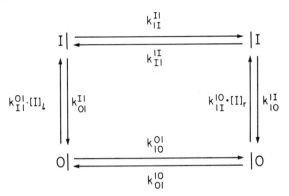

FIG. 11. Schematic representation of the interconnections among the four kinetic states in a one-site carrier. The symbol | denotes that access toward that side is blocked.

associated with the scheme have been solved for the stationary[22,72-76] and the transient[77-80] cases. Hladky[81] should be consulted for an incisive review. The stationary state probabilities are listed in Section 3 in the appendix. The net flux is equal to the net flux between any two adjacent states, e.g., between I| and |I:

$$j = k_{II}^{II}W(I|) - k_{II}^{II}W(|I)$$
$$= (k_{II}^{OI}k_{II}^{II}k_{IO}^{II}k_{OI}^{IO}[I]_l - k_{II}^{IO}k_{II}^{II}k_{OI}^{II}k_{IO}^{OI}[I]_r)/DD_3 \qquad (75)$$

where W is defined in Eqs. (A3.4) and (A3.5) and the denominator DD_3 is defined in Eq. (A3.6). As for the channel, Eq. (75) can be written in terms of experimental parameters:

$$j = (k_l^l[I]_l - k_r^r[I]_r)/(1 + [I]_l/K_l^l + [I]_r/K_r^r + [I]_l[I]_rX_I) \qquad (76)$$

where k_l^l and k_r^r denote rate coefficients for ion movement mediated by the carrier from left to right and right to left, respectively; K_l^l and K_r^r are the concentrations for half-maximal flux when the trans concentration is zero;

[72] V. S. Markin, L. I. Krishtalik, Y. A. Liberman, and V. P. Topaly, *Biofizika (Engl. Transl.)* **14**, 272 (1969).

[73] P. Läuger and G. Stark, *Biochim. Biophys. Acta* **211**, 458 (1970).

[74] S. B. Hladky, *J. Membr. Biol.* **10**, 67 (1972).

[75] S. M. Ciani, G. Eisenman, R. Laprade, and G. Szabo, *in* "Membranes: Lipid Bilayers and Antibiotics" (G. Eisenman, ed.), Vol. 2, pp. 61-177. Dekker, New York, 1973.

[76] W. R. Lieb and W. D. Stein, *Biochim. Biophys. Acta* **373**, 178 (1974).

[77] G. Stark, B. Ketterer, R. Benz, and P. Läuger, *Biophys. J.* **11**, 981 (1971).

[78] V. S. Markin and Y. A. Liberman, *Biofizika (Engl. Transl.)* **18**, 475 (1973).

[79] R. Laprade, S. M. Ciani, G. Eisenman, and G. Szabo, *in* "Membranes" (G. Eisenman, ed.), Vol. 3, pp. 127-214. Dekker, New York, 1975.

[80] S. B. Hladky, *J. Membr. Biol.* **46**, 213 (1979).

[81] S. B. Hladky, *Curr. Top. Membr. Transp.* **12**, 53 (1979).

and X_I is a cross coefficient. The carriers' stationary permeability characteristics are described by these parameters, which can be expressed in terms of the underlying rate constants:

$$k_I^l = k_{II}^{OI} k_{II}^{II} k_{IO}^{II} k_{OI}^{IO} / ((k_{IO}^{OI} + k_{OI}^{IO})(k_{II}^{II} k_{IO}^{II} + k_{II}^{II} k_{OI}^{II} + k_{OI}^{II} k_{IO}^{II})) \tag{77a}$$

$$k_I^r = k_{II}^{IO} k_{II}^{II} k_{OI}^{II} k_{IO}^{OI} / ((k_{IO}^{OI} + k_{OI}^{IO})(k_{II}^{II} k_{IO}^{II} + k_{II}^{II} k_{OI}^{II} + k_{OI}^{II} k_{IO}^{II})) \tag{77b}$$

$$K_I^l = \frac{(k_{IO}^{OI} + k_{OI}^{IO})(k_{II}^{II} k_{IO}^{II} + k_{II}^{II} k_{OI}^{II} + k_{OI}^{II} k_{IO}^{II})}{(k_{II}^{OI}\{k_{OI}^{IO}(k_{II}^{II} + k_{II}^{II} + k_{IO}^{II}) + k_{II}^{II} k_{IO}^{II}\})} \tag{78a}$$

$$K_I^r = \frac{(k_{IO}^{OI} + k_{OI}^{IO})(k_{II}^{II} k_{IO}^{II} + k_{II}^{II} k_{OI}^{II} + k_{OI}^{II} k_{IO}^{II})}{(k_{II}^{IO}\{k_{IO}^{OI}(k_{II}^{II} + k_{II}^{II} + k_{OI}^{II}) + k_{II}^{II} k_{OI}^{II}\})} \tag{78b}$$

$$X_I = k_{II}^{OI} k_{II}^{IO}(k_{II}^{II} + k_{II}^{II}) / ((k_{IO}^{OI} + k_{OI}^{IO})(k_{II}^{II} k_{IO}^{II} + k_{II}^{II} k_{OI}^{II} + k_{OI}^{II} k_{IO}^{II})) \tag{79}$$

As for the channel, k_I^l and k_I^r are related through the constraint imposed by detailed balance, Eq. (A3.6): $k_I^l / k_I^r = \exp\{z \, \Delta U\}$. Stationary net flux experiments at a given ΔV are thus described by four independent parameters, while the ion translocation kinetics depend on seven independent rate constants!

The important qualitative difference between the simple carrier and channel models, the return of the "empty" carrier, is formally reflected in the appearance of the cross-term, $[I]_l[I]_r X_I$, in the denominator in Eq. (76). Other differences and similarities can be ascertained by comparing Eqs. (56) and (57) and (77)–(79). It should, in particular, be noted that the concentration for half-maximal conductance no longer is simply related to the translocator's equilibrium ion-binding characteristics.

At low permeant ion concentrations, when $[I]_l/K_I^l$ and $[I]_r/K_I^r$ (and $[I]_l[I]_r X_I$) are much less than 1, Eq. (76) reduces to $j = k_I^l[I]_l - k_I^r[I]_r$ (cf. the simple channel model). It is again possible to define a limiting permeability coefficient, p_I^o, equal to k_I^l (and k_I^r) when $\Delta V = 0$. The maximal fluxes when the trans concentrations are zero, $j_{max}^{\rightarrow'}$ and $j_{max}^{\leftarrow'}$, are equal to $p_I^o K_I^l$ and $p_I^o K_I^r$, respectively, and the Haldane equation [Eq. (58)] is recovered.

The small-signal conductance per carrier molecule is obtained as for the simple channel, and the conductance–concentration, g–$[I]$, relation in symmetrical solutions is

$$g = [(ze)^2/kT] p_I^o[I]/(1 + [I]/K_I + [I]^2 X_I) \tag{80}$$

where $K_I = K_I^l K_I^r/(K_I^l + K_I^r)$. The associated net permeability coefficient, $p_I \, (= [kT/(ze)^2]g/[I])$, is

$$p_I = p_I^o/(1 + [I]/K_I + [I]^2 X_I) \tag{81}$$

The conductance is *not* a monotonically increasing function of [I], but will

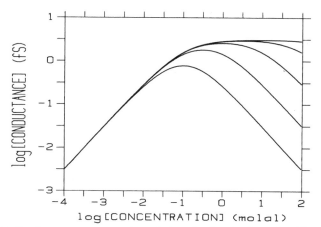

FIG. 12. Conductance–concentration relations for the simple carrier. The curves were calculated using Eq. (82), with the following parameters held constant, $K_{II} = 1$ m, $\kappa_{IO}^{II} = 10^7$ sec^{-1}, and $\kappa_{II}^{II} = 10^3$ sec^{-1}. κ_{OI}^{IO} was 10^3 sec^{-1} (lower curve), 10^4 sec^{-1}, 10^5 sec^{-1}, 10^6 sec^{-1}, and 10^7 sec^{-1} (upper curve). The shape is determined by the ratio $\kappa_{II}^{II}/\kappa_{OI}^{IO}$, and the concentration for half-maximal conductance varies as a function of κ_{OI}^{IO}.

go through a maximum at $[I] = (X_I)^{-0.5}$ and decrease as $1/[I]$ as $[I] \to \infty$ (see Fig. 12). The conductance decrease results from a competition between the two possible events that follow the dissociation of an ion from the carrier at the trans side: "return" of the empty carrier, and association of an ion with the carrier at the trans side (see Fig. 11). As $[I]$ increases, the latter process becomes increasingly likely. This behavior is most clearly illustrated for a symmetrical carrier, $\kappa_{II}^{OI} = \kappa_{II}^{IO}$, etc., where the g–$[I]$ relation becomes

$$g = [(ze)^2/kT]\kappa_{II}^{II}\kappa_{IO}^{II}[I]/\{2(K_{II} + [I])(2\kappa_{II}^{II} + \kappa_{IO}^{II} + (\kappa_{II}^{OI}\kappa_{II}^{II}/\kappa_{OI}^{IO})[I])\} \quad (82)$$

where $K_{II} = \kappa_{OI}^{II}/\kappa_{II}^{OI}$. Equation (82) should be compared with the corresponding expression for the simple channel:

$$g = [(ze)^2/kT]\kappa_{OI}^{IO}\kappa_{OO}^{OI}[I]/\{(K_{IO} + 2[I])(2\kappa_{OI}^{IO} + \kappa_{OO}^{OI})\} \quad (83)$$

As $\kappa_{II}^{II}/\kappa_{OI}^{IO} \to 0$, Eq. (82) will increasingly resemble a simple saturating function (Fig. 12), and the carrier becomes (almost) indistinguishable from a channel. The limiting form of Eq. (82), where $\kappa_{II}^{II}/\kappa_{OI}^{IO} = 0$, differs from Eq. (83) only by the factor of 2 preceding K_{II}. This is the formal equivalent of the qualitative criterion for carrier- and channel-mediated ion movement[25]: an empty carrier is accessible from only one solution, while an empty channel is accessible from both. (Likewise, the condition $\kappa_{II}^{II}/\kappa_{OI}^{IO} = 0$ denotes that the carrier has complete amnesia.)

As for the simple channel, the unidirectional flux determined with a finite trans concentration of the solute differs from the net flux determined when the trans concentration is zero. It is again necessary to examine this situation by introducing two permeant species, A and B (see Fig. 17 and Section 4 in the appendix). The net flux of A, j_A, is given by

$$j_A = k_{IA}^{AI} W(A|) - k_{AI}^{IA} W(|A) \tag{84}$$

$$= \{E_B(k_{OI}^{IO}k_{AI}^{OI}k_{IA}^{AI}k_{IO}^{IA}[A]_l - k_{IO}^{OI}k_{IA}^{IO}k_{AI}^{IA}k_{OI}^{AI}[A]_r)$$
$$+ k_{AI}^{OI}[A]_l k_{IA}^{AI}k_{IO}^{IA}k_{IB}^{IO}[B]_r k_{BI}^{IB}k_{OI}^{BI}$$
$$- k_{IA}^{IO}[A]_l k_{AI}^{IA}k_{OI}^{AI}k_{BI}^{OI}[B]_l k_{IB}^{BI}k_{IO}^{IB}\}/DD_4 \tag{85}$$

$$= (k_{OI}^{IO} + k_{IO}^{OI})(k_A^l[A]_l W(O|) - k_A^r[A]_r W(|O)) \tag{86}$$

where E_B is defined as in Eq. (A4.2) and DD_4 is defined in Eq. (A4.8). Correspondingly, the net flux of B, j_B, is given by

$$j_B = (k_{OI}^{IO} + k_{IO}^{OI})(k_B^l[B]_l W(O|) - k_B^r[B]_r W(|O)) \tag{87}$$

When A and B are chemically indistinguishable, such that $k_{AI}^{OI} = k_{BI}^{OI}$ (= k_{II}^{OI}), etc., and if $[A]_r = [B]_l = 0$, the denominator in Eq. (85) is equal to the denominator in Eq. (76) with $[I]_l = [A]_l$ and $[I]_r = [B]_r$. The net flux of A is again the unidirectional (tracer) flux of I from left to right:

$$j^{\to} = \frac{k_I^l[I]_l\{1 + k_I^r[I]_r(k_{OI}^{IO} + k_{IO}^{OI})/(k_{OI}^{IO}k_{IO}^{OI})\}}{(1 + [I]_l/K_I^l + [I]_r/K_I^r + [I]_l[I]_r X_I)} \tag{88}$$

In contrast to the conductance, the unidirectional flux is a simple saturating function of $[I]_l$:

$$j^{\to} = j_{max}^{\to}[I]_l/(K_I' + [I]_l) \tag{89}$$

where

$$j_{max}^{\to} = k_I^l K_I^l\{1 + k_I^r[I]_r(k_{IO}^{IO} + k_{IO}^{OI})/(k_{OI}^{IO}k_{IO}^{OI})\}/(1 + [I]_r K_I^l X_I) \tag{90}$$

and

$$K_I' = K_I^l(1 + [I]_r/K_I^r)/(1 + [I]_r K_I^r X_I) \tag{91}$$

with corresponding expressions for $j^{\gets}$. [If A and B are chemically dissimilar, Eqs. (85)–(87) describe the countertransport of A and B (e.g., Ref. 13). If $\kappa_{OI}^{IO} = 0$, the equations describe the homoexchange or heteroexchange of solutes across the membrane (e.g., Refs. 82 and 83). These permeation modes will not be discussed further.]

[82] J. O. Wieth, O. S. Andersen, J. Brahm, P. J. Bjerrum, and C. L. Borders, Jr., *Philos. Trans. R. Soc. London, Ser. B* **229**, 383 (1982).
[83] O. Fröhlich and R. B. Gunn, *Biochim. Biophys. Acta* **864**, 169 (1986).

The qualitatively different concentration dependencies for the conductance and the unidirectional flux results because net ion movement eventually becomes limited by the $|O \leftrightarrow O|$ transition, while the unidirectional flux is the sum of two components: the net flux component as well as an exchange flux component that results from $|I \leftrightarrow I|$ transitions and regeneration of $O|$ from $I| \rightarrow O|$ transitions. The $[I]_r$ dependence of $j_{\max}^{\rightarrow}$ and K_I' reflects the importance of the return via the $|I \leftrightarrow I|$ transition and the distribution of the carrier among the various states.

The tracer permeability coefficient, p_I^*, is again defined as $j^{\rightarrow}/[I]_l$ when $\Delta V = 0$. In symmetrical solutions,

$$p_I^* = p_I^0(1 + p_I^0[I](k_{OI}^{IO} + k_{IO}^{OI})/(k_{OI}^{IO}k_{IO}^{OI}))/(1 + [I]/K_I + [I]^2X_I) \quad (92)$$

and the correlation factor is *greater* than 1.0:

$$f = 1 + p_I^0[I](\kappa_{OI}^{IO} + \kappa_{IO}^{OI})/(\kappa_{OI}^{IO}\kappa_{IO}^{OI}) \quad (93a)$$
$$= 1 + \kappa_{II}^{OI}\kappa_{II}^{II}\kappa_{IO}^{II}/(\kappa_{IO}^{OI}(\kappa_{II}^{II}\kappa_{IO}^{II} + \kappa_{II}^{II}\kappa_{OI}^{II} + \kappa_{OI}^{II}\kappa_{IO}^{II}) \quad (93b)$$

If the dissociation rate constants are much larger than the translocation rate constants (κ_{II}^{II}, $\kappa_{II}^{II} \ll \kappa_{IO}^{II}$, κ_{OI}^{II}), Eq. (93) simplifies considerably: $f \simeq 1 + (\kappa_{II}^{II}/\kappa_{IO}^{OI})[I]/K_{II}$. Note that $f \rightarrow 1$ as $\kappa_{II}^{II}/\kappa_{OI}^{IO} \rightarrow 0$, and the carrier becomes qualitatively similar to the simple channel [cf. Fig. 12 and Eqs. (82) and (83)].

That f is greater than 1.0 implies that the exponent, n', in the generalized flux ratio equation

$$j^{\rightarrow}/j^{\leftarrow} = ([I]_l/[I]_r)n' \exp\{n'z \Delta U\} \quad (94)$$

is less than 1.0, since

$$f = 1/n' \quad (95)$$

Equation (95) follows from Eqs. (5) and (94) by an argument similar to that relating Eqs. (6) and (13).

As for the simple channel, a carrier's ability to discriminate between two permeant species, A and B, is quantified by the permeability ratio, p_A/p_B. For ions of the the same valency, the current zero condition is again that $j_A = -j_B$ or, after substitution and rearrangement of Eqs. (86) and (87):

$$k_A^l[A]_l + k_B^l[B]_l = k_A^r[A]_r + k_B^r[B]_r \quad (96)$$

Invoking the detailed balance constraint [cf. the derivation of Eq. (73)], Eq. (96) becomes

$$U_{rev} = 1/z \ln\{((k_A^l/k_B^l)[A]_r + [B]_r)/((k_A^l/k_B^l)[A]_l + [B]_l)\} \quad (97)$$

Equation (97) is similar to Eq. (72) [and Eq. (14b)], and k_A^l/k_B^l can be

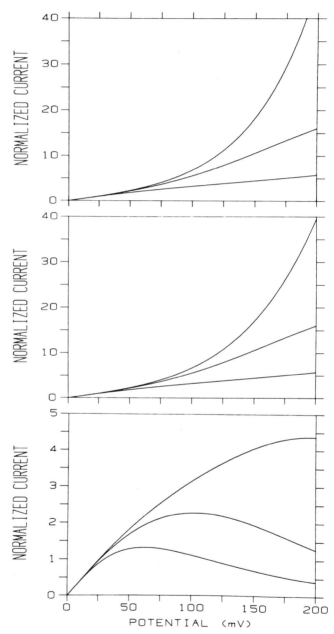

FIG. 13. Current–voltage relations for the simple carrier. The curves were calculated using Eq. (98) and plotted as i_n versus ΔV. The following parameters were held constant: $K_{II} = 0.1$ m; $z = 1$; $\kappa_{OI}^{IO} = \kappa_{II}^{II} = 10^3$ sec^{-1}, $\kappa_{OO}^{OI} = 10^7$ sec^{-1}; and $\delta_1 = 0.05$, $\delta_2 = 0.1$, $\delta_3 = 0.35$.

interpreted as a permeability ratio, p_A/p_B. The term $k_{OI}^{IO}/(k_{OI}^{IO} + k_{IO}^{OI})$ [cf. Eq. (77a)] factors out in the ratio k_A^I/k_B^I, and the same comments apply to Eq. (97) as to Eq. (72).

For the simple carrier, stationary net flux experiments at a given ΔV can be used to evaluate four experimental parameters, but the ion translocation kinetics is determined by seven independent rate constants (three parameters and four rate constants for a symmetrical carrier at $\Delta V = 0$). A fifth parameter can be determined from tracer flux measurements [cf. Eq. (93a)]. For a symmetrical carrier, it is thus possible to determine *all* the underlying rate constants from stationary flux measurements at $\Delta V = 0$. Generally, it is necessary to complement these flux measurements with stationary $i-V$ and current–relaxation measurements.[77,79,80] The principles underlying the analysis of the stationary $i-V$ relations can again be illustrated by noting how the shape of the $i-V$ relation vary as a function of the permeant concentration. For a symmetrical carrier, with simple exponential voltage dependence of the rate constants, Eq. (27), the explicit expression for the $i-V$ relation is

$$i = \frac{ze\kappa_{II}^{II}[I] \sinh\{(\delta_1 + \delta_2 + \delta_3)z \, \Delta U\}}{\begin{aligned} &K_{II} \cosh\{\delta_3 z_c \, \Delta U\}((\kappa_{II}^{II}/\kappa_{IO}^{II}) \cosh\{(\delta_2 z + \delta_3(z + z_c))\Delta U\} + 0.5) \\ &+ [I] \cosh\{((\delta_1 + \delta_2)z - \delta_3 z_c)\Delta U\} \\ &+ [I](\kappa_{II}^{II}/\kappa_{IO}^{II})(\cosh\{(\delta_1 + \delta_3)z \, \Delta U\} + \cosh\{(\delta_1 - \delta_3)z \, \Delta U\}) \\ &+ [I](\kappa_{II}^{II}/\kappa_{OI}^{IO})(\cosh\{((\delta_1 + \delta_2)z + \delta_3(z + z_c)\Delta U\} \\ &+ ([I]/K_{II}) \cosh\{\delta_3(z + z_c)\Delta U\}) \end{aligned}} \tag{98}$$

where δ_1, δ_2, and δ_3 denote the fractions of ΔU that affect κ_{II}^{OI}, κ_{OI}^{II}, and κ_{II}^{II} (and κ_{OI}^{IO}), respectively: $\delta_1 + \delta_2 + \delta_3 = 0.5$; z_c denotes the valence of the "empty" carrier. For any set of voltage dependencies (δ_1, δ_2, and δ_3) and valencies (z and z_c), the $i-V$ relations vary as a function of κ_{II}^{II}, $\kappa_{II}^{II}/\kappa_{IO}^{II}$, $\kappa_{II}^{II}/\kappa_{OI}^{IO}$, and $[I]/K_{II}$. These variations can again be displayed in $i-V$ plots. The importance of the carrier's valence is illustrated by comparing the concentration dependence of the shape of the i_n-V relations when $z = 1$ and z_c is either 0, -1, or -2, while other parameters remain unchanged (Fig. 13). When $z_c = 0$ (Fig. 13, top), one obtains a set of curves that is familiar from studies on mobile carrier-mediated ion movement through

---

(Top) $z_c = 0$, $[I] = 0.0001$ m (upper curve), 0.01 m (middle curve), and 1 m (lower curve). $i_n - V$ relations at $[I] < 0.0001$ m and at $[I] > 1$ m are similar to the upper and lower curves, respectively. (Middle) $z_c = -1$, $[I] = 10$ m (upper curve), 1 m (middle curve), and 0.1 m (lower curve). $i_n - V$ relations at $[I] > 10$ m and at $[I] < 0.1$ m are similar to the upper and lower curves, respectively. (Bottom) $z_c = -2$, $[I] = 10$ m (upper curve), 1 m (middle curve), and 0.1 m (lower curve). $i_n - V$ relations at $[I] > 10$ m and at $[I] < 0.1$ m are similar to the upper and lower curves, respectively.

planar lipid bilayers[84,85]: when $[I]/K_I \ll 1$, the $i-V$ relations are *superlinear;* when $[I]/K_I \gg 1$, the $i-V$ relations are *sublinear.* The concentration dependence of the shape changes is opposite to that for the simple channel. When $z_c = -1$ (Fig. 13, middle), the opposite pattern is observed. This situation is qualitatively similar to that observed for the simple channel. When $z_c = -2$ (Fig. 13, bottom), the currents go through a maximum and decrease as ΔV increases. The basis for this behavior is that the applied potential will bias the $I| \leftrightarrow |I$ and $|O \leftrightarrow O|$ transitions in the same direction. Only if the valence is known *a priori* can the qualitative concentration variation in the shape of the $i-V$ relations be used to distinguish among qualitatively different permeation schemes. The concentration-dependent shape changes tend to be larger than for the simple channel (cf. Figs. 10 and 13), but the same general comments apply to the carrier as to the channel.[53]

Complex Kinetic Models

The two basic schemes described above can be extended to incorporate the simultaneous binding of two or more solute particles (cotransport)[13,86-88] and single-file movement,[57,63,89-93] and allosteric control (slow conformational transitions) and permeant ion-induced conformational fluctuations.[94-96] The ensuing kinetic equations become increasingly complex and most models have quite complicated properties.[25] A general feature is, however, that there is no simple relation between equilibrium ion binding and ion permeation. This will be illustrated for single-filing channels.

If ions can enter a channel that is already occupied, permeant ions can interact *within* the channel. These ion–ion interactions have consequences for the channel's permeability function. The situation is particularly interesting when the permeating ions for structural or energetic reasons cannot

[84] G. Stark and R. Benz, *J. Membr. Biol.* **5**, 133 (1971).

[85] S. B. Hladky, *Biochim. Biophys. Acta* **352**, 71 (1974).

[86] D. Sanders, U.-P. Hansen, and C. L. Slayman, *J. Membr. Biol.* **77**, 123 (1984).

[87] D. Sanders, *J. Membr. Biol.* **90**, 67 (1986).

[88] P. Läuger and P. Jauch, *J. Membr. Biol.* **91**, 275 (1986).

[89] A. L. Hodgkin and R. D. Keynes, *J. Physiol. (London)* **128**, 61 (1955).

[90] B. Hille and W. Schwarz, *J. Gen. Physiol.* **72**, 409 (1978).

[91] B. W. Urban and S. B. Hladky, *Biochim. Biophys. Acta* **554**, 410 (1979).

[92] J. Sandblom, G. Eisenman, and J. Hägglund, *J. Membr. Biol.* **71**, 61 (1983).

[93] D. G. Levitt, *Curr. Top. Membr. Transp.* **21**, 181 (1984).

[94] P. Läuger, W. Stephan, and E. Frehland, *Biochim. Biophys. Acta* **602**, 167 (1980).

[95] P. Läuger, *Curr. Top. Membr. Transp.* **21**, 309 (1984).

[96] G. Eisenman, *in* "Ion Transport through Membranes" (K. Yagi and B. Pullman, eds.), pp. 101–129. Academic Press, Tokyo, 1986.

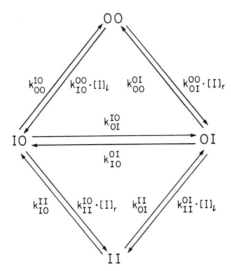

Fig. 14. Schematic representation of the interconnections among the four kinetic states in a two-site channel with double ion occupancy.

pass each other in the channel interior, in which case the ion permeation proceeds as a single-file process.[57,63,89-93] If the channel is *approximated* as having two binding sites (such that at most two ions simultaneously can reside in the channel), ion movement through the channel can be described as transitions among four states (Fig. 14): empty, OO; an ion in the left binding site, IO; an ion in the right binding site, OI; and ions in both binding sites, II. The kinetic equations associated with the scheme can be solved as described for the simple channel. The stationary state probabilities are listed in Section 5 in the appendix. The net flux through the channel is

$$j = k_{OI}^{IO} W(IO) - k_{IO}^{OI} W(OI)$$
$$= \{(k_{OI}^{II} + k_{IO}^{II})(k_{IO}^{OO} k_{OI}^{IO} k_{OO}^{OI}[I]_l - k_{OI}^{OO} k_{IO}^{OI} k_{OO}^{IO}[I]_r)$$
$$+ (k_{IO}^{OO}[I]_l + k_{OI}^{OO}[I]_l)(k_{II}^{OI} k_{IO}^{II} k_{OI}^{OI}[I]_l - k_{II}^{IO} k_{OI}^{II} k_{OI}^{IO}[I]_r)\}/DD_5 \quad (99)$$

where W is defined in Eqs. (A5.2) and (A5.3), and the denominator DD_5 is defined in Eq. (A5.5). The net flux through the channel can be depicted as the sum of two components: one due to transitions among OO, IO, and OI, and one due to transitions among OI, II, and OI. At low concentrations, the flux is predominantly through the sequence OO ↔ IO ↔ OI ↔ OO; as the concentration increases the flux will increasingly be through the sequence OI ↔ II ↔ IO ↔ OI. At very high concentrations, the net flux will go through a maximum and eventually decrease as 1/[I], because the IO ↔ OI transitions decrease in frequency as the rate of ion entry into a singly occupied channel, $k_{II}^{IO}[I]_r W(IO)$, increases and eventually exceeds the rate of ion translocation through the channel interior, $k_{OI}^{IO} W(IO)$ (and vice versa for ion movement in the opposite direction).

In symmetrical solutions, the small-signal single-channel conductance is, from Eqs. (99), (A1.5), and (A5.6),

$$g = \{(\kappa_{OI}^{II} + \kappa_{IO}^{II})\kappa_{IO}^{OO}\kappa_{OI}^{IO}\kappa_{OO}^{OI}[I] + (\kappa_{IO}^{OO} + \kappa_{OI}^{OO})\kappa_{II}^{OI}\kappa_{IO}^{II}\kappa_{IO}^{OI}[I]^2\}/DD_5 \quad (100)$$

The complete forms of Eqs. (99) and (100) are cumbersome, albeit straightforward. Hille and Schwarz[90] and Urban and Hladky[91] should be consulted for details about the general case.

A major simplification occurs if the channel is symmetrical, $\kappa_{IO}^{OO} = \kappa_{OI}^{OO}$, etc., where Eq. (100) reduced to (e.g., Ref. 57)

$$g = \frac{(ze)^2}{kT} \cdot \frac{2[I]}{K_{IO} + 2[I] + [I]^2/K_{III}} \cdot \frac{\kappa_{OI}^{IO}(\kappa_{OO}^{OI} + \kappa_{II}^{OI}[I])}{2(2\kappa_{OI}^{IO} + \kappa_{OO}^{OI} + \kappa_{II}^{OI}[I])} \quad (101)$$

where $K_{IO} = \kappa_{OO}^{IO}/\kappa_{IO}^{OO}$ and $K_{III} = \kappa_{OI}^{II}/\kappa_{II}^{OI}$. The second ratio term in Eq. (101) denotes the probability of finding the channel singly occupied, while the third keeps track of transitions in singly occupied channels. Many fairly different combinations of the underlying rate constants produce essentially indistinguishable $g-[I]$ relations (cf. Table III in Hladky and Haydon[97]). As for the simple channel, it will generally be necessary to incorporate additional information, e.g., from $i-V$ relations and permeability ratios, in the data analysis.[68]

As for the one-ion channel, the tracer flux differs from the net flux determined when the trans concentration is zero. In contrast to the one-ion channel, however, the tracer permeability coefficient, p_I^*, will generally differ from the net permeability coefficient, p_I, because of the interactions among ions *within* the channel. The analysis involves again two permeant species, A and B (see Fig. 18, in the appendix). The net flux of A, j_A, is

$$j_A = k_{OA}^{AO} W(AO) - k_{AO}^{OA} W(OA) \quad (102)$$

with a similar expression for the net flux of B, j_B. Here W denotes the probability of finding the channel in the indicated states [see Eqs. (A6.3)–(A6.6) and (A6.12)–(A6.16)].

The essential feature of ion–ion interactions vis-à-vis their effects on the flux is that the productive motion of an ion from left to right through the lower cycle in Fig. 14 (OI ↔ II ↔ IO ↔ OI) occurs only when the ion in the right binding site originated from the left solution. The conditional probability for this event, w, can be obtained from Eqs. (A6.12)–(A6.15). For a symmetrical channel at $\Delta V = 0$, with A and B chemically indistin-

[97] S. B. Hladky and D. A. Haydon, *Curr. Top. Membr. Transp.* **21**, 327 (1984).

guishable and $[A]_l = [B]_r (= [I])$, and $[A]_r = [B]_l = 0$, the expression for w is[57]

$$w = 2\kappa_{OI}^{IO}/(2(2\kappa_{OI}^{IO} + \kappa_{OO}^{OI}) + \kappa_{II}^{OI}[I]) \tag{103}$$

The net flux A is the product of two terms: the probability of finding the channel singly occupied by A, which is $[I]/(K_{IO} + 2[I] + [I]^2/K_{III})$; and $\kappa_{OI}^{IO}[(1 - w) - w] = \kappa_{OI}^{IO}(1 - 2w)$, the distribution of A between AO and OA times the rate constant for IO $\leftrightarrow$ OI transition. The resulting expression for the net flux of A, the unidirectional flux of I, is

$$j^{\rightarrow} = \frac{2[I]}{K_{IO} + 2[I] + [I]^2/K_{III}} \cdot \frac{\kappa_{OI}^{IO}(2\kappa_{OO}^{OI} + \kappa_{II}^{OI}[I])}{2[2(2\kappa_{OI}^{IO} + \kappa_{OO}^{OI}) + \kappa_{II}^{OI}[I]]} \tag{104}$$

The net permeability coefficient, p_I, is defined as $[kTk/(ze)^2]g/[I]$, and the tracer permeability coefficient, p_I^*, is defined as $j_I^{\rightarrow}/[I]$. As a consequence of single-file movement, $p_I^* \le p_I$ [cf. Eqs. (101) and (104)]. The correlation factor is

$$f = \frac{2\kappa_{OO}^{OI} + \kappa_{II}^{OI}[I]}{\kappa_{OO}^{OI} + \kappa_{II}^{OI}[I]} \cdot \frac{2\kappa_{OI}^{IO} + \kappa_{OO}^{OI} + \kappa_{II}^{OI}[I]}{2(2\kappa_{OI}^{IO} + \kappa_{OO}^{OI}) + \kappa_{II}^{OI}[I]} \tag{105}$$

Only rate constants relating to transitions in a singly occupied channel enter into the expression for f. Since $f \le 1$, the flux-ratio exponent, n' $(= 1/f)$, will be larger than 1. For a two-site channel, the flux-ratio exponent can vary between 1 and 2. Generally, the flux-ratio exponent can vary between 1 and m, where m is related to the number of sites that simultaneously can be occupied, n: $n - 3 \le m \le n - 1$ when $n \ge 4$.[98]

There is no simple correspondence between equilibrium ion binding and either the single-channel conductance or the flux-ratio exponent. This can be demonstrated by comparing $g-[I]$ and $n'-[I]$ relations for three different cases where κ_{IO}^{OO}, κ_{OI}^{OI}, and κ_{OO}^{OI} (and thus K_{IO}) are held constant, and κ_{II}^{OI} and κ_{IO}^{II} vary proportionally such that K_{III} is held fixed (see Fig. 15). In particular, it should be noted that n' can approach 2 for average ion occupancies ranging anywhere between 1 (when $x_{OO}^{OI} \ll \kappa_{II}^{OI}[I] \ll \kappa_{OI}^{OI}$, κ_{IO}^{II}) and 2 (when $\kappa_{OO}^{OI} < \kappa_{OI}^{II} \ll \kappa_{II}^{OI}[I] \ll \kappa_{OI}^{IO}$).[91] In the former case, the conductance is a linear function of $[I]$, but the major flux contribution nevertheless arises from transitions around the lower loop in Fig. 14 (OI $\leftrightarrow$ II $\leftrightarrow$ IO $\leftrightarrow$ OI). It is readily appreciated that the one-ion assumption in the simple channel model means not only that the maximal occupancy is 1.0, but also that a second ion, for structural or energetic reasons, does not enter an already occupied channel—at least not at a rate that is sufficient to have measurable consequences.

[98] H.-H. Kohler and K. Heckmann, *Membr. Sci.* **6**, 45 (1980).

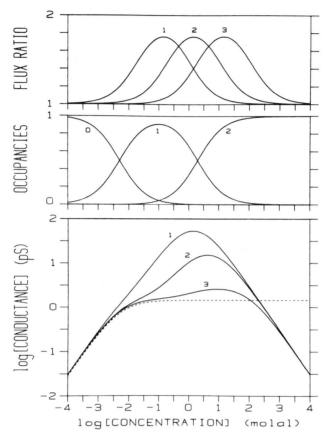

FIG. 15. Single-channel conductance (bottom), ion occupancy (middle), and flux-ratio exponent (top) as a function of permeant concentration. The curves were calculated using Eqs. (101) and (105) with the following parameters held constant: $K_{IO} = 0.01$ m, $K_{III} = 1$ m, $\kappa_{OI}^{IO} = 5 \times 10^7$ sec^{-1}, $\kappa_{OO}^{OI} = 10^6$ sec^{-1} (and $\kappa_{IO}^{OO} = 10^8$ $m^{-1} \cdot$ sec^{-1}). In the middle panel, the curves labeled 0, 1, and 2 refer to the probability of finding the channel occupied by 0, 1, and 2 ions. In the upper and lower panels, the curves with similar labels belong together: (1), $\kappa_{IO}^{II} = 10^8$ sec^{-1} (and $\kappa_{II}^{OI} = 10^8$ $m^{-1} \cdot$ sec^{-1}); (2), $\kappa_{IO}^{II} = 10^7$ sec^{-1} (and $\kappa_{II}^{OI} = 10^7$ $m^{-1} \cdot$ sec^{-1}); (3), $\kappa_{IO}^{II} = 10^6$ sec^{-1} (and $\kappa_{II}^{OI} = 10^6$ $m^{-1} \cdot$ sec^{-1}). For comparison, the interrupted curve in the lower panel denotes the $g - $[I] relation for a singly occupied channel (setting $\kappa_{II}^{OI} = 0$).

Ion–ion interactions within the channel imply further that the channel's ability to discriminate among permeant ions will vary as a function of the permeant ion concentrations. That is, the permeability ratio is concentration dependent because it depends not only on p^o but also on equilibrium ion binding in the channel.[90–92,99]

[99] H.-H. Kohler and K. Heckmann, *J. Theor. Biol.* **85**, 575 (1980).

Appendix: Stationary State Probabilities

This appendix contains expressions for the stationary probability of finding a carrier or a channel in different occupancy states. For the carrier, there is a single binding site, exposed toward the left or the right aqueous phase; the probability that an occupied or an unoccupied site is facing toward the left is denoted $W(I|)$ and $W(O|)$, etc. For the channel, there are two binding sites, exposed toward the left and the right aqueous phases, respectively; the probability of finding the channel unoccupied is denoted $W(OO)$, and the probability of finding the left binding site occupied is denoted $W(IO)$, etc. The rate constants are designated as $k_{\text{final state}}^{\text{initial state}}$; k_{IO}^{OO} denotes, for example, the association rate constant for an unoccupied channel at the left entrance, while the dissociation rate constant is denoted k_{OO}^{IO}, and the corresponding rate constants for a carrier are $k_{I|}^{O|}$ and $k_{O|}^{I|}$. Ion concentrations in the left and right aqueous phases are denoted by the subscripts "l" and "r," respectively.

The equations are written in a form that should facilitate comparisons among the various schemes, and ease their implementation in numerical simulations and data analysis.

1. *Two-site–one-ion channel, one permeant ion (Fig. 9)*. The kinetic equations can be solved algebraically or by the King and Altman[100] diagrammatic method (e.g., Ref. 26):

$$W(OO) = \{k_{OI}^{IO}k_{OO}^{OI} + k_{IO}^{OI}k_{OO}^{IO} + k_{OO}^{IO}k_{OO}^{OI}\}/DD_1 \qquad (A1.1)$$

$$W(IO) = \{k_{IO}^{OO}[I]_l(k_{IO}^{OI} + k_{OO}^{OI}) + k_{OI}^{OO}[I]_r k_{OI}^{OI}\}/DD_1 \qquad (A1.2)$$

$$W(OI) = \{k_{OI}^{OO}[I]_r(k_{OI}^{IO} + k_{OO}^{IO}) + k_{IO}^{OO}[I]_l k_{OI}^{IO}\}/DD_1 \qquad (A1.3)$$

where

$$
\begin{aligned}
DD_1 = {}& k_{OI}^{IO}k_{OO}^{OI} + k_{IO}^{OI}k_{OO}^{IO} + k_{OO}^{IO}k_{OO}^{OI} \\
& + k_{IO}^{OO}[I]_l(k_{OI}^{IO} + k_{IO}^{OI} + k_{OO}^{OI}) \\
& + k_{OI}^{OO}[I]_r(k_{OI}^{IO} + k_{IO}^{OI} + k_{OO}^{IO})
\end{aligned}
\qquad (A1.4)
$$

The rate constants are interrelated by the constraint imposed by detailed balance (e.g., Ref. 37):

$$k_{IO}^{OO}k_{OI}^{IO}k_{OO}^{OI}/(k_{OI}^{OO}k_{IO}^{OI}k_{OO}^{IO}) = \exp\{z\,\Delta U\} \qquad (A1.5)$$

which can be rewritten in terms of equilibrium constants:

$$K_{IO}/K_{OI} = L_I \exp\{-z\,\Delta U\} \qquad (A1.6)$$

---

[100] E. L. King and C. Altman, *J. Phys. Chem.* **60**, 1375 (1956).

where

$$K_{IO} = k_{OO}^{IO}/k_{IO}^{OO}, \qquad K_{OI} = k_{OO}^{OI}/k_{OI}^{OO}, \qquad \text{and} \qquad L_I = k_{OI}^{IO}/k_{IO}^{OI}$$
(A1.7a–c)

The detailed balance constraint implies that only five of the six rate constants and only two of the three equilibrium constants are independent.

2. *Two-site–one-ion channel, two permeant ions (Fig. 16).* The kinetic equations are most conveniently solved by the King and Altman method:

$$W(OO) = E_A E_B / DD_2$$
(A2.1)

where

$$E_A = k_{OA}^{AO} k_{OO}^{OA} + k_{AO}^{OA} k_{OO}^{AO} + k_{OO}^{AO} k_{OO}^{OA}$$
(A2.2)

with a corresponding expression for E_B, and

$$W(AO) = E_B\{k_{AO}^{OO}[A]_l(k_{AO}^{OA} + k_{OO}^{AO}) + k_{OA}^{OO}[A]_r k_{AO}^{OA}\}/DD_2$$
(A2.3)

$$W(OA) = E_B\{k_{OA}^{OO}[A]_r(k_{OA}^{AO} + k_{OO}^{OA}) + k_{AO}^{OO}[A]_l k_{OA}^{AO}\}/DD_2$$
(A2.4)

$$W(BO) = E_A\{k_{BO}^{OO}[B]_l(k_{BO}^{OB} + k_{OO}^{BO}) + k_{OB}^{OO}[B]_r k_{BO}^{OB}\}/DD_2$$
(A2.5)

$$W(OB) = E_A\{k_{OB}^{OO}[B]_r(k_{OB}^{BO} + k_{OO}^{OB}) + k_{OB}^{OO}[B]_l k_{OB}^{BO}\}/DD_2$$
(A2.6)

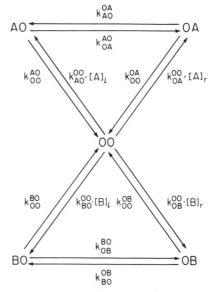

FIG. 16. Schematic representation of the interconnections among the five kinetic states in a two-site channel with single occupancy in the presence of two permeant ions.

where

$$DD_2 = E_A E_B$$
$$+ E_B\{k_{AO}^{OO}[A]_l(k_{OA}^{AO} + k_{AO}^{OA} + k_{OO}^{AO}) + k_{OA}^{OO}[A]_r(k_{OA}^{AO} + k_{AO}^{OA} + k_{OO}^{OA})\}$$
$$+ E_A\{k_{BO}^{OO}[B]_l(k_{OB}^{BO} + k_{BO}^{OB} + k_{OO}^{BO}) + k_{OB}^{OO}[B]_r(k_{OB}^{BO} + k_{BO}^{OB} + k_{OO}^{OB})\} \quad \text{(A2.7)}$$

For each ion, the detailed balance constraint is similar to Eq. (A1.5) [and Eq. (A1.6)].

3. One-site carrier, one permeant ion (Fig. 11). The kinetic equations are conveniently solved by the King and Altman method:

$$W(O|) = \{k_{O|}^{|O}(k_{||}^{||}k_{|O}^{||} + k_{||}^{||}k_{O|}^{||} + k_{O|}^{||}k_{|O}^{||}) + k_{||}^{|O}[I]_r k_{||}^{||}k_{O|}^{||}\}/DD_3 \quad \text{(A3.1)}$$

$$W(|O) = \{k_{|O}^{O|}(k_{||}^{||}k_{|O}^{||} + k_{||}^{||}k_{O|}^{||} + k_{O|}^{||}k_{|O}^{||}) + k_{||}^{O|}[I]_l k_{||}^{||}k_{|O}^{||}\}/DD_3 \quad \text{(A3.2)}$$

$$W(I|) = \{k_{O|}^{|O}k_{||}^{O|}[I]_l(k_{|O}^{||} + k_{||}^{||}) + k_{|O}^{O|}k_{||}^{|O}[I]_r k_{||}^{||}$$
$$+ k_{||}^{O|}[I]_l k_{||}^{|O}[I]_r k_{||}^{||}\}/DD_3 \quad \text{(A3.3)}$$

$$W(|I) = \{k_{|O}^{O|}k_{||}^{|O}[I]_r(k_{O|}^{||} + k_{||}^{||}) + k_{O|}^{|O}k_{||}^{O|}[I]_l k_{||}^{||}$$
$$+ k_{||}^{|O}[I]_r k_{||}^{O|}[I]_l k_{||}^{||}\}/DD_3 \quad \text{(A3.4)}$$

where

$$DD_3 = (k_{O|}^{|O} + k_{|O}^{O|})(k_{||}^{||}k_{|O}^{||} + k_{||}^{||}k_{O|}^{||} + k_{O|}^{||}k_{|O}^{||})$$
$$+ k_{||}^{O|}[I]_l\{k_{O|}^{|O}(k_{||}^{||} + k_{||}^{||} + k_{|O}^{||}) + k_{||}^{||}\}k_{|O}^{||}$$
$$+ k_{|O}^{O|}[I]_r\{k_{O|}^{O|}(k_{||}^{||} + k_{||}^{||} + k_{O|}^{||}) + k_{||}^{||}\}k_{O|}^{||}$$
$$+ k_{||}^{O|}[I]_l k_{||}^{|O}[I]_r(k_{||}^{||} + k_{||}^{||}) \quad \text{(A3.5)}$$

From detailed balance,

$$k_{||}^{O|}k_{||}^{||}k_{|O}^{||}k_{O|}^{|O}/(k_{||}^{|O}k_{||}^{||}k_{O|}^{||}k_{|O}^{O|}) = \exp\{z\,\Delta U\} \quad \text{(A3.6)}$$

which can be rewritten in terms of equilibrium constants

$$K_{I|}/K_{|I} = (L_I/L_O)\exp\{-z\,\Delta U\} \quad \text{(A3.7)}$$

where

$$K_{I|} = k_{O|}^{||}/k_{||}^{O|}, \qquad K_{|I} = k_{|O}^{||}/k_{||}^{|O}, \qquad L_I = k_{||}^{||}/k_{||}^{||}, \qquad \text{and } L_O = k_{|O}^{O|}/k_{O|}^{|O}$$
$$\text{(A3.8a–d)}$$

The detailed balance constraint implies that only seven of the eight rate constants and only three of the four equilibrium constants are independent.

4. One-site carrier, two permeant ions (Fig. 17). The kinetic equations are conveniently solved by the King and Altman[100] method:

$$W(O|) = \{E_A E_B k_{O|}^{|O} + E_B k_{|A}^{|O}[A]_r k_{A|}^{|A}k_{O|}^{A|} + E_A k_{|B}^{|O}[B]_r k_{B|}^{|B}k_{O|}^{B|}\}/DD_4 \quad \text{(A4.1)}$$

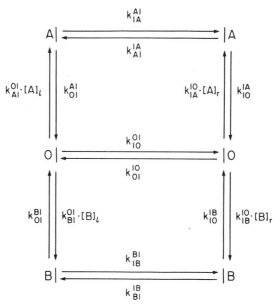

Fɪɢ. 17. Schematic representation of the interconnections among the six kinetic states in a one-site carrier in the presence of two permeant ions.

where

$$E_A = k_{IA}^{AI}k_{IO}^{IA} + k_{AI}^{IA}k_{OI}^{AI} + k_{OI}^{AI}k_{IO}^{IA} \tag{A4.2}$$

with a corresponding expression for E_B, and

$$W(|O) = \{E_A E_B k_{IO}^{OI} + E_B k_{AI}^{OI}[A]_l k_{IA}^{AI}k_{IO}^{IA} + E_A k_{BI}^{OI}[B]_l k_{IB}^{BI}k_{IO}^{IB}\}/DD_4 \tag{A4.3}$$

$$\begin{aligned} W(A|) = \{&E_B(k_{OI}^{IO}k_{AI}^{OI}[A]_l(k_{AI}^{IA} + k_{IO}^{IA}) + k_{IO}^{OI}k_{IA}^{IO}[A]_r k_{AI}^{IA}) \\ &+ E_B k_{AI}^{OI}[A]_l k_{IA}^{IO}[A]_r k_{AI}^{IA} + k_{AI}^{OI}[A]_l k_{IB}^{IO}[B]_r k_{BI}^{IB}k_{OI}^{BI}(k_{AI}^{IA} + k_{IO}^{IA}) \\ &+ k_{IA}^{IO}[A]_r k_{BI}^{OI}[B]_l k_{AI}^{IA}k_{IB}^{BI}k_{IO}^{IB}\}/DD_4 \end{aligned} \tag{A4.4}$$

$$\begin{aligned} W(|A) = \{&E_B(k_{IO}^{OI}k_{IA}^{IO}[A]_r(k_{IA}^{AI} + k_{OI}^{AI}) + k_{OI}^{IO}k_{AI}^{OI}[A]_l k_{IA}^{AI}) \\ &+ E_B k_{IA}^{OI}[A]_r k_{AI}^{OI}[A]_l k_{IA}^{AI} + k_{IA}^{IO}[A]_r k_{BI}^{OI}[B]_l k_{IB}^{BI}k_{IO}^{IB}(k_{IA}^{AI} + k_{OI}^{AI}) \\ &+ k_{AI}^{OI}[A]_l k_{IB}^{IO}[B]_r k_{IA}^{AI}k_{IB}^{IB}k_{OI}^{BI}\}/DD_4 \end{aligned} \tag{A4.5}$$

$$\begin{aligned} W(B|) = \{&E_A(k_{OI}^{IO}k_{BI}^{OI}[B]_l(k_{BI}^{IB} + k_{IO}^{IB}) + k_{IO}^{OI}k_{IB}^{IO}[B]_r k_{BI}^{IB}) \\ &+ E_A k_{BI}^{OI}[B]_l k_{IB}^{IO}[B]_r k_{BI}^{IB} + k_{BI}^{OI}[B]_l k_{IA}^{IO}[A]_r k_{AI}^{IA}k_{OI}^{AI}(k_{BI}^{IB} + k_{IO}^{IB}) \\ &+ k_{IB}^{IO}[B]_r k_{AI}^{OI}[A]_l k_{BI}^{IB}k_{IA}^{AI}k_{IO}^{IA}\}/DD_4 \end{aligned} \tag{A4.6}$$

$$\begin{aligned} W(|B) = \{&E_A(k_{IO}^{OI}k_{IB}^{IO}[B]_r(k_{IB}^{BI} + k_{OI}^{BI}) + k_{OI}^{IO}k_{BI}^{OI}[B]_l k_{IB}^{BI}) \\ &+ E_A k_{IB}^{OI}[B]_r k_{BI}^{OI}[B]_l k_{IB}^{BI} + k_{IB}^{IO}[B]_r k_{AI}^{OI}[A]_l k_{IA}^{AI}k_{IO}^{IA}(k_{IB}^{BI} + k_{OI}^{BI}) \\ &+ k_{BI}^{OI}[B]_l k_{IA}^{IO}[A]_r k_{IB}^{BI}k_{AI}^{IA}k_{OI}^{AI}\}/DD_4 \end{aligned} \tag{A4.7}$$

where

$$DD_4 = E_A E_B(k_{IO}^{OI} + k_{OI}^{IO})$$
$$+ E_B k_{AI}^{OI}[A]_l\{k_{OI}^{IO}(k_{IA}^{AI} + k_{AI}^{IA} + k_{IO}^{IA}) + k_{IA}^{AI}k_{IO}^{IA}\}$$
$$+ E_B k_{IA}^{IO}[A]_r\{k_{IO}^{OI}(k_{IA}^{AI} + k_{AI}^{IA} + k_{OI}^{AI}) + k_{IA}^{IA}k_{OI}^{AI}\}$$
$$+ E_A k_{BI}^{OI}[B]_l\{k_{OI}^{IO}(k_{IB}^{BI} + k_{BI}^{IB} + k_{IO}^{IB}) + k_{IB}^{BI}k_{IO}^{IB}\}$$
$$+ E_A k_{IB}^{IO}[B]_r\{k_{IO}^{OI}(k_{IB}^{BI} + k_{BI}^{IB} + k_{OI}^{BI}) + k_{BI}^{IB}k_{OI}^{BI}\}$$
$$+ E_B k_{AI}^{OI}[A]_l k_{IA}^{IO}[A]_r(k_{IA}^{AI} + k_{AI}^{IA})$$
$$+ E_A k_{BI}^{OI}[B]_l k_{IB}^{IO}[B]_r(k_{IB}^{BI} + k_{BI}^{IB})$$
$$+ k_{AI}^{OI}[A]_l k_{IB}^{IO}[B]_r\{k_{IB}^{BI}k_{AI}^{IA}(k_{OI}^{AI} + k_{IO}^{IB})$$
$$+ k_{IA}^{AI}k_{IO}^{IA}(k_{IB}^{BI} + k_{BI}^{IB}) + k_{BI}^{IB}k_{OI}^{BI}(k_{IA}^{AI} + k_{AI}^{IA})\}$$
$$+ k_{BI}^{OI}[B]_l k_{IA}^{IO}[A]_r\{k_{IB}^{IB}k_{IA}^{AI}(k_{IO}^{IA} + k_{OI}^{BI})$$
$$+ k_{AI}^{IA}k_{OI}^{AI}(k_{IB}^{BI} + k_{BI}^{IB}) + k_{IB}^{BI}k_{IO}^{IB}(k_{IA}^{AI} + k_{AI}^{IA})\} \qquad (A4.8)$$

For each ion, the constraint imposed by detailed balance is similar to that in Eq. (A3.6) [and Eq. (A3.7)]. Importantly, since L_O in this model is independent of ion type, one has generally that

$$K_{AI}/(K_{IA}L_A) = (K_{BI}/(K_{IB}L_B)) \exp\{-(z_A - z_B)\Delta U\} \qquad (A4.9)$$

Equation (A4.9) denotes a thermodynamic constraint on ion permeation (simple ion translocation, countertransport, or exchange diffusion) mediated by a simple carrier.

5. *Two-site–two-ion channel, one permeant ion (Fig. 15).* The kinetic equations are conveniently solved by the King and Altman method:

$$W(OO) = \{(k_{IO}^{II} + k_{OI}^{II})(k_{OI}^{IO}k_{OO}^{OI} + k_{IO}^{OI}k_{OO}^{IO} + k_{IO}^{IO}k_{OO}^{OI})$$
$$+ k_{II}^{OI}[I]_l k_{IO}^{II}k_{OO}^{IO} + k_{II}^{IO}[I]_r k_{OI}^{II}k_{OO}^{OI}\}/DD_5 \qquad (A5.1)$$

$$W(IO) = \{(k_{IO}^{II} + k_{OI}^{II})(k_{IO}^{OO}[I]_l(k_{IO}^{OI} + k_{OO}^{OI}) + k_{OI}^{OO}[I]_r k_{IO}^{OI})$$
$$+ k_{II}^{OI}[I]_l k_{IO}^{II}(k_{IO}^{OO}[I]_l + k_{OI}^{OO}[I]_r)\}/DD_5 \qquad (A5.2)$$

$$W(OI) = \{(k_{IO}^{II} + k_{OI}^{II})(k_{OI}^{OO}[I]_r(k_{OI}^{IO} + k_{OO}^{IO}) + k_{IO}^{OO}[I]_l k_{OI}^{IO})$$
$$+ k_{II}^{OI}[I]_r k_{OI}^{II}(k_{OI}^{OO}[I]_r + k_{IO}^{OO}[I]_l)\}/DD_5 \qquad (A5.3)$$

$$W(II) = \{(k_{IO}^{OO}[I]_l + k_{OI}^{OO}[I]_r(k_{II}^{OI}[I]_l k_{II}^{IO}[I]_r + k_{II}^{OI}[I]_l k_{OI}^{IO} + k_{II}^{IO}[I]_r k_{IO}^{OI})$$
$$+ k_{IO}^{OO}[I]_l k_{OO}^{OI}k_{II}^{IO}[I]_r + k_{OI}^{OO}[I]_r k_{OO}^{IO}k_{II}^{OI}[I]_l\}/DD_5 \qquad (A5.4)$$

where

$$DD_5 = (k_{IO}^{II} + k_{OI}^{II})(k_{OI}^{IO}k_{OO}^{OI} + k_{IO}^{OI}k_{OO}^{IO} + k_{OO}^{OI}k_{OO}^{IO})$$
$$+ (k_{IO}^{II} + k_{OI}^{II})(k_{IO}^{OO}[I]_l(k_{OI}^{IO} + k_{IO}^{OI} + k_{OO}^{OI})$$
$$+ k_{OI}^{OO}[I]_r(k_{OI}^{IO} + k_{IO}^{OI} + k_{OO}^{IO}))$$
$$+ k_{II}^{OI}[I]_l\{k_{IO}^{OO}[I]_l(k_{OI}^{IO} + k_{IO}^{II}) + k_{IO}^{II}k_{OO}^{IO}\}$$

$$+ k_{II}^{IO}[I]_r(k_{OI}^{OO}[I]_r(k_{IO}^{OI} + k_{OI}^{II}) + k_{OI}^{II}k_{OO}^{OI})$$
$$+ k_{IO}^{OO}[I]_l k_{II}^{IO}[I]_r(k_{OI}^{OI} + k_{OI}^{II} + k_{OO}^{OI} + k_{II}^{OI}[I]_l)$$
$$+ k_{OI}^{OO}[I]_r k_{II}^{OI}[I]_l(k_{OI}^{IO} + k_{IO}^{II} + k_{OO}^{IO} + k_{II}^{IO}[I]_r) \qquad (A5.5)$$

There are two constraints from detailed balance. For transitions among OO, IO, and OI, the constraint is given by Eqs. (A1.5) and (A1.6). For transitions among IO, II, and OI, and constraint is

$$k_{II}^{OI}k_{OI}^{II}k_{OI}^{IO}/(k_{II}^{IO}k_{OI}^{II}k_{IO}^{OI}) = \exp\{z \, \Delta U\} \qquad (A5.6)$$

or, expressed in terms of equilibrium constants,

$$K_{III}/K_{III} = L_I \exp\{-z \, \Delta U\} \qquad (A5.7)$$

where L_I is defined in Eq. (A1.7b) and

$$K_{III} = k_{OI}^{II}/k_{II}^{OI} \quad \text{and} \quad K_{III} = k_{IO}^{II}/k_{II}^{IO} \qquad (A5.8a,b)$$

Because the right-hand sides of Eqs. (A1.6) and (A5.7) are identical, one finally has that $K_{III}/K_{III} = K_{IO}/K_{OI}$, which can be rewritten as $K_{OI}K_{III} = K_{IO}K_{III}$. The detailed balance constraints imply that only eight of the ten rate constants and only three of the five equilibrium constants are independent.

6. Two-site–two-ion channel, two permeant ions (Fig. 18). The kinetic equations for this case are very complex, and the King and Altman method is less useful for analytical work. The general solution given here is adapted from Urban and Hladky.[91] The notational complexity makes it advantageous to introduce a set of auxiliary functions, given in a form differing slightly from that of the original authors':

$$F_A^l = k_{AO}^{OA} + k_{AA}^{OA}[A]_l k_{AO}^{AA}/(k_{AO}^{AA} + k_{OA}^{AA}) \qquad (A6.1a)$$
$$F_A^r = k_{OA}^{AO} + k_{AA}^{AO}[A]_r k_{OA}^{AA}/(k_{AO}^{AA} + k_{OA}^{AA}) \qquad (A6.1b)$$

with corresponding expressions for B, and

$$G_A^l = k_{AB}^{OB}[A]_l k_{AO}^{AB}/(k_{AO}^{AB} + k_{OB}^{AB}) \quad \text{and} \quad G_A^r = k_{BA}^{BO}[A]_r k_{OA}^{BA}/(k_{BO}^{BA} + k_{OA}^{BA}) \qquad (A6.1c,d)$$

$$G_B^l = k_{BA}^{OA}[B]_l k_{BO}^{BA}/(k_{BO}^{BA} + k_{OA}^{BA}) \quad \text{and} \quad G_B^r = k_{AB}^{AO}[B]_r k_{OB}^{AB}/(k_{AO}^{AB} + k_{OB}^{AB}) \qquad (A6.1e,f)$$

$$H_A^l = F_A^l + G_B^l + k_{OO}^{OA} \quad \text{and} \quad H_A^r = F_A^r + G_B^r + k_{OO}^{AO} \qquad (A6.1g,h)$$

$$H_B^l = F_B^l + G_A^l + k_{OO}^{OB} \quad \text{and} \quad H_B^r = F_B^r + G_A^r + k_{OO}^{BO} \qquad (A6.1i,j)$$

$$I_A^l = 1 + k_{AA}^{OA}[A]_l/(k_{OA}^{AA} + k_{AO}^{AA}) + k_{BA}^{OA}[B]_l/(k_{BO}^{BA} + k_{OA}^{BA}) \qquad (A6.1k)$$

$$I_A^r = 1 + k_{AA}^{AO}[A]_r/(k_{OA}^{AA} + k_{AO}^{AA}) + k_{AB}^{AO}[B]_r/(k_{AO}^{AB} + k_{OB}^{AB}) \qquad (A6.1l)$$

$$I_B^l = 1 + k_{BB}^{BO}[B]_l/(k_{OB}^{BB} + k_{BO}^{BB}) + k_{AB}^{OB}[A]_l/(k_{OB}^{AB} + k_{AO}^{AB}) \qquad (A6.1m)$$

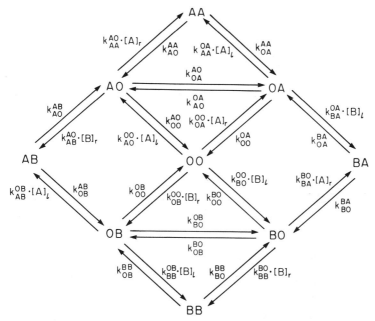

FIG. 18. Schematic representation of the interconnections among the nine kinetic states in a two-site channel with double occupancy in the presence of two permeant ions.

$$I_B^r = 1 + k_{BB}^{OB}[B]_r/(k_{OB}^{BB} + k_{BO}^{BB}) + k_{BA}^{BO}[A]_r(k_{BO}^{BA} + k_{OA}^{BA}) \tag{A6.1n}$$

The state probabilities can then be expressed as

$$W(OO) = D_6(O, O)/DD_6 \tag{A6.2}$$

$$W(AO) = D_6(A, O)/DD_6 \quad \text{and} \quad W(OA) = D_6(O, A)/DD_6 \tag{A6.3, A6.4}$$

$$W(BO) = D_6(B, O)/DD_6 \quad \text{and} \quad W(OB) = D_6(O, B)/DD_6 \tag{A6.5, A6.6}$$

$$W(AA) = \{(k_{AA}^{AO}[A]_r D_6(A, O) + k_{AA}^{OA}[A]_l D_6(O, A))/(k_{AO}^{AA} + k_{OA}^{AA})\}/DD_6 \tag{A6.7}$$

$$W(BB) = \{(k_{BB}^{BO}[B]_r D_6(B, O) + k_{BB}^{OB}[B]_l D_6(O, B))/(k_{BO}^{BB} + k_{OB}^{BB})\}/DD_6 \tag{A6.8}$$

$$W(AB) = \{(k_{AB}^{AO}[B]_r D_6(A, O) + k_{AB}^{OB}[A]_l D_6(O, B))/(k_{AO}^{AB} + k_{OB}^{AB})\}/DD_6 \tag{A6.9}$$

$$W(BA) = \{(k_{BA}^{BO}[A]_r D_6(B, O) + k_{BA}^{OA}[B]_l D_6(O, A))/(k_{BO}^{BA} + k_{OA}^{BA})\}/DD_6 \tag{A6.10}$$

where

$$D_6(OO) = (H_A^l H_A^r - F_A^l F_A^r)(H_B^l H_B^r - F_B^l F_B^r) + G_A^l G_A^r G_B^l G_B^r$$
$$- (F_A^l F_B^l G_A^r G_B^r + F_A^r F_B^r G_A^l G_B^l + G_A^l G_B^r H_A^l H_B^r + G_A^r G_B^l H_A^r H_B^l)$$

$$\text{(A6.11)}$$

$$D_6(AO) = (k_{OA}^{OO}[A]_r F_A^l + k_{AO}^{OO}[A]_l H_A^l)(H_B^r H_B^l - F_B^l F_B^r)$$
$$+ k_{OA}^{OO}[A]_r G_A^l F_B^r G_B^l + F_A^l G_A^r (k_{OB}^{OO}[B]_r F_B^l + k_{BO}^{OO}[B]_l H_B^l)$$
$$+ G_A^l H_A^l (k_{OB}^{OO}[B]_r H_B^r + k_{BO}^{OO}[B]_l F_B^r)$$
$$- k_{AO}^{OO}[A]_l G_A^r G_B^l H_B^r - k_{OB}^{OO}[B]_r G_A^l G_A^r G_B^l$$

$$\text{(A6.12)}$$

$$D_6(OA) = (k_{AO}^{OO}[A]_r F_A^r + k_{OA}^{OO}[A]_r H_A^l)(H_B^l H_B^r - F_B^l F_B^r)$$
$$+ k_{AO}^{OO}[A]_l G_A^r F_B^l G_B^r + F_A^r G_A^l (k_{BO}^{OO}[B]_l F_B^r + k_{OB}^{OO}[B]_r H_B^r)$$
$$+ G_A^r H_A^r (k_{BO}^{OO}[B]_l H_B^l + k_{OB}^{OO}[B]_r F_B^l)$$
$$- k_{OA}^{OO}[A]_r G_A^l G_B^r H_B^r - k_{BO}^{OO}[B]_l G_A^l G_A^r G_B^r$$

$$\text{(A6.13)}$$

$$D_6(BO) = (k_{OB}^{OO}[B]_r F_B^l + k_{BO}^{OO}[B]_l H_B^r)(H_A^r H_A^l - F_A^l F_A^r)$$
$$+ k_{OB}^{OO}[B]_r G_B^l F_A^r G_A^l + F_B^l G_B^r (k_{OA}^{OO}[A]_r F_A^l + k_{AO}^{OO}[A]_l H_A^l)$$
$$+ G_B^l H_B^l (k_{OA}^{OO}[A]_r H_A^r + k_{AO}^{OO}[A]_l F_A^r)$$
$$- k_{BO}^{OO}[B]_l G_B^r G_A^l H_A^l - k_{OA}^{OO}[A]_r G_B^l G_B^r G_A^l$$

$$\text{(A6.14)}$$

$$D_6(OB) = (k_{BO}^{OO}[B]_l F_B^r + k_{OB}^{OO}[B]_r H_B^r)(H_A^l H_A^r - F_A^l F_A^r)$$
$$+ k_{BO}^{OO}[B]_l G_B^r F_A^l G_A^r + F_B^r G_B^l (k_{AO}^{OO}[A]_l F_A^r + k_{OA}^{OO}[A]_r H_A^r)$$
$$+ G_B^r H_B^r (k_{AO}^{OO}[A]_l H_A^l + k_{OA}^{OO}[A]_r F_A^l)$$
$$- k_{OB}^{OO}[B]_r G_B^l G_A^r H_A^r - k_{AO}^{OO}[A]_l G_B^l G_B^r G_A^r$$

$$\text{(A6.15)}$$

and

$$DD_6 = D_6(O, O) + I_A^r D_6(A, O) + I_A^l D_6(O, A) + I_B^r D_6(B, O) + I_B^l D_6(O, B)$$

$$\text{(A6.16)}$$

For each permeant ion, the detailed balance constraints are similar to those in Eqs. (A5.8)–(A5.10). In addition, one has the constraints

$$k_{AO}^{OO} k_{AB}^{AO} k_{OB}^{AB} k_{OO}^{OB} = k_{OB}^{OO} k_{AB}^{OB} k_{AO}^{AB} k_{OO}^{AO}$$

$$\text{(A6.17)}$$

$$k_{OA}^{OO} k_{BA}^{OA} k_{BO}^{BA} k_{OO}^{BO} = k_{BO}^{OO} k_{BA}^{BO} k_{OA}^{BA} k_{OO}^{OA}$$

$$\text{(A6.18)}$$

or $K_{ABI} K_{OB} = K_{IAB} K_{AO}$, and $K_{BAI} K_{OA} = K_{IBA} K_{BO}$, where $K_{ABI} = k_{OB}^{AB}/k_{AB}^{OB}$, $K_{IAB} = k_{AO}^{AB}/k_{AB}^{AO}$, $K_{BAI} = k_{OA}^{BA}/k_{BA}^{OA}$, and $K_{IBA} = k_{BO}^{BA}/k_{BA}^{BO}$.

Acknowledgments

It is a pleasure to thank L. D. Chabala, J. T. Durkin, R. S. Eisenberg, D. C. Gadsby, W. N. Green, E. Jakobsson, E. J. Narcessian, and B. W. Urban for helpful discussions and critical reading of drafts of this chapter. This work was supported by NIH Grant GM 21342.

[5] Inhibition Kinetics of Carrier Systems

By Rosa Devés and R. M. Krupka

Because transport is a vectorial process involving two different compartments, the kinetics of transport inhibition has distinctive features not ordinarily encountered in enzyme kinetics. For example, it can be shown that the effect of an inhibitor on a transport system depends not only on the inhibition mechanism, but also on various other factors such as the relative location of the inhibitor and substrate, the affinity of the inhibitor for carrier sites on the inner and outer surfaces of the membrane, the transport properties of the substrate, and the transport mechanism. Failure to be aware of these complications can easily lead to mistakes in the interpretation of experimental results, because the same inhibitor could produce entirely different patterns of inhibition, depending on the conditions of the experiment.

The purpose of this chapter is to discuss the kinetics of transport inhibition from a practical point of view, drawing on the underlying kinetic theory in order to show how experiments may be designed and intepreted, but not providing a comprehensive treatment of transport kinetics. A general analysis of the kinetics of inhibition has been given elsewhere.[1] Strategies for determining the mechanism of inhibition are considered, for both reversible and irreversible inhibitors, and, in this connection, two questions are addressed: (1) the nature of the interaction between the substrate, the inhibitor, and the transport protein; that is, whether the binding of the substrate and inhibitor is either mutually exclusive, independent, or mutually dependent; and (2) the sidedness of the inhibition, that is, whether the inhibitor adds to a binding site on the outward-facing or the inward-facing carrier. It should be pointed out that, as transport proteins generally span the membrane and are exposed on both sides, it would be possible for an inhibitor site in a given carrier form to be exposed on the side of the membrane opposite from the substrate site; in such a case, an inhibitor adding to the inward-facing carrier form, for example, could actually add to the carrier on the outer surface of the membrane. In our analysis of the kinetics of transport, this complication will be ignored, and it will be assumed that a substrate and inhibitor adding to the same carrier form act on the same side of the membrane. The analysis of the kinetics of

---

[1] R. M. Krupka and R. Devés, *Int. Rev. Cytol.* **84**, 303 (1983).

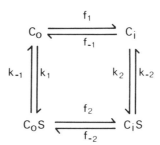

FIG. 1. Kinetic scheme for the carrier model. The conformation of the carrier alternates between inward-facing and outward-facing forms, C_o and C_i. Substrate molecules in the external or internal solutions, S_o and S_i, respectively, add to these two forms of the carrier; the corresponding substrate dissociation constants are given by $K_{S_o} = k_{-1}/k_1$; $K_{S_i} = k_{-2}/k_2$.

inhibition does reveal the carrier form attacked by the inhibitor, however, and if the actual location of the inhibitor or of the inhibitor site is determined, then it can be decided whether a site in the inner or outer carrier form can indeed be approached from the inner or outer compartment, respectively.

The problem of transport inhibition will be dealt with mainly in terms of the classical carrier model,[2-4] represented in the kinetic scheme in Fig. 1, because, despite its seeming simplicity, the model has proved capable of explaining a great variety of experimental observations. Its power derives from the fact that in it the movement of the substrate across the membrane depends on several different steps: addition of a substrate molecule to the carrier (k_1), reorientation of the carrier–substrate complex in the membrane (f_2), dissociation of the substrate from the carrier (k_{-2}), and finally the return of the free carrier to its original position (f_{-1}). The character of the model changes drastically, depending on which step is rate limiting. To take just one example, if the rate of reorientation of the free carrier (f_1, f_{-1}) is very rapid compared with the rate of reorientation of the substrate complex (f_2, f_{-2}), the scheme is reduced to three main intermediates — a single form of the free carrier, and an inner and an outer substrate complex. This is equivalent to the three-state channel model commonly invoked to explain ion movements,[5,6] and gives rise to kinetic rate equations of exactly the same form as does such a channel.

In the present treatment, the kinetic equations for the rates of substrate transport in the presence of competitive, noncompetitive, or uncompeti-

[2] W. F. Widdas, *J. Physiol. (London)* **118,** 23 (1952).

[3] H. G. Britton, *J. Physiol. (London)* **170,** 1 (1964).

[4] D. M. Regen and H. E. Morgan, *Biochim. Biophys. Acta* **79,** 151 (1964).

[5] R. Coronado, R. L. Rosenberg, and C. Miller, *J. Gen. Physiol.* **76,** 425 (1980).

[6] X. Cecchi, O. Alvarez, and R. Latorre, *J. Gen. Physiol.* **78,** 657 (1981).

tive inhibitors will be given for the case of rapid substrate dissociation, i.e., the case where the rate-limiting step in transport is reorientation of the carrier — either the free form or the substrate complex. The case of rate-limiting substrate dissociation (when the scheme reduces to one form of complex and one form of free carrier) need not be dealt with here, because the behavior is always simple: competitive inhibitors produce competitive behavior in all experiments, noncompetitive inhibitors produce noncompetitive behavior, and uncompetitive inhibitors produce uncompetitive behavior.

Reversible Inhibitors

A reversible inhibitor is one whose action is reversible by dilution. It may bind either at the substrate site, without itself undergoing transport, or outside the substrate site; in either case it interferes with the translocation of the substrate.

Observed Pattern of Inhibition

The study of reversible inhibition involves determining the effect of the substrate concentration on the transport rate at a fixed concentration of the inhibitor. Companion experiments are run in the presence or absence of the inhibitor, and the data are analyzed on the basis of the Michaelis–Menten equation. In the present Chapter, the analysis will be given in terms of the double-reciprocal (Lineweaver–Burk) plot, the most familiar and most widely used method, though other methods, such as the half-reciprocal plot of $[S]/v$ against $[S]$, or the direct linear plot, are preferred in statistical analysis.[7] In each experiment, two substrate parameters can be determined, a half-saturation constant and a maximum velocity. In the presence of an inhibitor, two inhibition constants may also be determined, corresponding to the substrate-dependent and substrate-independent components of the inhibition (the competitive and uncompetitive constants, respectively).

The various patterns of inhibition that could be produced by reversible inhibitors are shown in Fig. 2, as well as the procedure for calculating the competitive and uncompetitive inhibition constants. It may be seen that, depending on the properties of the system, increasing substrate concentrations could overcome the effect of the inhibitor (competitive inhibition), could be without effect (noncompetitive inhibition), or could even increase the effect of the inhibitor (uncompetitive inhibition).

[7] C. W. Wharton, *Biochem. Soc. Trans.* **11**, 817 (1983).

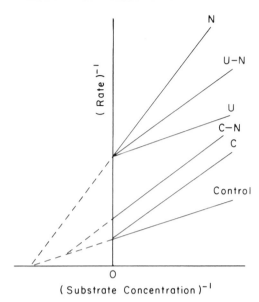

($ \text{Substrate Concentration} $)$^{-1}$

FIG. 2. Patterns of inhibition seen in double-reciprocal plots: control, rates in the absence of inhibitor; C, competitive inhibition; N, noncompetitive inhibition; U, uncompetitive inhibition; C–N, mixed competitive and noncompetitive inhibition; U–N, mixed uncompetitive and noncompetitive inhibition. In each case, two experimental half-saturation constants can be calculated. A substrate-dependent, or competitive, constant is calculated from the ratio of slopes in the presence and absence of a given concentration of inhibitor I:

$$K = [I] \left/ \left\{ \frac{\text{slope} \, (+I)}{\text{slope} \, (-I)} - 1 \right\} \right.$$

A substrate-independent, or uncompetitive, constant is calculated in a similar way from the ratio of intercepts:

$$K = [I] \left/ \left\{ \frac{\text{intercept} \, (+I)}{\text{intercept} \, (-I)} - 1 \right\} \right.$$

Transport Experiment. The rates of substrate transport can be measured in several types of experiments, each of which differ from one another in the initial distribution of substrates across the cell membrane. Indeed, the possibilities are so diverse that it was necessary to invent a rather elaborate terminology to describe the conditions of an experiment. By convention, the terms cis and trans refer to the relative locations of substrates or inhibitors across the membrane; cis designates the compartment occupied by the reference substrate, the one whose transport is followed, while trans designates the compartment on the opposite side of the membrane. For example, a cis inhibitor is present in the same compartment as the reference substrate, whereas a trans inhibitor is present in the opposite compartment.

In this chapter, we will deal with two different types of transport experiment: (1) zero-trans experiments, where the substrate is initially present in only one compartment and its flux across the membrane is determined; and (2) equilibrium exchange experiments, where the substrate is present in both compartments but only one is labeled, and the unidirectional flux of labeled substrate is measured.

Interaction between Inhibitor, Substrate, and Transport Protein

In discussing this topic, we will first assume the carrier model to be symmetrical in one respect: the rates of movement of the free carrier and the carrier–substrate complex are taken to be identical ($f_1 = f_2$ and $f_{-1} = f_{-2}$). A system with this property has a flux ratio equal to unity, which is to say that it exhibits neither accelerated exchange nor trans inhibition by the substrate. Later, the consequences of having an asymmetrical system will be considered.

Systems with Equal Rates of Reorientation of Free Carrier and Carrier–Substrate Complex. For the sake of clarity we will deal with two types of inhibitor separately: (1) inhibitors that are unable to cross the cell membrane and are therefore restricted to one compartment and (2) inhibitors that penetrate the membrane, for example, by simple diffusion, and are therefore present in both compartments.

Inhibitors restricted to one compartment. In this case, the inhibition mechanism can be unambiguously determined, provided the experiment is designed so as to have the inhibitor and the substrate (whose concentration is varied) in the same compartment. This requirement may be understood with the aid of the transport schemes in Fig. 3. An entry experiment, where a competitive inhibitor is present in the external solution, is shown in Fig. 3A. The observed kinetic pattern should be competitive, because the inhibitor and the substrate add, one at a time, to the same carrier form, the outward-facing carrier, C_o. Quite clearly, the substrate at sufficiently high concentrations should displace the inhibitor and thus overcome the inhibition. The result would be different in an exit experiment, shown in Fig. 3B, where the inhibitor and the substrate add to different carrier forms: the inhibitor adds to the outward-facing carrier and the substrate adds to the inward-facing carrier. The substrate, even at high concentrations, now fails to displace the inhibitor, and the inhibition is noncompetitive. Obviously, the latter experimental arrangement, with the substrate and inhibitor in different compartments, would give rise to a misleading pattern of inhibition, obscuring the actual mechanism. On the other hand, the former experimental arrangement, with the substrate and inhibitor in the same compartment, can disclose the mechanism directly (Table I).

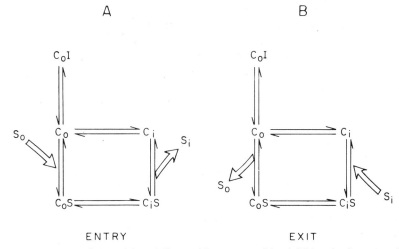

FIG. 3. The carrier model, as influenced by a competitive inhibitor in the suspending medium. (A) Substrate entry, where the observed inhibition would be competitive; (B) substrate exit, where the observed inhibition would be noncompetitive.

These patterns of inhibition in entry and exit experiments are predicted by the rate equations derived on the basis of the general inhibition scheme in Fig. 4. The definitions of the substrate and inhibitor constants which appear in the equations, and equivalent expressions in terms of individual rate constants in the transport scheme in Fig. 4, were given earlier.[1] Only the inhibitor constants need concern us here. They may be divided into two groups: those for addition to the free carrier,

$$\overline{K}_{I_o} = K_{I_o}(1 + f_1/f_{-1})$$

$$\overline{K}_{I_i} = K_{I_i}(1 + f_{-1}/f_1)$$

$$\tilde{K}_{I_o}^S = K_{I_o}(1 + f_1/f_{-2})$$

$$\tilde{K}_{I_i}^S = K_{I_i}(1 + f_{-1}/f_2)$$

and those for addition to the carrier–substrate complex,

$$\overline{K}_{S_oI_o} = K_{S_oI_o}(1 + f_2/f_{-1})$$

$$\tilde{K}_{S_oI_o}^S = K_{S_oI_o}(1 + f_2/f_{-2})$$

$$\overline{K}_{S_iI_i} = K_{S_iI_i}(1 + f_{-2}/f_1)$$

$$\tilde{K}_{S_iI_i}^S = K_{S_iI_i}(1 + f_{-2}/f_2)$$

Rate equations[8] for the transport of a substrate, S, in the presence of a competitive, noncompetitive, or uncompetitive inhibitor, I, will now be given for different experimental arrangements.

TABLE I
DETERMINATION OF THE MECHANISM OF INHIBITION FOR
REVERSIBLE OR IRREVERSIBLE INHIBITORS

	Experiment[a]	
Location of inhibitor[b]	Right	Wrong
A. Systems with equal rates of reorientation of free carrier and carrier–substrate complex		
I_o	Entry or exchange	Exit
I_i	Exit or exchange	Entry
I_o, I_i	Exchange	Entry or exit
B. Systems with unequal rates of reorientation of free carrier and carrier–substrate complex		
I_o	Exchange[c]	Entry or exit
I_i	Exchange	Entry or exit
I_o, I_i	Exchange	Entry or exit

[a] With an irreversible inhibitor, the experiment involves allowing the inhibitor to react with the system under the conditions of either entry (i.e., with the substrate outside), exit (i.e., with the substrate inside), or exchange (i.e., with the substrate on both sides); after removal of the unreacted inhibitor, the residual transport activity of the system is determined in any convenient transport assay.

[b] An inhibitor located in the solution outside the cell is termed I_o; an inhibitor inside is termed I_i.

[c] Equilibrium exchange experiments can generally be relied upon to reveal the mechanism of a noncompetitive inhibitor, except in the case of an extremely unsymmetrical transport system, where the substrate affinities on the two sides are very different; see the discussion in the text.

---

[8] The equations in the paper are written in terms of experimental parameters: K represents a half-saturation constant and V a maximum velocity. A nomenclature has been adopted [R. Devés and R. M. Krupka, Biochim. Biophys. Acta **556**, 533 (1979).] that is intended to suggest the significance of each constant directly: a subscript such as S_o, S_i, I_o, or I_i denotes the solute to which this constant applies and its location outside or inside the cell. Notation over parameters denotes the nature of the experiment in which the constant is determined: a single overbar refers to a zero-trans experiment, a double overbar refers to equilibrium exchange, and a tilde accompanied by a superscript "S" refers to an infinite-trans experiment (where there is saturating unlabeled substrate in the trans compartment). The equations are given in the reciprocal form, in order to make clear the predicted patterns of behavior in double-reciprocal plots.

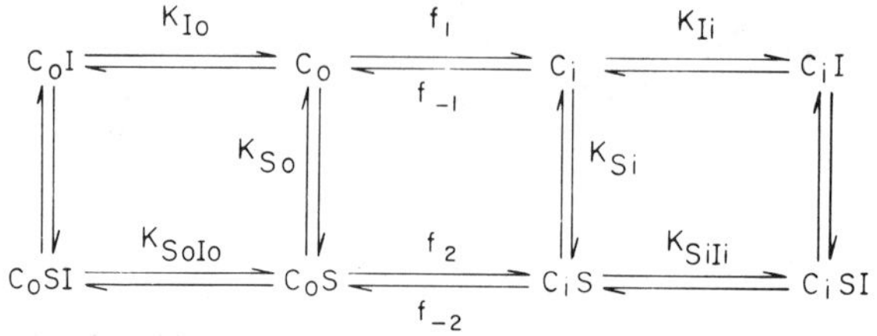

FIG. 4. Kinetic scheme for the carrier model, with an inhibitor (I) present on both sides of the membrane. The general case of inhibition is shown, in which the inhibitor adds to both the free carrier and the carrier–substrate complex.

1. The substrate and inhibitor are present in the same compartment (the right experiment).

  (a) Substrate influx, the inhibitor outside.

  Competitive inhibitor:

$$\frac{1}{v} = \frac{1}{\overline{V}_{S_o}}\left\{1 + \frac{\overline{K}_{S_o}}{[S_o]}\left(1 + \frac{[I_o]}{\overline{K}_{I_o}}\right)\right\} \qquad (1)$$

Noncompetitive inhibitor:

$$\frac{1}{v} = \frac{1}{\overline{V}_{S_o}}\left\{1 + \frac{[I_o]}{\overline{K}_{S_oI_o}} + \frac{\overline{K}_{S_o}}{[S_o]}\left(1 + \frac{[I_o]}{\overline{K}_{I_o}}\right)\right\} \qquad (2)$$

Uncompetitive inhibitor:

$$\frac{1}{v} = \frac{1}{\overline{V}_{S_o}}\left\{1 + \frac{[I_o]}{\overline{K}_{S_oI_o}} + \frac{\overline{K}_{S_o}}{[S_o]}\right\} \qquad (3)$$

  (b) Substrate efflux, the inhibitor inside.

  Competitive inhibitor:

$$\frac{1}{v} = \frac{1}{\overline{V}_{S_i}}\left\{1 + \frac{\overline{K}_{S_i}}{[S_i]}\left(1 + \frac{[I_i]}{\overline{K}_{I_i}}\right)\right\} \qquad (4)$$

Noncompetitive inhibitor:

$$\frac{1}{v} = \frac{1}{\overline{V}_{S_i}}\left\{1 + \frac{[I_i]}{\overline{K}_{S_iI_i}} + \frac{\overline{K}_{S_i}}{[S_i]}\left(1 + \frac{[I_i]}{\overline{K}_{I_i}}\right)\right\} \qquad (5)$$

Uncompetitive inhibitor:

$$\frac{1}{v} = \frac{1}{\overline{V}_{S_i}}\left\{1 + \frac{[I_i]}{\overline{K}_{S_iI_i}} + \frac{\overline{K}_{S_i}}{[S_i]}\right\} \qquad (6)$$

It can be seen that in these experiments there is agreement between the actual inhibition mechanism and the observed kinetics of inhibition. A competitive inhibitor produces competitive behavior; i.e., the inhibitor increases the slope of a double-reciprocal plot but not the intercept, or, put another way, sufficiently high substrate concentrations overcome the inhibition [Eqs. (1) and (4)]. A noncompetitive inhibitor produces noncompetitive behavior; both the slope and intercept of a double-reciprocal plot are increased, indicating that the substrate fails to overcome the inhibition [Eqs. (2) and (5)]. An uncompetitive inhibitor produces uncompetitive behavior; only the intercept increases, so that the inhibition becomes more severe as the substrate concentration is raised [Eqs. (3) and (6)].

2. The substrate and inhibitor are present in opposite compartments (the wrong experiment).

(a) Substrate influx, the inhibitor inside.

Competitive inhibitor:

$$\frac{1}{v} = \frac{1}{\overline{V}_{S_o}} \left\{ 1 + \frac{[I_i]}{\tilde{K}_{I_i}^S} + \frac{\overline{K}_{S_o}}{[S_o]} \left(1 + \frac{[I_i]}{\overline{K}_{I_i}} \right) \right\} \tag{7}$$

Noncompetitive inhibitor:

$$\frac{1}{v} = \frac{1}{\overline{V}_{S_o}} \left\{ 1 + \frac{[I_i]}{\tilde{K}_{I_i}^S} + \frac{\overline{K}_{S_o}}{[S_o]} \left(1 + \frac{[I_i]}{\overline{K}_{I_i}} \right) \right\} \tag{8}$$

Uncompetitive inhibitor:

$$\frac{1}{v} = \frac{1}{\overline{V}_{S_o}} \left\{ 1 + \frac{\overline{K}_{S_o}}{[S_o]} \right\} \tag{9}$$

(b) Substrate efflux, the inhibitor outside.

Competitive inhibitor:

$$\frac{1}{v} = \frac{1}{\overline{V}_{S_i}} \left\{ 1 + \frac{[I_o]}{\tilde{K}_{I_o}^S} + \frac{\overline{K}_{S_i}}{[S_i]} \left(1 + \frac{[I_o]}{\overline{K}_{I_o}} \right) \right\} \tag{10}$$

Noncompetitive inhibitor:

$$\frac{1}{v} = \frac{1}{\overline{V}_{S_i}} \left\{ 1 + \frac{[I_o]}{\tilde{K}_{I_o}^S} + \frac{\overline{K}_{S_i}}{[S_i]} \left(1 + \frac{[I_o]}{\overline{K}_{I_o}} \right) \right\} \tag{11}$$

Uncompetitive inhibitor:

$$\frac{1}{v} = \frac{1}{\overline{V}_{S_i}} \left\{ 1 + \frac{\overline{K}_{S_i}}{[S_i]} \right\} \tag{12}$$

With both competitive and noncompetitive inhibitors, the slopes and intercepts of the double-reciprocal plots are increased, indicating that the inhibition is noncompetitive; hence these experiments cannot be used to decide on the inhibition mechanism. The results are even more misleading with uncompetitive inhibitors, which now fail to inhibit.

Inhibitors present in both compartments. Some inhibitors, because they readily diffuse across the cell membrane, cannot be restricted to one compartment and are present on both sides of the membrane. Lipid-soluble substances such as alcohols and anesthetics are common examples. In cases such as this, it becomes obvious that the required experiment is one in which the substrate concentration is simultaneously varied on both sides of the membrane: i.e., equilibrium exchange (Table I). The kinetic equations written for the case where the external substrate pool is labeled and the internal pool is unlabeled are as follows.

1. The substrate and inhibitor are present in both compartments (the right experiment).
Equilibrium exchange, the inhibitor on both sides.
Competitive inhibitor:

$$\frac{1}{v} = \frac{1}{\overline{\overline{V}}_S} \left\{ 1 + \frac{\overline{\overline{K}}_{S_o}}{[S_o]} \left(1 + \frac{[I_o]}{\overline{K}_{I_o}} + \frac{[I_i]}{\overline{K}_{I_i}} \right) \right\} \tag{13}$$

Noncompetitive inhibitor:

$$\frac{1}{v} = \frac{1}{\overline{\overline{V}}_S} \left\{ 1 + \frac{[I_o]}{\tilde{K}^S_{S_oI_o}} + \frac{[I_i]}{\tilde{K}^S_{S_iI_i}} + \frac{\overline{\overline{K}}_{S_o}}{[S_o]} \left(1 + \frac{[I_o]}{\overline{K}_{I_o}} + \frac{[I_i]}{\overline{K}_{I_i}} \right) \right\} \tag{14}$$

Uncompetitive inhibitor:

$$\frac{1}{v} = \frac{1}{\overline{\overline{V}}_S} \left\{ 1 + \frac{[I_o]}{\tilde{K}^S_{S_oI_o}} + \frac{[I_i]}{\tilde{K}^S_{S_iI_i}} + \frac{\overline{\overline{K}}_{S_o}}{[S_o]} \right\} \tag{15}$$

In every case, the observed kinetic pattern corresponds to the actual mechanism of inhibition.

2. The substrate is present in one compartment, the inhibitor in both (the wrong experiment).
Substrate influx, the inhibitor on both sides.
Competitive inhibitor:

$$\frac{1}{v} = \frac{1}{\overline{V}_{S_o}} \left\{ 1 + \frac{[I_i]}{\tilde{K}^S_{I_i}} + \frac{\overline{K}_{S_o}}{[S_o]} \left(1 + \frac{[I_o]}{\overline{K}_{I_o}} + \frac{[I_i]}{\overline{K}_{I_i}} \right) \right\} \tag{16}$$

Noncompetitive inhibitor:

$$\frac{1}{v} = \frac{1}{\overline{V}_{S_o}}\left\{1 + \frac{[I_o]}{\overline{K}_{S_oI_o}} + \frac{[I_i]}{\tilde{K}_{I_i}^S} + \frac{\overline{K}_{S_o}}{[S_o]}\left(1 + \frac{[I_o]}{\overline{K}_{I_o}} + \frac{[I_i]}{\overline{K}_{I_i}}\right)\right\} \quad (17)$$

Uncompetitive inhibitor:

$$\frac{1}{v} = \frac{1}{\overline{V}_{S_o}}\left\{1 + \frac{[I_o]}{\overline{K}_{S_oI_o}} + \frac{\overline{K}_{S_o}}{[S_o]}\right\} \quad (18)$$

Here the same pattern of inhibition (noncompetitive) is found with both competitive and noncompetitive inhibitors; such an experimental arrangement, therefore, fails to reveal the actual mechanism. The same is true of substrate exit experiments, where the kinetic equations take the same forms as those above.

Systems with Unequal Rates of Reorientation of Free Carrier and Carrier–Substrate Complex. These systems exhibit either accelerated exchange or trans inhibition by the substrate. In terms of the carrier scheme in Fig. 1, accelerated exchange implies that the rate of reorientation of the carrier–substrate complex is faster than that of the free carrier ($f_2 > f_1$ and $f_{-2} > f_{-1}$); trans inhibition, on the other hand, implies that reorientation of the complex is the slower process ($f_2 < f_1$ and $f_{-2} < f_{-1}$).

When the rates of reorientation of the free carrier and its substrate complex are unequal, the addition of substrate to one side of the membrane alters the distribution of the carrier between its inner and outer forms. For example, if the complex moves across the membrane more rapidly than the free carrier returns, the presence of the substrate on one side causes the carrier to accumulate on the opposite side. On the other hand, if the complex moves more slowly than the free carrier, the carrier accumulates on the same side of the membrane as the substrate.

In an asymmetric system, the pattern of inhibition may therefore depend on the transport properties of the substrate, because, in shifting the carrier partition, the substrate would alter the accessibility of the carrier to an inhibitor that binds predominantly on one side of the membrane, and in this way would strengthen or weaken the inhibition. Consider, for example, the behavior of a noncompetitive inhibitor restricted to one compartment, in a system in which the carrier–substrate complex reorients faster than the free carrier. Because the substrate causes the carrier to accumulate on the opposite side of the membrane, it would tend to move the carrier out of reach of an inhibitor present in the same compartment as itself; the observed inhibition would therefore be either purely competitive or mixed competitive and noncompetitive, depending on the magnitude of the f_2/f_1 ratio. This is demonstrated by the kinetic equations, given above, for the case of a substrate and a noncompetitive inhibitor in the same compartment [e.g., Eq. (2)]. According to this equation, pure noncompetitive inhibition results if the competitive and uncompetitive constants are equal; i.e., if

$$\frac{\overline{K}_{S_oI_o}}{\overline{K}_{I_o}} = \frac{K_{S_oI_o}(1 + f_2/f_{-1})}{K_{I_o}(1 + f_1/f_{-1})} = 1 \tag{19}$$

Obviously, this ratio would equal unity with a purely noncompetitive inhibitor, for which $K_{S_oI_o} = K_{I_o}$, provided that $f_2 = f_1$. Where $f_2 \gg f_1$, however, the ratio becomes large and the inhibition becomes competitive. On the other hand, where $f_1 \gg f_2$, the ratio becomes small and the inhibition becomes partially uncompetitive.

How, then, may the inhibition mechanism be determined in an unsymmetrical system? Clearly, conditions are required in which the substrate does not alter the carrier distribution, and this usually may be achieved in equilibrium exchange, where the substrate adds on both sides of the membrane (Table I).

The rate of equilibrium exchange in the presence of a noncompetitive inhibitor outside is given by the following equation:

$$\frac{1}{v} = \frac{1}{\overline{\overline{V}}_S}\left\{1 + \frac{[I_o]}{\tilde{K}^S_{S_oI_o}} + \frac{\overline{\overline{K}}_{S_o}}{[S_o]}\left(1 + \frac{[I_o]}{\overline{K}_{I_o}}\right)\right\} \tag{20}$$

The pattern of inhibition is now independent of the f_2/f_1 ratio, as may be seen from the expression for the ratio of inhibition constants:

$$\frac{\tilde{K}^S_{S_oI_o}}{\overline{K}_{I_o}} = \frac{K_{S_oI_o}(1 + f_2/f_{-2})}{K_{I_o}(1 + f_1/f_{-1})} \tag{21}$$

An equilibrium exchange experiment should therefore reveal the actual mechanism even when f_1 and f_2 differ. There is one circumstance, however, in which it would fail to do so. This happens when the substrate dissociation constants on the two sides, K_{S_i} and K_{S_o}, are widely different, and where, in consequence, f_2/f_{-2} is very different from f_1/f_{-1}. In such a system, the substrate, even at equilibrium across the membrane, would alter the carrier distribution and could not be used to determine the inhibition mechanism. To avoid this difficulty, it would be necessary to use a substrate with a flux ratio near unity, for which the mobilities of the free carrier and the complex are similar.

Sidedness of Inhibition

While it is necessary for the substrate-binding site to become exposed on both membrane surfaces in order to bring about substrate translocation, an inhibitor-binding site may well be located on only one surface of the membrane, because an inhibitor, by definition, does not undergo transport. Considering this, it is of interest to determine the sidedness of inhibitor binding, particularly as inhibitors bound on only one side of the membrane can be important tools in the investigation of transport mechanisms.

Inhibitors That Can Be Restricted to One Compartment. Before the relative affinities of an impermeant inhibitor for the inward-facing and outward-facing carrier can be determined, it may be necessary to arrange to bring the inhibitor into contact with the cytoplasmic surface of the membrane. This can be achieved in various ways: (1) the membrane may be made temporarily permeable to the inhibitor (for example, by the addition of ionophores, or by preparing ghosts or vesicles in the presence of the inhibitor); (2) the inhibitor may be made temporarily permeant (for example, by changing its degree of ionization at an altered pH); or (3) sealed vesicles with an inverted orientation may be prepared. As these technical problems are beyond the scope of our discussion, they will not be dealt with here.

Provided the location of the inhibitor can be controlled, and it can be brought into contact with either the inner or outer surface of the membrane, the half-saturation constants on the two sides can be measured. It is of course necessary to determine comparable constants; for example, with a competitive or a noncompetitive inhibitor, the half-saturation constants for addition to the free enzyme, $\overline{K}_{I_o}$ and $\overline{K}_{I_i}$, may be determined from the slopes of double-reciprocal plots for either exit or entry (see Fig. 2). With an uncompetitive inhibitor, which does not add to the free enzyme, relative half-saturation constants can always be determined in equilibrium exchange experiments, though they may also be obtained in exit and entry, on the condition that the substrate and inhibitor are located in the same compartment.

Inhibitors Present in Both Compartments. The method for determining the sidedness of a permeant inhibitor, which is unavoidably present on both sides of the membrane, is not as obvious as in the case of a nonpenetrating inhibitor. The solution to the problem, however, emerges from a kinetic analysis of the carrier model.

Competitive inhibitors. The sidedness of inhibition can be determined from a comparison of the inhibition behavior in an entry and an exit experiment, provided the rate of reorientation of the carrier–substrate complex is not much slower than that of the free carrier (Table II). If the inhibitor binds symmetrically, and the carrier mechanism is also symmetrical, both experiments would give rise to apparent mixed competitive and noncompetitive behavior; if it binds exclusively to the outer carrier form, the inhibition of entry would be competitive and that of exit would be

TABLE II
DETERMINATION OF THE SIDEDNESS OF BINDING
FOR AN INHIBITOR PRESENT ON BOTH SIDES OF THE
MEMBRANE: THE REQUIRED FLUX RATIO OF THE
SUBSTRATE USED IN THE EXPERIMENT[a]

Type of inhibitor	Substrate flux ratio	
	Right	Wrong
Competitive	$\geqslant 1$	< 1
Noncompetitive	$< 1; > 1$	1
Uncompetitive	All values	—

[a] Sidedness is deduced from behavior in entry and exit experiments. The flux ratio is a measure of the effect of an unlabeled substrate, present in the trans compartment, on the unidirectional rate of transport of labeled substrate (homoexchange); it is equal to the rate of exchange transport divided by the rate of zero-trans transport. In the case of trans acceleration, the flux ratio is greater than unity; in the case of trans inhibition, it is less than unity. When substrate in the trans compartment has no effect on the unidirectional transport rate, the flux ratio is equal to unity.

TABLE III
INHIBITION PATTERNS PRODUCED BY REVERSIBLE OR
IRREVERSIBLE INHIBITORS PRESENT ON BOTH SIDES OF
THE MEMBRANE[a]

Site of inhibition	Substrate flux ratio	Experiment	
		Zero-trans exit	Zero-trans entry
Competitive inhibitors			
Outer	> 1	U–N	C
	1	N	C
	< 1	C	C
Inner	> 1	C	U–N
	1	C	N
	< 1	C	C
Both	> 1	N	N
	1	C–N	C–N
	< 1	C	C

TABLE III   *(continued)*

Noncompetitive inhibitors			
Outer	>1	U–N	C
	1	N	N
	<1	C	U–N
Inner	>1	C	U–N
	1	N	N
	<1	U–N	C
Both	>1	N	N
	1	N	N
	<1	N	N
Uncompetitive inhibitors			
Outer	>1	No inhibition	U
	1	No inhibition	U
	<1	No inhibition	U
Inner	>1	U	No inhibition
	1	U	No inhibition
	<1	U	No inhibition
Both	>1	U	U
	1	U	U
	<1	U	U

[a] For reversible inhibitors the abbreviations have the following meanings: C, competitive kinetics; N, noncompetitive kinetics; C–N, mixed competitive and noncompetitive kinetics; U–N, mixed uncompetitive and noncompetitive kinetics; U, uncompetitive kinetics. For irreversible inhibitors the abbreviations have meanings that are the counterparts of those for reversible inhibitors: C, full protection by the substrate; N, the substrate has no effect on the inactivation rate; C–N, partial protection by the substrate; U–N, the substrate accelerates the inactivation rate; U, inactivation only occurs in the presence of the substrate.

noncompetitive. The opposite would be true if the binding occurs exclusively inside. The inhibition patterns produced by competitive inhibitors present on both sides of the membrane, in relation to the sidedness of inhibition and the transport properties of the substrate, are given in Table III.

To find the relative affinities on the two sides, inhibition constants are calculated from the effect of the inhibitor on the intercept of a double-reciprocal plot [see Eq. (16)]; it may be noted that the constants calculated from the slope are independent of the sidedness of inhibition. The ratio of the constants in entry and exit experiments is

$$\frac{\tilde{K}_{I_i}^S}{\tilde{K}_{I_o}^S} = \frac{K_{I_i} (1 + f_{-1}/f_2)}{K_{I_o} (1 + f_1/f_{-2})} \tag{22}$$

The ratio is seen to depend not only on the inhibitor dissociation constants K_{I_i} and K_{I_o}, but also on the rate constants for movement of the free and complexed carrier; the true ratio of the affinities is obtained if $f_{-1}/f_2 = f_1/f_{-2}$.

Noncompetitive inhibitors. As in the case of competitive inhibitors, the sidedness of the binding of noncompetitive inhibitors may be inferred from their effects in entry and exit experiments. Here, however, a substrate must be used for which the rate of reorientation of the carrier complex is different from that of the free carrier ($f_2 \neq f_1$), because with a symmetrical system ($f_2, f_{-2} = f_1, f_{-1}$), entry and exit experiments would give the same pattern of inhibition (Tables II and III).

A substrate exhibiting trans acceleration ($f_2 > f_1$ and $f_{-2} > f_{-1}$), when present on one side of the membrane, displaces the equilibrium distribution of the carrier toward the opposite surface of the membrane. In this way the substrate alters the accessibility of the carrier to an inhibitor acting on only one side of the membrane. For example, if an inhibitor only binds outside, substrate in the outer compartment would tend to push the carrier out of reach, thus weakening the inhibition and making the inhibition pattern competitive; if the inhibitor only binds inside, on the other hand, the substrate would push the carrier toward the inhibitor and strengthen the inhibition, which would then be partially uncompetitive. The reverse would be true in exit experiments. On the other hand, an inhibitor bound equally on the two sides would be unaffected by the distribution of the carrier, and the inhibition would be noncompetitive in both exit and entry experiments.

By contrast, a substrate exhibiting trans inhibition ($f_2 < f_1$ and $f_{-2} < f_{-1}$), when present on only one side of the membrane, draws the carrier to the same side. The result is that if an inhibitor only binds outside, substrate in the outer compartment pulls the carrier toward the inhibitor, making the inhibition stronger; the inhibition pattern is then partially uncompetitive. If the inhibitor only binds inside, the substrate pulls the carrier away from the inhibitor, making the inhibition weaker and the inhibition pattern competitive. The reverse is true in exit experiments. Again, an inhibitor bound on both sides would be unaffected by the carrier distribution and would inhibit noncompetitively in both entry and exit. The predicted patterns of inhibition for a noncompetitive inhibitor present on both sides of the membrane are summarized in Table III.

It follows, then, that a large difference in the affinities on the two sides will be evident from the patterns of inhibition in exit and entry experiments. If the difference in affinities is less extreme, the relative values could be determined from the intercepts in double-reciprocal plots (see Fig. 2). From the equations for entry and exit experiments [see Eq. (17)], the ratio of the apparent inhibition constants is

$$\left(\frac{1}{\overline{K}_{S_oI_o}} + \frac{1}{\tilde{K}_{I_i}^S}\right) \bigg/ \left(\frac{1}{\overline{K}_{S_iI_i}} + \frac{1}{\tilde{K}_{I_o}^S}\right)$$

The individual half-saturation constants appearing in this expression were defined above. The ratio reduces to a simple form, which reflects the actual ratio of affinities, in two cases:

1. $f_2, f_{-2} \gg f_1, f_{-1}$

$$\frac{\dfrac{1}{K_{S_oI_o}}\left(\dfrac{f_{-1}}{f_2}\right) + \dfrac{1}{K_{I_i}}}{\dfrac{1}{K_{S_iI_i}}\left(\dfrac{f_1}{f_{-2}}\right) + \dfrac{1}{K_{I_o}}} \simeq \frac{K_{I_o}}{K_{I_i}} \tag{23}$$

2. $f_2, f_{-2} \ll f_1, f_{-1}$

$$\frac{\dfrac{1}{K_{S_oI_o}} + \dfrac{1}{K_{I_i}}\left(\dfrac{f_2}{f_{-1}}\right)}{\dfrac{1}{K_{S_iI_i}} + \dfrac{1}{K_{I_o}}\left(\dfrac{f_{-2}}{f_1}\right)} \simeq \frac{K_{S_iI_i}}{K_{S_oI_o}} \tag{24}$$

Uncompetitive inhibitors. The sidedness of action of an uncompetitive inhibitor may also be inferred from the behavior in entry and exit experiments (Table III). If the inhibitor binds exclusively to the outer carrier complex, the inhibition of entry would be uncompetitive; that is, the inhibition would become stronger at higher substrate concentrations. In an exit experiment, however, no inhibition would be observed, because an inhibitor of this kind does not bind to the free carrier, and the free carrier is the only form available to an inhibitor outside (assuming, as before, that substrate dissociation is rapid). The opposite behavior is expected of an inhibitor bound inside.

If the inhibitor is able to interact with both the internal and external carrier complexes, there would be uncompetitive inhibition in both exit and entry experiments, with the ratio of apparent inhibition constants given by

$$\frac{\overline{K}_{S_oI_o}}{\overline{K}_{S_iI_i}} = \frac{K_{S_oI_o}}{K_{S_iI_i}} \frac{(1 + f_2/f_{-1})}{(1 + f_{-2}/f_1)} \tag{25}$$

Irreversible Inhibitors

The experimental procedures used in studying reversible and irreversible inhibitors are different, but the principles underlying the behavior are the same, and therefore the same rules apply in interpreting experiments. With an irreversible inhibitor, the conclusions are based on measurements of the rates of inactivation of the transport system. These are obtained by incubating the system for varying periods of time with the inhibitor, and determining the residual transport activity after removal of the unreacted inhibitor. During the incubation, the inhibitor reacts progressively with the carrier, forming an inactive product:

$$\text{Active carrier} \xrightarrow{k} \text{Inactive carrier}$$

The data are plotted in accordance with a rate equation based on this scheme:

$$\ln (v/v_0) = kt \qquad (26)$$

where v is the transport rate for the system after treatment for a period of time t, and v_0 is the transport rate for the untreated system (all rates having been measured at the same substrate concentration). The constant k is the pseudo-first-order rate constant for the reaction and is a function of the inhibitor concentration. If the inhibitor reversibly forms a $1:1$ complex with the carrier, before reacting, the relationship would exhibit saturation behavior and take the form of the Michaelis–Menten equation. If the reversible binding of the inhibitor to form the initial complex is very weak, a linear relationship should be found between k and the inhibitor concentration. If the inhibition depends on reaction at two or more sites of similar reactivity, the inactivation rate constant would be proportional to the inhibitor concentration raised to the corresponding power.

Interaction between Inhibitor, Substrate, and Transport Protein

Substrates affect the rates of reaction of irreversible inhibitors in the same way that they affect the binding of reversible ones. The corresponding effects of the substrate on the two types of inhibition are as follows: (1) a reduction in the inactivation rate equivalent to competitive kinetics; (2) an increase in the inactivation rate, equivalent to an uncompetitive component in the kinetics; (3) the absence of any effect, equivalent to noncompetitive kinetics. The experimental procedure for deducing the actual inhibition mechanism (competitive, noncompetitive, or uncompetitive) is summarized in Table I.

Sidedness of Inhibition

The sidedness may be determined in experiments corresponding to those used with reversible inhibitors (Table II); the patterns of behavior expected in different cases are summarized in Table III.

Dependence of Kinetics of Inhibition on Transport Mechanisms

As was shown above [Eqs. (7) and (10)], the classical carrier model predicts that a competitive inhibitor in the trans compartment gives rise to noncompetitive inhibition. The same behavior, however, is not to be expected of all transport mechanisms, for example, the two models in Fig. 5. Either could represent a channel which is simultaneously accessible from both sides of the membrane and which can be occupied by only one substrate molecule at a time. The essential difference between these two types of channel is that in the model in Fig. 5A the rate-limiting step in transport is dissociation of the substrate from the complex, while in the model in Fig. 5B, the rate-limiting step is passage of the substrate through the channel.

It is interesting to note that the classical carrier model can give rise to either of the channel models in Fig. 5, depending on which steps are rate limiting. The model depicted in A results if substrate dissociation (k_{-1} and k_{-2} in Fig. 1) is rate limiting, and the model in B results if reorientation of the free carrier is very rapid compared with reorientation of the substrate complex ($f_1, f_{-1} \gg f_2, f_{-2}$). Therefore, it is possible to predict the inhibition behavior of these models by referring to the general kinetic treatment of the

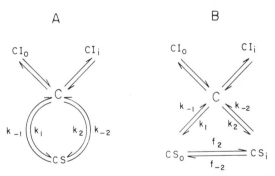

Fig. 5. Two-state and three-state channel models (A and B, respectively). In either, binding sites are simultaneously accessible from both sides of the membrane, but only one molecule is allowed into the channel at a time. (A) The rate-limiting step is dissociation of the substrate; (B) the rate-limiting step is movement of the ion within the channel. In A, different complexes must be allowed to form with an inhibitor in the internal and external solutions, because the formation of the same complex from either side would imply that the inhibitor undergoes transport (i.e., is a substrate).

carrier presented earlier.[1] As was seen above, the type of inhibition observed depends on the relative values of the competitive and uncompetitive inhibition constants [Eqs. (7) and (10)]. When the general expressions for these constants[1] are written, the ratio is found to be

$$\frac{\tilde{K}_{I_o}^S}{\overline{K}_{I_o}} = \frac{k_{-1}(f_1 + f_{-2}) + f_1(f_2 + f_{-2})}{k_{-1}f_{-2}(1 + f_1/f_{-1})} \tag{27}$$

If the dissociation step is rate limiting ($k_{-1}, k_{-2} \ll f_1, f_{-1}, f_2, f_{-2}$) as in Fig. 5A, the ratio becomes

$$\frac{\tilde{K}_{I_o}^S}{\overline{K}_{I_o}} = \frac{f_1(f_2 + f_{-2})}{k_{-1}f_{-2}(1 + f_1/f_{-1})} = \frac{f_1(1 + f_2/f_{-2})}{k_{-1}(1 + f_1/f_{-1})} \tag{28}$$

Because $k_{-1} \ll f_1$, the ratio is large and the inhibition is competitive. If, on the other hand, passage of the substrate through the channel is rate limiting ($k_{-1}, k_{-2} \gg f_1, f_{-1} \gg f_2, f_{-2}$), as in Fig. 5B, the ratio becomes

$$\frac{\tilde{K}_{I_o}^S}{\overline{K}_{I_o}} = \frac{f_1}{f_{-2}(1 + f_1/f_{-1})} \tag{29}$$

Because $f_1 \gg f_{-2}$, the ratio is large, and again the inhibition is competitive.

Concluding Remarks

In writing this chapter, one of our principal concerns has been to show that the elucidation of inhibition mechanisms, even in the simplest cases, requires a thorough understanding of the kinetic theory of transport inhibition, and that an uncritical application of enzyme kinetics to transport systems leads to error. The problem is that an inhibitor can produce various patterns of inhibition in different types of transport experiments; therefore, great care must be exercised in designing experiments, if valid conclusions are to be drawn. Once the subject is understood, however, discovering the effect of the substrate on the addition of the inhibitor to a given carrier form is simple enough. Further, the complicated behavior is not to be regarded as a mere obstacle, because with the inhibition mechanism known, this predictable diversity can be exploited, allowing the sidedness of the inhibition to be determined. Beyond this, inhibitors whose mechanism and sidedness have been established become useful tools in the investigation of the transport mechanism.

Acknowledgments

This work was supported in part by Grant B1540-8545 from the Departamento de Desarrollo de la Investigación, Universidad de Chile.

[6] Design of Simple Devices to Measure Solute Fluxes and Binding in Monolayer Cell Cultures

By JAYDUTT V. VADGAMA

A variety of techniques have been developed to measure solute fluxes across biological membranes (see the review in Ref. 1). Most of these techniques use methods suitable for studies with isolated tissues, cell suspensions, and vesicle preparations. These methods involve some means of separating cells or organelles from suspensions, for example, centrifugal sedimentation, centrifugal filtration, centrifugal layer filtration, and sieve filtration.[2-4] Most of these techniques have been well characterized and prove to be fairly reproducible.

One of the disadvantages of separating cells or organelles from a suspension is that only one sample can be assayed at a single time point. Modifications have been made to accommodate performing six simultaneous assays (see Fig. 13.7 in Ref. 1). This method was again slightly modified to study transport across red blood cells.[5] Similar adaptions have been made to measure bacterial iron transport (see Fig. 1 in Ref. 6). In this case the apparatus is constructed to hold eight funnel manifolds. With this adaptation, up to eight samples can be assayed at any given time. Much larger sample numbers can be accommodated if necessary. The apparatus can easily be adapted to measure transport by either vesicle or whole cell systems.

For adherent cells, tissue culture dishes or cover slips have been employed. The use of such techniques for measuring solute fluxes has been cataloged.[7-11] The main disadvantage of using cover slips or tissue culture

[1] H. N. Christensen, "Biological Transport," p. 405. Benjamin, New York, 1975.
[2] P. Ferdinanco and M. Klingenberg, this series, Vol. 56, p. 279.
[3] J. Lever, CRC Crit. Rev. Biochem., 187 (1980).
[4] J. D. Young and J. C. Ellory, in "Red Cell Membranes: A Methodological Approach" (J. C. Ellory and J. D. Young, eds.), p. 119. Academic Press, New York, 1982.
[5] J. V. Vadgama and H. N. Christensen, J. Biol. Chem. 260, 2912 (1985).
[6] H. Rosenberg, this series, Vol. 56, p. 388.
[7] D. Foster and A. Pardee, J. Biol. Chem. 244, 2685 (1979).
[8] A LeCam and P. Freychett, J. Biol. Chem. 252, 148 (1977).
[9] R. Bass and E. Englesberg, In Vitro 15, 829 (1979).
[10] R. Kletzien, M. Pariza, E. Becker, V. Porter, and F. Butcher, J. Biol. Chem. 251, 3014 (1976).
[11] L. Ø. Kristensen, J. Biol. Chem. 255, 5236 (1980).

dishes is that the assay methods are time consuming and somewhat cumbersome when experiments with multiple data points are needed.

In order to circumvent such problems, Gazzola et al.[12] rather ingeniously developed, initially in this laboratory, a simple set of devices based on modifications of the Costar 24-well tissue culture cluster dish. Members of this and other[13] laboratories have found that this technique revolutionizes our ability to study membrane transport for adherent cells under various experimental conditions, and with a high degree of reproducibility. The technique requires less separation time, is inexpensive, convenient, and allows assays for short (<10 sec) as well as for long terms.

In this report we will review (1) the construction of the essential devices, (2) their applications to study solute fluxes, and (3) the potentials of the method for studying other biochemical problems that would require multipoint assays, such as for hormone binding, drug toxicity, immunological studies, properties of cell surface proteins, and other metabolic studies.

Cell Culture

For details of culture conditions suitable for individual cell lines, see Refs. 14–19.

In general, medium 199 or minimum essential medium, Earle's (Gibco or KC biological)—supplemented with 5–10% fetal bovine serum, or, for hepatocytes in primary culture, ordinary bovine serum—has been found to be suitable. The cells are maintained at 37° in a humidified atmosphere of 5% CO_2/95% air. Stock cultures can be maintained in 75-cm^2 culture flasks (Corning 25110) and are prepared for transport by seeding approximately 4×10^4 cells into each 16-mm well of a Costar 24-well culture cluster dish (Costar 3524). Cells are allowed to grow to confluency in 1 ml of the appropriate medium for 3–5 days.

[12] G. C. Gazzola, V. Dall'Asta, R. Franchi-Gazzola, and M. F. White, *Anal. Biochem.* **115**, 368 (1981).

[13] G. C. Gazzola, Istituto di Patologia Generale, Universita di Parma, 43100 Parma, Italy; D. L. Oxender, Department of Biological Chemistry, University of Michigan, Ann Arbor, Michigan 48109; M. S. Kilberg, Department of Biochemistry and Molecular Biology, University of Florida, Gainesville, Florida 32610.

[14] M. F. White and H. N. Christensen, *J. Biol. Chem.* **257**, 4450 (1982).

[15] J. V. Vadgama and H. N. Christensen, *J. Biol Chem.* **258**, 6422 (1983).

[16] M. S. Kilberg, M. E. Handlogten, and H. N. Christensen, *J. Biol. Chem.* **255**, 4011 (1980).

[17] M. A. Shotwell and D. L. Oxender, *J. Biol. Chem.* **256**, 5422 (1981).

[18] P. Boerner and M. H. Saier, Jr., *J. Cell. Physiol.* **113**, 240 (1982).

[19] L. Weissbach, M. E. Handlogten, H. N. Christensen, and M. S. Kilberg, *J. Biol. Chem.* **257**, 12006 (1982).

Cells can be seeded at a density (3×10^5) sufficient to measure fluxes after 24 hr. Alternatively, the seeding density can be altered according to the purpose of the experiment. Freshly isolated fetal rat hepatocytes[15] or adult rat hepatocytes[16] are usually seeded at a density of 3×10^5 cells per well. Cells in primary culture are generally used after 24 hr of initial culture or after selected intervals (in the range of 4 to 72 hr) for studying in culture the changes in nutrient transport with time.[19]

Construction of Equipment

Wash Tray

Cells in monolayer cultures can be washed rapidly in the 24-well cluster tray using the device shown in Fig. 1. To construct this wash tray, take the top cover of a 24-well Costar dish (Costar 3524) and simply drill 24 12.5-mm holes through it. Smooth out the rough edges around the holes with a file or sandpaper. Now insert into each hole a plastic test tube, 12 mm in diameter and 75 mm in length (#2052, Falcon Plastics, Los Angeles, California). The tubes are inserted so that about 3.5 mm of the open end is exposed over the surface of the tray, and it is then glued into that position with polystyrene cement. Several wash trays should be constructed so that sequential washes are possible. The 24-well culture (as shown in Fig. 2a) is fitted over such a wash tray suitably loaded with test substrates. The whole assembly is inverted and the contents of the wash tray are rapidly transferred into each well of the Costar tray (Fig. 2b).

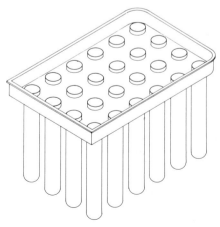

FIG. 1. Schematic diagram of a wash tray.

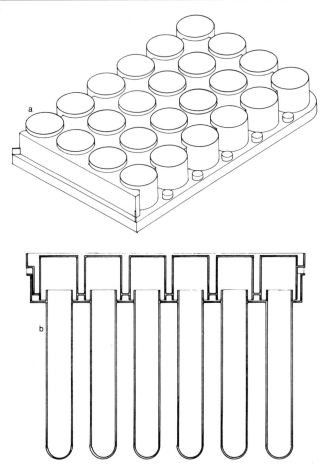

FIG. 2. Schematic diagram of (a) a 24-well Costar dish (Costar 3524) and (b) a longitudinal section of the 24-well Costar dish fitted over the wash tray.

Experimental or Uptake Tray

Here we begin with a Costar tray in which cells have been grown and are affixed in each well. In order to add a small volume of a solution of the test substrate (frequently isotopically labeled) to the monolayer cultures, an uptake tray is constructed as shown in Fig. 3a.

The construction procedure is similar to that of the wash tray, except that instead of the long polystyrene tubes, polypropylene embedding capsules (beam size 00, Polysciences, Inc., Warrington, Pennsylvania, No. 0224) are inserted into 9-mm holes drilled into another cover for the Costar 24-well cluster tray. The capsules can be attached with cyanoacry-

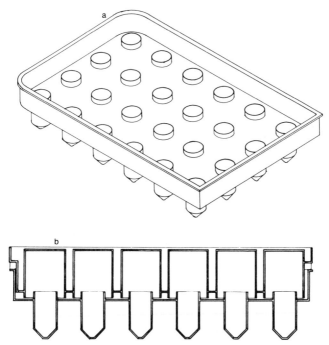

Fig. 3. Schematic diagram of (a) an experimental or uptake tray and (b) a longitudinal section of the 24-well Costar dish fitted over the uptake tray.

late adhesive (or any other suitable adhesive) so that the open end of each capsule extends about 3 or 4 mm inside. This arrangement ensures quantitative transfer, on subsequent inversion of the whole uptake assembly (Fig. 3b), of the contents of each capsule simultaneously onto the adherent cells in each of the 24 wells of the Costar dish. For preparing trays for the next experiment, each uptake tray is washed first with tap water, then with three cycles of distilled water, and is dried in an oven at 37–40°.

Assay Procedure

As the first step, a suitable volume (200 to 500 μl) of the test substrate solution is added to each of the polypropylene embedding capsules in the uptake tray. This uptake tray can be set up in a variety of ways, depending on the nature of the experiment. For instance, to determine the K_m of an amino acid for a given transport system specific to a cell type, one can systematically vary the substrate concentration from 0.001 mM in well #A1 (manufacturer's label) to 25 mM in well #D6. Alternatively, one can screen the inhibitory effect of eight amino acids, each with triplicate determinations, all in the 24 wells of a single tray. Or, the pH range can be varied from 4.5 to 8.0, e.g., with a total of eight different pH values, also securing triplicate analyses in the same tray. Several such trays can be set up for a given experiment. Typically, one can assay up to 144 data points with reasonable convenience in a single day's experiment.

Uptake measurements can be made in either sodium-containing or sodium-free Krebs–Ringer Phosphate buffer. The sodium-free medium is prepared by replacing sodium chloride and sodium hydrogen phosphate buffer with choline chloride and choline hydrogen phosphate, respectively. In order to minimize the so-called trans effects by analogous substrates already present, cells in culture are washed with 2 ml of choline media placed in each of the Falcon tubes in the wash tray, and then incubated in fresh 2-ml portions of the same buffer for 1 hr at 37° to reduce endogenous levels of such substances. This second depletion step may not be necessary when using monolayer cultures for purposes other than transport, such as hormone binding, drug sensitivity, etc. The wash trays can be rapidly filled with buffer solution using an appropriate Repipet (Labindustries, Berkeley, California).

To recapitulate, transport is initiated by adding simultaneously 0.25 to 0.50 ml of KRP media containing the labeled amino acid or other substrate plus any other test compounds to the monolayer cultures in the Costar 24-well cluster tray. Uptake is usually measured for 30 sec to 1 min at 37°, while gently shaking the cluster tray to avoid creating unstirred aqueous layers. Incubations can, however, be performed for any reasonable time intervals.

The uptake is stopped by pouring off the media into a receiving vessel for discard, followed by immediate rinsing with ice-cold phosphate-buffered saline (PBS), pH 7.4, using the wash trays. The cluster tray is drained or air dried and the labeled amino acid is extracted from the monolayer cell culture with 220 μl of 5% (w/v) trichloroacetic acid (TCA) for 1 hr at room temperature. The TCA is conveniently added to each well with the Distrivar Manual Dispenser (Gilson France S.A., Villiers-le Bel, France; or Ranin Instrument Co., Inc., Woburn, Massachusetts). The TCA-soluble radioactivity in a 200-μl extract is counted in 2 ml of the scintillant 3a70B (Research Product International) with a scintillation counter.

Protein Estimation

A further advantage of the cluster tray technique is that one can estimate protein content directly in each well, thus minimizing pipetting and dilution errors. The excess TCA (20 μl) is drained off from each well, and the precipitated protein is estimated using a modified Lowry procedure.[20,21] Add 100–200 μl of 1 N NaOH using a Distrivar Manual Dispenser. The protein is completely dissolved in about 30 min. Next, 1 ml of solution containing disodium cupric ethylenedinitrotetraacetate (0.25 g/liter), obtained from Eastman Kodak Co., Na_2CO_3 (20.0 g/liter), NaOH (4.0 g/liter), and 1% sodium dodecyl sulfate (SDS) is added. Occasionally, for cells in primary culture, such as fetal hepatocytes, it may be necessary to include SDS to disperse the cells. This addition does not alter the color yields.[22] Note that the 10 g of SDS should initially be dissolved in 200–300 ml of distilled water before adding it to the remaining components of the Lowry reagent. When SDS is included, it may be necessary to filter the Lowry reagent through a 0.45-μm filter to remove insoluble impurities.

Add 1 ml of the Lowry reagent to each well and gently swirl the tray. After 20 min, add to each well 0.1 ml of Folin–Ciocalteau reagent diluted with an equal volume of distilled water. Additions of the Folin reagent can be facilitated by means of a Distrivar dispenser, calibrated at 0.1 ml. Immediately swirl the cluster tray gently, so that the contents do not spill over. After 25 to 30 min, read the absorbance of the blue solution at 750 nm (Hitachi Spectrophotometer Model 100-60) if the protein content is below 100 μg, or at 500 nm if the concentration is above 100 μg. Usually the protein content ranges from 25 (human fibroblast) to 120 μg (HTC and hepatocytes). Bovine serum albumin is used as the standard. This assay yields a linear relationship up to 120 μg protein.

The Lowry determination can be adapted to an automated procedure which uses 96-well microtest plates.[23] In this case, protein samples from the 24-well Costar dish need to be transferred into the 96-well microtest plates, because automatic readers are not yet available for the 24-well tray. Protein solutions of 50 μl or less can be read very rapidly (<2 min) with such 96-well microtest plates. This equipment has a widespread use in routine serological procedures in which qualitative end points usually suffice. A Model MR-600 microplate reader, an Autodilutor II, and a Microshaker II can be purchased from Dynatech Laboratories. The Autodilutor II, composed of a Gilson Repetman and Hamilton syringes fitted to repeating dispensers, is used to dispense small volumes.

Calculations

Amino acid uptake expressed as nanomoles · milligram of protein^{-1} · min^{-1} is calculated from the primary data with a BASIC computer

[20] O. Lowry, J. Rosebrough, N. Farr, and R. Randall, *J. Biol. Chem.* **193**, 265 (1957).
[21] L. Lefferet, K. Koch, T. Moran, and M. Williams, this series, Vol. 58, p. 536.
[22] M. Lees and S. Paxman, *Anal. Biochem.* **7**, 184 (1972).
[23] C. Shopsis and G. Mackay, *Anal. Biochem.* **140**, 104 (1984).

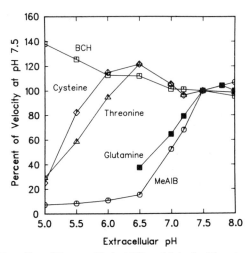

FIG. 4. The pH profiles of Systems N, A, ASC, and L in fetal hepatocytes. The uptake of 0.05 mM glutamine (System N), MeAIB (System A), cysteine and threonine (System ASC), and 0.1 mM BCH (System L) was measured for 1 min at 37° in KRP buffer at the indicated pH values. For Systems N, A, and ASC, the results show the Na^+-dependent rates corrected for velocities in Na^+-free buffer at the same pH values. For System L, the results show Na^+-independent rates corrected for a small nonsaturable component at the same pH. The Na^+-dependent rates (nanomoles · milligram of protein^{-1} · min^{-1}) at pH 7.5 were 0.50 ± 0.008 for glutamine, 0.052 ± 0.0007 for MeAIB, 0.58 ± 0.0164 for threonine, and 0.76± 0.008 for cysteine. For BCH, the Na^+-independent rate at pH 7.5 was 0.63 ± 0.006. All values are averages of three determinations with their SE. The uptake of cysteine was measured in the presence of 10 mM dithiothreitol.

program on a Radioshack TRS-80 computer. Secondary kinetic parameters (±SE) are determined by best computer fit to FORTRAN programs[24] using a Gauss–Newton least squares analysis.

Results and Discussion

Some of the results obtained by applying the cluster tray assay for measuring solute fluxes are discussed below.

Figure 4 shows pH profiles of the different transport systems in fetal hepatocytes. Uptake of five different amino acids was measured at different external pH values. For each amino acid, two 24-well Costar trays were used, one measuring the uptake of the given amino acid at eight different pH conditions in Na^+-containing medium, and the second measuring the uptake in the same pH range but in the absence of sodium ion. Uptake at each pH value was measured in triplicate. The difference between the

[24] W. Cleland, this series, Vol. 63, p. 103.

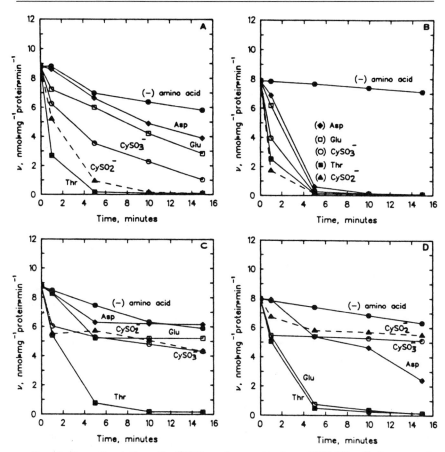

FIG. 5. Transstimulation of L-[³H]threonine exodus from CHO-K1 cells by external amino acids. The details of the loading procedure are described in the text. Exodus of L-[³H]threonine was measured at 0-, 1-, 5-, 10-, and 15-min intervals in the presence or absence of the indicated 10 mM external amino acid into Na⁺-containing medium at pH 7.4 (A) or 5.5 (B) and into Na⁺-free buffer at pH 7.4 (C) or 5.5 (D).

Na⁺-containing and the Na⁺-free medium measures the Na⁺-dependent velocities. The results in Fig. 4 were obtained with nine 24-well Costar trays, a total of 216 data points. The entire experiment took 2 to 3 hr.

Influx measurements were compared for various amino acids into CHO cells with two different assay procedures.[12] The results showed that the relative standard deviation decreased by a factor of 3 when the cluster tray assay was substituted for the 35 mM tissue culture dishes.[12,25]

Figure 5 measures trans stimulation of threonine exodus from

[25] G. Cecchini, M. Lee, and D. L. Oxender, *J. Supramol. Struct.* **4,** 441 (1976).

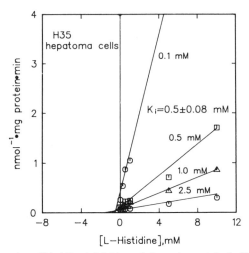

FIG. 6. A Dixon plot of histidine inhibition of glutamine uptake in H35 cells. Uptake of 0.1, 0.5, 1.0, and 2.5 mM L-[U-^{14}C]glutamine from KRP was measured for 1 min at 37°. The results shown are the Na$^+$-dependent velocities obtained after correcting for their respective rates in Na$^+$-free buffer. An additional point for 0.1 mM glutamine lies on the extended line at $v = 4.4$.

CHO-K1 cells by different amino acids. The monolayers were first washed with 2 ml of choline KRP, pH 7.4, and then incubated for 90 min with a fresh 2-ml aliquot of the same medium at 37° to deplete the intracellular pool of amino acids. These steps can be performed using the wash tray apparatus. Next, by using the uptake trays, cells were loaded with labeled [^{3}H]threonine for 30 min in Na$^+$-containing KRP. After the loading period, the monolayers were washed for less than 10 sec with choline KRP at pH 7.5 or 5.5. The exodus of the labeled amino acid into Na$^+$ or choline KRP was observed at the same pH values. After the indicated time interval, the cells were rapidly washed with ice-cold PBS, and the amount of radioactive label remaining in the monolayers was determined. Using this equipment we could conveniently manipulate the various conditions and were able to study the effect of six amino acids during five different time intervals and in four different media in a single experiment.

Figure 6 measures the K_i of histidine or glutamine uptake by H35 hepatoma cells. Typically in this case we might examine the uptake of 0.1, 0.5, 1.0, and 2.5 mM L-[U-^{14}C]glutamine in the presence or absence of four inhibitor concentrations of histidine. With triplicate data points one can obtain adequate K_i values from two 24-well trays.

Table I shows the inhibitory effect of 12 amino acids on the uptake of glutamine and histidine in the presence and absence of Na$^+$ ions and into four different cell types. These and other possible arrangements render the

TABLE I

INHIBITION OF 0.05 mM GLUTAMINE AND HISTIDINE UPTAKE INTO EHRLICH, FETAL, AND HEPATOMA CELLS HTC AND H35[a]

Inhibitor (10 mM)	Ehrlich cells		Fetal hepatocytes		HTC		H35	
	Na$^+$ dependent	Na$^+$ independent	Na$^+$ dependent	Na$^+$ independent	Na$^+$ dependent	Na$^+$ independent	Na$^+$ dependent	Na$^+$ independent
L-Glutamine uptake								
None	100 (2.24)	100 (0.23)	100 (1.68)	100 (0.06)	100 (1.22)	100 (0.08)	100 (1.24)	100 (0.09)
L-Glutamine	9	12.4	12.9	64.6	5.2	46.3	5.6	37.6
L-Histidine	21	7.3	8.7	29.5	22.2	25	4.8	26.9
L-1-Me-histidine	65.4	9.9	43.1	40	63.7	17.5	21	17.2
L-3-Me-histidine	38.2	9.4	90.8	52.1	31.2	26.3	60	29
L-Asparagine	6.6	29.2	34.3	59.1	12.5	63.8	32.8	51.6
MeAIB	23.2	102	81.4	102.3	16.9	80	100	90
L-Threonine	34.9	16.7	74.6	49.6	32	72.5	81	43
L-Arginine	69.5	86.3	66.1	98.8	73.3	75	49	108
L-Homoarginine	77.8	89.7	54.7	97.7	ND	ND	77	83
BCH	89.5	6.9	84.4	42.1	73	26.3	94	16.1
MPA	82.8	9.9	ND	ND	75.3	27.5	ND	ND
L-Histidine uptake								
None	100 (0.87)	100 (1.74)	100 (0.5)	100 (0.105)	100 (0.54)	100 (2.82)	2.82 ± 0.091	2.79 ± 0.18
L-Glutamine	13.7	3.62	9.8	42	0	27.5	0.24 ± 0.007	0.18 ± 0.005
L-Histidine	23.3	1.55	0.6	33.3	26.6	2.66	0.07 ± 0.003	0.04 ± 0.002
L-1-Me-histidine	58.9	1.03	32.2	25.7	73.2	1.84	0.10 ± 0.0018	0.03 ± 0.0018
L-3-Me-histidine	18.4	1.72	46.4	45.7	18.1	4.36	0.15 ± 0.011	0.05 ± 0.0003
L-Asparagine	20.7	8.97	25.8	124.8	0	43.8	0.53 ± 0.018	0.49 ± 0.038
MeAIB	14.1	93	120.6	154.3	57	80	3.38 ± 0.165	2.99 ± 0.389
L-Threonine	21	3.68	84.0	38.1	23.2	19	0.33 ± 0.0003	0.16 ± 0.013
L-Arginine	75.9	71.3	23.4	110.5	148.2	83.7	2.65 ± 0.088	2.43 ± 0.005
L-Homoarginine	90	78.2	27.2	114.3	ND	ND	2.69 ± 0.079	2.42 ± 0.023
BCH	53	2.64	117.4	27.6	73	14.2	ND	0.03 ± 0.002
MPA	26.9	1.09	87.6	25.7	34.9	5.63	0.16 ± 0.0097	0.04 ± 0.00008

[a] The uptake of 0.05 mM L-[U-^{14}C]glutamine or L-[U-^{14}C]histidine into various cells was measured for 1 min at 37°. The data are in each case the averages of three determinations and are expressed as percentages of the rates of the Na$^+$-dependent and Na$^+$-independent uptake seen in the absence of an inhibitor. The 100% control rates are given in parentheses as nanomoles · milligram of protein^{-1} · min^{-1}. For L-histidine uptake into H35 cells, the absolute rates in Na$^+$ and choline buffers are given since the difference between the two is very small. ND, Not determined. MPA, 4-Amino-1-methylpiperidine-4-carboxylic acid.

technique very useful for qualitative and quantitative analysis. The reproducibility from one tray to the other, or from one day to another, is typically within 10%.

This apparatus can be used for a variety of other problems such as immunological studies, hormone binding or function, properties of cell surface proteins, and metabolic studies. For instance, one can study the time course of specific binding of [^{125}I]monoiodoinsulin isomers to cells in culture using the cluster tray technique. Conventionally the extent of binding is assessed by rapidly separating cells from the incubation media by centrifugation using a silicon oil gradient system.[26] This method is somewhat cumbersome, because one needs to remove the oil before the cells can be extracted with TCA to separate protein-bound hormones.

Lectins are often used as molecular probes to detect topological alternations of cell surface sugar moieties due to their ability to bind specific saccharide receptors. A variety of lectins could be screened rapidly by using the cluster tray technique.

Isolab Inc. (Ohio) has designed an apparatus to transfer 96 specimens at once using a Pegasus Transfer System (VP#400). This system consists of a gang of 96 plastic syringes held together in an anodized aluminum frame. The plungers of the syringes are connected in another frame, which is used to raise and lower the plungers in unison. This system is suggested to be ideal for adding cells to multiwell units for RNA–DNA hydridization studies. In our opinion the cluster tray apparatus would be just as convenient and less expensive for such studies.

Acknowledgment

I am grateful to Satish Vekaria for preparing the illustrations, Jacqueline Benson for her excellent typing skills, Dr. Halvor Christensen for his comments on this manuscript, and for support from Grant HDO1233 from the Institute for Child Health and Human Development, National Institutes of Health, United States Public Health Service.

[26] D. Peavy, J. Abram, J. Ferroni, C. Hooker, B. Frank, and W. Duckworth, in "Isolation, Characterization and Use of Hepatocytes" (R. Harris and N. Cornell, eds.), p. 261. Elsevier, Amsterdam, 1983.

[7] Utilization of Binding Energy and Coupling Rules for Active Transport and Other Coupled Vectorial Processes

By WILLIAM P. JENCKS

Coupled vectorial processes bring about the interconversion of chemical energy and work, as in the ATP-driven mechanical work of muscle contraction or work against a gradient of chemical potential in active transport [Eq. (1)].

$$\text{ATP} \rightleftharpoons \text{ADP} + \text{P}_i + \text{work} \tag{1}$$

These processes are often reversible, as in the synthesis of ATP that is driven by proton transport across a gradient during oxidative phosphorylation and photophosphorylation. The problem is how to describe and think about processes such as those described by Eq. (1). Coupled vectorial systems have little or no relationship to steam engines or Carnot cycles.

It is useful to consider separately two components of coupled vectorial processes, the *utilization of binding energy* and the *mechanism of coupling* between chemical energy and the production or utilization of work. Both of these components are important in coupled vectorial processes, but there is not necessarily a direct connection between them. Much of the confusion surrounding the mechanisms of coupled vectorial processes has arisen from a failure to distinguish between the roles of binding energy and coupling.

The primary role of binding energy is kinetic. Binding energy is *utilized* in order to avoid high- or low-energy intermediates along the reaction path under physiological conditions. High-energy intermediates are likely to introduce a barrier to rapid turnover of the system (Fig. 1A), while low-energy intermediates cause the enzyme to fall into an energy well from which escape is difficult (Fig. 1B). The energies of the intermediates in the reaction cycle of the Ca^{2+}-ATPase from sarcoplasmic reticulum illustrate how the energies are balanced in a system of this kind in order to avoid high- or low-energy intermediates (Fig. 1C).[1] This diagram is based on physiological concentrations of reactants and a concentration gradient of 10^4 for calcium ions across the membrane of the vesicle. The utilization of

---

[1] C. M. Pickart and W. P. Jencks, *J. Biol. Chem.* **259,** 1629 (1984).

METHODS IN ENZYMOLOGY, VOL. 171

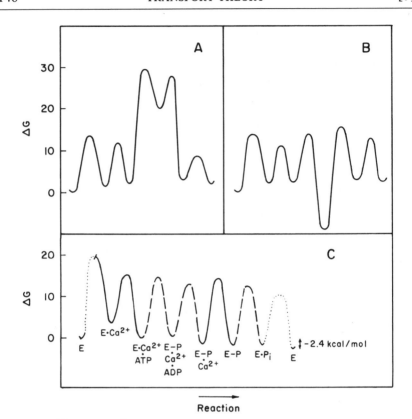

Reaction

FIG. 1. Gibbs energy diagram to show how high-energy (A) or low-energy (B) intermediates along a reaction pathway cause kinetic barriers for reaction; in B the enzyme will accumulate in the energy well so that there is a large barrier for further reaction. (C) The Gibbs energy profile for the Ca^{2+}-ATPase at pH 7.0, based on standard states corresponding to approximate physiological concentrations of 8 mM ATP, 40 μM ADP, 5 mM Mg^{2+}, 0.1 μM external Ca^{2+}, and 1 mM internal Ca^{2+}.[1]

binding energies to avoid high- or low-energy intermediates is described by interaction energies.[1-6]

It should be noted that there is no "energized state" or particular step in which "energy transfer" occurs. The Gibbs energies of all of the states of the system are similar. Individual steps involve the binding and dissociation of ligands that have high or low free energies of hydrolysis and may have high or low chemical potentials under physiological conditions. These

[2] J. Wyman, Jr., *Adv. Protein Chem.* **19**, 223 (1964).
[3] G. Weber, *Adv. Protein Chem.* **29**, 1 (1975).
[4] T. L. Hill, "Free Energy Transduction in Biology." Academic Press, New York, 1977.
[5] W. D. Stein and B. Honig, *Mol. Cell. Biochem.* **15**, 27 (1977).
[6] W. P. Jencks, *Adv. Enzymol. Relat. Areas Mol. Biol.* **51**, 75 (1980).

differences are balanced in the different states, so that turnover can occur easily.

The coupling process is described by a set of rules. These rules generally represent enzyme specificities for catalysis that *change* in different states of the enzyme. Enzymes that catalyze vectorial processes differ from most enzymes in their ability to change specificity in different steps of the reaction cycle. The specificity can be for chemical catalysis or for some vectorial property. Specificities for catalysis can give tighter coupling than binding specificities, because they are usually much larger; the rate constants k_{cat} and k_{cat}/K_m for specific and nonspecific substrates can differ by as much as 10^{10}, or more, whereas binding specificities are usually smaller. The problem of understanding the mechanism of coupling is to identify the rules for a particular system and to characterize the mechanisms and structures that are responsible for the operation of these rules.[6,7]

Utilization of Binding Energy

Reactions that are effectively irreversible in solution are often readily reversible at the active sites of enzymes that catalyze coupled vectorial processes. For example, the hydrolysis of ATP is reversible in the active sites of myosin,[8] dynein,[9] and the F_1- and CF_1-ATPases,[10-12] and the hydrolysis of an acyl phosphate is reversible at the active sites of the Ca^{2+}- and Na^+-transporting ATPases.[13-15] This behavior can be described by the thermodynamic box of Eq. (2) and the Gibbs free energy diagram of Fig. 2 for myosin ATPase.[6,7,16-19]

$$
\begin{array}{ccc}
\text{ATP} + \text{M} \xrightleftharpoons{K_3} & \text{M}\cdot\text{ATP} & \\
K_1 \big\Vert & K_4 \big\Vert & (2a) \\
\text{ADP} + \text{P}_i + \text{M} \xrightleftharpoons{K_2} & \text{M}\cdot^{\text{ADP}}_{\text{P}_i} &
\end{array}
$$

[7] W. P. Jencks, *Proc. XVIIIth Solvay Conf. Chem., Brussels 1983,* pp. 59–80. Springer-Verlag, Berlin and New York, 1986.

[8] C. R. Bagshaw and D. R. Trentham, *Biochem. J.* **133**, 323 (1973).

[9] K. A. Johnson, *Annu. Rev. Biophys. Biophys. Chem.* **14**, 161 (1985).

[10] J. Rosing, C. Kayalar, and P. D. Boyer, *J. Biol. Chem.* **252**, 2478 (1977).

[11] C. Grubmeyer, R. L. Cross, and H. S. Penefsky, *J. Biol. Chem.* **257**, 12092 (1982).

[12] R. I. Feldman and D. S. Sigman, *J. Biol. Chem.* **257**, 1676 (1982).

[13] R. L. Post and S. Kume, *J. Biol. Chem.* **248**, 6993 (1973).

[14] T. Kanazawa and P. D. Boyer, *J. Biol. Chem.* **248**, 3163 (1973).

[15] H. Masuda and L. de Meis, *Biochemistry* **12**, 4581 (1973).

[16] C. R. Bagshaw and D. R. Trentham, *Biochem. J.* **141**, 331 (1974).

[17] D. R. Trentham, J. F. Eccleston, and C. R. Bagshaw, *Q. Rev. Biophys.* **9**, 217 (1976).

[18] R. S. Goody, W. Hofmann, and H. G. Mannherz, *Eur. J. Biochem.* **78**, 317 (1977).

[19] W. P. Jencks, *Proc. Natl. Acad. Sci. U.S.A.* **78**, 4046 (1981).

$$\Delta G_1^0 + \Delta G_2^0 = \Delta G_3^0 + \Delta G_4^0 \qquad (2b)$$

$$\Delta G_1^0 - \Delta G_4^0 = \Delta G_3^0 - \Delta G_2^0 = \Delta G_I \qquad (2c)$$

The large change in the equilibrium constant for ATP hydrolysis at the active site of the enzyme (K_4) compared with that in solution (K_1) requires that ATP bind very tightly (K_3) compared with ADP and P_i (K_2). The difference between the standard Gibbs free energy change for the hydrolysis of ATP in solution and at the active site is $\Delta G_1^0 - \Delta G_4^0$ in Eq. (2). The thermodynamic box requires that this difference in Gibbs energy be equal to the difference in the Gibbs energy for binding of the reactants and the products, $\Delta G_3^0 - \Delta G_2^0$. This difference is the *interaction* energy, ΔG_I. This is shown in Eqs. (2b) and (2c).

It is useful to illustrate the energetics of a thermodynamic box by a bar graph, such as that shown in Fig. 2. The graph shows that the synthesis of ATP in solution is very unfavorable, whereas the equilibrium constant for synthesis at the active site is close to unity.[6,7,16-19] It is immediately apparent that this difference requires that the binding of ATP to the enzyme must be more favorable than the binding of ADP and P_i. Note that these Gibbs energy changes are defined in terms of some standard state, which is 1 M in this case.

The difference in the ΔG^0 values on any two opposite sides of the thermodynamic box or the diagram is the interaction energy, ΔG_I. In the diagram this represents the amount by which the two opposite lines connecting the energy bars are not parallel. Parallel lines are shown by the dashed lines in Fig. 2, and the interaction energy is indicated. The interaction energy is the amount by which ATP binds tighter than ADP and P_i,

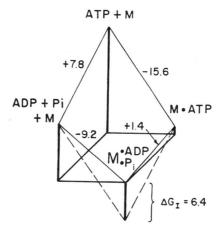

FIG. 2. Gibbs free energy diagram for the binding of ATP, ADP, and P_i to myosin.[7,16-19]

and it is the difference between the Gibbs energies for the hydrolysis of ATP in solution and at the active site of the enzyme. The value of ΔG_I is approximately 6.4 kcal mol^{-1} for a standard state of 1 M.

Two requirements must be met in order to make possible the facile synthesis of ATP at the active site. First, the binding of ATP to the enzyme must be very strong. This pulls the equilibrium toward ATP synthesis. The strong binding may be regarded as an expression of a large *intrinsic binding energy* between ATP and the active site. Second, the binding of ADP and P_i must be relatively very weak, in spite of the fact that ADP and P_i contain almost the same chemical structures for binding to the enzyme that are found in ATP. This weak binding produces a relatively high Gibbs energy of the bound ADP and phosphate in the active site, so that they can react easily to form ATP. The binding energy of ATP is *expressed,* to give the observed strong binding, while the binding energy of ADP and phosphate is *utilized,* to overcome destabilization and induce intramolecularity, so that the observed binding is weaker. The interaction energy is the difference between the strong binding, or stabilization, of ATP and the relatively weak binding, or destabilization, of ADP and P_i at the active site. The system will work only if there are *both* very strong binding of ATP and weak binding of ADP and P_i; i.e., there must be an interaction energy. If there were no interaction energy, the Gibbs energies would be described by the dashed lines in Fig. 2 and ATP synthesis at the active site would be just as difficult as it is in solution.

The relatively weak binding of ADP and P_i, as reflected in the interaction energy, is the result of two factors:

1. A *destabilization* of bound ADP and P_i, which may be described by ΔG_D. This may result from compression, desolvation, strain, and unfavorable electrostatic interactions. Relief of these unfavorable interactions when ATP is synthesized favors ATP synthesis. The destabilization may be so large that the binding of the substrates is incomplete; i.e., part of the ADP molecule may not be bound in its binding site or may bind nonproductively in the ground state of the system.

2. A decrease in the *entropy* of bound ADP and P_i at the active site, compared with ADP and P_i in solution at a standard state of 1 M. The most obvious, and possibly the most important, difference between the enzymatic and nonenzymatic synthesis of ATP in this system is that the enzymatic reaction is intramolecular and the nonenzymatic reaction is bimolecular. Intramolecular reactions are known to be more favorable than bimolecular reactions by factors of up to 10^8 for 1 M solutions, in the absence of strain. Formation of a new covalent bond is improbable; it requires a large loss of entropy because of loss of translational and rota-

tional freedom of the reactants. This loss is much smaller and the reaction is much more probable if the reactants are already tightly bound in exactly the correct position to react. A negative value of $T \Delta S$ corresponds to a positive ΔG, so that loss of entropy contributes to the interaction energy.[20,21]

More binding energy is required to bring about binding if there is strain, or some other kind of destabilization, and if there is a loss of entropy when ADP and P_i bind. The observed binding is therefore weaker, by the amount of the destabilization and entropy loss, which is equal to the interaction energy.

The relative importance of the contributions of destabilization and etropy loss is not known. The $T \Delta S$ term alone could account for the entire advantage of the intramolecular reaction. However, it is virtually certain that destabilization also contributes to the interaction energy because it is not possible to hold reactants very tightly, with a large loss of entropy, unless they are compressed so that their freedom of movement and low-frequency vibrations are restricted.[21]

Unfortunately, the relative contributions of the ΔG_D and $- T \Delta S$ terms to ΔG_I cannot be determined by measuring the thermodynamic parameters of the reaction. Compression, strain, and other components of ΔG_D do not necessarily appear as an unfavorable ΔH term, because the observed thermodynamic parameters include large compensating enthalpy and entropy terms that result from hydrophobic and electrostatic interactions, changes in protein conformation and solvent "structure," and other factors. Some of these effects cause large changes in ΔH and $T \Delta S$ with little change in ΔG for the system. It is difficult or impossible to separate the contributions of these different factors for the binding of small molecules to proteins in aqueous solution.[21,22]

Figure 2 is based on a standard state of 1 M. However, it is often of interest to consider the energetics of a system under physiological conditions. If the diagram is drawn for more nearly physiological standard states of 5 mM for ATP, 0.05 mM for ADP, and 1 mM for P_i,[23] the interaction energy is 13 kcal mol^{-1}. This value is much larger because it includes the large loss of entropy for the synthesis of ATP from dilute reactants at physiological concentrations. Values of ΔG under physiological conditions have been called "basic" Gibbs energy changes.[24]

[20] M. I. Page and W. P. Jencks, *Proc. Natl. Acad. Sci. U.S.A.* **68,** 1678 (1971).
[21] W. P. Jencks, *Adv. Enzymol.* **43,** 219 (1975).
[22] E. Grunwald, *J. Am. Chem. Soc.* **106,** 5414 (1984).
[23] J. A. Sleep and S. J. Smith, *Curr. Top. Bioenerg.* **11,** 239 (1981).
[24] R. M. Simmons and T. L. Hill, *Nature (London)* **263,** 615 (1976).

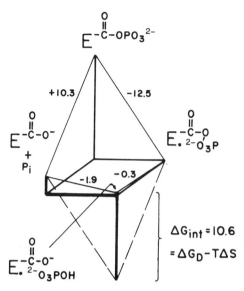

FIG. 3. Gibbs energy diagram to show how an acyl phosphate is formed spontaneously from the Ca^{2+}-ATPase and P_i. Noncovalent binding interactions between the phosphoryl group and the active site of the enzyme are indicated by the heavy dot. The dashed lines show that acyl phosphate formation at the active site would not occur if there were no interaction energy.[1]

The energetics of ATP hydrolysis at the active site are quite similar for dynein[9] and for the F_1-ATPase of mitochondria.[11]

The Gibbs energy diagram is slightly different for ATPases that form a covalent acyl phosphate intermediate in a reversible reaction, because one of the reacting molecules is part of the enzyme. An energy diagram for formation of the acyl phosphate intermediate, $E-P$, from the Ca^{2+}-ATPase and inorganic phosphate [Eq. (3)] is shown in Fig. 3 (pH 7.0, 25°, 100 mM KCl, 5 mM MgSO$_4$).[1]

$$E + P_i \rightleftharpoons E \cdot P_i \rightleftharpoons E-P \qquad (3)$$

The left-hand side of the diagram shows the unfavorable Gibbs energy change that would be observed if the acyl phosphate were formed at the active site of the enzyme with the same ΔG^0 as the nonenzymatic reaction in solution, i.e., with the phosphate extending out into the solvent and without noncovalent binding of the phosphate group into its complementary binding site on the enzyme. In fact, inorganic phosphate binds readily to the enzyme to form a noncovalent complex, $E \cdot P_i$, and this complex goes on to form the covalent enzyme–phosphate intermediate with an equilibrium constant close to 1.

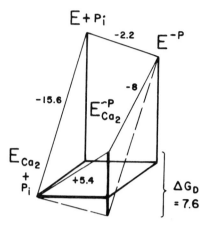

FIG. 4. Energy bar diagram to describe the binding of calcium and of phosphate to the Ca^{2+}-ATPase and their mutual destabilization in $E \sim P \cdot Ca_2$. The dashed lines show the binding that would occur if there were no interaction energy that causes destabilization.[1]

The back side of the diagram shows that this spontaneous formation of an acyl phosphate is made possible by the intrinsic binding energy of the phosphate moiety of the acyl phosphate, which binds to the active site in a strongly downhill reaction. The strong binding *pulls* the reaction toward acyl phosphate synthesis. One reason why this binding is so strong is that it is intramolecular, so that it does not require loss of translational and overall rotational entropy. The binding site includes a magnesium ion that contributes to this strong binding.

Again, in order for the reaction to proceed, the Gibbs energy of the noncovalent complex, $E \cdot P_i$, must be increased by an interaction energy, ΔG_{int}, so that it has a high enough Gibbs energy to form the covalent $E-P$ species spontaneously. This involves a destabilization energy, ΔG_D, and a loss of entropy upon binding, $-T \Delta S$, because the bimolecular reaction in solution is converted to an intramolecular reaction at the active site, as in the synthesis of ATP at the active site of myosin.

The second curious property of the Ca^{2+}-ATPase is that the low-energy $E-P$ intermediate, which is at equilibrium with P_i in solution, is converted into the high-energy intermediate $E \sim P \cdot Ca_2$, which is at equilibrium with ATP, upon binding two Ca^{2+} ions.[25] The Na^+,K^+-ATPase shows very similar behavior upon binding three Na^+ ions.[26] The energetics of these reactions for the calcium enzyme can be illustrated by Fig. 4.[1,6] Starting with the back of the diagram, two Ca^{2+} ions bind much more strongly to

[25] A. F. Knowles and E. Racker, *J. Biol. Chem.* **250**, 1949 (1975).
[26] K. Taniguchi and R. L. Post, *J. Biol. Chem.* **250**, 3010 (1975).

the free enzyme, to form $E \cdot Ca_2$, than to $E-P$, to form $E \sim P \cdot Ca_2$. The difference corresponds to an interaction energy of $\Delta G_I = \Delta G_D = 7.6$ kcal mol^{-1}. The same interaction energy of $\Delta G_I = 7.6$ kcal mol^{-1} describes the difference in ΔG^0 for the hydrolysis of $E-P$ and $E \sim P \cdot Ca_2$. The formation of $E-P$ from E and P_i is favorable by 2.2 kcal mol^{-1} (Fig. 3). The relatively weak binding of Ca^{2+} to $E-P$ requires that the synthesis of $E \sim P \cdot Ca_2$ from $E \cdot Ca_2$ is unfavorable by 5.4 kcal mol^{-1}. Thus, the interaction energy increases the energy of $E \sim P \cdot Ca_2$ and makes it an unstable species with respect to the transfer of both calcium and the phosphoryl group. It accounts for the large negative free energy of hydrolysis of $E \sim P \cdot Ca_2$ and brings this species to a high energy level, so that it can readily transfer phosphate to ADP to form ATP.

The interaction energy of $E \sim P \cdot Ca_2$ represents a mutual destabilization of Ca^{2+} and bound P_i at the active site that increases the Gibbs energy and the escaping tendency of both species. It is important that there be a large escaping tendency for bound Ca^{2+} in this species, because Ca^{2+} dissociates from $E \sim P \cdot Ca_2$ to the inside of the vesicle, where the concentration of Ca^{2+} is very high; if Ca^{2+} were bound tightly to the enzyme it could not dissociate and the enzyme could not turn over. It is equally important for turnover of the enzyme that Ca^{2+} bind very tightly to E, from the outside of the vesicle, so that it can be transported from the cytoplasm at physiological concentrations of $10^{-6} - 10^{-7}$ M. Thus, these changes in the stability of bound Ca^{2+} and P allow turnover to occur at a useful *rate* by avoiding very high-energy or very low-energy intermediates along the reaction path. They do not, however, explain how the hydrolysis of ATP is coupled to the transport of Ca^{2+} across the membrane of the vesicle.

It is possible that $E \sim P \cdot Ca_2$ is actually a rapidly equilibrating mixture of two species, with a high group potential or escaping tendency for phosphate transfer in one species and for calcium in the other, or that there are two different interconverting species of very different energy.[1,6,27-29] However, there is no evidence for these possibilities. It should be noted that although the hydrolysis of $E \sim P \cdot Ca_2$ is thermodynamically downhill, it is blocked kinetically. The species $E \sim P \cdot Ca_2$ will transfer P only to ADP, not to HOH, as a result of the changing catalytic specificity of the enzyme that is described below.

There is an analogous interaction energy that represents a mutual destabilization of ATP and actin bound to myosin in the myosin ATPase system[18,30]; very similar behavior is found in the dynein–tubulin system.[9]

[27] L. de Meis and A. L. Vianna, *Annu. Rev. Biochem.* **48**, 275 (1979).
[28] C. Tanford, *Proc. Natl. Acad. Sci. U.S.A.* **79**, 6161 (1982).
[29] T. L. Hill and E. Eisenberg, *Q. Rev. Biophys.* **14**, 463 (1981).
[30] B. Finlayson, R. W. Lymn, and E. W. Taylor, *Biochemistry* **8**, 811 (1969).

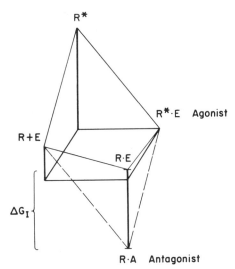

FIG. 5. Gibbs energy diagram to show how an interaction energy is needed for a receptor–effector interaction.[7] An active hormone, drug, or neurotransmitter, E, uses its intrinsic binding energy to bind strongly to the activated form of the receptor, R*, and increases the concentration of this species; it binds relatively weakly to the inactive receptor, R, because of the interaction energy ΔG_I. A compound, A, which binds tightly to R with no interaction energy, is an antagonist.

The binding of ATP decreases the affinity of myosin for actin and greatly increases its rate of dissociation from actomyosin, to give M·ATP. There is a similar interaction energy between the binding of actin and of ADP and P_i to myosin. After the hydrolysis of ATP, the strong binding of actin to myosin provides a strong thermodynamic driving force to pull the cycle in the forward direction and carry out mechanical work as ADP and P_i dissociate.[17] A large fraction of the overall free energy of hydrolysis of ATP is released in this step, because two molecules dissociate from myosin, with a large increase in translational and rotational entropy.

The intrinsic binding energy between ligands and macromolecules is utilized in a similar way in many biological processes. It may not be a gross exaggeration to suggest that anything interesting that happens in biology involves interaction energies. The extraordinary catalytic activity of enzymes arises, in large part, from the expression of a large intrinsic binding energy in the transition state but not in the enzyme–substrate complex.[7,21,31,32] The interaction energy describes the utilization of intrinsic binding energy to induce strain, destabilization, and loss of entropy of the

[31] L. Pauling, *Am. Sci.* **36**, 58 (1948).
[32] A. Fersht, "Enzyme Structure and Mechanism," 2nd Ed. Freeman, San Francisco, California, 1985.

bound substrates, so that the enzyme – substrate complex is raised to a high enough Gibbs energy that it can reach the transition state easily at the active site. If there were no interaction energy the rate constants would be the same for reaction at the active site and for the nonenzymatic reaction in solution.

Receptors for many hormones, neurotransmitters, and drugs act by expressing the intrinsic binding energy of the ligand in the active state of the receptor, but much less in the ground state.[7,33] This is illustrated by Fig. 5, in which the energy of the ground state, R, is much lower than that of the active state, R* in the absence of an effector, E. The difference between the binding of E to the ground state and the active state is the interaction energy, which provides the driving force for conversion of the receptor into the active state, R*E, upon binding the effector. This interaction energy can increase the rate constant as well as the equilibrium constant for formation of the active state. If there is no such interaction energy, the receptor is not converted to the active state and the ligand is an antagonist instead of an agonist. This is shown by $R \cdot A$ in Fig. 5.[7]

In all of these examples the interaction energy results in an observed binding for a ligand that is much weaker than the maximum binding that would be observed in the absence of interaction energy. It follows that the most active substrates, hormones, and drugs may not be those that show the strongest binding to an enzyme or receptor. This is because the strong binding, the intrinsic binding energy, must be manifested in the active state or transition state, not the ground state.

Coupling

The mechanism of coupling can be described by a set of *rules*. These rules usually represent enzyme specificities that change during the reaction cycle. Enzymes that catalyze coupled vectorial processes differ from most enzymes in that there is a change in their specificity for chemical catalysis or some vectorial reaction as the enzyme switches from one state to another during the reaction cycle. The reaction may be described as a *switch cycle*. A fully coupled reaction takes place when the reaction proceeds exactly as described by the chemical equations of the complete reaction cycle. To the extent that the specificity rules are not followed exactly the reaction becomes partially or completely uncoupled.[6,7] It is important to realize that there is no one step in the cycle that is responsible for coupling; failure of a coupling rule in any step will lead to uncoupling.[6,29]

[33] B. Katz and S. Thesleff, *J. Physiol. (London)* **138**, 63 (1957).

An understanding of the mechanism and nature of these rules and changes in specificity represents an understanding of the coupling mechanism. Ideally, we would like to understand the basis for these changing specificities at the molecular and structural level. In the absence of such understanding, a first step is to characterize the rules and specificities as completely as possible.

The switch cycle can often be described by a sort of matrix of changing specificities, as shown in Eq. (4), in which E_1 and E_2 represent one pair of different specificities and E^0 and E^* represent a different pair of changing specificities. It is important to define the specificities of the different states

$$E_1^0 \rightleftharpoons E_1^* \leftarrow \text{specificity 1}$$
$$\Updownarrow \quad \Updownarrow$$
$$E_2^0 \rightleftharpoons E_2^* \leftarrow \text{specificity 2} \qquad (4)$$
$$\uparrow \quad \uparrow$$
$$0 \quad *$$
$$\text{specificity}$$

in such a cycle as clearly as possible. There is a danger that without a clear definition, terms such as E_1 and E_2 can take on a life of their own, hindering more than helping the understanding of coupled vectorial processes.

The changes in specificity that are responsible for coupling are best explained by example. We will describe coupling rules for the Ca^{2+}-transporting ATPase of sarcoplasmic reticulum, a hypothetical set of rules to illustrate a mechanism for phosphorylation coupled to proton transport in oxidative phosphorylation, and rules that are concerned only with the control of sequential binding and dissociation steps for muscle contraction.

Ca^{2+}-ATPase

Scheme 1 is a simplified reaction cycle that illustrates a set of coupling rules for the Ca^{2+}-ATPase of sarcoplasmic reticulum.[1] The rules are based on well-known properties of different chemical states of the enzyme.[27]

The *chemical specificity* for catalysis by $E \cdot Ca_2$ is for phosphorylation of the enzyme by ATP and for transfer of phosphate to ADP from $E \sim P \cdot Ca_2$ in the reverse direction. When calcium dissociates from the enzyme, the specificity of the enzyme changes to catalysis of the phosphorylation of the enzyme by P_i and for transfer of phosphate from $E-P$ to water in the reverse direction. If either of these rules is violated, the system would hydrolyze ATP without transporting calcium and would be uncoupled. This would happen if $E \sim P \cdot Ca_2$ reacted with water to give P_i, or if E reacted with ATP to give $E-P$, which would then hydrolyze to P_i.

The *vectorial specificity* of the free enzyme, E, is that it will bind and dissociate calcium only on the outside (the cytoplasmic side) of the mem-

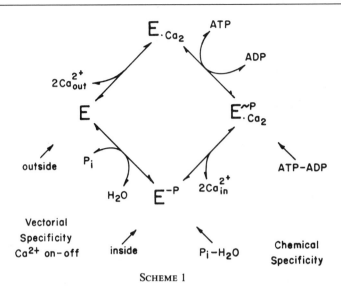

SCHEME 1

brane, whereas the phosphorylated enzyme, E–P, will bind and dissociate calcium only on the inside of the vesicle. If either of these rules is violated the system will become uncoupled because calcium would leak out of the loaded vesicle.

If these specificity rules are followed rigorously, two calcium ions will be transported across the membrane for each molecule of ATP that is hydrolyzed or synthesized in one turn of the reaction cycle (Scheme 1); failure of one of the rules leads to partial or complete uncoupling.

This switch cycle works because it follows an *ordered kinetic mechanism;* it represents one of the few instances in biochemistry in which it is essential that a particular kinetic mechanism be followed for the successful functioning of an enzyme. In order for Ca^{2+} to dissociate to the outside of the vesicle from $E \sim P \cdot Ca_2$, it is required that phosphate first be transferred to ADP to make ATP. In order for phosphate to be transferred to water in the other arm of the cycle, it is necessary that calcium first dissociate from $E \sim P \cdot Ca_2$ to the inside of the vesicle. Failure of this ordered sequence violates a rule and would lead to uncoupling.

The ordered mechanism divides the chemical reaction and the vectorial reaction into two parts, which are sandwiched between each other. Therefore, neither reaction can be completed without the other taking place. ATP hydrolysis is divided into phosphorylation of the enzyme and hydrolysis of the phosphoenzyme; calcium transport is divided into binding on one side of the membrane and dissociation on the other side.

The ordered mechanism also provides a connection between the energetics and the coupling in this system. Both the $\sim P$ and the Ca^{2+} of $E \sim P \cdot Ca_2$ are effectively "energy rich"; they have a high escaping tendency. An energy-rich $\sim P$ is needed to form ATP at a useful rate and weakly bound Ca^{2+} is needed in order to dissociate calcium into the high concentration of calcium that exists inside the vesicle. The ordered kinetic mechanism ensures that the high-energy species dissociates in a readily reversible process; its failure would presumably relieve the mutual destabilization of bound $\sim P$ and Ca^{2+} in $E \sim P \cdot Ca_2$, so that a subsequent step would become so unfavorable energetically that it could not proceed at a useful rate.

The Na^+,K^+-ATPase follows a very similar set of coupling rules, with binding of three sodium ions (instead of two calcium ions) to the enzyme, a resulting change in specificity for catalysis of phosphoryl transfer, and then dissociation of the sodium ions from the phosphoenzyme on the other side of the membrane. The switch cycle follows an ordered kinetic mechanism, with binding of Na^+ to the enzyme before phosphorylation (by ATP) on one side and dissociation of Na^+ before dephosphorylation (by water) on the other side of the membrane. An analogous ordered mechanism in the reverse direction applies to the back transport of K^+.[6,34,35] A proton-transporting ATPase from gastric mucosa and related enzymes from other sources appear to follow a similar mechanism.[36]

Proton-Transporting ATPases of Mitochondria and Chloroplasts

The rules that describe the coupling mechanism of the ATPases of oxidative phosphorylation and photophosphorylation are not known, but some of their properties are partially understood. In order that ATP synthesis may be coupled to proton transfer, catalysis of the synthesis and hydrolysis of ATP must be dissociated from the binding and dissociation steps.[37-43] If these processes are not separated, the enzyme would catalyze

---

[34] L. C. Cantley, *Curr. Top. Bioenerg.* **11**, 201 (1981).
[35] W. P. Jencks, *Curr. Top. Membr. Transp.* **19**, 1 (1983).
[36] G. Sachs, H. R. Koelz, T. Berglindh, E. Rabon, and G. Saccomani, *in* "Membranes and Transport" (A. N. Martonosi, ed.), Vol. 1, p. 633. Plenum, New York, 1982.
[37] C. Kayalar, J. Rosing, and P. D. Boyer, *J. Biol. Chem.* **252**, 2486 (1977).
[38] G. Rosen, M. Gresser, C. Vinkler, and P. D. Boyer, *J. Biol. Chem.* **254**, 10654 (1979).
[39] R. L. Hutton and P. D. Boyer, *J. Biol. Chem.* **254**, 9990 (1979).
[40] R. L. Cross, C. Grubmeyer, and H. S. Penefsky, *J. Biol. Chem.* **257**, 12101 (1982).
[41] C. C. O'Neal and P. D. Boyer, *J. Biol. Chem.* **259**, 5761 (1984).
[42] S. D. Stroop and P. D. Boyer, *Biochemistry* **24**, 2304 (1985).
[43] H. S. Penefsky, *J. Biol. Chem.* **260**, 13735 (1985).

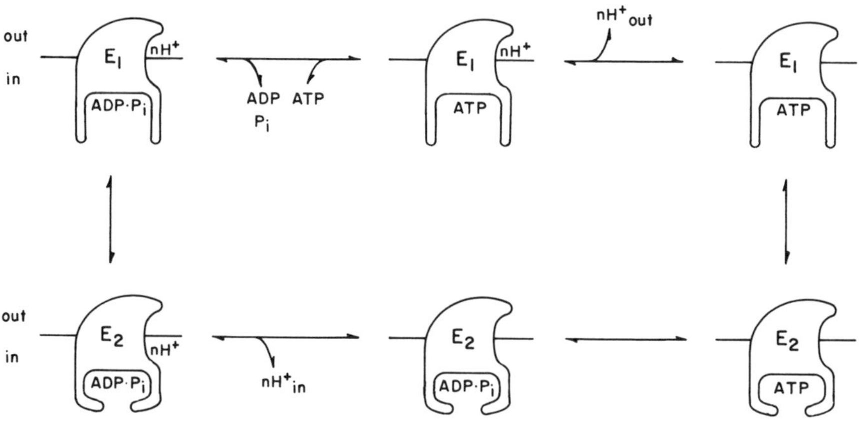

SCHEME 2

ATP hydrolysis that is not coupled to the transport of protons. The kind of mechanism that could bring this about is shown in Scheme 2.[6,7,44] The enzyme in Scheme 2 represents the F_0F_1 complex; however, it is possible that the proton transport does not involve F_1 directly and that coupling to proton transport is mediated through F_0, presumably by a conformational change.

The coupling rules for Scheme 2 are as follows:

1. The binding and dissociation of ATP, ADP, and P_i occur only with the E_1 state of the enzyme, which binds and dissociates protons only on one side of the membrane, whereas catalysis of the reversible cleavage and synthesis of ATP occurs only with the E_2 state of the enzyme, which binds and dissociates protons only on the other side of the membrane. The proton-binding site is alternately exposed to the two sides of the membrane, presumably through one or more channels.

2. Proton transfer across the membrane occurs only with $EH_n^+ \cdot ADP \cdot P_i$, whereas interconversion of unprotonated E_1 and E_2 occurs only with $E \cdot ATP$.

These rules prevent the uncoupled hydrolysis of ATP by requiring transfer of protons across the membrane before the dissociation of ADP and P_i. They prevent leakage of protons by requiring net hydrolysis or synthesis of ATP for each cycle that transports n protons. A favorable proton gradient will drive the cycle in the direction of ATP synthesis on the enzyme by trapping the E_2 species as the transported protons dissociate at the relatively high pH inside the mitochondrion; dissociation of ATP will

---

[44] P. D. Boyer, *FEBS Lett.* **58,** 1 (1975).

be favored by trapping the $E_1 \cdot ATP$ species as protons bind at the relatively low pH outside the mitochondrion.

A difference in the affinity of the E_1 and the E_2 states for protons is not required, in principle. However, the enzyme binds ATP very strongly and a decrease in this affinity must be brought about to drive its dissociation.[37-43] This may be mediated directly or indirectly by the binding of protons through an interaction energy, with a corresponding change in proton affinity, or it could simply reflect a shift in the $E_1 - E_2$ equilibrium that is driven by the proton gradient. A pH gradient across the mitochondrial membrane will increase the relative concentration of $E_1 \cdot ATP$ by protonating this species, and will therefore increase the rate constant for the dissociation of ATP. It has been shown directly that substrate oxidation can decrease the affinity of the mitochondrial ATPase for nucleotides and cause the dissociation of ATP.[43] It is known that nucleotides on other β subunits of the F_1-ATPase can facilitate nucleotide dissociation by decreasing the affinity for ATP and ADP.[39-42] The relationship between the effects of substrate oxidation, the proton gradient, and nucleotide-mediated subunit interactions is still not understood.

The assignment of the rules for the ATPase could be reversed. For example, E_1 could catalyze the chemical reaction and E_2 could allow binding and dissociation. The important points are that (1) there is a switch in the specificities for these two processes in different states of the enzyme and (2) proton transfer across the membrane drives the cyclic interconversion of these states. The conversion to $E_1 \cdot H_n^+ \cdot ATP$ that is driven by the proton is analogous to the role of actin in the actomyosin system; in both cases binding of the ligand induces nucleotide dissociation.

Cotransport of Ions and Molecules

The coupling rules are simpler for systems in which the driving force for the transport of one ion or molecule uphill, against a concentration gradient, is provided by the transport of another ion or molecule downhill, following a concentration gradient. The rule for coupled transport in such symport or antiport systems is simply that the transport of one ion or molecule can occur *only* when the other ion or molecule is transported, in the same or the opposite direction and in the same or an alternating step. However, the structural basis for coupling in these systems is not yet known.

Muscle Contraction

In the actomyosin system of muscle contraction (Scheme 3) the vectorial requirement for coupling is that a myosin head, M, that is attached to

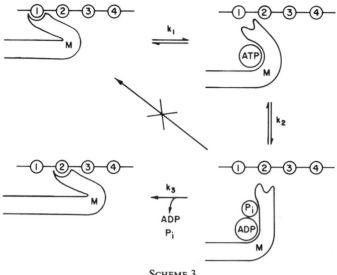

SCHEME 3

the thick filament must dissociate from a binding site on one actin mono-mer in the thin filament, A_1, and bind to a different actin monomer, A_2, when ATP is hydrolyzed (these sites are usually not on adjacent actin molecules of the thin filament). The chemical specificity requires that ATP must not be hydrolyzed to *free* ADP and P_i until this movement has taken place. These requirements must be satisfied in order that useful work may be performed when ATP is hydrolyzed. If ATP hydrolysis is followed by return of the myosin head to the original actin residue, A_1, with release of ADP and P_i, no work is performed and the system is uncoupled. This occurs in isometric contraction.[4,6,17,45]

The exact nature of the specificity rules that define the coupling is still not entirely clear, but the basic requirements for coupling are described by four simple rules for the specificity of association–dissociation steps. (1) The dissociation of nucleotides and P_i from myosin, and their binding to myosin, occur only in the presence of actin. (2) The binding of actin to myosin and the dissociation of actomyosin occur only in the presence of nucleotides. (3) The binding (and dissociation) of ATP occurs only with A_1M. (4) The dissociation (and binding) of ADP and P_i occurs only with A_2M. These rules are summarized in Scheme 4.

If these rules are followed, ATP hydrolysis will be coupled to the performance of mechanical work; to the extent that they are not followed the efficiency of contraction will decrease or ATP hydrolysis will be un-coupled.

[45] H. E. Huxley, *Science* **164**, 1356 (1969).

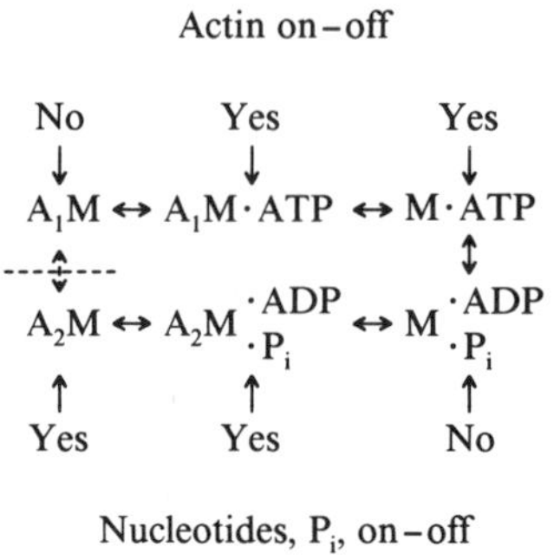

Actin on–off

No　　　　　Yes　　　　　Yes
↓　　　　　　↓　　　　　　↓
A_1M ↔ $A_1M \cdot ATP$ ↔ $M \cdot ATP$

--- ⇕ ---　　　　　　　　　　⇕

A_2M ↔ $A_2M \begin{matrix}\cdot ADP \\ \cdot P_i\end{matrix}$ ↔ $M \begin{matrix}\cdot ADP \\ \cdot P_i\end{matrix}$

↑　　　　　　↑　　　　　　↑
Yes　　　　　Yes　　　　　No

Nucleotides, P_i, on–off

SCHEME 4. Binding–dissociation rules.

Again, ATP hydrolysis is divided into two steps that are separated by two vectorial steps. ATP hydrolysis is controlled by the binding step and the product dissociation step; these are separated by the actin binding and dissociation steps, which represent the vectorial components of the switch cycle.

The hydrolysis of ATP on myosin is rapid and reversible; it can occur in either the presence or absence of actin and is not involved directly in the coupling rules.[16-18] Note that the *rate constants* for the binding–dissociation steps define the specificity rules; the changes in rate constants may or may not be accompanied by changes in equilibrium constants. In fact, the increases in the rate constants for dissociation of actin in the presence of ATP, and for dissociation of ADP and P_i in the presence of actin, are accompanied by changes in equilibrium constants for dissociation that serve to balance the concentrations of the intermediate species under physiological conditions so as to avoid the situations shown in Fig. 1 (A and B).

The critical rule for vectorial coupling is rule 4, which ensures that $M \cdot ADP \cdot P_i$ binds to A_2, to give movement, rather than to A_1, before the dissociation of products. There are three possible ways by which this could be brought about. (a) The hydrolysis of $M \cdot ATP$ to $M \cdot ADP \cdot P_i$ can be associated with a conformational change that moves the myosin head enough so that it can no longer bind to A_1 but can bind to A_2.[6,46] An extreme example of such movement is shown in Scheme 3. (b) Sliding of the thick and thin filaments relative to each other may be brought about by conformational changes in other AM residues that pull the detached $M \cdot ADP \cdot P_i$ toward A_2.[47] One myosin molecule that is attached to actin

[46] R. W. Lymn and E. W. Taylor, *Biochemistry* 10, 4617 (1971).

[47] W. P. Jencks, *in* "Membranes and Transport" (A. N. Martonosi, ed.), Vol. 1, p. 515. Plenum, New York, 1982.

may exert a "power stroke" after product dissociation that pulls the thick filament containing an unattached myosin–products complex into a position in which it can only bind to a new actin monomer, A_2. (c) The binding of ATP to actomyosin produces a high-energy conformational state that leads to rapid actomyosin dissociation. Relaxation of this strained conformation after dissociation may give enough movement to cause preferential binding of myosin to A_2.[9,46,48]

The change in chemical specificity that couples ATP hydrolysis to work in this system occurs only indirectly, through vectorial control of product dissociation. Although the hydrolysis step is readily reversible and occurs rapidly in either $M \cdot ATP$ or $A_1 \cdot M \cdot ATP$, the dissociation of myosin from A_1 in the presence of ATP is much faster than the hydrolysis and dissociation of hydrolysis products from $A_1M \cdot ADP \cdot P_i$.[30]

There is a possibility that $M \cdot ATP$ will bind to A_2 to give $A_2 \cdot M \cdot ATP$ and then $A_2 \cdot M$, in violation of rule 3. However, this binding will ordinarily be followed by fast dissociation, as with $A_1 \cdot M \cdot ATP$. If ATP should dissociate to give A_2M and the development of tension, another ATP molecule would immediately bind to this species and cause dissociation to $A_2 + M \cdot ATP$. The binding of ATP in this situation is favored by the resulting relief of tension. Although the binding of ADP and P_i to AM might be expected to cause dissociation, this binding is too weak to have a significant effect. This is a case in which a coupling rule does depend on binding strength for its specificity.

The coupling rules for the dynein–tubulin system appear to be almost identical to those for the actin–myosin system.[9]

Other Coupled Processes

An important function of nucleotide triphosphate hydrolysis is to provide directionality and control of metabolic reactions, including control of the duration of the effects of messenger molecules such as hormones, that might be thought of as directed movement in time. The mechanisms of a number of these processes involve a coupling of nucleotide triphosphate hydrolysis to directionality that is closely related to the behavior of coupled vectorial processes and can be described in the same way.[7,48]

Several steps in the pathway of protein synthesis require the hydrolysis of GTP. This may be coupled to movement, as in the translocation of peptidyl-tRNA and aminoacyl-tRNA binding sites on the ribosome relative to mRNA. In other reactions, such as the binding of aminoacyl-tRNA to ribosomes, the hydrolysis of GTP is utilized to ensure directionality in time and to increase the specificity of the reaction.[49-51] These reactions proceed through a series of ordered steps that are coupled to nucleotide triphosphate hydrolysis in order to drive the reaction in a particular direction.[52]

Hydrolysis of GTP in the reaction cycles initiated by hormones or drugs provides a directionality and time dependence, as in the action of receptors coupled to adenylate cyclase that leads to the release of cAMP.[53] Inhibition of this hydrolysis can lead to continued activity of the receptor, as in the action of cholera toxin.[54]

The changes in specificity, activity, and binding energy in these systems are analogous to those in more complex coupled vectorial processes. Their mechanism of action can be defined by analysis of the role of the utilization of binding energies to effectors and nucleoside triphosphates in different states and by elucidation of the rules and changes in specificity that are responsible for the coupling of nucleoside triphosphate hydrolysis to the directionality of the process. Several of these systems are becoming well enough characterized that it will be possible to understand at least the general nature of the mechanisms by which their switch cycles produce coupling.

[49] R. C. Thompson and P. J. Stone, *Proc. Natl. Acad. Sci. U.S.A.* **74**, 198 (1977).

[50] R. C. Thompson, D. B. Dix, R. B. Gerson, and A. M. Karim, *J. Biol. Chem.* **256**, 81 (1981).

[51] J. L. Yates, *J. Biol. Chem.* **254**, 11550 (1979).

[52] Y. Kaziro, *Biochim. Biophys. Acta* **505**, 95 (1978).

[53] A. Levitzky, "Receptors: A Quantitative Approach." Benjamin/Cummings, Menlo Park, California, 1984.

[54] D. Cassel and Z. Selinger, *Proc. Natl. Acad. Sci. U.S.A.* **74**, 3307 (1977).

[8] Generation of Steady-State Rate Equations for Enzyme and Carrier-Transport Mechanisms: A Microcomputer Program

By KEITH R. RUNYAN and ROBERT B. GUNN

Introduction

The rate equation for an enzymatic reaction describes mathematically the net rate of formation of a product or disappearance of a substrate in terms of rate constants and the concentrations of substrates and products. Steady state and rapid equilibrium, or combinations of both assumptions, are used in the derivation of the rate equations as well as in the design of kinetic experiments to which the rate equations are applied. By comparing the behavior of many rate equations with experimental data obtained under comparable conditions, one may eliminate certain enzymatic reaction mechanisms from consideration. Hence, the ability to generate rate equations for proposed models is a powerful tool for the kineticist.

The only method for generating steady-state rate equations prior to 1956 was the determinant method. This method is a very time-consuming and error-prone process when large mechanisms are considered. In 1956, King and Altman[1] developed a graphical method for the derivation of rate equations. Their method greatly simplifies the process but is tedious with large mechanisms. Many modifications and extensions of their method have been developed. These include Wong and Hanes,[2] Volkenstein and Goldstein,[3] Cha,[4] Fisher and Hoagland,[5] Fromm,[6] and Orsi.[7] Other graphical and algebraic methods include those of Lam and Priest,[8] Indge and Childs,[9] Cornish-Bowden,[10] and Chou and Forsen.[11] Several computer programs have been written using the determinant method or the graphical and algebraic methods mentioned above. These computer programs are listed in Table I. In addition, a BASIC program designed for microcomputer use has been developed previously in this laboratory.[12] These existing programs have several pitfalls. First, most are written for mainframe computers, which are not as readily available as microcomputers. Second, most of the programs are not written to allow irreversible reaction steps. Third, the output from these programs is largely unsorted and difficult to interpret. Finally, only one of the programs[13] directly generates the rate equation, namely the difference between the forward and reverse reaction rates; the remainder generate only the distribution equation (i.e., the steady-state fractional concentration) for each enzyme species. However, the FORTRAN 77 program presented by Herries[13] generates the rate equation for mechanisms containing at most 10 enzyme forms, 10 reactants, or 28 rate constants. In addition, a BASIC version of their program that runs on a

[1] E. L. King and C. Altman, *J. Phys. Chem.* **60**, 1375 (1956).

[2] J. T. F. Wong and C. S. Hanes, *Can. J. Biochem. Physiol.* **40**, 463 (1962).

[3] M. V. Volkenstein and B. N. Goldstein, *Biochim. Biophys. Acta* **115**, 471 (1966).

[4] S. Cha, *J. Biol. Chem.* **243**, 820 (1968).

[5] J. R. Fisher and V. D. Hoagland, *Adv. Biol. Med. Phys.* **12**, 163 (1968).

[6] H. J. Fromm, *Biochem. Biophys. Res. Commun.* **40**, 692 (1970).

[7] B. A. Orsi, *Biochim. Biophys. Acta* **258**, 4 (1972).

[8] C. F. Lam and D. G. Priest, *Biophys. J.* **12**, 248 (1972).

[9] K. J. Indge and R. E. Childs, *Biochem. J.* **155**, 567 (1976).

[10] A. Cornish-Bowden, *Biochem. J.* **165**, 55 (1977).

[11] K. C. Chou and S. Forsen, *Biochem. J.* **187**, 829 (1980).

[12] V. K. Gottipaty and R. B. Gunn, *J. Gen. Physiol.* **84**, 32a (1984).

[13] D. G. Herries, *Biochem. J.* **223**, 551 (1984).

TABLE I

COMPUTER PROGRAMS GENERATING STEADY-STATE EQUATIONS[a]

Method	Computer	Program language	Largest models[b]	Reference
Determinant	PDP6	MACRO 6	NG	Rhoads and Pring[c]
Determinant	IBM 360	FORTRAN IV	9	Hurst[d]
Connection matrix	IBM 7040	FORTRAN IV	NG	Fisher and Schulz[e]
Determinant	IBM 360	PL/1	13	Rudolph and Fromm[f]
Algebraic	NG	NG	8	Lam and Priest[g]
Chart method	NG	NG	NG	Seshagiri[h]
Algebraic	Nova 820	ALGOL/BASIC	NG/9	Indge and Childs[i]
Algebraic	IBM 370	FORTRAN IV	NG	Cornish-Bowden[j]
Determinant	IBM 360	PL/1	NG	Fromm[k]
Algebraic	Amdahl 470 V/7	FORTRAN 77	10	Herries[l]

[a] Authors of programs which generate steady-state equations, and the method, computer, and programming language used, are listed.

[b] The largest number of enzyme species in a model that can be solved by a given computer program is listed. When a number was not given (NG), the largest number of nodes allowed was a function of the available computer memory.

[c] D. G. Rhoads and M. Pring, *J. Theor. Biol.* **20,** 297 (1968).

[d] R. O. Hurst, *Can. J. Biochem.* **47,** 941 (1969).

[e] D. D. Fisher and A. R. Schulz, *Math. Biosci.* **4,** 189 (1969).

[f] F. B. Rudolph and H. J. Fromm, *Arch. Biochem. Biophys.* **147,** 515 (1971).

[g] C. F. Lam and D. G. Priest, *Biophys. J.* **12,** 248 (1972).

[h] N. Seshagiri, *J. Theor. Biol.* **34,** 469 (1972).

[i] K. J. Indge and R. E. Childs, *Biochem. J.* **155,** 567 (1976).

[j] A. Cornish-Bowden, *Biochem. J.* **165,** 55 (1977).

[k] H. J. Fromm, this series, Vol. 63, p. 84.

[l] D. G. Herries, *Biochem. J.* **223,** 551 (1984).

BBC microcomputer was developed. However, it has much longer computation times than the FORTRAN 77 version, which runs on the Amdahl 470 V/7 mainframe computer.

In this work, we present a new computer program based on the algorithm on Indge and Childs[9] that overcomes these difficulties. The program generates the full rate equation as well as the distribution equation for each enzyme form. Rate equations can be derived for the overall rate of a reaction mechanism or for the rate through just one step in the reaction sequence. Reaction mechanisms may contain irreversible reaction steps and up to 20 enzyme forms, 40 reactants, and 198 rate constants. The rate equation is displayed in either coefficient form, with concentration terms sorted alphabetically, or in complete form, with rate constant terms sorted and grouped by concentration term. Reaction mechanisms are easily entered and altered through the use of menus, and are solved faster than previously published programs that run on mainframe computers.

Formulation of the Rate Equation

The overall rate of a steady-state reaction can be equivalently expressed as (1) the net rate of disappearance of a substrate, (2) the net rate of appearance of a product, or (3) the rate of transformation(s) in the forward direction of an obligatory intermediate.[2] For example, if enzyme species E_i reacts with substrate S to form ES_j,

$$E_i + S \underset{k_{ji}}{\overset{k_{ij}}{\rightleftharpoons}} ES_j$$

the net rate of disappearance of S can be expressed as

$$v_{net} = \sum (V_i - V_j) \tag{1}$$

where v_{net} is expressed as a specific velocity ($=$ net rate/e_0, where e_0 is the total enzyme concentration), V_i and V_j are the rates of the forward reaction, ij, and the reverse reaction, ji, respectively, and the summation sign indicates that the rate is the sum of all such reaction steps involving S. This rate can also be expressed as

$$v_{net} = \sum (k_{ij}[S]N_i - k_{ji}N_j)/D \tag{2}$$

where k_{ij} and k_{ji} are the rate constants for the conversion of E_i to ES_j and vice versa, [S] is the concentration of substrate S, N_i and N_j are the numerators of the distribution equations for E_i and ES_j, respectively, and D represents the denominator of the rate equation and distribution equations (i.e., $D = \sum_{i=1}^{k} N_i$, where k = total number of enzyme forms). Mathematically, the rate equation consists of a numerator with positive (e.g., $k_{ij}[S]N_i$) and negative (e.g., $k_{ji}N_j$) terms composed of the product of rate constants and concentration terms, while the denominator, D, consists of similar positive terms. The properties of the terms of the rate equation have been thoroughly examined by Wong and Hanes,[2] and are explored in more detail below.

Program Description

Computer Hardware and Software

The steady-state rate equation generator (SSREG.BAS) program was written in BASIC on an IBM personal computer (PC). The program can be entered into an IBM PC or compatible microcomputer equipped with at

```
1 : ENTER NEW OR CORRECT OLD MODEL
2 : SOLVE ONE MODEL
3 : ENTER NEW OR CORRECT OLD MODEL LIST
4 : SOLVE MULTIPLE MODELS
5 : END PROGRAM
ENTER CHOICE ?
```

FIG. 1. The main menu used to select the functions of the program.

least 64K memory and one floppy disk drive either from the listing of SSREG.BAS in the appendix herein or from a floppy disk available from the authors.[14] It can be executed using either the IBM PC Basic Interpreter (version 2.10), which is supplied with the computer, or the IBM PC Basic Compiler (version 1.00). The IBM PC BASIC Compiler speeds execution time considerably and is recommended for models containing more than six enzyme species. For mechanisms containing more than ten enzyme species, a 5- or 10-megabyte hard disk drive is recommended. If the IBM PC BASIC Interpreter is used, then the command >BASICA/F:4 must be entered to start the interpreter so that four disk files can be used.

Program Execution

Because the terms of the rate equation are stored on the floppy or hard disk, when the program is executed the user is first required to enter the amount of available space on the disk (in bytes). This number is displayed when the directory command, >DIR, is entered ("243,712 bytes free," for example). From the amount of available disk space, the program calculates the maximum allowable number of denominator terms (MND) using Eq, (3), where N is the total number of enzyme forms.

$$MND = \text{Disk space (bytes)}/(4N) \qquad (3)$$

This number of terms is adjusted so that it does not exceed 32,760, the maximum number of terms allowed. During generation of the denominator terms, this number is checked to ensure that the available disk space will not be exceeded. Next, a menu is presented to the user, as shown in Fig. 1. Choice 1 allows a new reaction mechanism to be entered or an old one to be altered. Choice 2 is used to solve one model. Choice 3 allows a new model list (i.e., a list of input file names) to be created or an old one to be altered. Choice 4 allows multiple models to be solved using the model list created previously in choice 3.

---

[14] An executable compiled version of SSREG.BAS and DETAA.BAS along with the source code and example input files will be provided on request from Robert B. Gunn (Department of Physiology, Emory University School of Medicine, Atlanta, Georgia 30322) if a $5\frac{1}{4}$-inch double-sided, double-density floppy disk and two mailing labels are provided.

```
1 : GET MODEL FROM INPUT FILE
2 : ENTER RATE CONSTANTS SURROUNDING THE NODES
3 : ENTER RATE CONSTANTS LEAVING THE NODES
4 : ENTER RATE EQUATION FORMULA
5 : ENTER PSEUDO-FIRST ORDER RATE CONSTANTS
6 : OMIT NODES OR RATE CONSTANTS
7 : STORE MODEL IN INPUT FILE
8 : DISPLAY MODEL
9 : RETURN TO MAIN MENU
ENTER CHOICE ?
```

FIG. 2. The input file menu used to enter, alter, display, and store a model.

Entry of Reaction Mechanisms

In order to enter a model into the computer, one first writes down the model, including all concentration terms, rate constants, and the directions of the reactions in graphical form. Each reaction must have a different rate constant and the numerical subscript of all rate constants must range from -99 to 99, not including zero. The rate constants of forward and reverse reactions must differ only by a minus sign. If two or more rate constants in a mechanism are known to have equal magnitudes, they still must be represented here by distinct rate constants until the rate equation is generated. Next, one transforms the model to a form suitable for program entry by assigning a node number to each enzyme species (node is the term given to an enzyme species in topological graph theory). Although the order of assignment of node numbers does not affect the solution, node numbers are assigned in a specific order to speed computation time. The highest node number is assigned to the enzyme species with the largest number of rate constants entering and leaving it. The remaining node numbers are best assigned in decreasing order to adjacent nodes. Because all of the reactions must be first-order reactions, second-order rate constants are replaced by pseudo-first-order rate constants. Rate constants (e.g., k_1, k_{-2}, k_{10}) are indicated by their numerical subscript (e.g., 1, -2, 10), and pseudo-first-order rate constants are indicated by the numerical subscript of the second-order rate constant multiplied by its associated concentration term. For example, in the reaction

$$E + S \underset{k_{-1}}{\overset{k_1}{\rightleftharpoons}} ES$$

k_1 will become 1S and k_{-1} will become -1. Choice 1 from the main menu (Fig. 1) is then selected, which displays the input file menu shown in Fig. 2. Using the second through fifth choices of this menu the model is entered into the computer. First, the rate constants surrounding each node (i.e., the rate constants for reactions directed toward or away from an enzyme species) are entered (maximum of 20 nodes allowed, 16 rate constants per node). Only the numerical portion of the pseudo-first-order rate constants is entered at this point; the concentration portion is entered later. The rate constants surrounding each node can be entered in an abbreviated format if all of the reaction steps in the model are reversible. In the abbreviated mode, only the positive rate constants (half of the total number of reaction steps entering and leaving that node) are entered; the negative rate constants are inserted by the program. Then, the rate constants leaving each node (i.e., the rate constants for reactions directed away from an enzyme species) are entered (maximum of eight rate constants per node) and, from these, forbidden combinations of rate constants are generated by the program. Again, only the numerical portion of the pseudo-first-order rate constants is entered at this point. Next, the rate equation formula is entered which consists of a pair of numbers (maximum of 16 pairs) for each term in Eq. (2). The first number in each pair has the sign of the corresponding term and value of the subscript of the N in that term. The second number in each pair corresponds to the numerical subscript (sign and integer value) of the k in that term of Eq. (2). Next, the pseudo-first-order rate constants (e.g., 1S) are entered as pairs (e.g., 1, S; maximum of 40 pairs) of second-order rate constants along with their associated concentration terms (each consisting of at most six characters). After entering the model, choice 7 in the input file menu (Fig. 2) *must* be selected to store the model as an input file on the disk. Failure at this point to name the model and store it results in loss of all input and the program terminates. Input files are given names of at most eight characters, and the file name extension ".KIN" (for *kin*etics) is added by the program. This input file can then be recalled and used to generate the rate equation or can be altered and stored again.

Alteration of Reaction Mechanisms

Previously stored models can be read into memory by selecting choice 1 of the input file menu (Fig. 2). Choices 2 through 5 can be selected to alter a portion of the input file. In addition, one or more nodes or rate constants can be omitted from a model by selecting choice 6. In this way, reversible reactions can be made irreversible, or certain substrate or product concentrations can be fixed to the value zero, but creation of a node that does not contain at least one rate constant directed toward and one rate constant directed away from it is improper. Similarly, one or more nodes can be eliminated from the model to remove, for example, a dead-end inhibitor. This allows the rate equation for many related mechanisms to be generated without entering each mechanism separately.

Generation of the Rate Equation

After the model is entered and stored, choice 2 in the main menu (Fig. 1) can be selected to generate the rate equation. The user selects where and in what form the rate equation is to be displayed as discussed below. The terms of the rate equation and distribution equations consist of products of rate constants and concentration terms (e.g., 1S − 2 3P 4 − 5, where S and P represent the concentration of substrate S and product P in this term, composed of five rate constants). The program first generates the rate constant portion of each term (e.g., 1 − 2 3 4 − 5) and will be referred to as a rate constant term, and, from this, generates the concentration portion of each term (e.g., SP), referred to as a concentration term; finally both are displayed together as a term of the rate equation of distribution equations. Potential rate constant terms of the denominator of the rate equation or distribution equations are formed by alphanumerically multiplying the rate constants surrounding each of the first $N - 1$ nodes ($N =$ number of nodes in the model), thus forming terms composed of the product of $N - 1$ rate constants. A given potential term is eliminated if it contains two or more rate constants whose absolute values are equal (representing repeated or directly opposing rate constants) according to the Wang algebra product rule,[9,15] or if it contains one or more forbidden pairs of rate constants (since no terms in the steady-state rate equation may contain two or more rate constants leaving the same node).[9] If the term is not eliminated, it is stored on the disk until all of the potential rate constant terms have been either eliminated or stored. These potential denominator terms are then sorted on the disk and all pairs of identical terms are eliminated according to the Wang algebra summation rule,[9,15] since they represent invalid cycles. The rate constant terms remaining constitute the rate constant portion of the denominator of the rate equation as well as of the distribution equations.

The starting and ending date and time are displayed, the difference being the time required to generate and sort the denominator terms. The numerator of the distribution equation for each node is generated (and displayed optionally) by selecting those denominator terms which do not contain a rate constant directed away from the given node. Then, the denominator of the rate equation and of the distribution equations is displayed as described below. The positive and negative numerator terms of the rate equation are generated according to Eq. (2). Potential positive numerator rate constant terms are formed by alphanumerically multiplying k_{ij} and N_i and eliminating any terms containing two or more rate constants whose absolute values are equal according to the Wang algebra product rule. The concentration of the substrate, [S], is included in the rate

[15] R. J. Duffin, *Trans. Am. Math. Soc.* **93,** 114 (1959).

equation by associating it with its corresponding rate constant, k_{ij}. Similarly, potential negative numerator rate constant terms are formed according to Eq. (2) by alphanumeric multiplication of k_{ji} and N_j, with elimination of terms by the Wang algebra product rule. Thus, each numerator rate constant term is a product of N rate constants. Any rate constant terms common to both the positive and negative numerators are eliminated (i.e., identical rate constant terms having opposite signs are canceled). The program displays the positive numerator terms and then the negative numerator terms as described below.

Form of the Rate Equation

The numerators and denominator of both the rate equation and distribution equations can be displayed on the screen or printer in complete form or in coefficient form. In either form, the program selects the concentration portion of a term (e.g., S and P) from the pseudo-first-order rate constants (e.g., 1S and 3P) whose second-order rate constants match those in the term to be displayed (e.g., 1 and 3 in the term $1 - 2 3 4 - 5$). Then, the unique concentration terms are selected from all of the terms to be displayed and are sorted alphabetically. In the complete form of display, each unique concentration term is displayed, *followed by* the rate constant terms associated with that concentration term. Each rate constant term is displayed as the product of N or $N - 1$ rate constants (sorted by their absolute value in ascending order), separated by a space to indicate summation. These rate constant terms are sorted numerically in ascending order. Rate constant terms without associated concentration terms are displayed first after the word "constant." In the coefficient form of display, only the unique concentration terms are displayed, separated by a space to indicate summation. Each unique concentration term is implicitly multiplied by a constant which represents one, or the sum of more than one, rate constant terms. The presence of rate constant terms without associated concentration terms is indicated by the word "constant." Hence, the coefficient form of the rate equation or distribution equations describes the rate of reaction or distribution of enzyme species as a function of concentrations of substrates and products without regard to individual rate constants.

Generating Multiple Rate Equations

Multiple models can be solved in sequence without further input by first selecting choice 3 of the main menu (Fig. 1). This produces another menu which allows a list of input file names (each composed of at most

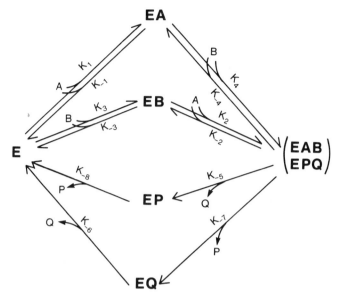

FIG. 3. Random Bi Bi mechanism: initial rate mechanism in the absence of products P and Q.

eight characters) to be entered, altered, and saved. The rate equation for each of these models is generated by selecting choice 4 of the main menu.

Application of the Program to a Specific Problem

An Enzyme Example

The generation of a rate equation using SSREG.BAS will be demonstrated using the random Bi Bi mechanism in the absence of product as shown in Fig. 3. The model is transformed into a form suitable for program entry by assigning node numbers as shown in Fig. 4. In order to speed computation time, the highest node number, 6, can be assigned to either enzyme species E or EAB–EPQ, because both have six rate constants surrounding them (i.e., six branches identified by six rate constants directed toward or away from the node). Once node 6 is assigned to enzyme species EAB–EPQ, then remaining node numbers are best assigned in decreasing order to adjacent nodes. This model is entered into the computer (after executing SSREG.BAS and entering the amount of available memory as previously described) by selecting choice 1 in the main menu (Fig. 1), which displays the input file menu (Fig. 2). Choice 2 is selected to enter the rate constants surrounding each node. Since irreversible reactions

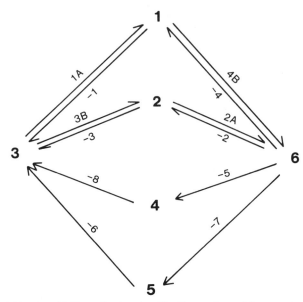

FIG. 4. Random Bi Bi mechanism (see Fig. 3) transformed for program entry.

are present in this model, the abbreviated mode of entry cannot be used, and 2 is entered in response to "ABBREVIATED MODE OF ENTRY, YES (1), NO (2) ?" The number of nodes, 6, is entered, the number of rate constants surrounding each node is entered, and the rate constants surrounding each node are entered one by one. For node 1, 4 is entered (the number of rate constants surrounding node 1), and the following rate constants are entered: $1, -1, 4,$ and -4, one by one in response to separate questions. The rate constants for the remaining nodes are entered in an analogous way. Choice 3 of the menu (Fig. 2) is selected to enter the rate constants for branches leaving each node. Again, the number of nodes, 6, is entered, the number of rate constants leaving each node is entered, and the rate constants for branches leaving each node are entered one by one. For node 1, 2 is entered (the number of rate constants leaving node 1), and the rate constants -1 and 4 are entered one by one. The rate constants for branches leaving each of the remaining nodes are entered in turn. Next, choice 4 allows one to enter the rate equation formula. The rate can be expressed as the net rate of disappearance of substrate A according to Eq. (2). It has four terms,

$$\text{Rate} = k_1[\text{A}]N_3 - k_{-1}N_1 + k_2[\text{A}]N_2 - k_{-2}N_6 \qquad (4)$$

and is entered as four pairs of numbers. The first number in each pair has the sign of the corresponding term and value of the subscript of the N in that term. The second number in each pair corresponds to the numerical subscript of the k in that term of Eq. (4). The number of pairs is entered (4 in this example), followed by each pair on a separate line with the numbers in the pair separated by a comma. In this example, one enters: 3,1; $-1, -1$; 2,2; and $-6, -2$. Then, the pseudo-first-order rate constants are entered as pairs by selecting choice 5. The number of pairs is entered and each pair is entered on a different line as follows: 1,A; 2,A; 3,B; 4,B.

Choice 7 is selected to store the model on the disk as an input file with the name RBB, for random Bi Bi. The file name extension ".KIN" is added by the program so that the model will be stored on the disk under the file name RBB.KIN. This completes the entry of the random Bi Bi model (RBB.KIN) shown in Fig. 3. The model can be displayed on the screen or printer by entering choice 8 (Fig. 2). The results are shown in Fig. 5.

```
FILE NAME IS RBB.KIN

NODE #      K's SURROUNDING NODE
  1            1-1  4-4
  2            2-2  3-3
  3            1-1  3-3-6-8
  4           -5-8
  5           -6-7
  6            2-2  4-4-5-7

NODE #   K's LEAVING NODE      FORBIDDEN COMBINATIONS
  1        -1  4                   -1  4,
  2         2-3                     2-3,
  3         1  3                    1  3,
  4        -8
  5        -6
  6        -4-2-5-7           -4-2, -4-5, -4-7, -2-5, -2-7, -5-7,

RATE EQUATION FORMULA
  3              1
 -1             -1
  2              2
 -6             -2

PSEUDO-FIRST ORDER RATE CONSTANTS
  1              A
  2              A
  3              B
  4              B
```

Fɪɢ. 5. Program output showing the input file, RBB.KIN, for the random Bi Bi mechanism in Fig. 4.

```
DISPLAY DISTRIBUTION EQUATION NUMERATORS, YES(1), NO(2)? 2
COMPLETE (1), OR COEFFICIENT (2) FORM OF DISPLAY? 1
OUTPUT TO SCREEN(1) OR PRINTER(2)? 2
ENTER NAME OF MODEL? RBB
```

FIG. 6. Program options selected for the generation of the rate equation for the random Bi Bi mechanism shown in Fig. 7.

Choice 9 is entered to return to the main menu (Fig. 1). The rate equation is generated by selecting choice 2 (Fig. 1). The form of the rate equation was selected for the random Bi Bi mechanism as shown in Fig. 6. These selections produce the complete form of the rate equation, shown in Fig. 7. The equation for the random Bi Bi mechanism in the absence of product consists of 8 positive numerator terms representing three unique concentration terms and 40 denominator terms representing one constant term and seven unique concentration terms. The coefficient form of this rate equation is given by

$$v = \frac{c_1 A^2 B + c_2 AB + c_3 AB^2}{c_4 + c_5 A + c_6 A^2 + c_7 A^2 B + c_8 AB + c_9 AB^2 + c_{10} B + c_{11} B^2} \qquad (5)$$

where

```
START = 02-12-1986 21:54:36,   END = 02-12-1986 21:54:40

constant
 -1-3-6-7-8 -1-3-5-6-8 -1-3-4-6-8 -1-2-3-6-8
A
 -1 2-6-7-8 -1 2-5-6-8 -1 2-4-6-8  1-3-6-7-8  1-3-5-6-8  1-3-4-6-8  1-2-3-6-8
AA
  1 2-6-7-8  1 2-5-6-8  1 2-4-6-8
AAB
  1 2 4-7-8  1 2 4-6-8  1 2 4-5-6
AB
 -1 2 3-7-8 -1 2 3-6-8 -1 2 3-5-6  1-3 4-7-8  1-3 4-6-8  1-3 4-5-6  1-2 4-6-8
  2 3-4-6-8  2 4-6-7-8  2 4-5-6-8
ABB
  2 3 4-7-8  2 3 4-6-8  2 3 4-5-6
B
 -3 4-6-7-8 -3 4-5-6-8 -2-3 4-6-8 -1-2 3-6-8 -1 3-6-7-8 -1 3-5-6-8 -1 3-4-6-8
BB
 -2 3 4-6-8  3 4-6-7-8  3 4-5-6-8
DENOMINATOR HAS 40 TERMS AND 7 UNIQUE CONCENTRATION TERMS.

AAB
  1 2 4-6-7-8  1 2 4-5-6-8
AB
 -1 2 3-6-7-8 -1 2 3-5-6-8  1-3 4-6-7-8  1-3 4-5-6-8
ABB
  2 3 4-6-7-8  2 3 4-5-6-8
THERE ARE 8 POSITIVE NUMERATOR TERMS AND 3 UNIQUE CONCENTRATION TERMS.
```

FIG. 7. Program output showing the complete form of the denominator and numerator of the rate equation for the random Bi Bi mechanism in Fig. 4.

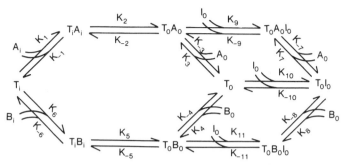

FIG. 8. A Ping-Pong model of obligatory exchange of anions across the human erythrocyte membrane in the presence of an impermeant external inhibitor, I_o. T_i and T_o represent the intracellularly and extracellularly facing conformations, respectively, of the anion transport site. T_i can bind to the intracellular anions A_i and B_i (forming T_iA_i and T_iB_i, respectively), while T_o can bind to the extracellular anions A_o, B_o, and I_o (forming T_oA_o, T_oB_o, and T_oI_o, respectively), I_o can bind to T_oA_o and T_oB_o, and A_o or B_o can bind to T_oI_o. No complexes involving I_o can undergo the conformational reactions (2 and 5) that result in the transport of anions.

$$c_1 = k_1k_2k_4k_{-6}k_{-7}k_{-8} + k_1k_2k_4k_{-5}k_{-6}k_{-8}$$

$$c_2 = k_{-1}k_2k_3k_{-6}k_{-7}k_{-8} + k_{-1}k_2k_3k_{-5}k_{-6}k_{-8}$$
$$+ k_1k_{-3}k_4k_{-6}k_{-7}k_{-8} + k_1k_{-3}k_4k_{-5}k_{-6}k_{-8}$$

$$c_3 = k_2k_3k_4k_{-6}k_{-7}k_{-8} + k_2k_3k_4k_{-5}k_{-6}k_{-8}$$

$$c_4 = k_{-1}k_{-3}k_{-6}k_{-7}k_{-8} + k_{-1}k_{-3}k_{-5}k_{-6}k_{-8}$$
$$+ k_{-1}k_{-3}k_{-4}k_{-6}k_{-8} + k_{-1}k_{-2}k_{-3}k_{-6}k_{-8}, \text{ etc.}$$

A Transport Example

The utility of the coefficient form of display is demonstrated in the generation of the rate equation for a Ping-Pong mechanism, which is of interest to our laboratory (Fig. 8). This is a model of anion transport in human red blood cells facilitated by the integral membrane protein, band 3. The transport mechanism has previously been shown to involve binding of either an intracellular or extracellular anion to a single anion-binding site, a conformational change of the band 3 protein causing movement of the anion to the opposite side of the membrane, and unloading of the anion (almost no conformational changes occur without transport of anions).[16] The model shown in Fig. 8 is an extension of this anion transport model in which an impermeant inhibitor, I_o, is present in the extracellular space. The binding of this inhibitor while allowing binding of

[16] R. B. Gunn and O. Frohlich, in "Chloride Transport in Biological Membranes" (J. A. Zadunaisky, ed.), p. 33. Academic Press, New York, 1982.

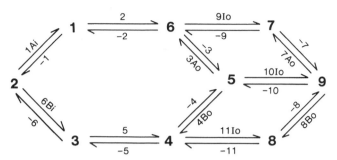

FIG. 9. Ping-Pong model of transport (see Fig. 8) transformed for program entry.

anions to the transporter prevents the conformational change of band 3 and hence slows the measured influx of extracellular anions. The rate equation for this model is generated by first transforming the model for program entry by assigning node numbers as previously described (see Fig. 9). SSREG.BAS is executed and the model is entered by selecting choice 1 of the main menu (Fig. 1). By selecting choice 2 of the input file menu (Fig. 2), the rate constants surrounding each node are entered. Because all reactions in this model are reversible, the abbreviated mode of entry is used to save time. In the abbreviated mode, only the positive rate constants are entered. Hence, for node 1, 4 is entered (the number of rate constants surrounding node 1), but only the two positive rate constants, 1 and 2, are entered. The -1 and -2 are inserted by the program. The rate constants leaving each node, rate equation formula (expressed as the net rate of disappearance of substrate A_i), and pseudo-first-order rate constants are entered as previously described. The resulting input file for this model is shown in Fig. 10. The coefficient form of the rate equation as generated by SSREG.BAS is shown in Fig. 11. The rate equation consists of 15 positive and negative numerator terms and 720 positive denominator terms. In coefficient form, the numerator consists of 7 positive and negative concentration terms, and the denominator consists of 54 concentration terms. Special experimental conditions can simplify the rate equation. One may calculate the initial rate of disappearance of A_i (which equals the initial rate of appearance of B_i) when the concentrations of A_o and B_i are zero. To do this, the rate constants associated with these concentration terms, namely, 3 and 7 for A_o and 6 for B_i, are deleted from the model as previously described; and A_i is held constant by not associating A_i with k_1 (i.e., by not including the pseudo-first-order rate constant $1A_i$ in the input file). The resulting rate equation in coefficient form is shown in Fig. 12. In this equation, some of the coefficients of the concentration terms are implicit functions of the initial concentration of A_i. Thus, the rate equation takes the form of a polynomial in B_o (the transported species) as predicted by Wong and Hanes.[2]

```
FILE NAME IS PINGPONG.KIN

NODE #      K's   SURROUNDING NODE
 1                1-1  2-2
 2                1-1  6-6
 3                5-5  6-6
 4                4-4  5-5  11-11
 5                3-3  4-4  10-10
 6                2-2  3-3  9-9
 7                7-7  9-9
 8                8-8  11-11
 9                7-7  8-8  10-10

NODE #   K's LEAVING NODE      FORBIDDEN COMBINATIONS
 1    -1  2                  -1  2,
 2     1  6                   1  6,
 3     5-6                    5-6,
 4    -4-5  11               -4-5, -4  11, -5  11,
 5     3  4  10               3  4,  3  10,  4  10,
 6    -2-3  9               -2-3, -2  9, -3  9,
 7    -7-9                   -7-9,
 8    -8-11                  -8-11,
 9     7  8-10               7  8,  7-10,  8-10,

RATE EQUATION FORMULA
 2                1
-1               -1

PSEUDO-FIRST ORDER RATE CONSTANTS
 1                Ai
 3                Ao
 7                Ao
 6                Bi
 4                Bo
 8                Bo
 9                Io
10                Io
11                Io
```

Fig. 10. Program output showing the input file, PINGPONG.KIN, for the Ping-Pong model of transport shown in Fig. 9.

Verification of the Program

The validity of the program has been tested by two independent methods. First, the rate equation generated by the program was compared to several previously published kinetic models. The denominators of the 11 ordered and random mechanisms of Plowman[17] agreed with those generated by SSREG.BAS (except for a total of 10 minor errors in 4 of the 11 mechanisms given in Appendix B of Plowman[17]). Similarly, identical rate

[17] K. M. Plowman, "Enzyme Kinetics." McGraw-Hill, New York, 1972.

```
START = 02-12-1986 21:49:26,  END = 02-12-1986 21:51:39

Ai AiAo AiAoAo AiAoAoIo AiAoAoIoIo AiAoBo AiAoBoIo AiAoBoIoIo AiAoIo AiAoIoIo
AiAoIoIoIo AiBo AiBoBo AiBoBoIo AiBoBoIoIo AiBoIo AiBoIoIo AiBoIoIoIo AiIo
AiIoIo AiIoIoIo Ao AoAo AoAoBi AoAoBiIo AoAoBiIoIo AoAoIo AoBi AoBiBo AoBiBoIo
AoBiBoIoIo AoBiIo AoBiIoIo AoBiIoIoIo AoBo AoBoIo AoIo AoIoIo Bi BiBo BiBoBo
BiBoBoIo BiBoBoIoIo BiBoIo BiBoIoIo BiBoIoIoIo BiIo BiIoIo BiIoIoIo Bo BoBo
BoBoIo BoIo BoIoIo
DENOMINATOR HAS 720 TERMS AND 54 UNIQUE CONCENTRATION TERMS.

AiAoBo AiAoBoIo AiBo AiBoBo AiBoBoIo AiBoIo AiBoIoIo
THERE ARE 15 POSITIVE NUMERATOR TERMS AND 7 UNIQUE CONCENTRATION TERMS.

AoAoBi AoAoBiIo AoBi AoBiBo AoBiBoIo AoBiIo AoBiIoIo
THERE ARE 15 NEGATIVE NUMERATOR TERMS AND 7 UNIQUE CONCENTRATION TERMS.
```

FIG. 11. Program output showing the denominator and numerator of the rate equation in coefficient form for the Ping-Pong model of transport shown in Fig. 9.

equations were generated for the two mnemonical enzyme models of Ricard et al.,[18] and 10 Ping-Pong and ordered mechanisms of Cleland.[19] The numerators of the rate equations for the random Bi Bi and general two-substrate, two-product mechanism given by Chou,[20] using a graphical method to generate only the nonzero numerator terms, were also in agreement with that generated by SSREG.BAS. Second, the number of denominator terms in the rate equation for more than 50 mechanisms (containing no irreversible reactions) of interest to our laboratory has agreed with that predicted by another computer program developed by the authors, called DETAA.BAS, based on the method of Lam and Priest.[8] DETAA.BAS uses the same model input file created by SSREG.BAS. However, all reactions must be reversible and the absolute value of the rate constants must start at 1 and increase consecutively.

Program Computation Time

The computation time of SSREG.BAS for several ordered and branched mechanisms was compared with that of a FORTRAN IV program, which uses the determinant method.[21] As shown in Table II, SSREG.BAS generates the rate equation denominator faster than the program of Hurst,[21] which ran on an IBM 360 mainframe computer. Also shown in Table II are the computational times for several mechanisms of interest to our laboratory.

[18] J. Ricard, J. C. Meunier, and J. Buc, Eur. J. Biochem. 49, 195 (1974).
[19] W. W. Cleland, Biochim. Biophys. Acta 67, 104 (1963).
[20] K. C. Chou, J. Theor. Biol. 89, 581 (1981).
[21] R. O. Hurst, Can. J. Biochem. 47, 941 (1969).

START = 02-12-1986 21:55:43,   END = 02-12-1986 21:56:16

constant Bo BoBo BoBoIo BoBoIoIo BoIo BoIoIo BoIoIoIo Io IoIo IoIoIo
DENOMINATOR HAS 192 TERMS AND 10 UNIQUE CONCENTRATION TERMS.

Bo BoBo BoBoIo BoIo BoIoIo
THERE ARE 12 POSITIVE NUMERATOR TERMS AND 5 UNIQUE CONCENTRATION TERMS.

FIG. 12. Program output showing the denominator and numerator of the rate equation in coefficient form for the Ping-Pong model of transport (see Fig. 9), with the concentration of A_o and B_i fixed to the value zero, and A_i held constant and suppressed.

Concluding Remarks

The microcomputer program presented in this work, SSREG.BAS, is superior to previously published computer programs, making it a valuable tool for the enzyme and transport kineticist. SSREG.BAS generates the rate equation for large reaction mechanisms containing reversible as well as irreversible reaction steps. It allows easy input, alteration, and storage of reaction mechanisms through the use of menus. Finally, it rapidly generates the complete steady-state rate equation and distribution equations in complete or coefficient form sorted by concentration term.

TABLE II
COMPARISON OF COMPUTATION TIMES

Mechanism	Number of nodes	Number of rate constants	ND[b]	IBM 360[a] (sec)	IBM PC compiler[c]
Ordered Bi Bi	4	8	16	0.74	1 sec
Ordered with 1 dead-end complex	5	10	20	5.30	1 sec
Ordered with 2 dead-end complexes	6	12	24	25.5	2 sec
Ordered with 3 dead-end complexes	7	14	28	16.3	3 sec
Ordered with 4 dead-end complexes	8	16	32	32.6	4 sec
Ordered Bi random Bi	5	12	60	35.2	4 sec
Random Bi Bi	6	16	192	23.7	15 sec
Ping-Pong (see Fig. 8)	9	22	720	—	2 min
Modified Divalent I mechanism[d]	8	22	1280	—	4 min
Ping-Pong—3 inhibitors	12	26	4032	—	35 min
Ping-Pong—3 inhibitors, slippage	12	32	9312	—	1 hr 13 min
Ping-Pong—2 mixed inhibitors	12	32	12600	—	2 hr 57 min
Ping-Pong—4 inhibitors	14	36	17920	—	4 hr 54 min
Ping-Pong—6 inhibitors	20	48	25600	—	17 hr 38 min

[a] From R. O. Hurst, *Can. J. Biochem.* **47**, 941 (1969).
[b] ND, The number of denominator terms in the rate equation.
[c] From this work; using IBM PC compiler.
[d] From J. T. F. Wong, *Biochim. Biophys. Acta* **94**, 102 (1965).

Appendix

```
10 REM *************** STEADY-STATE RATE EQUATION GENERATOR ***************
20 REM *******                                                     *******
30 REM *******     written by    KEITH R. RUNYAN       Jan. 1986   *******
40 REM *******                                                     *******
50 REM *******  This program generates steady-state rate equations *******
60 REM *******  using the algorithm of J. Indge and Robert E. Childs, *******
70 REM *******  BIOCHEM. J. 155, (1977), pp. 567-570.              *******
80 REM *******                                                     *******
90 REM * Define all variables beginning with the letters E-H as string, A-D
100 REM * and I-Y as integer, and Z as single precision real.
110 DEFSTR E-H :DEFINT A-D,I-Y :DEFSNG Z   :OPTION BASE 1
120 REM * Dimension arrays :A(20,16) - rate constants surrounding each node
130 REM * (max. of 20 nodes, 16 rate constants around each node), NA(20) -
140 REM * # of rate consts. around each node, B(20,8) - rate consts. leaving
150 REM * each node, NB(20) - # rate consts. leaving each node, E(1502) -
160 REM * used in sorting denominator and numerator terms, NC(10), C(10) -
170 REM * store node # and multiplying rate const. respectively, used in
180 REM * defining the rate equation, RD(40), HD(40) - stores rate constants
190 REM * and conc. terms that make up the pseudo-first order rate constants.
200 REM * HU(40) - stores a single character string that represents conc. terms
210 REM * I(20),R(20),S(20) - I(20) is used to serve as pointer to the rate
220 REM * constants in A(,), R(20),S(20) both are used to store integer forms
230 REM * of terms, H(4) - are buffers for 4 files used to store terms.
240 REM * The remainder of the dimensioned arrays are used in sorting terms.
250 DIM NA(20),A(20,16),NB(20),B(20,8),E(1502),NC(16),C(16),RD(40),HD(40)
260 DIM HU(40),I(20),R(20),S(20),H(4),NF(4),AF(4),T(4),TA(2),TB(2),TC(2),V(4)
270 REM * The following variables are assigned as maximum values of array
280 REM * dimensions and must be adjusted if array dimensions are changed.
290 MNA=20 :MSN=16 :MNC=16 :MLN=8 :MRD=40 :MX=1500 :MX2=MX/2
300 INPUT "ENTER AVAILABLE DISK MEMORY (in bytes)";ZMND
310 CLS :PRINT "Steady-State Rate Equation Generator" :PRINT
320 PRINT "1 : ENTER NEW OR CORRECT OLD MODEL" :PRINT "2 : SOLVE ONE MODEL"
330 PRINT "3 : ENTER NEW OR CORRECT OLD MODEL LIST"
340 PRINT "4 : SOLVE MULTIPLE MODELS" :PRINT "5 : END PROGRAM"
350 INPUT "ENTER CHOICE";I :IF I < 1 OR I > 5 GOTO 350
360 ON I GOSUB 580,1940,380,1830,5580
370 GOTO 310 '****** MODEL LIST MANIPULATOR ***********
380 CLS :PRINT "Model List Menu" :PRINT
390 PRINT "1 : GET MODEL LIST FROM FILE" :PRINT "2 : DISPLAY MODEL NAMES"
400 PRINT "3 : INSERT MODEL IN LIST" :PRINT "4 : DELETE MODEL FROM LIST"
410 PRINT "5 : CREATE NEW MODEL LIST" :PRINT "6 : STORE MODEL LIST IN FILE
420 PRINT "7 : RETURN TO MAIN MENU"
430 INPUT "ENTER CHOICE";I :IF I < 1 OR I > 7 GOTO 430
440 ON I GOTO 460,480,490,520,540,560,450
450 RETURN   '**** RETURN TO MAIN MENU ****
460 INPUT "ENTER NAME OF MODEL LIST FILE";F:OPEN "I",#1,F+".KIN":INPUT #1,NMTS
470 FOR I=1 TO NMTS :INPUT #1,E(I) :NEXT I :CLOSE #1 :GOTO 380
480 FOR I=1 TO NMTS :PRINT I,E(I) :NEXT I :INPUT I :GOTO 380
490 INPUT "ENTER : NUMBER IN LIST THAT MODEL SHOULD APPEAR, MODEL NAME";J,G
500 IF J>NMTS THEN NMTS=NMTS+1 :E(NMTS)=G :GOTO 380
510 NMTS=NMTS+1:FOR I=NMTS TO J+1 STEP -1 :E(I)=E(I-1):NEXT I:E(J)=G :GOTO 380
520 INPUT "ENTER MODEL NUMBER TO DELETE";J :NMTS=NMTS-1 :IF J=NMTS GOTO 380
530 FOR I=J TO NMTS :E(I)=E(I+1) :NEXT I :GOTO 380
540 INPUT "ENTER NUMBER OF MODELS IN LIST";NMTS :FOR I1=1 TO NMTS
550 PRINT "ENTER NAME OF MODEL NUMBER";I1; :INPUT E(I1) :NEXT I1 :GOTO 380
560 INPUT "ENTER NAME OF MODEL LIST FILE";F:OPEN "O",#1,F+".KIN":WRITE #1,NMTS
570 FOR I1=1 TO NMTS :WRITE #1,E(I1) :NEXT I1 :CLOSE #1 :GOTO 380
580 CLS :PRINT "Input File Menu"  '**** ENTER AN INPUT FILE ****
590 PRINT :PRINT "1 : GET MODEL FROM INPUT FILE"
600 PRINT "2 : ENTER RATE CONSTANTS SURROUNDING THE NODES"
610 PRINT "3 : ENTER RATE CONSTANTS LEAVING THE NODES"
620 PRINT "4 : ENTER RATE EQUATION FORMULA"
630 PRINT "5 : ENTER PSEUDO-FIRST ORDER RATE CONSTANTS"
640 PRINT "6 : OMIT NODES OR RATE CONSTANTS" :PRINT "7 : DISPLAY MODEL"
650 PRINT "8 : STORE MODEL IN INPUT FILE" :PRINT "9 : RETURN TO MAIN MENU"
```

```
660 INPUT "ENTER CHOICE";I :IF I < 1 OR I > 9 GOTO 660
670 ON I GOTO 700,720,870,960,1040,1120,690,1590,680
680 RETURN '********* PRINT MODEL *********
690 INPUT "OUTPUT TO SCREEN(1), PRINTER(2)";P :GOSUB 4010 :GOTO 580
700 GOSUB 1690 :GOTO 580 '****** GET MODEL FROM INPUT FILE ******
710 REM ******** ENTER RATE CONSTANTS SURROUNDING THE NODES ********
720 INPUT "ABBREVIATED MODE OF ENTRY, YES(1), NO(2)";P :IF P<1 OR P>2 GOTO 720
730 INPUT "ENTER # OF NODES";N :N1=N-1 :IF N<3 OR N>20 GOTO 730
740 FOR I=1 TO N
750 PRINT "ENTER # OF RATE CONSTANTS SURROUNDING NODE #";I;:INPUT NA(I)
760 IF NA(I) < 2 OR NA(I) > MSN GOTO 750
770 IF P=1 THEN NA(I)=NA(I)\2
780 FOR J=1 TO NA(I)
790 PRINT "ENTER RATE CONSTANT #";J; :INPUT A(I,J)
800 IF A(I,J)=0 OR ABS(A(I,J)) > 99 GOTO 790
810 NEXT J
820 IF P=2 GOTO 850 'In abbrev. mode, -k is added to A(,) where k=rate const.
830 I1=NA(I) :NA(I)=NA(I)+NA(I) :FOR J=NA(I) TO 1 STEP -2 :A(I,J)=-A(I,J-I1)
840 A(I,J-1)=A(I,J-I1) :I1=I1-1 :NEXT J
850 NEXT I :GOTO 580
860 REM ********** ENTER RATE CONSTANTS LEAVING THE NODES *********
870 INPUT "ENTER # OF NODES";N :N1=N-1 :IF N<3 OR N>20 GOTO 870
880 FOR I=1 TO N
890 PRINT "ENTER # OF RATE CONSTANTS LEAVING NODE #";I; :INPUT NB(I)
900 IF NB(I) < 1 OR NB(I) > MLN GOTO 890
910 FOR J=1 TO NB(I)
920 PRINT "ENTER RATE CONSTANT #";J; :INPUT B(I,J)
930 IF B(I,J)=0 OR ABS(B(I,J)) > 99 GOTO 920
940 NEXT J,I :GOTO 580
950 REM ********* ENTER RATE EQUATION FORMULA **********
960 INPUT "ENTER # OF RATE EQUATION PAIRS";NNC
970 IF NNC < 0 OR NNC > MNC GOTO 960
980 FOR I1=1 TO NNC
990 PRINT "FOR PAIR #";I1;"ENTER : + OR - NODE #,";
1000 INPUT "RATE CONSTANT";NC(I1),C(I1)
1010 IF NC(I1)=0 OR ABS(NC(I1)) > 20  OR C(I1)=0 OR ABS(C(I1)) > 99 GOTO 990
1020 NEXT I1 :GOTO 580
1030 REM ********* PSEUDO-FIRST ORDER RATE CONSTANTS *********
1040 INPUT "ENTER # OF PSEUDO-FIRST ORDER RATE CONSTANTS";NRD
1050 IF NRD < 0 OR NRD > 40 GOTO 1040
1060 FOR I1=1 TO NRD
1070 PRINT "FOR PAIR #";I1;"ENTER - ";
1080 INPUT "RATE CONSTANT, CONCENTRATION TERM";RD(I1),HD(I1)
1090 IF RD(I1)=0 OR ABS(RD(I1)) > 99 GOTO 1070
1100 NEXT I1 :GOTO 580
1110 REM *********      OMIT NODES OR RATE CONSTANTS      *********
1120 CLS :PRINT "1 : OMIT NODES" :PRINT "2 : OMIT RATE CONSTANTS"
1130 INPUT "ENTER CHOICE";I :IF I<1 OR I>2 GOTO 1130
1140 ON I GOSUB 1150,1310 :GOTO 580
1150 INPUT "ENTER # OF NODES TO REMOVE FROM MODEL";NN
1160 IF NN < 1 OR NN > N-3 THEN RETURN
1170 FOR I5=1 TO NN
1180 PRINT "ENTER NODE NUMBER #";I5; :INPUT I(I5)
1190 IF I(I5) < 1 OR I(I5) > N GOTO 1180
1200 NEXT I5 '** SORT NODES TO BE DELETED IN DESCENDING ORDER **
1210 FOR I5=1 TO NN-1 :FOR I6=I5+1 TO NN :IF I(I5)<I(I6) THEN SWAP I(I5),I(I6)
1220 NEXT I6,I5 :FOR I5=1 TO NN
1230 N=N1:N1=N-1:NR=NA(I(I5)):FOR I1=1 TO NA(I(I5)):S(I1)=A(I(I5),I1):NEXT I1
1240 IF I(I5)>N GOTO 1270
1250 FOR I1=I(I5) TO N :NA(I1)=NA(I1+1) :NB(I1)=NB(I1+1) :FOR I2=1 TO NA(I1)
1260 A(I1,I2)=A(I1+1,I2) :B(I1,I2)=B(I1+1,I2) :NEXT I2,I1
1270 IF NR>0 THEN GOSUB 1370
1280 FOR I1=1 TO NNC
1290 IF I(I5) < ABS(NC(I1)) THEN NC(I1)=SGN(NC(I1))*(ABS(NC(I1))-1)
1300 NEXT I1,I5 :RETURN '**** OMIT RATE CONSTANTS ****
1310 INPUT "ENTER # OF RATE CONSTANTS TO REMOVE FROM MODEL";NR
```

```
1320 IF NR < 1 OR NR > 20 THEN RETURN
1330 FOR I1=1 TO NR
1340 PRINT "ENTER RATE CONSTANT #";I1; :INPUT S(I1)
1350 IF S(I1)=0 OR ABS(S(I1)) > 99 GOTO 1340
1360 NEXT I1
1370 FOR I1=1 TO N :FOR I2=1 TO NA(I1) :FOR I3=1 TO NR
1380 IF A(I1,I2)<>S(I3) GOTO 1420
1390 IF NA(I1)<3 THEN PRINT "MUST HAVE >1 RATE CONSTANT AROUND A NODE":RETURN
1400 NA(I1)=NA(I1)-1 :IF I2=NA(I1)+1 GOTO 1370
1410 FOR I4=I2 TO NA(I1) :A(I1,I4)=A(I1,I4+1) :NEXT I4 :GOTO 1370
1420 NEXT I3,I2 :FOR I2=1 TO NB(I1) :FOR I3=1 TO NR
1430 IF B(I1,I2)<>S(I3) GOTO 1470
1440 IF NB(I1)<2 THEN PRINT "MUST HAVE >0 RATE CONSTS. LEAVING A NODE":RETURN
1450 NB(I1)=NB(I1)-1 :IF I2=NB(I1)+1 GOTO 1370
1460 FOR I4=I2 TO NB(I1) :B(I1,I4)=B(I1,I4+1) :NEXT I4 :GOTO 1370
1470 NEXT I3,I2,I1 :IF NNC=0 GOTO 1530
1480 FOR I1=1 TO NNC :FOR I2=1 TO NR
1490 IF C(I1)<>S(I2) GOTO 1520
1500 NNC=NNC-1 :IF I1=NNC+1 GOTO 1480
1510 FOR I3=I1 TO NNC :C(I3)=C(I3+1) :NC(I3)=NC(I3+1) :NEXT I3 :GOTO 1480
1520 NEXT I2,I1
1530 IF NRD=0 THEN RETURN
1540 FOR I1=1 TO NRD :FOR I2=1 TO NR :IF RD(I1)<>S(I2) GOTO 1570
1550 NRD=NRD-1 :IF I1=NRD+1 GOTO 1540
1560 FOR I3=I1 TO NRD :RD(I3)=RD(I3+1) :HD(I3)=HD(I3+1) :NEXT I3 :GOTO 1540
1570 NEXT I2,I1 :RETURN
1580 REM ********* STORE MODEL IN FILE **************
1590 INPUT "ENTER FILE NAME TO STORE MODEL";E1
1600 OPEN "O",#1,E1+".KIN" :WRITE #1,N
1610 FOR I=1 TO N :WRITE #1,NA(I) :FOR J=1 TO NA(I) :WRITE #1,A(I,J) :NEXT J,I
1620 FOR I=1 TO N :WRITE #1,NB(I) :FOR J=1 TO NB(I) :WRITE #1,B(I,J) :NEXT J,I
1630 WRITE #1,NNC,NRD :IF NNC=0 GOTO 1650
1640 FOR I=1 TO NNC :WRITE #1,NC(I),C(I) :NEXT I
1650 IF NRD=0 GOTO 1670
1660 FOR I=1 TO NRD :WRITE #1,RD(I),HD(I) :NEXT I
1670 CLOSE #1 :GOTO 580
1680 REM ********* GET OLD MODEL FROM FILE **************
1690 INPUT "OUTPUT TO SCREEN(1) OR PRINTER(2)";P :IF P < 1 OR P > 2 GOTO 1690
1700 INPUT "ENTER NAME OF MODEL";E1
1710 OPEN "I",#1,E1+".KIN" :INPUT #1,N :N1=N-1
1720 FOR I=1 TO N :INPUT #1,NA(I) :FOR J=1 TO NA(I) :INPUT #1,A(I,J) :NEXT J,I
1730 FOR I=1 TO N :INPUT #1,NB(I) :FOR J=1 TO NB(I) :INPUT #1,B(I,J) :NEXT J,I
1740 INPUT #1,NNC,NRD :IF NNC=0 GOTO 1760
1750 FOR I=1 TO NNC :INPUT #1,NC(I),C(I) :NEXT I
1760 IF NRD=0 THEN Q3=Q2 :Q2=1 :GOTO 1810
1770 FOR I=1 TO NRD :INPUT #1,RD(I),HD(I) :NEXT I
1780 FOR I=1 TO NRD-1 :FOR J=I+1 TO NRD
1790 IF HD(I) > HD(J) THEN SWAP HD(I),HD(J) :SWAP RD(I),RD(J)
1800 NEXT J,I
1810 CLOSE #1 :IF Q1<>2 THEN GOSUB 4010
1820 RETURN '************ SOLVE MULTIPLE MODELS **************
1830 INPUT "ENTER NAME OF MODEL LIST";FMODEL :FMODEL=FMODEL+".KIN"
1840 INPUT "DISPLAY INPUT FILES, YES(1),NO(2)";Q1 :IF Q1<1 OR Q1>2 GOTO 1840
1850 INPUT "DISPLAY DISTRIBUTION EQUATION NUMERATORS, YES(1), NO(2)";Q
1860 IF Q<1 OR Q>2 GOTO 1850
1870 INPUT "COMPLETE (1), OR COEFFICIENT (2) FORM OF DISPLAY";Q2
1880 IF Q2 < 1 OR Q2 > 2 GOTO 1870
1890 INPUT "OUTPUT TO SCREEN(1) OR PRINTER(2)";P :IF P < 1 OR P > 2 GOTO 1890
1900 OPEN "I",#1,FMODEL :INPUT #1,NMTS :CLOSE #1
1910 FOR I0=1 TO NMTS '******* LOOP THROUGH ALL MODELS IN LIST ********
1920 OPEN "I",#1,FMODEL:INPUT #1,I:FOR I=1 TO I0:INPUT #1,E1:NEXT I:CLOSE #1
1930 GOSUB 1710 :GOSUB 2010 :NEXT I0 :RETURN
1940 REM ************ SOLVE ONE MODEL ************
1950 INPUT "DISPLAY DISTRIBUTION EQUATION NUMERATORS, YES(1), NO(2)";Q
1960 IF Q<1 OR Q>2 GOTO 1950
1970 INPUT "COMPLETE (1), OR COEFFICIENT (2) FORM OF DISPLAY";Q2
```

```
1980 IF Q2 < 1 OR Q2 > 2 GOTO 1970
1990 Q1=1 :GOSUB 1690
2000 REM ************* SUBROUTINE TO SOLVE ONE MODEL *******************
2010 IF Q1=2 AND P=1 THEN PRINT :PRINT "FILE NAME IS ";E1;".KIN"
2020 IF Q1=2 AND P=2 THEN LPRINT :LPRINT "FILE NAME IS ";E1;".KIN"
2030 E1=DATE$+" "+TIME$ :J=2 :K=2 :ND=0 :B4=1
2040 OPEN "FILE1" AS #1 LEN=N :FIELD #1,N AS H(1)
2050 OPEN "FILE2" AS #2 LEN=N :FIELD #2,N AS H(2)
2060 OPEN "FILE3" AS #3 LEN=N :FIELD #3,N AS H(3)
2070 OPEN "FILE4" AS #4 LEN=N :FIELD #4,N AS H(4)
2080 IF NRD=0 GOTO 2130 ELSE NU=1 :HU(NU)=CHR$(NU+132)
2090 FOR I1=2 TO NRD :IF HD(I1)=HD(I1-1) THEN HU(I1)=HU(I1-1) :GOTO 2110
2100 NU=NU+1 :HU(I1)=CHR$(NU+132) 'elements in the HU array are single char.
2110 NEXT I1 'equivalents of the concentration terms in the HD array.
2120 REM * Calculate MND (max. # of denominator terms)
2130 IF ZMND/(4*N) > 32762 THEN MND=32762 ELSE MND=ZMND/(4*N)
2140 FOR A1=1 TO NA(1)    :I(1)=A1 'Begin generation of denominator terms. A
2150 FOR A2=1 TO NA(2)    :I(2)=A2  :IF N1=2 GOTO 2330 'potential denominator
2160 FOR A3=1 TO NA(3)    :I(3)=A3  :IF N1=3 GOTO 2330 'term is generated by
2170 FOR A4=1 TO NA(4)    :I(4)=A4  :IF N1=4 GOTO 2330 'selecting one rate
2180 FOR A5=1 TO NA(5)    :I(5)=A5  :IF N1=5 GOTO 2330 'constant surrounding
2190 FOR A6=1 TO NA(6)    :I(6)=A6  :IF N1=6 GOTO 2330 'each of the first N-1
2200 FOR A7=1 TO NA(7)    :I(7)=A7  :IF N1=7 GOTO 2330 'nodes in the model.
2210 FOR A8=1 TO NA(8)    :I(8)=A8  :IF N1=8 GOTO 2330 'By nesting N-1 For/Next
2220 FOR A9=1 TO NA(9)    :I(9)=A9  :IF N1=9 GOTO 2330 'loops where each loop
2230 FOR A10=1 TO NA(10)  :I(10)=A10 :IF N1=10 GOTO 2330 'goes through each rate
2240 FOR A11=1 TO NA(11)  :I(11)=A11 :IF N1=11 GOTO 2330 'constant surrounding
2250 FOR A12=1 TO NA(12)  :I(12)=A12 :IF N1=12 GOTO 2330 'the first N-1 nodes in
2260 FOR A13=1 TO NA(13)  :I(13)=A13 :IF N1=13 GOTO 2330 'the model, all
2270 FOR A14=1 TO NA(14)  :I(14)=A14 :IF N1=14 GOTO 2330 'possible potential
2280 FOR A15=1 TO NA(15)  :I(15)=A15 :IF N1=15 GOTO 2330 'denominator terms are
2290 FOR A16=1 TO NA(16)  :I(16)=A16 :IF N1=16 GOTO 2330 'generated. Each
2300 FOR A17=1 TO NA(17)  :I(17)=A17 :IF N1=17 GOTO 2330 'potential term is
2310 FOR A18=1 TO NA(18)  :I(18)=A18 :IF N1=18 GOTO 2330 'pointed to in the A(,)
2320 FOR A19=1 TO NA(19)  :I(19)=A19 'array by I1 and I(I1), i.e. A(I1,I(I1)).
2330 IF J<K THEN K=J '**** see Note 1 below ****
2340 FOR I1=J TO N1 :FOR I2=I1-1 TO 1 STEP -1
2350 IF ABS(A(I1,I(I1)))=ABS(A(I2,I(I2))) GOTO 2580
2360 NEXT I2,I1 '* ELIMINATE TERMS WITH FORBIDDEN COMBINATIONS see Note 2 below
2370 FOR I1=K TO N1 :K=I1 :FOR I2=I1-1 TO 1 STEP -1
2380 FOR I3=1 TO N :FOR I4=1 TO NB(I3)-1 :FOR I5=I4+1 TO NB(I3)
2390 IF A(I1,I(I1))=B(I3,I4) GOTO 2410
2400 IF A(I2,I(I2))=B(I3,I4) GOTO 2420 ELSE GOTO 2440
2410 IF A(I2,I(I2))=B(I3,I5) GOTO 2580 ELSE GOTO 2430
2420 IF A(I1,I(I1))=B(I3,I5) GOTO 2580
2430 NEXT I5
2440 NEXT I4,I3,I2,I1
2450 REM ******  ADD TERM TO DENOMINATOR  ********
2460 ND=ND+1 :FOR I1=1 TO N1 :S(I1)=A(I1,I(I1)) :NEXT I1
2470 FOR I1=1 TO N1-1 :FOR I2=I1+1 TO N1 '*** SORT RATE CONSTANTS IN A TERM ***
2480 IF ABS(S(I1)) > ABS(S(I2)) THEN SWAP S(I1),S(I2)
2490 NEXT I2,I1 :G="" :FOR I1=1 TO N1 :G=G+CHR$(S(I1)+132) :NEXT I1
2500 IF ND > MX GOTO 2520 ELSE E(ND)=G :I1=N1 :IF ND < MX GOTO 2580
2510 FOR I1=1 TO ND :LSET H(B4)=E(I1) :PUT #B4,I1 :NEXT I1 :I1=N1 :GOTO 2580
2520 LSET H(B4)=G :PUT #B4,ND
2530 IF ND=MND THEN D=K :GOSUB 4980 :GOSUB 5470 ELSE I1=N1 :GOTO 2580
2540 K=D :I1=N1 :IF ND < MND GOTO 2580
2550 IF P=1 THEN PRINT "DISK SPACE MAY BECOME EXCEEDED, STOP AFTER";ND;"TERMS"
2560 IF P=2 THEN LPRINT "DISK SPACE MAY BECOME EXCEEDED, STOP AFTER";ND;"TERMS"
2570 IF P=1 THEN INPUT I :GOTO 3990 ELSE GOTO 3990
2580 ON I1 GOTO 2760,2760,2750,2740,2730,2720,2710,2700,2690,2680,2670,2660,2650
,2640,2630,2620,2610,2600,2590
2590 J=19 :NEXT A19 'Note 1: The potential term is tested for the Wang product
2600 J=18 :NEXT A18 'rule by checking for any two rate constants whose absolute
2610 J=17 :NEXT A17 'value is equal. The variable J is used so that no two rate
2620 J=16 :NEXT A16 'constants are checked for the Wang product rule more than
```

```
2630 J=15 :NEXT A15 'once. The order of checking is designed to be able to find
2640 J=14 :NEXT A14 'equal rate constants in adjacent nodes quickly, hence the
2650 J=13 :NEXT A13 'reason for assigning a specific order to the assignment of
2660 J=12 :NEXT A12 'node numbers to a model.
2670 J=11 :NEXT A11
2680 J=10 :NEXT A10
2690 J=9  :NEXT A9  'Note 2: The potential term having passed the Wang product
2700 J=8  :NEXT A8  'rule must be tested for forbidden combinations. The
2710 J=7  :NEXT A7  'variable K is used in the same way as variable J above,
2720 J=6  :NEXT A6  'ie. it is used to ensure that no two rate constants are
2730 J=5  :NEXT A5  'tested as a forbidden combination more than once. The
2740 J=4  :NEXT A4  'order of checking is the same as above but does not speed
2750 J=3  :NEXT A3  'the process of checking for forbidden combinations.
2760 J=2  :NEXT A2,A1
2770 GOSUB 4980 '****        SORT DENOMINATOR USING MERGE SORT        ****
2780 GOSUB 5470 '**** ELIMINATE REPEATED TERMS BY WANG SUMMATION RULE ****
2790 IF NRD=0 THEN Q2=Q3 '**** NOW SORTED LEGAL TERMS ARE IN FILE # B4 ****
2800 IF P=1 THEN PRINT "START = ";E1;",   END = ";DATE$+" "+TIME$
2810 IF P=2 THEN LPRINT "START = ";E1;",   END = ";DATE$+" "+TIME$
2820 IF P=1 THEN PRINT ELSE LPRINT
2830 REM ******  GENERATE AND PRINT OUT EACH NODE NUMERATOR  ********
2840 REM Node Sorting is carried out by looping through each node, and for
2850 REM each node, looping through all denominator terms. Each denominator
2860 REM term is checked to see if it belongs with the given node (ie. if the
2870 REM term contains no rate constants leaving the node). Then, each of the
2880 REM concentration terms associated with each selected term is determined,
2890 REM and a list of the unique conc. terms is formed in the E( ) array. When
2900 REM all terms for that node have been processed, the list of unique conc.
2910 REM terms is sorted. Then a loop is started through all the unique conc.
2920 REM terms; it is first printed, and if selected the rate constant terms
2930 REM are printed as follows. All the denominator terms are searched again,
2940 REM the terms for the given node selected, the conc. term determined and
2950 REM is compared with the conc. term currently being printed. If it matches
2960 REM the rate constant portion is printed. The process of printing the
2970 REM denominator is similar, except there is no need to loop through all
2980 REM the nodes to select terms for that given node.
2990 IF Q=2 GOTO 3240 '** SKIP THIS SECTION IF SELECTED **
3000 IF P=1 THEN PRINT "*****  NODE SORTING  *****" :PRINT
3010 IF P=2 THEN LPRINT "*****  NODE SORTING  *****" :LPRINT
3020 NE=N1 :FOR I1=1 TO N :IF NRD=0 THEN NU=1 :E(NU)=" " :GOTO 3080
3030 NU=0 :FOR I4=1 TO ND :GET #B4,I4 :G=H(B4) :FOR I5=1 TO N1
3040 S(I5)=ASC(MID$(G,I5,1))-132
3050 FOR I6=1 TO NB(I1) :IF S(I5)=B(I1,I6) GOTO 3070
3060 NEXT I6,I5 :GOSUB 4350
3070 NEXT I4 :K=NU :GOSUB 4600
3080 NN=0 :FOR I2=1 TO NU :GOSUB 4520    '*** PRINT UNIQUE CONC. TERM ***
3090 IF Q2=2 GOTO 3150
3100 FOR I4=1 TO ND :GET #B4,I4 :G=H(B4) :FOR I5=1 TO N1
3110 S(I5)=ASC(MID$(G,I5,1))-132
3120 FOR I6=1 TO NB(I1) :IF S(I5)=B(I1,I6) GOTO 3140
3130 NEXT I6,I5 :GOSUB 4380
3140 NEXT I4
3150 NEXT I2 :IF P=1 THEN PRINT TAB(1)" "; ELSE LPRINT TAB(1)" ";
3160 G="UNIQUE CONCENTRATION TERMS."
3170 IF Q2=2 GOTO 3200
3180 IF P=1 THEN PRINT "NODE #";I1;"HAS";NN;"TERMS AND";NU;G
3190 IF P=2 THEN LPRINT "NODE #";I1;"HAS";NN;"TERMS AND";NU;G
3200 IF Q2=2 AND P=1 THEN PRINT "NODE #";I1;"HAS";NU;G
3210 IF Q2=2 AND P=2 THEN LPRINT "NODE #";I1;"HAS";NU;G
3220 IF P=1 THEN INPUT I ELSE LPRINT
3230 NEXT I1        '******** PRINT DENOMINATOR WITH CONCENTRATION TERMS *******
3240 NE=N1 :IF NRD=0 THEN NU=1 :E(NU)=" " :GOTO 3270
3250 NU=0 :FOR I4=1 TO ND :GET #B4,I4 :G=H(B4) :FOR I5=1 TO N1
3260 S(I5)=ASC(MID$(G,I5,1))-132 :NEXT I5 :GOSUB 4350 :NEXT I4 :K=NU:GOSUB 4600
3270 NN=0 :FOR I2=1 TO NU :GOSUB 4520    '*** PRINT UNIQUE CONC. TERM ***
3280 IF Q2=2 GOTO 3310
```

```
3290 FOR I4=1 TO ND :GET #B4,I4 :G=H(B4) :FOR I5=1 TO N1
3300 S(I5)=ASC(MID$(G,I5,1))-132 :NEXT I5 :GOSUB 4380 :NEXT I4
3310 NEXT I2 :IF P=1 THEN PRINT TAB(1)" "; ELSE LPRINT TAB(1)" ";
3320 G="UNIQUE CONCENTRATION TERMS."
3330 IF P=1 THEN PRINT "DENOMINATOR HAS";ND;"TERMS AND";NU;G
3340 IF P=2 THEN LPRINT "DENOMINATOR HAS";ND;"TERMS AND";NU;G
3350 IF P=1 THEN INPUT I ELSE LPRINT
3360 IF NNC=0 GOTO 3990 '**** NOW GENERATE RATE EQUATION NUMERATORS ****
3370 REM * Generating the rate equation numerator is similar to node sorting
3380 REM * except an added step is needed. Here the first loop is not through
3390 REM * all nodes, but through all nodes listed in section 3 of the input
3400 REM * file. Node sorting for the nodes listed is carried out and in
3410 REM * addition, the multiplier rate constant is multiplied by the term if
3420 REM * it does not contain any rate constants whose absolute value equals
3430 REM * the absolute value of the multiplier rate constant.Once the new term
3440 REM * now containing N rate constants is formed, it is stored in one file
3450 REM * if the sign before the node # in the input file is +, and in another
3460 REM * file if it is negative. After all terms are generated, the files are
3470 REM * checked to make sure no identical terms are present in both the +
3480 REM * and - term file. Then the terms are printed, sorted by conc. term
3490 REM * as described above.
3500 T(1)=2:T(2)=3 :TC(1)=0 :IF T(1)=B4 THEN T(1)=4 ELSE IF T(2)=B4 THEN T(2)=4
3510 TC(2)=0 :FOR I3=1 TO NNC :I2=ABS(NC(I3)) :IF NC(I3) ) 0 THEN K=1 ELSE K=2
3520 FOR I4=1 TO ND :GET #B4,I4 :G=H(B4)
3530 FOR I5=1 TO N1 :S(I5)=ASC(MID$(G,I5,1))-132
3540 FOR I6=1 TO NB(I2) :IF S(I5)=B(I2,I6) GOTO 3610
3550 NEXT I6,I5 '**** ELIMINATE TERMS CONTAINING MULTIPLIER RATE CONSTANT ****
3560 FOR I5=1 TO N1 :IF ABS(S(I5))=ABS(C(I3)) GOTO 3610
3570 NEXT I5 :S(N)=C(I3) :FOR I5=1 TO N1:FOR I6=I5+1 TO N '* MULT. RATE CONST.
3580 IF ABS(S(I5)) ) ABS(S(I6)) THEN SWAP S(I5),S(I6)
3590 NEXT I6,I5 :G="":FOR I5=1 TO N:G=G+CHR$(S(I5)+132):NEXT I5:TC(K)=TC(K)+1
3600 LSET H(T(K))=G :PUT #T(K),TC(K)
3610 NEXT I4,I3 :B9=TC(1) :IF B9 ( TC(2) THEN B9=TC(2) ELSE IF B9=0 GOTO 3990
3620 B1=1 :IF TC(1)=0 OR TC(2)=0 GOTO 3720 '*** ELIMINATE TERMS COMMON TO ***
3630 IF NNC=2 AND C(1)=-C(2) GOTO 3700     '*** BOTH + AND - NUMERATORS ***
3640 FOR I3=B1 TO TC(1) :GET #T(1),I3 :G=H(T(1)) :FOR I4=1 TO TC(2)
3650 GET #T(2),I4 :IF G()H(T(2)) GOTO 3690
3660 GET #T(1),TC(1) :G=H(T(1)) :TC(1)=TC(1)-1 :LSET H(T(1))=G :PUT #T(1),I3
3670 GET #T(2),TC(2) :G=H(T(2)) :TC(2)=TC(2)-1 :LSET H(T(2))=G :PUT #T(2),I4
3680 B1=I3 :GOTO 3640
3690 NEXT I4,I3 '*** NOW + TERMS IN B4 AND - TERMS TO FILE 1 FOR SORTING ***
3700 FOR I3=1 TO TC(2) :GET #T(2),I3:G=H(T(2)):LSET H(1)=G:PUT #1,I3+B9:NEXT I3
3710 B4=T(1) :GOTO 3740 '*** NOW + OR - TERMS ARE IN FILE B4 FOR SORTING ***
3720 IF TC(1)=0 THEN K=2 ELSE K=1
3730 B4=T(K) '    **** SORT + AND/OR - NUMERATORS IN FILE B4 ****
3740 IF TC(1)=0 THEN B8=2 ELSE B8=1
3750 ND=TC(B8) :IF TC(B8) ) MX GOTO 3770
3760 FOR I3=1 TO ND :GET #B4,I3 :E(I3)=H(B4) :NEXT I3
3770 GOSUB 4980 :IF TC(1)=0 OR TC(2)=0 THEN T(B8)=B4 :TB(B8)=0 :GOTO 3840
3780 IF B4=1 THEN B8=2 ELSE B8=1
3790 FOR I3=1 TO TC(2) :GET #1,I3+B9 :G=H(1) :LSET H(B8)=G :PUT #B8,I3 :NEXT I3
3800 FOR I3=1 TO TC(1) :GET #B4,I3 :G=H(B4) :LSET H(1)=G :PUT #1,I3+B9 :NEXT I3
3810 B4=B8 :ND=TC(2) :IF TC(2) ) MX GOTO 3830
3820 FOR I3=1 TO ND :GET #B4,I3 :E(I3)=H(B4) :NEXT I3
3830 GOSUB 4980 :T(1)=1 :TB(1)=B9 :T(2)=B4 :TB(2)=0 '* FIND UNIQUE TERMS *
3840 NE=N :FOR I1=1 TO 2 :IF TC(I1)=0 GOTO 3980
3850 IF NRD=0 THEN NU=1 :E(NU)=" " :GOTO 3880
3860 NU=0 :FOR I2=1 TO TC(I1) :GET #T(I1),TB(I1)+I2 :G=H(T(I1)) :FOR I3=1 TO N
3870 S(I3)=ASC(MID$(G,I3,1))-132 :NEXT I3 :GOSUB 4350 :NEXT I2 :K=NU:GOSUB 4600
3880 NN=0 :FOR I2=1 TO NU :GOSUB 4520     '*** PRINT UNIQUE CONC. TERM ***
3890 IF Q2=2 GOTO 3920
3900 FOR I3=1 TO TC(I1) :GET #T(I1),TB(I1)+I3 :G=H(T(I1)) :FOR I4=1 TO N
3910 S(I4)=ASC(MID$(G,I4,1))-132 :NEXT I4 :GOSUB 4380 :NEXT I3
3920 NEXT I2 :IF P=1 THEN PRINT TAB(1)" "; ELSE LPRINT TAB(1)" ";
3930 IF I1=1 THEN G="POSITIVE NUMERATOR " ELSE G="NEGATIVE NUMERATOR "
3940 E1="UNIQUE CONCENTRATION TERMS."
```

```
3950 IF P=1 THEN PRINT "THERE ARE";TC(I1);G;"TERMS AND";NU;E1
3960 IF P=2 THEN LPRINT "THERE ARE";TC(I1);G;"TERMS AND";NU;E1
3970 IF P=1 THEN INPUT I ELSE LPRINT
3980 NEXT I1
3990 CLOSE #1,#2,#3,#4  :KILL "FILE1" :KILL "FILE2" :KILL "FILE3"
4000 KILL "FILE4"   :RETURN '**** SUBROUTINE TO PRINT INPUT DATA ****
4010 IF P=1 THEN PRINT "FILE NAME IS ";E1;".KIN" :PRINT
4020 IF P=2 THEN LPRINT "FILE NAME IS ";E1;".KIN" :LPRINT
4030 IF P=1 THEN PRINT "NODE #    K's  SURROUNDING NODE"
4040 IF P=2 THEN LPRINT "NODE #    K's  SURROUNDING NODE"
4050 FOR I2=1 TO N :IF P=1 THEN PRINT I2, ELSE LPRINT I2,
4060 FOR I3=1 TO NA(I2)
4070 IF P=1 THEN PRINT STR$(A(I2,I3)); ELSE LPRINT STR$(A(I2,I3));
4080 NEXT I3 :IF P=1 THEN PRINT ELSE LPRINT
4090 NEXT I2 :IF P=1 THEN PRINT :INPUT J ELSE LPRINT
4100 IF P=1 THEN PRINT "NODE #   K's LEAVING NODE     FORBIDDEN COMBINATIONS"
4110 IF P=2 THEN LPRINT "NODE #   K's LEAVING NODE     FORBIDDEN COMBINATIONS"
4120 FOR I2=1 TO N :IF P=1 THEN PRINT I2;"   "; ELSE LPRINT I2;"   ";
4130 FOR I3=1 TO NB(I2) :IF P=1 THEN PRINT STR$(B(I2,I3));
4140 IF P=2 THEN LPRINT STR$(B(I2,I3));
4150 NEXT I3 :IF P=1 THEN PRINT TAB(25);"  "; ELSE LPRINT TAB(25);"  ";
4160 IF NB(I2)=1 THEN IF P=1 THEN PRINT :GOTO 4210 ELSE LPRINT :GOTO 4210
4170 FOR I3=1 TO NB(I2)-1 :FOR I4=I3+1 TO NB(I2)
4180 IF P=1 THEN PRINT STR$(B(I2,I3))+STR$(B(I2,I4));",";
4190 IF P=2 THEN LPRINT STR$(B(I2,I3))+STR$(B(I2,I4));",";
4200 NEXT I4,I3 :IF P=1 THEN PRINT ELSE LPRINT
4210 NEXT I2 :IF P=1 THEN PRINT :INPUT J ELSE LPRINT
4220 IF NNC=0 GOTO 4280
4230 IF P=1 THEN PRINT "RATE EQUATION FORMULA"
4240 IF P=2 THEN LPRINT "RATE EQUATION FORMULA"
4250 FOR I2=1 TO NNC
4260 IF P=1 THEN PRINT NC(I2),STR$(C(I2)) ELSE LPRINT NC(I2),STR$(C(I2))
4270 NEXT I2 :IF P=1 THEN PRINT :INPUT J ELSE LPRINT
4280 IF NRD=0 GOTO 4340
4290 IF P=1 THEN PRINT "PSEUDO-FIRST ORDER RATE CONSTANTS"
4300 IF P=2 THEN LPRINT "PSEUDO-FIRST ORDER RATE CONSTANTS"
4310 FOR I2=1 TO NRD
4320 IF P=1 THEN PRINT STR$(RD(I2)),HD(I2) ELSE LPRINT STR$(RD(I2)),HD(I2)
4330 NEXT I2 :IF P=1 THEN PRINT :INPUT J ELSE LPRINT
4340 RETURN '****** SUBROUTINE TO FIND UNIQUE CONCENTRATION TERMS ********
4350 NU=NU+1 :GOSUB 4430 :E(NU)=G '**** ADD TERM IF UNIQUE ****
4360 FOR I5=1 TO NU-1 :IF E(I5) = E(NU) THEN NU=NU-1 :RETURN
4370 NEXT I5 :RETURN '* SUBROUTINE TO PRINT THE RATE CONSTANT PORTION OF TERM *
4380 IF NRD=0 GOTO 4400
4390 GOSUB 4430 :IF E(I2)<>G THEN RETURN
4400 G=" " :FOR I5=1 TO NE :G=G+STR$(S(I5)) :NEXT I5
4410 NN=NN+1 :IF P=1 THEN PRINT G; ELSE LPRINT G;
4420 RETURN '* SUBROUTINE USED BY ABOVE 2 SUBRS. TO FIND CONC. PORTION OF TERM *
4430 I=0 :FOR I5=1 TO NE :FOR I6=1 TO NRD
4440 IF S(I5)=RD(I6) THEN I=I+1 :R(I)=I6 :GOTO 4460
4450 NEXT I6
4460 NEXT I5 :IF I=0 THEN G=" " ELSE G="" '**** SORT CONC.'S IN THIS TERM ****
4470 FOR I5=1 TO I-1 :FOR I6=I5+1 TO I
4480 IF R(I5) > R(I6) THEN SWAP R(I5),R(I6)
4490 NEXT I6,I5 '* CONVERT INDICES OF HD( ) IN R( ) TO CHARACTER EQUIVALENTS *
4500 FOR I5=1 TO I :G=G+HU(R(I5)) :NEXT I5 :RETURN
4510 REM ******* SUBROUTINE TO PRINT CONC. TERMS    ********
4520 G=" " :FOR I4=1 TO LEN(E(I2)) :FOR I5=1 TO NRD
4530 IF MID$(E(I2),I4,1)=HU(I5) THEN G=G+HD(I5) :GOTO 4550
4540 NEXT I5
4550 NEXT I4 :IF G=" " THEN NU=NU-1 :G="constant"
4560 IF Q2=1 THEN IF P=1 THEN PRINT TAB(1)G ELSE LPRINT TAB(1)G
4570 IF Q2=2 THEN IF P=1 THEN PRINT G; ELSE LPRINT G;
4580 RETURN '********* SUBROUTINE TO QUICK SORT TERMS INTERNALLY **********
4590 REM * The Quick Sort is called QUICKERSORT written by
4600 J=K:I=1:L=1 '-) R.S. Scowen, Algorithm 271, Comm. ACM 8 (Nov. 1965), 669.
```

```
4610 IF J-I <= 1 GOTO 4720
4620 B7 = (J+I)/2 :G=E(B7) :E(B7)=E(I) :B5=J
4630 FOR B6=I+1 TO B5 :IF E(B6) <= G GOTO 4680
4640 FOR B5=B5 TO B6 STEP -1
4650 IF E(B5) < G THEN SWAP E(B6),E(B5) :B5=B5-1 :GOTO 4680
4660 NEXT B5
4670 B5=B6-1 :GOTO 4690
4680 NEXT B6
4690 E(I)=E(B5) :E(B5)=G
4700 IF 2*B5 > I+J THEN R(L)=I :S(L)=B5-1 :I=B5+1 ELSE R(L)=B5+1:S(L)=J:J=B5-1
4710 L=L+1 :GOTO 4610
4720 IF I < J THEN IF E(I) > E(J) THEN SWAP E(I),E(J)
4730 L=L-1 :IF L > 0 THEN I=R(L) :J=S(L) :GOTO 4610
4740 RETURN '    **********    END OF QUICKERSORT    ***********
4750 REM ** Subroutine to sort terms internally and write to file 1 or 2 **
4760 REM ** K=# of terms to sort, first term is B1=I2-K+1, last term is B5=I2
4770 GOSUB 4600   '** when I1=1 or B1=I2-K+1-NF(1) and B5=I2-NF(1) when I1=2.
4780 B1=I2-K+1 :B5=I2 :K=0 :IF I1=2 THEN B1=B1-NF(1) :B5=B5-NF(1)
4790 FOR I3=B1 TO B5 :K=K+1 :LSET H(I1)=E(K) :PUT #I1,I3 :NEXT I3 :K=0 :RETURN
4800 REM ******** SUBROUTINE TO READ IN SECTIONS OF SORTING FILES ********
4810 REM * This subroutine is used to mimic buffers for the merge sort files
4820 REM * since by default there exists one input and output file buffer.
4830 REM * This is described in more detail with the merge sort subroutine.
4840 REM * Receives J=1,2,3 = which V(J), T(K)=which file to read from,
4850 REM * AF(T(K))=which term to read, Returns V(J)=index of E( ) with term
4860 REM * Uses TA( ) which indicates the source file #.
4870 IF TA(1)=T(K) OR TA(2)=T(K) GOTO 4920
4880 IF NF(T(K))-AF(T(K)) < MX2 THEN B7=NF(T(K)) ELSE B7=AF(T(K))+MX2-1
4890 TA(K)=T(K) :TB(K)=AF(T(K)) :IF K=1 THEN B5=0 ELSE B5=MX2
4900 FOR I1=TB(K) TO B7:B5=B5+1:GET #T(K),I1:E(B5)=H(T(K)):NEXT I1:GOTO 4870
4910 REM  *****  LOCATE DESIRED TERM  ******
4920 V(J)=AF(T(K))-TB(K)+1 :IF K=2 THEN V(J)=V(J)+MX2
4930 IF V(J)<>MX2 AND V(J)<>MX GOTO 4960
4940 I=V(J) :IF K=1 THEN V(J)=MX+1 ELSE V(J)=MX+2
4950 TA(K)=0 :E(V(J))=E(I)
4960 IF AF(T(K))=NF(T(K)) THEN TA(K)=0
4970 RETURN '**** MERGE SORT SUBROUTINE TO SORT TERMS IN E( ) OR FILE B4 ****
4980 IF ND > MX GOTO 5000   '****   SORTED FILE WILL BE FILE #B4   ****
4990 I1=2 :K=ND :I2=ND :GOSUB 4770 :B4=I1 :RETURN
5000 NF(1)=ND/2 :NF(2)=ND-NF(1)
5010 REM ******* PARTIALLY SORT TERMS INTO FILES 1 AND 2 *******
5020 FOR I1=1 TO 2 :K=0 :IF I1=1 THEN B2=1 :B3=NF(1) ELSE B2=NF(1)+1 :B3=ND
5030 FOR I2=B2 TO B3 :K=K+1 :GET #B4,I2 :E(K)=H(B4) :IF K=MX THEN GOSUB 4770
5040 NEXT I2 :I2=B3 :IF K>1 THEN GOSUB 4770 ELSE IF K=1 THEN GOSUB 4780
5050 NEXT I1 :TA(1)=0   '****** MERGE THE PARTIALLY SORTED FILES ******
5060 REM * This merge sort is designed to sort a large number of items in a
5070 REM * relatively small amount of internal memory space.  The terms
5080 REM * are partially sorted using the Quickersort algorithm above,
5090 REM * creating two files, each partially sorted.  The merge sort then
5100 REM * compares items from each file, placing the smaller in a third file.
5110 REM * When the first partially sorted list in each file has been trans-
5120 REM * fered to the third file, the next partially sorted lists are moved
5130 REM * to the fourth file.  This process continues, the result being that
5140 REM * the new files contain half as many partially sorted lists as before.
5150 REM * On the last pass,the two files are merged into one completely sorted
5160 REM * list.  The terms are also in order on the disk so that each GET
5170 REM * issued to get a term from the file does not necessarily access the
5180 REM * disk.   The speed of reading from the disk is improved by using an
5190 REM * array in internal memory to act as a file buffer.  This is done by a
5200 REM * subroutine which uses the first half of the E( ) array for one input
5210 REM * and the second half of E( ) for the other input file.
5220 TA(2)=0 :NF(3)=0 :NF(4)=0 :FOR I1=1 TO 4 :T(I1)=I1 :AF(I1)=1 :NEXT I1 :L=3
5230 K=1:J=1 :GOSUB 4870 :K=2 :J=2 :GOSUB 4870 'Read first two terms, each pass
5240 IF E(V(1)) < E(V(2)) THEN K=1 ELSE K=2 '* Compare two terms in two files
5250 AF(T(K))=AF(T(K))+1 :NF(T(L))=NF(T(L))+1 'Write smaller term to other file
5260 LSET H(T(L))=E(V(K)) :PUT #T(L),NF(T(L))
```

```
5270 IF AF(T(K)))NF(T(K)) THEN AF(T(K))=1:NF(T(K))=0:GOTO 5300 'Done with file?
5280 J=3:GOSUB 4870 'Read in next term from list where smaller term was found
5290 IF E(V(3)))=E(V(K)) THEN SWAP V(K),V(3):GOTO 5240'Still in sorted section?
5300 IF K=1 THEN K=2 ELSE K=1 'Transfer part of file until end of file or
5310 J=K :GOSUB 4870          'new sorted segment is reached.
5320 NF(T(L))=NF(T(L))+1 :AF(T(K))=AF(T(K))+1
5330 LSET H(T(L))=E(V(K)) :PUT #T(L),NF(T(L))
5340 REM ***** CHECK TO SEE IF AT END OF TRANSFER SOURCE FILE  ******
5350 IF AF(T(K)))NF(T(K)) THEN AF(T(K))=1:NF(T(K))=0:GOTO 5380
5360 J=3 :GOSUB 4870 'Read in next term in the transfer list.
5370 IF E(V(3)) )= E(V(K)) THEN SWAP V(K),V(3) :GOTO 5320
5380 IF L=3 THEN L=4 ELSE L=3 '* Now at end of transfer list or new segment.
5390 IF AF(T(K))=1 GOTO 5420
5400 IF K=1 THEN K=2 ELSE K=1 'At end of a segment in both files if AF(T(K))<>1
5410 IF AF(T(K))<>1 GOTO 5230 ELSE 5300 'Else if AF(T(K))=1 then at end of
5420 IF K=1 THEN K=2 ELSE K=1 'segment in transfer file + end of other file.
5430 IF AF(T(K))=1 GOTO 5440 ELSE 5310 'At end of both files, ELSE end of
5440 IF NF(T(L))<>0 THEN SWAP T(1),T(3) :SWAP T(2),T(4) :GOTO 5230
5450 IF L=3 THEN L=4 ELSE L=3 'transfer file, and end of segment in other file.
5460 B4=T(L) :RETURN '******* Subroutine to eliminate repeated terms by *******
5470 B1=B4 :IF B1=1 THEN B4=2 ELSE B4=1 '****  Wang Summation Rule  ****
5480 K=1 :B3=0 'transfer terms from file B1 to B4 except all pairs of = terms
5490 K=1 :I1=I1+1 :GET #B1,I1 :E(K)=H(B1) :IF I1=ND GOTO 5560
5500 IF K=1 THEN K=2 ELSE K=1
5510 I1=I1+1 :GET #B1,I1 :E(K)=H(B1)
5520 IF E(1)=E(2) THEN IF I1=ND GOTO 5570 ELSE GOTO 5490
5530 IF K=1 THEN K=2 ELSE K=1
5540 B3=B3+1 :LSET H(B4)=E(K) :PUT #B4,B3 :IF I1 < ND GOTO 5510
5550 IF K=1 THEN K=2 ELSE K=1
5560 B3=B3+1 :LSET H(B4)=E(K) :PUT #B4,B3
5570 ND=B3 :RETURN '*******   now sorted legal terms are in file #B4  *******
5580 END '******************   END   OF   SSREG.BAS   *******************
```

Section II

Model Membranes and Their Characteristics

[9] Liposome Preparation and Size Characterization

By MARTIN C. WOODLE and DEMETRIOS PAPAHADJOPOULOS

Introduction

Liposomes are rapidly becoming accepted as pharmaceutical agents which may improve the therapeutic activity of a wide variety of compounds. Consequently, there is an increasing need for easily reproducible and efficient preparation methods at both a laboratory and an industrial scale. A number of reviews have described studies of liposome production and properties, their use as carriers for drugs, and their interaction with a variety of cell types.[1-7] Many studies have shown that liposome behavior can vary dramatically, depending on several properties, including particle-size distribution, number of lamellae, chemical composition, and amount of trapped volume. Therefore, it is essential to use methods producing well-defined liposomes, with respect to these parameters, in order to understand the impact of each property on liposome function. Reliable characterization of liposomes with respect to these parameters is critical and especially so for particle-size distribution.

Recent reviews have focused on the therapeutic properties of liposomes without consideration of the differences produced by current production methods, which may greatly affect the behavior of liposomes.[6,8-11] In this review we will present a scheme categorizing all liposome preparation methodologies along with a detailed description of the most useful methods. Principal concerns and problems for producing well-defined

[1] F. C. Szoka and D. Papahadjopoulos, *Annu. Rev. Biophys. Bioeng.* **9,** 467 (1980).

[2] D. Deamer and P. S. Uster, *in* "The Liposomes" (M. J. Ostro, ed.), p. 27. Dekker, New York, 1983.

[3] G. Gregoriadis (ed.), "Liposome Technology," Vols. 1-3. CRC Press, Boca Raton, Florida, 1984.

[4] P. R. Cullis, M. J. Hope, M. B. Bally, T. D. Madden, L. D. Mayer, and A. S. Janoff, *in* "Liposomes from Biophysics to Therapeutics" (M. J. Ostro, ed.), p. 39. Dekker, New York, 1987.

[5] D. Lichtenberg and Y. Barenholz, *Methods Biochem. Anal.* **33,** 337 (1988).

[6] J. H. Senior, *Crit. Rev. Ther. Drug Carrier Syst.* **3,** 123 (1987).

[7] P. Machy and L. Leserman (eds.), "Liposomes in Cell Biology and Pharmacology," p. 6. Libbey, London, 1987.

[8] J. N. Weinstein, *Cancer Treat. Rep.* **68,** 127 (1984).

[9] G. Lopez-Berestein, *Antimicrob. Agents Chemother.* **31,** 675 (1987).

[10] R. Kirsh and G. Poste, *Adv. Exp. Med. Biol.* **202,** 171 (1986).

[11] J. Sunamoto and K. Iwamoto, *Crit. Rev. Ther. Drug Carrier Syst.* **2,** 117 (1986).

METHODS IN ENZYMOLOGY, VOL. 171

liposomes will also be discussed. Finally, the existing methods for size characterization will be reviewed. The usefulness of these methods as applied to liposomes will be described.

Conceptual Mapping of Liposome Preparation Methods

The procedures for liposome preparation can be generalized and divided into three stages: first, preparation of the aqueous and lipid phases; second, primary processing involving lipid hydration; and third, secondary processing steps essential in some cases and optional in others. The secondary processing steps can be used separately or combined, in some cases. Based on these three stages it is possible to categorize all existing preparation methodologies. Such a conceptual mapping helps to clarify the relationship between methods and thus the choices that must be made when choosing one. A categorization is presented below with notes on the critical elements involved in each step and the specific method to which it relates.

Preparation of Molecular Mixtures of Components

Preparation of Aqueous Phase. Aqueous buffer, drug, chelator, osmolarity, ionic strength, pH, and concentration of material to be encapsulated have to be adjusted in relation to the medium in which the liposomes will be suspended after formation. For most detergent solubilization methods, detergent is added to the aqueous phase.

Preparation of Molecular Mixture of Lipids. The lipid material is dissolved in a suitable (organic) solvent or mixture which can be either miscible or immiscible with the aqueous phase, depending on the method to be used.

For many methods, a dry lipid mixture is produced by removing the solvent. The form of the dry lipid (powder, film, etc.) affects the liposomes formed and is determined by both the solvent removal process and the surface used for drying the lipid. The solvent can be removed either by solvent vaporization [classic multilamellar vesicle (MLV)], by solvent sublimation (freeze drying), or by spray drying.[12-14]

[12] F. J. T. Fildes, *in* "Research Monographs in Cell and Tissue Physiology. Liposomes: From Physical Structure to Therapeutic Applications" (C. G. Knight, ed.), Vol. 7, p. 465. Elsevier, Amsterdam, 1981.

[13] T. E. Thompson, B. R. Lentz, and Y. Barenholz, *FEBS Symp.* **42,** 47 (1985).

[14] G. Redziniak and A. Meybeck, U.S. Patent 4,508,703 (1985).

Primary Processing (Hydration of Lipids: Liposome Formation)

Hydration of Dry Powder or Film with Aqueous Phase. After removing the solvent, dry lipid mixtures can be dispersed into the aqueous phase by shaking or vortexing, forming the now classic MLV (see *Hydration of Dry Lipid Films*). Large (LUV) or small (SUV) unilamellar vesicles can be formed spontaneously by controlling the lipid components and/or the form of the dry lipid mixture. This is discussed in detail below (see *Spontaneous Vesiculation*). Detergent solubilization of lipid, either dry mixture or previously hydrated, produces mixed micelles of detergent and lipids. Controlled selective removal of the detergent results in lipid reorganization into unilamellar vesicles. Details of the controlling parameters are discussed below (see *Detergent Solubilization and Removal*).

Replacement of Immiscible Organic Solvent with Aqueous Phase. When lipids are dissolved in aqueous immiscible solvents they can form an emulsion in the aqueous phase. Removal of the organic solvent under the proper conditions forms LUVs. This is called the reverse phase evaporation (REV) procedure which is discussed below with several modified versions (see *Reverse Phase Evaporation and Solvent Vaporization;* see also *Large Unilamellar Vesicles, Reverse Phase Evaporation*).

Hydration by injection of the lipid solution into the aqueous phase, maintained under a reduced pressure to vaporize the solvent, produces oligolamellar vesicles (OLVs), LUVs, or MLVs, depending on the parameters. This technique will be described in more detail below (see *Reverse Phase Evaporation and Solvent Vaporization;* see also *Volatile Solvent Injection*).

Replacement of Miscible Organic Solvent with Aqueous Phase. Injection of aqueous miscible lipid solutions, typically ethanol, into the aqueous phase under the proper conditions will form SUVs. Strategies for removal of the solvent include dialysis, ultrafiltration, and gel permeation chromatography. This method will be discussed in greater detail [see *Solvent (Ethanol) Injection*].

Secondary Processing (Size Changes after Liposome Formation)

Mechanical Fragmentation. Many methods exist for fragmentation of MLVs into particles that are smaller and/or unilamellar. Ultrasonication of an MLV dispersion converts the lipid into minimally sized SUVs. Passage through the French Press cell also converts MLVs into SUVs. Homogenization of MLVs can form either SUVs or OLVs, depending on conditions and number of passes used. Two types of equipment, microfluidizers available from BDC or Microfluidics and traditional homogenizers from Gaulin, have been used. Extrusion of MLV dispersions through filters with defined pores can be used to reduce the particle size to approximately that of the pore and to reduce the number of lamellae. LUVs can be formed in some cases. [See *Description of Specific Methods* for detailed descriptions of these processes (pp. 203 and 208).]

Physicochemically Induced Fragmentation. By temporarily altering the chemical nature of the aqueous phase, the physical properties of the head group are changed, dramatically altering the organization of the lipid dispersions. One manifestation of such a method is based on hydration changes occurring on protonation or deprotonation of the lipid head group. Addition of sufficient base to alter the protonation of head groups followed by a return to neutral pH will convert MLVs containing an acidic lipid into LUVs. The details of this procedure are described below [see *Description of Specific Methods* (p. 207)].

Fusion-Mediated Size Increase of SUVs. SUVs are unstable, spontaneously increasing in size during storage. This results in LUVs in some cases and large aggregates in others. Freezing and thawing SUVs (or LUVs) will induce the release of entrapped aqueous contents. Simultaneously, larger particles are formed which are typically multilamellar. Addition of ions or other materials can induce aggregation and fusion, resulting in LUVs in some cases. Most commonly, divalent cations are added to formulations with acidic lipids, resulting in cochleated cylinders which convert to LUVs upon addition of chelators. Addition of detergents and subsequent removal can convert SUV to LUV. [See *Description of Specific Methods* for details (p. 208).]

Dehydration–Rehydration–Induced Size Increases (MLVs). Dehydration of the lipid head groups can be achieved by evaporation, lyophilization, or even freezing. Reconstitution (rehydration of the dry lipid in the case of evaporation, and lyophilization and thawing in the case of freezing) results in substantial changes in the particle structure. Typically, SUVs increase in size, forming OLVs or MLVs. Freeze–thaw cycling of MLVs increases the aqueous entrapment, but with only minimal changes in overall size distribution. [See *Description of Specific Methods* for details (pp. 199 and 200).]

Guide for Choice of Method

The methodology for liposome preparation has evolved rapidly during the last few years as a response to the need for preparing well-defined liposomes for specific applications. As a consequence, the value of each method or each combination of methods depends entirely on the intent of the investigator and the requirements of their intended use. It follows that, depending on the specification at hand, some methods are much more suitable than others. The critical appraisal that follows is based on a broad categorization of various types of liposomes and some specific usages. Specific details and references for each method will be given in the following section. As we discuss the methods of choice, we will also define whether they are particularly suitable for laboratory scale (1.0–100 ml of 10–100 mg/ml), for the relatively larger scale required for clinical testing (up to 10 liters of 100–300 mg/ml), or for industrial-scale production (more than 10 liters of up to 400 mg/ml).

Multilamellar Liposomes

If there is no restriction or specific preference concerning geometry or size, the MLV form of liposomes is especially advantageous when lipophilic material is to be encapsulated, or if there is no concern with the efficiency of aqueous encapsulation. For laboratory-scale production, the classic "thin film hydration" method is adequate. Its usefulness can be extended by subsequent "extrusion" through defined pore membranes, which produces a reasonably narrow size distribution. For a larger scale, other strategies should be considered: alternative drying methods, such as lyophilization, spray drying, or solvent injection and vaporization, all of which can be optimized for large-scale production.

If the efficiency of encapsulation is a concern, and there is a desire for high encapsulation of water-soluble solutes, the following methods should be considered: (1) reverse phase evaporation; (2) one of the freeze–thaw or dehydration–rehydration methods followed by extrusion; or (3) extrusion, homogenization, or solvent injection and vaporization under conditions giving high final lipid concentrations. The latter methods (3) are particularly well-suited for preparations of relatively large amounts of liposomes, such as 0.1- to 1.0-liter volumes, while the former methods (1 and 2) are best for laboratory scale.

Small Unilamellar Liposomes

The small size, relatively narrow size distribution, and high surface-to-volume ratio of SUVs makes them a good choice for some applications, especially if there is no need for efficient encapsulation of water-soluble solutes. For laboratory-scale preparation, sonication, French Press, or ethanol injection methods should be considered. The choice between these depends on the importance of the specific limitations of each, such as exposure to ultrasonic irradiation, residual ethanol, or limited access to equipment (such as a French Press). For larger scale, up to a few liters, homogenization procedures produce fairly small liposomes after repeated passage. These may not be unilamellar, but can be made at high enough lipid concentrations to achieve reasonable capture of water-soluble solutes. Alternatively, ethanol injection on a large scale should be considered when an ethanol content of up to 10% is acceptable. In fact, this level of ethanol may be of interest as a preservative in some situations where the resulting liposome properties, such as increased bilayer fluidity and aqueous leakage, are not a problem.

Large Unilamellar Liposomes

These liposomes are particularly well suited for efficient encapsulation of water-soluble solutes because of their favorable aqueous volume-to-surface ratio. They are also most desirable for membrane protein reconstitution and other studies of membrane permeability and fusion. The detergent-removal methods are most commonly used for membrane protein reconstitution, while the reverse phase evaporation is the most efficient procedure for high degrees of encapsulation of water-soluble solutes and macromolecules. Convenient commercial apparatus for laboratory-scale preparations has recently been introduced for the detergent dialysis method (Dupont, Wilmington, Delaware). Extrusion is recommended for secondary processing to achieve a narrower size distribution. When extrusion is used under high pressure, with high lipid concentration and small pore-size membranes, it produces liposomes with both high capture and a very homogeneous size distribution.

Although good methods for large-scale production of LUVs particularly at the industrial scale, are currently lacking, all the possibilities have not been exhausted and solutions are expected, perhaps even in the near future. For example, modification of existing methods, such as reverse phase evaporation, into continuous processes, rather than simply increasing the batch size, have yet to be fully explored. Other strategies for modifying existing methods or development of entirely new methods may provide larger scale LUV production.

Description of Specific Methods

Multilamellar Vesicles

Hydration of Dry Lipid Films. Liposomes prepared by shaking or vortexing dried lipids with the aqueous phase were the first to be characterized as closed vesicles composed of multiple concentric lipid bilayers (MLVs).[15-17] These liposomes are very heterogeneous in size[18] and have a

[15] A. D. Bangham, M. M. Standish, and J. C. Watkins, *J. Mol. Biol.* **13,** 238 (1965).
[16] A. D. Bangham, J. de Gier, and G. D. Greville, *Chem. Phys. Lipids* **1,** 225 (1967).
[17] A. D. Bangham, M. W. Hill, and N. G. A. Miller, *Methods Membr. Biol.* **1,** 1 (1974).

low level of aqueous encapsulation. However, because the method is exceptionally simple and the MLVs form spontaneously, it is used extensively as the hydration step, frequently followed by one or more secondary processing steps such as extrusion or sonication.

Many studies have demonstrated that lipid composition and concentration as well as the mode of hydration affect the final product. The mode of hydration may be the most important factor.[17,18] The usual approach has been to use a large surface area to dry the lipid and hence to reduce the thickness of the lipid film. The method for drying the lipid, such as spray drying, also has shown that the exact form of the dried lipid can influence the resulting liposomes.[12-14] Nevertheless, the full extent of the role played by the form of the dry lipid has yet to be determined. Many other factors are also important, such as initial moisture content, aqueous phase used, final lipid concentration, lipid composition, the actual hydration process, and time for hydration.[18] The principal problem in producing MLVs by shaking or vortexing is reproducible control of all the parameters. Often the form of dry lipid and the hydration process vary uncontrollably, making it difficult to achieve a well-defined preparation.

Dehydration–Rehydration. An approach which improves entrapment of aqueous solutes by MLV preparations is to dry the lipid in the presence of the aqueous solute to be entrapped, thus forming a mixed lipid film with solute trapped between layers.[19,20] This mixed film is then hydrated gradually with a minimum of aqueous phase. Interestingly, the mixed film hydrates differently when certain aqueous solutes are used.[20] The result is a convenient method for producing MLVs with a higher percentage of entrapped aqueous phase. In the usual procedure, which is denoted dehydration–rehydration, preformed liposomes are added to an aqueous solution of the solute and the mixture is either lyophilized[19] or evaporated[20] with similar results. SUVs may be preferable to MLVs, but both have been used. Subsequent rehydration forms MLVs with a high level of entrapment of the aqueous solutes added. Consequently, this method is useful for entrapping a wide variety of aqueous components, especially proteins which can be freeze-dried, without the need for exposure of aqueous solutes to organic solvents or detergents. Another manifestation of this technique is to add the aqueous solution to an ethanolic solution of the lipid, which is then dried.[21]

---

[18] F. Olson, C. A. Hunt, F. C. Szoka, W. J. Vail, and D. Papahadjopoulos, *Biochim. Biophys. Acta* **557**, 9 (1979).

[19] C. J. Kirby and G. Gregoriadis, *in* "Liposome Technology" (G. Gregoriadis, ed.), Vol. 1, p. 19. CRC Press, Boca Raton, Florida, 1984.

[20] R. L. Shew and D. W. Deamer, *Biochim. Biophys. Acta* **816**, 1 (1985).

[21] S. M. Gruner, R. P. Lenk, A. S. Janoff, and M. J. Ostro, *Biochemistry* **24**, 2833 (1985).

In most cases, gradual controlled rehydration of the resulting material with a minimal aqueous volume gives a heterogeneous population of MLVs but with a higher percentage of entrapped aqueous solutes. Subsequent processing, such as extrusion,[18] is recommended to reduce the size of the particles and improve the homogeneity.

Freezing alone may essentially dehydrate the lipid head groups,[22] with rehydration occurring upon thawing. Therefore, freeze–thaw cycles are effectively a form of dehydration–rehydration. This method is discussed in the next section.

Freeze–Thaw Processing of MLVs. Repeated freezing followed by thawing is a convenient method for increasing the trapped volume of MLV preparations or inducing leakage of SUVs and LUVs.[23-27] The bilayer structure is dramatically altered as determined by freeze-fracture electron microscopy (EM) and nuclear magnetic resonance (NMR) studies.[26] A likely explanation may involve a combination of both head group dehydration and physical disruption of the multilamellar structure by ice crystals formed during freezing. These result in improved hydration of the lipid upon thawing. However, despite the appearance of an isotropic component in NMR spectra attributed to small liposomes, the overall size distribution is not significantly decreased.[27] Both the lipid and aqueous compositions influence the process such that the compositions which reduce membrane damage on freezing also diminish the effects of the freeze–thaw procedure. The presence of cryoprotectants and nonionic aqueous solutes reduce membrane damage.[28,29] Interestingly, entrapped glycoprotein from serum also exhibits protection during freezing. Low salt concentrations and cholesterol are also stabilizing.[29] Therefore, each formulation may behave differently and this must be considered and tested.

Reverse Phase Evaporation and Solvent Vaporization. Both of these general procedures were developed to produce LUVs,[30,31] but under many

[22] S. W. Hui, T. P. Stewart, L. T. Boni, and P. L. Yeagle, *Science* **212**, 921 (1981).

[23] R. Rendi, *Biochim. Biophys. Acta* **135**, 333 (1967).

[24] D. Siminovitch and D. Chapman, *FEBS Lett.* **16**, 207 (1971).

[25] U. Pick, *Arch. Biochem. Biophys.* **212**, 186 (1981).

[26] L. D. Mayer, M. J. Hope, P. R. Cullis, and A. S. Janoff, *Biochim. Biophys. Acta* **817**, 193 (1985).

[27] M. J. Hope, M. B. Bally, L. D. Mayer, A. S. Janoff, and P. R. Cullis, *Chem. Phys. Lipids* **40**, 89 (1986).

[28] L. M. Crowe, C. Womersley, J. H. Crowe, D. Reid, L. Appell, and A. Rudolph, *Biochim. Biophys. Acta* **861**, 131 (1986).

[29] G. Strauss, *in* "Liposome Technology" (G. Gregoriadis, ed.), Vol. 1, p. 197. CRC Press, Boca Raton, Florida, 1984.

[30] D. Deamer and A. D. Bangham, *Biochim. Biophys. Acta* **443**, 629 (1976).

[31] F. Szoka and D. Papahadjopoulos, *Proc. Natl. Acad. Sci. U.S.A.* **75**, 4194 (1978).

conditions MLVs are actually formed.[21,32] The specific conditions under which these methods can be used to prepare LUVs are discussed below (see *Large Unilamellar Vesicles*). Nevertheless, these methods are very useful for MLV production, either as the hydration step to be followed by secondary processing or for situations where heterogeneous MLVs are adequate.

Solvent injection and vaporization are especially good methods for situations requiring large-scale hydration of lipid and a relatively large percentage of aqueous entrapment, even though OLVs or MLVs may be formed.[33-36] The reverse phase evaporation method is most useful on the laboratory scale.[31] Claims of unique stability have been made[21] for MLVs produced by a specific modification of the REV technique, and to indicate this they have been given a special name, stable plurilamellar vesicles (SPLVs). The SPLV method consists of bath sonicating an emulsion of aqueous phase in an ether solution of lipid while evaporating the ether.[21] Another method using phase reversal produces multivesicular liposomes.[37,38]

Unilamellar Vesicles

Unilamellar vesicles are liposomes with a single bilayer surrounding the entrapped aqueous phase. The term small unilamellar vesicles is used for minimally sized liposomes. The minimum size (200–250 Å diameter), usually attributed to a maximum of curvature allowed by the packing of lipids in the bilayer, results in a low amount of entrapped aqueous volume per lipid. As the size increases, the amount of entrapped aqueous phase per mole of lipid increases.[39] Stability of size also increases with size, usually attributed to a decrease in the amount of curvature and resulting strain.[39-42] Unilamellar liposomes with diameters of 100 nm and larger are generally referred to as large unilamellar vesicles. In practice, however,

[32] C. Pidgeon, F. McNeely, T. Schmidt, and J. Johnson, *Biochemistry* **26**, 17 (1987).
[33] D. Deamer, in "Liposome Technology" (G. Gregoriadis, ed.), Vol. 1, p. 29. CRC Press, Boca Raton, Florida, 1984.
[34] D. S. Cafiso, H. R. Petty, and H. M. McConnell, *Biochim. Biophys. Acta* **649**, 129 (1981).
[35] H. Schieren, S. Rudolph, M. Finkelstein, P. Coleman, and G. Weissmann, *Biochim. Biophys. Acta* **542**, 137 (1978).
[36] F. J. Martin and G. West, U.S. Patent Appl. Serial No. 908765 (1986).
[37] S. Kim and S. B. Howell, *Cancer Treat. Rep.* **71**, 705 (1987).
[38] S. Kim, M. S. Turker, E. Y. Chi, S. Sela, and G. M. Martin, *Biochim. Biophys. Acta* **728**, 339 (1983).
[39] D. Lichtenberg, E. Freire, C. F. Schmidt, Y. Barenholz, P. L. Felgner, and T. E. Thompson, *Biochemistry* **20**, 3462 (1981).
[40] M. P. Sheetz and S. I. Chan *Biochemistry* **11**, 4573 (1972).
[41] C. Huang and J. T. Mason, *Proc. Natl. Acad. Sci. U.S.A.* **75**, 308 (1978).
[42] B. R. Lentz, T. J. Carpenter, and D. R. Alford, *Biochemistry* **26**, 5389 (1987).

oligolamellar vesicles, which have a large entrapped volume per lipid but with several bilayers widely separated, are often described as unilamellar due to a difficulty in distinguishing between the two. In some procedures,[43,44] unilamellar vesicles with diameters as large as 10 μm have been made and are called giant unilamellar vesicles (GUVs). Despite variation in terminology in the literature, the distinctions between SUVs, LUVs, OLVs, and GUVs are becoming standardized.[5,7,45]

Small Unilamellar Vesicles

Sonication. Sonication of MLVs or other aqueous dispersions of phospholipids produces small particles[46,47] that have been shown to be SUVs that entrap an aqueous space.[48] Extensive characterization of these minimally sized SUVs[49] has greatly enhanced the understanding of lipid bilayers, liposomes, and biological membranes in general.[17,50] Because of their minimal size, these liposomes are a relatively homogeneous population of unilamellar vesicles, but they are strained[40,41,51] and are likely to undergo spontaneous slow transformations to larger vesicles.[40,44,52-55] High-speed centrifugation has been used as a simple but effective procedure to remove any larger liposomes.[56] The principal disadvantages of SUVs in general are a relatively low percentage of aqueous entrapment and an incompatibility for sonication with many compounds, including some lipids.[57] Sonication under an inert atmosphere has been used to reduce lipid damage,[49] although this is not always successful, especially with probe sonicators.[57] Milder sonication using lower power bath sonicators may

[43] S. Kim and G. M. Martin, *Biochim. Biophys. Acta* **646**, 1 (1981).
[44] N. Oku and R. C. MacDonald, *Biochemistry* **22**, 855 (1983).
[45] D. Papahadjopoulos and W. J. Vail, *Ann. N.Y. Acad. Sci.* **308**, 259 (1978).
[46] L. Saunders, J. Perrin, and D. B. Gammack, *J. Pharm. Pharmacol.* **14**, 567 (1962).
[47] M. B. Abramson, R. Katzman, and H. P. Gregor, *J. Biol. Chem.* **239**, 70 (1964).
[48] D. Papahadjopoulos and N. Miller, *Biochim. Biophys. Acta* **135**, 624 (1967).
[49] C. Huang, *Biochemistry* **8**, 344 (1969).
[50] L. B. Margolis, *Biochim. Biophys. Acta* **779**, 161 (1984).
[51] T. E. Thompson, C. Huang, and B. J. Litman, *in* "The Cell Surface in Development" (A. A. Moscona, ed.), p. 1. Wiley, New York, 1974.
[52] J. Suurkuusk, B. R. Lentz, Y. Barenholz, R. L. Biltonen, and T. E. Thompson, *Biochemistry* **15**, 1393 (1976).
[53] A. L. Larrabee, *Biochemistry* **18**, 3321 (1979).
[54] M. Wong, F. H. Anthony, T. W. Tillack, and T. E. Thompson, *Biochemistry* **21**, 4126 (1982).
[55] R. A. Parente and B. R. Lentz, *Biochemistry* **23**, 2353 (1984).
[56] Y. Barenholz, D. Gibbes, B. J. Litman, J. Goll, T. E. Thompson, and E. D. Carlson, *Biochemistry* **16**, 2806 (1977).
[57] H. Hauser, *Biochem. Biophys. Res. Commun.* **45**, 1049 (1971).

have an advantage of reducing lipid damage.[58,59] However, the bath sonicators have several disadvantages: they require much longer times; sample sizes, containers, and placement in the bath are often critical so that reproducibility is hard to achieve; and the residual concentration of large particles is greater.[1] The model G112SP1T bath sonicator (Laboratory Supplies Co., Hicksville, New York) has an excellent reputation, but in our experience the actual bath component has a short lifetime. Finally, for large-scale production, sonication is severely limited by difficulties in achieving uniform results and in scale up.

French Press. Extrusion of MLV dispersions through a French Press is a gentle method for producing SUVs.[60-63] The average size of the resulting particles is slightly larger and the heterogeneity of the preparations with respect to size is greater than that obtained by sonication. Thus the liposomes may well be more stable but the reproducibility of the method has yet to be established.

Homogenization. Homogenization of MLVs or other lipid dispersions also produces SUVs in some circumstances. Several different types of equipment have been used: microfluidizers such as those supplied by Microfluidics and Biotechnology Development Corporation[64] and homogenizers such as those supplied by Gaulin.[65] Unfortunately, little data have been published and much of that is only limited data in support of a patent. The principal consideration is the shear force generated, temperature of operation, and lipid fluidity at that temperature.[64,66] The resulting SUVs are larger than the minimal size produced by sonication and thus may be more stable. However, significant amounts of larger particles are also present.

In the case of the microfluidizer developed by Biotechnology Development Corporation, the instrument is based on a high-pressure collision of two feed streams containing MLVs. Despite only limited data, the resulting

[58] D. Papahadjopoulos and H. K. Kimelberg, *in* "Progress in Surface" (S. G. Davidson, ed.), Vol. 4, Part 2, p. 141. Pergamon, Oxford, 1973.

[59] E. Racker, *Biochem. Biophys. Res. Commun.* **55**, 224 (1973).

[60] Y. Barenholz, S. Amselem, and D. Lichtenberg, *FEBS Lett.* **99**, 210 (1979).

[61] R. L. Hamilton, J. Goerke, L. S. S. Guo, M. C. Williams, and R. J. Havel, *J. Lipid Res.* **21**, 981 (1980).

[62] R. L. Hamilton and L. S. S. Guo, *in* "Liposome Technology" (G. Gregoriadis, ed.), Vol. 1, p. 37. CRC Press, Boca Raton, Florida, 1984.

[63] P. I. Lelkes, *in* "Liposome Technology" (G. Gregoriadis, ed.), Vol. 1, p. 51. CRC Press, Boca Raton, Florida, 1984.

[64] E. Mayhew, R. Lazo, W. J. Vail, J. King, and A. M. Green, *Biochim. Biophys. Acta* **775**, 169 (1984).

[65] R. J. Klimchak and R. P. Lenk, *Biopharmacology* **1**, 18 (1988).

[66] R. C. Gamble, Eur. Patent Appl. 0-190-050 (1986).

size clearly depends on the pressure used and the number of passes through the device.[64] However, even after 40 recycles at the maximum pressure, the resulting liposomes contained a large quantity of larger particles. Whether the size distribution was bimodal or a single population skewed toward large particles was not determined.[5] The microfluidizer can be easily scaled up to an industrial scale, but it is only applicable to situations where SUVs are desirable and even then a second processing step may be necessary if the presence of larger particles cannot be tolerated.

Solvent (Ethanol) Injection. Injection of lipid solutions in a solvent miscible with the aqueous phase is a simple method for preparing SUVs. However, it is limited by the need for subsequent processing to remove the solvent, the inevitability of residual solvent, and the low solubility of some lipid components in aqueous miscible solvents such as ethanol. Ethanol was the first solvent used for this method and is still the most common.[67,68] The size of the resulting liposomes is dependent on several parameters: relatively homogeneous SUVs can be produced by keeping a low lipid concentration, a fast rate of injection, and keeping the aqueous phase above the T_c phase transition of the lipid.[68] Removal of ethanol from the resulting liposomes by gel permeation chromatography may be more effective than dialysis, but still does not remove it completely.[69] For some applications, the residual solvent may not be a limitation, as, for example, when using ethanol with very low levels of impurities.

Spontaneous Vesiculation. Small unilamellar vesicles can also be formed spontaneously during the hydration of certain lipids, such as a mixture of short- and long-chain lecithins[70,71] or lecithin with lysolecithin.[72] A recent report also indicates that the surface on which the phospholipids are deposited during dehydration can contribute to spontaneous formation of SUVs upon hydration, without sonication or application of shearing forces.[73] In a related procedure, micelles are homogeneously converted to SUVs by acyl transfer to lysophosphatidylcholine (lyso-PC).[74]

[67] S. Batzri and E. D. Korn, *Biochim. Biophys. Acta* **298**, 1015 (1973).
[68] J. M. H. Kremer, M. W. J. Esker, C. Pathmamanoharan, and P. H. Wiersema, *Biochemistry* **16**, 3932 (1977).
[69] J. R. Norlund, C. F. Schmidt, S. N. Dicken, and T. E. Thompson, *Biochemistry* **20**, 3237 (1981).
[70] N. E. Gabriel and M. F. Roberts, *Biochemistry* **23**, 4011 (1984).
[71] N. E. Gabriel and M. F. Roberts, *Biochemistry* **25**, 2812 (1986).
[72] H. Hauser, *Chem. Phys. Lipids* **43**, 283 (1987).
[73] D. D. Lasic, J. Kidric, and S. Zagorc, *Biochim. Biophys. Acta* **896**, 117 (1987).
[74] V. Gavino and D. Deamer, *J. Bioenerg. Biomembr.* **14**, 513 (1982).

Small or Large Unilamellar Vesicles

Detergent Solubilization and Removal. The method of detergent solubilization and removal was originally developed for purification and reconstitution of membrane proteins giving rise to particles or vesicles that were eventually described as liposomes.[75-78] With this procedure, either dried lipid mixtures or preformed liposomes are solubilized with the detergent-containing aqueous phase. In either case the critical step is the detergent removal from the mixed micelles. By controlling the rate of detergent removal, the size and structure of the resulting liposomes can be controlled, but the actual dependence on lipids, detergent(s), and conditions are not clear. Despite advances in development of the theory,[5] the current approach is largely empirical.[79-82] Nevertheless, by carefully controlling the detergent removal, SUVs as small as 60 nm or LUVs of 100 nm can be produced.[79,83] Recent introduction of a commercial apparatus by Dupont may facilitate use of this method on the laboratory scale.

One of the principal problems with this method is removal of residual detergent.[7,84] Several methods have been explored based on differences between lipids and detergents. Detergent dialysis is based on a substantial concentration of detergent monomers, approximately the CMC, compared to a very low concentration of lipid monomers. Therefore, dialysis of the detergent is considerably faster than that of the lipid.[77-79,85] The residual detergent is determined by the dialysis conditions used as well as the detergent and lipids. Dilution is based on the same difference between detergent and lipid monomer concentration. When mixed micelles are diluted below the CMC for the detergent, all the detergent leaves the micelles, forcing the lipid to reorganize into bilayers. However, the detergent monomer still must be removed. A couple of removal strategies are

[75] S. Fleischer and H. Klouwen, *Biochem. Biophys. Res. Commun.* **5**, 378 (1961).

[76] M. Kasahara and P. Hinkle, *Proc. Natl. Acad. Sci. U.S.A.* **73**, 396 (1976).

[77] Y. Kagawa and E. Racker, *J. Biol. Chem.* **246**, 5477 (1971).

[78] S. Razin, *Biochim. Biophys. Acta* **265**, 241 (1972).

[79] H. G. Weder and O. Zumbuehl, *in* "Liposome Technology" (G. Gregoriadis, ed.), Vol. 1, p. 79. CRC Press, Boca Raton, Florida, 1984.

[80] R. E. Stark, G. J. Grosselin, J. M. Donovan, M. C. Carey, and M. F. Roberts, *Biochemistry* **24**, 5599 (1985).

[81] S. Almog, T. Kushnir, S. Nir, and D. Lichtenberg, *Biochemistry* **25**, 2597 (1986).

[82] P. Schurntenberger, R. Bertani, and W. Kanzig, *J. Colloid Interface Sci.* **114**, 82 (1986).

[83] M. H. Milsmann, R. A. Schwendener, and H. G. Weder, *Biochim. Biophys. Acta* **512**, 147 (1978).

[84] T. M. Allen, *in* "Liposome Technology" (G. Gregoriadis, ed.), Vol. 1, p. 109. CRC Press, Boca Raton, Florida, 1984.

[85] V. Rhoden and S. Goldin, *Biochemistry* **18**, 4173 (1979).

based on selective detergent removal. In one application, the mixed micelle solution is passed over a gel permeation column. As the particles pass down the column, the detergent monomers are retained, depleting the detergent and eventually forming liposomes, which are eluted.[86-88] In another method, specific absorption of detergent to beads is used.[88-91]

Recently, a different strategy, using detergents, was reported[92] for preparation of homogeneous LUVs containing membrane proteins. Well-defined preformed liposomes are added to solutions of detergent-solubilized protein, but without an excess of detergent. Presumably, the liposomes are not completely solubilized and their homogeneity is maintained. The preliminary data reported for bacteriorhodopsin indicate that not only were the resulting liposomes very homogeneous, but also a very high degree of specificity for orientation was found.

Large Unilamellar Vesicles

Reverse Phase Evaporation. The reverse phase evaporation technique[31] has become a widely accepted method for producing LUVs with a high percentage of aqueous phase entrapment on a laboratory scale.[7] The actual type and size of liposomes produced depends on a number of factors, such as choice of lipids (percentage cholesterol and charged lipids), lipid concentration used in the organic solvent, choice of solvent, ratio of organic to aqueous volumes in the emulsion, rate of evaporation, and ionic strength of the aqueous phase.[1] By controlling these parameters it is possible to produce LUVs or MLVs (or SPLVs) reproducibly.[32,93-96] By combining the procedure with extrusion, the final size can be selected over a wide range.[96,97] However, scale up of the method for LUVs is limited due

[86] J. Bruner, P. Skrabal, and H. Hauser, *Biochim. Biophys. Acta* **455**, 322 (1976).

[87] H. G. Enoch and P. Strittmatter, *Proc. Natl. Acad. Sci. U.S.A.* **76**, 145 (1979).

[88] T. M. Allen, A. Y. Romans, H. Kercret, and J. P. Segrest, *Biochim. Biophys. Acta* **601**, 328 (1980).

[89] P. W. Holloway, *Anal. Biochem.* **53**, 304 (1973).

[90] W. J. Gerritsen, A. J. Verkleij, R. F. Zwaal, and L. L. M. Van Deenen, *Eur. J. Biochem.* **85**, 255 (1978).

[91] P. S. J. Cheetam, *Anal. Biochem.* **92**, 447 (1979).

[92] M. Paternostre and J. L. Rigaud, *Biophys. J.* **53**, 504a (1988).

[93] J. N. Weinstein, R. L. Magin, M. B. Yatvin, and D. S. Zaharko, *Science* **204**, 188 (1979).

[94] C. G. Knight (ed.), "Research Monographs in Cell and Tissue Physiology. Liposomes: From Physical Structure to Therapeutic Applications," Vol. 7. Elsevier, Amsterdam, 1981.

[95] F. C. Szoka and D. Papahadjopoulos, *in* "Research Monographs in Cell and Tissue Physiology. Liposomes: From Physical Structure to Therapeutic Applications" (C. G. Knight, ed.), Vol. 7, p. 51. Elsevier, Amsterdam, 1981.

[96] N. Duzgunes, J. Wilschut, K. Hong, R. Fraley, C. Perry, D. S. Friend, T. L. James, and D. Papahadjopoulos, *Biochim. Biophys. Acta* **732**, 289 (1983).

[97] F. Szoka, F. Olson, T. Heath, W. Vail, E. Mayhew, and D. Papahadjopoulos, *Biochim. Biophys. Acta* **601**, 559 (1980).

to a requirement for a small surface-area-to-volume ratio. Other limitations of the method include sonication of lipids during emulsion formation, limited solubility of lipids and instability of some proteins in the solvent, and a need to eliminate the peroxide impurities in ether that may degrade the lipids. These factors are not necessarily limiting for every situation and can be circumvented in some cases. For example, sonication to form the emulsion has been replaced with other methods.[98]

Volatile Solvent Injection. Injection of solutions of lipid in volatile solvents into an aqueous phase under reduced pressure is a very convenient method for controlled hydration of lipids even on larger scales.[30,34-36,99,100] In general, the aqueous phase is maintained above the phase transition of the lipids (T_c) and at a reduced pressure to vaporize the lipid solvent during the injection. This method can produce LUVs with a relatively high percentage of entrapped aqueous volume per mole of lipid.[27,33] However, OLVs and MLVs can also form, depending on the conditions. A slow rate of injection combined with rapid solvent removal and low lipid concentration in the solvent favors LUV formation. The total percentage of aqueous volume entrapped can be as high as 65% when the final aqueous lipid concentration is high.[35] Unfortunately, a complete understanding of the relationship between process conditions and the lamellarity of the resulting liposomes is lacking. Solubility of the lipids in the solvent can be a severe limitation for some lipids. The solvent used originally, ethyl ether, is also limited by its tendency to decompose into peroxides, which are not volatile and induce lipid damage. For these reasons use of other volatile solvents has been explored.[34-36] When an appropriate solvent is found the method has many advantages. Even when mostly OLVs or MLVs are formed, the preparation can be extruded at high lipid concentration to produce liposomes with a high aqueous entrapment efficiency and with the size controlled over a wide range.[27,97]

Transient pH Jump. A recently developed method for reorganizing MLVs into LUVs takes advantage of lipid hydration and structural changes following deprotonation of acidic lipids.[101-104] Currently this

[98] P. Machy and L. D. Leserman, *in* "Liposome Technology" (G. Gregoriadis, ed.), Vol. 1, p. 221. CRC Press, Boca Raton, Florida, 1984.

[99] M. J. Ostro, D. Giacomoni, D. Lavelle, W. Paxton, and S. Dray, *Nature (London)* **274**, 921 (1978).

[100] M. J. Ostro, D. Lavelle, W. Paxton, B. Matthews, and D. Giacomoni, *Arch. Biochem. Biophys.* **201**, 392 (1980).

[101] H. Hauser and N. Gains, *Proc. Natl. Acad. Sci. U.S.A.* **79**, 1683 (1982).

[102] N. Gains and H. Hauser, *Biochim. Biophys. Acta* **731**, 31 (1983).

[103] T. S. Aurora, W. Li, H. Z. Cummins, and T. H. Haines, *Biochim. Biophys. Acta* **820**, 250 (1985).

[104] W. Li and T. H. Haines, *Biochemistry* **25**, 7447 (1986).

method has been demonstrated to work with only a few acidic lipids, but this restriction may decrease with further studies.[72] Characterization of the dependence of size and distribution on many parameters has yet to be completed, but some control may be possible.

Fusion. A number of fusion processes can alter the size of liposomes but most give rise to poorly defined particles and little control is possible. SUV are relatively unstable, especially when produced by sonication, and can spontaneously, but slowly, grow in size, at best becoming LUV[42] and at worst forming large aggregates. Several techniques to fuse SUV are available such as addition of specific mono or divalent ions to acidic liposomes,[105-109] incubation at temperatures at or below the T_c of the lipids,[54] or addition and subsequent removal of detergents.[81,87,110] However, with some exceptions[54,107,111] these methods are difficult to control and often the resulting liposomes are poorly defined. Another related method uses freeze thaw techniques similar to those applied to MLV (p. 200) to fuse SUV, forming LUV and GUV up to 10 μm in diameter.[112]

Unilamellar or Oligolamellar Vesicles

Extrusion through Filters with Well-Defined Pores. Extrusion of MLV dispersions through polycarbonate membranes with defined diameter pores results in a decrease in particle size and polydispersity.[18] The characteristics of the final liposomes depend on many factors, including lipid composition and charge, solvent composition, filter pore diameter, the pressure used for the extrusion, temperature, and number of passes used.[18,113] The influence of all the parameters is not clearly understood but the following conclusions can be made. With a pore size of 0.2 μm and larger, the resulting liposomes are not unilamellar nor is the population very homogeneous. However, by repeatedly extruding the sample the num-

[105] D. Papahadjopoulos, W. J. Vail, K. Jacobson, and G. Poste, *Biochim. Biophys. Acta* **394**, 483 (1975).

[106] D. Papahadjopoulos, W. J. Vail, C. Newton, S. Nir, K. Jacobson, G. Poste, and R. Lazo, *Biochim. Biophys. Acta* **465**, 579 (1977).

[107] J. Wilschut, N. Duzgunes, and D. Papahadjopoulos, *Biochemistry* **20**, 3126 (1981).

[108] N. Duzgunes and D. Papahadjopoulos, *in* "Membrane Fluidity in Biology" (R. I. Aloia, ed.), Vol. 2, p. 187. Academic Press, New York, 1983.

[109] N. Duzgunes, R. M. Straubinger, P. A. Baldwin, D. S. Friend, and D. Papahadjopoulos, *Biochemistry* **24**, 3091 (1985).

[110] P. Schurntenberger, N. Mazer, and W. Kanzig, *J. Phys. Chem.* **89**, 1042 (1985).

[111] S. Gould-Fogerite and R. J. Mannino, *Anal. Biochem.* **148**, 15 (1985).

[112] R. I. MacDonald and R. C. MacDonald, *Biochim. Biophys. Acta* **735**, 243 (1983).

[113] L. D. Mayer, M. J. Hope, and P. R. Cullis, *Biochim. Biophys. Acta* **858**, 161 (1986).

ber of lamellae can be reduced to less than five.[114,115] With repeated extrusion the heterogeneity of the size distribution also decreases, but this effect is diminished as the filter size increases. By using high pressure, up to 15 atm, MLV samples can be extruded directly through 0.1- or 0.05-μm filters to form LUVs. Interestingly, combining this technique with two stacked filters may improve the size homogeneity.[27] An onionskin model has been proposed to explain the results of the extrusion method.[116]

The use of extrusion as a secondary processing technique has many advantages. MLVs can be converted to well-defined oligolamellar vesicles over a large range of sizes (0.2 to 0.5 μm) controlled by the pore size of the filter used, or to LUVs from 0.1 to 0.05 μm. Extrusion can be performed on a large scale, limited primarily by the size of filters and the availability of apparatus which can withstand high pressures. Identifying filter failure often presents a difficulty, especially during the final passes when only a small percentage of large particles remain.

Critical Issues and Problems to Avoid

Aqueous Phase

The choice of aqueous phase must take into consideration several factors that depend on the desired final product: osmolarity and ionic strength, pH, choice of buffer and concentration, any aqueous component to be entrapped, and the maximum levels of contaminants such as metal ions and dust. Liposomes are osmotically active; they shrink or swell with changes in osmolarity.[16] In extreme cases this can result in loss of entrapped aqueous markers and large changes in size distribution. Minimally sized SUVs seem to be an exception to this in that they are not sensitive to osmolarity changes. Many ions, and most notably divalent cations, can interact specifically with the head group of negatively charged lipids, resulting in changes in zeta potential, which can lead to aggregation, fusion, or affect the interaction between liposomes and other surfaces, such as cells. Likewise, the pH can alter the ionization of lipids, giving rise to similar effects or even completely destabilizing the bilayer structure in some cases, such as is observed for pure phosphatidylethanolamine

[114] D. Lichtenberg and T. Markello, *J. Pharm. Sci.* **73**, 122 (1984).
[115] H. Talsma, H. Jousma, K. Nicolay, and D. J. A. Crommelin, *Int. J. Pharm.* **37**, 171 (1987).
[116] H. Jousma, H. Talsma, F. Spies, J. G. H. Joosten, H. E. Junginger, and D. J. A. Crommelin, *Int. J. Pharm.* **35**, 263 (1987).

(PE).[48,117-119] Furthermore, most lipids are subject to acid- or base-catalyzed hydrolysis. Lipid oxidation is also enhanced by aqueous components, especially some metal ions. Consequently, it is essential to control the osmolarity, ionic strength, and pH of the aqueous phase. In addition, antioxidants such as tocopherol, aqueous antibiotics or bacterial inhibitors, and metal chelators are often added to the aqueous phase to improve the overall stability of liposomes.

Organic Solvent

The choice of an organic solvent for solubilizing the lipids is based on two overriding considerations. First, all the lipid components must be soluble at the desired concentration so that a molecular mixture can be formed. Second, removal of the solvent from the lipids must be uniform so that the lipid components are uniformly distributed throughout the particles.[120-122] For example, many lipid components crystallize before others during the drying process. When this happens the film will be heterogeneous and in some cases the liposome preparations will contain heterogeneities or even residual crystals. In addition, common impurities or degradation products of some solvents are concentrated upon solvent removal and can lead to lipid degradation. This is especially true of peroxides formed by many ethers.

Lipid Choice

Although the choice for lipids depends on the particular application, the following guidelines generally relate to the factors that enter in choosing a particular lipid over another. Although liposomes can be made with any phospholipid, most work on liposomes has been done with phosphatidylcholine (PC) derived from hen egg yolks. This material is available in both large quantities and high purity and is probably the best characterized phospholipid. It is usually mixed with cholesterol at a 2 : 1 or 1 : 1 mole

[117] R. J. Y. Ho, B. T. Rouse, and L. Huang, *Biochem. Biophys. Res. Commun.* **138**, 931 (1986).

[118] H. Ellens, J. Bentz, and F. C. Szoka, *Biochemistry* **25**, 285 (1986).

[119] H. Ellens, J. Bentz, and F. C. Szoka, *Biochemistry* **25**, 4141 (1986).

[120] T. M. Estep, D. B. Mountcastle, R. L. Biltonen, and T. E. Thompson, *Biochemistry* **17**, 1984 (1978).

[121] T. M. Estep, D. B. Mountcastle, Y. Barenholz, R. L. Biltonen, and T. E. Thompson, *Biochemistry* **18**, 2112 (1979).

[122] D. Bach, Y. Barenholz, T. E. Thompson, and I. R. Miller, *Thermochim. Acta,* in press (1989).

ratio for increasing the stability and decreasing the permeability of liposomes both in buffer and in the presence of plasma proteins.[64,123-125]

A negatively charged phospholipid is added in many preparations in the form of phosphatidylglycerol (PG) or phosphatidylserine (PS) at a 10–30% mole ratio to PC. The inclusion of the acidic lipid decreases the tendency of PC liposomes to aggregate and can increase the efficiency of encapsulation of certain drugs either by improving the aqueous entrapment[16] or by specific charge interactions.[94] The negative charge also increases the efficiency of the uptake of liposomes by cells *in vitro*,[126,127] and *in vivo* decreases their half-life $(t_{1/2})$ in blood following iv injection.[128] The specific lipid used to impart the negative charge is important for some applications. Recent work indicates that when either ganglioside G_{M1} or phosphatidylinositol (PI) is added for the negative charge, the resulting liposomes tend to have long $t_{1/2}$ in blood.[129,130]

The fluidity of the liposome bilayer can be controlled by the choice of lipid, in terms of the number of double bonds and the length of the acyl chains. Synthetic PCs with long, saturated acyl chains, such as dipalmitylphosphocholine (DPPC) and distearylphosphocholine (DSPC) (with and without cholesterol) have been used to produce liposomes which are "rigid" at ambient temperatures. These show much lower permeability to various solutes, and increased stability and circulation time in blood.[124,128,131] However, such liposomes have to be prepared at a temperature above the solid-to-fluid phase transition (T_c) for each lipid (42 and 54°, respectively, for DPPC and DSPC). Once formed and annealed at high temperature (including during extrusion and other secondary processing), they can then be used at low temperature. Sphingomyelin is a natural phospholipid that has a high transition temperature[132] and also forms very stable liposomes.[133]

[123] D. Papahadjopoulos, M. Cowden, and H. K. Kimelberg, *Biochim. Biophys. Acta* **330**, 8 (1973).
[124] C. J. Kirby, J. Clarke, and G. Gregoriadis, *Biochem. J.* **186**, 591 (1980).
[125] L. S. S. Guo, R. L. Hamilton, J. Goerke, J. N. Weinstein, and R. J. Havel, *J. Lipid Res.* **21**, 993 (1980).
[126] J. Damen, J. Regts, and G. Scherphof, *Biochim. Biophys. Acta* **665**, 538 (1981).
[127] R. Fraley, R. M. Straubinger, G. Rule, L. Springer, and D. Papahadjopoulos, *Biochemistry* **20**, 6978 (1981).
[128] J. Senior, G. Gregoriadis, and K. A. Mitropoulos, *Biochim. Biophys. Acta* **760**, 111 (1983).
[129] T. M. Allen and A. Chonn, *FEBS Lett.* **223**, 42 (1987).
[130] A. Gabizon and D. Papahadjopoulos, *Proc. Natl. Acad. Sci. U.S.A.* **85**, 6949 (1988).
[131] J. Senior and G. Gregoriadis, *Life Sci.* **30**, 2123 (1982).
[132] M. Shinitzky and Y. Barenholz, *J. Biol. Chem.* **249**, 2652 (1974).
[133] T. M. Allen, *Biochim. Biophys. Acta* **640**, 385 (1981).

Liposomes become very permeable at temperatures close to the T_c[134] and this has been used to create temperature-sensitive liposomes for drug delivery to hyperthermic tumors.[84] The presence of cholesterol at high mole ratio abolishes the T_c and also inhibits the formation of a highly permeable state.[134-136]

For most phospholipids, liposomes can be formed at a wide pH range of 6-9 in the presence of 0.1 M monovalent ions. When liposomes are made of acidic phospholipids such as PS, PG, PI, or phosphatidic acid (PA), care should be taken to avoid divalent or polyvalent metals, which tend to bind strongly to negatively charged groups on the lipid, thus producing aggregation and fusion.[108] Addition of metal chelators can reduce this problem.

PE, when pure, is the only phospholipid that does not form closed vesicles at neutral pH and physiological ionic strength.[137] It will form liposomes however, at low ionic strength and high pH[137] or in combination with another component.[138-140] This property has been utilized to produce pH-sensitive liposomes, by incorporating PE or other amphiphilic molecules that have a carboxyl group that can be protonated as the pH drops from 7.0 to 6.0,[109,118,119] or to produce target-sensitive liposomes.[117]

Although most of the work with liposomes has been done with pure lipids or mixtures of pure lipids, certain impurities can be tolerated while still maintaining acceptable stability. For example, the small percentage of lyso-PC or fatty acids that accumulates during hydrolysis would be acceptable if there were no other interference with any of the specific functions contemplated. However, storage conditions at either low (≤ 6.5) or high pH (≥ 8.5) may generate enough hydrolytic products to destabilize liposomes, depending on the time and temperature. Oxidation is another difficult problem that results in unstable liposomes.[141-144] Since oxidation is an

[134] D. Papahadjopoulos, K. Jacobson, S. Hir, and T. Isac, *Biochim. Biophys. Acta* **311**, 330 (1973).

[135] E. Oldfield and D. Chapman, *FEBS Lett.* **23**, 285 (1972).

[136] M. K. Jain, *Curr. Top. Membr. Transp.* **6**, 1 (1975).

[137] D. Papahadjopoulos and J. C. Watkins, *Biochim. Biophys. Acta* **135**, 639 (1967).

[138] P. R. Cullis and B. DeKruijff, *Biochim. Biophys. Acta* **559**, 399 (1979).

[139] R. J. Y. Ho and L. Huang, *J. Immunol.* **134**, 4035 (1985).

[140] T. F. Tarashi, T. M. Van Der Steen, B. DeKruijff, C. Tellier, and A. J. Verkeij, *Biochemistry* **21**, 5756 (1982).

[141] T. Nakazawa and S. Nagatsuka, *Int. J. Radiat. Biol.* **38**, 537 (1980).

[142] D. A. Barrow and B. R. Lentz, *Biochim. Biophys. Acta* **645**, 17 (1981).

[143] T. Nakazawa, S. Nagatsuka, and T. Sakurai, *Int. J. Radiat. Biol.* **40**, 365 (1981).

[144] F. Ianzini, L. Guidoni, P. L. Indovina, V. Viti, G. Erriu, S. Onnis, and P. Randaccio, *Radiat. Res.* **98**, 154 (1984).

autocatalytic process accelerated by metal ions, both the purity of the starting material and the storage conditions (including antioxidants, chelation, and nitrogen or argon atmosphere) are to be monitored and controlled. This is particularly important for polyunsaturated fatty acids present in both egg yolk and soybean PC. In this respect, synthetic palmityloleylphosphatidylcholine (POPC), DPPC, and DSPC are much more stable to oxidative degradation and, if this is a crucial parameter, their use is recommended. In terms of stability to hydrolysis, sphingomyelins and ether analogs of PC (instead of the diacyl-PC) are much more stable than egg or soybean PC.

Liposomes can also be made of single-chain amphophiles, such as mixtures of long-chain fatty acids and alcohols[145] or nonphospholipid synthetic ether compounds,[146] which may have some advantages for certain applications that require increased stability to hydrolysis.

Finally, when liposomes are made to encapsulate lipophilic drugs that tend to associate with the bilayer, the stability and physicochemical characteristics of the system may be altered depending on the interaction between lipid and the drug.[94]

Methods for Particle-Size Determination

Particle-size distribution is an especially important parameter for liposomes. For example, it clearly influences the distribution of the liposomes within the body when they are administered parenterally.[147-151] However, despite the existence of many methods, there is no single method capable of accurate and reliable size distribution analysis of liposome samples. Except for electron microscopy, none of the methods is capable of resolving the distribution over the entire range of particle size required for liposomes, from 10 nm to 50 μm. In fact, many methods only determine the average size and provide little or no indication of the distribution. Particle shape can also be important, especially for larger particles, those greater than 1 μm. Therefore, comparison and compilations of results from

[145] W. R. Hargreaves and D. Deamer, *Biochemistry* **17**, 3759 (1978).
[146] A. J. Baillie, A. T. Florence, L. R. Hume, G. T. Muirhead, and A. Rogerson, *J. Pharm. Pharmacol.* **37**, 863 (1985).
[147] R. Juliano and D. Stamp, *Biochem. Biophys. Res. Commun.* **63**, 651 (1975).
[148] R. M. Abra and C. A. Hunt, *Biochim. Biophys. Acta* **666**, 493 (1981).
[149] T. M. Allen and J. M. Everest, *J. Pharmacol. Exp. Ther.* **226**, 539 (1983).
[150] Y. Sato, H. Kiwada, and Y. Kato, *Chem. Pharm. Bull.* **34**, 4244 (1986).
[151] F. C. Szoka, D. Milholland, and M. Barza, *Antimicrob. Agents Chemother.* **31**, 421 (1987).

many methods are necessary to fully characterize the sizes of liposomes within a particular preparation.

When choosing size measurement methods, several factors must be taken into consideration. Because liposomes are populations of particles, there is a distribution of size, rather than a single value, which can be expressed in a variety of ways. For instance, the percentage of mass, volume, or even surface area in each resolved size class can be determined. However, each of these may have a very different shape and calculated mean, even for the same distribution.[152]

While a single, absolute liposome size-distribution measurement does not exist, a variety of useful and reliable sets of information can be obtained. When a more specific question is asked, then the method or methods can be chosen more effectively. A few generalized situations have been addressed in the following discussion. Large liposomes, those greater than 1 μm, can be adequately measured by light microscopy and the Coulter counter. For very small liposomes, those smaller than 0.2 μm, several methods are very useful: gel permeation chromatography (GPC), dynamic light scattering (DLS), and EM. The major difficulty is in the range between these two extremes. Apart from EM, which is cumbersome, modifications of one or more methods, such as centrifugation, sedimentation field flow fractionation, or flow cytometry, may improve size characterization of liposomes in this intermediate range. Alternatively, new methods may be developed to fill this need.

So far, this discussion has assumed that absolute size distributions are to be measured. Nevertheless, a number of situations can be satisfied by relative measurements or indications of change. For example, control of process parameters that may affect size distribution or identification of component failures can be accomplished by measurements of change without absolute measures. In such cases, measurements limited to the mean size, such as DLS and turbidity, can be quite useful.

Single-Particle Properties

Microscopic methods have been used extensively to characterize liposomes, not only for size but also for shape and structure. However, the advantage of measurements on individual particles is offset by a need to count a large number of particles for a good representation of the entire sample. In addition, quantitation of images is difficult and tedious, especially for a large number of particles. Image-processing techniques can be

---

[152] C. Orr and E. Y. H. Keng, *in* "Handbook on Aerosols" (R. L. Dennis, ed.), p. 114. Energy Res. Dev. Admin., G. C. A. Corp., Bedford, Massachusetts, 1976.

applied to facilitate the data collection and analysis but have not been standardized.

Light microscopy has the advantage of determining particle shape and to some extent the amount of multilamellarity. The principal limitation is the lack of resolution below 0.5 μm with the best of optics, although the use of lasers may extend the range even further.[153] Another problem is that visualization by light microscopy of truly unilamellar liposomes of any size is difficult, if not impossible, without the use of encapsulated fluorescent markers.

Electron microscopy has a much smaller minimum resolution but the samples must be fixed. The fixation can introduce artifacts, especially with negative staining. Consequently, quantitation of images from negative stain are of very limited use. Freeze-fracture has less of a tendency for artifacts but must be corrected for deviations of the fracture plane from the midplane of the particle.[154] Freeze etching minimizes this problem.[1,18,155]

The Coulter counter is very useful for determination of the distribution by particle volume but only for particles of 1 μm and larger. It is also dependent to a lesser extent on particle asymmetry. Because of the range limitation it is best for determination of the number of particles present above 1 μm or for use in combination with another method.

Hydrodynamic Radius Using Gel Permeation Chromatography

This method fractionates the sample by hydrodynamic radius, allowing a determination of the amount of phospholipid in each size class resolved (fraction), but with some limit to the actual resolution.[49] When the column is calibrated the amount of lipid (or drug) mass versus size can be determined.[156,157] However, the currently available columns are only capable of resolving particle sizes smaller than 0.2 μm. All the larger particles are lumped together in the void volume fraction. Thus, use of GPC in combination with the Coulter counter can provide size characterization from 10 nm to more than 1 μm, but with a gap from 0.2 to 1 μm. Recent procedures using high-performance liquid chromatography (HPLC) columns can be automated for routine analyses.[158]

[153] W. Jiskoot, T. Teerlink, E. C. Beuvery, and D. J. A. Crommelin, *Pharm. Weekbl. Sci. Ed.* **8**, 259 (1986).
[154] R. Van Venetie, J. Leunissen-Bijvelt, A. J. Verkleij, and P. H. Ververgaert, *J. Microsc.* **118**, 401 (1980).
[155] P. Guiot and P. Baudhuin, *in* "Liposome Technology" (G. Gregoriadis, ed.), Vol. 1, p. 163. CRC Press, Boca Raton, Florida, 1984.
[156] Y. Nozaki, D. D. Lasic, C. Tanford, and J. A. Reynolds, *Science* **217**, 366 (1982).
[157] J. A. Reynolds, Y. Nozaki, and C. Tanford, *Anal. Biochem.* **130**, 471 (1983).
[158] M. Ollivon, A. Walter, and R. Blumenthal, *Anal. Biochem.* **152**, 262 (1986).

Light Scattering

The turbidity of liposome samples depends on the particle-size distribution, among other parameters. Turbidity has been used to characterize the average size but it is severely limited by the many other factors which also affect turbidity.[159] The primary usefulness of turbidity is in detecting formation of particles, or a change in particle size, as processing takes place.

Dynamic light scattering [also referred to as photocorrelation spectrometry (PCS) and quasielastic light scattering (QELS)] has been used extensively to characterize liposome size distribution but is reliable only for homogeneous samples. The reliability and accuracy deteriorates rapidly with increasing heterogeneity.[160] In addition, the method is limited to sizes smaller than 3 μm and its ability to resolve 1-μm particles in the presence of smaller ones is questionable. Nevertheless, DLS can be very useful to characterize liposome samples, especially SUVs and small LUVs, giving an average diameter and polydispersity. These values can be used to detect relative differences or changes occurring on processing or swelling.[161-163] The mathematical techniques required to calculate the size distribution operate by fitting a summation of exponentials to the autocorrelation function. The effect is that the distribution analyses can often be misleading. By use of multiple mathematical procedures on a single autocorrelation function this problem can be minimized but not eliminated.[160]

Density

Centrifugation can be used to fractionate liposome preparations by differences in sedimentation rate; the resultant separation may be due to either differences in size or some other factor affecting liposome density, such as number of lamellae.[164] In order for the method to be applicable, there must be sufficient density differences between the particles and the solvent. For liposomes, ultracentrifugation or alteration of the buffer density may be required, especially for smaller particles. By use of analytical ultracentrifuge techniques the distribution of scattered intensity versus density can be calculated.[165] If the particle density is homogeneous, and

[159] D. A. Barrow and B. R. Lentz, *Biochim. Biophys. Acta* **597,** 92 (1980).
[160] T. Prouder, *ACS Monogr.,* 332 (1987).
[161] E. Hantz, A. Cao, J. Escaig, and E. Taillandier, *Biochim. Biophys. Acta* **862,** 379 (1986).
[162] W. Li, T. S. Aurora, T. H. Haines, and H. Z. Cummins, *Biochemistry* **25,** 8220 (1986).
[163] M. S. McCracken and M. C. Sammons, *J. Pharm. Sci.* **76,** 56 (1987).
[164] E. Goormaghtigh and G. A. Scarborough, *Anal. Biochem.* **159,** 122 (1986).
[165] H. Coll and C. G. Searles, *J. Colloid Interface Sci.* **115,** 121 (1987).

known, the distribution with size can be calculated. Since the density of liposomes may vary, this presents a limitation of the method's accuracy.

Sedimentation field flow fractionation (SFFF) suffers from the same limitations as centrifugation, because the determination is also based on density differences between the particles and the solvent. Initial testing of SFFF for liposomes was promising but no further reports have appeared.[166]

Others

A variety of other methods have been used for characterization of liposome size based on parameters influenced by size, in addition to other properties: NMR, DSC, zeta potential, flow cytometry, etc. In most cases, the dependence of the method on size varies and the extent of the influence of other parameters makes the results uncertain. Nevertheless, the results from each of these methods can provide important alternative information for a complete characterization of the liposomes. In the case of flow cytometry, it is conceivable that fluorescent measurements of liposomes may be possible down to 0.1 μm by optimization for much smaller particle size. By a combination of lipid-soluble dyes and aqueous entrapped dyes, simultaneous measurements of both lipid and aqueous mass per particle are conceivable, allowing determination of the lamellarity of each particle. Unfortunately, none of the commercial instruments has been designed for this purpose.

[166] J. J. Kirkland, W. W. Yau, and F. C. Szoka, *Science* **215**, 296 (1982).

[10] Preparation of Microcapsules from Human Erythrocytes: Use in Transport Experiments of Glutathione and Its S-Conjugate

By TAKAHITO KONDO

Sealed inside-out microvesicles have been used in the characterization of transport systems in erythrocyte membranes.[1-5] The method for preparing the microvesicles was originally reported by Steck and Kant,[6,7] using

[1] H. Sze and A. K. Solomon, *Biochim. Biophys. Acta* **550**, 393 (1979).
[2] J. E. Mollman and D. E. Pleasure, *J. Biol. Chem.* **255**, 569 (1980).
[3] J. M. Gimbel, *J. Biol. Chem.* **257**, 10781 (1982).
[4] A. J. Caride, A. F. Rega, and P. J. Garahan, *Biochim. Biophys. Acta* **734**, 363 (1983).
[5] J. S. Lolkema and G. T. Robillard, *Eur. J. Biochem.* **147**, 69 (1985).
[6] T. L. Steck, R. S. Weinstein, J. H. Straus, and D. F. H. Wallach, *Science* **168**, 255 (1970).
[7] T. L. Steck and J. A. Kant, this series, Vol. 31, p. 172.

ultracentrifugation in a dextran density gradient. An alternative purification method has been developed which relies on the observation that glycoproteins are located only on the external surface of erythrocyte membranes. Glycoprotein sugar chains of the erythrocyte membranes extend outward into the plasma.[8] The concanavalin A – cellulose described here has a binding affinity for these sugars on the external surface of right-side-out vesicles,[9] whereas inside-out vesicles do not interact with concanavalin A.

Preparation of Concanavalin A – Cellulose. The coupling of concanavalin A to insoluble polysaccharides is performed by the method described by Sanderson and Wilson.[10] α-Cellulose (Sigma) is washed with distilled water and excess water is decanted. Oxidation is carried out by adding 1 volume of 0.1 M sodium periodate in water for 1 hr at room temperature. The oxidized α-cellulose is collected on a coarse sintered-glass funnel G3 and washed with 1 liter of distilled water. Concanavalin A (Sigma) employed for the coupling is 30 mg per 1 g α-cellulose. Concanavalin A in 3 volumes of phosphate-buffered saline (PBS, 9 parts of 0.154 M NaCl and 1 part of 0.1 M sodium phosphate, pH 8.0) is added to the moist α-cellulose and the mixture is gently rotated using a rotary mixer for 24 hr at 4°. The particles are then washed with 1 liter of phosphate-buffered saline and treated with 20 ml of 100 mM NaBH$_4$ in water for 15 min. The particles are washed again with phosphate-buffered saline and suspended in 0.5 mM Tris – HCl, pH 8.0 containing 1 mM each CaCl$_2$ and MnCl$_2$. Under the conditions employed the coupling yield is approximately 18 mg concanavalin A per 1 g cellulose.

Preparation of Erythrocytes. Venous blood is freshly collected in 1 mg of EDTA/ml and is freed of leukocytes and most platelets by passing through a column of α-cellulose and microcrystalline cellulose (Sigma) (1 : 1, w/w)[11] which has been previously washed with phosphate-buffered saline (9 parts of 0.154 M NaCl and 1 part of 0.1 M sodium phosphate, pH 7.4). Usually the bed volume of the column is equal to that of the blood sample. The column is washed with the buffer until colorless, and the eluted fraction is washed three times with the buffer to remove plasma.

Preparation of Ghosts. Hemolysis is performed by thoroughly mixing the erythrocytes with 25 volumes of ice-cold 10 mM Tris – HCl, pH 7.0. The buffer should be kept at 0° to rapidly remove the contaminated hemoglobin in the membranes. The mixture is centrifuged at 22,000 g for

[8] T. L. Steck, *J. Cell Biol.* **62**, 1 (1974).
[9] N. Sharon and H. Lis, *Science* **177**, 949 (1972).
[10] C. J. Sanderson and D. V. Wilson, *Immunology* **120**, 1061 (1971).
[11] E. Beutler, C. West, and K. G. Blume, *J. Lab. Clin. Med.* **88**, 328 (1976).

TABLE I
PREPARATION OF INSIDE-OUT VESICLES[a]

Preparation	Volume (ml)	Protein (mg/ml)	%	Glyceraldehyde-phosphate dehydrogenase (I.U./mg protein)	Acetylcholinesterase (I.U./mg protein)
Unsealed membranes	10	1.34	100	2.90	1.82
Homogenized with needle	10	1.32	100	3.77	0.76
Concanavalin A–cellulose treated	5.5	1.75	71.8		
Without detergent	—	—	—	5.55	0.38
With detergent[b]	—	—	—	6.01	2.10
Accessibility (%)	—	—	—	92.3	18.1

[a] Accessibility was calculated as (activity without detergent/activity with detergent) × 100. Reprinted with permission from T. Kondo et al., Biochim. Biophys. Acta 602, 127 (1980).
[b] Triton X-100 in a final concentration of 0.02% was added as a detergent.

10 minutes at 0°, and membranes are rinsed and washed with the same buffer until white.

Preparation of Sealed Vesicles. Vesicles are prepared from the erythrocyte membranes as described by Steck and Kant.[7] The collected white ghosts are suspended in 40 volumes of ice-cold 0.5 mM Tris–HCl, pH 8.0, for 2 hr, washed twice with the same buffer, and centrifuged at 28,000 g for 30 min at 0°. The pellet is then vesiculated by passing through a 27-gauge needle five times; this produces a mixture of right-side-out and inside-out vesicles.

Preparation of Sealed Inside-Out Microvesicles. Homogenized vesicles are added to 1 volume of a 20% (v/v) suspension of concanavalin A–cellulose. After 10 min at room temperature, the concanavalin A–cellulose is poured onto a coarse sintered-glass funnel G3 and washed with 5 volumes of 0.5 mM Tris–HCl, pH 8.0. The vesicles which do not bind to the mesh are concentrated by centrifugation at 26,000 g for 20 min at 4°.

Determination of the Sideness. The sideness of these vesicles is determined by measurement of glyceraldehyde-phosphate dehydrogenase as a marker of the erythrocyte inner surface and acetylcholinesterase as a marker of the erythrocyte outer surface.[7,12] As a control, the vesicles are dissolved in Triton X-100 to allow assay of these total; enzyme activities (Table I).

Comments. The resealed microvesicles obtained do not contain intrinsic proteins such as spectrin. Because these proteins seem not to participate in the transport systems of the erythrocyte membranes, spectrin-deleted

[12] E. Beutler, in "Red Cell Metabolism: A Manual of Biochemical Methods," 3rd Ed., pp. 51 and 103. Grune & Stratton, Orlando, Florida, 1985.

membranes would be applicable for the transport experiments. On the other hand, treatment of the erythrocyte ghosts with Mg^{2+}-ATP before the vesiculation results in the formation of inside-out vesicles containing fully intact spectrin components,[13] and these vesicles may be applicable for the study of functions of contractile proteins. In preliminary studies, the use of commercially available concancavalin A–Sepharose was ineffective in preparation of vesicles. The preparation of concanavalin A–cellulose is simple and allows repeated usage merely by washing with 0.1 M α-methylmannoside. The method for the preparation of vesicles is comparably rapid and efficient. The purity of the vesicles determined on the basis of the specific gravity is as same as the proposed method described by Steck and Kant.[7]

Examples of Application of Inside-Out Vesicles in Transport Experiments. Inside-out microvesicles from human erythrocytes have been found to be useful in the study of transport systems for glutathione and its compounds.[14-16] Since the intracellular membrane proteins can be exposed to the external environment, the effects of substrates, activators, and inhibitors on the transport systems are easily studied.

Glutathione in human erythrocytes is synthesized by glutamate–cysteine ligase (γ-glutamyl cysteine synthetase) and glutathione synthase; it protects erythrocyte membranes, hemoglobin, and other cell constituents against oxidative damage by free radicals and peroxides. Glutathione is transported outside of the cells in the oxidized form (GSSG). The ATP-dependent transport systems for GSSG was first described by Srivastava and Beutler.[17] Further studies demonstrated that the rate of GSSG transport out of cells is sufficient to account for the physiological turnover rate of glutathione.[18] Glutathione forms S-conjugates with a number of electrophilic compounds, including various drugs and xenobiotics.[19] Glutathione S-transferase, which catalyzes the reaction, exists in human erythrocytes,[20] and glutathione S-conjugates (GSH S-conjugates) formed in the erythrocytes are eliminated from the cells.[21] The transport system for GSSG and

---

[13] W. Birchmeier, J. H. Lanz, K. H. Winterhalter, and M. J. Conrad, *J. Biol. Chem.* **254**, 9298 (1979).

[14] T. Kondo, G. L. Dale, and E. Beutler, *Proc. Natl. Acad. Sci. U.S.A.* **177**, 6359 (1980).

[15] T. Kondo, G. L. Dale, and E. Beutler, *Biochim. Biophys. Acta* **645**, 132 (1981).

[16] T. Kondo, M. Murao, and N. Taniguchi, *Eur. J. Biochem.* **125**, 551 (1982).

[17] S. K. Srivastava and E. Beutler, *J. Biol. Chem.* **244**, 9 (1969).

[18] G. Lunn, G. L. Dale, and E. Beutler, *Blood* **54**, 238 (1979).

[19] W. B. Jacoby, W. H. Habig, J. H. Keen, J. N. Ketley, and M. J. Pabst, *in* "Glutathione: Metabolism and Function" (I. M. Arias and W. B. Jacoby, eds.), p. 189. Raven, New York, 1976.

[20] C. J. Marcus, W. H. Habig, and W. B. Jacoby, *Arch. Biochem. Biophys.* **188**, 287 (1979).

[21] P. G. Board, *FEBS Lett.* **124**, 163 (1981).

GSH S-conjugates is characterized in detail from data obtained utilizing the inside-out microvesicles.

Preparation of [³H]GSSG. [³H]GSSG is prepared by the incubation of the ³H-reduced form of glutathione (GSH) with a 10 molar excess of H_2O_2 at pH 8.0 in a Dubnoff shaker for 20 min at 37°.[22] After the incubation, residual H_2O_2 was removed by the addition of 18 units of bovine catalase (Sigma).

Purification of Glutathione S-Transferase. An anionic form of glutathione S-transferase is purified from human liver as described,[23] using an affinity chromatography column prepared by coupling glutathione to epoxy-activated Sepharose 4B.[24]

Preparation of GSH S-Conjugate. GSH S-conjugate, S-(2,4-dinitrophenyl) glutathione (GS-N_2ph), is prepared enzymatically. The incubation system consists of 3 mM [³H]GSH, 4 mM 1-chloro-2,4-dinitrobenzene, and 1 unit of glutathione S-transferase, in a final volume of 3.2 ml in 50 mM phosphate buffer, pH 6.5. After 3 hr at 37° the system is cooled to 0° to terminate the reaction. The formation of GS-N_2ph is estimated photometrically at 340 nm as described.[25] GS-N_2ph is purified at room temperature on a column of Dowex-1 (formate form, 100–200 mesh, Bio-Rad) employing the following method.[18] The assay mixture is diluted with 10 volumes of water to decrease the ionic strength and is applied to a 1.0 × 15-cm column of Dowex-1. The column is washed with 10 ml of water and elution is performed with a linear gradient established between 25 ml of water and 25 ml of 4 M formic acid. GS-N_2ph emerges at approximately 3.5 M. The eluate collected is essentially free of other compounds and is evaporated using a rotary evaporator. Traces of formic acid are removed by suspending the sediments in 0.5 ml of cold acetone and 2 ml of hexane followed by centrifugation at 2000 g for 10 min. The supernatant is dried under N_2 gas and the conjugate is stored at −20° until use. The concentration of [³H]GS-N_2ph is determined by liquid scintillation counting.

Analytical Methods. Protein concentrations are estimated by the method of Lowry *et al.*,[26] with bovine serum albumin as a standard. GSSG is determined as reported.[27]

[22] J. Prchal, S. K. Srivastava, and E. Beutler, *Blood* **46**, 111 (1975).
[23] Y. C. Awasthi, D. D. Dat, and R. P. Saneto, *Biochem J.* **191**, 1 (1980).
[24] P. C. Simons and D. L. Vander Jagt, *Anal. Biochem.* **82**, 334 (1977).
[25] A. Wallander and H. Sies, *Eur. J. Biochem.* **96**, 441 (1979).
[26] O. H. Lowry, N. J. Rosebrough, A. L. Farr, and R. J. Randall, *J. Biol. Chem.* **193**, 265 (1951).
[27] E. Beutler "Red Cell Metabolism: A Manual of Biochemical Methods," 3rd Ed., p. 134. Grune & Stratton, Orlando, Florida, 1985.

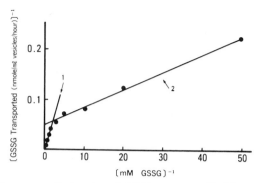

FIG. 1. A Lineweaver–Burke plot of GSSG transport. Vesicles were incubated with various concentrations of GSSG, 2 mM ATP, and 10 mM MgCl$_2$ in 250 μl phosphate-buffered saline, pH 7.4 for 2 hr at 37°. Vesicles without ATP served as a blank. Line 1: $K_m = 7.3$ mM and $V = 210$ nmol/ml/hr. Line 2: $K_m = 0.1$ mM and $V = 20$ nmol/ml/hr. (Reprinted with permission from Kondo et al.[15])

Active Transport of GSSG. Unless otherwise indicated, the assay system for glutathione transport consists of 50 μl of inside-out vesicles, 2 mM MgCl$_2$, and 20 μM or 5 mM [^{3}H]GSSG, in a final volume of 250 μl phosphate-buffered saline, pH 7.4. A blank assay mixture contains no ATP. After a 2-hr incubation at 37°, the reaction is stopped by the addition of 40 volumes of ice-cold 30 mM NaF in 0.5 mM Tris–HCl, pH 8.0. The vesicles are washed two times with the same solution by centrifuging at 25,000 g for 20 min. Water (1 ml) is added to the vesicles, and the mixture is then frozen and thawed. The lysed vesicles are centrifuged to remove membrane fragments, and the radioactivity in the supernatant is determined. The lysed membrane fragments are removed to prevent the contamination of [^{3}H]GSSG that reacts with sulfhydryl groups on the membrane surface. When the vesicles are incubated in the absence of ATP, no GSSG is transported into the vesicles. In the presence of ATP, GSSG is transported at a rate linear with time. The transport rate increases as the concentration of GSSG increases. The dependence of GSSG transport on GSSG concentration measured over a 10 μM to 10 mM range of initial extravesicular GSSG concentrations gives a biphasic plot on a Lineweaver–Burke analysis (Fig. 1). One phase has a high affinity for GSSG, with an apparent K_m of 0.1 mM GSSG and a low transport velocity with an apparent V of 20 nmol of GSSG of vesicles/milliliter of vesicles/hour. The other phase has a low affinity for GSSG, with an apparent K_m of 7.3 mM GSSG and a high transport velocity with an apparent V of 210 nmol of GSSG/milliliter of vesicles/hour. This may indicate either the existence of two independent transport systems that have different kinetic parameters or a negative cooperativity in a single transport system. In

TABLE II

COMPARISON OF TWO DIFFERENT GSSG TRANSPORT PROCESSES IN HUMAN ERYTHROCYTES[a]

Property or assay condition	Unit	Low K_m process	High K_m process
Apparent K_m (GSSG)	mM	0.1	7.3
V	nmol/ml vesicles/hr	20	210
K_m (Mg-ATP)	mM	0.63	1.25
Transport rate	nmol/ml/hr	3.90/1.20/(20)[b]	71.0/9.70/(20)
pH optimum	—	6.5	pH 7.2
Temperature	—	Plateau at 25–37°	Maximal activity at 37°
Erythrocyte age			
Top fraction (young cells)	nmol/ml/hr	3.82/0.84/(4)	70.8/1.8/(4)
Bottom fraction (old cells)	nmol/ml/hr	2.84/0.37/(4)	73.5/2.5/(4)
Nucleotide requirement			
2 mM ATP	nmol/ml/hr	3.1/1.1/(4)	75.1/6.3/(4)
2 mM GTP	nmol/ml/hr	10.5/3.3/(4)	88.6/7.4/(4)
Stability			
0 day	nmol/ml/hr	3.89/0.17/(4)	71.2/1.4/(4)
2 days	nmol/ml/hr	2.50/0.76/(4)	73.1/3.3/(4)
5 days	nmol/ml/hr	1.51/0.05/(4)	67.5/3.8/(4)
Effect of thiols and thiol reagents (%)			
1 mM dithiothreitol	—	94[c]	170
0.1 mM p-chloromercuribenzoate	—	101	49
1 mM N-ethylmaleimide	—	94	75
1 mM iodoacetamide	—	81	66
Control	—	100	100

[a] The low K_m process and the high K_m process were monitored with 20 μM GSSG and 5 mM GSSG, respectively. See *Methods* for details of individual reactions. Reprinted with permission from Kondo *et al.*[15]

[b] These values are mean 1 SD (number of determinations).

[c] Averages of two experiments.

order to determine these two possibilities, the effect of modifiers on the two phases of the transport curve was studied, choosing two GSSG concentrations to represent the two phases of the transport; 20 μM GSSG for the low K_m system and 5 mM GSSG for the high K_m system. Table II illustrates the comparison of the two phases of GSSG transport processes in the vesicles. The K_m for Mg^{2+}-ATP in the transport of GSSG was 0.63 mM when the GSSG concentration was 20 μM and 1.25 mM when the GSSG concentration was 5 mM. At the low K_m process, the GSSG transport was most active at an acidic pH and at 25°. This transport activity was easily lost during the storage of vesicles at 4°. Guanosine triphosphate substituted for ATP. Neither dithiothreitol nor thiol reagents affected this transport activity. At the high K_m process, the transport was most active at pH 7.2 and 37°, and this transport activity was stable during storage of vesicles at 4°

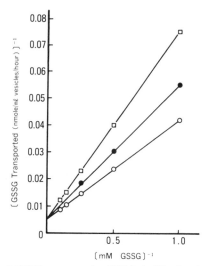

FIG. 2. Inhibition of GSSG transport by GS-N$_2$ph. The incubation mixture comprised 25 μl of inside-out vesicles, various concentrations of GSSG, 2 mM ATP, and 10 mM MgCl$_2$, in the presence (●, 0.1 mM; □, 0.5 mM) or absence (○) of GS-N$_2$ph, in 125 μl of phosphate-buffered saline, pH 7.4. Vesicles were incubated for 2 hr at 37°. GS-N$_2$ph transported was estimated based on a lysed solution of vesicles.

for several days. Guanosine triphosphate substituted for ATP with no change in activity. Dithiothreitol increased the activity, and thiol reagents such as p-chloromercuribenzoate inhibited the transport. Erythrocytes were separated into two age groups on the basis of specific gravity according to the method described by Murphy.[28] Inside-out vesicles were prepared from these cells. At the low K_m process, the transport activity decreased slightly during cell aging, but the transport activity at the high K_m process did not change. These results suggest that there are two independent transport processes for GSSG in human erythrocytes.

The effect of GS-N$_2$ph on each GSSG transport process was examined. As shown in Fig. 2, the high K_m process was inhibited by various concentrations of GS-N$_2$ph. On the other hand no apparent inhibitory effect of the conjugate was observed at the low K_m process.

Active Transport of Glutathione S-Conjugate. The assay system for glutathione conjugate transport consists of 25 μl of inside-out vesicles, 2 mM ATP, 10 mM MgCl$_2$, and 1 mM [^{3}H]GSH-Dnp, in a final volume of 125 μl of phosphate-buffered saline, pH 7.4. Vesicles are incubated with 1 mM GS-N$_2$ph in the presence of ATP at 37°; no transport activity is observed in the absence of ATP. The transport rate is dependent on the concentration of GS-N$_2$ph, with an apparent K_m of 0.94 mM GS-N$_2$ph

[28] J. R. Murphy, *J. Lab. Clin. Med.* **82**, 334 (1973).

and an apparent V of 183 nmol of GS-N$_2$ph/milliliter of vesicles/hour. The transport is dependent on Mg^{2+}-ATP, with an apparent K_m value of 0.76 mM at 1 mM GS-N$_2$ph. It is suggested that glutathione S-conjugates are transported out of the erythrocytes via the same transport system for GSSG elevated under oxidative stress.

[11] Planar Lipid–Protein Membranes: Strategies of Formation and of Detecting Dependencies of Ion Transport Functions on Membrane Conditions

By Hansgeorg Schindler

Strategies to Form Planar Lipid Membranes

Molecular understanding of ion channel function involves assignment of channel transport activities to constituents, to their arrangements, and to membrane conditions. One major strategy for identifying such structure–function relationships is to reconstitute isolated channel proteins into planar lipid membranes. From the different approaches used in the past (see reviews by Montal *et al.*[1] and by Miller[2]) two principal strategies have emerged. They both use lipid–protein vesicles as starting material, either native membrane vesicles or reassembled lipid–protein vesicles (see Fig. 1). In the first strategy,[3] vesicle–bilayer fusion (VBF)[3a] vesicles are fused to preformed planar lipid membranes. Transfer of chan-

---

[1] M. Montal, A. Darszon, and H. Schindler, *Q. Rev. Biophys.* **14**, 1 (1981).

[2] C. Miller, *Physiol. Rev.* **63**, 1209 (1983).

[3] C. Miller and E. Racker, *J. Membr. Biol.* **30**, 283 (1976).

[3a] Abbreviations and symbols: VBF (Vesicle–bilayer fusion): technique to incorporate lipid–protein vesicles into preformed planar lipid bilayers. SVB (Septum-supported vesicle-derived bilayer): technique to form planar membranes from lipid–protein vesicles. PVB (Pipet-supported vesicle-derived bilayer): technique to form lipid–protein bilayers at the tip of a glass micropipet. PC, Patch-clamp technique; A_{bil}, area of bilayer, set equal to the area of the aperture in the Teflon septum within which the bilayer is formed according to the SVB technique—i.e., $A_{bil} = \pi d^2/4$, where d is the diameter of the aperture; N_L, number of lipids per vesicle; $\eta_{v,b}$, relates molar ratio of protein to lipid in the bilayer, $(p/l)_{bil}$, to that in the vesicles, $(p/l)_{ves}$, used to form the bilayer—i.e., $\eta_{v,b} = (p/l)_{bil}/(p/l)_{ves}$; π_e, vesicle monolayer equilibrium surface pressure, used to assay the lateral pressure in the planar bilayer; n_s, number of proteins per unit area of bilayer; t_c, average time for first collisions of proteins in planar bilayers, initially distributed at random; D_{lat}, lateral diffusion of membrane proteins.

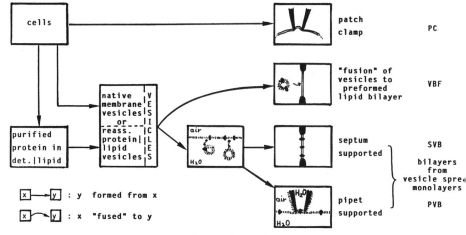

FIG. 1. Schematic diagram of strategies to form planar lipid–protein membranes. Straight arrows indicate that the membrane material is transformed from one configuration to another. The curved arrow indicates combination or "fusion" of two configurations of vesicles to a preformed planar bilayer. Abbreviations and references: VBF,[3] vesicle–bilayer fusion; SVB,[4] septum-supported, vesicle-derived bilayer; PVB,[9-11] pipet-supported, vesicle-derived bilayer. The patch clamp (PC[12]) has been included, because it provides reference data for reconstitution by either the VBF method or the SVB or PVB methods, where vesicles are used as starting materials in all three techniques.

nel proteins from vesicles to planar membranes by this method requires the presence of calcium ions, of osmotic gradients across the vesicles and the planar membrane, and of negatively charged lipids. In the second strategy, planar bilayers are formed from vesicles.[4] The technique is based on the finding that monolayers spontaneously form at the air–water interface of any vesicle suspension.[5-7] There are two distinct ways of combining two such monolayers (normally one lipid–protein and one lipid monolayer) to form a bilayer. In the first method, the monolayers are apposed within an aperture in a thin Teflon septum[4] [septum-supported vesicle-derived bilayer (SVB)]. By raising the water levels in both compartments above the aperture, the two monolayers cover the Teflon septum and combine to a bilayer within the aperture, in close analogy to bilayer formation from solvent-spread monolayers at lens pressure.[8] In the alternate method, the support is not a septum between the two monolayers but

[4] H. Schindler, *FEBS Lett.* **122**, 77 (1980).
[5] F. Pattus, P. Desnuelle, and R. Verger, *Biochim. Biophys. Acta* **507**, 62 (1978).
[6] H. Schindler, *Biochim. Biophys. Acta* **555**, 316 (1979).
[7] T. Schürholz and H. Schindler, *Eur. Biophys. J.* (in press).
[8] M. Montal and P. Mueller, *Proc. Natl. Acad. Sci. U.S.A.* **69**, 3561 (1972).

a glass pipet used to bring two monolayers into bilayer contact [pipet-supported vesicle-derived bilayer (PVB)].[9-11]

Both strategies, vesicle–bilayer fusion and vesicle-derived bilayers, have yielded satisfactory recovery of channel functions for several different channels, starting from isolated proteins. Comparison of channel characteristics obtained from patch clamp (PC) studies[12] on whole cells and reconstitution data allowed for clear identification of particular protein species as primary constituents of electrophysiologically observed channel conductances. Further structure–function assignments via reconstitution would require at least some knowledge of physical and structural conditions in the membrane while observing channel activities. More specifically, it would be desirable to relate channel conductances (1) to the number and stoichiometry of components; (2) to their lateral and transverse distributions, such as modes of association; and (3) to variables characterizing the overall physical state of the membrane. Channel reconstitution based on physical control of bilayers needs further study. The vesicle–bilayer fusion method is not particularly suited for this, since number, arrangement, and environmental conditions of inserted channel proteins are not sufficiently known to infer relationships between membrane structure aspects and observed channel properties. The second principal strategy, vesicle-derived bilayers, has been implemented for the control of at least a few basic structural parameters of planar membranes containing channel proteins. It is the goal of this chapter to give a detailed account of how structural control is achieved in SVB, with all necessary technical information to build a planar bilayer stand, to form SVB, and to analyze channel conductances under variable conditions, with references given to published examples.

Experimental Setup

Electrical Equipment

As schematically shown in Fig. 2, one compartment of the bilayer cell (cf. Fig. 4) is connected to an input source and the other compartment to an amplifier (current-to-voltage converter). Different types of input signals are used, dc, sine, and ramp. A dc signal of high stability is absolutely

[9] V. Wilmsen, C. Methfessel, N. Hanke, and G. Boheim, in "Physical Chemistry of Transmembrane Ion Motions" (G. Spach, ed.), p. 479. Elsevier, Amsterdam, 1983.

[10] T. Schürholz and H. Schindler, FEBS Lett. **152**, 187 (1983).

[11] R. Coronado and R. Latorre, Biophys. J. **43**, 231 (1983).

[12] O. P. Hamill, A. Marty, E. Neher, B. Sakman, and F. J. Sigworth, Pfluegers Arch. **391** (1981).

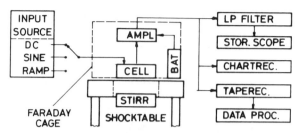

FIG. 2. Electrical equipment for planar bilayer reconstitution work. CELL: planar bilayer cell with two electrodes, one connected to the voltage INPUT SOURCE (DC, SINE, or RAMP) and the other connected to a current–voltage converter (AMPL) supplied with voltage by batteries (BAT). The output signal from the amplifier is connected to a storage scope (STOR. SCOPE) via a low-pass filter (LP FILTER), to a chart recorder (CHARTREC.), and to a tape recorder (TAPEREC.) for later processing of data (DATA PROC.). STIRR: stirring magnets, one for each compartment of the bilayer cell.

required. (A circuit diagram of a simple, low-cost battery-driven dc source is shown in Fig. 7a. Its noise is sufficiently low, and voltage values can be conveniently changed between preset values.) A sine wave, 50 Hz and 16 mV peak to peak, is used to monitor membrane formation. For this purpose, sine waveform distortions and noise are of no importance, so that it is sufficient to simply transform line power as shown in Fig. 7b. The amplifier (Fig. 7c) is also easy to build using the particular elements given in *Appendix A*. A voltage ramp signal is not absolutely required, but is convenient for monitoring current–voltage relations of channel conductances. A simple scheme to generate triangular waveforms is described in Ref. 13, p. 120, with additional information on electrical equipment for planar bilayer work, including fabrication of the Ag–AgCl electrodes. The other devices in Fig. 2 are standard commercial units. The low-pass filter, normally an 8-pole Bessel filter, should have settings of (1 and 3) $\times$ 10^nHz, for $n = 0$, 1, 2, and 3. A digital storage scope with rolling data mode is of advantage. The chart recorder is used only to continuously protocol experiments, so any simple one-channel recorder may be used. The nonfiltered data are stored on tape. There are two possibilities. The expensive version employs a FM four-channel tape recorder (up to 40 kHz resolution) such as from RACAL Ltd., Southampton, England, type store 4D. It has the advantage that data can be replayed at up to 64 times slower speed. The other version employs any video recorder with a processor that can be modified to accept dc signals. Its resolution is even higher than that of the FM tape recorder, and its dynamic range is just appropriate (up to

[13] O. Alvarez, *in* "Ion Channel Reconstitution" (C. Miller, ed.), p. 115. Plenum, New York, 1986.

10 khz). Unfortunately, there is no version yet on the market which allows for replay slower than real time. This, however, may not be a serious disadvantage if a data processing unit is used which can store channel data in real time. One of the less expensive versions, but well adapted in its software for bilayer work, is the B-Scope offered by med-NATIC (Munich, Germany). The + 15 and − 15 V supply for the amplifier is provided by 20 long-life batteries with 1.5 V each, assembled in a box within the Faraday cage. For stirring we use two small magnets, one for each compartment, fixed to the axes of two small motors, which can be separately regulated by an external unit to revolutions in the range of 1 to 10 Hz.

Mechanical Parts

The mechanical hardware of the bilayer stand we use is shown in Fig. 3. Dimensions are not critical and may be estimated from Fig. 3, except for the bilayer cell and Teflon septum, which are described in Figs. 4a and 8, respectively. The shock table consists of a framework of rectangular steel tubing on four shock feet (used for workshop machines of similar weight). A wooden plate (70 × 70 cm) is tightly screwed to the steel stand. The top plate is of iron (90 kg), which rests on three rubber stoppers.; The iron and wooden plates each have a circular hole (see Fig. 3b) of about 6 cm in diameter for installment of the stirring motors and magnets. The two magnets are positioned below the two compartments of the bilayer cell and vertically just below the bottom plate of the Faraday cage. The motors are in a plastic holder which is fixed from below the wooden plate. The Faraday cage is made from 2-mm sheet iron and has a sliding door at the front and BNC input–output connections at the rear. The cage should be tightly fixed to the iron plate. Inside the cage (see Fig. 3b) are the battery box (at the rear), a solid holder for the electrodes, and the amplifier box (small cylindrical part at the front, right side). The holder is worked from a brass pipe (10 cm in diameter, 6-mm wall thickness, and 15 cm in length). Its lower rim fits exactly to the cell holder, so that there is no play when the amplifier and electrode holder are lowered (compare Fig. 3c and d). The bilayer cell consists of two half-cells, between which the Teflon septum is positioned. They are clamped together by a tapered metal ring and tight-ened by hand. For the type of septum used, this gives reliable electrical tightness.

The critical aspects of the whole design are the shape of the compart-ments in the Teflon cell (addressed in the legend to Fig. 4), fabrication of the Teflon septum (described in *Appendix B*), but especially the punching of the aperture within which the planar bilayers are to be formed. Because

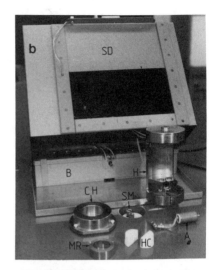
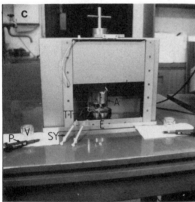

FIG. 3. Photographs showing mechanical parts of a planar bilayer stand. (a) Faraday cage (F), iron plate (I), wooden plate (W), steel stand (S), and shock feet (SF). (b) Sliding door (SD), battery box (B), holder (H) for amplifier (A) and electrodes (E), stirring magnets (SM), half-cell (HC), metal ring (MR), and cell holder (CH). (c) Amplifier (A), electrodes (E), syringes (SY) with teflon tubings (TT) inserted into the two compartments, and pipet (P) with a special glass tip and vial (V) for vesicle suspension. (d) Amplifier and electrode holder in raised position.

we know that success in bilayer work depends on the septum and aperture, and that fabrication is not trivial even for a good machine shop, we have decided to offer some help for those who want to use this technique (see last section, *Start-Up Kit for Users*).

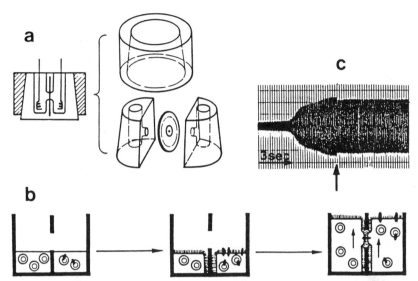

FIG. 4. (a) Drawing of the assembly of a bilayer cell. Dimensions: compartments are 16 mm deep and 12 mm in diameter. The common hole of the half-cells is 4 mm in diameter; its center is 9 mm above the bottom of each compartment. The height of the Teflon cell is 20.5 mm. Upper and lower outer diameters of the tapered Teflon cell are 32.5 and 35.5 mm, respectively. Each half-cell has one 2-mm hole from the top surface to the bottom of the compartment (not shown) for insertion of the Teflon tubing, which is connected to a syringe used to add vesicle suspension to each compartment. The tapered metal ring is 18 mm high; the outer diameter is 48 mm and the inner diameters are 33 and 35 mm at the upper and lower ends, respectively. The common hole (4 mm in diameter) should enlarge in a smooth curvature to join the compartments without edges. Dimensions of the Teflon septum are given in Fig. 8. (b) Schematic illustration of three states during planar bilayer formation. State 1 (left): vesicles have just been added. State 2 (middle): monolayers have spontaneously formed from vesicles. State 3 (right): a planar bilayer has been formed by raising the water levels, thus apposing the two monolayers within the aperture in the Teflon septum (for details and mechanism see the text). (c) Electrical capacitance signal during bilayer formation (see the text).

Principle of Planar Membrane Formation from Vesicles

This technique[4] makes use of physical forces[14] leading to spontaneous (re)organization of membrane material at interfaces and between interfaces. During membrane reorganization, three distinct situations lead to formation of planar membranes from vesicles, shown schematically as three bilayer cells in Fig. 4b. In the first situation, vesicle suspensions have been added to the two compartments of the bilayer cell to just below the

[14] H. Schindler and T. Schürholz, *Eur. Biophys. J.* (in press).

aperture (cf. Fig. 4a for details). In the second situation a spontaneous process occurs in the bilayer cell, the self-assembly of lipid–protein mono-layers at the interfaces.[5-7] Typically within 10 sec the surface pressure approaches its equilibrium value,[6] characteristic for the lipids and aqueous phases used. Monolayer compositions closely mimic those of the overall vesicle suspensions[7] just after monolayer formation. The two monolayers, being in equilibrium with membranes, spontaneously (re)combine with a membrane when they are apposed within the aperture, after raising the water levels above the aperture (the third bilayer cell). The whole proce-dure takes about 1 min.

Appropriate Vesicle Suspensions

Spontaneous formation of monolayers and recombination of two monolayers to a bilayer impose certain constraints on vesicle samples to be used for planar bilayer formation. They all relate to the equilibrium pres-sure, π_e, which should be reached for successful bilayer formation. De-pending on the type of lipid used and on the type and strength of ions in solution, measured π_e values range between about 20 and 45 mN/m.[15] Bilayer formation becomes impracticable for π_e values below 24 mN/m. For practical purposes of bilayer formation, however, it is not required to know or measure the π_e value of the vesicle sample used. The surface pressure will adjust to the proper π_e value provided the following condi-tions are fulfilled.

1. Vesicle diameter should exceed 100 nm. For smaller vesicles of the same composition, considerably lower "apparent" π_e values are mea-sured,[4] reflecting an unfavorable energy state due to limited size (caused by energy input, such as sonication, during vesicle preparation). Therefore, vesicle preparation methods should be used which require little mechanical or chemical energy input, yielding vesicles larger than 100 nm (for tech-niques, see *Appendix C*). The time approach to π_e should be sufficiently fast (up to 1 min) for the purpose of bilayer formation.

2. Vesicle concentration should be about 1 mg lipid/ml.[4] Although π_e is independent of vesicle concentration,[6] the approach to π_e becomes controlled by diffusion flux of vesicles to the surface when vesicle concen-tration is, for example, below 0.3 mg/ml for vesicles 120 nm in diameter.

3. Buffers should contain at least 10 mM univalent ions or at least 1 mM calcium or magnesium ions. At lower concentrations, electrostatic

[15] H. Schindler, unpublished observations (1988).

barriers, especially when using negatively charged lipids, impede monolayer formation[14] (see below).

4. The following restriction is more severe. It was found[14] that pure lipid vesicles (with protein or proteolipid impurities less than 10^{-5} mol per mole lipid) do not spread to monolayers or do so only extremely slowly. However, when only one protein or proteolipid is present per vesicle (about 10^5 lipids), the monolayer formation rate increases by orders of magnitude to values practical for bilayer formation. Therefore, when using very pure synthetic lipids we add trace amounts of proteolipid isolated from soybean lipid[14] to facilitate monolayer formation. Commercially available lipids, obtained from lipid extracts, have thus far yielded sufficiently fast monolayer formation, probably due to residual nonlipid components.

5. Considerable vesicle aggregation should be avoided, because the presence of aggregates reduces monolayer formation rates. This is interpreted in the following way. Analysis of vesicle–monolayer equilibration showed[6] that equilibration of the system can be regarded as a sequence of preequilibration steps between monolayer and individual vesicles (net flow of a few lipid molecules from the vesicle outer layer has adjusted to that of lipids in the monolayer). There are many more such elementary steps of lipid flow required for system equilibration (zero net flow) than even one densely packed layer of vesicles could provide, so that preequilibrated vesicles have to make room for vesicles which have not yet interacted. Aggregates, however, tend to adhere more strongly to the region below the monolayer than do single vesicles, due to increased van der Waals forces, and therefore impede the final approach to system equilibrium, so that the monolayer pressure remains stationary at a value below the equilibrium value π_e. When membranes are not associated with aggregates of single-walled vesicles but with oligolamellar vesicles, reduction of monolayer formation rates is also found, but only by a factor of 2 to 3. It should be added that adhesion is generally strengthened by high salt concentrations, such as 1 M KCl. Minimizing adverse effects of adhesion and satisfying aspect (3) above leaves an optimal salt range of 100 to 200 mM univalent salt.

Protocol for Planar Membrane Formation

Preparatory Procedures

1. Vesicle samples. For one set of experiments prepare about 20 ml of lipid–vesicle suspension at 1 mg/ml (see *Appendix C*). For bilayers asym-

metric with respect to lipid, two such samples are prepared. Two glass vials (shown in Fig. 3c or d) are filled with 2 ml of suspension, one vial for the rear and one for the front compartment of the bilayer cell. Lipid–protein vesicles, prepared by dialysis or the fast dilution technique *(Appendix C)*, are added in a defined aliquot to one of the vials and are mixed by shaking. The vials should be sterilized before use and used only once.

2. Ag–AgCl electrodes (stored in distilled water) are connected to the amplifier holder.

3. A Teflon septum with optional aperture diameter (we use apertures of 160, 100, 80, or 60 μm, depending on application) is briefly sonicated in a bath in chloroform–methanol–water (2:1:3), washed in distilled water, and stored dry between Kleenex sheets.

4. The cell: For one bilayer setup, or for each project, at least six cells should be available. They are recycled by cleaning in warm 1% detergent solution with stirring inside the compartments for 2 hr, followed by the same procedure with several washes in distilled water, and stored in a dust-tight box. It is advisable to take a fresh cell for any change of conditions. Before reusing the same cell it should be rinsed with water by repeatedly filling with water and clearing by aspiration (no hydrocarbon solvents should be used for cleaning, because their evaporation is slow from the capillary Teflon surfaces, and they may partition into the lipid–protein monolayer during the next experiment).

5. For addition of vesicle suspensions to the compartments, two syringes and two pipets are used (see Fig. 3c and d). The former are assembled from sterile 1-ml syringes, 1-mm (o.d.) hypodermic needles, and Teflon tubing about 8 cm long (all used as disposable elements). The Teflon tubing has been cleaned inside and outside by a flush of detergent solution and distilled water, each for 2 hr. Tubing is stored in a dust-tight box with one end sticking out. Fixed 200-μl capillary glass pipets with a Teflon-tipped plunger are used as pipets. Glass capillaries are pulled to small openings to deliver smaller droplets; the pulled part ($\sim$2 cm in length) is bent by about 30° for comfortable delivery of droplets to the compartments (see below). These capillaries may be reused after cleaning in hexane when dry and dust-free.

6. For septum coating use hexadecane in hexane (5 μl/ml) in an absolutely clean small glass vial with an air-tight Teflon cover. A capillary glass pipet with one end pulled to a small opening is used to apply this solution to the septum (see note 5 above). A closed-end rubber tube is fixed to the other end of the capillary tube and is used for filling the glass tube and for delivering tiny droplets. This pipet should be carefully cleaned in hexane before loading with hexadecane solution.

7. For monitoring planar bilayer formation a sine wave input voltage is

applied. The output signal, V_{out}, is then related to the input, V_{in}, by

$$V_{out} = R_f \left(\frac{V_{in}}{R_m} + C \frac{dV_{in}}{dt} \right) \tag{1}$$

with $V_{in} = \frac{1}{2} V_{in}^{p\text{-}p} \sin 2\pi \nu t$. At a sine wave frequency $\nu = 50$ Hz, the capacitive current $[C(dV_{in}/dt)]$ is larger by orders of magnitude than the noncapacitive current (V_{in}/R_m) even in the presence of several open channels. Under this assumption the peak-to-peak output voltage is

$$V_{out}^{p\text{-}p} \approx R_f \, V_{in}^{p\text{-}p} 2\pi \nu C \tag{2}$$

for $R_m \gg (2\pi \nu C)^{-1}$. There are three distinct stationary states of the output signal, referred to as S_{sep}, S_{bil}, and S_{sat}.

S_{sep} is the capacitive current contribution of the septum. The septum is about half-covered by the aqueous phases during bilayer formation, corresponding to capacitance values of 30 and 15 pF for the two Teflon thicknesses used, 6 and 12 μm, respectively. Example: $S_{sep} = 1.5$ V for $R_f = 10^{10}$ Ω, $V_{in}^{p\text{-}p} 2\pi = 100$ mV, $\nu = 50$ Hz, using 6-μm-thick Teflon.

S_{bil} is the capacitive current across the completely formed planar lipid bilayer, including S_{sep}. Bilayer capacitance is proportional to bilayer area (A_{bil}):

$$C_{bil} = C_{spec} A_{bil} \tag{3}$$

Example: $S_{bil} = 9$ V for the same values as in the above example, and for 160-μm aperture size and $C_{spec} = 0.75$ μF/cm^2 (value for soybean lipid bilayer). $C_{bil} = 150$ pF (it would be 20 pF for a 60-μm aperture). Capacitive resistance $(2\pi \nu C)^{-1}$ is 2×10^7 Ω.

S_{sat} is the current across the aperture when the membrane is broken which is always high enough to drive the amplifier into saturation.

Step-by-Step Guide for Bilayer Formation

Step 1. Using a capillary pipet, a tiny droplet (~ 0.5 μl) of the hexadecane solution is placed on the septum at the aperture. The droplet will spread to some extent (it should not roll away from the aperture, indicating adverse surface properties of this particular septum) yet and should cover the hole during hexane evaporation. This is repeated on the other side of the septum (the droplets should not flow through the hole, which again would indicate inappropriate surface conditions on the Teflon; if this occurs after washings, the septum should be discarded). To remove or smooth the excess hexadecane, the septum is placed between soft, dust-free tissue and is lightly pressed. An alternate way of coating is to devise an evaporation chamber, where hexadecane (or squalene) is delivered in definable amounts to only the aperture, without using solvent. In our laboratory this elaborate procedure had no advantage either in bilayer formation or in observing channels (compared with acetylcholine receptor channels). It is essential that the annulus along the perimeter of the hole, provided by hexadecane coating, is fluid, because the contact angle between bilayer and annulus needs to self-adjust to a particular value[16] for a stable bilayer to form. Hexadecane is, therefore, inappropriate for coating below 16°, where it freezes.

Step 2. The coated septum is placed on one half-cell and the second half-cell is placed on top. After setting the cell down, the tapered metal ring is pressed lightly downward on the tapered cell, which reliably ensures watertight contact between septum and cell compartment. The septum is centered in the hole of the cell and the cell is inserted into the cell holder in the Faraday cage with compartments at the front and the rear. The stage with amplifier and electrodes is lowered for a tight fit over the cell holder, (cf. Fig. 3c and d). The electrodes should not touch the bottom or side walls to avoid microphonics. Add stirring bars if needed.

Step 3. The two syringes are filled with vesicle suspensions and inserted into the cell, and about 0.8 ml is placed in each half-cell.

Step 4. The two pipets are filled with 200 μl of the corresponding vesicle suspensions. Add suspension to the rear compartment by continuously forming small droplets, which fall to the interface. The water level should now be above the aperture (within 1 mm above the middle of the 4-mm hole as judged by visual inspection); if it is still too low, add more droplets; if too high, remove suspension with the syringe. The syringe is removed from the rear compartment (to lower pickup noise during monitoring of bilayer formation).

Step 5. In the same way, droplets are added to the front compartment. The scope should be watched during this step. Stop adding droplets when the sine wave signal jumps from S_{sep} to S_{sat}, which normally occurs between addition of droplets.

Step 6. Immediately after the signal jumps to S_{sat}, the front water level is lowered until the signal returns to S_{sep}. Because the water surface tends to adhere slightly to the septum surface, this lowering should be done by a few short and stepwise pulls on the syringe. The water level is now just below the aperture and only a little additional volume should be required to form the bilayer.

Step 7. The front water level is again raised, this time using the syringe in a slow and continuous fashion, during which the capacitive signal should develop as shown in Fig. 4c, from S_{sep} to S_{bil} in 1–5 sec. Toward the

[16] S. H. White, D. C. Peterson, S. Simon, and M. Yafuso, *Biophys. J.* **16**, 481 (1976).

end of formation, the signal rises slightly beyond the expected value for a bilayer, but then suddenly drops to S_{bil} (see arrow in Fig. 4c). This decrement reflects the "locking in" of the bilayer into its minimum area configuration within the hole. Remove the syringe, close the door of the Faraday cage. Switch input signal to dc.

Criteria

The planar membrane is normally stable and acceptable for quantitative channel studies provided the following three criteria are fulfilled: the bilayer is locked in to S_{bil}, bilayer formation succeeds at the first try, and no conductance artifacts are observed.

Comments

1. Bilayers which do not lock in to S_{bil} are prone to break or to show artifacts due to physical mismatch and folding at the boundary.

2. Formation at the first try ensures that the protein-to-lipid ratio is conserved during bilayer formation. Repetitive tries (of *Steps 6* and *7* above) lead to increasing deposition of material around the aperture, which contributes to the bilayer composition and which also impedes locking in.

3. First-try and locked-in bilayers may still show conductance artifacts (for possible reasons see below). The third criterion reads in practical terms: switch the sine wave off and apply a constant voltage of 100 mV. Reverse polarity about every 5 sec. If no conductance events in excess of capacitive spikes or expected channel conductances occur during 1 min, it is unlikely that artifacts will appear later even when the voltage is increased to 200 mV.

4. The rationale for addition of droplets is as follows. During the elevation of the water levels the total surface area to be covered by the monolayer (air–water interface and Teflon walls) increases. This leads to a transient reduction of the surface pressure π to below π_e. (See below the remark regarding lipid absorption by fresh cells.) This is compensated for by adding droplets. Because droplets are formed from vesicle suspensions, coverage of them by a monolayer begins during their formation. During passage of the droplet across the interface, part of this loosely packed monolayer is integrated into the monolayer at the interface. This temporarily raises the surface pressure (even above π_e), facilitating bilayer formation.[4]

Criteria Not Fulfilled (Troubleshooting Guide)

Locking in Not Occurring or Too Slow or S Values $\neq S_{bil}$. (1) This is most often caused by the presence of too much hexadecane, so that the annulus is too thick for the bilayer to find its minimum area configuration. (2) This also occurs when the vesicles show a considerable degree of aggregation. (3) For locking in the diameter of the aperture should be at least 10 times the septum thickness.

Bilayer Breaks during Formation. (1) This is most often caused by surface pressure ($\pi < \pi_e$) that is too low. This relates to the inappropriate vesicle sample conditions mentioned above. One more technical cause should be added. The rough surface of fresh, dry Teflon cells adsorbs a considerable amount of lipid when vesicle sample is added. This slows down the approach to π_e, which becomes critical under conditions where the monolayer formation rate is already slow. In this case it helps to preincubate the cell with vesicle suspension before use. (2) Check the hexadecane solution for dust or replace it. (3) Replace the septum or check the septum used for irregularities around the perimeter of the hole. (4) Coating may be imperfect, i.e., the Teflon surface at the hole is not homogeneously coated with hexadecane. (5) Lytic materials are present (see below, *Conductance Artifacts*).

No Onset of Bilayer Formation. This is mostly caused by imperfect apertures as well as the same reasons listed above for breakage.

Conductance Artifacts. When bilayers have formed at the first try and are locked in properly but show artifactual conductances, and this occurs after replacing septum and coating solution, this is not due to the bilayer formation procedure but rather the presence of artifact-inducing materials (AIMs). There are two ways to proceed: identification of origin and measures to avoid AIMs. Knowledge is required of possible origins, at least of classes of origins, which is a nontrivial task because some materials (see below) induce artifacts at extremely low concentrations. AIMs may be found in the buffer, the lipid preparation, the coating solution, the Teflon cell, glassware, pipets, syringes, Teflon tubing, and paper tissue.

Avoidance of AIMs. (1) Glassware, before being washed, should be cleansed of any tape residue or colors from marking pens, because these materials often are or contain very potent ionophoric or lytic materials. (2) Plasticizers also are dangerous. Plastic tips for pipets, plastic dispensers, or containers, or any plastic which comes into intimate contact with solutions for bilayer work, should be avoided or carefully checked. (3) Other major sources of AIMs are microorganisms and their metabolites. Filtering buffers with sterile 0.2-μm filters are often not sufficient. Metabolites have been found to survive even double distillation. Crucial glassware (for vesicles and buffers) should be sterile at the start and water should be as clean as possible; techniques such as serial deionization, activated charcoal filtering, sterile filtering, and double distillation should be used. If facilities are available, stock buffers should be autoclaved.

Avoidance and Identification of AIMs. This necessitates rigid control of all preliminary and peripheral components of bilayer work. The following strategy is easy to establish and has proved valuable. For all solutions and materials used, one should save (under stable conditions) controls that reliably yield AIMs-free bilayers and thus can be used for cross-checks to identify origins of AIMs. For example, coating solution should be prepared in a large batch, divided into about 100 ampules of 1 ml each (in a dust-free hood), and stored at $-20°$. Before the batch is used up, prepare a new one. This procedure should be applied to lipid samples [ampules with different kinds of lipids or common lipid mixtures in solvent, directly used for vesicle formation (see *Appendix C*)]; buffers (bottles with 20 ml of standard buffers used for vesicle formation), Teflon tubing for 1-ml syringes (100 m cleaned and stored, 20×5 m in airtight sacs, at $-20°$); paper tissue, which should be purchased in large stock and dated; Teflon sheets, Teflon cells, and dishwashing detergent.

Electrical Characterization of Bilayer and Channels

Lipid Bilayer

Specific capacitance has been described earlier. Specific resistance values are in the range of $1-2 \times 10^8$ Ωcm^2 at 100 mV applied voltage for typical lipid mixtures used for channel studies. The stability against voltage is measured by the breakdown voltage, which is defined by the voltage value where the membrane breaks during a voltage ramp of 10 mV/sec. For soybean lipid with cholesterol [12:1 (w:w) 120 mM KCl, 0.1 mM CaCl$_2$, buffered to pH 7)], this is typically 370 mV. Estimated lifetimes at continuous voltage are ≥ 1 min (250 mV), $\geq 10-15$ min (150 mV), and only weak correlation between lifetime and voltage is found below 100 mV. These estimates do not apply to bilayers just after formation.

The capacitive noise of a planar bilayer sets a limit for channel resolution. It is proportional to the bilayer area, i.e., it decreases with decreasing aperture diameter. However, at about a 60-μm diameter the capacitive noise has decreased to the noise level of the best-equipped current-to-voltage converters. For the equipment described here, total rms current noise at 1 kHz filtering and 10^{10} Ω feedback resistor is 1, 0.4, and 0.2 pA for holes of diameter 160, 100, and 60 μm, respectively.

Channels

There is abundant literature on electrical characterization of channel conductances,[13,17] to which the reader is referred because such details are

[17] B. Hille, "Ion Channels of Excitable Membranes." Sinauer, Sunderland, Massachusetts, 1984.

outside the scope of this chapter. Two basic terms are defined below.

1. Channel current, i, or conductance, λ, is given by Eq. (4)

$$V_{out} = iR_f = V_{in}\lambda R_f \qquad (4)$$

where V_{in} is the dc input voltage, V_{out} is the output voltage of the amplifier, and R_f is the feedback resistance (see *Appendix A*).

2. Time and amplitude resolution: A required time resolution is determined by the time the channel to be studied is open. Choose the highest R_f value which just allows for this time resolution, thus optimizing amplitude resolution at this particular time resolution. For the amplifier described (see *Appendix A*) suggested choices are $R_f = 10^9$, 10^{10}, and 10^{11} Ω for time resolutions of 0.2, 1, and >4 msec. If this optimized amplitude resolution is not sufficient to actually resolve the channel, use smaller aperture sizes (see *Lipid Bilayer*). For instance, a channel of 2 pS conductance and 10 msec mean open time can be resolved at 100 mV using a 60-μm aperture (channel current to noise ratio 1 : 1 at a time resolution 10 times the mean open time), but is not resolvable using 160-μm apertures. However, choice of optimal aperture size also depends on other independent factors outlined in the next section.

Assays to Relate Channel Function to Membrane Structure

The main advantage to forming planar bilayers from vesicle-derived monolayers is the control of several structural and physical variables of membranes to which channel activities can then be related.

Channel Proteins

Number of Proteins per Bilayer. The number of proteins per bilayer is approximately equal to the number of lipids per bilayer times the molar ratio of protein to lipid in the overall vesicle suspension. The conservation of the protein-to-lipid ratio (p/l) during bilayer formation is evident in the first step — monolayer formation from vesicles — for several representative vesicle suspensions using radioactive tracer methods.[7,18]

$$(p/l)_{mon} = (p/l)_{ves}\eta_{v,m} \qquad (0.7 < \eta_{v,m} < 1.4) \qquad (5)$$

This result applies to times up to 10 min after monolayer formation, a time much longer than is needed for bilayer formation. At longer times η has been found to increase.

Conservation of (p/l) ratios during bilayer formation from monolayers is expected from the process of apposition, but is difficult to assess. For

[18] H. Schindler and U. Quast, *Proc. Natl. Acad. Sci. U.S.A.* **77**, 3052 (1980).

matrix protein channels almost quantitative recovery has been shown.[19] The number of observed channels divided by the estimated number of lipids (l) in one layer of the bilayer ($l =$ area of bilayer divided by area per lipid, about 0.65 nm^2) gave a value which was 0.5–1 times the (p/l) ratio in the vesicle suspension. The same study revealed, however, that this conservation of (p/l) ratios was no longer guaranteed when bilayers were formed after the first attempt (cf. *Channels,* comment 2 above). This is another reason for suggesting the "first-try criterion."

For a channel study at a more qualitative level it can be assumed from the above reasoning that

$$(p/l)_{\text{bil}} = (p.l)_{\text{ves}}\eta_{\text{v,b}} \qquad (0.5 < \eta_{\text{v,b}} < 2) \qquad (6)$$

provided bilayers are formed at the first try. This should be confirmed if accuracy is required, as it is for assaying lateral distributions of proteins (described below).

For single- or multiple-channel traces there should be about 10–100 channel proteins present in the planar bilayer. The number of lipids per bilayer, assuming 0.65 nm^2 area per lipid, is about 3×10^{10} (160-μm hole) and 4×10^9 (60-μm hole) for each of the two layers. Practical lipid-to-protein ratios in vesicle suspensions used for channel studies range between $10^6 < (l/p)_{\text{ves}} < 10^{10}$. If, for some reason, one wants to have as many proteins as possible in the planar bilayer, there is an upper limit at about $(l/p) \approx 10^4$. At higher protein content artifacts appear, probably related to condensation of protein at the annulus, which creates instabilities at the bilayer boundary.

Transverse Distribution of Proteins (Sidedness). For those channels for which sidedness could be assayed by ligand binding,[18,20–25] complete functional sidedness was found when vesicles with protein were present only in one compartment. This, however, is not an obvious expectation and it remains uncertain whether this generally applies. The kind and degree of sidedness for various proteins may differ from that in vesicles to that in the

[19] H. Schindler and J. P. Rosenbusch, *Proc. Natl. Acad. Sci. U.S.A.* **78**, 2302 (1981).

[20] N. Mazurek, H. Schindler, T. Schürholz, and I. Pecht, *Proc. Natl. Acad. Sci. U.S.A.* **81**, 6841 (1984).

[21] H. Schindler, F. Spillecke, and E. Neumann, *Proc. Natl. Acad. Sci. U.S.A.* **81**, 6222 (1984).

[22] N. Mazurek, T. Schürholz, H. Schindler, V. Rulich, I. Pecht, and B. Rivnay, *Immunol. Lett.* **12**, 37 (1986).

[23] L. Hymel, J. Striessnig, H. Glossman, and H. Schindler, *Proc. Natl. Acad. Sci. U.S.A.* **85**, 4290 (1988).

[24] L. Hymel, M. Inui, S. Fleischer, and H. Schindler, *Proc. Natl. Acad. Sci. U.S.A.* **85**, 441 (1988).

[25] L. Hymel, H. Schindler, M. Inui, and S. Fleischer, *Biochem. Biophys. Res. Comm.* **152**, 308 (1988).

monolayer. So far, this has not been investigated, because sidedness could be imposed by asymmetric vesicle addition, as reported both in cited and in several unpublished studies. It should be added that for generating protein sidedness the first-try criterion is again important.

Lateral Distribution of Protein. Control of the initial distribution of proteins in the bilayer plane turns out to be a most informative tool in structure–function investigations.[19-25] Random or nonrandom distributions simply result from appropriate choices of vesicle suspensions.

1. Random initial distribution of single proteins. At the high lipid-to-protein ratios (typically 10^8) used for single-channel studies, proteins occur singly in vesicles. More precisely (for the techniques in *Appendix C*), vesicle diameters *(d)* range between 100 and 300 nm, which correspond to a range of lipids per vesicle from 10^5 to 10^6 [$N_L = 2\pi d^2/A_L = 2\pi d^2/0.65$ nm$^2 \approx 10d^2$ (nm)^2]. Provided proteins occur in the original detergent solution as monomers, and provided that they do not aggregate during detergent dialysis for vesicle formation (avoided by the fast dilution technique), one may safely conclude that proteins occur single in the vesicles. Because there are 100–1000 vesicles per protein, the probability of finding two proteins in the same vesicle is negligible. Consequently, in monolayers generated from such vesicles, proteins will be distributed singly at random, and likewise in bilayers formed from such a monolayer and a lipid monolayer.

Starting from this initially random distribution, proteins will undergo lateral diffusion, with occasional encounters. If encounters lead to a change in channel function, this is detected by following channel activity with time, as shown in Fig. 5. Matrix protein trimers[19] are present in soybean lipid vesicles at a ratio of 10^8 lipids per trimer. The expected number of trimers per bilayer is 300 (160-μm aperture, 3×10^{10} lipids per layer, $\eta_{v,b} \approx 1$). After bilayer formation no conductance events are observed at 180 mV constant applied voltage. After about 2 min, triple-channel events appear (arrows in Fig. 5), characteristic for matrix protein channel conductances (cf. legend to Fig. 5 and Ref. 19 for details). It is concluded that aggregation of matrix protein trimers is required for channel activation. The time lag for onset of observing channels should relate to the mean collision time t_c. This time should be in the range of several minutes (time counts begin from monolayer formation, and it is about 0.5 to 1 min, the time of bilayer formation, before analysis can be started). It can always be adjusted to within this practical range by choosing appropriate values of n_s, the number of proteins per unit area of bilayer (in the present case, $n_s \approx 10^6$ trimers/cm^2). Assuming binary collisions upon random walk,[26] an

26 E. W. Montroll, *J. Math. Phys.* **10**, 753 (1969).

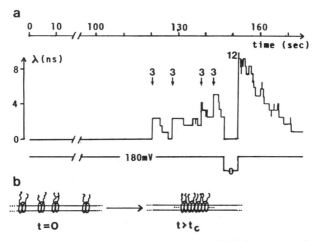

Fig. 5. (a) Channel conductance trace observed for an initially random distribution of about 300 matrix protein trimers in a planar bilayer. Zero time refers to the time just after bilayer formation and application of 180 mV. The time between monolayer formation and bilayer formation was about 30 sec, which has to be added to the time indicated for estimating collision times (see the text). The number "3" over each arrow above the channel trace indicates simultaneous activation of three channels (cf. Refs. 19 and 25 for details), which then start to close due to the high value of applied voltage. Four activation events of triple channels are shown. The applied voltage is then set to zero for a short period. This causes the activated 12 channels to reopen, as is evident from renewed application of 180 mV, followed by the voltage-driven closing of the 12 channels. The data shown represent the onset of activation; activation continues over a period of 15 min. (b) The conclusion drawn from the data in a. Activation of channels requires association of trimers. This is inferred from the absence of channel events in the first 2 min after having formed the planar bilayer under conditions where trimers are expected to be distributed at random. After a time that is consistent with estimates for the time required for aggregate formation (first collision time, t_c; see the text) via lateral diffusion, channels are observed (for details see the text). Aqueous phases were buffered to pH 7.4 and contained 1 M NaCl and 1 mM CaCl$_2$.

approximate relation between t_c, and the lateral diffusion constant D_{lat} is $t_c = k(n_s D_{lat})^{-1}$. For initial estimates one may assume k to be 0.05 (empirically determined for the matrix protein) and $D_{lat} = (1-5) \times 10^{-10}$ cm^2/sec, typical values for single integral membrane proteins. Obviously, collisions or associations of channel proteins may have effects on channel function different than the one found for matrix protein (see, for instance, Refs. 20, 21, 23, and 24). The above relation for t_c provides a detailed test, since the product $t_c n_s$ should be constant for collisions or associations to be the cause of observed changes of channel function. Also, this provides an *a posteriori* test for the assumption made, that proteins had been single in detergent.

2. Nonrandom initial distributions of proteins. Such distributions are generated by preparing lipid–protein vesicles with, on the average, several to many proteins per vesicle and mixing a small aliquot of these vesicles with lipid vesicles. For example, vesicles are prepared at an $l/p \approx 10^4$, so that 10–100 proteins per vesicle are expected. Of this a 1-μl sample is added to 10 ml of lipid vesicle sample (both at 1 mg lipid/ml), yielding an overall ratio $l/p \approx 10^8$. Using this vesicle preparation for bilayer formation results in initial occurrence of protein clusters. This assay complements results from random single-protein distributions. For example, when applied to matrix proteins, channels are immediately observed and the number of channels neither decreases nor increases with time showing that association, which activates channels, is irreversible.

3. Codistributions of different constituents. Functional interactions can be analyzed between different kinds of proteins, or proteins and lipids, via generation of initial codistributions of the components in question (for first examples, see Refs. 19, 20, and 22). Consider two different components (indices a and b), available in purified form. One component is a channel protein, the other may be either another protein (for example, see Refs. 20 and 22) or a particular lipid (see Ref. 19). Seven different kinds of vesicle samples can be prepared by incorporating the components separately or together into lipid vesicles. Separate incorporations yield four different samples: V_{a1}, V_{b1}; and V_{am}, V_{bn}. The first two samples contain per vesicle at most one copy of component a (V_{a1}) or of component b (V_{b1}); the latter two samples are prepared at conditions of, on the average, m and n copies per vesicles of the components a and b, respectively. Simultaneous incorporation of the two components yields three additional kinds of vesicle samples: $V_{a1,bn}$; $V_{am,b1}$; and $V_{am,bn}$.

Mixing these vesicle preparations in different combinations, together with lipid vesicles ($V_{0,0}$), presents different kinds of initial codistributions of the two components in the planar membranes (where α, β, and γ are molar ratios of the different vesicles). (1) Uncorrelated distributions: (a) both components at random ($\alpha V_{a1} + \beta V_{b1} + \gamma V_{0,0}$),[22] (b) one component at random, the other nonrandom ($\alpha V_{am} + \beta V_{b1} + \gamma V_{0,0}$, or $\alpha V_{a1} + \beta V_{bn} + \gamma V_{0,0}$),[19,20] (c) both components nonrandom ($\alpha V_{am} + \beta V_{bn} + \gamma V_{0,0}$). (2) Correlated distributions: (a) one component at random, the other nonrandom ($\alpha V_{a1,bn} + \beta V_{0,0}$, or $\alpha V_{am,b1} + \beta V_{0,0}$),[19] (b) both components nonrandom ($\alpha V_{am,bn} + \beta V_{0,0}$).

Differences in channel activities observed for these different types of initial codistributions, together with the development of channel activities with time, will provide detailed insight into roles of protein–protein or protein–lipid interactions in channel function. (For details on lipid–protein interactions see the following section, *Lipids.*)

Lipids

Lipid Composition and Asymmetry. There are no restrictions on the type of lipid mixture applicable for planar bilayer formation, provided the lipids form stable vesicles of sufficient size (≥ 100 nm in diameter) and provided they generate monolayers sufficiently fast (requirements have been detailed in the section *Appropriate Vesicle Suspensions*). Mixtures of phospholipids with various polar groups, acyl chain lengths, and degrees of saturation, like most native lipid extracts, generally form stable vesicles and planar bilayers rather than single types of phospholipids. This is probably due to a local structural and energetic match of components which alone may not form lamellar phases (e.g., phosphatidylethanolamine with saturated chains) or may form less stable lamellar phases. As discussed for proteins, compositions of monolayers were found to be practically identical to those of vesicles used to form the monolayers.[7] Apposition of two differently composed monolayers results in a planar membrane with an initial lipid asymmetry, which, in time, may randomize for particular types of lipids, such as sterols. Unadulterated initial asymmetry implies first-try membranes.

Initial Lateral Distributions of Lipid. Concomitant with protein distributions, one may impose lipid distributions, initially in the bilayer plane, in order to identify lipid–protein interactions[19] and to investigate whether association is irreversible[19] or reversible. Possible codistributions of a particular type of protein and a particular type of lipid were listed in the preceding section. The first of the possibilities (see 1a) appears, however, unrealizable for lipid–protein interaction studies, in contrast to protein–protein interaction studies. Also, in estimating times for association from noncorrelated initial distributions or randomization from initially correlated distributions, one has to take into account the relatively fast lateral diffusion of lipids ($D_{lat} \approx 10^{-8}$ cm^2/sec).

Lateral Membrane Pressure

The equilibrium pressure π_e of a monolayer in equilibrium with vesicles (see *Principle of Planar Membrane Formation from Vesicles*) can be regarded as a measure for the physical state of the vesicle membrane and of the planar bilayer formed. In other words, π_e values that are measured for particular vesicle and aqueous phase compositions are a measure for the lateral pressure values in planar bilayers with the same compositions. π_e values depend on almost any parameter of membranes and aqueous phases, most notably cholesterol, calcium ions, and degrees of unsaturation of acyl chains.[15] The significance of this concept has been shown for the acetylcholine receptor channel,[21] where channel aggregation was found only above a critical π_e value, irrespective of how π_e values were adjusted via membrane and aqueous phase conditions. Likewise, for colicin A[15] the unique dependence of functional parameters on lateral pressure can be found. In comparison with other already mentioned adjustable membrane parameters, quantitative control of lateral membrane pressure is less advanced,[15] but it has been included here because lateral pressure is a membrane state variable of principal importance.

Aqueous Phases

There is direct access to both aqueous phases, although changes at the high ohmic side (connected to the amplifier) are increasingly delicate for increasing values of the feedback resistor (grounding of solutions and glass pipet before addition is advantageous). In the front chamber (connected to the input signal) complete replacement of aqueous phases is routinely achieved without membrane breakage using coupled in–out syringes.

General Comments on Structure–Function Assignments

The assays thus far described address relations between particular adjustable membrane parameters and channel function. However, observed effects of a particular parameter may be either direct or indirect because of other structural parameters. For example, if a particular lipid is found to induce changes in channel activity, at least four different modes of action may be distinguished.

Indirect coupling modes. The lipid may exert its effect indirectly by changing the physical state of the membrane, assayed by the lateral pressure parameter π_e. Such changes of π_e may either induce changes of channel activities[15] directly or indirectly via pressure-induced aggregation or demixing of components.[21] Direct π_e effects *(mode 1)* can be identified by independent adjustments of π_e using other components. Effects of π_e-induced association *(mode 2)* can be identified using initial distribution assays.

Direct coupling modes. The lipid may exert its effect by directly interacting with the channel protein. This interaction mode can be inferred when the lipid is effective at constant π_e values (if the lipid has, in addition, an effect on π_e, this change of π_e has to be compensated by other means). Still, there are two modes to be distinguished for direct lipid–protein interactions. Such interactions may be restricted to single interacting units, i.e., one channel protein surrounded by lipids, where lipids may be either in partition with bulk lipid or rigidly bound[19] *(mode 3)*. Such interactions may further induce association to large clusters with repeating lipid–

protein units *(mode 4)*, as observed for lipopolysaccharide–matrix protein interaction.[19,27]

Perhaps the most interesting findings so far relate to the L-type Ca^{2+} channel of transverse tubule membranes and to the Ca^{2+} release channel of sarcoplasmic reticulum. Using initial distribution assays, both kinds of channels were recently shown to function as "oligochannels," where several channels in channel protein associates open and close synchronously, giving the appearance of one larger channel.[23–25]

Other findings can be treated analogously. A complete systematic account of identifiable coupling modes between adjustable parameters of membrane structure and.observed channel function would be beyond the scope of this technically oriented chapter.

Start-Up Kit for Users

We regard the information presented regarding this technique to be rather complete. In setting up the bilayer apparatus and getting used to the assays two difficulties are generally encountered. One relates to the fabrication of Teflon septa with precise apertures and of a cell with the appropriate shape (see *Mechanical Parts* and *Appendix B*). For starting up experimentation of this type we can provide a few septa with different aperture sizes and one bilayer cell (which we request be returned to us after being copied). Additional septa can be purchased from our institute for a reasonable price.

The second difficulty concerns acquiring expertise in performing channel analysis using the assays described. It is advisable to gain confidence by using a stable, nondelicate channel protein that allows for a representative screening of the assays described. For this purpose we suggest using two different types of channel proteins. One is a porin, PhoE (kindly provided as purified trimers by Dr. J. Rosenbusch, Biozentrum, University of Basel). Typical PhoE channel conductance traces are shown in Fig. 6a and b. Channel characteristics are almost identical to those of matrix protein channels[19,27] (see Fig. 5). Trimers form triple channels in associates of trimers, with a voltage-dependent open–close equilibrium. The protein is very stable and easy to handle. The other protein is colicin A [kindly provided in purified form by D. F. Pattus, European Molecular Biology Laboratory (EMBL) Heidelberg, Federal Republic of Germany]. Colicin A is a bacterial toxin that provides an example of channel formation upon

[27] A. Engel, A. Massalski, H. Schindler, D. L. Dorset, and J. P. Rosenbusch, *Nature (London)* **317**, 643 (1985).

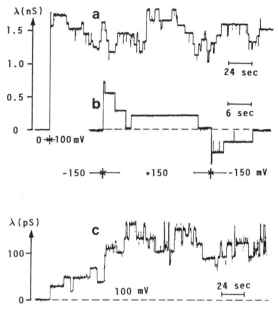

FIG. 6. PhoE channel traces. (a) Multiple-channel trace upon application of 100 mV. (b) Triple-channel events with voltage-induced closing at a constant 150 mV of both polarities. Voltage reversal after channel closing leads to transient channel opening, indicating two distinct closed states for either voltage polarity. Application of 100 mV (trace a) leads to only partial voltage-driven closing. These features (including others not shown) are the same as observed for matrix protein channels.[19,27] Lipids used were soybean lipid with cholesterol [12 : 1 (w : w)]. Aqueous phases contained 100 mM NaCl, 0.1 mM EGTA, and 0.2 mM CaCl$_2$ and were buffered to pH 7.4. (c) Colicin A channel trace. Colicin A was added to one side of a preformed planar membrane to a final concentration of 0.1 ng/ml. After ~ 20 sec, channel conductance events are observed, the number increasing in time to a (voltage-dependent) stationary value. Colicin A insertion to form active channels requires the presence of voltage across the bilayer, with more positive potential at the side of colicin addition. Once activated, the channels show a voltage-dependent open–closed equilibrium.[28,29] Soybean lipid was used. Aqueous phases contained 1 M KCl and were buffered to pH 6.1.

protein insertion in the membrane from the aqueous phase.[28,29] A typical channel trace is shown in Fig. 6c. Channel formation upon insertion requires application of voltage with certain polarity,[28] the equilibrium between open and closed states is highly voltage-dependent and the single channel conductance is dependent on pH[28] and π_e.[15] These dependencies, as well as others not mentioned, will provide a detailed exercise to gain confidence in channel characterization.

[28] S. J. Shein, B. L. Kagan, and A. Finkelstein, *Nature (London)* **276**, 159 (1978).
[29] F. Pattus, D. Cavard, R. Verger, C. Lazdunski, J. P. Rosenbusch, and H. Schindler, *in* "Physical Chemistry of Transmembrane Ion Motions" (G. Spach, ed.), p. 407. Elsevier, Amsterdam, 1983.

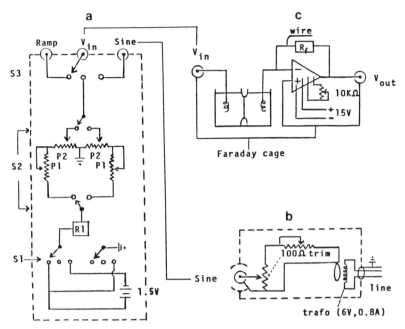

FIG. 7. Circuit diagrams to build a dc source (a), a sine wave source (b), and a current-to-voltage converter (c). trafo, Transformer.

Appendix A

Figure 7 shows a simple design to generate input signals [dc signal (a); sine wave (b)] and to convert membrane current to a low ohmic voltage signal (c).

The elements shown in (a) are mounted on a metal box (about $10 \times 12 \times 6$ cm). Three switches (S1, S2, and S3) and two potentiometers (P2; these are 1-kΩ, 10-turn potentiometers with dials) are on top and three BNC jacks are at one side. Voltage is supplied to R1 (220 Ω) using a 1.5-V battery (9 Ahr), where switch S1 selects polarity and zero voltage (+ V, 0, – V). Switch S2 selects between two voltage dividers. This allows for voltage jumps between preset voltage values using the 10-turn potentiometers. The trim-potentiometers (P1, 500 Ω) are adjusted to 100 mV per turn of P2. Switch S3 selects between the three kinds of input sources used.

One loop around the transformer (depicted in Fig. 7b, "trafo") provides a sine wave at line frequency, the peak-to-peak value of which is adjusted by the trim-potentiometer to about 16 mV, used to monitor planar bilayer formation.

A small circuit board (about 2×5 cm) (depicted in Fig. 7c) carries the amplifier (type OPA 128LM, Burr Brown, Tucson, Arizona) the 10-kΩ trim-potentiometer, the feedback resistor R_f (ELTEC Instruments, Inc.,

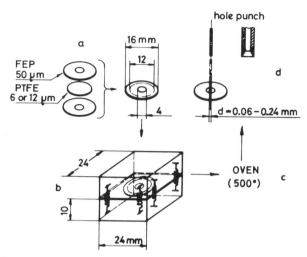

FIG. 8. Schematic representation of the assembly of the Teflon septum from three sheets (a) in a spring-loaded clamp (b), the FEP outer rings of which are fused together in an oven (c); punching a hole with a steel punch sharpened only on the inside (d).

Daytona Beach, Florida, Model 104), available with resistance values (1, 2, and 5×10^n, $n = \cdots 8, 9, 10, 11$), and a third wire which defines capacitance to 0.002 pF. The negative input of the amplifier should be shielded and directly connected with the Ag–AgCl electrode via a high-ohmic, low-capacitance jack. Voltage ($+15$ and -15 V) is supplied by 20 1.5-V batteries mounted in a box within the Faraday cage and the output signal V_{out} is connected to a BNC jack at the rear of the Faraday cage, as is the input signal. As a housing for the circuit board we use a cylindrical steel tube about 2 cm in diameter and 5 cm long.

Appendix B

Figure 8 illustrates the fabrication of Teflon septa. Sharp steel punches are used to obtain Teflon rings of 50-μm FEP Teflon with a 4-mm inner diameter and 16-mm outer diameter and 6- or 12-μm PTFE disks 12 mm in diameter (both types of Teflon are purchased from Saunders Corp., Los Angeles, California). The three sheets are sandwiched between stainless-steel blocks, both covered inside with smooth aluminum foil. The blocks are connected by springs (spring force about 0.5 N, which should not significantly drop during heating) and are heated for about 4 min in an oven at 500°. During this, the assembly is positioned between preheated ceramic plates to ensure even heat flow from top and bottom. After removing from the oven the assembly is shock-cooled by thrusting it into cold water. This procedure should yield a septum, the outer rim of which is fused where the FEP sheets are in contact. The PTFE sheet in the 4-mm inner part should be planar. The contact between FEP and PTFE should be tight (they are not fused together because the melting temperature of PTFE should not be reached during heating). To obtain such septa there are three adjustable parameters: time of heating, spring force, and weight of ceramic plate. The aluminum foil should be replaced if the septum starts to stick to it. Apertures are punched with sharpened stainless-steel tubes (sharpened only from the inside). Fabrication of punches is a rather involved procedure, requiring a special drilling tool with conical shape, triangular cross-section, and sharp tip. This is moved into a revolving stainless-steel tube with self-adjustment of the triangular tool to the inner rim of the steel tube (bending forces of the thin tube during sliding in of the tool should be minimized by control under a microscope). During sharpening by slowly moving the tool inward (use a hydraulic stage), a fine grindstone is held to the outer rim of the tube at one edge of the triangular tool in order to avoid bending out the rim and to polish the outside of the tube.

Appendix C

Lipid Vesicles

Lipid vesicles are generally used for diluting lipid–protein vesicles to appropriate overall lipid–protein ratios and for generating the different types of initial distributions of protein in the planar bilayer.

Preparation method[18]: 20 mg of lipid in 4 ml of solvent is added to a 1-liter round-bottom flask. The lipid is dried down to a thin film in the following way. A stream of nitrogen is applied via a Pasteur pipet inserted into the flask. The flask is rapidly swirled in such a way that the solvent remains spread as thinly as possible until it is evaporated. To remove residual solvent, nitrogen flow is continued for a few minutes. Buffer is always filtered through a 0.2-μm sterile filter directly into the flask. After adding about 30 glass beads (2 mm in diameter, carefully cleaned), the flask is rotated for 5 to 10 min such that rolling of the glass beads is smooth and extends over the whole lipid film. The resulting vesicle suspension is filtered through a 0.4-μm polycarbonate filter (Nuclepore) into a sterile flask. This simple and fast method yields vesicles which are well-suited for planar bilayer formation (cf. *Appropriate Vesicle Suspensions*) with, however, restrictions on the choice of lipids. The presence of at least a small percentage of negatively charged lipids appears to be required, otherwise multilamellar structures and aggregation dominate, reducing monolayer formation rates to below reasonable values (discussed earlier).

Lipid–Protein Vesicles

Fast dilution technique[21]: The procedure is the same as for lipid vesicles described above except for the addition of protein to the buffer. Since details are of importance the procedure will be described step by step. (1) Prepare a lipid film in a round-bottom flask with continuous nitrogen flow until step 5. (2) Prepare a crude suspension of the same lipid by adding about 5 mg of dry lipid to 2 ml of buffer, then vortex for 10 sec. (3) About 1 ml of this suspension is added to 19 ml of buffer by filtering it through an 80- to 200-nm polycarbonate filter (choose pore size as small as possible but with easy flow of suspension). (4) Immediately after step 3, add a small aliquot $(1-5\ \mu l)$ of protein sample (purified protein in detergent and, optionally, lipid) to the 20 ml of buffer during both shaking and mild bath sonication of the buffer. (5) The buffer is immediately added to the flask and the lipid film is resuspended with glass beads as described above.

This technique has been devised to ensure single protein incorporation into vesicles and to achieve much lower detergent-to-lipid ratios as in dialysis methods. The rationale is as follows. Protein is added in such a way (low amount in large volume during shaking and sonication) that the proteins are immediately dispersed. If they were single in detergent, they will still be single after the fast dilution technique, because encounters are highly improbable. The detergent added with the protein is also "fast diluted," but there is sufficient lipid present to which the protein can associate with during loss of detergent. More precisely, the described way of adding lipids ensures that there are prevesicular or nonclosed lipid structures offering sufficient association sites for the few proteins added. During resuspension of the lipid film, these structures are partly integrated into forming vesicles or partially complemented to vesicular structures. This is inferred from the relatively high recovery of the (p/l) ratio when monolayers are formed from these vesicles. For example, for the acetylcholine receptor the recovery was 70%, i.e., $\eta_{\text{fast dil}} = (p/l)_{\text{mon}}/(p/l)_{\text{bulk}} = 0.7$. Also, the detergent-to-lipid ratio may be lowered to 10^{-5} or 10^{-6} when fast dilution is followed by dialysis for 1 or 24 hr, respectively. During dilution the detergent molecules primarily dissolve into the buffer and can therefore be removed to a greater extent, as after vesicle formation by dialysis. Detergent is apparently trapped in lipid structures as inferred from the low detergent-to-lipid ration of 10^{-4} after exhaustive dialysis.

It should be added that this fast dilution technique is applicable only for incorporation of small amounts of protein; optimal conditions for step 4 are the addition of 2 μl of protein sample with about 10 μg/ml protein. The number of lipids in the film should be adjusted to yield appropriate (p/l) ratios for the particular assays to be employed.

Acknowledgments

The development of the reconstitution technique described in this paper has been carried out at the Biocentre of the University of Basel, Switzerland, and was supported by research grants from the Swiss National Science Foundation (Grants 3.390.78, 3.773.80, and 3.303.82).

[12] Spontaneous Insertion of Integral Membrane Proteins into Preformed Unilamellar Vesicles

By DAVID ZAKIM and ANTHONY W. SCOTTO

Complete understanding of the function of proteins that are integral components of membranes depends on the preparation of pure forms, which then can be studied in well-defined synthetic membranes. The model system for such studies usually is a unilamellar bilayer of phospholipids in the form of a vesicle (ULV). The pure proteins hence must be assembled into these vesicles. Many techniques have been described for achieving this end. Most depend on forming the ULVs in the presence of the protein of interest, as, for example, by cosonication of phospholipids and proteins of interest or by mixing phospholipids and membrane proteins in the presence of detergents, which then are removed by dialysis. Recently, however, techniques have been described by which pure membrane proteins can be inserted spontaneously into preformed ULVs.[1-4] These techniques are advantageous because of their rapidity and the mild conditions needed. In addition, they allow for direct study of the mechanisms by which integral membrane proteins may be assembled into lipid bilayer membranes, which is not possible if the bilayer is formed in the presence of proteins. We describe in this chapter simple methods for the spontaneous insertion of several integral membrane proteins into small sonicated ULVs of dimyristoylphosphatidylcholine (DMPC). Although the methods have been tested thoroughly with only three integral mem-

[1] A. W. Scotto and D. Zakim, Biochemistry 24, 4066 (1985).
[2] A. W. Scotto and D. Zakim, Biochemistry 25, 1555 (1986).
[3] N. A. Dencher, Biochemistry 25, 1195 (1986).
[4] N. A. Dencher, this volume [13].

brane proteins and bilayers comprising only one type of phospholipid (DMPC), there is good reason to believe that the methods will have general utility.

Principles of the Method

The underlying principle of the method appears to be that membrane proteins, even when present as large aggregates or sheets (as in the case of bacteriorhodopsin), insert spontaneously and rapidly into ULVs in which there are defects in the packing of the acyl chains.[1,2,5] There are, undoubtedly, many ways in which defects in packing can be induced. The methods described here are based on the presence of impurities in otherwise pure bilayers of DMPC. The method for reconstitution of the glucose transporter from red cell membranes,[6] in which vesicles are frozen and thawed in the presence of the protein, probably is a variation of this idea. The freeze–thaw cycle can induce transient defects in packing of acyl chains.[7,8] We have found, in fact, that membrane proteins will insert spontaneously into pure vesicles of DMPC that are sonicated below the temperature for the gel to liquid crystalline phase transition and that defects induced in this manner can be removed by annealing of vesicles.[9] Spontaneous reconstitution can be effected for DMPC in either gel or fluid states, but conditions for promoting spontaneous reconstitution depend on the phase of the vesicles. It is important to emphasize that the conditions promoting spontaneous insertion of membrane proteins into preformed ULVs also tend to promote fusion between ULVs,[1,2,10,11] enhanced rates of phospholipid flip-flop,[12] and leakiness of the ULVs to solutes trapped on the inside of the vesicles.[13] Thus, the presence of defects in packing has generalized effects on the characteristics of ULVs. It is possible, however, to observe spontaneous insertion of membrane proteins into ULVs under conditions that do not cause fusion between protein-free vesicles.[1,2] This observation indicates that the energy barrier constraining spontaneous insertion of membrane proteins into ULVs is lower than for fusion between ULVs.

[5] M. K. Jain, *Membr. Fluid. Biol.* **1**, 1 (1983).
[6] M. Kasahara and P. C. Hinkle, *J. Biol. Chem.* **252**, 7384 (1977).
[7] R. Lawaczeck, M. Kainosho, and S. I. Chan, *Biochim. Biophys. Acta* **443**, 313 (1976).
[8] N. J. Salsbury and D. Chapman, *Biochim. Biophys. Acta* **163**, 314 (1968).
[9] A. W. Scotto and D. Zakim, *J. Biol. Chem.* **263**, 18500 (1988).
[10] H. L. Kantor and J. H. Prestegard, *Biochemistry* **14**, 1790 (1975).
[11] M. Liao and J. H. Prestegard, *Biochim. Biophys. Acta* **599**, 81 (1980).
[12] S. Massari, P. Arslan, A. Nicolussi, and R. Colonna, *Biochim. Biophys. Acta* **599**, 118 (1980).
[13] M. A. Singer and M. K. Jain, *Can. J. Biochem.* **58**, 815 (1980).

Preparation of Lipid Vesicles

Purification of DMPC

The purity of DMPC used to prepare ULVs is critical. This is so because, as noted already, impurities in the lipids make ULVs fusogenic. Although this is the basis for the method of spontaneous insertion of membrane proteins into preformed bilayers, the presence of uncharacterized impurities in ULVs will mean that the conditions for reconstitution cannot be controlled rigorously. Among other difficulties posed by this lack of experimental control is that the presence of uncharacterized impurities make it extremely difficult to study the process of insertion per se.

The most common and obvious impurities in DMPC are colored compounds that impart a yellow tint to concentrated solutions (100 mg/ml) of the phospholipid in chloroform. A second category of impurities arises from improper storage of DMPC in the powder form. Presumably, there is hydrolysis of the phospholipid resulting in free myristate and lysomyristoylphosphatidylcholine. Storage of saturated phosphatidylcholines in chloroform is not recommended, however, since it leads to a more rapid degradation of the lipid versus storage of a desiccated powder at $-17°$. A third type of impurity is the isomer 1,3-DMPC, which is formed during synthesis of DMPC. This compound is difficult to separate completely from 1,2-DMPC. Routinely, we have been able to obtain a sufficiently pure DMPC (from Avanti, Birmingham, Alabama), which is designated as $\geq 99\%$. Numerous lots obtained from other suppliers have been too contaminated to use directly. These can be repurified, however, as described below. In addition, material from Avanti may deteriorate after lengthy periods of storage and may have to be repurified.

DMPC obtained from Avanti usually can be used without further purification. Radioactive [14C]DMPC can be obtained from Amersham, Arlington Heights, Illinois. DMPC can be repurified essentially according to Colacicco.[14] A necessary precaution, however, is that the silicic acid must be washed extensively. (Silicar silica gel 60–100 mesh type 60A, Mallinckrodt, Paris, Kentucky.) Silicic acid is packed in a column (2.5 × 150 cm) and washed with a gradient from pure chloroform to pure methanol (1 liter), and then with 500 ml of methanol : water (1 : 1). Finally, the silicic acid is washed with absolute methanol, and reactivated by heating at 95° for 1 hr. This treatment removes a visible yellow material and other noncolored substances that otherwise would coelute with DMPC. Then 1–2 g of DMPC in 25–50 ml of chloroform is applied to the column. DMPC is eluted in a stepwise fashion using reagent grade or freshly redis-

14 G. Colacicco, *Biochim. Biophys. Acta* **266**, 313 (1972).

tilled solvents. The column is washed with 500–1000 ml of 100% chloroform. Lipid then is eluted successively with 500 ml of 90, 80, 70, 60, and 50% chloroform in methanol (vol/vol). The first fraction contains 1,3-DMPC, which coelutes with the front of the 1,2-DMPC. The subsequent fractions eluted with 80 and 70% chloroform contain 1,2-DMPC. The fraction eluted with 60% chloroform contains DMPC of unknown purity. Each of the fractions eluted are saved, evaporated to dryness, resuspended in chloroform and quantitated. An aliquot of each is used to prepare vesicles, which are tested for their suitability. Monitoring of the elution of the DMPC can be done by thin-layer chromatography (TLC) or determination of phosphate. However, inclusion of 1.25–2.5 μCi of [14C]DMPC facilitates analysis and quantitation of the purified DMPC.

Preparation of Vesicles of DMPC

DMPC (70 mg) in chloroform is added to a stainless-steel tube. Solvent is removed under a stream of dry nitrogen and dispersed in 10 mM Tris or HEPES, pH 7.5, (30°) and 100 mM KCl. Alternatively, the lipids can be suspended in 100 mM KCl and the pH can be adjusted after hydration and/or after sonication. The mixture is sonicated at 30° in a stainless-steel tube suspended in a water bath. Sonication is carried out for 30 min with a Heat Systems W225 sonicator under a stream of dry argon. The standard tip is used at 30% of maximum power; the output is pulsed so that power is delivered for 60% of the total time of sonication. These precautions are taken to prevent degradation of the phospholipid during sonication. Samples should be clear at the end of sonication.

Vesicles comprising DMPC and myristate are prepared by cosonication. An ethanolic solution of sodium myristate (3 mg/ml) is added to DMPC in chloroform and the solvents are removed by drying under a stream of nitrogen. This must be followed by vacuum evaporation for 1 to 2 hr to ensure the complete removal of the chloroform. When vesicles containing cholesterol are prepared, DMPC is added to the tube and dissolved in chloroform. The cholesterol, which also is dissolved in chloroform (10 mg/ml), is added and the solutions are mixed before drying. Unlike fatty acids, cholesterol will not diffuse rapidly between membranes.[15-17] Therefore, it must be comixed with the phospholipid before hydration to ensure a homogeneous distribution in the ULVs. Solvent is removed as above, leaving a film of cholesterol and DMPC on the vessel wall. The dried residue is hydrated for 1 hr at 37° in buffer with

[15] C. Daniels, N. Noy, and D. Zakim, *Biochemistry* **24**, 3286 (1985).
[16] A. Jonas and G. T. Maine, *Biochemistry* **18**, 1722 (1979).
[17] L. R. McLean and M. C. Phillips, *Biochemistry* **20**, 2893 (1981).

intermittent mixing. ULVs are prepared by sonication, as described above. Sonicated vesicles are centrifuged for 30 min at 39,000 rpm in a Beckman #40 rotor to remove titanium particles and any multilamellar vesicles, which may remain at the end of sonication. Temperature during centrifugation is 37°.

Trace amounts of [^{14}C]DMPC can be added to simplify quantitation of the ratio of phospholipid to protein in final lipid–protein complexes. The concentration of DMPC is determined after centrifugation by assay of inorganic phosphorus.[18] The specific activity of the radiolabel is determined in separate aliquots by liquid scintillation counting. Vesicles must be maintained above the phase transition of the bilayer, preferably 37°, to prevent the fusion of the vesicles. ULVs can be kept for several days at 37° without significant degradation or increase in turbidity. However, with time, the vesicles form a white sediment on the bottom of the tube. These vesicles should not be used.

Preparation of Proteins

We have observed spontaneous insertion of the following preparations of proteins into ULV's of DMPC: bacteriorhodopsin,[19] the GT_{2P} form of microsomal UDPglucuronosyltransferase,[20] mitochondrial cytochrome oxidase,[21] and insulin receptor from human placenta.[22] As prepared in Ref. 19, bacteriorhodopsin is contaminated with residual phytolipids, but contains no detergent. GT_{2P} is purified in the presence of deoxycholate, and is dialyzed extensively prior to use.[20] This enzyme, however, is delipidated. The residual contamination with detergent and lipid of the cytochrome oxidase and insulin receptor is unstudied. Although residual membrane lipids, as contaminants of membrane proteins, do not appear to interfere with reconstitution, more than small amounts of detergent may inhibit spontaneous insertion of membrane proteins into preformed vesicles in the gel phase. Small amounts of deoxycholate, less than CMC, promoted spontaneous insertion of GT_{2P} into preformed vesicles of DMPC in a liquid crystalline state.[1] Only membrane proteins form complexes with ULVs under the conditions described. The several cytosolic proteins tested do not.[1]

[18] A. Chalvardjian and E. Rudniski, *Anal. Biochem.* **36**, 225 (1970).
[19] D. Oesterhelt and W. Stoeckenius, this series, Vol. 88, p. 667.
[20] Y. Hochman and D. Zakim, *J. Biol. Chem.* **258**, 4143 (1983).
[21] T. L. Mason, R. D. Poyton, D. C. Wharton, and G. Schatz, *J. Biol. Chem.* **248**, 1346 (1973).
[22] J. M. Maturo III, W. H. Shackelford, and M. D. Hollenberg, *Life Sci.* **23**, 2063 (1978).

TABLE I
TIME DEPENDENCE OF MYRISTATE-INDUCED
INCORPORATION OF BACTERIORHODOPSIN
INTO ULVs OF DMPC[a]

| Time (min) | R_f | Ratio phospholipid/protein | |
		w/w	mol/mol
0	1.00	0.06	12
0.5	0.47–0.60	1.36	51
1	0.42–0.54	1.86	68
2	0.34–0.49	2.30	84
4	0.32–0.49	2.06	77
15	0.33–0.48	3.14	117
30	0.24–0.49	4.30	162
60	0.24–0.44	4.70	175
120	0.24–0.29	7.37	277

[a] Bacteriorhodopsin was added to preformed ULVs of DMPC containing 11.3 mol% myristate. Samples were mixed at 30°, cooled to 18° for the time shown, and then returned to 30°. The mixtures were centrifuged at 30° on 10–60% glycerol gradients. R_f is the relative distance from the top of the gradient for complexes of bacteriorhodopsin and phospholipid versus the distance from the top of the gradient for essentially lipid-free bacteriorhodopsin.

Spontaneous Insertion of Membrane Proteins into ULVs of DMPC Plus Myristate

ULVs of DMPC and myristate may fuse in the gel phase, depending on the concentration of myristate.[10,11] ULVs containing about 10 mol% myristate fuse readily, but fusion cannot be detected by measurement of light scattering in ULVs containing 0.1 mol% myristate.[1] In addition, the rate of fusion is termperature dependent. It is maximal at 18°.[1,10] Fusion is hard to detect at 5° for ULVs with moderate concentrations of myristate, e.g., 3 to 5 mol%.

ULVs of DMPC and myristate must be kept above the phase transition temperature (24°) to prevent fusion prior to adding the membrane protein of interest. This is so because vesicles that have fused prior to addition of bacteriorhodopsin will not promote insertion of proteins. Thus, protein and ULVs are mixed at 30° and the temperature of the mixture then is lowered to 18°. The mixture at 18° will become turbid almost immedi-

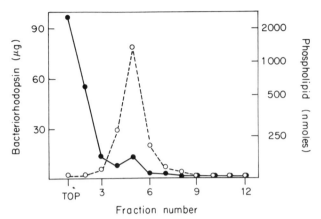

FIG. 1. Distribution of bacteriorhodopsin and ULVs of DMPC reconstituted for 30 sec at 18°. Bacteriorhodopsin (250 μg) was mixed at 30° with 2.5 mg of ULVs of DMPC containing 11.3 mol% myristate. The mixture was treated at 18° for 30 sec and then warmed to 30° prior to centrifugation. Centrifugation at 30° was for 17 hr at 39,000 rpm in a SW41 rotor using a 10–60% glycerol gradient containing 10 mM Tris–HCl, pH 7.5. Phospholipid (●); protein (○).

ately, which is an index that the proteins inserted into the ULVs and that fusion of ULVs occurred. The ratio of protein to lipid in an experiment can be variable; typically we use 10 mg total lipid per milligram protein. It is important to appreciate that the interaction of membrane proteins and vesicles, under these conditions, is a two-step process. The two steps consist of rapid insertion of protein into a small percentage of vesicles followed by growth, via fusion, between protein-free vesicles and lipid–protein complexes (proteoliposomes). The data in Table I indicate the course of a typical experiment with bacteriorhodopsin. With increasing time of treatment at 18°, there is a progressive increase in the ratio of lipid to protein in proteoliposomes isolated by density gradient centrifugation. Note, however, that lipid–protein complexes are formed almost instantly upon mixing ULVs and proteins, that is, within less than 0.5 min. We know, in fact, that proteoliposomes formed at 5 sec under the conditions in Table I will be active in proton pumping.

The growth of the proteoliposomes formed initially, i.e., at 0.5 min (Table I), can be stopped by raising the temperature above 24°. The concentration of protein in proteoliposomes (ratio protein/lipid) can be varied in this way, which also is illustrated by the data in Table I; protein–lipid complexes can be separated from protein-free vesicles by density gradient centrifugation. The distribution of phospholipid and bacteriorhodopsin, with conditions for a typical experiment, are illustrated in Fig. 1. The proteoliposomes in the gradient can be recovered easily by side puncture of the tube, especially when using bacteriorhodopsin or when they scatter light. When colorless proteins are reconstituted and the reconstituted proteoliposomes are small and not easily seen, the gradient must be fractionated. The most convenient method for doing so, in our experience, is to fractionate them from the top using an ISCO model 185 density gradient fractionator. The reconstituted vesicles can be localized then by measurement of the lipid and protein or the appropriate activity of the protein. One need only achieve minimal separation of the two components of the mixture in Fig. 1. This can be accomplished by centrifugation for as little as 1 hr in a SW41 rotor in the case of bacteriorhodopsin. Longer periods of centrifugation will be required for membrane proteins that are less dense than the crystalline arrays of bacteriorhodopsin. Centrifugation, however, must be carried out above the temperature for the phase transition of DMPC. This can be a problem in the reconstitution of temperature-labile proteins.

The rate of growth of proteoliposomes secondary to fusion with protein-free ULVs at 18° depends on the concentration of myristate.[1] Increasing concentrations of myristate, up to the highest concentration we have tested (11.4 mol%), speed up this rate. Even at this concentration, however, the fusion process is not complete until about 2 hr after mixing. Moreover, all the lipids never become associated with the protein. This result reflects that some of the ULVs undergo cycles of fusion with each other, which precludes their subsequent fusion with proteoliposomes. By contrast with the process of fusion between ULVs, the insertion of bacteriorhodopsin into ULVs is extremely rapid. It appears to be complete in less than 1 min, even at concentrations of myristate of 3 mol%. Concentrations of myristate as low as 0.18 mol% (the smallest concentration tested) also promote spontaneous insertion of bacteriorhodopsin, but the rate of this process has not been examined in vesicles containing 0.18 mol% myristate. We also do not know the exact rates for insertion of proteins, other than bacteriorhodopsin, into performed ULVs; but these events seem to be complete within minutes. The size of aggregated proteins, however, may influence spontaneous reconstitution in that sonicated preparations of bacteriorhodopsin but not sheets of purple membrane will insert spontaneously into ULVs that have fused with each other.[1,2]

Rapid cooling, from 30 to 5°, of ULVs comprising DMPC and variable concentrations of myristate is associated with relatively slow rates of fusion. Because fusion at 5° is relatively slow it is possible to insert bacteriorhodopsin — and presumably other membrane proteins — into preformed ULVs of DMPC and myristate at 5°. It is important in this context, however, not to let the ULVs "age" at 5° for more than a few minutes. The proteoliposomes formed spontaneously at 5°, between

DMPC and bacteriorhodopsin, fuse relatively rapidly with protein-free vesicles. Although we have not studied the time course of the insertion event for bacteriorhodopsin at 5°, insertion per se is completed at least within minutes of mixing. One minor shortcoming of reconstitution at 5° is that the secondary growth of proteoliposomes (by fusion with ULVs) cannot be controlled, except by warming the system to a temperature above the phase transition. On the other hand, it is likely that membrane proteins other than bacteriorhodopsin also will insert into vesicles under the above condition (5°). This means that proteins that are unstable at 18° probably can be incorporated spontaneously into preformed ULVs at 5°. Proteoliposomes with variable ratios of lipid to protein can be prepared at 5° by using different initial amounts of lipid and allowing the fusion process (secondary growth of proteoliposomes) to go essentially to completion. This should take no more than 2 hr. Subsequently, protein-free lipids can be separated from proteoliposomes by density gradient centrifugation, as outlined above.

Spontaneous Insertion of Membrane Proteins into ULVs of DMPC Plus Cholesterol

Conditions for inserting membrane proteins into ULVs containing DMPC and cholesterol are basically similar to those already discussed.[2] The phospholipids must be in a gel phase. There are, however, some important differences between using cholesterol or myristate as the "promotor" for spontaneous insertion. The optimum temperature for cholesterol-induced fusion of ULVs is 21°, not 18°. Insertion of a membrane protein into ULVs and subsequent secondary growth of the proteoliposomes by fusion with protein-free ULVs occurs most rapidly at 21°. The process of fusion can be stopped either by raising the temperature above T_c for DMPC or by lowering it. For example, bacteriorhodopsin will insert spontaneously into ULVs of DMPC and cholesterol at 5° but the proteoliposomes do not fuse with protein-free vesicles at this temperature.[2]

Concentrations of cholesterol as low as 0.1 mol% in ULVs of DMPC have been effective in promoting spontaneous insertion of bacteriorhodopsin. Increasing the concentration of cholesterol presumably does not enhance significantly the rate of the insertion event per se, but it does increase the rate of fusion between proteoliposomes and protein-free ULVs. There is a direct relationship between the amount of cholesterol in ULVs of DMPC and rates of fusion at 21°, up to 20 mol% cholesterol. ULVs with DMPC and cholesterol in excess of 20 mol% seem to fuse to a minimal extent and then to remain stable in size.[2] We do not know, however,

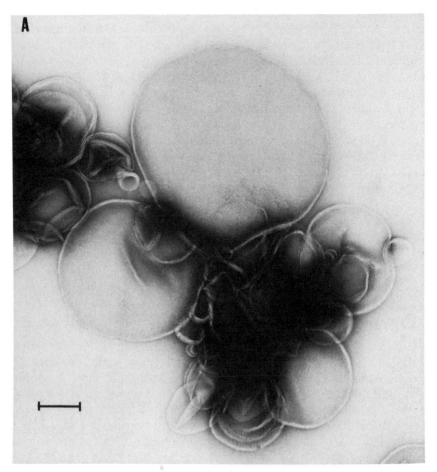

FIG. 2. Electron microscopy of proteoliposomes prepared from bacteriorhodopsin and ULVs of DMPC. The bacteriorhodopsin–DMPC complexes were isolated on a 10–40% glycerol gradient. (A) The ULVs of DMPC contained 11.3 mol% myristate. The mixture of bacteriorhodopsin and ULVs of DMPC were treated at 18° for 1 min prior to centrifugation. Bar, 0.25 μm. (B) Bacteriorhodopsin was mixed with ULVs of DMPC plus 12 mol% cholesterol. After 60 min at 21°, the protein–lipid complexes were separated from protein-free ULVs by gradient centrifugation as in Fig. 1. The protein–lipid complexes were obtained by side puncture of the tube. The fractions containing proteoliposomes were prepared for electron microscopy by negative staining (with 5% uranyl acetate) of samples adhered to glow-discharged carbon film grids. Bar, 0.25 μm.

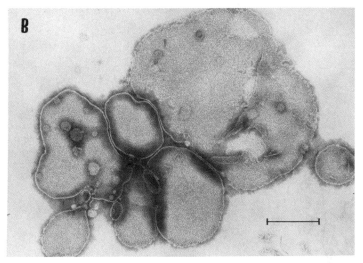

FIG. 2B.

whether concentrations of cholesterol in excess of 20 mol% still promote spontaneous insertion of membrane proteins into vesicles.

Properties of Reconstituted Lipid–Protein Vesicles

Electron microscopy of negatively stained preparations shows that proteoliposomes reconstituted by the above methods are large and unilamellar (Fig. 2). The vesicles made from DMPC and cholesterol, as compared with those made from DMPC and myristate, appear to be stiffer. This conclusion is based on the presence, in the former instance, of vesicles that primarily are large spheres (Fig. 2B), whereas many vesicles composed of protein, DMPC, and myristate have regions that are folded on each other (Fig. 2A). The lipid–protein vesicles prepared as above are more heterodisperse in size versus proteoliposomes prepared by detergent dialysis or cosonication of protein and phospholipids. Also, depending on how finely a gradient is fractionated, some preparations may contain vesicles with variable ratios of lipid to protein (Fig. 3). This, however, is not the cause for variability in the sizes of proteoliposomes, which occurs even in proteoliposomes with a constant ratio of lipid to protein. The size and heterodispersity of large proteoliposomes can be reduced by brief treatment with sonic energy.[1]

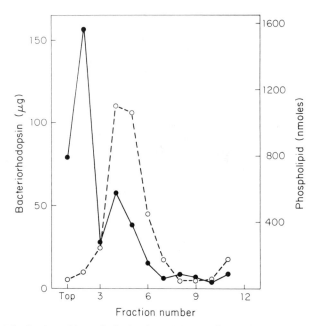

FIG. 3. Distribution of bacteriorhodopsin and ULVs of DMPC reconstituted for 90 min at 21°. Bacteriorhodopsin (250 µg) was mixed at 30° with 3.0 mg of ULVs of DMPC containing 12 mol% cholesterol. The mixture was treated at 21° for 90 min and then warmed to 30° prior to centrifugation. Centrifugation at 30° was for 17 hr at 39,000 rpm in a SW41 rotor using a 10–60% glycerol gradient containing 10 mM Tris–HCl, pH 7.5. Phospholipid (●); protein (○).

Spontaneous reconstitution appears to occur with the proteins in an outside-out orientation. For example, reconstituted bacteriorhodopsin pumps protons to the inside of the vesicles,[1] the active site region of GT_{2P} is exposed to the bulk aqueous phase,[23] and the insulin receptor binds insulin.[24]

[23] Y. Hochman, M. Kelley, and D. Zakim, *J. Biol. Chem.* **258,** 6509 (1983).
[24] J. M. Maturo III, A. W. Scotto, and D. Zakim, unpublished observations.

[13] Gentle and Fast Transmembrane Reconstitution of Membrane Proteins

By NORBERT A. DENCHER

Membrane biology and bioenergetics have gained enormous progress by the application of reconstituted protein–lipid vesicles. Reconstituted vesicles enable the study of structural and functional properties of isolated membrane proteins, the investigation of mechanisms involved in energy coupling, and analysis of lipid–protein and protein–protein interactions. Furthermore, reconstituted vesicles can be used for the formation of planar lipid bilayers with inserted transmembrane proteins. Although nowadays several approaches are successfully applied for reconstituting integral membrane proteins, still much effort is made to develop new procedures that allow easy and fast reconstitution but avoid the usually required harsh treatments such as prolonged sonication, solubilization by detergents, organic solvents, freezing and thawing. In addition, any new approach might give insight into the mechanisms involved in the *in vivo* insertion of membrane proteins in biological membranes. Based on the observation[1] that unilamellar lipid vesicles form spontaneously after mixing aqueous suspensions of long-chain lecithins with small amounts of micellar synthetic short-chain lecithins (fatty acid chain lengths of six to eight carbons), a technique was developed for gentle and rapid functional reconstitution of integral membrane proteins into lipid vesicles.[2] Here, the detailed protocol for spontaneous insertion of the light-energized H^+-pump bacteriorhodopsin (BR) into lipid vesicles composed of various long-chain phospholipids by aid of the short-chain phospholipid diheptanoylphosphatidylcholine (DHPC, Fig. 1) is described. BR has been selected as model membrane protein since it is one of the best characterized transmembrane proteins, and therefore has been used also in several other reconstitution procedures.[3-7] Furthermore, the purple membrane (PM) applied as starting material can be easily isolated in pure form without the use of deter-

[1] N. E. Gabriel and M. F. Roberts, *Biochemistry* **23**, 4011 (1984).
[2] N. A. Dencher, *Biochemistry* **25**, 1195 (1986).
[3] E. Racker, this series, Vol. 55, p. 699.
[4] P. W. M. van Dijck and K. van Dam, this series, Vol. 88, p. 17.
[5] M. P. Heyn and N. A. Dencher, this series, Vol. 88, p. 31.
[6] N. A. Dencher, P. A. Burghaus, and S. Grzesiek, this series, Vol. 127, p. 746.
[7] D. Zakim and A. W. Scotto, this volume [12].

METHODS IN ENZYMOLOGY, VOL. 171

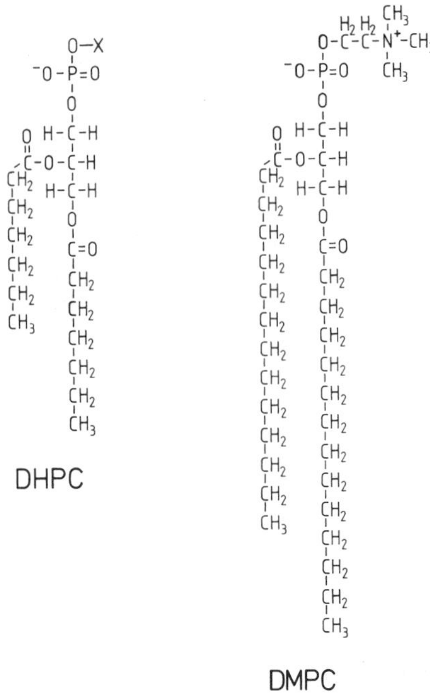

DHPC

DMPC

Fig. 1. Structural comparison of the short-chain phospholipid diheptanoylphosphatidyl-choline (DHPC, fatty acid chain length of 7 carbons) and the long-chain phospholipid dimyristoylphosphatidylcholine (DMPC, chain length of 14 carbons). The characteristic configuration around the glycerol backbone adopted in the lipid bilayer is shown.[8] The moiety X denotes the choline head group.

gents and contains, in addition to BR, only about seven native lipids per protein. The novel method presented will allow rapid, easy, and gentle functional reconstitution of other membrane systems and isolated membrane proteins as well. Recently this method has been successfully applied for the coreconstitution of BR and a H^+-ATPase of yeast plasma membrane (N. A. Dencher and A. Wach, unpublished experiments).

Materials

Long-chain phospholipids: dimyristoylphosphatidylcholine (DMPC, Fig. 1 and Ref. 8 therein) and dipalmitoylphosphatidylcholine (DPPC) were obtained from Fluka and showed only one spot in thin-layer chromatography. Soybean phospholipids (SBPL) were purchased from Sigma (St. Louis, Missouri) and purified by a stan-

[8] G. Büldt and G. H. de Haas, *J. Mol. Biol.* **158**, 55 (1982).
[9] Y. Kagawa and E. Racker, *J. Biol. Chem.* **246**, 5477 (1971).

dard procedure[9]; assumed molecular weight of 800 after purification.

Short-chain phospholipid: diheptanoylphosphatidylcholine (DHPC, Fig. 1) in chloroform from Sigma.

Purple membrane (PM) isolated from *Halobacterium halobium* S9 showed a single band in sodium dodecyl sulfate–polyacrylamide gel electrophoresis and an absorbance ratio A_{280}/A_{568} of 1.5–1.6 in light-adapted samples.

For all sonication steps a bath-type sonicator was used (Sonorex RK 100 H, Bendelin Electronic, Federal Republic of Germany; 35 khz).

Protocol for Reconstitution of Bacteriorhodopsin into Lipid Vesicles

Reconstitution proceeds via the following steps: long-chain and short-chain phospholipids are dissolved separately in a small volume of chloroform and spread as a thin film on the wall of two glass test tubes by evaporation of the solvent under a stream of nitrogen. Residual traces of solvent are removed under vacuum ($\leq 10^{-3}$ Torr, 5–15 hr) at room temperature. The dried lipid films are hydrated in an appropriate aqueous solvent, e.g., 150 mM KCl solution. This step can be assisted by 10 sec of sonication. The aqueous suspension of the respective long-chain phospholipid (DMPC, DPPC, or SBPL), at a final concentration of 20 mM after addition of all constituents, is mixed with a concentrated aqueous suspension of PM. The PM should be sonicated prior to mixing to reduce the size of the membrane sheets; however, sonication is not essential. Sonication of the PM for about 60 sec, depending on the power of the sonicator, is recommended to obtain an optimum in the extent of incorporation of BR as well as in functional activity. To this mixture of long-chain phospholipid and PM, the aqueous micellar 50 mM DHPC solution is added, yielding a final DHPC concentration of 5 mM (20 mol% of total lipid), and the sample is immediately vortexed. Finally, this lipid–PM mixture is sonicated in the bath sonicator for 30 sec. This sonication step is only required for SBPL and can be omitted with DMPC and DPPC. All manipulations should be performed at temperatures above the lipid phase transition of the respective long-chain phospholipids, e.g., room temperature for SBPL, 30° for DMPC, and 44° for DPPC. Transmembrane insertion of BR into the lipid vesicles occurs immediately on addition of DHPC (and 30 sec of sonication, in the case of SBPL). Usually, the samples are incubated at room temperature or 30° for 3–8 hr and thereafter subjected to density gradient centrifugation (e.g., linear 5–40%, w/w, sucrose gradient in 150 mM KCl; 141,000 g for 8 hr at 4°).

As an example, a specific reconstitution protocol for DMPC and BR is

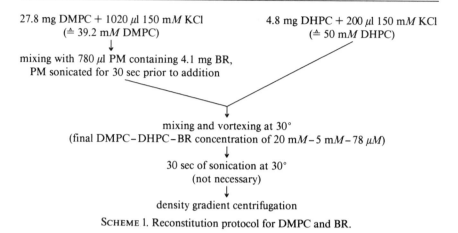

SCHEME 1. Reconstitution protocol for DMPC and BR.

given in Scheme 1 (2-ml sample volume, initial molar total lipid-to-BR ratio of 320, 20 mol% DHPC; depicted in Ref. 2 are the absorbance and light-scattering changes of this sample occurring during reconstitution). Under these conditions, complete reconstitution is obtained. Neither free lipid nor nonincorporated PM can be detected in the gradient. The vesicles in the main band have an average molar lipid-to-BR ratio of 366 and that of the minor band a ratio of 263.

Inspection of Vesicle Formation and Transmembrane Insertion of BR

Stable BR–lipid vesicles are spontaneously formed upon mixing all constituents. The large alterations in light-scattering of the sample, which can be followed even by visual inspection, allow detection of vesicle formation as illustrated in Fig. 2. The suspension of multibilayer DPPC (20 mM final concentration) and PM is extremely turbid (spectrum 1). The broad band around 560 nm represents the absorption maximum of BR and resides on a strong background of light scattering. Addition of DHPC (5 mM final concentration) at 23°, i.e., at a temperature far below the DPPC phase transition, leads to a drastic drop in the turbidity (spectrum 2). If this sample is warmed up to 46°, i.e., above the phase transition of DPPC, and vortexed again, a further pronounced reduction in light scattering occurs. The sample is now optically clear, also after recooling to room temperature or 4°. With DMPC or SBPL as long-chain phospholipid the same observations are made, however, in the case of SBPL the final 30-

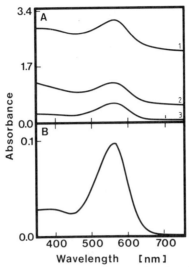

FIG. 2. (A) Formation of reconstituted DPPC–DHPC–BR vesicles monitored as changes in absorbance and light scattering. (1) Dispersion of DPPC and PM (PM sonicated for 30 sec prior to addition), (2) 3 min after addition of aqueous micellar 50 mM DHPC at 23° and vortexing, (3) after vortexing (2) at 46° and recooling to 23°. DPPC–DHPC–BR concentration of 20 mM–5 mM–78 μM, respectively. (B) Reconstituted vesicles, L/BR = 715, after density gradient centrifugation and storage at 8° for 48 hr. Spectrum recorded with an integrating sphere attachment. All spectra are measured at room temperature with a 1-mm path length cuvette.

sec sonication step is essential (compare Figs. 1 and 2 of Ref. 2). Control experiments show that in the absence of DHPC no significant change in turbidity is noticeable on vortexing and 30 sec of sonication of SBPL–PM, DMPC–PM, and DPPC–PM suspensions. Both the observed light-scattering changes and electron microscopy reveal that DHPC facilitates the formation of stable unilamellar vesicles from aqueous suspensions of PM and preexisting multibilayers of long-chain phospholipids. DHPC fails to decrease the turbidity of suspensions of branched diphytanoyl–PC and PM. Furthermore, no satisfactory results are obtained if a dried mixture of SBPL and DHPC is hydrated with an aqueous PM suspension.

In the linear 5–40% sucrose gradient, reconstituted BR–lipid vesicles can be easily identified by their color and position in the gradient and are clearly separated from pure lipid at the top of the gradient and nonincorporated PM at the bottom. The extent of reconstitution depends on both the long-chain phospholipid used and the duration of sonication, i.e., the size of PM. As discussed in the preceding paragraph, complete reconstitution is obtained with DMPC. With DPPC, again all lipid is reconstituted but not all PM. The amount of nonincorporated PM decreases with decreasing PM size. In the sample depicted in Fig. 2 with an initial molar lipid-to-BR ratio of 320 and with PM subjected to only 30 sec of sonication, a considerable proportion of the PM remains nonincorporated, and the majority of the reconstituted DPPC–DHPC–BR vesicles have a lipid-to-BR ratio of only 715. With PM sonicated for 420 sec and an initial lipid-to-BR ratio of 984, the reconstituted material in the band at lower and higher density in the gradient has lipid-to-BR ratios of 1149 and 627, respectively. Both the band pattern and the actual lipid-to-BR ratio are very similar in preparations with and without the final 30-sec sonication step. Contrary to DMPC and DPPC, a large amount of SBPL is not reconstituted but forms a band close below the surface of the gradient. Since the majority of PM is reconstituted, the final lipid-to-BR ratio is considerably lower than the initial one. Starting with an initial lipid-to-BR ratio of 984, the reconstituted vesicles have a ratio of 649 with PM sonicated for only 20 sec and of 178 with PM sonicated for 420 sec. An initial ratio of 320 and 180 sec of sonication of PM results in a preparation with a final ratio of 104.

Transmembrane insertion of BR into the lipid bilayer is verified both by circular dichroism spectroscopy and transport studies. The CD spectrum of the reconstituted DPPC–DHPC–BR vesicles collected from the density gradient (Fig. 2B) exhibits at 5.3° the same characteristic spectral features as observed for BR in the PM (solid curve in Fig. 3A). This indicates that the BR molecules in the reconstituted vesicles are arranged as trimers in a two-dimensional hexagonal lattice. Increasing the temperature from below to above (43.4°) the lipid phase transition temperature leads to an alteration of the pair of positive and negative exciton bands into a broad positive band typical for monomeric BR (dotted curve in Fig. 3A). The CD measurements on DMPC– and SBPL–DHPC–BR vesicles are in agreement with this result and can be explained by BR molecules being inserted into the vesicle bilayer membrane composed from long- and short-chain phospholipids. Changes in the physical state of this new lipid environment induce the observed reversible aggregation–disaggregation of BR. In contrast, PM incubated with 5 mM DHPC under the same conditions used for reconstitution, but omitting the long-chain phospholipids, reveals no pronounced change of the exciton CD spectrum in the temperature range between 5.0° (solid curve in Fig. 3B) and 45.0° (dotted curve). (The small alterations are due to the different temperature and light–dark adaptation of BR.)

The reconstituted BR–lipid vesicles are highly active in light-energized vectorial H$^+$ translocation. Steady-state values corresponding to a net inward transport of about 11 H$^+$/BR are observed for SBPL–DHPC–BR vesicles during illumination with light absorbed by BR, as compared to ≤0.16 H$^+$/BR in the absence of DHPC. The 70-fold enhancement of

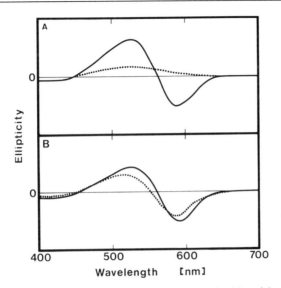

Fig. 3. (A) Changes in the CD spectrum of DPPC–DHPC–BR vesicles, molar lipid-to-BR ratio of 715, when the temperature is raised from 5.3° (solid line) to 43.4°(dotted line). (B) CD spectrum of a mixture of 78 μM PM and 5 mM DHPC at 5.0° (solid line) and after warming up to 45.0° (dotted line).

vectorial transport activity at otherwise identical conditions demonstrates best the efficiency of DHPC to reconstitute BR into SBPL vesicles. From the direction of light-induced proton translocation, i.e., into the vesicles, a net inside-out orientation of BR in the SBPL–DHPC vesicles has to be concluded. On the other side, with DMPC– and DPPC–DHPC–BR vesicles, both below and above the lipid phase transition temperature, light-induced fast acidification occurs corresponding to ≤ 1.3 H$^+$/BR released into the external medium. (The less efficient transport activity of DMPC– and DPPC–DHPC–BR vesicles as compared with the corresponding SBPL vesicles is an inherent property of the former long-chain phospholipids and is observed for detergent dialysis vesicles lacking DHPC as well.[10]) The net outside-out orientation of BR molecules in DMPC–DHPC and DPPC–DHPC vesicles deduced from the direction of the pH signal is in accordance with the outside-out orientation of the majority of the BR molecules determined by selective proteolysis and subsequent gel electrophoresis (P. A. Burghaus, unpublished observations). The fact that the net orientation of BR depends on the long-chain phospholipids used is an interesting feature of this reconstitution procedure. The efficient light-induced vectorial H$^+$ transport not only confirms transmembrane insertion of BR but also illustrates that DHPC is not deleterious for BR.

[10] N. A. Dencher and M. P. Heyn, *FEBS Lett.* **108**, 307 (1979).

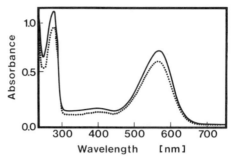

Fig. 4. Absorption spectrum of light-adapted PM before (solid line) and after incubation (dotted line) with 5 mM DHPC for 48 hr at room temperature. BR concentration: 78 μM after the 10% dilution by addition of the DHPC suspension; 1 mm optical path length.

The size of the reconstituted vesicles seems to depend on the long-chain phospholipids. According to electron microscopy, the majority of the SBPL–DHPC–BR vesicles have diameters between 35 and 100 nm, whereas the DMPC–DHPC–BR vesicles are much larger, ranging from 100 to 450 nm in diameter. Electron microscopic data, the clarity of the vesicle suspensions (Fig. 2), the extent of proteolytic digestion of BR, and the relatively fast rates of the passive back-diffusion of protons upon termination of illumination indicate that the reconstituted BR–lipid vesicles are predominantly unilamellar.

Interaction of DHPC with Bacteriorhodopsin and the Lipid Bilayer

Various experimental observations unambiguously indicate that there is no harmful interaction of DHPC with BR. This appears to be the predominant advantage of this reconstitution procedure, besides its easiness and quickness. As illustrated in Fig. 4, incubation of PM with 5 mM DHPC at room temperature for 48 hr (molar BR : native PM lipids : DHPC ratio of 1 : 7 : 64) does not affect the absorption spectrum. As compared to the absorption spectrum of untreated PM (Fig. 4, solid spectrum), the amplitude and position of both the chromophore absorption band at 568 nm and the aromatic amino acid region around 280 nm are not changed (dotted spectrum). (The small decrease in amplitude is solely due to the 10% dilution by addition of DHPC.) Also the CD spectrum of this sample reveals no DHPC-induced alterations. Even at 45.0° BR remains immobilized in the DHPC-treated PM (Fig. 3B). Absorption (Fig. 2) and CD (Fig. 3A) spectra of reconstituted BR also indicate no perturbation by DHPC. Storage of DPPC–DHPC–BR vesicles at 8° for 48 hr has no influence on the absorption spectrum (Fig. 2B), even after 9 days the amplitude decreased by less than 5%. All these unchanged spectral features

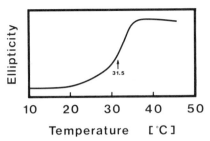

FIG. 5. Temperature dependence of the ellipticity at 585 nm, the wavelength of maximal difference between the CD spectra of aggregated and monomeric BR, for BR in DPPC–DHPC vesicles.

and the considerable H^+ transport activity illustrate best that DHPC has no harmful effect on BR.

Although DHPC does not affect BR itself, it influences the physical parameters of the lipid phase. This can be expected for such a short-chain component (Fig. 1) at this relatively high concentration (20 mol% of total lipid). The CD transition curve that monitors the reversible crystallization of BR induced by the lipid phase transition has its midpoint at 31.5° for DPPC–DHPC–BR vesicles (Fig. 5) as compared to 34.5° for BR in a pure DPPC bilayer. A similar alteration of the transition midpoint temperature is observed for DMPC–DHPC–BR vesicles, i.e., 9.9° instead of about 13–15°. Furthermore, in DMPC–DHPC–BR vesicles, a different cleavage pattern of BR by chymotrypsin is found as compared to BR in PM and in reconstituted DMPC vesicles (P. A. Burghaus, unpublished results). This might indicate a thinner or/and perturbed lipid bilayer. Replacement of DHPC by dioctanoyl–PC with its longer fatty acid chains (eight carbons), or a decrease in the relative proportion of DHPC, e.g., by a factor of two, might prevent or at least reduce the described effects on the bilayer. A fivefold reduction of the molar DHPC-to-SBPL ratio, however, leads to a considerably smaller extent of vesicle formation. By dilution of the reconstituted vesicles, part of the DHPC might be removed from the lipid bilayer.[11] Fortunately, at least with SBPL–DHPC–BR vesicles, the permeability of the lipid bilayer for protons and/or for the counterions potassium and chloride is not affected by the presence of DHPC, if compared with SBPL–BR detergent-dialysis vesicles. On the other side, the fusogen myristate that has also been used for reconstitution[7] increases the proton/hydroxide ion conductance by about 10-fold at 10–20 mol%.[12]

Transmembrane insertion of the PM into the lipid bilayer seems to proceed during intercalation of the short-chain DHPC into the preexisting

[11] N. E. Gabriel and M. F. Roberts, *Biochemistry* 25, 2812 (1986).
[12] J. Gutknecht, *Biophys. J.* 49, 516a (1986).

large, multilamellar vesicles of long-chain phospholipids. DHPC, possibly by the mismatch in fatty acyl chain lengths, perturbs the multibilayers and leads, similar to sonication, to a decrease in vesicle size, a conversion of the multibilayer to unilamellar vesicles, and a concomitant transmembrane integration of the protein. As indicated in Fig. 2A, DHPC interacts more readily with long-chain phospholipid bilayers in the liquid–crystalline rather than in the gel state. The decrease in the vesicle size during this process is demonstrated by the large decrease in turbidity of the sample and by electron microscopy. It should be mentioned that reconstitution of BR with the aid of DHPC can also be achieved with sonicated small (diameter of ~30 nm), unilamellar SBPL vesicles, at least in the presence of a small amount of detergent.

Because the procedure presented is easy, fast, and particularly has no harmful effect on the membrane protein, it might be a method of choice for functional reconstitution of other membrane systems (in particular, membranes enriched in one protein species) as well as isolated purified membrane proteins (e.g., in the presence of small amounts of detergents).

Acknowledgments

I thank P. A. Burghaus for performing proteolytic degradation experiments, R. Groll for determination of the size distribution of vesicles by electron microscopy, and I. Wallat for excellent technical assistance. This research was supported by the Deutsche Forschungsgemeinschaft (SFB 312/B4, Heisenberg Grant De 300/1).

[14] Ion Carriers in Planar Bilayers: Relaxation Techniques and Noise Analysis

By R. BENZ, H.-A. KOLB, P. LÄUGER, and G. STARK

Ion Carriers

A number of macrocyclic compounds such as valinomycin, enniatin A and B, beauvericin, and monactin (Fig. 1) have been shown to act as carriers for alkali ions in lipid bilayer membranes.[1] In the complex with the carrier the ion is located in a cage of oxygen ligands which replace the primary hydration shell of the solvated ion.

[1] W. Burgermeister and R. Winkler-Oswatitsch, *Top. Curr. Chem.* **69**, 91 (1977).

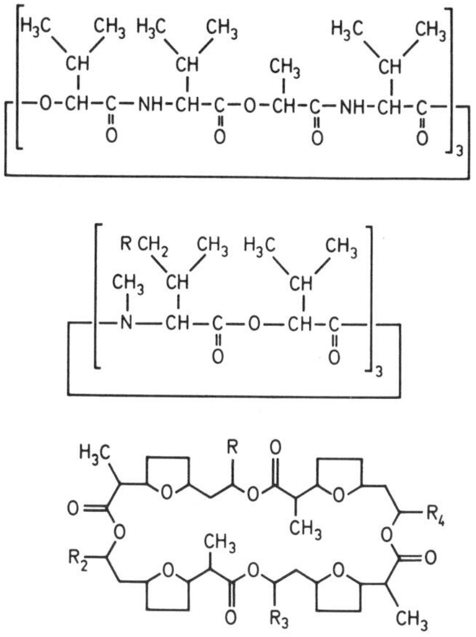

Valinomycin

$R = CH_3$ Enniatin A

$R = H$ Enniatin B

$R_1 = R_2 = R_3 = R_4 = CH_3$		Nonactin
$R_1 = R_2 = R_3 = CH_3$	$R_4 = C_2H_5$	Monactin
$R_1 = R_3 = CH_3$	$R_2 = R_4 = C_2H_5$	Dinactin
$R_1 = CH_3$	$R_2 = R_3 = R_4 = C_2H_5$	Trinactin
	$R_1 = R_2 = R_3 = R_4 = C_2H_5$	Tetranactin

FIG. 1. Structure of cation carriers.

Carrier-mediated ion transport involves four main reaction steps: (1) binding of the ion by the carrier at one membrane interface, (2) conformational change or translocation of the carrier and exposure of the binding site to the opposite interface, (3) release of the ion, and (4) transition back to the original orientation of the carrier (Fig. 2). The representation of carrier-mediated transport processes by a reaction scheme such as given in Fig. 2 involves a number of simplifications. In reality, the association in the interface is a multistep process in which the water molecules of the primary hydration shell are substituted one after the other by ligand groups of the carrier molecule. The association rate constant k_R describes the overall rate of this process, which may involve a conformational change of the carrier molecule as a rate-limiting step. In a similar way, the translocations of free carrier S and complex MS^+ are treated as monomolecular reaction steps (rate constants k_S and k_{MS}); this is a reasonable approximation for particles which spend most of the time in adsorption sites in the membrane–solution interface.

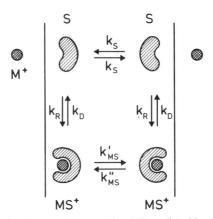

FIG. 2. Transport of a metal ion M^+ mediated by a carrier S.

Planar Bilayers as a Tool for Studying Carrier-Mediated Ion Transport

Planar lipid bilayers with areas up to several square millimeters may be formed in an aqueous medium from a variety of lipids. In the original version of the method[2] a lipid, such as dioleoyllecithin, is dissolved in an alkane solvent (e.g., n-decane), and a small amount of the solution is spread over a hole (diameter of ~ 1 mm) in a Teflon septum separating two aqueous electrolyte solutions. The lipid–alkane film spontaneously thins and eventually forms a lipid bilayer containing a residual amount of solvent. In an alternative method[3] a lipid monolayer is created at the air–water interface in a two-compartment trough. When first one and thereafter the other water level are raised over the hole in the Teflon septum separating the two compartments, a lipid bilayer forms over the hole in a zipperlike fashion.

Carrier molecules may be incorporated into the bilayer either by adding the compound to the membrane-forming lipid solution or to the aqueous phase.[4] When the carrier is added to the aqueous phase a partition equilibrium between membrane and water is set up which is strongly in favor of the membrane (in the case of valinomycin the partition coefficient is of the order of 10^4).

The electrical conductance of an undoped lipid bilayer in, say, a 1 M KCl solution is of the order $0.01 - 1$ $\mu S/cm^2$ (1 S = 1 ohm^{-1}). Incorporation of valinomycin (or similar compounds) into the membrane raises

[2] P. Mueller, D. O. Rudin, M. T. Tien, and W. D. Wescot, *Nature (London)* **194,** 979 (1962).

[3] M. Montal and P. Mueller, *Proc. Natl. Acad. Sci. U.S.A.* **69,** 3561 (1972).

[4] G. Stark and R. Benz, *J. Membr. Biol.* **5,** 133 (1971).

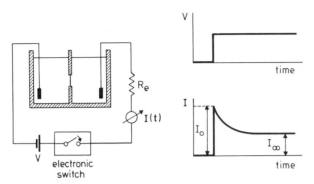

FIG. 3. Principle of the voltage-jump current-relaxation method. In the actual experiment the voltage step is applied from a signal generator with a rise time of 10 nsec or less. Silver/silver chloride electrodes or platinized platinum electrodes may be used. The current transient $I(t)$ is recorded with a digital storage oscilloscope. R_e is the total resistance of the external circuit (solutions plus electrodes plus input resistance of the amplifier). The initial current spike resulting from the charging of the membrane capacitance is not shown.

the conductance by several orders of magnitude. The conductance increase is linear with respect to valinomycin concentration and the concentration of the transported ion over several decades.[4,5] This finding is consistent with the assumption that charge is transported across the membrane by a 1:1 complex between valinomycin and the alkali ion.

Relaxation Techniques

From stationary conductance measurements only limited information on the carrier system can be obtained. For the evaluation of the rate constants of the single transport steps relaxation techniques may be used. In a relaxation experiment an external parameter, such as pressure, temperature, or voltage, is quickly changed; after the perturbation the approach of the system toward a new stationary state is followed.

Current Relaxation Following a Voltage Jump

The state of an ion transport system depends, in general, on voltage. If the voltage across the membrane is quickly changed (Fig. 3), the conductance (and thus the transmembrane current) relaxes toward a new stationary value.[6] For the generation of the voltage step, commercial signal generators are available with intrinsic rise times of 10 nsec or less. The time

[5] R. Laprade, S. Ciani, G. Eisenman, and G. Szabo, in "Membranes: A Series of Advances" (G. Eisenman, ed.), Vol. 3, p. 127. Dekker, New York, 1975.

[6] G. Stark, B. Ketterer, R. Benz, and P. Läuger, Biophys. J. 11, 981 (1971).

resolution of the voltage-jump method is limited by the charging time τ_c of the membrane. Under most conditions the membrane resistance is much higher than the external resistance R_e (solutions plus electrodes plus input resistance of the amplifier) so that τ_c is approximately given by $\tau_c = R_e C = (R_s + R_i)C$, where R_s is the resistance of solutions and electrodes, R_i the input resistance of the amplifier, and C the combined electrical capacitance of membrane and septum. Unless the electrodes are placed very close to the membrane, a lower limit of R_s is given by the two-sided convergence resistance of a circular opening of radius r: $R_s \approx \rho/\pi r$ (ρ is the specific resistance of the electrolyte solutions). As the capacitance of the septum is usually negligible, the relation $C \approx \pi r^2 C_m$ holds, where C_m is the specific membrane capacitance. This gives $R_s C > r\rho C_m \simeq 0.2$ μsec for a membrane of radius $r = 0.5$ mm and 1 M alkali–chloride solutions ($\rho \simeq 10$ Ωcm; $C_m \simeq 0.4$ μF/cm^2). Therefore, under conditions where R_i may be kept of the same order or smaller than R_s, charging times as low as 0.2 μsec can be achieved. The actual time resolution of the method is about 10 times this value (~ 2 μsec), since the charging current has to decline to a small fraction of the initial value before relaxation currents of low amplitude can be measured.

The existence of a finite circuit resistance $R_e = R_s + R_i$ which is in series with the membrane resistance R_m represents a possible source of systematic error in the analysis of voltage-jump relaxation experiments. A constant voltage V which is applied to the circuit is divided between R_m and R_e according to $V = J(R_m + R_e) = V_m + JR_e$. Thus, the voltage V_m across the membrane which determines the rate of ion transport is not strictly constant but varies with R_m. Since $V_m = V/(I + R_e/R_m)$, it is seen that R_e should be much smaller than R_m in order to minimize the time variation of V_m. For $|V_m| < 25$ mV the effect of series resistance can be accounted for by introducing a correction term into the calculation of rate constants.[7]

From the foregoing comments it is clear that the input resistance R_i of the amplifier has to be kept small in order to ensure a good time resolution. The sensitivity of the measurement is often limited by noise in the electronic circuit. In such cases the signal-to-noise ratio can be considerably increased by repetitive application of a square voltage and signal averaging using a fast transient recorder. Since a straightforward theoretical treatment of relaxation phenomena is considerably simplified for small perturbations, it is preferable to choose the amplitude of the voltage step as small as possible (of the order of 10 mV). Furthermore, in experiments with solvent-containing membranes, care should be taken to avoid electrostric-

[7] P. Jordan and G. Stark, *Biophys. Chem.* **10**, 272 (1979).

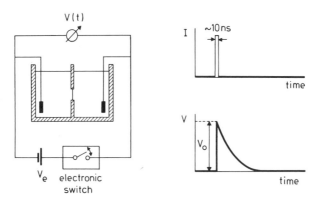

Fig. 4. Principle of the charge-pulse voltage-relaxation technique. The membrane capacitance is charged to an initial voltage V_0 by a brief current pulse, and at the end of the pulse the charging circuit is switched to virtually infinite resistance. For this purpose a fast FET switch may be used which has an impedance of $> 10^{12} \, \Omega$ in the "open" state and $< 1000 \, \Omega$ in the "closed" state. Alternatively, a commercial pulse generator with a short rise time and a low output impedance may be connected to the membrane cell by a diode with a reverse-voltage resistance of the order of $10^9 \, \Omega$. The decay of voltage which results from charge movement within the membrane is recorded with a storage oscilloscope. Matched pairs of silver/silver chloride or platinized platinum electrodes may be used. Electrode polarization effects may be minimized by measuring the voltage with a separate pair of electrodes.

tion of the lipid torus surrounding the membrane. Voltage-dependent changes of membrane geometry may be minimized by using membranes of large diameter (~ 3 mm) and by applying only small voltage steps.

The performance of the measuring circuit should be checked using test circuits with predictable relaxation behavior. For this purpose a membrane may be used which is formed in the usual way but without added ion transport system, and instead an external resistor (or a suitable combination of resistors and capacitors) is placed in parallel to the membrane capacitance in order to simulate an intrinsic conductance. Or the membrane may be entirely replaced by a combination of resistors and capacitors between electrodes and amplifier. Signal distortion from bandwidth limitation of amplifiers or from electrode polarization can be detected in this way.

Voltage Relaxation Following a Charge Pulse

The bandwidth limitation by the RC time of the circuit is circumvented in the so-called charge-pulse technique. The membrane capacitance is charged to an initial voltage (typically 10 mV or less) by a brief current pulse of 10–100 nsec duration, and at the end of the pulse the external charging circuit is switched to virtually infinite resistance (Fig. 4). The subsequent decay of the voltage which results from charge movement within the membrane contains kinetic information on the ion transport system under study. This method which has been developed for the study of fast electrode reactions was subsequently used in quasistationary conductance measurements with lipid bilayers.[8] A fast version of the method was applied to kinetic studies of ion carriers in planar bilayer membranes.[9]

The initial voltage V_0 is given by the time integral of the charging current and by the membrane capacitance C_M; if J is the average current during the charging pulse of length Δt, then $V_0 \simeq J \Delta t / C_M$. In principle, the charging time Δt may be very short if J is sufficiently large. The actual time resolution of the method is limited by other factors, namely, the bandwidth of the voltage amplifier and the stray capacitance of the circuit. Furthermore, during the charging pulse a voltage drop occurs between the electrodes, which tends to overload the amplifier. For this reason, an amplifier with a short recovery time should be used. By careful adjustment of the circuit parameters a time resolution of about 40 nsec has been achieved.[10]

Temperature-Jump Experiments

The two relaxation methods described above require that the state of the ion transport system under study can be perturbed to a sufficient extent by a change of electric field strength in the membrane. This condition is not always met, however. An alternative method consists in changing suddenly the temperature of the membrane and the adjacent aqueous solutions. The temperature jump may be generated by absorption of an intense flash of light.[11,12] If a flash lamp with main output in the visible range is used, the aqueous solution bathing the membrane has to contain an absorbing dye.[12] In an improved version of the method[13] the heating of the solution is brought about by an infrared laser pulse (Fig. 5). At the emission wavelength of the Nd–glass laser ($\lambda = 1.06\ \mu m$), the decadic absorption coefficient of water is $0.063\ cm^{-1}$. This permits temperature jumps of about $0.3 – 0.4$ K in the fixed-Q mode of the laser.

With the relatively low absorption coefficient of water at $\lambda = 1.06\ \mu m$, temperature gradients in the cell remain small. As the membrane is only about 5 nm thick, temperature equilibrium between membrane and solution is established within less than 1 nsec. In the fixed-Q mode of the method, the time resolution is limited by the duration of the laser pulse

[8] S. W. Feldberg and G. Kissel, *J. Membr. Biol.* **20**, 269 (1975).

[9] R. Benz and P. Läuger, *J. Membr. Biol.* **27**, 171 (1976).

[10] R. Benz and U. Zimmerman, *Biochim. Biophys. Acta* **597**, 637 (1980).

[11] H. Staerk and G. Czerlinski, *Nature (London)* **205**, 63 (1965).

[12] W. Knoll and G. Stark, *J. Membr. Biol.* **37**, 13 (1977).

[13] W. Brock, G. Stark, and P. Jordan, *Biophys. Chem.* **13**, 329 (1981).

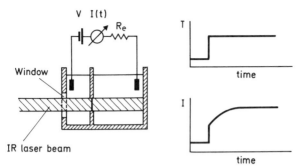

Fig. 5. Principle of the temperature-jump relaxation method. Membrane and adjacent solutions are heated up by a pulse from a high-intensity infrared laser. A voltage V is already applied to the membrane prior to the temperature jump. The temperature jump leads to a virtually instantaneous rise of membrane conductance (resulting from the temperature dependence of the rate constants), followed by slower relaxation processes. R_e is the total resistance of the external circuit.

(about 0.4 msec). The time resolution may be improved to 5 μsec by using the Q-switch mode of the laser (Awiszus and Stark, to be published). An upper limit of the time range in which temperature-jump relaxation experiments can be carried out is given by the temperature stability in the cell. It was found that with a diameter of 16 mm of the unfocused laser beam the temperature of the membrane remains stable for about 4 sec.[13]

In the usual version of the experiment a constant voltage V is already applied to the membrane, prior to the temperature jump (Fig. 5). The temperature jump leads to a virtually instantaneous increase of membrane conductance resulting from the temperature dependence of the elementary rate constants of the ion transport system. This steplike increase of membrane current is followed by slower relaxation processes associated with concentration changes and redistribution of membrane components.

Analysis of Relaxation Experiments

In the analysis of relaxation experiments with alkali ion carriers, it is usually assumed that the association and dissociation reactions in the interface are voltage independent and that the only transport step which is affected by the electric field in the membrane is the translocation of MS^+. Accordingly, the dependence of the translocation rate constants k'_{MS} and k''_{MS} on membrane voltage V_m is approximately given by

$$k'_{MS} \approx k_{MS}(1 + u/2) \tag{1}$$

$$k''_{MS} \approx k_{MS}(1 - u/2) \tag{2}$$

$$u \equiv V_m e_0/kT \tag{3}$$

where k_{MS} is the translocation rate constant for $V = 0$, k is Boltzmann's constant, T is the absolute temperature, and e_0 is the charge of the proton. Equations (1) and (2) hold for $|V_m| \ll RT/e_0 \simeq 25$ mV under the assumption of a symmetrical, narrow energy barrier.[14]

Denoting the concentrations of S and MS$^+$ at the left-hand and right-hand interface by N'_S, N''_S, N'_{MS}, N''_{MS} (referred to unit area), and the aqueous concentration of M$^+$ by c_M, the rate equations may be written as

$$dN'_S/dt = -k_R c_M N'_S + k_D N'_{MS} - k_S N'_S + k_S N''_S \qquad (4)$$

$$dN''_S/dt = -k_R c_M N''_S + k_D N''_{MS} - k_S N''_S + k_S N'_S \qquad (5)$$

$$dN'_{MS}/dt = k_R c_M N'_S - k_D N'_{MS} - k'_{MS} N'_{MS} + k''_{MS} N''_{MS} \qquad (6)$$

$$dN''_{MS}/dt = k_R c_M N''_S - k_D N''_{MS} - k''_{MS} N''_{MS} + k'_{MS} N'_{MS} \qquad (7)$$

Equations (4)–(7) are valid under the condition that the exchange of S and MS$^+$ between water and membrane is slow within the time scale of the relaxation experiment (microseconds to milliseconds).[14]

The transmembrane current I is given by

$$I = F(k'_{MS} N'_{MS} - k''_{MS} N''_{MS}) \qquad (8)$$

Solving Eqs. (4)–(7) yields the following expression for the current after a voltage jump of small amplitude ($|u| \ll 1$)[6]:

$$I(t) = I_\infty + (I_0 - I_\infty)[a_1 \exp(-t/\tau_1) + a_2 \exp(-t/\tau_2)] \qquad (9)$$

$$I_0 = e_0 N_{MS} k_{MS} u \qquad (10)$$

$$I_\infty = e_0 N_{MS} k_{MS} \frac{k_S k_D}{k_{MS}(c_M k_R + 2k_S) + k_S k_D} u \qquad (11)$$

$$a_1 = \frac{1}{2} + \frac{P(c_M k_R + 2k_S) + c_M k_R k_D}{2(c_M k_R + 2k_S)\sqrt{s}} \qquad (12)$$

$$a_2 = 1 - a_1 \qquad (13)$$

$$\tau_1 = \frac{1}{Q - \sqrt{s}}, \qquad \tau_2 = \frac{1}{Q + \sqrt{s}} \qquad (14)$$

$$P \equiv \tfrac{1}{2}(c_M k_R - k_D + 2k_S - 2k_{MS}) \qquad (15)$$

$$Q \equiv \tfrac{1}{2}(c_M k_R + k_D + 2k_S + 2k_{MS}) \qquad (16)$$

$$s \equiv P^2 + c_M k_R k_D \qquad (17)$$

$N_{MS} = c_M k_R N_S / k_D$ is the equilibrium concentration of complexes in one

[14] P. Läuger, R. Benz, G. Stark, E. Bamberg, P. C. Jordan, A. Fahr, and W. Brock, Q. Rev. Biophys. **14**, 513 (1981).

interface. Equations (9)–(11) show that the current declines from an initial value I_0 to a stationary value I_∞ which is always smaller than I_0. The origin of the current relaxation may be visualized in the following way. At equilibrium ($u = 0$) the concentration of charged complexes is the same in both interfaces. After the voltage jump a redistribution of MS^+ sets in, which gives rise to an initial current I_0. In this way a concentration difference $N'_{MS} - N''_{MS}$ is built up which counteracts the driving force of the voltage and diminishes the current.

In a similar way, the decay of transmembrane voltage V_m in a charge-pulse experiment is described by

$$V_m(t) = V_m^0 [A_1 \exp(-\lambda_1 t) + A_2 \exp(-\lambda_2 t) + A_3 \exp(-\lambda_3 t)] \quad (18)$$

$$A_1 + A_2 + A_3 = 1 \quad (19)$$

The relaxation times $\lambda_i^* = 1/\lambda_i$ and the relaxation amplitudes A_i may be expressed as functions of the rate constants and of the total carrier concentration $N_0 = N'_S + N''_S + N'_{MS} + N''_{MS}$, but the resulting equations are rather cumbersome. An alternative method is preferable which requires that all three relaxation times $1/\lambda_i$ and relaxation amplitudes A_i can be resolved in the experiment.[9] Defining the quantities

$$P_1 = \lambda_1 + \lambda_2 + \lambda_3 \quad (20)$$

$$P_2 = \lambda_1\lambda_2 + \lambda_1\lambda_3 + \lambda_2\lambda_3 \quad (21)$$

$$P_3 = \lambda_1\lambda_2\lambda_3 \quad (22)$$

$$P_4 = A_1\lambda_1 + A_2\lambda_2 + A_3\lambda_3 \quad (23)$$

$$P_5 = A_1\lambda_1^2 + A_2\lambda_2^2 + A_3\lambda_3^2 \quad (24)$$

it may be shown that

$$k_{MS} = \tfrac{1}{2}(P_5/P_4 - P_4) \quad (25)$$

$$k_D = 1/2k_{MS}[P_1 P_5/P_4 - P_2 + P_3/P_4 - (P_5/P_4)^2] \quad (26)$$

$$k_S = 1/2k_D(P_3/P_4) \quad (27)$$

$$c_M k_R = P_1 - P_4 - 2k_S - 2k_{MS} - k_D \quad (28)$$

$$N_0 = 2N^*(P_4/k_{MS})(1 + k_D/c_M k_R) \quad (29)$$

$N^* = kTC_m/e_0^2$ is the number of elementary charges (per unit area) required to charge the membrane capacitance C_m to a voltage $kT/e_0 \simeq$ 25 mV. In this way the four rate constants, k_R, k_D, k_S, and k_{MS}, as well as the total carrier concentration N_0, may be obtained from the five experimental quantities P_1, P_2, P_3, P_4, and P_5.

Noise Analysis

Theoretical Basis

On a microscopic scale, carrier-mediated ion transport across a membrane is a discontinuous process, consisting of single discrete events (association and dissociation processes in the interface, jumps of ion–carrier complexes across the dielectric barrier). Accordingly, the membrane current is not constant but fluctuates among a mean value $\bar{I}$:

$$I(t) = \bar{I} + \delta I(t) \tag{30}$$

The fluctuating part $\delta I(t)$ of the current contains information on the kinetic parameters of the transport system. Thus, analysis of current noise represents an alternative method for evaluating the rate constants of a carrier system.

A straightforward method of extracting information from $\delta I(t)$ consists in measuring the spectral intensity $S_I(f)$ of current noise as a function of frequency f. $S_I(f)$ is the value of $\overline{(\delta I)^2}$ per unit frequency interval. This means that in the frequency interval (f_1, f_2) the mean square of $\delta I(t)$ is given by

$$[\overline{(\delta I)^2}]_{f_1, f_2} = \int_{f_1}^{f_2} S_I(f)\, df \tag{31}$$

A simple situation is given when the system is in equilibrium, i.e., when the aqueous phases have identical composition and when no external voltage is applied. In this case the average current vanishes, but $\delta I(t)$ differs from zero, since charged ion–carrier complexes randomly jump across the membrane in either direction. According to Nyquist's theorem, the spectral intensity $S_I(f)$, under equilibrium conditions, is related to the frequency-dependent complex admittance of the membrane:

$$S_I(f) = 4kT\, \text{Re}[Y(f)] \tag{32}$$

where Re means "real part of." The (small-signal) admittance $Y(f)$ may be calculated in a straightforward way for the carrier model described above. The result reads ($\omega \equiv 2\pi f$)[15]:

$$S_I(\omega) = 4e_0^2 A_m N_{MS} k_{MS} \left(1 - \frac{\alpha_1}{1 + \omega^2 \tau_1^2} - \frac{\alpha_2}{1 + \omega^2 \tau_2^2} \right) \tag{33}$$

$$\alpha_1 = \frac{k_{MS}}{\sqrt{s}} \frac{P + \sqrt{s}}{Q - \sqrt{s}}; \qquad \alpha_2 = \frac{k_{MS}}{\sqrt{s}} \frac{\sqrt{s} - P}{\sqrt{s} + Q} \tag{34}$$

[15] H.-A. Kolb and P. Läuger, *J. Membr. Biol.* **41**, 167 (1978).

A_m is the membrane area; the quantities τ_1, τ_2, s, P, and Q are given by Eqs. (14)–(17). It is seen from Eq. (33) that $S_I(\omega)$ approaches frequency-independent ("white") noise both at low and high frequencies ω. Between $S_I(0)$ and $S_I(\infty)$ two dispersion regions are located at frequencies $\omega_1 = 1/\tau_1$ and $\omega_2 = 1/\tau_2$.

Current noise from carrier systems under nonequilibrium conditions may be analyzed by calculating the correlation function of current fluctuations.[16] In this case $S_I(\omega)$ contains three time constants (τ_1, τ_2, and τ_3) instead of two.

Measurement of Current Noise

For the analysis of current fluctuations, Ag/AgCl electrodes are introduced into the aqueous solutions bathing the membrane, and the electrodes are connected to a low-noise preamplifier which is located within the Faraday shield containing the membrane cell. The signal of the preamplifier is fed into the main amplifier which is used in the ac-coupled mode with appropriately selected lower and upper cutoff frequencies.[15,16] The spectral intensity $S_I(f)$ is calculated using a Fourier analyzer.

A common difficulty in the analysis of current fluctuations is the background noise of the measuring system. The main source of background noise is the voltage noise of the preamplifier, which is transformed into current noise at the membrane capacitance. The intrinsic noise of the preamplifier may be described to a first approximation by introducing at the input of the amplifier a voltage–noise source of spectral intensity S_v and an independent current–noise source of spectral intensity S_i. If the input of the preamplifier is connected to a source of impedance Z (the membrane), the spectral intensity $S_{I,b}$ of the background current noise generated by the preamplifier is given by[15]

$$S_{I,b} = S_i + S_v \left\{ \frac{1}{R_f^2} + \frac{1}{|Z|^2} \left[1 + 2\,\frac{\text{Re}(Z)}{R_f} \right] \right\} + \frac{4kT}{R_f} \tag{35}$$

R_f is the feedback resistance of the preamplifier. The total spectral intensity $S_{I,t}$ is the sum of the noise contrication of the source $S_{I,s}$ and of the preamplifier $S_{I,b}$:

$$S_{I,t} = S_{I,s} + S_{I,b} \tag{36}$$

The background noise and the frequency response of the measuring system should be carefully checked. For this purpose the membrane is to be replaced by a model circuit consisting of a capacitance C and a parallel

[16] H.-A. Kolb and E. Frehland, *Biophys. Chem.* **12**, 21 (1980).

resistance R; the values of C and R should match the membrane capacitance and the steady-state membrane resistance in the carrier experiment. The noise spectrum of the model circuit may be calculated from the Nyquist relation [Eq. (32)] and compared with the observed spectrum. Alternatively, an undoped (high-resistance) bilayer membrane in combination with an external parallel resistance may be used as a test circuit.

Comparison with Relaxation Methods

From the foregoing discussion it is clear that noise analysis and relaxation techniques yield the same information on the characteristic time constants of the carrier system. However, for a given system one of the relaxation amplitudes may become very small so that the corresponding time constant cannot be obtained from the relaxation experiment, whereas the same time constant may still appear in the noise spectrum. Values of the kinetic parameters of the valinomycin/Rb$^+$ system as determined from noise analysis agree, within the limits of experimental error, with the results of voltage-jump and charge-pulse relaxation experiments.[15] Noise analysis has the advantage that the measurement can be carried out without any external perturbation of the system. On the other hand, noise measurements require more sophisticated instrumentation; for this reason most kinetic studies of ion carriers have been done so far with relaxation techniques.

[15] Ion Interactions at Membranous Polypeptide Sites Using Nuclear Magnetic Resonance: Determining Rate and Binding Constants and Site Locations

By Dan W. Urry, T. L. Trapane, C. M. Venkatachalam, and R. B. McMichens

Introduction

The dominant considerations in this presentation of ion interaction at membranous polypeptide sites using nuclear magnetic resonance (NMR) are relaxation studies on the alkali metal ions. Such studies provide estimates of binding constants and rate constants. In addition ion resonance chemical shifts and carbon-13 chemical shifts of binding site carbonyls will be considered as complementary sources of binding constant data which aid in substantiating interpretation of the relaxation data. It is the same ion-induced carbonyl carbon chemical shifts that can be used to determine the location of the binding site. Thus it becomes possible to monitor ion–polypeptide interactions by looking at the binding site and by looking at the ion in complementary ways. The fact that the polypeptide–ion binding site is within a large membranous structure makes more difficult some studies and analyses, while simplifying others. Monitoring a site within the membrane is made difficult by the slow reorientation of the membrane; resonance lines, for example, from polypeptide backbone sites within the membrane are very broad, and increased temperature is necessary to increase mobility sufficiently to obtain an adequately defined resonance line. For interpreting ion correlation times, however, it is helpful when the reorientation correlation time of the site is long compared to the lifetime of the ion in the channel. As will be shown for the membranous gramicidin channel, the ion correlation time, derived from resonance linewidth analyses, can be interpreted as the ion occupancy time in the channel. The reciprocal of the ion occupancy time becomes the important off-rate constant for an ion leaving the channel.

For several reasons the gramicidin channel is a particularly favorable membranous peptide which can function as a proving ground for these NMR approaches to the characterization of ion interactions. (1) The magnitudes of the ion currents passing through the channel at unit ion activity are in the range of 10^7 ions/sec.[1] Since the voltage dependence for an ion entering or leaving the channel is weak, this sets an approximate lower value for the off-rate constant to be 10^7/sec. With nuclear magnetic resonance observation frequencies for the alkali metal ions being in the 10^7/sec range and for slowly reorienting sites, the frequency range is favorable for estimating off-rate constants. (2) The conductances (current/applied voltage) are well known for the alkali metal ions — Li^+, Na^+, K^+, Rb^+, and Cs^+ — with the spread of values being greater than a factor of 20.[2-5] This provides an opportunity to compare the experimental currents with the currents calculated for the several ions using the NMR-derived constants. The result provides a substantiation of the NMR analyses due to Bull, Forsén, and colleagues.[6-8] (3) The binding processes possible within the

[1] D. W. Urry, in "Topics in Current Chemistry" (F. L. Boschke, ed), p. 175. Springer-Verlag, Heidelberg, Federal Republic of Germany, 1985.

[2] S. B. Hladky and D. A. Haydon, Biochim. Biophys. Acta 274, 294 (1972).

[3] E. Bamberg, K. Noda, E. Gross, and P. Läuger, Biochim. Biophys. Acta 419, 223 (1976).

[4] G. Eisenman, J. Sandblom, and E. Neher, Biophys. J. 22, 307 (1978).

[5] S. B. Hladky and D. A. Haydon, Curr. Top. Membr. Transp. 21, 327 (1984).

[6] T. E. Bull, J. Magn. Reson. 8, 344 (1972).

[7] T. E. Bull, S. Forsén, and D. L. Turner, J. Chem. Phys. 70, 3106 (1979).

[8] For an excellent review of ion binding in biological systems with emphasis on quadrupolar nuclei, see S. Forsén and B. Lindman, in "Methods of Biochemical Analysis" (D. Glick, ed.), p. 289. Wiley, New York, 1981.

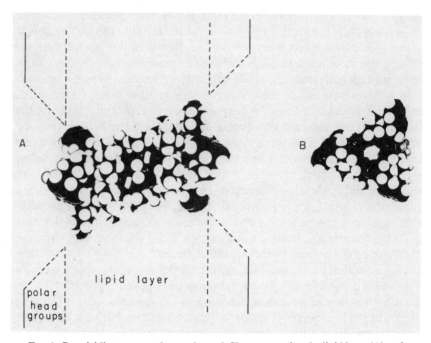

FIG. 1. Gramicidin transmembrane channel. Shown spanning the lipid layer (A) and seen in channel view (B). This channel is selective for monovalent cations with currents of 10^7 ions/sec or more.

channel are simplified by its twofold symmetry and by the structurally imposed single filing of water and ion.[9-11] (4) Finally, the ion translocation within the channel appears not to provide a significant fluctuating electric field gradient and is not rate limiting, such that the on- and off-rate constants are the dominant factors in determining current magnitude.[1] Thus the process of substantiating or validating the interpretation or analyses of NMR data on ion interaction with membranous polypeptide is facilitated by the particular characteristics of the gramicidin channel transport and structure.

Briefly, the gramicidin channel is a structure of two gramicidin molecules, e.g. (HCO-L-Val^1-Gly^2-L-Ala^3-D-Leu^4-L-Ala^5-D-Val^6-L-Val^7-D-Val^8-L-Trp^9-D-Leu^{10}-L-Trp^{11}-D-Leu^{12}-L-Trp^{13}-D-Leu^{14}-L-Trp^{15}-NHCH$_2$CH$_2$OH)$_2$ each in a single-stranded left-handed β-helical conformation containing 6.3

[9] D. W. Urry, in "The Enzymes of Biological Membranes" (A. N. Martonosi, ed.), p. 229. Plenum, New York, 1985.
[10] D. W. Urry, Proc. Natl. Acad. Sci. U.S.A. **68**, 672 (1971).
[11] D. W. Urry, M. C. Goodall, J. D. Glickson, and D. F. Mayers, Proc. Natl. Acad. Sci. U.S.A. **68**, 1907 (1971).

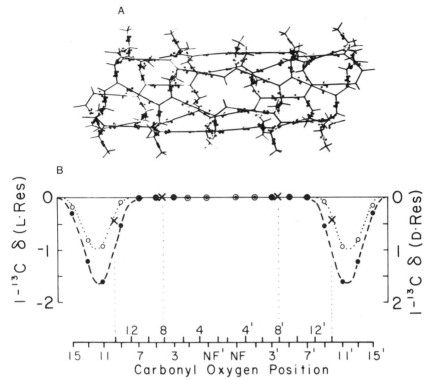

FIG. 2. (A) Wire model of the backbone and β-carbon atoms of the gramicidin trans-membrane channel. (B) Plotted with respect to the wire model are the ion-induced carbonyl carbon chemical shifts exhibited by 1-¹³C-enriched residues. The chemical shifts, whether induced by sodium ion (Na⁺) (O) or by thallium ion (Tl⁺) (●), demonstrate two binding sites related by twofold symmetry and occurring at the mouth of the channel. (Adapted with permission from Ref. 12.)

residues per turn and attached head to head (formyl end to formyl end) by means of six intermolecular hydrogen bonds.[9] A space-filling model of this structure is shown spanning a lipid layer in Fig. 1, which also shows the channel axis view. It is this slightly less than 4 Å diameter channel that can selectively pass monovalent cations at a rate of better than 10^7 ions/sec while excluding anions and divalent cations. Using cation-induced carbonyl carbon chemical shifts, two binding sites have been located just inside the ends of the channel as shown in Fig. 2.[12] It is the exchange of ions and the binding of ions in these membranous polypeptide sites that will be examined primarily by nuclear magnetic resonance relaxation approaches.

[12] D. W. Urry, J. T. Walker, and T. L. Trapane, *J. Membr. Biol.* **69**, 225 (1982).

General Consideration of Quadrupolar Ion
NMR Relaxation Methods

When certain nuclei are placed in a magnetic field their nuclear magnetic moments become quantized and align with components either parallel or antiparallel to the direction of the magnetic field.[13] The energy, E_0, derived by this orientation is $\gamma \hbar H_0 m$, where γ is the gyromagnetic ratio characteristic of the nucleus; $\hbar$ is Planck's constant, h, divided by 2π; H_0 is the external magnetic field; and m is the magnetic quantum number. The selection rules for the transition between parallel and antiparallel orientation are $\Delta m = \pm 1$; so for a nucleus with a spin quantum number, I, of $\frac{1}{2}$, the two states are $m = +\frac{1}{2}$ and $m = -\frac{1}{2}$. For nuclei with electric quadrupolar moments, i.e., quadrupolar nuclei such as ^{7}Li, ^{23}Na, ^{39}K, and ^{87}Rb with $I = \frac{3}{2}$, the states are $-\frac{3}{2}$, $-\frac{1}{2}$, $\frac{1}{2}$, and $\frac{3}{2}$. For ^{23}Na, $\gamma = 7076/G$-sec, such that at a 23.5-kG field with $\Delta E = \gamma \hbar H_0 = h\nu_0$, the difference between energy levels, ΔE, is 1.75×10^{-19} erg/nucleus, i.e., 0.00252 cal/mol, and the frequency ν_0 would be 26.4 MHz. This is a small energy difference but sufficient to give a substantial macroscopic magnetic moment, M_z, with the field direction taken along the z axis, and, of course, the frequency at which such transitions would be observed is within a few hundred parts per million (ppm) of 26.4×10^6 Hz for sodium ion nuclei. The chemical shift in ppm from a reference value (e.g., 10 mM NaCl in water) is due to the presence of local magnetic fields arising from electronic motion induced by the external magnetic field, H_0.

The NMR relaxation methods are a perturbation technique in which the macroscopic magnetic moment is displaced from equilibrium by a short intense radio frequency pulse causing the macroscopic magnetic moment to be rotated from along the field direction.[14] Commonly, pulses of sufficient intensity are given to rotate the macroscopic magnetic moment by 90 or 180°. With a 180° pulse the magnetic moment is reversed, i.e., M_z becomes $-M_z$; the NMR signal is inverted and the time characterizing its return to the equilibrium value is a characteristic time, T_1, of interest. In the relaxation methods the pulse sequence most commonly used to measure T_1 accurately is $(180° - \tau - 90°)$, which is referred to as the inversion recovery method. For this pulse sequence the intensity of M_z at

---

[13] For an excellent guidebook with graphic descriptions and practical experimental detail as well as a list of other review texts on NMR, see E. Fukushima and S. B. W. Roeder, "Experimental Pulse NMR: A Nuts and Bolts Approach." Addison-Wesley, London, 1981.

[14] For an elementary and readable review which is more mathematical than Ref. 13, see T. C. Farrar and E. D. Becker, "Pulse and Fourier Transform NMR." Academic Press, New York, 1971.

pulse interval, τ, is given by

$$M_z(\tau) = M_0[1 - 2\exp(-\tau/T_1)] \tag{1}$$

where M_0 is the resonance line intensity for infinite τ. For determining M_0 it is usually sufficient to use a delay time between sets of pulses that is $5 \times T_1$. Several sets of partially relaxed Fourier transform (PRFT) spectra with the τ values indicated are given in Fig. 3 at 30 mM ion chloride for cesium-133 and for sodium-23 in water and in the presence of 3 mM channels.[15] Plotted in Fig. 4 are the values for M_z (actually the normalized peak heights) as a function of the pulse interval where the values go from $-M_0$ to $+M_0$. A major point to be made by analysis of Fig. 4 is that the relaxation is seen to be very much faster in the presence of the membranous polypeptide binding site. The value of τ at which M_z is zero is called τ_{null}. By Eq. (1) it can be seen that the null point is a means of determining the value of T_1, i.e., $T_1 = \tau_{null}/\ln 2$. This approach is seen below to be particularly useful when there is more than one Lorentzian component to the resonance signal. In general T_1 is obtained from a plot of $\ln[M_0 - M_z(\tau)]$ versus τ wherein a straight line should be obtained with a slope of $-1/T_1$.

The magnitude of T_1, the longitudinal relaxation time (also called the spin-lattice relaxation time), depends on the environment or lattice surrounding the ion. The relaxation mechanism depends on the properties of the nucleus, the surrounding lattice, and the interactions between the two. For nuclei with electric quadrupole moments, the dominant relaxation mechanism depends on the magnitude of the electric field gradient, eq of the site(s) experienced by the nucleus. For quadrupolar nuclei the longitudinal relaxation time for the extreme narrowing condition (i.e., $\omega^2\tau_c^2 \ll 1$) is written

$$\frac{1}{T_1} = \frac{1}{T_2} = \frac{3(2I+3)}{40I^2(2I-1)}\left(1+\frac{\eta^2}{3}\right)\left(\frac{e^2qQ}{\hbar}\right)^2\tau_c \tag{2}$$

where I is the spin quantum number; η is the asymmetry parameter for the electric field gradient, eq; the quantity eQ is the electric quadrupole moment; $\hbar$ is as defined above; ω is the resonance frequency in radians per sec, i.e., $\omega = 2\pi\nu_0$; and τ_c is the correlation time. The quantity T_2 is called the transverse or spin–spin relaxation time and in the absence of complications such as significant resonance line broadening due to field inhomogeneity, it is equal to $(\pi\nu_{1/2})^{-1}$ where $\nu_{1/2}$ is the full resonance linewidth at half-intensity.

[15] D. W. Urry, T. L. Trapane, R. A. Brown, C. M. Venkatachalam, and K. U. Prasad, *J. Magn. Reson.* **65**, 43 (1985).

A

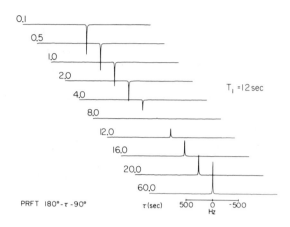

B

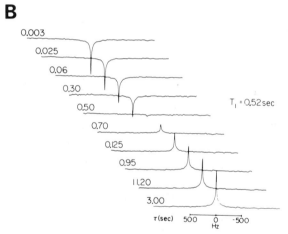

Fig. 3. Longitudinal relaxation data, partially relaxed Fourier transformed (PRFT) data, for (A) cesium-133 at 30°, 30 mM CsCl in water (^2H_2O), and in the presence of (B) 3 mM lysolecithin-packaged malonyl(gramicidin A)$_2$ channels and for (C) sodium-23 at 30°, 30 mM, NaCl in water (^2H_2O), and in the presence of (D) 3 mM lysolecithin-packaged malonyl(gramicidin A)$_2$ channels. Note that the change in T_1 due to the presence of channels is greater for Cs^+ than for Na^+. (Reproduced with permission from Ref. 15.)

Multiple Relaxations for Spin $\frac{3}{2}$ Nuclei Undergoing Exchange

For spin $\frac{3}{2}$ nuclei and for two-site exchange when one site (the hydrated ion in aqueous solution) is under the extreme narrowing condition and the other site (e.g., a membranous peptide binding site) is not, it is possible for

C

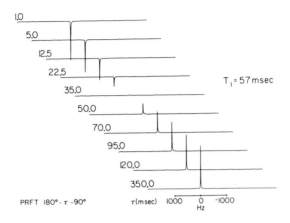

D

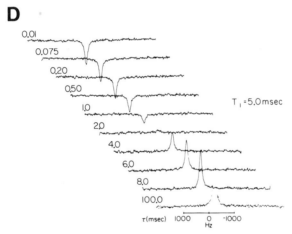

FIG. 3C and D.

two components to be observed in the longitudinal relaxation, i.e.,

$$M_z(t) = M_0[0.2 \exp(-t/T_1') + 0.8 \exp(-t/T_1'')] \tag{3}$$

where T_1' and T_1'' are the characteristic longitudinal relaxation times for the two relaxations.[8] With T_{1f} being the relaxation time of the free ion in solution, for a rate that an ion leaves the site that is rapid with respect to the relaxation rate at the site, for the case of axial symmetry, i.e. $(1 + \eta^2/3) = 1$, and with the quadrupolar coupling constant, χ, defined $e^2qQ/\hbar$, T_1' and

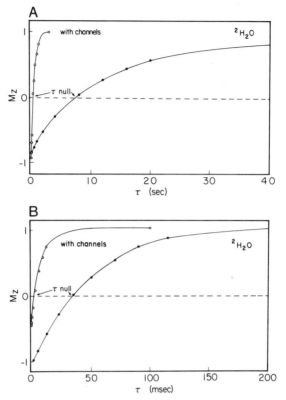

FIG. 4. Plots of recovery time of the inverted macroscopic magnetic moment, M_z, for the data in Fig. 3. This demonstrates the rapid relaxation of the inverted signal effected by the presence of channels and it allows comparison of (A) cesium-133 and (B) sodium-23 data. The time for which the intensity of M_z passes through zero is called the τ_{null} and is equal to $T_1/\ln 2$.

T_1'' can be written as

$$\frac{1}{T_1'} = \frac{1}{T_{1f}} + \frac{P_B\chi^2}{10}\left[\frac{\tau_c}{(1 + \omega^2\tau_c^2)}\right] \tag{4}$$

and

$$\frac{1}{T_1''} = \frac{1}{T_{1f}} + \frac{P_B\chi^2}{10}\left[\frac{\tau_c}{(1 + 4\omega^2\tau_c^2)}\right] \tag{5}$$

where P_B, the mole fraction of ions at the binding site, is very small with respect to 1, that is, P_B/P_A is essentially P_B because P_A, the mole fraction of ions in solution, is essentially 1.

For the same conditions as Eqs. (3)–(5), the time dependence of the transverse magnetization, $M_T(t)$, is also due to two relaxation processes

$$M_T(t) = M_T(0)[0.6 \exp(-t/T_2') + 0.4 \exp(-t/T_2'')] \tag{6}$$

where

$$\frac{1}{T_2'} = \frac{1}{T_{2f}} + \frac{P_B\chi^2}{20}\left[\tau_c + \frac{\tau_c}{(1 + \omega^2\tau_c^2)}\right] \tag{7}$$

and

$$\frac{1}{T_2''} = \frac{1}{T_{2f}} + \frac{P_B\chi^2}{20}\left[\frac{\tau_c}{(1 + 4\omega^2\tau_c^2)} + \frac{\tau_c}{(1 + \omega^2\tau_c^2)}\right] \tag{8}$$

Methods for Determining Relaxation Time Constants

Curve Resolution for Determining T_2' and T_2''. With $T_2' = (\pi\nu_{1/2}')^{-1}$ and $T_2'' = (\pi\nu_{1/2}'')^{-1}$, the experimental spectrum is examined for the presence of both narrow and broad Lorentzian curves. An experimental spectrum is shown in Fig. 5A for 350 mM NaCl in the presence 2.6 mM malonylgramicidin channels.[16] (Malonylgramicidin is the covalent dimer where the two formyl hydrogens have effectively been replaced by a bridging CH_2 moiety. In subsequent reference this covalent dimer is also referred to as malonyl[gramicidin]$_2$). In Fig. 5B the experimental data are shown resolved into two Lorentzian curves, a broad Lorentzian curve, called the fast component, representing 60% of the total signal intensity, and a narrow Lorentzian curve, called the slow component, representing 40% of the total signal intensity.[8]

The Null Method for Estimating T_1', T_1'', T_2', and T_2''. These two components can also be experimentally dissected by the null method. Initially the pulse interval is found near the elapsed time where the narrow component changes from being negative to positive. This leaves only the broad component from which $\nu_{1/2}'$ can be measured, and the τ_{null} for the narrow component allows determination of T_1'. At slightly longer values of τ the broad component is seen to be nulled. This leaves the narrow component alone from which $\nu_{1/2}''$ is estimated and the τ_{null} for the broad component allows determination of T_1''. The null method is demonstrated in the partially relaxed Fourier transform (PRFT) spectra of Fig. 6 for rubidium-87 as 2 M RbCl in the presence of 3.16 mM gramicidin channels.[17] At

[16] C. M. Venkatachalam and D. W. Urry, *J. Magn. Reson.* **41**, 313 (1980).

[17] D. W. Urry, T. L. Trapane, C. M. Venkatachalam, and K. U. Prasad, *J. Am. Chem. Soc.* **108**, 1448 (1986).

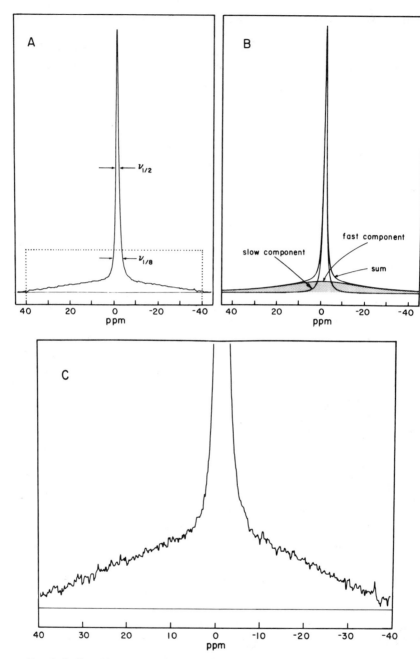

FIG. 5. Sodium-23 NMR spectrum of 350 mM NaCl in the presence of 2.6 mM malo-nyl(gramicidin A)$_2$ channels. From the experimentally observed lineshape (A) it is possible by means of curve resolution (B) to determine the linewidths at half-intensity for the broad ($v'_{1/2}$) and narrow ($v''_{1/2}$) components. The dotted boundary in A is shown enlarged (C), where both

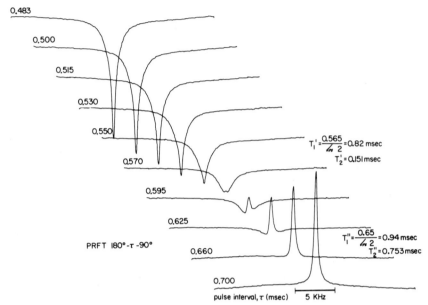

FIG. 6. Partially relaxed Fourier transform (PRFT) data through the null time regions of the narrow and broad components for 2 M RbCl in the presence of 3.16 mM channels. By noting the null time for the narrow component, T_1' may be calculated and from the linewidth at half-intensity for the remaining broad component, $T_1'' = (\pi v_{1/2}')^{-1}$ is calculated. At the null time for the broad component, T_2'' may be calculated and from the linewidth at half-intensity of the remaining narrow component, $T_2'' = (\pi v_{1/2}'')^{-1}$, can be calculated. By means of Eq. (9) these values are used to calculate the correlation time. (Reproduced with permission from Ref. 17.)

$\tau = 0.483$ msec the resonance is inverted and composed of broad and narrow components and the $v_{1/8}/v_{1/2}$ ratio is much larger than $\sqrt{7} = 2.65$, which is the value expected for a single Lorentzian. As the negative intensity decreases, it is seen at $\tau = 0.595$ msec that the narrow component has become positive while the broad component is yet negative. By extrapolation the narrow component is estimated to be nulled at 0.565 msec, which on division by ln 2 gives a T_1' of 0.82 msec. At 0.570 msec the $v_{1/8}/v_{1/2}$ ratio is near $\sqrt{7}$ and the linewidth at half-intensity provides a measure of $v_{1/2}'$, which gives a $T_2' = 0.151$ msec. When the pulse interval is 0.660 msec, the broad component is near its null point and the resulting resonance exhibits

broad and narrow components are readily apparent from these values; the transverse relaxation times T_2' and T_2'' are obtained by the relation $T_2 = (\pi v_{1/2})^{-1}$. The values are $T_2' = 0.25$ msec and $T_2'' = 8.6$ msec. These values may be used in Eq. (9) to calculate the correlation time. This represents the curve resolution method. (Reproduced with permission from Ref. 16.)

a $v_{1/8}/v_{1/2}$ ratio that is $\sqrt{7}$ within the measurable accuracy. This Lorentizan curve gives $v''_{1/2}$, which yields T''_2. The null point for the broad component is estimated to be 0.65 msec, which on division by $\ln 2$ gives a T''_1 of 0.94 msec. Thus the null method provides estimates for T'_1, T''_1, T'_2, and T''_2.

Spin-Echo Method with Curve Stripping for Estimating T'_2 and T''_2. Another means of evaluating T'_2 and T''_2 is by means of a spin-echo experiment. One common spin-echo experiment with a pulse sequence of $90°-(\tau-180°-\tau)_n$ is called the Carr–Purcell–Meiboom–Gill (CPMG) sequence. In this case the 180° pulses are phase shifted 90° with respect to the initial 90° pulse; that is, the 90° pulse is along the x' axis and the 180° pulses are along the y' axis, with the axes defined in the rotating frame.[13,14] With this approach the pulse sequence corrects for and utilizes field inhomogeneities to give rise to spin echoes which decrease in intensity with time. The transverse magnetization, M_T, is directly obtained as a function of time and, as shown in Fig. 7 when $\ln M_T$ is plotted versus time for sodium-23 nuclei in the presence of channels, the plot is nonlinear. There is an initial rapid loss of transverse magnetization followed by a slower linear rate of decline. By fitting the linear data at longer times to a straight line, the slow process is characterized by the slope which gives T''_2. By taking the difference between this fitted line and the experimental curve at short times, a second linear set of values is obtained, the slope of which gives the value of T'_2. This is the direct demonstration of the double exponential decay process of Eq. (6) and the intercepts give the relative intensities of the two components. The intercepts are 3.450 for the fast process with T'_2 as the relaxation time, and 3.075 for the slower process with T''_2 as the relaxation time. The ratio, $\exp(3.450)/\exp(3.075) = 1.46$, is very close to the ratio of theoretical coefficients of 0.6/0.4 in Eq. (6).

Thus there are three different ways to evaluate T'_2 and T''_2, and when one approach is compromised, or substantiation is desired, another may be used. To obtain T_1 a standard longitudinal relaxation experiment (as in Fig. 3) and Eq. (1) may be used, or the null method of Fig. 6 may be used, in which case, when present, both T'_1 and T''_1 may be obtained.

Approaches for Estimating Correlation Times: Four Experimental Cases

The relaxation data on quadrupolar nuclei can be considered in terms of four experimental cases which require different analyses in order to arrive at the experimental correlation time and finally at a rate constant. The correlation time, once obtained, must then be interpreted in terms of the molecular process or processes responsible for the nuclear relaxation. The four experimental cases are (1) the extreme narrowing condition with

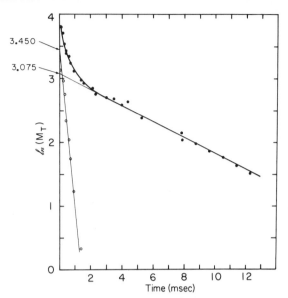

FIG. 7. Sodium-23 NMR spin-echo experiment (Carr–Purcell–Meiboom–Gill sequence) following the relaxation of the transverse magnetization of 350 mM NaCl in the presence of 2.5 mM malonyl(gramicidin A)$_2$ channels. The decay is clearly nonlinear or, more correctly, composed of two linear components. By fitting the data at long times to a straight line, the difference between the extrapolated straight line at short times and the experimental values (●) define a second straight line (○). From the slope at long times is obtained T_2'' and from the slope of the process occurring at short times, the value T_2' is obtained. This is called the spin-echo curve-stripping method for obtaining transverse relaxation times. (Reproduced with permission from Ref. 16.)

a Lorentzian-shaped resonance, (2) the observation of both narrow and broad components, (3) the observation of a non-Lorentzian lineshape, and (4) observation of only the narrow component with a broad component being too broad to observe in the experimental window or to determine due to instrumental limitations. These four cases are shown schematically in Fig. 8.

Case 1. Extreme Narrowing Condition. The extreme narrowing condition is for the case of very short correlation times. In particular, $\omega^2\tau_c^2 \ll 1$, as is the case for Eq. (2). With an observation frequency of $\omega = 2\pi \times 26.3 \times 10^6$ Hz $= 1.65 \times 10^8$/sec for sodium-23, this means that τ_c would be much less than 6×10^{-9} sec. Such a situation is recognized as shown in Eq. (2) by the experimental finding that $T_1 = T_2$ and further by the experimental observation that T_1 and T_2 are independent of the observation frequency, i.e., are field independent. Another characteristic of this situation, but not uniquely so, is that the resonance line is Lorentzian in shape. This may be assessed by taking the $\nu_{1/8}/\nu_{1/2}$ ratio, which is $\sqrt{7}$ for a

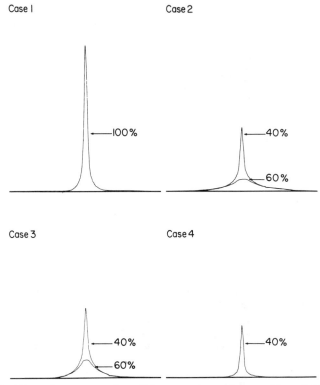

FIG. 8. Schematic representation of four experimental cases. (Case 1) The extreme narrowing condition, $\omega^2 \tau_c^2 \ll 1$ and $T_1 = T_2$, where a simple Lorentzian curve is obtained containing the total intensity. (Case 2) Both distinct broad and narrow components are readily apparent; $\omega^2 \tau_c^2 \gg 1$ and $T_1 \neq T_2$. Curve resolution, the null approach, or a spin-echo sequence can be used to obtain T_2' and T_2'' for calculation of the correlation time using Eq. (9). (Case 3) Non-Lorentzian (two components less differentiated) resonance signal; $T_1 \neq T_2$ and $\nu_{1/8}/\nu_{1/2}$ is greater than $\sqrt{7}$. With determination of T_1 from a PRFT experiment and of T_2, for example, from the linewidth at half-intensity, both in the presence of peptide binding site, and with the reference T_1 and T_2 determined in the absence of the binding site, Eq. (10) may be used to calculate the ion correlation time. (Case 4) Broad component too broad to observe. Integrated signal is 40% of the reference signal. Limits can be placed on the width of the nonobserved broad component, which, with the linewidth of the narrow component (essentially a Lorentzian curve), can put limits on the value of τ_c. More directly, when this single component shows exchange broadening, it has been used to calculate the rate constant for the exchange process.

Lorentzian line. Beyond placing a limit on the magnitude of τ_c, a more specific definition of τ_c for the extreme narrowing case requires knowledge of the magnitude of the quadrupolar coupling constant and of the symmetry of the site. Fortunately, for consideration of ion interactions with membranes and in particular of channel transport with currents of the order of 10^7 ions/sec, longer correlation times are often the experimental result, as discussed below.

Case 2. Observation of Both Narrow and Broad Components. Under certain favorable circumstances, as, for example, in Figs. 5 and 6, two components are apparent in the spectrum. This is accompanied by the experimental finding that $T_1 \neq T_2$ and is the result of $\omega^2\tau_c^2 > 1$. Under these conditions Eqs. (3)–(8) apply. By taking the ratio of Eqs. (4) and (5) or of Eqs. (7) and (8), it is possible to obtain an expression containing τ_c that depends otherwise on known quantities, as was first shown by Bull.[6] As is apparent in Fig. 6, T_1' and T_1'' are similar in magnitude, with T_1'' being only 15% larger than T_1'. On the other hand T_2'' is 500% larger than T_2'. In the case of T_1' and T_1'' errors in these values and the approximations of the expressions make Eqs. (4) and (5) of limited use, whereas the large differences in the values of T_2' and T_2'' make Eqs. (7) and (8) of much use. Following Bull[6] we write

$$\frac{1/T_2' - 1/T_{2f}}{1/T_2'' - 1/T_{2f}} \approx \frac{1 + (1 + \omega^2\tau_c^2)^{-1}}{(1 + 4\omega^2\tau_c^2)^{-1} + (1 + \omega^2\tau_c^2)^{-1}} = \Delta(R_2'/R_2'') \qquad (9)$$

For convenience this expression is referred to as the Bull ratio or $\Delta(R_2'/R_2'')$. A plot of $\Delta(R_2'/R_2'')$ as a function of correlation time, τ_c, is given in Fig. 9A for the alkali metal ions for a 23.5-kG magnet and in Fig. 9B for a set of observation frequencies. Actually, by translation of a curve along the x axis with the proper positioning of the observation frequency at the top of the figure, the curve is obtained for any frequency. Thus when T_2' and T_2'' are experimentally determined, it is possible to obtain a correlation time for any observation frequency using Fig. 9B.

Case 3. Observation of a Non-Lorentzian Lineshape. When $\omega^2\tau_c^2$ is of the order of one ($\simeq 1$), two components are not immediately apparent on inspection of the resonance line. On measuring the resonance linewidth at half-intensity and at one-eighth intensity, the ratio $\nu_{1/8}/\nu_{1/2}$ can be found to be significantly greater than $\sqrt{7}$, that is greater than 2.65. This is a non-Lorentzian lineshape. Also, experimentally it is found that $T_1 \neq T_2$ but curve resolution may not be unique for two components present in the signal and the spin-echo and null approaches may not be sufficient to separate out T_2' and T_2''. Under these circumstances it is possible to take ratios of approximate expressions due to Bull[6,18] as first suggested by Rose and Bryant[19]

[18] H. Gustavsson, B. Lindman, and T. E. Bull, *J. Am. Chem. Soc.* **100,** 4655 (1978).
[19] K. Rose and R. G. Bryant, *J. Magn. Reson.* **31,** 41 (1978).

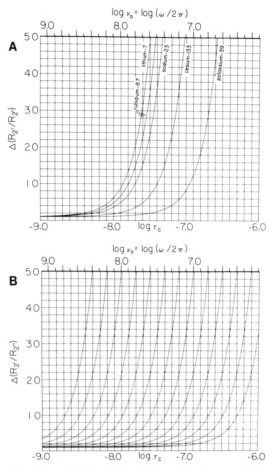

FIG. 9. Plots of Eqs. (9) and (10) for different observation frequencies. (A and B) The Bull ratio, $\Delta(R_2'/R_2'')$, plotted as a function of correlation time, τ_c, for the observation frequencies of each of the alkali metal ions (A) and for a set of observation frequencies in steps of log 0.2 (B). (C and D) The Rose and Bryant ratio, $\Delta(R_2/R_1)$, plotted as a function of correlation time, τ_c, for each of the Group IA ion observation frequencies (C) and for a set of observation frequencies in steps of log 0.2 (D). The values go up to 250. (E and F) The same as C and D with an expanded vertical scale. By these plots of Eqs. (9) and (10), experimental results may be used to estimate correlation times.

which utilize the known quantities T_1, T_2, T_{1f}, and T_{2f} as shown below:

$$\frac{1/T_2 - 1/T_{2f}}{1/T_1 - 1/T_{1f}} = \frac{0.6 + 0.4(1 + 4\omega^2\tau_c^2)^{-1} + (1 + \omega^2\tau_c^2)^{-1}}{1.6(1 + 4\omega^2\tau_c^2)^{-1} + 0.4(1 + \omega^2\tau_c^2)^{-1}} \qquad (10)$$
$$= \Delta(R_2/R_1)$$

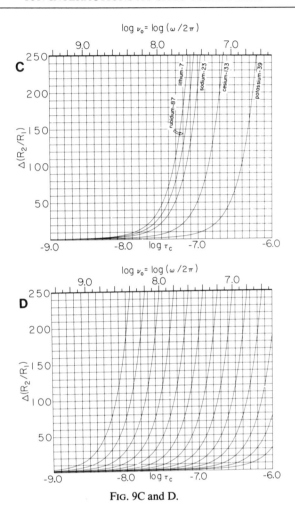

FIG. 9C and D.

This will be referred to as the Rose and Bryant ratio or as $\Delta(R_2/R_1)$. A plot of this equation is given in Fig. 9C for the alkali metal ions given a 23.5-kG magnetic field and in Fig. 9D are curves for the logarithm of the observation frequencies differing by units of ln 0.2. Expanded scales with $\Delta(R_2/R_1)$ ranging from 1 to 10 are given in Fig. 9E and F. Again, as with the Bull ratio, a careful translation of any curve along the x axis will allow it to be used for any observation frequency as marked at the top of the figures. In this case with $\omega\tau_c \simeq 1$, when the observation frequency is 4.65×10^6 Hz as for potassium-39, the correlation time would be of the order of 30×10^{-9} sec.

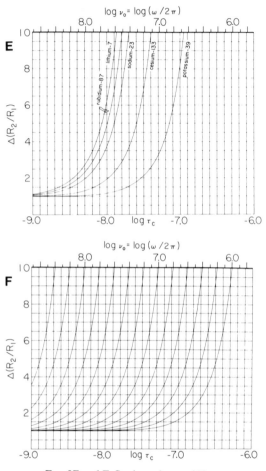

FIG. 9E and F. See legend on p. 302.

Case 4. Broad Component Too Broad to Observe. When the correlation time is sufficiently long that T_2' is too short to detect experimentally, the broad component is too broad to observe. For example, when the delay time between the end of the pulse and the beginning of data accumulation is the same magnitude or longer than T_2', then this fast relaxation will have occurred before data accumulation begins. A very short delay time is 20 μsec. If $T_2' = 0.02$ msec, then 63% of the relaxation of the fast component will have already occurred. Another perspective is that a very wide observation window would be required to observe the broad component with $\nu_{1/2}' = (\pi T_2')^{-1}$, i.e., the linewidth at half-intensity would be 16 kHz. This situation of a broad component being too broad to observe is diagnosed by either a field dependence (observation frequency dependence) of T_1 or by an integration of signal intensity and comparison to a signal of known 100% intensity. Since the remaining signal is the narrow component, i.e., T_2'', by Eq. (6) with a 0.4 coefficient for this term, an intensity as low as 40% of a reference intensity could be found. Care in this case should be taken not to use Eq. (10) even though T_1 may not equal T_2. When this situation is observed, the correlation time is so long and the exchange of nuclei between solution and binding site may be sufficiently slow that exchange broadening of the resonance line can begin to be apparent. Since the narrow component is due to the $+\frac{1}{2}$ to $-\frac{1}{2}$ transition, it has been treated as a simple spin $\frac{1}{2}$ case to obtain an estimate of an exchange rate (see below) of relevance to ion interaction with the gramicidin A transmembrane channel.[16,20,21]

Possible Contributions to the Correlation Time

Quadrupolar nuclei relax by means of the interactions of their nuclear quadrupole moments with fluctuating electric field gradients. Therefore interpretation of correlation times requires consideration of all of the processes which would result in a quadrupolar nucleus to experience a fluctuating electric field gradient (EFG). As it is the rate processes that are additive, one writes that the inverse of the correlation time is the sum of all of the inverse mean correlation times for each of the rate processes, τ_i, that would cause the nucleus to experience a fluctuating electric field gradient, i.e.,

$$\frac{1}{\tau_c} = \sum_i \frac{1}{\tau_i} \tag{11}$$

When the environment of the nucleus is not of cubic symmetry, the reorientation of a site in the magnetic field constitutes a fluctuation leading to relaxation.[22] This is referred to as the reorientation correlation time, τ_r. For a membranous site with dimensions of several hundred angstroms or more, this is a relatively long reorientation correlation time of the order of microseconds. For a membranous ion-binding site, an ion going from solution, where there is little or no EFG, to a binding site with a substantial EFG constitutes a fluctuating electric field gradient with a mean correlation time that would be the mean occupancy time, τ_b.[22] Also, an ion in the

[20] D. W. Urry, C. M. Venkatachalam, A. Spisni, P. Läuger, and M. A. Khaled, *Proc. Natl. Acad. Sci. U.S.A.* **77**, 2028 (1980).
[21] D. W. Urry, C. M. Venkatachalam, A. Spisni, R. J. Bradley, T. L. Trapane, and K. U. Prasad, *J. Membr. Biol.* **55**, 29 (1980).
[22] A. G. Marshall, *J. Chem. Phys.* **52**, 2527 (1970).

site could possibly experience internal motion at the binding site, e.g., vibrating ligands, that could constitute a fluctuating EFG. This is designated as a vibrational correlation time, τ_{vib}.[23] Finally, a channel with more than one binding site also represents a special case in which an intrachannel ion translocation could result in a fluctuating EFG even with identical symmetrically related binding sites. This is because the EFG of the ion in transit may be quite different than when bound at a site. The translocational correlation time is designated as τ_{tr}.[24] Accordingly, the experimental correlation time can be expressed as

$$\frac{1}{\tau_c} = \frac{1}{\tau_r} + \frac{1}{\tau_{vib}} + \frac{1}{\tau_b} + \frac{1}{\tau_{tr}} \tag{12}$$

Determination of Rate Constants

In order to utilize the correlation time data obtained as outlined above, it is necessary to determine which terms in Eq. (12) are contributing to τ_c^{-1}. Also, for the case where the broad component is too broad to observe (see experimental Case 4 above), the expressions for exchange broadening will be considered.

Interpretation of the Correlation Time

The experimental sodium-23 correlation time for 1 M sodium chloride interacting with 3 mM channels at 30° is 4.8×10^{-8} sec.[25] Since the reorientation correlation time of the membranous site is in the microsecond range, τ_r^{-1} of Eq. (12) is too small to contribute to τ_c^{-1}. By means of the temperature dependence of τ_c it is possible to assess whether $\Delta S^{\ddagger}$ and $\Delta H^{\ddagger}$ values are indicative of a vibrational process. This data are given in Fig. 10. The plot of $-\ln \tau_c$ versus $1/T$ (degrees K) yields a well-defined straight line with a slope giving a $\Delta H^{\ddagger}$ of 5.9 kcal/mol and an intercept giving a $\Delta S^{\ddagger}$ of -5.4 cal/mol-deg. A value of 5.9 kcal/mol appears to be high for a vibrational process. With respect to the entropy of activation it is possible to determine what vibrational frequency would give such an entropy by means of the statistical mechanical expression for entropy using a har-

[23] S. Forsén and B. Lindman, *Chem. Br.* **14**, 29 (1978).

[24] D. W. Urry, *in* "Ion Transport through Membranes" (K. Yagi and B. Pullman, eds.), p. 233. Academic Press, New York, 1987.

[25] D. W. Urry, T. L. Trapane, C. M. Venkatachalam, and K. U. Prasad, *Int. J. Quantum Chem., Quantum Biol. Symp.* **12**, 1 (1986).

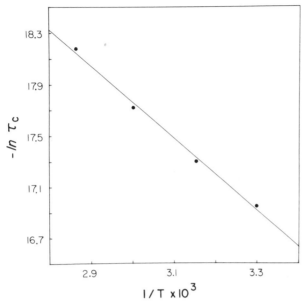

FIG. 10. Plot of $-\ln \tau_c$ versus $1/T$ (degrees K) for 1 M NaCl in the presence of 3 mM channels. The values of τ_c were estimated using the Bull ratio, Eq. (9). The linewidth is consistent with a simple barrier process. Using Eyring rate theory, with $\tau_c^{-1} = kT/he^{\Delta S^{\ddagger}/R}e^{-\Delta H^{\ddagger}/RT}$, resulted in values of 5.9 kcal/mol for $\Delta H^{\ddagger}$ and -5.4 cal/mol-deg for $\Delta S^{\ddagger}$. As discussed in the text, these are not values characteristic of ligand vibrations with 48 nsec correlation times, but rather are indicative of a barrier for an ion leaving the channel. (Adapted with permission from Ref. 25.)

monic oscillator partition function, i.e.,

$$S_i = R \ln(1 - e^{-h\upsilon_i/kT})^{-1} + \frac{Nh\upsilon_i}{T}(e^{h\upsilon_i/kT} - 1)^{-1} \tag{13}$$

With $\upsilon_i = (2\pi\tau_{\text{vib}})^{-1}$, an entropy of -5.4 cal/mol-deg defines a correlation time, τ_{vib}, of tenths of a picosecond rather than of 48 nsec. Because of this, it is argued that a vibrational process of the ligand field is not responsible for the experimental τ_c. This means that τ_c^{-1} is either τ_b^{-1} or τ_{tr}^{-1} or a combination of the two. To assess this situation, it is possible to compare channels with different intrachannel ion translocation rates. For the malonyl (head to head) covalent dimer of gramicidin the relaxation time (determined by dielectric relaxation studies) for the ion jump from one site to another at 25° is 1.2×10^{-7} sec.[26] For the gramicidin channel the relaxation time is more than an order of magnitude shorter (R. Henze and

[26] R. Henze, E. Neher, T. L. Trapane, and D. W. Urry, *J. Membr. Biol.* **64** (1982).

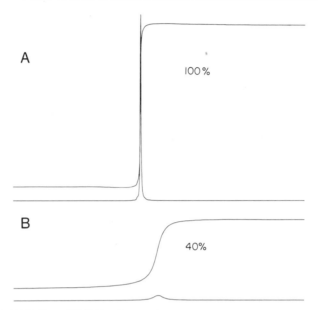

FIG. 11. (A) Sodium-23 NMR reference signal of 20 M NaCl in the presence of lysophosphatidylcholine. This resonance intensity is integrated and taken as 100%. (B) Resonance intensity of 20 mM NaCl in the presence of lysophosphatidylcholine on addition of 3 mM channels. The relative intensity on addition of channels, shown by the integration, is about 40%. Therefore, this is an example of Case 4 as shown in Fig. 8 and discussed in the text, where the broad component is too broad to observe.

D. W. Urry, unpublished observations), yet the NMR-derived correlation times for the two channels are essentially the same.[20,21] Since the NMR-derived correlation time is unchanged by a changing τ_{tr}, it can be argued that this possible EFG fluctuation is not contributing to τ_c^{-1}. Therefore it appears that the channel system provides a particularly favorable case for the application of this data to ion transport where $\tau_c^{-1} = \tau_b^{-1}$. Now the inverse of the occupancy time is the off-rate constant. Since this value of τ_c was obtained at high concentration where there is double occupancy, it represents exchange with the weak binding site, that is, $\tau_c^{-1} = k_{off}^w$.

Exchange Broadening of Narrow Component When Broad Component Is Too Broad to Observe

At low ion concentrations, e.g., 20 mM NaCl, in the presence of 3 mM channels at 30°, the sodium-23 resonance obtained is shown integrated in Fig. 11B. In the absence of channels with all other components present and under identical instrumental conditions, the resonance is shown in Fig.

11A. On comparison it is seen that the intensity is greatly reduced to about 40% in the presence of channels. This is the experimental Case 4 noted above where the broad component (60% of intensity) is too broad to observe. There is only the narrow component at low ion concentrations because the relaxation time T_2' is too short to observe and the correlation time is too long (i.e., the rate constant is too small) to determine as considered above. These are the conditions where there is at most one ion in the channel. This interpretation of single-ion occupancy is corroborated, by studies on other transported ions as well, when following the chemical shift of carbonyl carbons at the ion-binding site (see below). For the sodium ion at low concentrations this means that $\tau_c \gg 4.8 \times 10^{-8}$ sec, which is the value obtained at high ion concentration, e.g., 1 M NaCl, where the relaxation process is dominated by the second ion occupancy of the channel. The second ion occupancy at high ion concentration is referred to as the weak site and the first ion occupancy as the tight site. Furthermore this means that $k_{off}^t \ll k_{off}^w$, i.e., the off-rate constant for the singly occupied channel is very much smaller than the off-rate constant for the doubly occupied channel.

An estimate of k_{off}^t becomes possible by combining the chemical shift and linewidth dependencies on ion concentration at low ion concentration. Since k_{off}^t is small, it does not contribute measurably to the current passing through the channel. Therefore it is only important for understanding gramicidin channel transport to obtain an approximate or limiting value for k_{off}^t. A plot of the sodium-23 ion chemical shift, ν, at low ion concentration for the malonylgramicidin channel in phospholipid bilayers formed on heating malonylgramicidin with lysolecithin is given in Fig. 12. Under conditions of fast exchange the ion chemical shift, ν, is described by the expression

$$\nu = P_f \nu_f + P_t \nu_t \tag{14}$$

where P_f and P_t are the mole fractions of ions free in solution and bound as the first ion in the channel, respectively, and ν_f and ν_t, respectively, are the chemical shift for the ion free in solution and for the first ion in the channel, i.e., for the tight binding site. A fitting of Eq. (14) to the data in Fig. 12 gives $\nu_t = 1850$ Hz. The fast exchange condition assumed for Eq. (14), consistent with the observation of a single peak throughout the titration, means that $k_{off}^t > 10^4$/sec ($= 2\pi \times 1850$/sec) with ν_f taken as the reference of zero chemical shift. So far we have that 10^4/sec $< k_{off}^t = 10^7$/sec.

Also, plotted with respect to the right-hand ordinate in Fig. 12, is the concentration dependence of the linewidth at half-intensity, $\nu_{1/2}$. The ex-

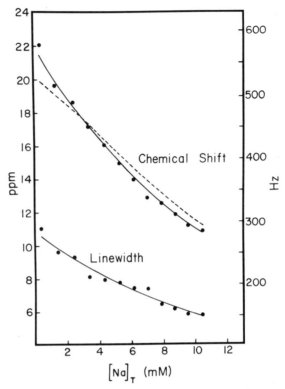

FIG. 12. Concentration dependence of chemical shift and linewidth at low ion concentration for sodium-23 nuclei in the presence of 3 mM malonyl(gramicidin)$_2$ channels. From the fitted curves $v_t = 1850$ Hz and $v^t_{1/2} = 900$ Hz. These values provide quantities required for estimating k^t_{off} by means of Eq. (16). (Reproduced with permission from Ref. 21.)

pression for the linewidth dependence may be written.[16,27,28]

$$v_{1/2} = v^f_{1/2} + P_t(v^t_{1/2} - v^f_{1/2}) + \frac{4\pi P_t(1 - P_t)^2 v^2_t}{k^t_{off}} \tag{15}$$

where $v^f_{1/2}$ and $v^t_{1/2}$ here designate the linewidths at half-intensity of the ion free in solution and in the singly occupied channel, respectively. Combining Eqs. (14) and (15) and eliminating P_t by setting $v_f = 0$ and $v/v_t = P_t$

[27] G. Binsch, *in* "Topics in Stereochemistry" (E. L. Eliel and N. L. Allinger, eds.), p. 97. Wiley, New York, 1968.

[28] J. Feeney, J. G. Batchelor, J. P. Albrand, and G. C. K. Roberts, *J. Magn. Reson.* **33,** 519 (1979).

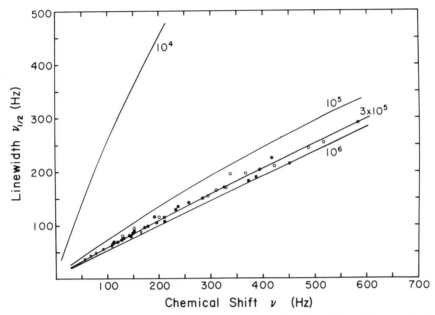

FIG. 13. Plot of linewidth at half-intensity $\nu_{1/2}$ versus chemical shift, ν. The calculated curves use Eq. (16) and the data of Fig. 12 with $\nu_{1/2} = 1850$ Hz and $\nu_{1/2}^t = 838$ Hz. A value for k_{off}^t of 3×10^{-5}/sec is seen to fit well with the experimental data. (Reproduced with permission from Ref. 21.)

gives[16,21]

$$\nu_{1/2} = \nu_{1/2}^f + \frac{\nu}{\nu_t}(\nu_{1/2}^t - \nu_{1/2}^f) + \frac{4\pi\nu(\nu_t - \nu)^2}{\nu_t k_{\text{off}}^t} \tag{16}$$

Using this equation, a replot of the data (and additional data) in Fig. 13 allows evaluation of k_{off}^t. The best fit gives $k_{\text{off}}^t \approx 3 \times 10^5$/sec.

Determination of Ion Binding Constants

Binding constants can be estimated a number of ways using NMR. Below are examples of using longitudinal relaxation rates, ion resonance chemical shifts, and site resonance chemical shifts as a function of ion concentration. The multiple approaches have obvious advantages. One method can be used to corroborate another method, and one method may have a limitation that can be overcome by a second method. For example, because the reorientation correlation time for a lipid bilayer membrane system is long, it is often necessary to use an elevated temperature to follow ion binding when monitoring an element within the membrane, e.g., when following the chemical shift of a nucleus within a membrane-bound moiety affected by ion binding. On the other hand the longitudinal relaxation rate approach, which can readily follow the ion, can be used at any temperature and therefore can substantiate the meaning of data obtained at elevated temperature by examining the binding behavior over the range from the elevated temperature to a lower temperature. Also the ion chemical shift approach has its favorable and less favorable situations. When there are two binding sites, a tight site and a weak site, and when the chemical shift at the weak site is the same magnitude or less than at the tight site, it is often difficult to characterize the weak site or even to recognize its presence. At low ion concentration there can be observed a large chemical shift because, following Eq. (14), the P_t/P_f ratio is a reasonable value. But at high ion concentration with a several orders of magnitude lower concentration of weakly binding sites, the P_w/P_f ratio becomes quite unfavorable. Under these circumstances the observed chemical shift would be relatively much smaller and could even be difficult to delineate from the tight binding process. For these reasons it is useful to begin with consideration of longitudinal relaxation rates.

Excess Longitudinal Relaxation Rates

The excess longitudinal relaxation rate (ELR) is defined as the difference in the observed longitudinal relaxation rate in the presence ($R_1 = 1/T_1$) and in the absence ($R_{1f} = 1/T_{1f}$) of binding site, i.e., the ELR is ($R_1 - R_{1f}$). Assuming conditions of fast exchange and considering only the tight binding site,

$$R_1 = P_f R_{1f} + P_t R_{1t} \tag{17}$$

the excess longitudinal relaxation rate is written

$$(R_1 - R_{1f}) = P_t(R_{1t} - R_{1f}) \tag{18}$$

For conditions where site concentration is small with respect to total ion concentration, James and Noggle[29] derived an expression using the reciprocal of the ELR, i.e.,

$$(R_1 - R_{1f})^{-1} = \frac{\mathrm{Na_T} + (K_b^t)^{-1}}{C_T(R_{1t} - R_{1f})} \tag{19}$$

where in this case $\mathrm{Na_T}$ and C_T are used as the total sodium ion and channel tight site concentrations, respectively. A plot of $(R_1 - R_{1f})^{-1}$ versus sodium

[29] T. L. James and J. H. Noggle, *Proc. Natl. Acad. Sci. U.S.A.* **62**, 644 (1969).

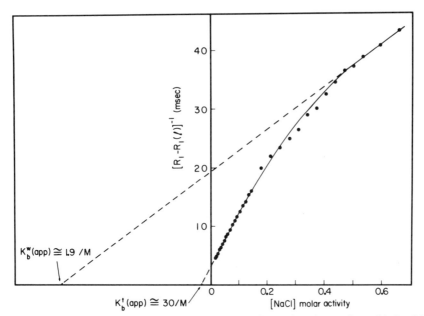

FIG. 14. Excess longitudinal relaxation rate plot for sodium interaction with 3 mM malonyl(gramicidin)$_2$ channels at 60°. The intercept for the low concentration data gives an apparent tight binding constant of 30/M, and that for the high concentration data gives an apparent weak binding constant of 1.9/M. The tight binding constant is the same as obtained for sodium ion interaction with gramicidin A channels and the weak binding constant differs by less than a factor of two (see Fig. 23B).

ion concentration is such that fitting the data to a straight line allows determination of K_b^t from the inverse of the negative x axis intercept. For the case of two binding sites the slope defined by the low ion concentration range can be used to estimate the tight binding constant, and the limiting slope at high ion concentration can be used to estimate the weak binding constant. More thorough analyses of multiple binding processes for the gramicidin transmembrane channel system and for the alkali metal ions have been presented elsewhere.[15,17,20,21,30,31] Some of the relevant data are given in the section below on alkali metal ions. An example of the ELR approach and of the determination of the apparent tight and weak binding constants is given in Fig. 14 for the malonylgramicidin channel at elevated temperatures and where $R_1(l)$ stands for the reference of lipid in the absence of channels. The apparent binding constants at 60° are $K_b^t(app) = 30/M$ and $K_b^w(app) = 1.9/M$.

[30] D. W. Urry, T. L. Trapane, C. M. Venkatachalam, and K. U. Prasad, *J. Phys. Chem.* **87,** 2918 (1983).
[31] D. W. Urry, T. L. Trapane, and C. M. Venkatachalam, *J. Membr. Biol.* **89,** 107 (1986).

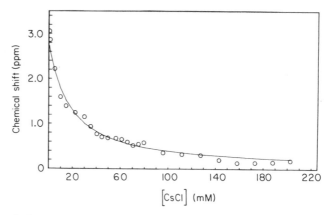

FIG. 15. Cesium-133 NMR chemical shift of CsCl in the presence of 3 mM malonyl(gramicidin)$_2$ channels. The magnitude of the positive chemical shift is quite small; it provides the possibility of estimating a tight binding constant ($\sim$ 50/M), but provides no evidence for a second weak binding constant that is clearly seen in Fig. 23E. (Reproduced with permission from Ref. 15.)

Ion Resonance Chemical Shift

The use of ion resonance chemical shift to obtain binding constants utilizes Eq. (14), recognizing, of course, that K_b^t is a function of P_t and P_f. An interesting example of following ion chemical shift as a function of ion concentration is given in Fig. 15 for cesium ion interaction with the malonylgramicidin channel. This ion is unique among the alkali metal ions in that it exhibits a downfield chemical shift rather than an upfield chemical shift on interaction with the channel binding site. The usually found upfield chemical shift is considered to arise from the channel site being a relatively electron-rich site. The surrounding electron cloud has a circular motion or current induced in it by the magnetic field. This produces a local magnetic field which opposes the external field and therefore the nucleus requires a higher field to come into resonance. The opposite chemical shift, exhibited by cesium ion in the binding site at the mouth of the channel, is suggested to arise from ion pairing with chloride ion just outside the mouth of the channel.[15] An interesting feature apparent in the cesium ion chemical shift data is that the chemical shift quickly becomes very small with little chemical shift above 100 mM CsCl such that any binding process at higher concentration cannot be observed and in fact would not be apparent. When following the site, two binding processes are easily recognized for Tl$^+$, Rb$^+$, and K$^+$ at elevated temperature (see Fig. 16), as they are for the cesium ion in the ELR plot of Fig. 23E (which see).

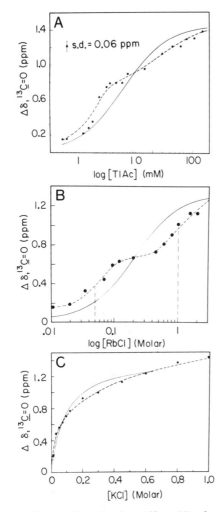

FIG. 16. Ion-induced carbonyl carbon chemical shifts at 70° of synthetic [1-^{13}C]Trp^{11}- or Trp^{13}-gramicidin A channels for thallium acetate (A), for rubidium chloride (B), and for potassium chloride (C). All three demonstrate two binding processes—a tight binding process on entry of the first ion into the channel and a weak binding process on entry of the second ion into the channel. (A and C are reproduced with permission from Ref. 33 and B is reproduced with permission from Ref. 17.)

In general, ion chemical shifts for the alkali metals do not give evidence for two binding processes, as seen by examining the data in Fig. 22. The ELR data for the alkali metal ions in Fig. 23 and the site chemical shift data of Fig. 16, on the other hand, both show two binding processes.

Site (Carbon-13) Chemical Shift

The phospholipid packaging process for obtaining the channel state is one of heat incubation of gramicidin with lysolecithin for 12 or more hours at 70°. The product is the channel state in a lipid bilayer, that is, the interaction of gramicidin with lysolecithin converts the lysolecithin from micelles to lipid bilayers.[32] Verification of the channel state is achieved primarily by means of the circular dichroism spectrum, which is entirely unique and which changes dramatically during the early hours of the heating process and then becomes constant and independent of further temperature variation. It has been verified that even a week of heating at 70° causes neither significant breakdown of the phospholipid nor of the channel.[33] That the incorporation is achieved at 70° and that there is good stability at this temperature is particularly fortunate for following a resonance at the membranous binding site. This is because the reorientation correlation time of the large membranous particles limits the reorientation correlation time of the ion-binding site and a high temperature (60–70°) is required to obtain sufficiently narrow carbonyl carbon resonance lines. Titrations have been carried out on 90% enriched [1-^{13}C]Trp^{13}- or Trp^{11}-gramicidin A using thallium acetate (Fig. 16A),[33] rubidium chloride (Fig. 16B)[17], and potassium chloride (Fig. 16C).[33] In the logarithmic plots of Fig. 16A and B, with a standard deviation of ±0.06 ppm having been established for the reproducibility of data points, it is quite apparent that there are two binding processes; they are seen as two separated sigmoid curves. Also included are single sigmoid curves which represent an effort to fit the data to a single binding process. Even the magnitude of change for each process is what is to be expected based on half of site occupancy for the first ion and complete occupancy of sites on addition of the second ion. This was established for a similar relative orientation of carbonyl and cation in the monovalent cation complexes of enniatin B.[33] The binding constants can be estimated from the midpoints of the sigmoid curves. The data for potassium chloride are plotted linearly with respect to concentration in Fig. 16C. The initial rapid rise and the slower subsequent rise require two binding constants, as shown by the calculated dashed curve. The solid curve represents the best fit for a single binding process in this nonlogarithmic plot. The estimated binding constants for K$^+$ are 28/M for the tight site and 2.4/M for the weak site. From a study of the temperature dependence of sodium binding using the ELR approach, the decrease in either of the

[32] I. Pasquali-Ronchetti, A. Spisni, E. Casali, L. Masotti, and D. W. Urry, *Biosci. Rep.* **3,** 127 (1983).

[33] D. W. Urry, T. L. Trapane, C. M. Venkatachalam, and K. U. Prasad, *Can. J. Chem.* **63,** 1976 (1985).

TABLE I
GENERAL NMR PROPERTIES OF ALKALI METAL IONS

Nucleus	Observation frequency, v_0 (23.5 kG)	Natural abundance (%)	Sensitivity relative to ^1H	Receptivity (vs ^{13}C)	Gyromagnetic ratio, γ (rad/G-sec)
^7Li	38.9 MHz	92.6	0.293	1.54×10^3	10396.4
^{23}Na	26.4 MHz	100	0.0925	5.25×10^2	7076.1
^{39}K	4.66 MHz	93.1	0.00051	2.69	1248.3
^{87}Rb	32.7 MHz	27.8	0.175	2.77×10^2	8753.2
^{133}Cs	13.1 MHz	100	0.0474	2.69×10^2	3508.7

binding constants on going from 30 to 75° is less than a factor of two.[25] For the potassium ion the binding constant for the weak site, estimated by the ELR approach at 30°, was found to be 8.3/M.[31] This is just over a factor of three from the value of 2.4/M obtained at 70° from Fig. 16C. Thus it is apparent that binding studies at elevated temperatures using site (carbon-13) chemical shifts are relevant to the binding processes observed at lower temperatures using ion chemical shift and excess longitudinal relaxation rate data.

The Alkali Metal Ions

General Considerations

The alkali metal ions as a group and particularly Na^+ and K^+ are of great interest in membrane transport; Na^+ and K^+ pass through membrane channels at rates of 10^7 ions/sec or greater. It is a favorable circumstance that the NMR frequencies of these nuclei are in the 10^7 Hz range, as shown in Table I, and it is further good fortune for researchers interested in transmembrane channel transport that the analysis of NMR-derived correlation times and rate constants is facilitated when the rate constants are of the magnitude of the observation frequency. This point was made in the section on the determination of rate constants. The natural abundance of the useful nuclei is good (see Table I, column 3), and while the sensitivity relative to the proton (^1H) is not so favorable, the product of abundance times relative sensitivity, that is, the receptivity, considered with respect to carbon-13 (column 5 of Table I), is very favorable. This means that the pulse methods so effective in studying naturally abundant carbon-13 could be expected to be effective in studying alkali metal ions in biological systems, and it means that one could expect to detect resonances and to follow resonance chemical shifts and exchange broadening as can be done for carbon-13.

TABLE II

QUANTITIES RELEVANT TO RELAXATION PROPERTIES OF ALKALI METAL IONS

Nucleus	Spin quantum number	Electric quadrupole moment, Q $(10^{-28}$ m$^2)$	Sternheimer antishielding factor, γ_∞ as $(1 - \gamma_\infty)$	$Q(1 - \gamma_\infty)$	$\omega = 2\pi\nu_0$ (rad/sec at 23.5 kG)
^{7}Li	$\frac{3}{2}$	-0.03	0.74	0.022	24.4×10^7
^{23}Na	$\frac{3}{2}$	0.14–0.15	5.1	0.714	16.6×10^7
^{39}K	$\frac{3}{2}$	0.11	18.3	2.013	2.93×10^7
^{87}Rb	$\frac{3}{2}$	0.13	48.2	6.266	20.5×10^7
^{133}Cs	$\frac{7}{2}$	-0.003	111	0.333	8.23×10^7

As seen in Table II, these alkali metal ions have spin quantum numbers of $\frac{3}{2}$ or greater, and this allows that they may have useful relaxation properties. In Eq. (2) it was seen that the magnitude of the relaxation times depended on the quadrupole coupling constant, which is a product of the nuclear electric quadrupole moment, eQ, times the electric field gradient, eq. In Table II the electric quadrupole moments in units of e, the unit charge, are seen to be of similar magnitude with the exception of cesium-133, which is an order of magnitude smaller. This in itself would suggest that cesium-133 may not be a favorable nucleus for relaxation studies, but it was shown in Figs. 3 and 4 that the change in relaxation time for cesium-133 on interaction with the channel was actually more favorable than for sodium-23. This is thought to be due in part to the ion pairing discussed above, but in larger measure it is due to the Sternheimer antishielding factor, γ_∞. This factor reflects the effect of the electrons surrounding the nucleus to enhance the electric field gradient of the environ-

TABLE III

COEFFICIENTS FOR CALCULATION OF ACTIVITY COEFFICIENTS FOR ALKALI METAL ION CHLORIDE SOLUTIONS[a]

Ion chloride	a	b	c	d
Li$^+$	0.51006	1.49	9.122×10^{-1}	4.766×10^{-3}
Na$^+$	0.50214	1.3496	2.85×10^{-2}	3.57×10^{-3}
K$^+$	0.51129	1.2998	9.85×10^{-4}	3.09×10^{-3}
Rb$^+$	0.54	1.32	2.267×10^{-3}	2.31×10^{-3}
Cs$^+$	0.4936	0.7748	1.272×10^{-2}	8.4×10^{-4}

[a] Coefficients to be used in Eq. (20).

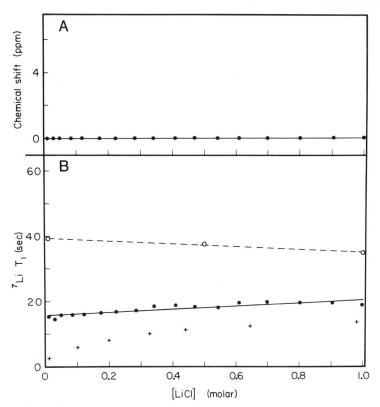

FIG. 17. (A) Lithium-7 NMR background data showing the absence of the concentration dependence of chemical shift in the presence of (●) lysolecithin micelles (lysophosphatidylcholine). This is the situation in water above and on addition of membranous channels (see Fig. 22). (B) The concentration dependence of longitudinal relaxation time, T_1, in (O) water (^2H_2O), in the presence of (●) lysophosphatidylcholine, and a few representative points on addition of (+) gramicidin channels. The fractional change in T_1 on addition of channels to phospholipid is the smallest of the alkali metal ions. Because of this, estimates of binding constants for lithium ion interaction with the channel are more approximate for this ion.

ment. Thus the product, $Q(1 - \gamma_\infty)$, more nearly reflects the relative use of these nuclei for relaxation studies.

The observation frequency in radians/second is included in Table II because of its relationship to the observable window of relaxation rates. By Eqs. (9) and (10) and the plotted values in Fig. 9, it is apparent that τ_c can be determined when $\omega \tau_c > 1$. But for sodium-23 interaction with the gramicidin channel, it was found that $\tau_c^{-1} = k_{off}^w$. Accordingly when $k_{off}^w \lesssim \omega$, its value can be estimated. For a channel to have currents of

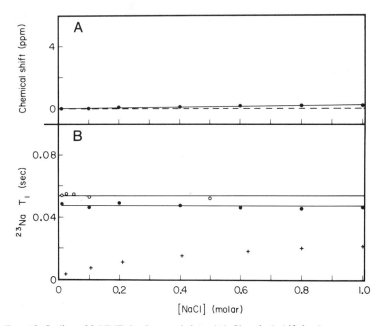

FIG. 18. Sodium-23 NMR background data. (A) Chemical shift in the presence of (●) lysolecithin micelles (lysophosphatidylcholine) is found to be small and to occur only at higher ion concentration. This is the NaCl concentration dependence of chemical shift. (B) Longitudinal relaxation time, T_1, NaCl concentration dependence for (O) ^2H_2O, in the presence of (●) lysolecithin micelles, and a few representative points on addition of (+)gramicidin channels.

10^7 ions/sec or more requires that the off-rate constant for the ion leaving the channel must be 10^7/sec or more. Therefore it is apparent that NMR relaxation studies on alkali metal ions provide just the appropriate window to evaluate one of the most important elemental rate processes for channel transport.

Additionally, since longitudinal relaxation rate studies can be used to estimate binding constants, the situation is also favorable, as shown above, to evaluate binding constants and to do so at the relatively high ion concentration of interest for the sodium and potassium ions. In order to properly compare binding constants of ions in the alkali metal ion series, it is necessary to make corrections for ion activities. The activity coefficients compiled by Harned and Owen[34] were fitted to a Guggenheim-like equa-

[34] H. S. Harned and B. B. Owen, "The Physical Chemistry of Electrolyte Solutions," 3rd Ed. Rheinhold, New York, 1967.

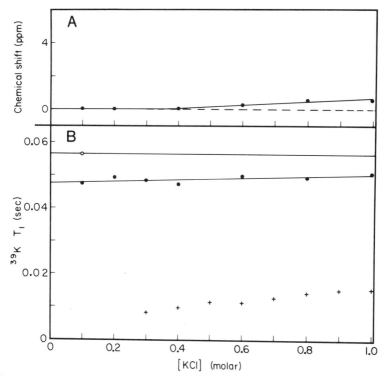

FIG. 19. Potassium-39 NMR background data. (A) Chemical shift in the presence of (●) lysolecithin micelles, which is a KCl concentration dependence to be expected in ^2H_2O alone. (B) Longitudinal relaxation time, T_1, as a function of KCl concentration in (○) ^2H_2O, in the presence of (●) lysolecithin micelles, and a few representative points on addition of (+) gramicidin channels.

tion of the form

$$\log(\text{activity coefficient}) = -a(m)^{1/2}/(1 + b(m)^{1/2}) + cm + dm^2 \quad (20)$$

The values of the a, b, c, and d coefficients of Eq. (20) are listed in Table III for each of the alkali metal ion chlorides.

Background Data for Studies of Interactions with Polypeptide Sites within Membranes

Before using NMR to characterize ion interactions with membranous polypeptide sites, data from a number of preliminary studies are required. These data are the ion chloride concentration dependence of chemical shift

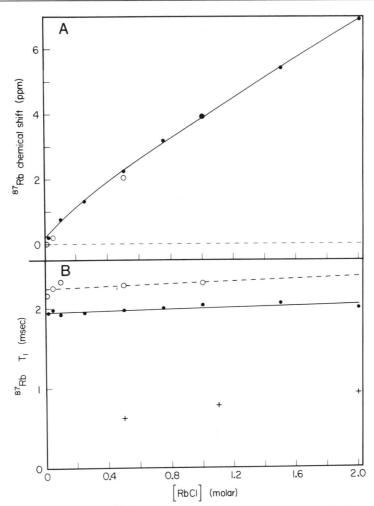

FIG. 20. Rubidium-87 NMR background data. (A) Chemical shift as a function of RbCl concentration in (O) ^2H_2O and in the presence of (●) lysolecithin micelles (lysophosphatidyl-choline). (B) Longitudinal relaxation time, T_1, as a function of RbCl concentration in (O) ^2H_2O, in the presence of (●) lysolecithin micelles, and a few representative points on addition of (+) gramicidin channels.

in water and in the presence of the phospholipid of interest and the ion chloride concentration dependence of the longitudinal relaxation time also in the presence of water and in the presence of the phospholipid of interest. Then preliminary experiments are necessary to determine the magnitudes

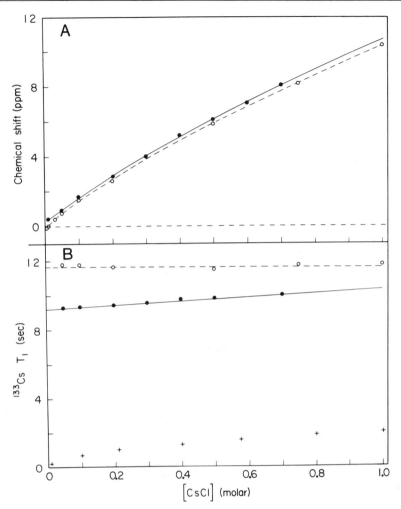

FIG. 21. Cesium-133 NMR background data. (A) Chemical shift dependence on CsCl concentration in (○) ^2H_2O and in the presence of (●) lysolecithin micelles. (B) Longitudinal relaxation time, T_1, dependence on CsCl concentration in (○) ^2H_2O, in the presence of (●) lysolecithin micelles, and a few representative points on addition of (+) 3 mM gramicidin channels.

of the chemical shifts and T_1 changes on adding the binding site of interest before knowing the extent to which either of these experimental quantities will be useful. Background data for lithium-7, sodium-23, potassium-39, rubidium-87, and cesium-133 are given in Figs. 17–21 using lysolecithin (lysophosphatidylcholine) micelles as the phospholipid. Checks with di-

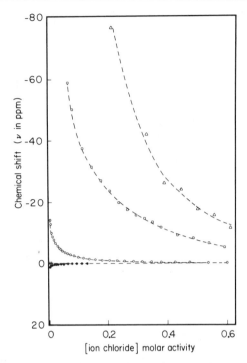

FIG. 22. Ion resonance chemical shift data as a function of ion chloride molar activity in the presence of 3 mM gramicidin channels. The lithium-7 resonance (+) shows no chemical shift. The sodium-23 resonance exhibits a substantial negative (upfield) chemical shift (O). The potassium-39 resonance exhibits the largest chemical shift (Δ), but the low sensitivity due to the low observation frequency and an exchange-broadened resonance makes it difficult to determine a chemical shift below 200 mM. The rubidium-87 resonance shows a large negative (upfield) chemical shift ($\square$). The cesium-133 resonance exhibits a very weak positive (downfield) chemical shift (●). This is different from the other alkali metal ions and it may be the result of ion pairing with chloride when the ion is in the binding site at the mouth of the channel.

myristoylphosphatidylcholine showed this phospholipid to give essentially the same results for sodium ions as lysophosphatidylcholine. The data in Figs. 17–21 are useful when considering a study of the interaction of the alkali metal ions with polypeptide binding sites in phosphatidylcholine membranes. Other lipids or mixtures of lipids should be examined similarly.

Ion Chemical Shift Data Due to Channel Interactions

Plotted in Fig. 22 as a function of ion chloride concentration are the ion resonance chemical shifts exhibited by the alkali metal ions on interaction with the channel binding site. The relationship among the Group IA ions with respect to magnitude of chemical shift exhibited is not simple. Lithium-7 shows little or no chemical shift; sodium-23 shows a substantial negative (upfield) chemical shift; potassium-39 (where obtained at high concentrations) shows large negative chemical shifts; rubidium-87 also exhibits large negative chemical shifts; cesium-133 exhibits an uncharacteristic weak positive (downfield) chemical shift. All but lithium-7 exhibit chemical shifts that are useful in characterizing ion binding. Care should be taken, however, on fitting ion chemical shift data when there is more than one binding process. Depending on the concentration range used, the calculated binding constant can vary. Because of the multiple binding processes, comparisons of binding constants derived from chemical shift data and relaxation data should utilize common ion activity ranges. When this is done the values compare favorably, generally well within a factor of two. The case for cesium ion is particularly favorable. The values calculated using curve e of Fig. 22 and the curve of Fig. 23E (low concentration range) are 60/M and 64/M, respectively.[15]

Excess Longitudinal Relaxation Rate Plots for Channel Interactions

The excess longitudinal relaxation rate plots for the interactions of each of the alkali metal ions with the gramicidin channel are given in Fig. 23. The excess rate is given with respect to the relaxation rates in the presence of the same concentration of phospholipid and for the correct ion activity using the data in Figs. 17–21. Complete titrations were possible for lithium-7, sodium-23, and cesium-133. The low sensitivity of potassium-39 at 4.65 MHz and the occurrence of broad lines at lower concentrations made it necessary to begin the titration at about 200 mM activity and to use 0.3 mM channels, whereas the other studies used 3 mM channels. Also, the very broad lines for rubidium-87, with very short relaxation times particularly at low ion concentrations, spread the resonance intensity over a wide spectral window such that data were not obtained much below an activity of 400 mM. Fortunately it was possible for all of the Group IA ions to estimate the weak binding constants from the negative x axis intercepts obtained from the limiting slopes at high ion concentration. For Li$^+$, Na$^+$, and Cs$^+$ the tight binding constants were obtained from straight lines fitted to the limiting slopes in the low concentration range. As will be shown below, it is the weak binding constant and the off-rate constant for double occupancy (i.e., for the weak site) that are the key constants for calculating the channel currents. For calculating the relative currents or the conductance ratios, the negative x axis intercepts provide a satisfactory means of obtaining good relative magnitudes of binding constants. The binding constants are given in Table IV.

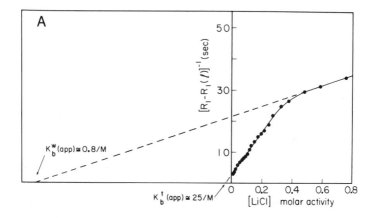

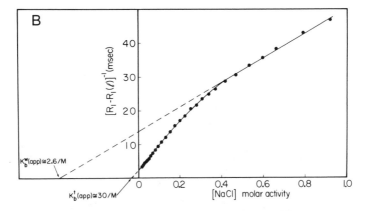

FIG. 23. Excess longitudinal relaxation rate (ELR) data for each of the alkali metal ions in the presence of 3 mM gramicidin channels with the exception of the KCl titration, which was at 0.3 mM gramicidin channels. $R_1(l)$ is the reciprocal of the T_1 values in the presence of lysolecithin in Figs. 17–21. All values are given in terms of molar ion activities. (A) Lithium-7 ELR plot giving an estimated tight binding constant of 25/M and weak binding constant of 0.8/M. Because of the relatively small difference in T_1 on interaction with the channel (see Fig. 17B and note the low values of the ordinate), these estimates are the most approximate. (Adapted with permission from Ref. 30.) (B) Sodium-23 ELR plot giving apparent binding constants of 30/M for the tight binding constant and 2.6/M for the weak binding constant. (Adapted with permission from Ref. 25.) (C) Potassium-39 ELR plot in the presence of 0.3 mM gramicidin channels giving an apparent weak binding constant of 8.3/M. (Adapted with permission from Ref. 31.) (D) Rubidium-87 ELR plot in the presence of 3 mM gramicidin channels giving an apparent weak binding constant of 3.9/M. (Adapted with permission from Ref. 17.) (E) Cesium-133 ELR plot in the presence of 3 mM gramicidin channel giving an apparent tight binding constant of 60/M and an apparent weak binding constant of 4.2/M. (Adapted with permission from Ref. 15.)

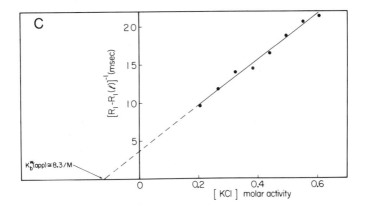

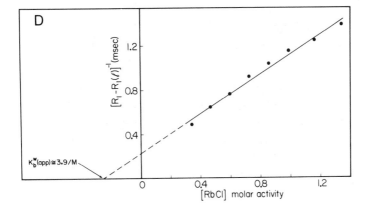

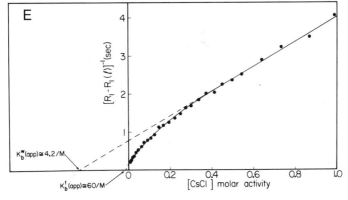

FIG. 23C–E.

TABLE IV

BINDING AND RATE CONSTANTS FOR ALKALI
METAL ION INTERACTIONS WITH THE GRAMICIDIN
TRANSMEMBRANE CHANNEL

Ion	K_b^l	K_b^w	k_{off}^w
Li$^+$	25/M	0.77/M	1.7×10^7/sec
Na$^+$	30/M	2.6/M	2.1×10^7/sec
K$^+$	~50/M	8.3/M	2.6×10^7/sec
Rb$^+$	~60/M	4/M	6.6×10^7/sec
Cs$^+$	60/M	4/M	5.9×10^7/sec[a]

[a] Treating Cs$^+$ as a spin $\frac{3}{2}$ nucleus.

While in Fig. 23 the intercepts were used to obtain apparent binding constants, more thorough analyses of these data have been published elsewhere.[15,17,25,30,31] The more complete analyses yield both binding constants and relaxation rates at the sites, e.g., $R_{lt} = 1/T_{lt}$ and $R_{lw} = 1/T_{lw}$. Knowledge of these values is useful for verifying that the conditions for the approximations leading to Eqs. (9) and (10) are satisfied. In order that the process of entering and leaving the channel provide a fluctuating electric field gradient, it is necessary that the occupancy time, τ_b, be very much shorter than the relaxation time for the site. This condition is not well met at low sodium ion concentrations and, in part because of this, the use of Eq. (10) based solely on the observation that $T_1 \neq T_2$ would result in an incorrect τ_c.

Rate Constant Data for Channel Interactions

Lithium-7. Because gramicidin channel current data are well known for the alkali metal ions and because the off-rate constant for double occupancy is the overwhelmingly dominant factor in determining the magnitude of the current, an estimate of k_{off}^w from the NMR data can be checked against the reality of the current data. This provides a guide to the correct approach or approaches, and can even encourage consideration of approaches that in the absence of a check may not have been considered. This is the case for lithium-7. The true linewidth for lithium-7 is extremely narrow. The linewidth expected in the extreme narrowing case of water or lipid alone is one to two orders of magnitude smaller than the experimental linewidth due to field inhomogeneity line broadening. Yet it is apparent when channels are present that the line broadens beyond that due to field inhomogeneity. An average value for the excess broadening due to chan-

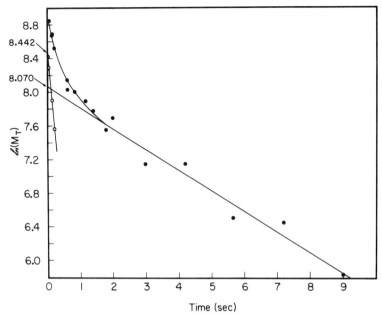

FIG. 24. Lithium-7 NMR spin-echo experiment using the Carr–Purcell–Meiboom–Gill (CPMG) pulse sequence for $1\ M$ LiCl in the presence of 3 mM gramicidin channels. This experiment is made difficult due to the long times involved and the data are quite scattered. The slope at long times gives a T_2'' of 4.0 sec and the stripped curve at short times gives a value for T_2' of 0.23 sec. The ratio of the intercepts for the two curves (exp 8.442/exp 8.070) is a reasonable value of 1.45.

nels at high ion concentration is 1.1 Hz. This gives an estimated value for T_2 of 0.29 sec. The experimental T_1 in the presence of channels is 13.6 sec and the reference value in the presence of lipid is 19.2 sec. Using the spin-echo approach $T_2(l)$ was found to be 12 sec and using these values as T_{1f} and T_{2f} in Eq. (10) gives a value for $\Delta(R_2/R_1)$ of 160. Using the curve for lithium-7 in Fig. 9E gives a τ_c of 5.9×10^{-8} sec. Taking the reciprocal to be $k_{\text{off}}^{\text{w}}$ gives 1.7×10^7/sec. Immediately this number is compromised by the small increase in linewidth due to the presence of channels and because the quadrupolar relaxation mechanism is sufficiently limited [note the value of $Q(1 - \gamma_\infty)$ for lithium-7 in Table II] for this ion that dipolar relaxation could possibly make a significant contribution to the relaxation rate. In order to overcome the line broadening due to field inhomogeneity, a spin echo experiment can be used to estimate T_2' and T_2'' directly, as was done in Fig. 7. This data is given in Fig. 24 for $1\ M$ LiCl in the presence of 3 mM channels. The values for T_2' and T_2'' are 0.23 sec and 4.0 sec, respec-

A

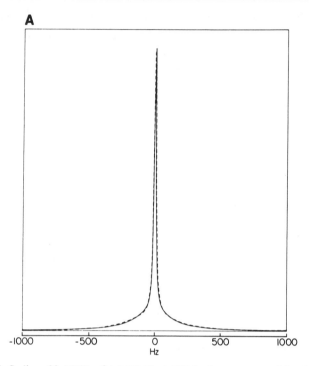

FIG. 25. Sodium-23 NMR of 1 M NaCl at 30° in the presence of 3 mM gramicidin channels results in a resonance with a broad and narrow component. Curve resolution with a 60% broad (fast) component and a 40% narrow (slow) component gives curves with linewidths at half-intensity of $v'_{1/2} = 325$ Hz and $v''_{1/2} = 16$ Hz. These values are used to calculate T'_2 and T''_2 and to estimate a correlation time by means of Eq. (9) with a value for $T_2(l) \simeq T_{2f}$ of 31.8 msec. The calculated value of τ_c is 4.8×10^{-8} sec. (Adapted with permission from Ref. 25.)

tively; the ratio of intercepts is 1.45 and the value for $T_2(l)$ is 12 sec. Using these values in Eq. (9) the calculated correlation time is 2.2×10^{-8} sec, the inverse of which is taken as the weak off-rate constant, $k^w_{off} = 4.5 \times 10^7/$ sec. Of these two approaches for estimating k^w_{off} the former will be used, as the data in Fig. 24 are quite scattered. It is interesting to see what relative channel current is obtained by the weak off-rate constant of 1.7×10^7/sec.

Sodium-23. The resonance line shape for sodium-23 in the presence of 3 mM gramicidin channels with 1 M NaCl at 30° is given in Fig. 25. By the indicated curve resolution the half linewidths for the two components are $v'_{1/2} = 325$ Hz and $v''_{1/2} = 16$ Hz. This gives values of 0.98 msec for T'_2 and of 19.9 msec for T''_2. With the transverse relaxation time of 31.8 msec in the absence of channels, this gives by way of Eq. (9) a correlation time of

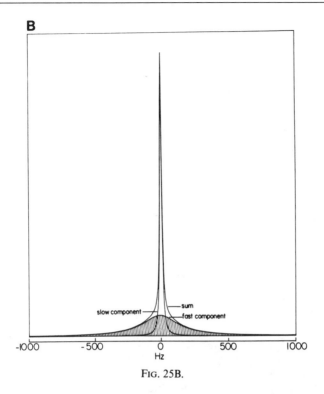

FIG. 25B.

4.8×10^{-8} sec. This is essentially the same value that was obtained for the malonylgramicidin channel (see Fig. 5) and it gives a weak off-rate constant of 2.1×10^{7}/sec, based on the preceding analyses for Na which demonstrated that $k_{off}^{w} = v_{c}^{-1}$.

Potassium-39. The T_1 and T_2 values for potassium-39 at high ion concentration (1 M KCl) in the presence of 0.3 mM channels are given in Fig. 26. Here T_1 is 15.3 msec and T_2 is 7.3 msec. The values for T_1 and T_2 in the absence of channels but with lipid present (see Fig. 19) are 55 and 50 msec, respectively. These values by way of Eq. (10) yield a value for τ_c of 3.89×10^{-8} sec. The reciprocal of τ_c, following the detailed arguments for sodium, gives the value for k_{off}^{w} of 2.6×10^{7}/sec.

Rubidium-87. The data for rubidium-87 are given in Fig. 6 for 2 M RbCl in the presence of 3.16 mM channels at 30°. Under these conditions T_2' is 0.151 msec and T_2'' is 0.753 msec. For the reference system in the absence of channels, $T_2 = 1.9$ msec. By means of Eq. (9) the value of τ_c is found to be 1.53×10^{-8} sec. Again following the arguments pursued in detail for the sodium ion, this gives a value for k_{off}^{w} of 6.6×10^{7}/sec.

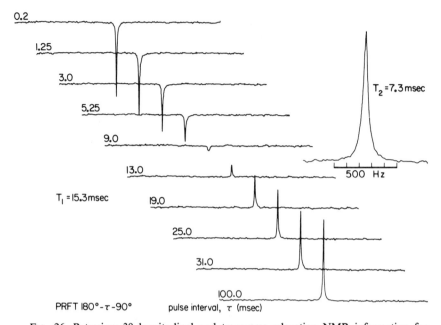

FIG. 26. Potassium-39 longitudinal and transverse relaxation NMR information for 1 M KCl in the presence of 0.3 mM lysolecithin-packaged gramicidin A channels at 30°. The partially relaxed Fourier transform data give a longitudinal relaxation time, T_1, of 15.3 msec, and from the linewidth at half-intensity of the peak, enlarged as the inset, a value for T_2 of 7.3 msec is obtained. In the absence of channels, i.e., in the presence of lysolecithin micelles, $T_1 = 55$ msec and $T_2 = 50$ msec. These values are used in Eq. (10) as plotted in Fig. 9E to calculate a correlation time of 3.89×10^{-8} sec. (Adapted with permission from Ref. 31.)

Cesium-133. As seen in Table II, cesium-133 is a spin $\frac{7}{2}$ nucleus whereas Eqs. (9) and (10) are approximate expressions for spin $\frac{3}{2}$ nuclei. More complex expressions, if they could be obtained in analytical form, would obtain for spin $\frac{7}{2}$ nuclei.[7] Consideration of this channel transport system where the current ratios are known for all of the alkali metal ions, including cesium, makes it possible to assume the spin $\frac{3}{2}$ formalism and to see just how much in error would be an estimate of correlation time when interpreting the inverse of this as an off-rate constant. Often, of course, for biological as well as other systems, it is useful information to obtain an order of magnitude value.

Using the spin-echo (CPMG) data shown in Fig. 27A, at 30 mM CsCl, 3 mM channels, and 30°, $T_2' = 1.2$ msec; $T_2'' = 23$ msec, and $T_2(l) \simeq T_{2f}$ was also found to be about 6 sec such that by Eq. (10) a value of 5.7×10^{-8} sec was calculated for τ_c. Using null data at 210 mM CsCl of Fig. 5 in Ref. 15, a correlation time of 4.0×10^{-8} sec is calculated by Eq. (10).[15]

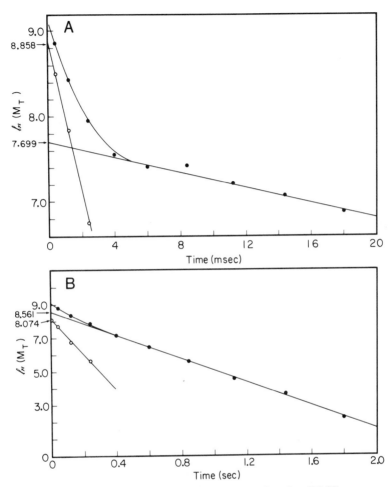

FIG. 27. Cesium-133 NMR spin-echo experiment using the CPMG sequence at 30 mM CsCl (A) and 1.5 M CsCl (B). In both cases, but most clearly at low concentration, the plot of $\ln M_T$ versus time is nonlinear. The data can be curve stripped to obtain the characteristic transverse relaxation times for two processes. At low ion concentration, the first relaxation is characterized with a time constant T_2' of 1.2 msec and the slow relaxation yields a T_2'' of 23 msec. At high ion concentration, the rapid relaxation is not as well delineated, with $T_2' = 98$ msec and $T_2'' = 286$ msec. In the absence of channels but in the presence of lysoleci-thin, the spin-echo data, while scattered, gave evidence only for a single relaxation with a value of 5–6 sec.

Using the spin-echo data in Fig. 27B for 1.5 M CsCl and 3 mM channels at 30°, $T_2' = 98$ msec, $T_2'' = 286$ msec, and $T_2(l) \simeq T_{2f} \simeq 5$ sec. Substituting these values in Eq. (9) gives a τ_c of 1.7×10^{-8} sec. Thus the apparent correlation time gets shorter as the concentration increases. As the recipro-

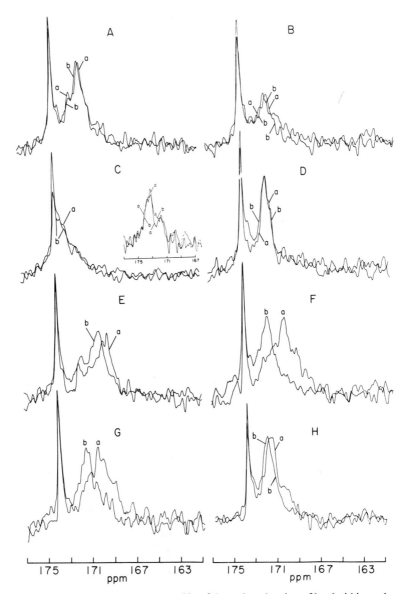

FIG. 28. Carbon-13 NMR spectra at 70° of the carbonyl region of lysolecithin-packaged synthetic gramicidin A's which have been individually 1-^{13}C-enriched at the peptide residue indicated as follows: (A) Val^1 $C = O$, (B) Ala^3 $C = O$, (C) Ala^5 $C = O$, (D) Val^7 $C = O$, (E) Trp^9 $C = O$, (F) Trp^{11} $C = O$, (G) Trp^{13} $C = O$, and (H) Trp^{15} $C = O$. Two spectra for each labeled position in the gramicidin channel are overlayed—one spectrum showing the chemical shift of the broad enriched carbonyl in the presence of 0.5 mM NaCl (a), which is taken as the reference spectrum, having no ion-induced chemical shift, and one spectrum observed in

cal is taken as the off-rate constant, the off-rate constant increases as the concentration increases, i.e., 1.75×10^7/sec at 30 mM, 2.5×10^7/sec at 210 mM, and 5.9×10^7/sec at 1.5 M. For cesium ion, single occupancy appears to result in an off-rate constant sufficient to contribute to the single-channel current but it is a relatively minor component of the current at high ion concentration. This is considered in more detail below.

Location of Binding Sites

Another way in which NMR can contribute to an understanding of ion interactions at a membranous polypeptide site is to identify or to locate the binding site or sites. For the channel system under consideration here the location of binding sites is necessary before the NMR-derived binding constants and rate constants can be used to calculate a current. Under the conditions in which the binding and rate constants were determined by NMR there is no net current through the channel. A net current through the channel results from an applied potential along the length of the channel, and the location of the binding site with respect to the transmembrane axis is necessary to consider how the applied electric field will modify the binding and rate constants. The location of the binding sites was shown in the *Introduction* (see Fig. 2) as part of the discussion as to why the gramicidin transmembrane channel was a favorable model system for these NMR studies. The data in Fig. 2 are based on the experimentally determined ion-induced carbonyl carbon chemical shifts exhibited by a series of synthetic gramicidin A molecules in which a single, different carbonyl carbon was 90% enriched in each synthesis. The ion-induced carbonyl carbon chemical shifts are shown in Fig. 28.[35] As discussed before, a temperature of 60 to 70° is necessary to get sufficiently narrow resonance lines from channel backbone nuclei. The lipid carbonyl carbon resonance is relatively sharp; its chemical shift is unchanged throughout the titration, and therefore this resonance provides a convenient chemical shift reference.

[35] D. W. Urry, T. L. Trapane, and K. U. Prasad, *Science* **221**, 1064 (1983).

the presence of 83 mM TlAc (b), which shows the change in the chemical shift of the labeled carbonyl upon binding of ion in the channel. The sharp downfield resonance at 174 ppm is due to the carbonyl of the lipid, which exhibits no shift upon addition of ion. It may be seen that residues 1, 3, 5, and 7 do not exhibit any ion-induced chemical shift upon binding of thallium ion, but that residues 9, 11, 13, and 15 do show substantial downfield shifts due to the presence of the ion at three positions in the channel. A plot of the observed magnitudes of these ion-induced chemical shifts are given in Fig. 2, which clearly locates the ion-binding sites in the gramicidin channel. (Adapted with permission from Ref. 35.)

The zero chemical shifts exhibited so clearly by the L-Val[1] and L-Val[7] carbonyl carbon resonances and also aparent in the L-Ala[3] and L-Ala[5] carbonyl carbon resonances indicate that there is no detectable ion occupation at these positions in the channel. If the ion had entered the channel and visited each position in the channel with equal frequency, i.e., had there been the same occupancy time at each position within the channel, then the carbonyl carbon resonance of each L-amino acid residue would have exhibited an essentially similar chemical shift. This is because the L-residue carbonyls are helically equivalent. Since this is strikingly not the case and since there are no detectable chemical shifts in the residue 1, 3, 5, and 7 carbonyl carbon resonances, it is easily concluded that an ion in the channel is very localized as shown by the plots in Fig. 2. From the ion titrations in Fig. 16 it is apparent that the tight binding site represents the entry of the first ion in the channel and the weak binding site represents the entry of the second ion in the channel. The difference in binding constants with the twofold symmetry of the sites is due to ion–ion repulsion. That the chemical shift for the L-Trp[15] carbonyl carbon resonance is small indicates that the ion has indeed entered the channel but as is apparent in comparing the chemical shifts with the wire model in Fig. 2, the ion is at each channel mouth where there would be only a small change in applied potential on going from solution just outside the mouth to the binding site. The process of introducing this small voltage dependence and voltage dependence in general has been outlined in detail elsewhere[1] and is briefly considered below.

Comparison of NMR-Derived and Experimental Channel Currents

Measurement of single-channel currents through the gramicidin transmembrane channel is achieved using an approach due to Mueller and Rudin.[36] The apparatus consists of an inert container containing a septum separating two chambers filled with the desired salt solution. The septum has a small hole in it; the hole is filled with a lipid, which on standing thins to a lipid bilayer. Gramicidin molecules associate with the lipid bilayer, transiently change into the channel conformation at the interface, and dip into the lipid layer by means of the formyl (head) end; when one such molecule pairs with a second from the opposite side of the membrane, head-to-head hydrogen bonding occurs and a continuous transmembrane channel is formed through which monovalent cations can pass. A pair of electrodes are used to apply a potential across the channel in the membrane and the current flowing across the membrane can be measured. As

[36] P. Mueller and D. O. Rudin, *Biochem. Biophys. Res. Commun.* **26**, 398 (1967).

first shown by Hladky and Haydon,[37] the technique can be made so sensitive that the current passing through a single channel can be measured. An example of the conductance (current/applied potential) steps as channels turn on and off is given in Fig. 29[38] for diphytanoylphosphatidylcholine membranes. The most probable single-channel conductances are in the range of 26 pS as shown in the histogram (Fig. 29A) for 1 M KCl at 30.7° and a 100-mV potential. On division by 1.6×10^{-19} C/ion and multiplication by 0.1 V, a single channel conductance of 26 pS is seen to be a current of 1.63×10^7 ions/sec. This is the current in ions/second for the potassium ion at 30°. An attempt to calculate this current and the currents for the other alkali metal ions will be made using the NMR-derived binding and rate constants.

Before proceeding to do so, however, additional data in Fig. 29 can be noted; that is, the data for des-L-Val[7]-D-Val[8]-gramicidin A. This is a shortened channel. On the basis of a transport mechanism dependent on diffusional length and with the shorter length over which the potential is applied, this shortened channel would be expected to have an approximate 40% increase in conductance over gramicidin A.[38] This would be 36 pS, whereas a much lower conductance is observed, approximately 16 pS. On the basis of this result, it is argued that length is not the limiting factor for the current of ions that pass through the channel but rather that a barrier perspective is appropriate. This is, of course, consistent with the binding and rate constants derived by the NMR methods discussed above. The question to be addressed now is whether the NMR-derived constants can be used to approximate the currents passing through the channel.

The complete expressions appropriate for calculating the currents passing through the channel based on a barrier perspective and suitable for utilizing the NMR-derived binding and rate constants have been presented elsewhere.[21] While complete calculations are possible for the alkali metal ions,[24] it is more instructive and almost as accurate to carry out approximate calculations. As we have seen for the sodium ion, the first ion that enters the channel does so with a stronger binding constant and importantly with an off-rate constant of 3×10^5/sec. For a channel with one ion in it, the contribution to the sodium ion current could be no greater than 3×10^5 ions/sec. In the preceding paragraph it was seen that the currents are of the order of 10^7 ions/sec or more. Thus the initial sodium ion occupancy does not contribute significantly to the sodium ion current. On the other hand, the off-rate constant for the doubly occupied state for the

[37] S. B. Hladky and D. A. Haydon, *Nature (London)* **225**, 451 (1970).

[38] D. W. Urry, S. A. Romanowski, C. M. Venkatachalam, T. L. Trapane, R. D. Harris, and K. U. Prasad, *Biochim. Biophys. Acta* **775**, 115 (1984).

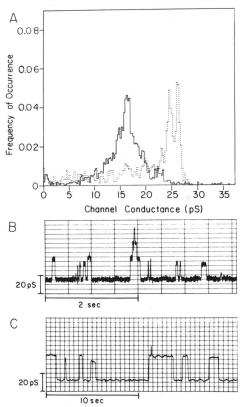

FIG. 29. (A) Conductance histograms at $\sim 31°$ for single-channel current events as measured by the planar lipid bilayer technique for gramicidin A (1547 channel events, 30.7°) ($\cdot \cdot \cdot \cdot \cdot$) and for synthetic des-L-Val^7-D-Val^8-gramicidin A (2040 channel events, 31.0°) (——) in diphytanoylphosphatidylcholine–n-decane bilayers at 1 M KCl and a 100-mV potential. Typical conductance traces for the shortened analog and for gramicidin A are given in B and C, respectively. The most probable conductance step for the full-length channel is on the order of 26 pS, whereas on decreasing the molecular length by two residues or the total channel length by four residues, as is the case for the des-L-Val^7-D-Val^8 analog, the most probable conductance decreases to approximately 16 pS. (Reproduced with permission from Ref. 38.)

sodium ion is 2.1 $\times$ 10^7/sec, which would give currents of the right magnitude. Considering only two states of the channel, singly occupied with a probability χ_s and doubly occupied with a probability of χ_d, $\chi_s + \chi_d = 1$, and $\chi_s = (1 + [\text{Me}^+]K_b^w)^{-1}$ and $\chi_d = K_b^w(1 + [\text{Me}^+]K_b^w)$. Taking the entry barrier to the channel to be rate limiting, the single-channel current, i, may be written as

$$i = \text{forward rate} - \text{backward rate} \tag{21}$$

$$i = [\text{Me}^+]k_{\text{on}}^w\chi_s e^{l_f zFE/2dRT} - k_{\text{off}}^w\chi_d e^{-l_b zFE/2dRT} \tag{22}$$

where $K_b^w = k_{\text{on}}^w/k_{\text{off}}^w$. At 1 M ion activity, i.e., $[\text{Me}^+] = 1$, this becomes

$$i = \frac{K_b^w k_{\text{off}}^w}{1 + K_b^w}(e^{l_f zFE/2dRT} - e^{-l_b zFE/2dRT}) \tag{23}$$

The term in parentheses is the Eyring factor, which introduces the voltage dependence. With the known structure of the channel and the known location of the binding sites, l is the distance from minimum to barrier ($l_f \simeq l_b \simeq 3$ Å); z is the charge on the ion; F is the Faraday (23 kcal/mol-V); E is the potential (0.1 V); $2d$ is the total length across the channel (30 Å); R is the gas constant (1.987 cal/mol-deg); and T is the absolute temperature (303 K). The Eyring factor ($e^{0.38} - e^{-0.38}$) becomes 0.78. Therefore, at 1 M ion activity the single-channel current in ions/second is simply approximated as $0.78 \times K_b^w k_{\text{off}}^w(1 + K_b^w)^{-1}$. Given for each of the alkali metal ions the value of K_b^w and k_{off}^w, it becomes possible to approximate the single-channel currents for 1 M ion activity. These values are given in Table V.

TABLE V

APPROXIMATE CALCULATION OF SINGLE-CHANNEL
CURRENTS AND CONDUCTANCES

Ion	Single-channel[a] current i (ions/sec)	Single-channel[b] conductance γ (pS)	$\gamma(\text{Me}^+)/\gamma(\text{Na}^+)$[c] Calc.	Expt.
Li$^+$	0.57×10^7	9	0.48	0.24
Na$^+$	1.2×10^7	19	1	1
K$^+$	1.8×10^7	29	1.5	~2
Rb$^+$	4.1×10^7	65	3.4	>3
Cs^{+d}	3.7×10^7	59	3.1	>3

[a] Single-channel current $i = 0.78\ K_b^w k_{\text{off}}^w/(1 + K_b^w)$ at 1 M ion activity.

[b] Single-channel conductance $\gamma = i \times 1.6 \times 10^{-19}$ C/V for 0.1 V applied potential.

[c] The conductance ratios vary with lipid and with ion activity. On the bases of Refs. 2–5 and 39, the values were chosen as listed. In general the values for Rb$^+$ and Cs$^+$ are 3.0 or greater, with those for Rb$^+$ being somewhat larger than for Cs$^+$. A complete set of values for phosphatidylcholine membranes at 1 M ion activity are not currently available.

[d] Treating Cs$^+$ as a spin $\frac{3}{2}$ nucleus.

For potassium ion the single-channel conductance calculates to be 29 pS, whereas in Fig. 29A the experimental determination of the most probable single-channel conductance was in the range of 26 pS. This is a remarkably similar result. Also included in Table V are the conductances similarly calculated for the other alkali metal ions. Of further interest is a comparison of the calculated (using NMR-derived binding and rate constants) and the measured (experimental) conductance ratios given, as is the common practice with respect to sodium ion.[2-5,39] Not only is the calculation of the conductance of a single alkali metal ion possible, all of the four spin $\frac{3}{2}$ alkali metal ions can be calculated to give values that are within a factor of two of the measured values.

For several reasons the cesium ion results require additional discussion. This is a spin $\frac{7}{2}$ nucleus whereas Eqs. (9) and (10), which are approximate expressions for spin $\frac{3}{2}$ nuclei, have been used to calculate the correlation time, the inverse of which is taken to be k_{off}^w. In evaluating this usage of Eqs. (9) and (10) it is helpful that rubidium ion and cesium ion give very nearly the same experimental single-channel currents and that their weak binding constants are essentially the same (see Fig. 23D and E and Table IV). By Eq. (23) this requires that their off-rate constants for the doubly occupied state of the channel, k_{off}^w, be essentially the same. At high ion concentration, (see Fig. 27B), the values for T_2' and T_2'' and $T_2(l)$ give a correlation time of 1.7×10^{-8} sec. This yields a value for k_{off}^w of 5.9×10^7/ sec for cesium ion at high concentration, whereas the value is 6.6×10^7/sec for rubidium ion at high concentration. This would be considered very satisfactory if the cesium nucleus were a spin $\frac{3}{2}$. Coming so close for a spin $\frac{7}{2}$ nucleus means that Eqs. (9) and (10) can, in practice, for binding site symmetries such as the gramicidin channel, be usefully applied to spin $\frac{7}{2}$ nuclei. This makes the analyses of Bull, Forsén, and their colleagues[6-8] all the more significant in practice.

A second feature demonstrated by the cesium ion is that values are obtained at low ion concentrations (30 mM) for k_{off}^t of 1.75×10^7/sec. This is much larger than estimated for the sodium ion (3×10^5/sec). The result is not inconsistent with the ion flux ratio data, which is 1.0 for sodium ion and 1.6 for the cesium ion.[40,41] We interpret this to mean that only one occupied state of the channel for sodium ion is conductive (the doubly occupied state), whereas both the singly and doubly occupied states contribute to the cesium ion conductances but not equivalently. It is interesting to note in this regard that the conductance for the singly occupied state $0.78K_b^t k_{off}^t/(1 + K_b^t)$, with $K_b^t = 60$/M and $k_{off}^t = 1.75 \times 10^7$/

[39] V. Myers and D. A. Haydon, *Biochim. Biophys. Acta* **274**, 313 (1976).

[40] J. Procopio and O. S. Andersen, *Biophys. J.* **25**, 8a (1979).

[41] A. Finkelstein and O. S. Andersen, *J. Membr. Biol.* **59**, 155 (1981).

sec, is 1.34×10^7 ions/sec and that of the doubly occupied state given in Table V is 3.7×10^7 ions/sec. At a concentration of 1 M CsCl, $\chi_s = 0.46$ and $\chi_d = 0.52$ such that the ratio of contribution of singly and doubly occupied states is $0.32:1$. This, of course, is approximate and needs, for example, to be corrected for the rate over the central barrier. Accordingly, we suggest that the flux ratio data be considered in this light. It would be of interest to pursue the same consideration for the rubidium ion, i.e., obtaining an estimate for k_{off}^t by means of Eqs. (9) and (10), but this is not possible with the present instrumentation due to the very short delay time required before the signal accumulation begins; that is, the linewidths for rubidium ion are already at the limit of broadness at high ion concentration and they would be broader yet at low ion concentrations where the transverse relaxation times are shorter.

When the correct single-channel currents were calculated for the sodium ion using NMR-derived rate constants,[20,21] this was quite satisfying. When the correct single-channel currents could be reasonably calculated for two different ions, the chance of a fortuitous result is greatly decreased. When the single-channel currents at high ion concentration for all four of the spin $\frac{3}{2}$ alkali metal ions can be calculated within a factor of two of the experimental currents, the result can be taken as a convincing demonstration of the NMR analyses. It would appear to have been demonstrated on the gramicidin transmembrane channel that NMR can be used with channels having currents in the 10^7 ions/sec range to derive binding and rate constant data relevant to the channel transport mechanism.

Conclusions

The ion interactions with membranous polypeptide sites using NMR, considered in this review, are alkali metal ion interactions with the gramicidin transmembrane channel. The NMR approaches involve primarily longitudinal and transverse relaxation studies of quadrupolar NMR supplemented by ion resonance and site resonance chemical shift studies. The gramicidin channel provides a unique opportunity to demonstrate and to develop the practical applications of these NMR methods. Longitudinal relaxation studies are shown to be a favorable means for estimating binding constants for complex binding processes when two binding constants differ by an order of magnitude. When the observation frequency, ν_0, for the nucleus exchanging between solution and binding site is within an order of magnitude of the off-rate constant for the ion leaving the channel, it can be possible to determine the value of k_{off} quite accurately by means of transverse relaxation studies, supplemented as necessary by longitudinal relaxation studies. Because the observation frequencies for the alkali metal ions are in the 10^7/sec range and the ionic currents which pass through the gramicidin channel, as with most physiological channels, are also in the range of 10^7 ions/sec, it is possible to calculate the currents passing through the gramicidin channel using NMR-derived binding and rate constants, not for just one or two of the alkali metal ions but rather for each of the five alkali metal ions. The calculated conductances (currents/applied potential) and the conductance ratios of the entire alkali metal ion series are within a factor of two of the electrically measured conductances. This is the case even for the spin $\frac{7}{2}$ cesium-133 nucleus when data are analyzed by the spin $\frac{3}{2}$ formalism due to Bull, Forsén, and colleagues. This demonstration with the gramicidin channel, verified by the capacity to calculate electrically measured currents, hopefully can provide a basis for the study of many ion interactions with membranous polypeptide sites and particularly for the characterization of ion interactions with ion-conducting channels.

Acknowledgment

This work was supported in part by NIH Grant GM-26898.

[16] Measuring Electrostatic Potentials Adjacent to Membranes

By David Cafiso, Alan McLaughlin, Stuart McLaughlin, and Anthony Winiski

Introduction

Figure 1 illustrates the electrostatic potential adjacent to a bilayer membrane formed from negative phospholipids such as phosphatidylserine (PS) or phosphatidylglycerol (PG) in 0.1 M NaCl. The fixed negative charges on the lipids attract sodium ions and produce an electrostatic "diffuse double layer" in the aqueous phase adjacent to the surface. This phenomenon was described first by Gouy[1] and Chapman,[2] who combined the Poisson equation with the Boltzmann relation. Stern[3] considered the possibility that the counterions, sodium ions in the example illustrated in Fig. 1, could bind to the membrane as well as exert a nonspecific "screening" effect in the diffuse double layer. Several introductory reviews discuss

[1] G. Gouy, *J. Phys.* **9**, 457 (1910).
[2] D. L. Chapman, *Philos. Mag.* **25**, 475 (1913).
[3] O. Stern, *Z. Elektrochem.* **30**, 508 (1924).

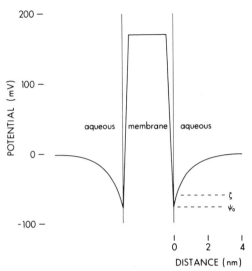

FIG. 1. The electrostatic potential adjacent to a bilayer membrane formed in 0.1 M NaCl from negative phospholipids such as PS or PG. The potential profile in the aqueous phase is drawn according to the Gouy–Chapman–Stern theory. The four major assumptions in this theory are (1) the surface charges are smeared uniformly over the membrane–solution interface, (2) the ions in the aqueous phase are point charges, (3) the dielectric constant of the aqueous phase is uniform up to the surface of the membrane, and (4) image charge effects are negligible. Sodium ions are assumed to bind to the negative lipids with an intrinsic association constant of 1 M^{-1}. Note the large positive potential within the membrane, which is due to molecular dipoles near the surface.

the Gouy–Chapman–Stern theory and refer to more detailed reviews and the original literature.[4-6] In 0.1 M NaCl, the potential at the surface of PS and PG membranes is determined experimentally to be about -80 mV (see below). This agrees with the value predicted from the Gouy–Chapman–Stern theory if we assume the intrinsic association constant of sodium with these negative lipids is 1 M^{-1}. The surface potential of membranes formed from PS or PG is more negative in KCl and less negative in LiCl solutions; monovalent cations bind to these lipids in the sequence Li > Na > K > Rb > Cs > TMA.[7] Note in Fig. 1 there is an approximately exponential decline in the potential with distance from the membrane: $\psi(x) \sim \psi_0 \exp(-\kappa x)$ where $1/\kappa$ is the Debye length, about 1 nm in 0.1 M NaCl. The potential at the surface of the membrane, ψ_0 (millivolts),

[4] R. Aveyard and D. A. Haydon, "An Introduction to the Principles of Surface Chemistry." Cambridge Univ. Press, London, 1973.
[5] S. McLaughlin, Curr. Top. Membr. Transp. 9, 71 (1977).
[6] J. N. Israelachvili, "Intermolecular and Surface Forces." Academic Press, New York, 1985.
[7] M. Eisenberg, T. Gresalfi, T. Riccio, and S. McLaughlin, Biochemistry 18, 5213 (1979).

is related to the net surface charge density, σ (electronic charges/Å^2), and the concentration C (molar), of $z:z$ electrolyte in the bulk aqueous phase by the Gouy equation ($T = 25°$):

$$\sinh(z\psi_0/51.4) = 136.6\sigma/(C)^{1/2} \tag{1}$$

When $z\psi_0 \gg 51$ mV, Eq. (1) predicts the potential changes 59 mV per decade change in monovalent salt concentration, 29 mV per decade change in divalent salt concentration, and 14 mV per decade change in tetravalent salt concentration. These predictions have been tested experimentally for monovalent,[5,8] divalent,[5,9] and tetravalent cations[10] and have been confirmed qualitatively.

Diffuse double-layer potentials influence many biological processes. For example, surface charges affect the conductance of ion channels in excitable membranes,[11-13] the affinity of enzymes for charged substrates,[14-18] and possibly the stacking of thylakoid membranes[19,20] and the flow of fluid in epithelia.[21]

Molecular dipoles near the surface of the membrane produce a large positive dipole potential within the membrane.[22-27] As illustrated in Fig. 1, the potential within the membrane is the sum of the diffuse double-layer potential and the dipole potential. Experiments with lipid-soluble ions suggest the dipole potential is about 250 mV for phosphatidylcholine (PC)

[8] F. Lakhdar-Ghazal, J.-L. Tichadou, and J.-F. Tocanne, *Eur. J. Biochem.* **134,** 531 (1983).

[9] S. McLaughlin, N. Mulrine, T. Gresalfi, G. Vaio, and A. McLaughlin, *J. Gen. Physiol.* **77,** 445 (1981).

[10] L. Chung, G. Kaloyanides, R. McDaniel, A. McLaughlin, and S. McLaughlin, *Biochemistry* **24,** 442 (1985).

[11] B. Hille, A. M. Woodhull, and B. I. Shapiro, *Philos. Trans. R. Soc. London* **270,** 304 (1975).

[12] T. Begenisich, *J. Gen. Physiol.* **66,** 47 (1975).

[13] J. Bell and C. Miller, *Biophys. J.* **45,** 279 (1984).

[14] L. Wojtczak and M. J. Nalecz, *Eur. J. Biochem.* **94,** 99 (1979).

[15] M. J. Nalecz and L. Wojtczak, *Eur. J. Biochem.* **112,** 75 (1980).

[16] R. Kramer, *Biochim. Biophys. Acta* **735,** 145 (1983).

[17] R. Gibrat, J.-P. Grouzis, J. Rigaud, and C. Grignon, *Biochim. Biophys. Acta* **816,** 349 (1985).

[18] R. Gibrat and C. Grignon, *Biochim. Biophys. Acta* **692,** 462 (1982).

[19] J. Barber, *Annu. Rev. Plant Physiol.* **33,** 261 (1982).

[20] K. R. Miller and M. K. Lyon, *Trends Biochem. Sci.* **10,** 219 (1985).

[21] S. McLaughlin and R. T. Mathias, *J. Gen Physiol.* **85,** 699 (1985).

[22] S. B. Hladky and D. A. Haydon, *Biochim. Biophys. Acta* **318,** 464 (1983).

[23] S. B. Hladky, *Biochim. Biophys. Acta* **352,** 71 (1974).

[24] O. S. Andersen and M. Fuchs, *Biophys. J.* **15,** 795 (1975).

[25] A. D. Pickar and R. Benz, *J. Membr. Biol.* **44,** 353 (1978).

[26] R. F. Flewelling and W. L. Hubbell, *Biophys. J.* **47,** 249a (1985).

[27] R. F. Flewelling and W. L. Hubbell, *Biophys. J.* **41,** 541 (1986).

bilayers.[24-27] Experiments with carriers suggest the dipole potential is very similar for PS and PG membranes (unpublished results). The origin of the dipole potential is unknown. It could arise from oriented carbonyl oxygens or water molecules. Moreover, the biological role of the dipole potential in transport processes is unclear. The dipole potential does not affect the concentration of ions adjacent to the membrane, as expected theoretically,[5] and should have little effect on the kinetics of enzymes that interact with ions. Furthermore, it has little effect on the conductance produced by gramicidin,[28] which forms channels in bilayers; dipole potentials affect only the conductance of channels that are long, narrow, and have their main energy barrier near the center of the membrane.[29]

We routinely examine the electrostatic potential adjacent to phospholipid bilayer membranes by making two different types of measurements. First, we estimate the zeta potential, ζ, the potential at the hydrodynamic plane of shear (see Fig. 1), by measuring the electrophoretic mobility of a vesicle and applying the Helmholtz–Smoluchowski equation. These measurements require no probe molecules and can be made on vesicles of any size and shape, including biological membranes. The results obtained with biological membranes, however, are difficult to interpret. Second, we estimate the surface potential, ψ_0 (see Fig. 1), by measuring the adsorption of ions to membranes and applying the Boltzmann equation; we have developed four different techniques, all of which utilize ions as "probes." (1) We make ^{31}P NMR measurements using a paramagnetic cation, such as Mn, as a probe. The NMR measurements sense the potential at a well-defined site, the phosphodiester group at the membrane surface, but the technique requires small sonicated vesicles and relatively large amounts of them. (2) We make fluorescence measurements using 2-(p-toluidinyl)naphthalene 6-sulfonate (TNS) and (3) electron paramagnetic resonance (EPR) measurements using spin labels as probes. The anion TNS and the charged spin labels are amphipathic molecules that adsorb hydrophobically to the membrane–solution interface. TNS and spin labels can be used to probe the surface potential of biological membranes, but they permeate some membranes[30] and are sensitive to many parameters other than the electrostatic potential. Fluorescent probes, for example, respond to the dielectric constant. (4) We measure the conductance of planar membranes using nonactin K or FCCP as cationic or anionic probes. These molecules respond to the potential in the center of a bilayer membrane, and sense both the dipole and the diffuse double-layer potentials illustrated in Fig. 1. The

[28] O. S. Andersen, in "Renal Function" (G. H. Giebisch and E. Purcell, eds.), p. 71. Independent Publ., Port Washington, New York, 1978.

[29] P. C. Jordan, Biophys. J. **41,** 189 (1983).

[30] H. Rottenberg, this volume [17].

potentials can be examined over a wide range of values, but the conductance produced by these carriers is difficult to measure on biological membranes. In summary, we can measure the potential at the hydrodynamic plane of shear, at the surface of the membrane, and within the bilayer.

Many other techniques have been used to measure the electrostatic potential adjacent to surfaces. A vibrating plate or ionizing electrode, for example, can be used to measure directly the potential above a monolayer.[8,22] Ehrenberg[31] reviewed the use of resonance Raman spectroscopy to estimate the surface potential. One advantage of the techniques we discuss below is that we recently tested all our probe molecules on membranes of known positive and negative surface charge density and are therefore aware of the artifacts associated with each of them.

Electrophoretic Mobility

Principles

Electrophoresis (movement of a charged particle in an electric field) and electroosmosis (field-induced movement of fluid adjacent to a fixed, charged surface) are manifestations of the same phenomenon. Consider the cylindrical tube illustrated in Fig. 2A, which is filled with fluid. The ends are open and are maintained at atmospheric pressure. If the walls of the tube bear a fixed negative charge, imposition of an electric field parallel to the wall of the pipe will cause the fluid to move, a phenomenon termed electroosmosis. Helmholtz[32] correctly described the phenomenon in 1879. Electroosmosis occurs because the fixed charges on the surface attract counterions (see insert of Fig. 2A), which "form an electric double layer along their boundary surface. . . . This layer has an extraordinarily small, but not disappearing, thickness. The side of this layer adjacent to the boundary surface clings immovably to the wall . . . a potential gradient directly propels the positively charged wall layer of the liquid. However, because of the internal friction of the liquid, the entire cross section of the tube assumes the same motion if no hydraulic back pressure results." Thus, the fluid velocity is constant throughout the tube, except for an extremely thin region adjacent to the wall (see insert of Fig. 2A). The quantitative description of the electroosmotic fluid velocity is due to Helmholtz[32] and von Smoluchowski[33]; detailed modern descriptions of

[31] B. Ehrenberg, this series, Vol. 127, p. 678.

[32] H. L. Helmholtz, *Ann. Phys. (Leipzig)* **7**, 337 (1879). [For an English translation entitled "Studies of Electric Boundary Layers," see *Eng. Res. Bull.* **33**, 5 (1951).]

[33] M. von Smoluchowski, *in* "Handbuch der Elektrizitat und des Magnetismus" (W. Graetz, ed.), Vol. 2, p. 366. Barth, Leipzig, Germany, 1921.

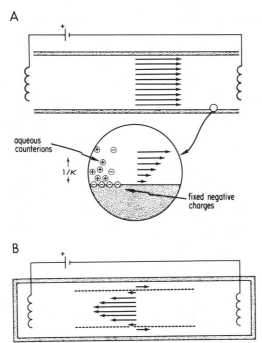

FIG. 2. (A) The velocity profile that results when fluid is driven through a rigid cylindrical tube of length l and radius r ($l \gg r$) by a gradient of voltage. A voltage is applied down the axis of the pipe by two electrodes. The potential has a value of $+\psi$ at $x = 0$ and a value of 0 at $x = l$. The ends of the pipe at $x = 0$, l are exposed to infinite reservoirs of fluid maintained at atmospheric pressure. The inset illustrates that the velocity, represented by the length of the arrows, is zero at the surface of the pipe, but increases to a maximum value within a few Debye lengths ($1/\kappa \sim 1$ nm for [NaCl] = 0.1 M). For a tube with $r \gg 1/\kappa$, the velocity profile is constant throughout most of the pipe. (B) The velocity profile that results when the ends of the pipe are closed. The dashed lines indicate the location of the "stationary layer." See the text for details.

electroosmosis[34-36] and brief derivations of the Helmholtz–Smoluchowski equation[4,37,38] are in the literature. The equation

$$u_{eo} = (d\psi/dx)(\epsilon_r \epsilon_0 \zeta/\eta) \tag{2}$$

relates the fluid velocity, u_{eo}, to the electric field along the axis of the tube, $-d\psi/dx$, the viscosity, η, the dielectric constant ϵ_r, the permittivity of free

[34] J. T. G. Overbeek and P. H. Wiersema, *in* "Electrophoresis" (M. Bier, ed.), Vol. 2, p. 1. Academic Press, New York, 1967.
[35] S. S. Dukhin and B. V. Derjauguin, *Surf. Colloid Sci.* **7**, 49 (1974).
[36] R. W. O'Brien and L. R. White, *J. Chem. Soc., Faraday Trans. 2* **74**, 1607 (1978).
[37] D. J. Shaw, "Electrophoresis." Academic Press, New York, 1969.
[38] R. J. Hunter, "Zeta Potential in Colloid Science." Academic Press, New York, 1981.

space, ϵ_0, and the zeta potential, ζ, the potential at the hydrodynamic plane of shear, which is 0.2 nm from the surface of a phospholipid bilayer exposed to a decimolar monovalent salt solution.[7,39,40]

Most tubes used for measuring the electrophoretic mobility of charged particles are closed when the electrodes are inserted. If the tube illustrated in Fig. 2 is closed, the movement of the fluid to the right under the influence of the electric field produces a pressure gradient within the tube that causes fluid to move to the left; the pressure gradient produces the conventional Poiseuille parabolic velocity profile. The net velocity profile, which is the sum of the parabolic profile due to pressure and the constant profile due to electroosmosis, is illustrated in Fig. 2B. The fluid velocity is zero at the "stationary layer" (dashed lines in Fig. 2B). The experimenter focuses on a particle at the stationary layer, measures its velocity per unit field, u, and calculates the zeta potential from the Helmholtz–Smoluchowski equation:

$$\zeta = u\eta/\epsilon_r\epsilon_0 \tag{3}$$

Note that Eq. (3) predicts the electrophoretic mobility independent of the size and shape of the particle. In fact, particles of different size and shape made of the same material all move with the same velocity in an electric field.

The main assumptions in deriving this equation are that the radius of the vesicle is much larger than the Debye length, the charges are at the membrane–solution interface, and the surface is smooth.[34-38] These assumptions are valid for multilamellar phospholipid vesicles, but not for biological cells such as erythrocytes. It is intuitively apparent why the Helmholtz–Smoluchowski equation cannot be used successfully to calculate the zeta potential from the measured electrophoretic mobility of a biological cell. Any sugars or proteins that protrude from the bilayer–solution interface of the cell will exert hydrodynamic drag that will decrease the electrophoretic mobility. On the other hand, the sialic acid residues responsible for most of the fixed charge on an erythrocyte are located a significant distance from the bilayer–solution interface, which will tend to increase the magnitude of the electrophoretic mobility. These effects are not small.[41-43] The electrophoretic mobility of an erythrocyte is about half the value predicted from the known charge density and the

[39] E. K. Rooney, J. M. East, O. T. Jones, J. McWhirter, A. C. Simmonds, and A. G. Lee, *Biochim. Biophys. Acta* **198**, 159 (1983).

[40] O. Alvarez, M. Brodwick, R. Latorre, A. McLaughlin, S. McLaughlin, and G. Szabo, *Biophys. J.* **44**, 333 (1983).

[41] E. Donath and V. Pastuschenko, *Bioelectrochem. Bioenerg.* **6**, 543 (1979).

[42] R. W. Wunderlich, *J. Colloid Interface Sci.* **88**, 385 (1982).

[43] S. Levine, M. Levine, K. A. Sharp, and D. E. Brooks, *Biophys J.* **42**, 127 (1983).

Helmholtz–Smoluchowski equation.[43] We studied these effects by forming vesicles from mixtures of phospholipids, such as phosphatidycholine, and gangliosides, such as G_{M1}.[44] X-Ray diffraction measurements demonstrate that in PC:G_{M1} membranes the head group of G_{M1} is fully extended and protrudes about 2 nm from the surface of the bilayer.[45] A combination of the nonlinear Poisson–Boltzmann and the Navier–Stokes equations predicts, and experiments confirm, that the electrophoretic mobility of these PG:G_{M1} vesicles is only about half the mobility of a phospholipid vesicle of equivalent charge density.[46]

We recently measured the zeta potential of vesicles formed from mixtures of PC and either negative (PG, PS) or positive (didodecyldimethylammonium bromide; DDDAB) lipids. The Gouy–Chapman–Stern theory describes the data in 0.1 and 0.01 M NaCl well if one assumes sodium ions bind to the phosphate group of the negative lipids and chloride ions bind to the amine group of the positive lipids with an intrinsic association constant of 1 M^{-1} (unpublished observations). These vesicles should be useful for testing other techniques used to estimate the surface potential of membranes.

Experimental Considerations

Ware and Flygare[47] developed an elegant new method for determining the electrophoretic velocity. They measured the Doppler shift of laser light scattered from the particles. The main advantage of this technique is that measurements can be made on particles too small to measure with direct microscopic observation or even with dark-field illumination, e.g., plasma proteins. One also obtains rapidly a spectrum of the entire distribution of electrophoretic mobilities. The disadvantages other than cost are discussed in other reviews.[38,48,49] It is our contention that a simple glass tube, a water bath, a pair of electrodes, an inexpensive, traveling microscope, and a stopwatch yield equally accurate results for most applications, including measurements on biological cells and lipid vesicles. The reason is simple. The main source of error is not the accuracy with which the velocity can be measured, but the uncertainty in the location of the stationary layer, particularly when a few particles have settled onto the surface of the tube.

[44] R. McDaniel, A. McLaughlin, S. McLaughlin, A. Winiski, and M. Eisenberg, *Biochemistry* **23**, 4618 (1984).

[45] R. McDaniel and T. McIntosh, *Biophys. J.* **49**, 94 (1985).

[46] R. McDaniel, K. Sharp, A. McLaughlin, A. Winiski, D. Brooks, D. Cafiso, and S. McLaughlin, *Biophys. J.* **49**, 741 (1986).

[47] B. R. Ware and W. H. Flygare, *Chem. Phys. Lett.* **12**, 81 (1971).

[48] B. A. Smith and B. R. Ware, *Contemp. Top. Anal. Clin. Chem.* **2**, 29 (1978).

[49] V. A. Bloomfield and T. K. Lim, this series, Vol. 48, p. 415.

Consider the economical microelectrophoresis machines available from Rank Bros. (Bottisham, Cambridge, England). In our opinion, the original Mark I machine, based on a design of Bangham et al.,[50] is superior to the Mark II machine for routine measurements, if only because it is less expensive and more rugged (one student broke four cylindrical Mark II tubes before obtaining any reasonable measurements). Particles falling to the bottom of cylindrical cells affect the location of the relevant stationary layer less in the Mark I than in the Mark II machine. If settling of particles is a serious problem, however, a flat cell can be used with the Mark II apparatus, which is designed to accommodate both types of cells. The dark-field illumination of the Mark II also allows measurements on particles the size of a single vesicular stomatitis virus particle (unpublished observations). However, sonicated phospholipid vesicles (radius ~ 15 nm) do not scatter sufficient light to be resolved with this machine.

We use the following procedure when inserting a new cylindrical tube into a Mark I or Mark II machine: clean the tube in chromic acid to remove all traces of organic material and wash it thoroughly (for about a day); calibrate the eyepiece reticule under water with a stage micrometer; align the tube in the holder and measure its diameter at the spot where measurements will be taken with the microscope. We then determine the effective length of the tube by filling it with solutions of known conductivity (e.g., 1, 10, and 100 mM NaCl), applying a known voltage, and measuring the current. It is important to check that no electrode polarization occurs in the solution of highest salt concentration used for electrophoretic mobility measurements.

Hunter[38] deals with the "measurement of electrokinetic parameters" and includes references to many original and review articles as well as much practical advice, as does an article by James[51] and a book by Shaw.[37] We shall not repeat here the detailed suggestions contained in these sources. In brief, focus on the inner surface of the cylindrical tube, then dial in the correct distance for the stationary layer (for a cylindrical tube this is 0.293r where r is the radius). Corrections are necessary for the planoconvex lens effect[52,53] on the Mark I (thick Mattson type of cell) but not the Mark II (thin van Gils type of cell) apparatus. Measure, with a stopwatch, the time a vesicle takes to move a known distance, which is determined with the calibrated eyepiece reticule. Average the results of at least 10 measurements with normal and reversed electrode polarity, and

[50] A. D. Bangham, R. Flemans, D. H. Heard, and G. V. F. Seaman, Nature (London) 182, 642 (1958).
[51] A. M. James, Surf. Colloid Sci. 11, 121 (1979).
[52] D. C. Henry, J. Chem. Soc. p. 997 (1938).
[53] E. S. Hall, Nature (London) 203, 1371 (1964).

calculate the zeta potential from Eq. (3). It is tedious to record the times manually and analyze the written data. We use an inexpensive system to automate these tasks: a Hewlett–Packard 41C programmable calculator–timer interfaced to a Brothers electronic typewriter.

We check that the microscope is focused on the stationary layer by measuring the electrophoretic mobility of egg PC vesicles. The phospholipid is obtained from Avanti (Birmingham, Alabama) and the vesicles are formed according the established procedures.[54] These vesicles have zero electrophoretic mobility in 1, 10, and 100 mM NaCl solutions.[7] We then check the other calibrations (effective tube length, applied voltage, and distance between lines of eyepiece reticule) by measuring the electrophoretic mobility of brain PS (Avanti) vesicles in 0.1 M NaCl. These vesicles have a zeta potential of -59 ± 2 mV.[55]

We calibrate a new tube with PC vesicles to ensure the velocity profile is parabolic (see Fig. 2B). The entire profile can be obtained on the thin-walled cylindrical van Gils cell in the Mark II apparatus. With the thick-walled Mattson cell in the Mark I apparatus, measurements can be made only up to the axis of the cell. Even with the Henry correction,[52] a complex form of astigmatism arises in this type of cell.[53]

The voltage electrodes embedded in the van Gils cell supplied with the Mark II apparatus make the tube very difficult to clean. Strong cleaning agents, such as chromic acid, weaken the bond between the glass and the embedded wires and ruin the tube. If requested, Rank Bros. will supply a tube without embedded voltage electrodes.

The zeta potential, calculated from Eq. (3), is the potential at the hydrodynamic surface of shear. The surface of shear "is an imaginary surface which is considered to lie close to the solid surface and within which the fluid is stationary." [38] Where is the surface of shear? It is 0.2 nm from the surface of a phospholipid vesicle in a 0.1 M salt solution.[7,39,40] We estimated this distance by independently measuring both the surface potential (see conductance section below) and the zeta potential, then assuming the potential profile is given correctly by the Gouy–Chapman theory (Fig. 1). The work of Parsegian, Rand and co-workers[56] and Israelachvili and co-workers[6] indicates this assumption is correct, at least for distances greater than 2 nm from the membrane. The theory is untested for distances closer to the bilayer than 1 nm, however, and the exact location of the plane of shear must be regarded as somewhat uncertain.

[54] A. D. Bangham, M. W. Hill, and N. G. A. Miller, *Methods Membr. Biol.* 1, 1 (1974).
[55] A. McLaughlin, W.-K. Eng, G. Vaio, T. Wilson, and S. McLaughlin, *J. Membr. Biol.* 76, 183 (1983).
[56] A. C. Cowley, N. L. Fuller, R. P. Rand, and V. A. Parsegian, *Biochemistry* 17, 3163 (1978).

Nuclear Magnetic Resonance

Principles

Paramagnetic divalent transition metal ions have large and well-understood effects on the narrow ^{31}P NMR signals from sonicated phospholipid bilayer vesicles,[57-61] and these effects can be used to estimate the surface potential of the membrane. The four common paramagnetic divalent transition metal cations are manganese, cobalt, nickel, and copper. Manganese exerts the largest effects on the ^{31}P NMR signal and is the most useful for surface potential measurements, although the original experiments were done with cobalt.[59]

The effect of manganese on the linewidth of the ^{31}P NMR signal from phospholipids in sonicated vesicles is proportional to the number of phosphodiester groups bound in inner sphere[62] complexes with manganese. In our experiments, less than 0.1% of the phosphodiester groups are bound to manganese ions. The concentration of bound complexes is thus proportional to the Boltzmann factor, $\exp(-2F\psi_P/RT)$, where ψ_P is the electrostatic potential in the aqueous phase adjacent to the phosphodiester group. The approach is perhaps best illustrated by a specific example: consider the effect of incorporating a charged lipid, PG, into a bilayer membrane that consists initially of the zwitterionic lipid, PC. We can calculate the change in potential at the phosphodiester group, $\Delta\psi_P$, using the following equation

$$[1/T_{2P}]/[1/T_{2P}^0] = \exp[-2F\Delta\psi_P/RT] \tag{4}$$

where $1/T_{2P}$ is the effect of a known free concentration of manganese on the transverse relaxation rate of the ^{31}P NMR signal from PC:PG vesicles and $1/T_{2P}^0$ is the effect of the same free concentration of manganese on the transverse relaxation rate of the ^{31}P NMR signal from PC vesicles. $1/T_{2P}$ and $1/T_{2P}^0$ are calculated from the effects of manganese on the observed linewidths, $\Delta\nu_P$ and $\Delta\nu_P^0$, using the relations $1/T_{2P} = \pi\Delta\nu_P$ and $1/T_{2P}^0 = \pi\Delta\nu_P^0$. The potential at a specific site at the membrane surface, in this case the phosphodiester group, is usually termed the "micropotential" to distinguish it from the average value of the potential at the surface of the membrane.

[57] A. C. McLaughlin, C. Gratwohl, and S. McLaughlin, *Biochim. Biophys. Acta* **513**, 338 (1978).
[58] A. C. McLaughlin, C. Grathwohl, and R. E. Richards, *J. Magn. Reson.* **31**, 283 (1978).
[59] A. Lau, A. C. McLaughlin, and S. McLaughlin, *Biochim. Biophys. Acta* **645**, 279 (1981).
[60] A. C. McLaughlin, *J. Magn. Reson.* **49**, 246 (1982).
[61] A. C. McLaughlin, *Biochemistry* **21**, 4879 (1982).
[62] H. C. Freeman, *in* "Inorganic Biochemistry" (G. Eichorn, ed.), Vol. 1, p. 121. Elsevier, Amsterdam, 1973.

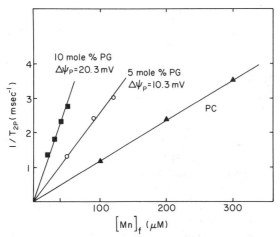

FIG. 3. The effect of manganese on the transverse relaxation rate ($1/T_{2P}$) of the ^{31}P NMR signal from phosphatidylcholine (PC) molecules in the outer monolayer of sonicated vesicles. The vesicles were formed from PC, or mixtures of PC and phosphatidylglycerol (PG). The aqueous solutions contained 0.1 M sodium chloride buffered to pH 7.4 with 5 mM MOPS. $T = 25°$. Changes in the micropotential at the phosphodiester group of PC, $\Delta\psi_P$, were calculated using Eq. (4).

Figure 3 illustrates the effect of manganese on the transverse relaxation rate of ^{31}P NMR signals from PC vesicles and PC:PG vesicles containing 5 or 10 mol% PG. The change in the micropotential was calculated from the ratio of the slopes of the lines using Eq. (4). The linear dependence of $1/T_{2P}$ on $[Mn]_f$ confirms that manganese does not alter the surface potential at the concentrations used in these experiments. This approach has been used to measure the effect of PS, PG, and gangliosides on the surface potential of PC membranes.[46] It can also be used to calculate the effect of diamagnetic cations on the surface potential of membranes formed from charged phospholipids. We used this approach to study the effect of calcium on the surface potential of PG membranes,[59] and the effect of calcium, organic divalent cations, and antibiotics on the surface potential of PS membranes.[10,40,55] We recently measured the surface potential of positively (PC:DDDAB) and negatively (PC:PS and PC:PG) charged vesicles in 0.1 and 0.01 M NaCl. The surface potentials determined from NMR measurements agree quite well with the theoretical predictions of the Gouy–Chapman–Stern theory and the zeta potential measurements.

Experimental Considerations

Sonicated unilamellar vesicles are prepared by suspending approximately 20 mg of dried lipid in 2 ml of buffered sodium chloride solution, and sonicating in an ice bath for 30 min (using a 30% duty cycle) under a stream of hydrated argon. We centrifuge the sample to remove titanium particles, and use the supernatant. We control the free manganese concentration by passing the sonicated vesicle samples over Sephadex columns (G-200, fine) loaded with buffered sodium chloride solutions containing known manganese concentrations. This procedure does not equilibrate the intravesicular space with manganese because the membrane is relatively impermeable to divalent cations.[63] The observed ^{31}P NMR spectrum is thus the superposition of a broad component from phospholipid molecules on the outer monolayer, which are exposed to manganese, and a narrow component from phospholipid molecules on the inner monolayer, which are not exposed to manganese. The narrow signal can, however, be nulled using a $(\pi, \tau, \pi/2)$ radio frequency pulse sequence that utilizes the difference in longitudinal relaxation times for the narrow and broad signals.[59] Thus we can determine the effect of manganese on the linewidth of the ^{31}P NMR signal from phospholipids in the outer monolayer. We note that a fresh sample of sonicated vesicles must be employed for each experimental data point and manganese must not be added to the membrane sample before sonication.

The experiments can be performed at any reasonable magnetic field strength, but higher magnetic field strengths increase the signal-to-noise ratio and decrease the time required for data acquisition. We use a Bruker WH-360 NMR spectrometer with a 15-mm sample tube containing 5 ml of the peak fractions from the Sephadex column. If the original sample contains 20 mg of phosphatidycholine, we can acquire sufficient data in 1–4 hr (depending on the ^{31}P NMR linewidth). The effect of manganese on the linewidth of the ^{31}P NMR signal from phosphodiester groups is very sensitive to temperature,[64] and precise temperature control ($\pm 0.5°$) is critical. Measurements with membranes containing PS must be done at pH 5.0 to supress "chelate" complexes that form between transition metal ions and amino acids at higher pH values.[46,61]

Fluorescence

Principles

The anionic fluorescent molecule 2-(p-toluidinyl)naphthalene 6-sulfonate has a number of properties that make it a good probe of the electrostatic potential at the membrane–solution interface. First, TNS adsorbs

[63] H. Degani, S. Simon, and A. McLaughlin, *Biochim. Biophys. Acta* **646,** 320 (1981).
[64] R. G. Shulman, H. Sterlicht, and B. J. Wyluda, *Chem. Phys.* **43,** 3116 (1965).

hydrophobically to bilayers formed from different neutral and zwitterionic lipids.[65,66] The adsorption is independent of the dipole potential and the nature of the head group. The charge on an adsorbed TNS molecule is presumably located outside the region where a large change in dipole potential occurs (see Fig. 1). Second, the adsorption coefficient ($\sim 2 \times 10^{-3}$ cm, which corresponds to an apparent association constant of $\sim 10^4 \, M^{-1}$ with the lipids) and the ratio of the quantum yield of the probe adsorbed onto the membrane to the quantum yield of the probe in the aqueous phase (700) are both large. Thus it is easy to distinguish adsorbed TNS molecules from those in the aqueous phase. Third, fluorescence measurements are sufficiently sensitive that one can make measurements at low probe concentrations where TNS does not itself perturb the surface potential. Fourth, the fluorescence characteristics of the adsorbed TNS molecules (excitation and emission spectra, quantum yield, and lifetime) are essentially independent of the type of lipids present in the membrane, at least for PC, PS, and PG.[7]

These properties imply that the observed net fluorescence, f [see Eq. (7)], is proportional to the surface concentration of adsorbed TNS, $\{TNS\}$, which is proportional to the aqueous concentration of TNS immediately adjacent to the membrane, $[TNS]_0$. This concentration is related to the electrostatic potential in the aqueous phase adjacent to the membrane, ψ_0, and the concentration of TNS in the bulk aqueous phase, $[TNS]$, by the Boltzmann relation:

$$(f/[L]) = \alpha\{TNS\} = \beta[TNS]_0 = \beta[TNS] \exp(F\psi_0/RT) \qquad (5)$$

where α and β are proportionality constants, F, R, and T have their usual meanings, and $[L]$ is the lipid concentration. The potential in the aqueous phase adjacent to a vesicle formed from a zwitterionic lipid such as PC is zero because the zeta potential is zero.[7] Using Eq. (5), we calculate the surface potential of vesicles formed from charged lipids, L:

$$f(L)[PC]/f(PC)[L] = \exp(F\psi_0/RT) \qquad (6)$$

where $f(L)$ and $f(PC)$ are the net fluorescence intensities of L and PC. Here $[L]$ and $[PC]$ are the molar concentrations of the charged lipids and PC. The net fluorescence, f, is calculated by subtracting the appropriate backgrounds from the measured intensity:

$$
\begin{aligned}
f = f(+TNS, +ves) &- f(-TNS, +ves) \\
&- [f(+TNS, -ves) - f(-TNS, -ves)] \qquad (7)
\end{aligned}
$$

[65] C. Huang and J. P. Charlton, *Biochemistry* **11**, 735 (1972).
[66] S. McLaughlin and H. Harary, *Biochemistry* **15**, 1941 (1976).

where each term indicates the observed fluorescence in the presence (+) or absence (−) of either TNS or vesicles. Above optical densities of about 0.08 OD unit, one must also correct for the inner filter effect.[67]

Experimental Considerations

It is important to determine the linear ranges for lipid and TNS concentrations. A linear relationship between net fluorescence and lipid concentration demonstrates that TNS adsorption onto the vesicles does not significantly lower the aqueous [TNS]. A linear relationship between net fluorescence and [TNS] demonstrates that the adsorbed TNS ions produce a negligible change in the surface potential. For egg PC sonicated vesicles, the net fluorescence versus lipid concentration relationship is linear up to ~0.1 mg/ml and the net fluorescence versus TNS concentration relationship is linear up to ~2 μM.[7] The ranges are higher for negatively charged vesicles and lower for positively charged vesicles (unpublished observations).[7] We have used this method to measure surface potentials of vesicles formed from mixtures of PC and either PS,[44,46] PG,[46] gangliosides,[44,46] or DDDAB (unpublished observations.).

We prepare sonicated vesicles by the following procedure. Lipids are dried down from their stock organic solvent solutions under nitrogen and then placed under vacuum overnight to remove any residual solvent. The dried lipids are resuspended in a buffered sodium chloride solution and sonicated (Branson sonifier) under nitrogen for about 30 min in an ice bath (or ~20° for mixtures containing bovine brain PS, which has a broad phase transition centered at 5°).[68] Ultracentrifugation at greater than 100,000 g for 90 min removes the multilamellar lipid vesicles and titanium fragments.[69] Phosphate analyses are used to determine the final lipid concentrations,[70] which are necessary for an accurate measure of ψ_0 [see Eq. (6)].

We add the fluorescent probe TNS (Sigma, St. Louis, Missouri) to the vesicle preparations from an aqueous stock solution. Teflon stir bars are not used when the vesicle concentration is low (<0.03 mg/ml) because lipids adsorb to them. We measure fluorescence intensities with a Spex Fluorocomp fluorometer (Spex Industries, Metuchen, New Jersey) equipped with a RCA 31034 photon-counting emission photomultiplier. The photon-counting mode has the sensitivity to measure the small fluo-

[67] J. R. Lakowicz, "Principles of Fluorescence Spectroscopy." Plenum, New York, 1983.
[68] K. Jacobson and D. Papahadjopoulos, *Biochemistry* **14**, 152 (1975).
[69] Y. Barenholz, D. Gibbes, B. J. Littman, J. Goll, T. E. Thompson, and F. D. Carlson, *Biochemistry* **16**, 2806 (1977).
[70] G. Rouser, S. Fleischer, and A. Yamamoto, *Lipids* **5**, 494 (1970).

rescence signals that are often encountered; the dark counts are typically less than 10 counts/sec. The intensity of the 150-W xenon light source is monitored through a rhodamine B filter by a reference photomultiplier. We use an excitation wavelength of 321 nm with a 2.5-nm bandwidth and an emission wavelength of 446 nm with a 5.0-nm bandwidth.

Egg PC bilayers or bilayers composed of mixtures of egg PC and anionic lipids are essentially impermeable to TNS. [Addition of TNS to a solution of sonicated vesicles produces an essentially instantaneous rise in fluorescence, with no further increase in TNS fluorescence for many hours. In contrast, we see a steady increase in fluorescence with a time constant of 30 min at 22° when sonicated egg PC vesicles are exposed to a membrane-permeable analog of TNS, 1-anilinonaphthalene 8-sulfonate (ANS).] Therefore, TNS probes the surface potential of the outer monolayer of the vesicle. TNS fluorescence increases with a time constant of 30 min at 22° with egg PC vesicles containing 5 mol% DDDAB (unpublished observations), suggesting that the probe leaks into these positively charged vesicles. However, we can determine the ψ_0 of the outer monolayer of these vesicles by measuring fluorescence intensities immediately after TNS addition (~2 min).

Two complications arise in determining ψ_0. First, since we measure the amount of adsorbed TNS, we must determine the amount of accessible lipid (i.e., that in the outer monolayer). Phosphate analyses measure only the total amount of phospholipid. To calculate ψ_0 from Eq. (6), we assume that the fraction of lipid in the outer monolayer is identical for neutral and charged vesicles. NMR studies on egg PC and egg PG vesicles support this assumption.[71] Second, there may be an asymmetric distribution of charged lipids in sonicated vesicles formed from a mixture of lipids.[71-75]

Work with TNS in model membranes reveals several problems that may arise in using it to study biological membranes. First, one must know the area of membrane exposed to TNS to calculate the value of ψ_0. It is difficult to measure this area and other approaches may be necessary (see later). Second, TNS tends to underestimate the surface potential of vesicles containing low concentrations of negative lipids.[16,44,46] The cause of this phenomenon is not known, but it will complicate measurements on biological membranes. Finally, membrane proteins have extra binding sites for both ANS[76] and TNS.[77] Some investigators have analyzed fluorescence

[71] D. M. Michaelson, A. F. Horowitz, and M. P. Klein, *Biochemistry* **12**, 2637 (1973).
[72] D. M. Michaelson, A. F. Horowitz, and M. P. Klein, *Biochemistry* **13**, 2605 (1974).
[73] J. A. Berden, R. W. Barker, and G. K. Radda, *Biochim. Biophys. Acta* **375**, 186 (1975).
[74] S. Masari, D. Pascolini, and G. Gradenigo, *Biochemistry* **17**, 4465 (1978).
[75] B. R. Lentz, D. R. Alford, and F. A. Dombrose, *Biochemistry* **19**, 2555 (1980).
[76] J. Slavik, *Biochim. Biophys. Acta* **694**, 1 (1982).
[77] J. H. Easter, R. P. Detoma, and L. Brand, *Biochim. Biophys. Acta* **508**, 27 (1978).

decay curves to distinguish between fluorophores bound to lipids and proteins.[78] Kramer discusses the use of TNS to estimate surface potentials in reconstituted membranes.[79]

Electron Paramagnetic Resonance

Principles

There are several advantages to using paramagnetic amphiphiles to estimate surface potentials: a wide range of probes can be synthesized, the sensitivity of the technique permits measurements on relatively small samples, and the turbidity of the samples is irrelevant. Figure 4 illustrates the four paramagnetic amphiphiles we used to measure membrane surface potentials. The distribution of these probes between the membrane and aqueous phases, which can be followed easily using EPR spectroscopy, depends on the membrane surface potential. The synthesis and detailed behavior of these probes in model and native membrane systems are described elsewhere.[80-84]

To estimate surface potentials using probes I–IV, we must quantify their phase partitioning in membrane suspensions. We have taken several approaches to obtain a value for λ, the ratio of label in the membrane and aqueous phases. One approach uses computer methods and our knowledge of bound and aqueous lineshapes to deconvolute a composite spectrum into a ratio of populations.[84,85] A simpler approach that is often adequate uses the fact that the resonance amplitude of the first-derivative EPR spectrum is proportional to the number of spins. The proportionality constant depends on the width of the resonance line. For the high-field nitroxide resonance, probes bound to the membrane have a broader resonance amplitude than probes in the aqueous phase. The amplitude of this resonance provides an excellent measure of the concentration of probe in the aqueous phase and only a minor correction is required for the bound

[78] K. Zierler and E. Rogus, *Biochim. Biophys. Acta* **514**, 37 (1978).
[79] R. Krämer, this volume [19].
[80] W. L. Hubbell, J. C. Metcalfe, S. M. Metcalfe, and H. M. McConnell, *Biochim. Biophys. Acta* **219**, 415 (1970).
[81] B. J. Gaffney and R. J. Mich, *J. Am. Chem. Soc.* **98**, 3044 (1976).
[82] J. D. Castle and W. L. Hubbell, *Biochemistry* **15**, 4818 (1976).
[83] R. J. Melhorn and L. Packer, this series, Vol. 56, p. 515.
[84] D. S. Cafiso and W. L. Hubbell, *Annu. Rev. Biophys. Bioeng.* **10**, 217 (1981).
[85] L. Berliner (ed.), "Spin Labeling Theory and Applications," Vols. 1 and 2. Academic Press, New York, 1979.

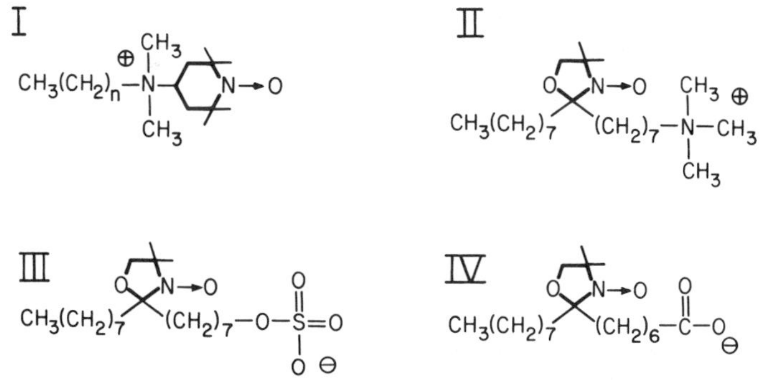

FIG. 4. Structures of four EPR probes. The synthesis and use of the alkylammonium I[82,83] and II,[81] alkyl sulfate,[84] and alkyl carboxylate probes are described elsewhere.

probe. In any membrane sample, the partitioning can be determined by calibrating this resonance amplitude in terms of the nanomoles of aqueous spin. These procedures are discussed elsewhere in this volume.[86]

Membranes have a low permeability to probes I–IV, which should therefore partition exclusively into the external membrane surface. We discuss exceptions to this assumption below. The electrostatic potential difference between the outer membrane surface and the bulk aqueous phase ψ_0 (see Fig. 1) is related to the phase partitioning of the probe, λ, by $\lambda = \lambda_0 \exp(-zF\psi_0/RT)$, where λ_0 is the partitioning of the probe in the absence of a potential. For studies on many model membrane systems, λ_0 is relatively easy to determine, because it is possible to form membranes from lipids with no net charge.[81,82] This approach assumes the concentration of lipid in the two samples is identical and all changes in the binding of the amphiphile to charged versus uncharged membranes are due to membrane surface potential.

In practice, it may be more accurate to determine ψ_0 by measuring λ as a function of the mass of lipid per unit volume (m_1/V_t). A plot of λ^{-1} versus V_t/m_1 yields a straight line of slope α. If α_0 is the slope in the absence of a potential, then $\psi_0 = -(RT/zF) \ln(\alpha/\alpha_0)$.[82]

It may be difficult to determine λ_0 (or α_0) in biological membranes, but several approaches are possible. If the native membranes are stable at high ionic strength, where ψ_0 is small, λ will be approximately equal to λ_0. Alternatively, we could determine λ over a range of ionic strengths.[82] If the assumptions inherent in the Gouy–Chapman theory are valid, the slope $(d\lambda/dC)$ can be used to deduce the surface charge density.

[86] D. S. Cafiso, this series, Vol. 172, p. 331.

Experimental Considerations

One can carry out partitioning experiments under a wide range of experimental conditions. One may vary both the vesicle concentration and the partition coefficient of the probe. We discussed elsewhere how the sensitivity of the technique depends on these parameters.[84] A major consideration is that the charged probes should not alter significantly the surface charge density. With this limitation in mind, we choose the length of the alkyl chain on the probe so that significant populations of both aqueous and membrane-bound probe are present. Finding a suitable vesicle concentration and probe partition coefficient is not difficult, because most EPR spectrometers easily detect 2 nmol of probe. In our measurement, probe:lipid ratios range from 1:1000 to 1:100 with 100-μl sample volumes.

When using the above procedures, we assume the probe does not cross the bilayer during the time course of the experiment. In model membrane systems this is usually the case. Probe I(8) permeates egg PC vesicles with a $t_{1/2} > 9$ hr at $0°$.[82] Probe IV (Fig. 4) should be used with caution since it may be partially protonated and this form can cross bilayers rapidly. In biological membranes, rapid transmembrane migration of these probes can occur. For example, the alkylammonium nitroxide I(8) equilibrates across photoreceptor disk membranes in several minutes.[84] We determined the initial phase partitioning of the probe by fitting a rapid mixing device to the EPR spectrometer.[87] The initial and equilibrium values for phase partitioning can be used to estimate both the internal and external surface potentials.[88,89]

Complications arise both when the charge is not distributed uniformly over the surface of the membrane and when the probe binds to proteins as well as to the lipid bilayer. In these cases, the lineshape of the bound probe can provide useful information about its distribution. Specific binding of the probe to a membrane protein, for example, is immediately apparent in the bound lineshape.[86]

Probes I–IV (Fig. 4) have been calibrated in model membrane systems under a range of conditions. Surface potentials were measured using probes I and II in sonicated egg PC vesicles containing phosphatidic acid, PA, and PS and in multilamellar egg PC vesicles containing cardiolipin.[46,81,82] These potentials were identical, within experimental error, to the potentials predicted by the Gouy equation [Eq. (1)]. We also measured potentials that agreed with the theory when we used probe I in positively

[87] D. S. Cafiso and W. L. Hubbell, *Biophys. J.* **39**, 263 (1982).
[88] S. A. Sundberg and W. L. Hubbell, *Biophys. J.* **41**, 192a (1983).
[89] S. A. Sundberg and W. L. Hubbell, *Biophys. J.* **49**, 553 (1986).

charged sonicated vesicles (unpublished observations). The negatively charged sulfate probe III and fatty acid probe IV, however, underestimated the potentials in PC:PS and PC:cardiolipin vesicles.[46,81] The reasons for this discrepancy, which we also observe with the negatively charged TNS label, are not understood.

Conductance Measurements

Principles

A phospholipid bilayer formed in a 0.1 M NaCl or KCl solution has a very low conductance (10^{-8} S/cm^2) because the Born energy required to move a sodium, potassium, or chloride ion from the aqueous to the membrane phase is very high.[90] An ionophore like nonactin functions as a carrier by solubilizing a potassium ion in the low dielectric interior of the bilayer. It does this in two ways. First, it increases the size of the ion (from about 0.2 to 0.5 nm), which reduces the Born energy. Second, it places a hydrophobic coat on the ion, which also increases its solubility in the bilayer. The anionic, A^-, forms of the weak acids that act as proton ionophores are soluble in the bilayer for the same two reasons: the charge is distributed over π electrons, which increases the effective radius of the ion, and the molecules contain hydrophobic moieties.

The detailed mechanism of action of the potassium[91-94] and proton[95,96] ionophores is considered elsewhere. The mechanism by which FCCP acts as a proton ionophore is illustrated in Fig. 5. We note here that for most solvent-free or decane-containing bilayer membranes, the interfacial reactions by which a proton combines with FCCP (Fig. 5) or a potassium ion combines with nonactin remain at equilibrium when a voltage is applied to the membrane. The rate-limiting step is the movement of the charged species across the membrane ($k_R[H^+]$, k_D, $k_{HA} > k'_A$ in Fig. 5). Thus the conductance is proportional to, and can be used to measure, the equilibrium concentration of ions in the membrane. For example, the conductance due to FCCP is proportional to both the mobility, m and the equilibrium concentration of anions in the membrane. The concentration of

[90] A. Parsegian, *Nature (London)* **221**, 844 (1969).
[91] S. G. A. McLaughlin, G. Szabo, and G. Eisenman, *J. Gen. Physiol.* **58**, 667 (1971).
[92] S. McLaughlin and M. Eisenberg, *Annu. Rev. Biophys. Bioeng.* **4**, 335 (1976).
[93] G. Szabo, G. Eisenman, S. G. A. McLaughlin, and S. Krasne, *Ann. N.Y. Acad. Sci.* **195**, 273 (1973).
[94] D. A. Haydon and V. B. Myers, *Biochim. Biophys. Acta* **307**, 429 (1973).
[95] S. G. A. McLaughlin and J. P. Dilger, *Physiol. Rev.* **60**, 825 (1980).
[96] R. Benz and S. McLaughlin, *Biophys. J.* **41**, 381 (1983).

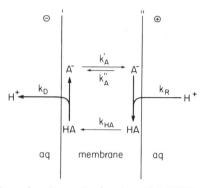

FIG. 5. Diagram illustrating the mechanism by which FCCP transports protons across bilayer membranes. As indicated by the circled positive and negative signs, a voltage exists across the membrane. The rate constants k_R and k_D refer to the heterogeneous reactions whereby a proton from the aqueous phase either recombines with or dissociates from an anion adsorbed to the membrane, the voltage-dependent rate constants k'_A and k''_A refer to the movement of A^- from the prime to the double prime and from the double prime to the prime interfaces, respectively, and the rate constant k_{HA} refers to the movement of HA between the two interfaces. (Reproduced from the *Biophysical Journal*, 1983, Vol. 41, pp. 381–398 by copyright permission of the Biophysical Society.)

anions in the membrane is equal to the product of the A^- concentration in the bulk aqueous phase, $[A^-]$, the partition coefficient of A^- into a neutral membrane, k, and the Boltzmann factor $\exp(F\psi/RT)$, where ψ is the potential in the center of the membrane, measured with respect to the bulk aqueous phase. Thus:

$$G = (F^2/d)mk[A^-] \exp(F\psi/RT) \tag{8}$$

where d is the thickness of the membrane.

Consider a specific example: measuring the effect of calcium on the surface potential of PS membranes.[40,91] If addition of calcium does not change the values of k and m, we can calculate the change in surface potential, $\Delta\psi$, from the measured ratio of the nonactin conductance in the presence, G', and absence, G, of calcium:

$$G'/G = \exp(-F\Delta\psi/RT) \tag{9}$$

Of course, calcium may change the value of k or m, and it is important to use both cations and anions as conductance probes.[91,93] Only changes in electrostatic potential change the conductance due to anions (e.g., FCCP) and cations (e.g., nonactin-K) symmetrically in opposite directions. Changes in other parameters (e.g., thickness, dielectric constant, fluidity of the membrane) change the conductance due to anions and cations in the same direction.[93]

Equation (9) illustrates that we can measure the change in the potential in the center of a membrane by measuring the ratio of the conductances; we can also determine the absolute value of the surface potential by combining conductance and electrophoretic mobility measurements. We measure the change in potential that occurs upon addition of ions that reduce the electrophoretic mobility and surface potential to zero. For example, 0.05 M calcium reduces the electrophoretic mobility of a PS vesicle formed in 0.1 M CsCl to zero, and higher concentrations make the vesicle positive.[40] Thus the surface potential of a PS planar bilayer membrane exposed to 0.05 M calcium is zero. The surface potential of a PS planar bilayer membrane in 0.1 M CsCl changes 105 mV upon addition of 0.05 M calcium.[40] Thus we conclude that the surface potential of a PS membrane is 0.1 M CsCl is -105 mV. This agrees with the measured value of the zeta potential, -75 mV, if the hydrodynamic plane of shear is 0.2 nm from the surface (see Fig. 1).[40]

Similar conductance measurements demonstrate the surface potential of PS membranes is -100 mV in 0.1 M KCl; analogous monolayer measurements indicate it is -80 mV in 0.1 M NaCl.[7] These surface potential results agree qualitatively with the zeta potential measurements of -70 mV (0.1 M KCl) and -60 mV (0.1 M NaCl) for PS vesicles, if the plane of shear is 0.2 nm from the surface.[7] The simplest interpretation of these results is that the monovalent cations bind to PS in the sequence Na > K > Cs.[7]

There is good agreement between the conductance measurements, which examine the potential within membranes, and the results obtained using the techniques described above, which estimate the potential at the surface and at the hydrodynamic plane of shear. We have used conductance measurements to examine the effect of salicylates,[97] local anesthetics,[98] aminoglycosides,[10] spermine,[10] and gangliosides[44,46] on the surface potential of membranes.

Experimental Considerations

We normally use a 1–2% solution of phospholipid in n-decane to form the membranes across a hole in the partition separating two aqueous compartments in a Teflon chamber. The bilayers are more stable if the hole is "prepainted" with a few microliters of a 1% solution of the lipid in chloroform and the organic solvent is allowed to dry under vacuum for 20 min. The chamber is then filled with an aqueous solution, less than 10 μl of

[97] S. McLaughlin, *Nature (London)* **243**, 234 (1973).
[98] S. McLaughlin, *in* "Progress in Anesthesiology" (B. R. Fink, ed.), Vol. 1, p. 193. Raven, New York, 1975.

the decane solution of lipid is applied to the hole with a Pasteur pipet, and the membrane is formed by sweeping an air bubble (on the end of the pipet) across the hole. The volume of the aqueous phases should be adequate to provide a sufficient reservoir of FCCP or nonactin, which are added to the aqueous phases in the form of concentrated ethanolic solutions. (The partition coefficients of FCCP and nonactin into decane are less than 10^4, so if 2 μl of decane solution is deposited on the hole, an aqueous phase of 20 ml provides a sufficient reservoir of carrier.)

The aqueous solutions must be well stirred and the membrane radius must be significantly larger than 0.1 mm, the thickness of the Nernstian unstirred layers. It would take many hours for nonactin to diffuse through these unstirred layers and equilibrate with the excess decane in the torus. This equlibration does not normally occur during an experiment and the torus acts as a sink for nonactin. On the other hand, the thin bilayer equilibrates with nonactin in the aqueous phase within a few minutes, as expected theoretically. The effect of stirring and membrane radius are discussed by Hladky.[99]

The equipment required for these measurements is simple and inexpensive. A high-impedance electrometer, a dc voltage source, a Teflon chamber, and a microscope to observe the membrane and measure its area are sufficient to make accurate measurements.

Acknowledgments

This work was supported by NIH Grants GM24971 and GM33837 and by NSF Grants BNS 85-01456 and BNS 86-04101.

[99] S. B. Hladky, *Biochim. Biophys. Acta* **307**, 261 (1973).

[17] Determination of Surface Potential of Biological Membranes

By HAGAI ROTTENBERG

Introduction

Biological membranes carry discrete electrical charges at their surface. These charges are carried by acidic phospholipids, gangliosides, and by membrane proteins, most of which are also negatively charged. Another significant contribution to the membrane surface charge, which tends to reduce the net negative charge, arises from the surface binding and absorption of various charged molecules and ions (mostly divalent cations). These discrete charges induce electrostatic potentials in the aqueous phase adjacent to the membrane surface.[1] Biological fluids and the cytoplasm contain relatively high concentrations of mono- and divalent cations, which would largely screen these charges and limit the effective surface potential to a distance of only a few nanometers. Nevertheless, this potential is believed to play a role in regulating a variety of important membrane processes.[1-3] The determination of the surface potential of artificial phospholipid membranes, both planar and vesicular, is described in great detail by Cafiso *et al.* in this volume [16]. The same principles and many of the same techniques are also used to determine surface potentials in biological membranes. However, because of the greater complexity and severe limitations imposed by the nature of biological membranes, a careful selection of an appropriate technique and a more prudent interpretation of the results are necessary.

To begin with, the distribution of phospholipids, proteins, gangliosides, and other charged species between the two membrane surfaces is asymmetric. Hence, the density and type of charges on one surface of the membrane are, in general, different from the opposite surface. It is, therefore, necessary to select methods that allow an estimation of the potential on each separate membrane surface. Second, in some biological membranes, protein clustering, aggregation, and lateral segregation of phospholipids may result in nonuniform distribution of charges. In such cases, a method which yields a measurement of an average delocalized surface potential may not be informative. Third, the charges on biological membranes may be located at various distances from the phospholipid head groups, which form the surface separating the membrane from the aqueous phase. Gangliosides and membrane proteins project into the aqueous phase up to a distance of several nanometers. A charge located at a distance from the phospholipid head group produces a potential which would be sensed differently by a probe located near the phospholipid head groups or on an adjacent site. For membranes in which a major fraction of their surface is occupied by proteins (e.g., the inner membrane of the mitochondria), the definition of a surface becomes blurred and the measured potentials would strongly depend on the localization of the probes. Finally, it must be remembered that in intact biological membranes, various processes, such as enzymatic reactions, active ion transport, and the existence of membrane potentials and pH gradients may greatly distort the information obtained in measurement of surface potentials. This may pose a severe

[1] B. H. Honig, W. L. Hubbell, and R. F. Flewelling, *Annu. Rev. Biophys. Biophys. Chem.* **15**, 163 (1986).

[2] J. Barber, *Biochim. Biophys. Acta* **594**, 253 (1980).

[3] C. M. Armstrong, *Physiol. Rev.* **61**, 44 (1981).

TABLE I
SURFACE POTENTIALS IN RAT LIVER MITOCHONDRIA

Method	Probe	External pH	ψ_s (mV)	Ref.
Electrophoretic mobility		7.4	-21	12
Fluorescence	ANS	7.2	-12	10
Phosphorescence	Tb^{3+}	7.2	-5 to -7	19
ESR	Cat_{12}	7.2	-20	24
ESR	Cat_{12}	7.2	-15	22

limitation since it is often desired to measure changes in surface charge in relation to membrane processes.

The application of these techniques to biological membranes, particularly as applied to the purple membrane of *Halobacterium halobium*, is discussed in some detail elsewhere in this series.[4] In this review, we emphasize the techniques which were used to study surface potentials in mitochondria.

Methods

Electrophoretic Mobility. This method, which depends on the measurement of the velocity of movement of charged particles in an electric field, is quite useful in model systems of uniform phospholipid particles. However, the method is of limited use with intact biological membranes. This is because the charge distribution and charge location in relation to the membrane–solution interface are not sufficiently uniform and the surface is not sufficiently smooth to satisfy the assumptions of the (Helmholtz–Smoluchowski) theory upon which the calculation of the ζ potential is based. (For further discussion, see Cafiso *et al.* [16], this volume.) The calculated ζ potential from measurement of electrophoretic mobility of mitochondria is compared with other methods in Table I.

Most other methods for the measurement of surface potential (and the calculation of surface charge density) depend on the potential-induced accumulation of probe ions near the charged surface. These methods are more applicable to biological membranes, but careful choice of the appropriate method is required. The most common difficulty in applying these methods to biological membranes is the rather high permeability of most biological membranes to charged molecules. A common error is to assume that a probe that was shown to be relatively impermeable in liposomes is

[4] B. Ehrenberg, this series, Vol. 127, p. 678.

similarily impermeable in all membranes. In general, biological membranes are orders of magnitude more permeable to ions and charged amphipilic molecules than liposomes. This is probably the result of the effect of integral membrane proteins on the membrane dielectric constant, which reduces the potential barrier in the core of the membrane, thus facilitating charge transfer.[5] This increased permeability is particularly problematic if it is desired to measure the surface potential in a system where transmembrane potentials, pH gradients, or other ion gradients exist or are induced during the measurements. In such cases, the measured response is often a result of the latter processes, which may affect both the probe distribution across the membrane and its binding to the membrane. Several cases of such complications are described below.

Methods for Measuring Directly the Accumulation of Probe Ion near the Membrane Surface. The electrostatic potential induced by the membrane surface charge leads to accumulation of ions of opposite charge according to the induced field. When the ions are at electrochemical equilibrium their excess concentration near the surface is related to the surface potential by the Boltzmann relationship:

$$C_s/C_0 = \exp[-zF\psi_s/RT] \qquad (1)$$

where C_0 is the concentration in the bulk phase, C_s is the concentration at the surface, z is the ion charge, ψ_s is the surface potential, F is the Farady constant, R is the gas constant, and T is the absolute temperature. Because the direct measurement of C_s/C_0 is usually quite difficult, these methods are not widely used.

Resonance raman dye probe. In this method the accumulation of a positively charged styryl dye (WW-638), which aggregates when present at high concentration near the membrane surface, is estimated from the aggregation-induced changes in its resonance raman spectrum. The method has been used to estimate the surface charge of purple membranes and liposomes reconstituted with bacteriorhodopsin.[4,6] Although the dye has been shown to be relatively impermeable in liposomes, its permeability in intact biological membranes is unknown and should be assessed before application (see below). The method is described in detail elsewhere in this series.[4]

Fluorescent acridines. The fluorescence of positively charged acridines is quenched when aggregated near the surface. This method is in

[5] J. P. Dilger, S. G. A. McLaughlin, T. J. McIntosh, and S. A. Simon, *Science* **206**, 1196 (1979).
[6] B. Ehrenberg and Y. Berezin, *Biophys. J.* **45**, 663 (1984).

principle identical to the method described above. 9-Aminoacridine quenching was used to probe surface potential in chloroplasts.[7,8] This probe should be used with the utmost caution. First, the probe permeates the membrane readily in its neutral form and distributes according to the pH gradient.[9] The accumulation of the probe in a compartment of low pH also results in fluorescence quenching. Thus the finding that the addition of salts to a chloroplast suspension increases fluorescence may be due, in part, to the effect of salts on the membrane Donnan potential, which is responsible for the existence of a substantial pH gradient in chloroplasts even in the dark.[9] Also, it is now recognized that 9-aminoacridine binds to membranes, which also induces quenching and modulates the surface charge, thus further complicating the interpretation of these results.

Methods Which Depend on the Enhanced Membrane Binding of a Charged Probe Due to Its Potential-Induced Accumulation Adjacent to the Membrane Surface. These methods are based on the same principle as the methods just described, except that instead of a direct determination of the probe accumulation near the surface, one measures the enhancement of probe binding that results from the increased accumulation. When the binding characteristic is analyzed, the surface potential can be estimated from its effect on the apparent dissociation constant of the probe

$$K_d = K_d^0 \exp(zF\psi_s/RT) \tag{2}$$

hence

$$\psi_s = (RT/zF) \ln(K_d/K_d^0) \tag{3}$$

where K_d is the apparent binding constant in the presence of surface charge (ψ_s). Depending on the membrane polarity (which is usually negative, ψ_s) and the probe charge (either $+z$ or $-z$), one would get enhancement of binding (lower K_d) or attenuation (higher K_d), respectively. This analysis holds for those probes that bind to saturable binding sites. In practice it is not always necessary to determine the values of K_d. Provided that this value is known approximately, and that the assay is carried out at a concentration well below the K_d, the extent of binding is proportional to the K_d. Hence, ψ_s can be calculated from the extent of binding (m) in the presence and absence (m_0) of surface potential

$$\psi_s = (RT/zF) \ln(m/m_0) \tag{4}$$

Of course, the latter equation can also be used with probes which do not

[7] W. S. Chow and J. Barber, *Biochim. Biophys. Acta* **591**, 82 (1980).

[8] R. W. Mansfield, H. Y. Nakatani, J. Barber, S. Mauro, and R. Lannoye, *FEBS Lett.* **137**, 133 (1982).

[9] H. Rottenberg, T. Grunwald, and M. Avron, *Eur. J. Biochem.* **25**, 54 (1972).

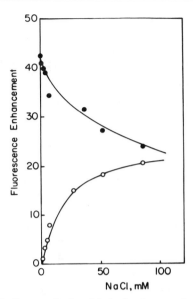

Fig. 1. ANS⁻ binding to mitochondria in the absence of surface potential.

show saturation binding, since in this case the extent of binding is proportional to the partition coefficient. Commonly, the determination of probe binding in the absence of surface charge (m_0 or K_d^0) is accomplished by the addition of high concentrations of salts which efficiently screen the surface charge and abolish surface potential. However, with many biological membranes this technique presents a problem since many ions, particularly cations, bind to the membrane surface, and thus directly modulate the surface charge. If the binding parameters of the screening ion are known, a correction can be made. A simple, but not a perfect, alternative way to overcome this problem is to perform a "null-point" titration of surface potential.[10] Figure 1 shows such a titration for the estimation of ANS⁻ binding to mitochondria in the absence of surface potential. The endogenous surface potential, which is negative, is first titrated with a monovalent salt solution, which reduces the negative charge of the membrane and enhances ANS⁻ binding (and fluorescence). Then the membrane charge is inverted by addition of tightly bound cations (Ca^{2+} or La^{3+}) and the positive surface potential is also titrated with the same salt. Since monovalent cation would not bind to the positively charged membrane, the asymptotic value of fluorescence approached in this titration more accurately reflects the binding of ANS⁻ in the absence of surface potential.

[10] D. E. Robertson and H. Rottenberg, *J. Biol. Chem.* **258**, 11039 (1983).

Another common complication in the application of these methods to biological membranes is the fact that the binding of the charged probe itself modulates the potential.[11] This can be taken into account when calculating charge density since the binding is usually measured. However, it is preferable, whenever possible, to choose a probe that can be measured with high sensitivity at surface concentration which is negligible when compared to the membrane charge density.

Nuclear magnetic resonance. In this technique, one measures the potential-induced binding of paramagnetic transition metal ions to the phospholipid head group from their effects on the ^{31}P NMR signal (this method is described in great detail by Cafiso et al. [16], this volume). The limitations of this technique, when applied to intact biological membranes, are considerable. First, the narrow ^{31}P NMR signal is obtained only in small sonicated vesicles. Second, some of the ions used are either transported by the Ca^{2+} carrier (i.e., manganese) or strongly inhibit various membrane processes, which may lead to further complications.

Fluorescent and phosphorescent probes. In this method, the binding of charged probes to the membranes is detected by the fluorescence (or phosphorescence) of the bound probe. To be useful it should be possible to distinguish clearly between the bound and free species. A large difference in quantum efficiency, emission wavelength, or both enables one to monitor changes in the amount of bound probe, free probe, or both.

1. *ANS and TNS.* These anionic fluorescent molecules (1-anilinonaphthalene 8-sulfonate and 2-(p-toluidinyl)naphthalene 6-sulfonate are the most widely used probes for surface potential determination[10-14] (see also Cafiso et al. [16], this volume). The hydrophobic anion fluorescence is greatly enhanced and exhibits a strong blue shift on binding to hydrophobic sites, thus enabling easy determination of the amount of bound probe. The use of TNS for surface potential determination in artificial phospholipid membrane is described elsewhere in this volume (Cafiso et al. [16]). Although these probes are also widely used in biological membranes, there are many complications which often result in erroneous determination. Unlike artificial phospholipid membranes, there are high-affinity sites for probe binding which are located on membrane proteins. These sites, which often contribute most of the fluorescence at low probe concentrations, may be located at various distances from the membrane surface, either in the

[11] S. McLaughlin and H. Harary, *Biochemistry* **15**, 1941 (1976).
[12] T. Aiuchi, N. Kamo, K. Kurihara, and Y. Kobatake, *Biochemistry* **16**, 1626 (1977).
[13] D. H. Haynes and H. Staerk, *J. Membr. Biol.* **17**, 313 (1974).
[14] R. Gibrat, C. Romieu, and C. Grignon, *Biochim. Biophys. Acta* **736**, 196 (1983).

membrane core or projecting into the aqueous bulk. Moreover, these sites are often heterogeneous, with different affinities and extremely different quantum yields. Some may not contribute at all to the fluorescence but strongly affect the binding parameters.[10] Finally, although these probes are relatively impermeable in liposomes they are quite permeable in biological membranes, and therefore respond to the membrane potential. Indeed, these same probes are used to estimate the transmembrane potential in vesicular membrane systems having a positive membrane potential[10,15] (see Rottenberg, this series, Vol. 172 [5]). Nevertheless, with caution, surface potentials can be estimated with these probes. A brief description of the estimation of surface potential in mitochondria with ANS is given below.

Method: mitochondria (2 mg protein/ml) are incubated in low-salt medium (0.25 M sucrose, 1 mM Tris–Cl, pH 7.2), rotenone (1 μM) is added to inhibit endogenous respiration, thus reducing the magnitude of the membrane potential which interferes with the measurement. The background fluorescence (F_b) of the mitochondrial suspension is measured first (excitation 400 nm, emission 470 nm). Then ANS (8 μM) is added and the fluorescence recorded again (F). This concentration of ANS is sufficient to give a reasonable signal although it is well below the determined K_d (47 μM); hence, the fluorescence is proportional to the binding. Moreover, since only a small fraction of the binding sites will be occupied at this concentration (much less than 1 nmol/mg protein), the binding of ANS itself should not affect the surface charge significantly. Next 150 mM NaCl is added to screen the surface charge, and the fluorescence is recorded again (F_0). The surface potential is calculated from these measurements as follows: (at room temperature) $\psi_s = 59 \log[(F\text{-}F_b)/F_0\text{-}F_b)]$. The value determined by this method in mitochondria at pH 7.2 is -12 mV.

Remarks: (i) a more accurate determination of F_0 can be accomplished by the null-point titration as described above (Fig. 1), although the difference is not always significant. (ii) The fluorescence of ANS increases slowly up to 20 min (at room temperature) after the addition of ANS. The slow component of the fluorescence is due to permeation of ANS into the mitochondrial matrix and binding to internal sites.[10] This leads to underestimation of the surface potential on the external surface. It is possible to overcome this problem by performing the assay at low temperatures (below 10°), which slows down ANS permeation considerably.[10] (iii) The mitochondrion also has an outer membrane. Thus, the surface potential determination is a weighted average of the two membranes. However, most of the high-affinity binding sites are on the inner membranes, hence the

[15] A. A. Jasaitis, V. V. Kuliene, and V. P. Skulachev, *Biochim. Biophys. Acta* **234**, 177 (1971).

contribution from the outer membrane is negligible.[10] A more accurate determination of the surface potential on the cytoplasmatic surface of the inner membrane can be performed on mitoplasts in which the outer membranes have been removed. (iv) In our view it is not possible to estimate surface potential changes from ANS fluorescence changes whenever a large modulation of membrane potential occurs simultaneously,[12] because fluorescence changes due to probe transport across the membrane are very large. However, when the surface potential was determined by the procedure described above, in a state having a large but constant transmembrane potential, no significant difference between the two states was observed.[10]

2. Tb^{3+}. This phosphorescent cation of the lanthanide series can also be used to measure surface potentials.[16] It binds with high affinity to high-affinity Ca^{2+} binding sites. This probe also binds to membrane proteins so that the same reservations discussed above in regard to ANS apply here. However, in mitochondria, at least, there are fewer types of binding sites and these sites are easily separated. The protein-bound probe is estimated from the phosphorescence, which is excited by fluorescence energy transfer from the protein. With this probe it is therefore possible to measure surface potential at specific sites. This may result in more interesting and biologically relevant information. The main drawback of this probe is its avidity in forming insoluble complexes with a variety of anions, particularly inorganic phosphate and phosphate esters. This can present a serious problem, as it severely limits the choice of medium composition. In mitochondria, the problem is particularly acute since the matrix contains large amounts of phosphate, which slowly leak out during the course of the measurement. We have solved this problem by briefly treating the mitochondria with N-ethylmaleimide (NEM) (40 nmol/mg protein, 1 min at room temperature) to inhibit phosphate transport. Alternatively, this problem can be minimized by performing the assay quickly at low temperatures (below 10°), or by depleting the mitochondria of phosphate before the determination. The problem does not exist with submitochondrial particles or other membrane preparations that do not leak phosphate. In mitochondria there are two sites for Tb^{3+} binding. One with high affinity (1.5 μM) and low capacity and the other with relatively lower affinity (29 μM) and higher capacity. We found the low-affinity site more suitable for surface potential determination. The sites are easily distinguishable because of different phosphorescence decay times. With a commercial phosphorescence spectrometer (Perkin–Elmer LS5), it is possible to select a window in which

[16] K. Hashimoto and H. Rottenberg, *Biochemistry* 22, 5738 (1983).

only the phosphorescence of the slow decay site is detected. We describe briefly the method as applied to surface potential determination in submitochondrial particles (SMPs).

Method: 3 ml of SMP suspension (0.1 mg protein/ml) containing 0.25 M sucrose, 10 mM MOPS–Na$^+$ (pH 7.2), and 1 μM rotenone in a quartz curvette is placed in the spectrofluorometer (Perkin–Elmer LS5). The instrument is used in its phosphorescence detection mode. Delay time (t_d) is set to 1 msec and gating time (t_g) is also 1 msec. Excitation is at 280 nm and the emission at 543 nm. After measuring the background phosphorescence (which is subtracted from all further measurements), 2 μM TbCl$_3$ is added and the phosphorescence measured again (P). Because the concentration is very low compared to the K_d of this site, only a very small fraction of the cation is bound. Thus the phosphorescence is proportional to the binding and the effect on surface charge at these sites is minimal. Finally, 150 mM NaCl is added and the phosphorescence is read again (P_0); surface potential is calculated from the following equation (at room temperature):

$$\psi_s = (-59/3) \log (P/P_0) = -19.7 \log P/P_0 \tag{5}$$

The surface potential measured in submitochondrial particles under these conditions is -6.5 mV.

Remarks: notice that since the Tb^{3+} cation is triply charged, a threefold change in phosphorescence is expected (per millivolt) when compared to ANS$^-$. This makes this probe more sensitive to small changes in surface potentials. The lower potential measured, compared with other probes, probably indicates that the Tb^{3+} is bound to a protein which projects out of the phospholipid surface. The method used for mitochondria is identical, except for the pretreatment with NEM as described above.

ESR probes. In this method a charged aliphatic compound which is labeled with a spin moiety (usually doxyl) is used to measure the free and bound species from the differences in their ESR spectra (the application of this method to liposomes is described in detail elsewhere in this volume; see Cafiso *et al.* [16]). The most serious problem in applying these probes to biological systems is the difficulty in finding a probe which would be sufficiently hydrophobic to allow detection of the signal from the bound species at practical membrane concentrations and at the same time be impermeable. Since permeability increases with partition coefficient, and the permeability of biological membranes is much greater than that of liposomes, it has not been possible, as yet, to identify with certainty such probes. The widely used Cat$_{12}$ probe (4-N,N,N-dimethyl-N-dodecyl)ammonium-2,2,6,6-tetramethylpiperidine-1-oxyl), which is sufficiently impermeable in liposomes,[17] was found to be permeable in red blood cells[18] and even more so in mitochondria.[19,20] Thus its use for determination of putative potential induced changes in surface charge[21,22] could not be justified, since the probe clearly responds to the induced membrane potential.[19] Another drawback of the spin probe technique is its relatively low signal sensitivity as compared to fluorescent probes. Finally, the nitroxy group is reduced quickly[23] if the membrane contains even very weak redox activity. The latter problem can be partially solved by lowering the temperature, and the addition of specific enzyme inhibitors and strong oxidants such as ferricyanide. An advantage of this probe over the above fluorescent probe is that it partitions into the lipid phase, thus allowing a determination of the surface potential adjacent to the phospholipid head group. A brief description of the method as applied to mitochondria is described below.

Method: mitochondria (8 protein mg/ml) are suspended in a medium composed of 0.25 M sucrose, 1 mM Tris–Cl (pH 7.2), 4 μM rotenone, 0.3 μg/ml antimycin, 1 mM $K_3Fe(CN)_6$), and 50 μM Cat$_{12}$ ($T = 10°$). Another sample is incubated in an identical medium with the addition of 150 mM NaCl. The ESR spectra of the two samples are taken. The apparent partition coefficient (P) of the probe in the low-salt medium and the high-salt medium (P_0) is calculated from the ratio of the amplitude (or integral) of the signal of the bound probe (h_H) and the free probe (h_P) (Fig. 2). Thus $P = h_{H/h_p}$ and hence, at 10°

$$\psi_s = -57 \log(P/P_0) \tag{6}$$

Surface pH indicators. This method is based on the fact that the existence of a surface charge would change the concentration of protons near the surface. A pH indicator (which binds protons) that is located at or near the surface can report these changes. The pH indicator -4-heptadecyl-4-hydroxycoumarin (pK 8.65) was used for the determination of surface potential in liposomes[24,25] and stomatitis virus.[26]

[17] D. J. Castle and W. L. Hubbell, *Biochemistry* **15**, 4818 (1976).
[18] G. S. B. Lin, R. I. Macey, and R. J. Mehlhorn, *Biochim. Biophys. Acta* **732**, 683 (1983).
[19] K. Hashimoto, P. Angiolillo, and H. Rottenberg, *Biochim. Biophys. Acta* **764**, 55 (1984).
[20] L. Wojtczak and A. Szewczyk, *Biochem. Biophys. Res. Commun.* **136**, 941 (1986).
[21] A. T. Quintanilha and L. Packer, *FEBS Lett.* **78**, 161 (1977).
[22] A. T. Quintanilha and L. Packer, *Arch. Biochem. Biophys.* **190**, 206 (1978).
[23] A. Quintanilha and L. Packer, *Proc. Natl. Acad. Sci. U.S.A.* **71**, 570 (1977).
[24] P. Fromherz, *Biochim. Biophys. Acta* **323**, 326 (1973).
[25] M. S. Fernandez, *Biochim. Biophys. Acta* **646**, 23 (1981).
[26] R. Pal, W. A. Petri, Jr., Y. Barenholz, and R. R. Wagner, *Biochim. Biophys. Acta* **729**, 185 (1983).

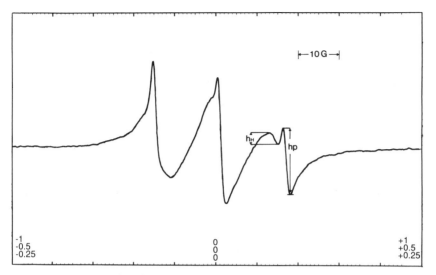

FIG. 2. ESR spectra of bound and free probe.

To demonstrate the various results that can be obtained by the use of these various methods in a single membrane preparation, we show in Table I (p. 366) estimates of surface potential in rat liver mitochondria obtained by different methods. As discussed above, the various estimates which result from different locations of the probe may be useful in establishing the heterogeneous distribution of surface charge on the mitochondrial membrane surface. Thus, the significantly lower potential detected by Tb^{3+} suggests a protein binding site which projects out of the membrane surface. Alternatively, lateral protein and lipid segregation may lead to uneven distribution of charges at the surface. These alternatives are being explored by various means and may lead to better understanding of the topology of the mitochondrial inner membrane. The application of these probes to the measurements of mitochondrial surface charge in the presence and absence of membrane potential suggests that there is no significant change in surface charge due to energization. The reported changes[15,21] were shown to be due to the effect of membrane potential on the transport and distribution of surface charge probes.[27]

Acknowledgment

Supported by PHS Grant #GM-28173.

[27] H. Rottenberg, in "Ion Interactions in Energy Transfer Biomembranes" (G. C. Papageorgiou, J. Barber, and S. Papa, eds.), p. 29. Plenum, New York, 1986.

[18] Lipid Coumarin Dye as a Probe of Interfacial Electrical Potential in Biomembranes

By PETER FROMHERZ

Introduction

An electrical potential Ψ_0 at the surface of a biomembrane affects the concentration c_i of ions at the membrane [Eq. (1a)], it controls the electrical field E_m in the membrane [Eq. (1b)], and it determines the repulsive force F_{el} between two membranes [Eq. (1c)] (c_w, bulk concentration; z, charge number; e_0, proton charge; kT, thermal energy; Ψ_{0e} and Ψ_{0i}, external and internal surface potential; d_m, membrane thickness; V_m, membrane potential; F_{el}^0, maximal force; d_w, membrane distance; $1/\kappa$, Debye length). Thus the surface potential Ψ_0 modulates important cellular processes such as ion transport, gating of pores, membrane contact, and fusion.

$$c_i = c_w \exp(-ze_0\Psi_0/kT) \tag{1a}$$

$$E_m = (V_m + \Psi_{0i} - \Psi_{0e})/d_m \tag{1b}$$

$$F_{el} = F_{el}^0 \, tgh^2(e_0\Psi_0/4kT) \exp(-\kappa d_w) \tag{1c}$$

We describe a method which provides a reliable estimate of the surface potential. It is derived from the changed interfacial concentration of the proton [Eq. (1a)]. This shift of "local pH" leads to a corresponding shift of the "apparent pK" of a pH indicator bound to the surface.[1-3] As a suitable probe we have introduced 4-alkyl-7-hydroxycoumarin (Fig. 1).[4] Ψ_0 is obtained from the shift of pK_{app} as referred to the pK_{app}^0 in a membrane of vanishing potential [Eq. (2)].[4-7]

$$\Psi_0 = -2.3kT/e_0(pK_{app} - pK_{app}^0) \tag{2}$$

In fact, however, the method does not rely on such a dubious concept as a local concentration. It is related directly to the physical notion of the interfacial potential as "the work of transfer per unit charge from bulk water to the interface in the limit of an infinitesimally small probe charge."

[1] G. S. Hartley and J. W. Roe, *Trans. Faraday Soc.* **36**, 101 (1940).
[2] P. Mukerjee and K. Banerjee, *J. Phys. Chem.* **68**, 3567 (1964).
[3] M. Montal and C. Gitler, *Bioenergetics* **4**, 363 (1973).
[4] P. Fromherz, *Biochim. Biophys. Acta* **323**, 326 (1973).
[5] P. Fromherz and B. Masters, *Biochim. Biophys. Acta* **356**, 270 (1974).
[6] M. S. Fernandez and P. Fromherz, *J. Phys. Chem.* **81**, 1755 (1977).
[7] P. Fromherz and R. Kotulla, *Ber. Bunsenges. Phys. Chem.* **88**, 1106 (1984).

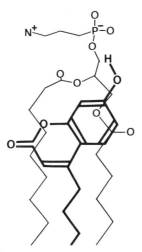

Fig. 1. 4-Alkyl-7-hydroxycoumarin in its approximate position relative to the head group of dipalmitoylphosphatidylcholine. The binding site of the proton probes the electrical potential. It is located close to the phosphate, in front of the glycerol backbone, behind the choline moiety. The position is invariant for substituents from nonyl to nonadecyl.

In the first part of the present chapter we discuss the theory of the pH indicator method, we derive the prerequisites of a potential probe, and we consider in detail the features of the lipoid coumarin. In the second part we describe the application of the method: fluorescence titration, evaluation of the surface potential, and interpretation.

Theory

Let us consider the cyclic process of phase transfers and chemical reactions sketched in Fig. 2. The total work vanishes according to the second law of thermodynamics [Eq. (3a)]: *Step 1:* We transfer a negative base from the bulk aqueous phase to the interface (work W_θ). *Step 2:* We discharge the base in the membrane by protonation from bulk water (work $- W_{mw}$). *Step 3:* We transfer the neutral acid into a reference interface without surface potential (work W_0). *Step 4:* We recharge the acid in the reference interface by deprotonation (work W_i). *Step 5:* We transfer the negative base from the reference interface to bulk water (work $- W_\theta^r$).

$$W_{mw} - W_i = W_\theta - W_\theta^r + W_0 \tag{3a}$$

$$\Delta G_{mw}^0 - \Delta G_i^0 = W_\theta - W_\theta^r + W_0 \tag{3b}$$

$$2.3kT(pK_{mw} - pK_i) = W_\theta - W_\theta^r + W_0 \tag{3c}$$

$$2.3kT(pK_{mw} - pK_i) = e_0\Psi_0 + (W_{pol} - W_{pol}^r) + (W_{mol} - W_{mol}^r) + W_0 \tag{3d}$$

$$2.3kT(pK_{mw} - pK_i) = -e_0\Psi_0 \tag{3e}$$

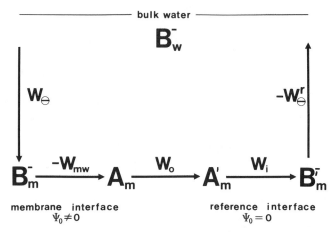

FIG. 2. Thermodynamic cycle connecting processes of charging/discharging (by deprotonation/protonation) and of phase transfer for acid A and base B^- of a pH indicator in a membrane interface (A_m, B_m^-) with an electrical potential Ψ_0, in a reference interface (A_m', B_m') with $\Psi_0 = 0$ and in bulk water (B_w^-). The cycle defines a measuring process for the electrical potential Ψ_0. For details see the text.

We relate the left-hand side of Eq. (3a) (work of charging/decharging) to pK values and the right-hand side of Eq. (3a) (work of transfer) to the surface potential.

The work of charging in the two interfaces is evaluated as the free energy of deprotonation of the neutral acid molecules. It is given by the standard Gibbs energy ΔG_{mw}^0 characterizing a two-phase dissociation of membrane-bound acid into membrane-bound base and proton in bulk water $A_m = B_m + H_w$.[6] With $\Delta G_{mw}^0 = \Delta G_i^0$ for the reference interface, the difference $\Delta G_{mw}^0 - \Delta G_i^0$ reflects the difference of work of charging if we consider the same standard state for both interfaces [Eq. (3b)].

The standard free energy ΔG_{mw}^0 is related to the equilibrium constant K_{mw} of two-phase dissociation, with $K_{mw} = K_i$ for the reference interface [Eq. (3c)]. It is given by the activities of the reactants,[6] i.e., by the pH_w in bulk water and by the degree of dissociation β_m of the probe in the interface at vanishing density [Eq. (4)].

$$pK_{mw} = pH_w - 2.3kT \log \beta_m/(1 - \beta_m) \tag{4}$$

We consider now the work of transfer. We partition W_θ into three components [Eq. (3d)]. (1) The negative probe is transferred against the electrostatic potential (work $- e_0\Psi_0$). (2) The finite charge of the probe induces a changed dielectric polarization of water and interface (work

W_{pol}). (3) The probe exhibits different molecular interactions with water and interface (work W_{mol}). A similar consideration holds for W_{θ}^r, where the electrostatic potential is omitted per definition.

We define now a reference interface such that the pK shift reflects exclusively the potential [Eq. (3e)]: (1) Its chemical structure must resemble that of the interface probed, such that the molecular interaction with the charged and with the uncharged probe is identical ($W_{mol} = W_{mol}^r$, $W_0 = 0$). (2) Its dielectric polarization by the finite charge of the probe must resemble that of the interface probed ($W_{pol} = W_{pol}^r$).

In summary, for a pH indicator bound to a membrane the observable apparent pK is a genuine pK_{mw} of a two phase dissociation, i.e., Eqs. (2) and (3e) are identical. To derive the interfacial potential from the pK shift we need two prerequisites: (1) a molecular probe fixed to the interface with sensitive detection of its degree of dissociation and (2) a reference interface of vanishing potential with dielectric polarizability and molecular interactions identical to the probed interface. For a unique interpretation we need further a defined location of the probe in the interface with minimal perturbation of the surround.

The Probe

Binding

The distribution coefficient of 4-heptyl-7-hydroxycoumarin in the system lecithin/water is 13,500.[8] It increases per addtional CH_2 by a factor of 10.[8] The pK_{mw} in lecithin/water is independent on the chain length above nine carbons.[8] Thus substituents such as nonyl to nonadecyl are suitable for a unique interfacial binding.

Location

The binding site of the proton in 4-alkyl-7-hydroxycoumarin is located close to the phosphate of lecithin (Fig. 1) as determined by NMR, evaluating the shielding of the lipid protons by the ring current of the dye.[8] Independent of the chain length, the probe is located beyond the electrical dipoles of the glyceride, within the dipole of phosphatidylcholine for nonyl to nonadecyl substituents. This assignment is confirmed by the change of pK_{mw} with changed packing of phosphatidylcholine[4,8,9] and by the invariance of pK_{mw} after a replacement of the esters by ether groups.[8] We

---

[8] P. Fromherz and A. Haase, manuscript in preparation.
[9] J. Teissie, J.-F. Tocanne, and W. G. Pohl, *Ber. Bunsenges. Phys. Chem.* **82**, 875 (1978).

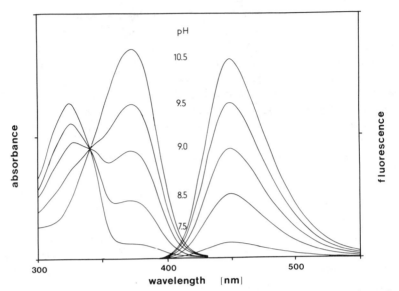

FIG. 3. Spectra of absorbance and fluorescence (excitation at 370 nm, uncalibrated) of 4-undecyl-7-hydroxycoumarin in micelles of Triton X-100 (5 mM) as a function of pH$_w$ of bulk water (5 mM Tris). Molar ratio of probe to surfactant, 1 : 100.

assume a similar position in charged membranes with the probe located in the plane of interfacial charges.

Perturbation

A local perturbation of the interface by the chromophore is unavoidable, although the coumarin is a fluorescent probe of minimal size. The potential determined refers to the geometry of the interfacial structure around the probe. The Coulomb potential of the finite probe charge itself does not introduce an intrinsic perturbation, as it is eliminated in the procedure of the measurement [Eq. (3)]. A collective effect of the probe molecules by their charge is avoided by low dotation.

pK Value

7-Hydroxycoumarin in water is deprotonated with a pK_w = 7.75.[10-12] In its acidic form it absorbs light around 325 nm, in its basic form around 370 nm (Fig. 3).[10,11] Both species emit the fluorescence of the base around 450 nm because of the strong acidity of the excited state (pK_w^* = 0.45).[12]

[10] R. F. Chen, *Anal. Lett.* **1**, 423 (1968).
[11] G. J. Yakatan, R. J. Juneau, and S. G. Schulman, *Anal. Chem.* **44**, 1044 (1972).
[12] V. Mikes, *Collect. Czech. Chem. Commun.* **44**, 508 (1979).

Under usual conditions the excitation spectrum reflects the acid–base equilibrium of the ground state. Exciting the base near 370 nm the fluorescence intensity normalized to the intensity at high pH_w reflects the dissociation degree (Fig. 3). The spectroscopic features of the coumarin (spectrum, lifetime, quantum yield) are rather insensitive to the environment.[13,14]

Reference Interface

We cannot assume for every neutral membrane a vanishing surface potential nor an identity of polarity and molecular interactions as compared to a charged interface, as seen by the following examples. (1) In lecithin the pK_{mw} of the coumarin is shifted to higher values by enhanced packing[4,8,9] due to a negative potential within the oriented head groups.[8] (2) The potential in micelles of lauryl sulfate evaluated relative to Triton X-100 varies from −125 to −300 mV, depending on the lipid pH indicator used,[8] as the interaction of the two interfaces with the different probes is not identical.

As a standard for the evaluation of the surface potential in biomembranes we suggest $pK_i = 8.8$ for two reasons: (1) For micelles of lauryl sulfate we have avoided the use of reference interface by applying two indicators, the 7-hydroxycoumarin with a negative base and a 7-aminocoumarin with a positive acid.[6] The potential determined is identical to that obtained with the hydroxycoumarin alone taking Triton X-100 as a reference.[6] The agreement indicates a similar nature of the highly charged and the neutral interface with respect to this indicator type. (2) The pK_{mw} in a series of neutral systems is found rather invariant for the hydroxycoumarin between 8.6 and 9.0. The systems studied include micelles and foam films of Triton X-100,[6,7] monolayers of methyl stearate and cholesterol,[4] and bilayers of phosphatidylglycerol[8] and egg lecithin.[15-17] Here $pK_i = 8.8$ is a representative average. Note the exception of dipalmitoylphosphatidylcholine.[4,8,18] The use of a standard $pK_i = 8.8$ may introduce an error of 5–10 mV [Eq. (3e)], as the actual pK_i for different neutral systems varies within 0.1–0.2 units and a similar variation may be hidden for charged systems.

[13] G. W. Pohl, *Z. Naturforsch.* **C31**, 575 (1976).
[14] R. Pal, W. A. Petri, V. Ben-Yashar, R. R. Wagner, and Y. Barenholz, *Biochemistry* **24**, 573 (1985).
[15] M. S. Fernandez, *Biochim. Biophys. Acta* **646**, 23 (1981).
[16] W. A. Petri, R. Pal, Y. Barenholz, and R. R. Wagner, *Biochemistry* **20**, 2796 (1981).
[17] R. Krämer, *Biochim. Biophys. Acta* **735**, 145 (1983).
[18] S. Lukacs, *J. Phys. Chem.* **87**, 5045 (1983).

Experimental

4-Heptadecyl-7-hydroxycoumarin is available from Molecular Probes Inc. (Junction City, Oregon). It is dissolved in ethanol or tetrahydrofuran. The dye is added to the sample at a ratio of dye/lipid below 1 : 1000.

Fluorescence Titration

The pH of the sample is adjusted by HCl and NaOH, avoiding dilution and change of ionic strength. A weak buffer is required. The pH is measured by a glass electrode, carefully calibrated under the conditions of pH adjustment (e.g., stirring) relative to reference buffers around the pK value expected.

The fluorescence intensity at 450 nm is observed, with excitation around 370 nm. To scale the fluorescence signal a preliminary sample is adjusted to pH 12. The background is determined at pH 2. The titration is started at the lowest pH tolerated by the sample, with the pH changed in steps of a few tenths of a pH unit up to the highest pH tolerated by the sample. The titration should be completed within a short time without changing any parameter of the fluorometer. Automatic titration has been applied successfully.[8] The wider the range of titration, the more precise is the pK_{mw} evaluated and the better is the control for a single acid–base equilibrium.

Evaluation of Titration

We obtain the dissociation degree β_m at a certain pH$_w$ from the fluorescence intensity I by subtracting the signal I_0 in the limit of low pH as a background and by normalization to the signal I_M in the limit of high pH [Eq. (5)].

$$\beta_m = (I - I_0)/(I_M - I_0) \tag{5}$$

The data of β_m are plotted versus pH$_w$ and fitted according to Eq. (4). Often we cover a limited range of pH, such that I_0 and I_M are not measured. In that case we plot log $\beta_m/(1 - \beta_m)$ versus pH$_w$ with best estimates of I_0 and I_M. We adjust them I_0 and I_M until the data are matched by line of unity slope [Eq. (4)]. Using a microcomputer the procedure is fast and reliable if the data are part of an ideal titration curve.

The determination of a pK value is not trivial. The values reported for 7-hydroxycoumarin in water and for 4-alkyl-7-hydroxycoumarin in several lipoid systems, respectively, differ by 0.2 pH units, and in some cases even more. Whether the calibration of the glass electrode is an art or whether the impurity of the material plays a role is unclear. In our laboratory the pK of the lipid coumarin in micelles of Triton X-100 is found to be pK_{mw} = 8.85 ± 0.05 relative to NBS buffers, constant over the years and independent of electrodes and persons involved.[6-8]

Excitation Spectra

A complete titration of a single sample provides the best determination of a pK. However, sometimes the sample has to be changed within a titration or a sample is too unstable for titration. In that case we determine in a first step a sequence of excitation spectra (range from 275 to 425 nm) of the coumarin in a model system as micelles of Triton X-100 (Fig. 3). From the fluorescence at 450 nm, at an excitation of 370 nm, the dissociation degrees are assigned as described. The dissociation degree is plotted versus the ratio of excitation around 370 nm (base) and excitation around 325 nm (acid). (Without spectral correction this calibration curve is valid only for the particular spectrometer used.) In the second step the excitation spectrum of the sample is measured at one pH$_w$ selected not too far from the pK expected. Interpolation of the calibration spectra provides an estimate of the dissociation degree β_m and of pK_{mw} [Eq. (4)].

Interpretation

Interfacial potentials arise from oriented molecular dipoles and from electrical double layers of interfacial charges, with their counterions being chemically bound, physically adsorbed, or diffusely accumulated.[19] Because the potential probe is located in the plane of the phosphates of the usual lipids (Fig. 1),[8] the molecular dipoles of glycerides are irrelevant. We have to consider only effects of the electrical double layer.

Lecithin

Pure lecithin forms electrically neutral membranes. In bilayers and monolayers of dipalmitoylphosphatidylcholine, a pK value of pK_{mw} = 10.2 – 10.5 is reported.[4,8,18] Increasing the salt concentration from 1 to 100 mM leads to a pK_{mw} drop from 10.2 to 9.7.[8] The pK_{mw} is unchanged for ether analogs, where the possible contribution by traces of negatively

[19] S. McLaughlin, *Curr. Top. Membr. Transp.* **9**, 71 (1977).

charged acid formed by hydrolysis is excluded.[8] Addition of cholesterol to membranes leads to $pK_{mw} = 9.0$.[4,8] Reduction of packing in monolayers leads to $pK_{mw} = 9.0$.[9] For membranes of egg lecithin with its unsaturated chains a pK value of $pK_{mw} = 8.65 - 8.8$ is reported.[15-17]

We assign the shift from the standard $pK_i = 8.8$ to $pK_{mw} = 10.2$ in dipalmitoylphosphatidylcholine to an electrical potential $\Psi_0 = -80$ mV caused by the double layer of the phosphatidylcholine head groups inclined with respect to the interface. We estimate the dislocation d of choline and phosphate normal to the plane using a condensator model [Eq. (6)].

$$d = \epsilon_0 \epsilon \Psi_0 a_M / e_0 \tag{6}$$

With an area per molecule of $a_M = 50$ Å^2 using the permittivity ($\epsilon_0 \epsilon$) of water we obtain $d = 2$ Å. A more involved procedure should account for the actual molecular packing of the interface made of head groups, water, and electrolyte. A preliminary treatment of the difficult problem leads to a similar result.[8]

Charged Interfaces

In negatively and positively charged biomembranes, lipid bilayers, monolayers, micelles, and foam films, considerable shifts $pK_{mw} - pK_i$ are reported as being sensitive to salt concentration.[4-7,15-18] The electrical potential is assigned to the electrical double layer spanned by the interfacial charge and the diffuse cloud of counterions in the adjacent electrolyte. Using the Poisson–Boltzmann equation (Gouy–Chapman model), an interfacial charge density σ_{GC} is assigned to the measured potential Ψ_0 [Eq. (7)].[20] This structural interpretation of the surface potential is consistent if σ_{GC} is independent of the concentration c of $1 - 1$ electrolyte.

$$\sigma_{GC} = \sqrt{8\epsilon_0 \epsilon kTc} \, \sinh(e_0 \Psi_0 / 2kT) \tag{7}$$

The charge density σ_{GC} may be compared with the charge density as calculated from the chemical composition or the physical structure of the interface.[5,7,15,21] The ratio of the measured and calculated charge density — the dissociation degree of the interface — is close to 1 in some monolayers and bilayers.[5,15] In surfactant systems dissociation degrees distinctly below 1 are found.[6,7,21]

[20] E. J. W. Verwey and J. T. G. Overbeek, "Theory of the Stability of Lyophobic Colloids." Elsevier, New York, 1948.

[21] J. Frahm, S. Dieckmann, and A. Haase, *Ber. Bunsenges. Phys. Chem.* **84**, 566 (1980).

The consistent interpretation of surface potentials in membranes by the Gouy–Chapman model is well established,[19] but is still astonishing. The deviations expressed by a low dissociation degree may be assigned to a penetration of the diffuse double layer into the plane of the head groups rather loosely packed in micelles and foam films.[7,22]

A contribution of a potential drop in a double layer of adsorbed counterions (Helmholtz potential) has never been observed in interfaces of lipids or surfactants. A dislocation of fixed charges and of "condensed" counterions is not visible. Only with macromolecular counterions such as proteins do the coumarin probes detect a potential within a "Helmholtz layer."[23]

Problems

In the inhomogeneous system of a biomembrane the lateral location of the probe is unknown. Although it may be assumed that the dye is localized in patches of lipid, an unspecific or specific adsorption to membrane proteins is not excluded. No information is available on that point. As biologically sensitive functions are localized in the core of membrane proteins, the correlation of the probed potential and the potential relevant for function may be indirect.

In a mixed membrane the negative probe may accumulate positive components in a kind of two-dimensional Debye–Hückel cloud leading to an enhanced local positive charge density. This finite charge effect is not compensated by the reference to the neutral interface [Eq. (3)]. With negative components an analogous reduction of the local negative charge density is expected to be less pronounced.

The surface potential of a membrane is pH dependent if interfacial acids and bases are present.[24] If their pK_{mw} is in the range of the pK_{mw} of the probe, the titration curve is the superposition of a spectrum of ideal titrations with defined pK_{mw}. A check for such a modulated titration allows the detection of pH-sensitive membrane functions.

Summary

The coumarin method has been applied for the determination of surface potentials in a few biological membranes, such as thylakoids,[25] elec-

[22] P. Fromherz, *Ber. Bunsenges. Phys. Chem.* **85**, 891 (1981).
[23] P. Fromherz, *in* "Electron Microscopy at Molecular Dimensions" (W. Baumeister and W. Vogell, eds.), pp. 338. Springer-Verlag, Berlin and New York, 1980.
[24] T. A. J. Payens, *Philips Res. Rep.* **10**, 425 (1955).
[25] A. Haase, Ph.D. thesis. University of Giessen, 1980.

troplax,[26] vesicular stomatitis virus,[27] and mycoplasms.[28] Furthermore, it has been used in the reconstituted system of the ATP/ADP transport protein of mitochondria.[17] As shown in the last example the method provides best values of surface potential in relation to the Gouy–Chapman model.

With respect to a general application of the method in biomembranes, there are two basic obstacles: (1) The location of the probe in a inhomogeneous membrane made of lipids and proteins is unknown. (2) The insight with respect to an actual effect of the average surface potential on membrane transport, pore gating, and membrane contact [Eq. (1)] is limited at the present time.

The practical problems are hardly more involved than for other techniques: (1) In some systems the blue fluorescence may be superimposed on some background. (2) The pK_{mw} in negative membranes is rather high. Up to now it has been impossible to design other lipid pH indicators (carbonylcoumarin,[29] dansyl,[30,31] acridine,[31] chinin,[31] and naphthoic acid[32]) with $pK_i = 6-7$ and red fluorescence, conserving the crucial feature of the coumarin and its insensitivity to details of the membrane surface.

The main success of the method up to now has been in the field of pure colloidal systems. Surface potential and charge density have been characterized in insoluble monolayers,[4,5] in micelles,[6,21] in lipid vesicles,[8,15,18] and in foam films.[7,33] These results are the basis for studies of native and of reconstituted biomembranes and of the formation of membrane junctions.

Some problems of the method remain to be solved: (1) The validity of the thermodynamic cycle has to be checked experimentally by relating the probed potentials with macroscopically determined electrical potentials in insoluble monolayers and in solvent-free bilayers. (2) The interaction of the probe with the lipid matrix has to be studied in more detail by NMR techniques and by theoretical simulation. (3) The polarization of a mixed matrix and the interaction with membrane proteins has to be investigated. (4) The effect of neutral interfaces on the pK has to be characterized such that for every charged system the appropriate pK_i is chosen.

[26] J. M. Gonzalez-Ros, M. Lanillo, A. Paraschos, and M. Martinez-Carrion, *Biochemistry* **21**, 3467 (1982).
[27] R. Pal, W. A. Petri, Y. Barenholz, and R. R. Wagner, *Biochim. Biophys. Acta* **729**, 185 (1983).
[28] U. Schummer and H. G. Schiefer, *FEMS Microbiol. Lett.* **23**, 143 (1984).
[29] H. Alpes and G. W. Pohl, *Naturwissenschaften* **65**, 652 (1978).
[30] W. L. C. Vaz, A. Niksch, and F. Jähnig, *Eur. J. Biochem.* **83**, 299 (1978).
[31] P. Fromherz, unpublished observations (cf. Ref. 25).
[32] B. Lovelock, F. Grieser, and T. W. Healy, *J. Phys. Chem.* **89**, 501 (1985).
[33] P. Fromherz and R. Kotulla, *Proc. 31st IUPAC Congr., Sofia* **7**, 239 (1987).

The lipid coumarin is a probe of interfacial polarity as emphasized previously[4,6] and as illustrated recently for a glycosurfactant.[34] Finally, it should be noted that the coumarin may be used not only as an analytical tool for probing surface potentials but as a synthetic tool for the design of sensors probing interfacial effects induced by agonists, enzymes, or cells, with direct transduction of the fluorescence signal to a semiconductor.[35]

Acknowledgments

We thank the Deutsche Forschungsgemeinschaft and the Fonds der Chemischen Industrie for their generous support.

[34] C. J. Drummond, G. G. Warr, F. Grieser, B. W. Ninham, and D. F. Evans, *J. Phys. Chem.* **89**, 2103 (1985).
[35] P. Fromherz and W. Arden, *J. Am. Chem. Soc.* **102**, 6211 (1980).

[19] Modulation of Membrane Protein Function by Surface Potential

By REINHARD KRÄMER

Introduction

Nearly all biomembranes bear a net negative surface charge which gives rise to a negative surface potential. This will affect the concentration of cations and anions in the region adjacent to the membrane and may thus influence kinetic parameters of enzymes and carriers located in the membrane. In a few cases the influence of surface charge and surface potential on transport activity has been analyzed *in situ*, i.e., with membrane proteins in their natural surroundings, e.g., in bacteria,[1,2] in mitochondria,[3,4] and in chloroplasts.[5] Furthermore, experiments have been reported where the surface charge was artificially changed within the natural membrane, for instance by adding charged surfactants to mitochondria.[6,7]

[1] G. W. F. H. Borst-Pauwels, *Biochim. Biophys. Acta* **650**, 88 (1981).
[2] A. P. R. Theuvenet and G.W. F. H. Borst-Pauwels, *Biochim. Biophys. Acta* **734**, 62 (1983).
[3] H. Rottenberg, *Mod. Cell Biol.* **4**, 47 (1985).
[4] I. M. Møller, C. J. Kay, and J. M. Palmer, *Biochem. J.* **223**, 761 (1984).
[5] J. Barber, *Annu. Rev. Plant Physiol.* **33**, 261 (1982).
[6] J. Duszynski and L. Woitczak, *FEBS Lett.* **40**, 72 (1974).
[7] L. Woitczak, *FEBS Lett.* **44**, 25 (1974).

However, in all these cases severe complications arise due to the high protein content of the membrane and due to the possibility of an asymmetrical charge distribution. This has eventually led to the reduction of experimental complexity, i.e., to the use of reconstituted systems.

Under the less complex conditions prevailing in the reconstituted system, the influence of surface potential on membrane-bound enzymes has been extensively analyzed.[8,9] In these studies, membrane-bound enzymes catalyzing reactions with charged substrates were influenced by the surface potential, which modulated the apparent K_m values but did not change the V_{max}. This is in agreement with the simple assumption that variation in the surface charge density changes the concentration of charged substrates at the active site of the enzyme close to the surface of the membrane.

In the case of carrier proteins the situation is more complicated. It is not only the diffusion step of charged substrates from the surrounding bulk phase to the binding site that may be influenced by surface charges, but also essential conformational change(s) of the carrier/substrate complex during the reaction cycle of the carrier protein. Additionally, it has to be remembered that carriers, in contrast to membrane-bound enzymes, are necessarily influenced by surface charges at both of the two different membrane sides.

This is even more significant when considering that biological membranes are in general asymmetric with respect to their phospholipid and charge distribution.[10]

The reconstituted ADP–ATP carrier from the inner mitochondrial membrane provides an ideal tool for studying all the factors of the basic question—the influence of surface charges and surface potential on the carrier function—for three main reasons: (1) the adenine nucleotide carrier transports the highly negatively charged substrates, ADP and ATP. (2) The natural surroundings of this membrane protein, the inner mitochondrial membrane, contain a high amount of negatively charged phospholipids, especially cardiolipin. (3) The inner mitochondrial membrane is asymmetric with respect to the distribution of surface charges.[10] These facts show clearly that the reconstituted ADP–ATP carrier is an ideal model system for analyzing the modulating influence of surface charge and surface potential on the kinetic parameters of membrane-bound enzymes and transport systems.

[8] L. Woitczak and M. J. Nalecz, *Eur. J. Biochem.* **94**, 99 (1979).

[9] M. J. Nalecz, J. Zborowski, K. S. Famulski, and L. Woitczak, *Eur. J. Biochem.* **112**, 75 (1980).

[10] J. A. F. Op den Kamp, *Annu. Rev. Biochem.* **48**, 47 (1979).

Experimental

The purpose of this chapter is not only to describe how carrier function can be modulated by surface potential, but to correlate quantitatively a certain surface potential to its effect on definite kinetic parameters of membrane-bound enzymes or carriers. Thus, apart from the preparation of suitable reconstituted proteoliposomes forming the basis for the following experiments, both the appropriate determination of surface potential or surface charge and the analysis of the characteristic kinetic constants of the enzymatic or transport process will be described.

Preparation of Proteoliposomes

For this type of experiment it is necessary to use a definite lipid composition. However, this prerequisite is in many cases not easy to fulfill. The phospholipids commonly used for physicochemical studies of membrane surface potential, e.g., phosphatidylcholine, phosphatidylethanolamine, and phosphatidylserine with saturated fatty acids, do not usually provide a suitable membrane environment for studying the function of carrier proteins. Highly unsaturated phospholipids that are in general necessary for the activity of these membrane proteins are both expensive and difficult to handle. We generally use purified egg yolk phospholipids[11] for pilot studies, since this material has a very low content of negatively charged lipids. The surface charge of liposomes prepared from this material prove to be reproducibly low and easy to determine.[12] For more elaborate experiments we isolate pure phosphatidylcholine and phosphatidylethanolamine from egg yolks by column chromatography.[13] In contrast to many preparations of commercially available phospholipids from natural sources, liposomes prepared from this material prove to be nearly completely free from net negative charges at the surface.[12]

Liposomes can be prepared by a number of different methods[14]; however, for the experiments discussed here, we are limited by the necessity to incorporate functionally active proteins into these membranes.[15] On the one hand, we are restricted to certain detergent removal techniques, such as dialysis,[15] column procedures,[14] and removal by hydrophobic absorption,[14,16] or, on the other hand, to the freeze/thaw/sonication tech-

[11] M. A. Wells and D. J. Hanahan, this series, Vol. 14, p. 178.
[12] R. Krämer, *Biochim. Biophys. Acta* **735**, 145 (1983).
[13] G. B. Ansell and J. N. Hawthorne, "Phospholipids." Elsevier, Amsterdam, 1964.
[14] F. Szoka, Jr. and D. Papahadjopoulos, *Annu. Rev. Biophys. Bioeng.* **9**, 467 (1980).
[15] E. Racker, this series, Vol. 55, p. 699.
[16] M. Ueno, C. Tanford, and J. A. Reynolds, *Biochemistry* **23**, 3070 (1984).

nique.[17,18] The reconstitution method for the model system described here, i.e., the ADP–ATP carrier from mitochondria, has already been described in detail.[19]

Variation in the surface charge of the reconstituted proteoliposomes can be achieved in two ways. First, charged lipids can be included in the lipid mixture before the preparation of proteoliposomes, leading to a more or less uniform distribution of the charged molecules in the bilayer of the proteoliposomes. This is the method of choice if negatively charged phospholipids, e.g., phosphatidylserine or cardiolipin, are used to establish negative surface potentials. The second method for changing the surface charge consists of an addition of lipophilic molecules to the preformed proteoliposomes, leading to an asymmetric distribution of the charged substances located mainly at the outside of the bilayer. This method is preferentially used with low-molecular-weight substances such as dicetyl phosphate, free fatty acids (negatively charged), or alkyltrimethylammonium bromide (positively charged). These compounds can be added either as micellar solution in aqueous buffers, or dissolved in small amounts of organic solvent; their effect on the reconstituted liposomes must be tested separately. The rate of flip-flop of these molecules to the inner half of the bilayer of the liposomes can easily be tested by measurement of surface potential.[12]

At this point there is an important fact to consider regarding the methods of variation of surface potential in natural membranes. This variation is usually achieved by changing the ionic strength in the surroundings of the (negatively) charged biomembrane. It has to be kept in mind that this procedure leads only to a variation of the actual surface potential in the phase adjacent to the membrane, which thus modulates the concentration of charged ligands. The intrinsic surface charge density of the membrane which may directly influence the inserted membrane proteins is in general not changed by this method (if we neglect direct binding to the surface charges), since usually the concentration of monovalent ions is varied. However, a will be seen in the following, it is in fact a variation in the surface charge density that may be significant for the elucidation of the influences on particular functions of the membrane protein.

In order to measure significant effects of surface potential on membrane proteins in the case of a prevailing low density of surface charges, i.e., less than 10% negatively charged lipids, one has to reduce the ionic strength outside the liposomes to relatively low levels. This is sometimes rather

[17] M. Kasahara and P. Hinkle, *J. Biol. Chem.* **252**, 7384 (1977).
[18] R. Krämer and M. Klingenberg, *FEBS Lett.* **82**, 363 (1977).
[19] R. Krämer, this series, Vol. 125, p. 610.

difficult to achieve, since it is in many cases not possible to prepare the proteoliposomes in low-salt buffers. Especially in the case of exchange carriers, the appropriate activating ions and substrates have to be present in the interior phase of the liposomes in sufficiently high concentrations and thus have to be added from the start. For the separation of the proteoliposomes from the surrounding ions ultracentrifugation is usually not a convenient method, due to the insufficient sedimentation behavior of liposomes. Our experience is that prolonged dialysis of proteoliposomes in low-salt buffer in general leads to aggregation of the vesicles and to inactivation of the reconstituted protein. We prefer to use gel filtration on Sephadex G-75 (fine) in buffers composed mainly of sucrose or mannitol, the concentrations of which balance the internal osmolarity. The volume of the sample applied to the column has to be relatively high, i.e., it has to amount to at least 10% of the bed volume of the column, in order to prevent peak broadening and tailing due to the high viscosity of the proteoliposome dispersion.

Measurement of Surface Potential

Many methods for measuring surface potential have been described.[20] Since membrane proteins and especially carrier proteins in proteoliposomes are not of unlimited stability, the method here should be both fast and not harmful to the incorporated proteins. We have found two methods suitable for this purpose.

The use of 2-p-toluidinylnaphthalene 6-sulfate (TNS) as charged fluorescent distribution probe is convenient for determining the surface potential of proteoliposomes. This method has been excellently described in the literature.[21-23] Although it leads to a possible underestimation of the true surface potential,[12,20] it proved to be the most accurate method of those applicable in the minute time range. In order to avoid artifacts and to increase the reliability of this method. we routinely determine the surface potential of a particular proteoliposome preparation several times, varying the ionic strength (monovalent ions). Then we calculate the corresponding surface charge density which gives rise to the surface potentials observed under the varying ion concentrations.[12,20,24] Only when the calculated surface charge is found to be constant over a considerable range of ionic strength values can the measurement be accepted.

[20] D. Cafiso, A. McLaughlin, S. McLaughlin, and A. Winiski, this volume [16].
[21] S. McLaughlin, Curr. Top. Membr. Transp. 9, 1977.
[22] M. Eisenberg, T. Gresalfi, T. Riccio, and S. McLaughlin, Biochemistry 18, 5213 (1979).
[23] H. Rottenberg, this volume [17].
[24] R. Krämer and G. Kürzinger, Biochim. Biophys. Acta 765, 353 (1984).

The second method we use for the determination of surface potential is based on the pK shift of the lipophilic pH indicator alkylumbelliferone. This method was developed by Fromherz[25] and provides a very accurate measurement of surface potential. For the determination we add the dye to the preformed proteoliposomes, thereby monitoring the surface potential of the outer half of the membrane.[12] The measurement of surface potential by alkylumbelliferone is more complicated than the use of the fluorescent distribution probe TNS, since it requires a simultaneous determination of fluorescence and pH. Furthermore, it cannot be applied to highly negatively charged membranes, due to the strong pK shift into the alkaline region that results under these circumstances. This method is, however, not only more accurate than the determinations using distribution probes, but also less prone to artifacts caused by adsorption of the probe to proteins.

Determination of Kinetic Constants

The kinetic analysis of the activity of the reconstituted membrane protein has to be carried out using the same preparation of proteoliposomes as that used for the determination of surface potential, since the reproducibility of exactly the same reconstitution conditions is in general not very high.

The most obvious parameter modulated by surface potential is the apparent dissociation constant, K_d, of a charged ligand dissociating from its protein target in the neighborhood of the membrane surface. Classically, according to the Boltzmann equation, a change in the K_d of an ion with the charge z can be described by the equation

$$K_{d, app} = K_d \exp(zq\psi_0/kT)$$

where q is the electronic charge.[12,26]

The influence of surface potential on the apparent K_m of a membrane-bound enzyme or on the transport affinity (K_m or K_t) of a carrier protein is in general much more complex. An excellent theoretical description of this complex situation has been published recently.[27] In some cases, in which the influence of surface charges on the apparent K_m has been quantitatively analyzed, linear reciprocal plots (i.e., Michaelis–Menten kinetics) and a clear correlation between surface potential and kinetic properties have been found, in spite of these difficulties. In other words, a relationship is found that is similar to that expressed in the above equation for the

[25] P. Fromherz, this volume [18].
[26] E. Katchalski, I. Silman, and R. Goldman, Adv. Enzymol. 34, 445 (1971).
[27] P. W. J. A. Barts and G. W. F. H. Borst-Pauwels, Biochim. Biophys. Acta 813, 51 (1985).

dependence of the apparent K_d on surface potential. These favorable circumstances held true not only for membrane-bound enzymes incorporated into liposomes,[8,9] but also, for instance, for NADH dehydrogenase in native plant mitochondria[4] and for the reconstituted ADP–ATP carrier from the inner mitochondrial membrane.[12]

It is important to consider a further possible effect of the surface potential on carrier kinetics. A change in the surface potential may introduce a simultaneous change in the difference of the potential between the two membrane interfaces, i.e., the membrane potential.[21] Especially when electrogenic carriers are involved, this may lead to additional complexity in the experimental situation. Thus, in order to simplify the analysis of surface potential effects on carrier kinetics, it is usually of benefit to cancel out an influence of the membrane potential by using symmetric membranes or by providing an equivalent flux of monovalent cations, e.g., K^+, through the membrane.

Although only the relatively simple conditions of a reconstituted system permit a thorough study of the more or less complete spectrum of modulating parameters with respect to the influence of surface charge, the problem of protein orientation arises in the experiments with reconstituted transmembranous proteins. In some cases, the orientation of the incorporated protein is easy to determine, due to the accessibility of its substrates only from one side of the membrane. In general, however, the protein orientation has to be determined for every reconstitution procedure. In the case of the adenine nucleotide carrier, for instance, this can be achieved by the use of the side-specific inhibitors carboxyatractylate and bongkrekate.[28] Especially in experiments where surface charges have been introduced only at the outer side of the reconstituted proteoliposomes by adding charged lipophilic compounds to the preformed vesicles (see above), the orientation of the membrane protein has to be considered.

The influence of surface potential on the enzyme or transport activity, i.e., the V_{max}, should also always be analyzed, although in principle this may not seem to be very obvious (see below). Again, the reconstituted ADP–ATP carrier provides a good example where the influence on V_{max} can be clearly discriminated from a K_m effect.[12,24]

Correlation of Kinetic Parameters with Surface Charge

To make a quantitative correlation of various kinetic parameters with the observed surface potential, one must always be aware of the complexity of possible interactions and influences at certain steps within a complete reaction cycle of an enzyme or carrier system. As mentioned before, the

[28] R. Krämer and M. Klingenberg, *Biochemistry* **18**, 4209 (1979).

most basic and simple approach is that of determining the dependence of pure K_d values on a variation of surface potential.

When a real transport function is investigated, one can in general discriminate between influences of surface potential on affinity parameters (transport affinity K_m) and those on velocity parameters (V_{max}). In these cases we have to take into account that the complex equation for the apparent transport affinity may also include rate constants (velocity parameters) and that the equation for V_{max} may include dissociation constants (affinity parameters), as shown in numerous analysis of transport kinetics.[29-31]

However, at least a qualitative but nevertheless important statement can be made. If the charged membrane surface predominantly modulates the affinity parameters of the particular membrane protein, one can attribute this to a simple surface *potential* effect. The most convincing example of this situation is achieved by modulating the kinetic properties of membrane-bound enzymes,[8,9] where no effect at all on V_{max} is found. If, on the other hand, a definite effect on the V_{max} of the protein function is observed, one can correlate this with an influence of the surface *charge*. In other words, a direct interaction of the charged lipid molecule at the membrane surface with the membrane protein under investigation can be assumed in the latter case, in contrast to the unspecific and indirect effect of the surface potential generated by these surface charges.

[29] R. J. Turner, *J. Membr. Biol.* **76**, 1 (1983).
[30] D. Sanders. U.-P. Hansen, D. Gradmann, and C. L. Slayman, *J. Membr. Biol.* **77**, 123 (1974).
[31] R. Krämer and M. Klingenberg, *Biochemistry* **21**, 1082 (1982).

Section III

Isolation of Cells

[20] Use of Nonequilibrium Thermodynamics in the Analysis of Transport: General Flow–Force Relationships and the Linear Domain

By Daniela Pietrobon and S. Roy Caplan

Introduction

The first serious attempt to apply the concepts of nonequilibrium thermodynamics to biological (and hence, by definition, enzymatic) transport was made by Kedem and Katchalsky.[1,2] Their pioneering efforts sparked off a great deal of work in a variety of membrane systems ranging from amphibian epithelia to mitochondria and chloroplasts.[3–5] Modern techniques in nonequilibrium thermodynamics are now playing an increasing role in the analysis of the coupled processes which constitute the basis of biological energy transduction, since the all-important interdependence of kinetic and thermodynamic parameters is not only stressed but displayed explicitly. In fact, nonequilibrium thermodynamics relates flows, e.g., chemical reactions and transmembrane ionic fluxes, to thermodynamic forces, e.g., the reaction affinity (the instantaneous negative Gibbs free energy change) and the electrochemical potential difference of the transported ion.

A major limitation of the classical Michaelis–Menten description of a transporting enzyme is its failure to account for the thermodynamic forces controlling the flows. Furthermore, since the algebraic complexity increases in multisubstrate reactions and even minimal models of cotransport or active transport involve six-state cyclic reaction schemes such as those depicted in Fig. 1, conventional Michaelis–Menten kinetic analyses invariably make a physically doubtful simplifying assumption: transmembrane translocation steps are rate limiting while ligand-binding reactions are at equilibrium. The dangers and limitations of this assumption have

[1] O. Kedem and A. Katchalsky, *Biochim. Biophys. Acta* **27**, 229 (1958).

[2] A. Katchalsky and P. F. Curran, "Nonequilibrium Thermodynamics in Biophysics." Harvard Univ. Press, Cambridge, Massachusetts, 1965.

[3] H. Rottenberg, *Biochim. Biophys. Acta* **549**, 225 (1979).

[4] S. R. Caplan and A. Essig, "Bioenergetics and Linear Nonequilibrium Thermodynamics: The Steady State." Harvard Univ. Press, Cambridge, Massachusetts, 1983.

[5] K. van Dam and H. V. Westerhoff, *in* "Bioenergetics" (L. Ernster, ed.), p. 1. Elsevier, Amsterdam, 1984.

METHODS IN ENZYMOLOGY, VOL. 171

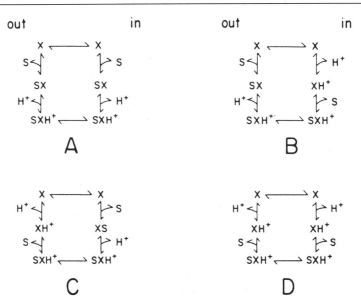

FIG. 1. (A–D) Four alternative models for cotransport of a solute (S) with a driver ion (H⁺), mediated by a carrier (X). All models depict transport of positive charge on the loaded carrier. (From Sanders et al.[6])

been pointed out by Slayman and co-workers.[6] They demonstrated that without it, using minimal models of the kind shown in Fig. 1, one can describe most of the kinetic diversity observed in cotransport or countertransport systems merely by assigning appropriate values to the rate constants. As emphasized by these workers, many more carrier states than those considered in Fig. 1 may exist in an experimental system, some of which may not even have been identified by the experimenter. Nevertheless, it turns out that, under steady-state conditions, such "hidden" states within a cyclic scheme do not affect the overall kinetic description of its behavior.[7] The models of Fig. 1, which include a single transmembrane electrical potential gradient dependent step, are therefore more general than one might think at first glance.

A procedure using "lumped" voltage-independent reaction constants to account for concealed states was used very successfully by Slayman and co-workers to study the current–voltage relationships of such models. By means of this procedure, all single cycle schemes, whatever the number of

[6] D. Sanders, U.-P. Hansen, D. Gradmann, and C. L. Slayman, *J. Membr. Biol.* **77**, 123 (1984).
[7] U.-P. Hansen, D. Gradmann, D. Sanders, and C. L. Slayman, *J. Membr. Biol.* **63**, 165 (1981).

states involved, may be reduced to pseudo-two-state schemes, or, where necessary, to pseudo-n-state schemes, where n is a small number. The simplified model can then describe rather well many of the observed properties of electrogenic pumps. Although this is a powerful approach, lumping cannot be resorted to when branching occurs, except in a restricted sense such that the number of cycles is not reduced. Current–voltage relationships have been derived (without using the lumping procedure) for a six-state model of Na^+,K^+-ATPase by Chapman et al.,[8] and for a generalized channel model of a proton pump by Läuger.[9] They also studied the current–voltage relationships in the case of more than one voltage-dependent step. In addition to these studies, nonclassical model simulations relating flows to the chemical potential difference (rather than the electrical potential difference) of the transported ion have been described.[10,11]

In all these studies the kinetic behavior was examined in terms of one thermodynamic force only. Moreover, the methods used to derive the relevant kinetic equations were rather cumbersome and, as in the case of the more conventional Michaelis–Menten treatment, essentially restricted to single cycle reaction schemes. This is especially true of the lumping procedure: branched schemes, such as those involving random binding of ligands, are impractical to deal with because of the vastly increased algebraic complexity. In particular, the method cannot readily handle the phenomenon of "slip," i.e., intrinsic uncoupling, which may be far more common in enzymes mediating multiple fluxes than hitherto assumed.[12–15]

The most convenient way of obtaining steady-state solutions to the kinetic equations of a branched or multicyclic system in terms of the thermodynamic forces and flows is to use the diagram method of Hill.[16] This is a massive extension of the original diagram method of King and Altman,[17] which was comparatively limited in scope. The Hill method

[8] J. B. Chapman, E. A. Johnson, and J. M. Kootsey, *J. Membr. Biol.* **74,** 139 (1983).

[9] P. Läuger, *Biochim. Biophys. Acta* **779,** 307 (1984).

[10] P. Gräber and E. Schlodder, *in* "Photosynthesis II: Electron Transport and Phosphorylation" (G. Akoyunoglou, ed.), p. 867. Balaban Int. Sci. Serv., Philadelphia, Pennsylvania, 1981.

[11] C. Tanford, *Proc. Natl. Acad. Sci. U.S.A.* **79,** 6161 (1982).

[12] D. Pietrobon, G. F. Azzone, and D. Walz, *Eur. J. Biochem.* **117,** 389 (1981).

[13] A. A. Eddy, *Adv. Microb. Physiol.* **23,** 1 (1982); see also S. J. D. Karlish and W. D. Stein, *J. Physiol. (London)* **328,** 295 (1982).

[14] O. Frölich, C. Leibson, and R. B. Gunn, *J. Gen. Physiol.* **81,** 127 (1983).

[15] H. V. Westerhoff and Z. Dancsházy, *Trends Biochem. Sci.* **9,** 112 (1984).

[16] T. L. Hill, "Free Energy Transduction in Biology." Academic Press, New York, 1977.

[17] E. L. King and C. Altman, *J. Phys. Chem.* **60,** 1375 (1956).

considers cycles, cycle fluxes, and coupling between fluxes explicitly, offering a graphical algorithm from which one can obtain the steady-state fluxes either as functions of the forces or, if desired, as functions of the substrate concentrations. Being a combination of kinetics and thermodynamics, it deals directly with the *energetics* of a given system. Moreover, it is far less cumbersome to handle than the methods used in all the modeling investigations referred to above, gives greater insight, requires no assumptions, and is eminently suited for the treatment of branched or multicyclic schemes. In this chapter we shall describe the force–flow relationships which emerge from such an approach, and the application of these relationships to the problems of bioenergetics.

Basic Concepts

The Dissipation Function

The dissipation function, Φ, is fundamental to any nonequilibrium thermodynamic description of a system and may be derived rigorously by applying the Gibbs equation to all its parts.[2,4] However, in most cases of the kind considered here the dissipation function may be written down immediately by the following procedure:

1. Decide which elements of the system (e.g., whole cell, vesicle, membrane) can be regarded as being in a steady state during the experiment.
2. Choose such an element, and enumerate *all* the processes occurring on it or in it during the steady state.
3. Express each process as a product of its conjugate thermodynamic force and flow (see below), whereupon the dissipation function is given by the sum of all such products.

Some elaboration of these steps is now required.

What Do We Mean by a Steady State? By definition this is a state in which none of the thermodynamic parameters of the element under consideration vary with time. These parameters include temperature, pressure, concentrations, and electrical potentials, and they have time-independent values irrespective of the fact that reactions and flows may be taking place. In practice we often deal with *pseudostationary* states, i.e., states whose parameters may be changing, but on a time scale slow enough relative to that of the phenomena under study that the variation may be ignored.

What Processes Are We Talking About? In the present context these will always be reactions occurring in or on a membrane, and transmembrane flows. It is important to realize that the processes to be considered in the dissipation function must be complete as seen from the compartments on either side of the membrane, and not partial processes involving the membrane itself. Thus the hydrolysis of ATP or the transport of a ligand are admissible, but not the binding of these species to an integral protein within the membrane.

What Are the Conjugate Forces and Flows? The flows are either reaction rates or net material fluxes. They may be given per unit area of membrane, or per total membrane area or some other index of membrane area (e.g., per milligram of protein). The conjugate forces are the reaction affinity, A, and the electrochemical potential difference, $\Delta\tilde{\mu}$ (for uncharged species the latter reduces to the chemical potential difference, $\Delta\mu$). The affinity A is identical to the negative of the instantaneous Gibbs free energy difference for the reaction, the quantity well known in classical thermodynamics as $-\Delta G$, providing the latter is understood to indicate a partial derivative, at constant temperature and pressure, with respect to the extent of reaction ξ (i.e., $A = -\Delta G \equiv -(\partial G/\partial\xi)_{T,p}$).[18] When A is positive, the reaction occurs spontaneously. For practical purposes A is given by

$$A = RT \ln(K/\Pi_i \, a_i^{\nu_i}) \tag{1}$$

where K is the equilibrium constant of the reaction, and a_i and ν_i are the activity and stoichiometric coefficient of the ith participating species, respectively. The stoichiometric coefficient is conventionally assigned a positive sign for products and a negative sign for reactants. The electrochemical potential difference of the jth species is given by

$$\Delta\tilde{\mu}_j = RT\Delta \ln a_j + z_j F\Delta\psi \tag{2}$$

where z_j is the charge number of the species, F is the faraday constant, and $\Delta\psi$ is the electrical potential difference between the compartments. Usually $\Delta\tilde{\mu}_j = \tilde{\mu}_{j(1)} - \tilde{\mu}_{j(2)}$ is taken to be positive for the spontaneous flow of the jth species from compartment 1 to compartment 2.

Once the forces and flows are identified, the dissipation function is given, in the case of r processes, by

$$\Phi = J_1 X_1 + J_2 X_2 + \cdots + J_r X_r \tag{3}$$

where each term represents a separate process (reaction or flow), and the J values denote fluxes and the X values denote forces. The dissipation function represents the rate of dissipation of free energy in isothermal systems: if we divide it by the absolute temperature we get the rate of entropy production by all the irreversible processes occurring in the steady-state element. It can be shown that the second law of thermodynamics imposes

[18] E. A. Guggenheim, "Thermodynamics." North-Holland Publ., Amsterdam, 1957.

the following condition:

$$\Phi \geq 0 \qquad (4)$$

In other words the dissipation function will always be positive providing the system is not at equilibrium. It is important to note that this does not require individual terms to be positive, so long as the sum is greater than zero.

Simple Linear Phenomenological Relations

Once the dissipation function is established for a system, the simplest possible phenomenological description that can be given is to suppose all flows to be linear functions of all the forces. This takes into account all conceivable coupling phenomena. Consider, for example, a membrane which mediates the occurrence of three processes. The dissipation function is

$$\Phi = J_1 X_1 + J_2 X_2 + J_3 X_3 \qquad (5)$$

and it generates the following set of linear phenomenological relations:

$$J_1 = L_{11} X_1 + L_{12} X_2 + L_{13} X_3$$
$$J_2 = L_{21} X_1 + L_{22} X_2 + L_{23} X_3 \qquad (6)$$
$$J_3 = L_{31} X_1 + L_{32} X_2 + L_{33} X_3$$

The L_{ij} values in this set of equations represent conductance coefficients relating each force to each flow. The three so-called "straight" coefficients (L_{11}, L_{22}, and L_{33}) must have positive values, but the six "cross"-coefficients (or coupling coefficients) may be positive, negative, or zero. This representation is obviously an idealization; on *a priori* grounds its range of applicability would be expected to be confined to very small values of the forces, i.e., to a domain very near equilibrium. It will be seen that this domain has considerable importance as a reference point even when we wish to discuss conditions far from equilibrium. If the system is sufficiently close to equilibrium so that Eq. (6) holds rigorously, an important theorem due to Onsager indicates that

$$L_{ij} = L_{ji} \qquad (7)$$

Thus the cross-coefficients are reduced in number—from nine to six in this example. We shall call the region of validity of Eq. (7) (which is generally referred to as Onsager's symmetry or reciprocity relation), the

"Onsager domain." A further condition which holds in this domain is that

$$L_{ij}^2 \le L_{ii}L_{jj} \tag{8}$$

This condition is a consequence of Eq. (4).

In most experimental studies we manipulate the forces and measure the flows. In some cases a force is not subject to experimental control but reaches a steady-state value dictated by the other constraints on the system. If so, its net conjugate flow may well become zero ("stationary state of minimal entropy production" [2]). If, for example, in the case corresponding to Eq. (5), X_3 were free to vary, reaching a steady-state value such that J_3 became zero, Eq. (5) would be reduced to two terms and instead of Eq. (6) we would have

$$
\begin{aligned}
J_1 &= L_{11}'X_1 + L_{12}'X_2 \\
J_2 &= L_{21}'X_1 + L_{22}'X_2
\end{aligned}
\tag{9}
$$

where the L_{ij}' values are simply related to the L_{ij} values of Eq. (6).

It is important to realize that although the validity of a linear formalism such as Eqs. (6) and (9) is confined to the Onsager domain, the validity of the dissipation function is unrestricted. This gives the dissipation function a special utility, which will be discussed later. It is also worth noting that for many transport processes involving only diffusion, the linear domain extends far from equilibrium.

Energy Transduction

In this section we shall take as an example the case of oxidative phosphorylation. Caplan and Essig showed that an adequate dissipation function to describe the steady state of the inner mitochondrial membrane during oxidative phosphorylation is, under certain conditions,[19,20]

$$\Phi = J_p A_p^{ex} + J_H \Delta \tilde{\mu}_H + J_o A_o^{ex} \tag{10}$$

where the subscripts p and o denote phosphorylation and oxidation, respectively, and the affinities and reaction rates are evaluated in the external medium. J_H is the net outward proton flux when $\Delta \tilde{\mu}_H = \tilde{\mu}_H^{in} - \tilde{\mu}_H^{ex}$. If the mitochondria as a whole reach a steady state with no externally imposed constraints on $\Delta \tilde{\mu}_H$, the net proton flux must be zero and the dissipation

[19] S. R. Caplan and A. Essig, *Proc. Natl. Acad. Sci. U.S.A.* **64**, 211 (1969).

[20] H. Rottenberg, S. R. Caplan, and A. Essig, *in* "Membranes and Ion Transport" (E. E. Bittar, ed.), Vol. 1, p. 165. Wiley (Interscience), London, 1970.

function reduces to

$$\Phi = J_p A_p^{ex} + J_o A_o^{ex} \tag{11}$$

If we choose to consider the steady state corresponding to the dissipation function of Eq. (11), we may immediately write the phenomenological relations for the Onsager domain in an abbreviated notation as follows:

$$
\begin{aligned}
J_p &= L_p A_p^{ex} + L_{po} A_o^{ex} \\
J_o &= L_{po} A_p^{ex} + L_o A_o^{ex}
\end{aligned}
\tag{12}
$$

Two-term dissipation functions such as that displayed in Eq. (11) frequently describe energy transduction, and of course mitochondria offer a classical example of this. To the extent that relations such as Eq. (12) apply, they provide a "black box" representation characterized by the three phenomenological coefficients. As we shall see, certain combinations of these coefficients are conceptually advantageous. Whether or not we are in the Onsager domain, the dissipation function enables us to write down the correct expression for the thermodynamic efficiency at sight. As alluded to earlier, the two terms on the right-hand side of Eq. (11) need not both be positive. Indeed, in actively phosphorylating mitochondria (i.e., mitochondria in "state 3") the term $J_p A_p^{ex}$ will be negative. If by convention A_p^{ex} is the affinity and J_p the velocity of the phosphorylation reaction, A_p^{ex} has a negative value, since the only spontaneous process which the $F_0 F_1$ complex (or indeed any other ATPase) can bring about under state 3 conditions is ATP hydrolysis. If by convention A_p^{ex} is the affinity of the hydrolysis reaction, the rate J_p is positive for ATP hydrolysis and negative for ATP synthesis. In contrast, the term $J_o A_o^{ex}$ is positive. Oxidation is occurring spontaneously, but the energy expended by the oxidation reaction is not being merely dissipated. Part of it is being converted into another form and stored. Hence $J_o A_o^{ex}$ measures the input to the transducer, and $-J_p A_p^{ex}$ measures the output (these quantities, like the dissipation function, have the dimensions of power). The efficiency, η, is therefore given by[21]

$$\eta = \frac{\text{output}}{\text{input}} = -\frac{J_p A_p^{ex}}{J_o A_o^{ex}} \tag{13}$$

The above definition of the efficiency holds universally. Added insight into the nature of the efficiency may be obtained by examining its behavior in the Onsager domain. For this it is useful to consider two new parameters:

[21] O. Kedem and S. R. Caplan, *Trans. Faraday Soc.* **61**, 1897 (1965).

Degree of Coupling, q. This quantity is defined, in terms of the quantities appearing in Eq. (12), by

$$q = L_{po}/\sqrt{L_p L_o} \tag{14}$$

On referring to Eq. (8), it is seen that in general q lies between 1 and -1. The sign of q has significance in cotransport systems — thus a positive value of q would indicate symport, a negative value antiport. However, in this article we shall restrict ourselves to positively coupled systems. Clearly, if $q = 0$ we have complete intrinsic uncoupling and both processes can only take place spontaneously. When $q = 1$, the two flows have a fixed stoichiometric ratio whatever the values of the forces.

Phenomenological Stoichiometry, Z. In terms of Eq. (12) this is defined by

$$Z = \sqrt{L_p/L_o} \tag{15}$$

It will be shown that in fully coupled systems only, i.e., systems in which $q = 1$, Z is identical to the mechanistic stoichiometry, n. The relationship of Z to n (and indeed the meaning of n) in incompletely coupled systems will become clear when we consider kinetic diagrams. The flow ratio can now be expressed as

$$\left(\frac{J_p/J_o}{Z}\right) = \frac{(ZA_p^{ex}/A_o^{ex}) + q}{1 + q(ZA_p^{ex}/A_o^{ex})} \tag{16}$$

where the quantities in parentheses are dimensionless, and the efficiency as

$$\eta = -\frac{(ZA_p^{ex}/A_o^{ex}) + q}{\dfrac{1}{Z(A_p^{ex}/A_o^{ex})} + q} \tag{17}$$

Hence, when $q = 1$, $J_p/J_o = Z$ and $\eta = -Z(A_p^{ex}/A_o^{ex})$. Under these circumstances $Z \equiv n$.

Energy transduction takes place when the output process is "driven" by the input process, i.e., η lies between zero and one. The condition for this to occur is

$$-1 \le \frac{ZA_p^{ex}/A_o^{ex}}{q} \le 0 \tag{18}$$

and it is seen from Eq. (17) that at either limit $\eta = 0$. Within this range η passes through a maximum. The maximum efficiency in the linear regime depends only on q:

$$\eta_{max} = \frac{q^2}{(1 + \sqrt{1 - q^2})^2} \tag{19}$$

The phenomenological coefficients treated above, and the thermodynamic parameters derived from them are, of course, functions of state, but in principle are independent of the forces. This is only true in the Onsager domain. Outside this domain, even though we may be in a linear regime, the utility of these quantities must be established either empirically or by kinetic modeling, as will be discussed below. Despite this limitation, it can be seen that we are dealing with an analytical tool which expressly brings out the interplay between two processes and highlights the energy transaction involved.

Level Flow and Static Head

In the previous section attention was drawn to two states in which $\eta = 0$. These are of sufficient physiological importance to merit special consideration.[21]

Level flow occurs when the output force vanishes. For example, individual ion pumps are readily studied under conditions where the electrochemical potential gradient of the pumped ion is zero.

Static head occurs when the output flow vanishes. This is the well-known "state 4" condition in mitochondria. If the input force is constant, and no restrictions are placed on the output force, the latter will increase in absolute magnitude during the process of energy transduction until it is sufficient to bring the output flow to a halt.

Clearly energy must be expended to maintain these states, even though output is zero. An exception is the case of static head when $q = 1$, as will be seen below. The occurrence of these states is, of course, not restricted to the Onsager domain.

Completely Coupled Systems

The thermodynamic implications of complete coupling are highly important. The flows are now "locked together," and hence the dissipation function contracts to a single term. Returning to Eq. (11), if $q = 1$, $J_p/J_o = Z = n$ and hence

$$\Phi = J_o(nA_p^{ex} + A_o^{ex}) \quad (q = 1) \tag{20}$$

Equation (20) is universally valid. We see that since the two forces can be independently varied, the composite force can be made as small as desired. Thus Φ may become vanishingly small compared with the input and output. The significance of this is that the system approaches reversible equilibrium as the rate tends to zero, and indeed in such systems static head corresponds to reversible equilibrium. Thus, if we know our system to be fully coupled, n can be estimated from static head measurements, for example

$$n = -A_o^{ex}/A_p^{ex} \quad (q = 1) \tag{21}$$

Linearized Analysis Outside the Onsager Domain

It has been observed empirically that many processes are linear over a surprisingly wide range outside the Onsager domain, and accordingly a number of workers have attempted to use the above formulation without modification to describe processes of this kind (for example, Stucki,[22] Essig and Caplan,[23] and Pietrobon *et al.*[24]). The significance of such an analysis will be examined later and specific instances will be discussed in the section *Applications.*

Proportionality. This term was introduced by Westerhoff and van Dam[25] to emphasize a feature of phenomenological equations such as Eqs. (6) and (9): if all the forces except one are reduced to zero, the fluxes are not only linearly related to the remaining force but are also proportional to it. As we shall see, outside the Onsager domain this need not necessarily be the case even if an experimental range is found far from equilibrium in which both linearity and symmetry are observed. However, it is the simplest possible tentative assumption.

Nonproportionality. The concept of linearity without proportionality originated in the work of Rottenberg.[26,27] Rottenberg considered a Michaelis–Menten-like carrier reaction under certain boundary conditions, and showed that the flux of transported ligand could be approximated as a linear function of its chemical potential difference over a considerable range of the latter. This occurred in the vicinity of an inflection point in the sigmoid flux–force curve, which could be situated far from equilibrium. Whenever such a linear range is *not* centered at equilibrium, the corresponding flux–force relation is no longer proportional. Similar considerations could be applied to enzyme-catalyzed reactions,

[22] J. W. Stucki, *Eur. J. Biochem.* **109,** 269 (1980).
[23] A. Essig and S. R. Caplan, *Am. J. Physiol.* **236,** F211 (1979).
[24] D. Pietrobon, M. Zoratti, G. F. Azzone, J. W. Stucki, and D. Walz, *Eur. J. Biochem.* **127,** 483 (1982).
[25] H. V. Westerhoff and K. van Dam, *in* "Membranes and Transport" (A. N. Martonosi, ed.), Vol. 1, p. 341. Plenum, New York, 1982.
[26] H. Rottenberg, *Biophys. J.* **13,** 503 (1973).
[27] H. Rottenberg, *Prog. Surf. Membr. Sci.* **12,** 245 (1978).

leading to flux–force relations of the type $J = LX + K$, K being an additive constant. This led Rottenberg to propose an empirical extension of the concept to coupled processes, giving rise to the following modification of Eq. (12):

$$J_p = L_p A_p^{ex} + L_{po} A_o^{ex} + K_p$$
$$J_o = L_{op} A_p^{ex} + L_o A_o^{ex} + K_o$$

(22)

In Eq. (22) symmetry has not been automatically assumed, since we are manifestly outside the Onsager domain. Similar considerations to those of Rottenberg were given by van der Meer et al.[28] and have been extensively exploited by van Dam and Westerhoff.[5] The latter workers, in an analogous treatment of completely coupled pumps, assumed explicitly that symmetry cannot be invoked. In other words the system is expected to manifest kinetic inequivalence in the sense that equivalent changes in the forces have different impacts on the flows. This was expressed by introducing a so-called "activity" factor multiplying one of the forces[5,29] (see below).

Model Systems

In order to study the general pattern of flow–force relationships that a representative two-flow coupled process can generate, and to analyze the properties of domains of approximate linearity far from equilibrium and the meaning of parameters such as q, Z, and η_{max} in these regions, one must turn to kinetic models. Recently we investigated the behavior of a six-state model of an electrogenic proton pump with the object of addressing several important questions not hitherto posed: the dependence of the fluxes on reaction affinity as well as $\Delta\tilde{\mu}_H$, the influence of intrinsic uncoupling, and the nature of domains of linearity.[30] It should be emphasized that in these studies, the methodology of which is to be outlined in the following sections, the proton pump is chosen merely as an example of an ion pump. It has the advantage of having been studied extensively in our laboratories, but the methods could equally well have been applied to other pumps or transport systems with appropriate modifications.

[28] R. van der Meer, H. V. Westerhoff, and K. van Dam, *Biochim. Biophys. Acta* **591**, 488 (1980).
[29] K. van Dam, H. V. Westerhoff, K. Krab, R. van der Meer, and J. C. Arents, *Biochim. Biophys. Acta* **591**, 240 (1980).
[30] D. Pietrobon and S. R. Caplan, *Biochemistry* **24**, 5764 (1985).

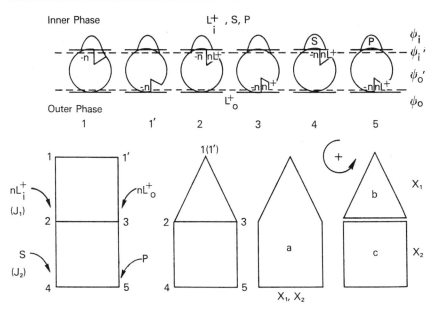

FIG. 2. Model for active ion transport. The univalent cation L^+ binds to an internal or external site on the transporting enzyme bearing a charge $-n$. Binding is assumed to occur in successive steps with transient intermediates that need not be specified. Six states of the enzyme are shown; each may represent the "reduction" of several states in sufficiently rapid equilibrium. This reduction is made explicit for states 1 and $1'$ in the accompanying Hill diagrams. Binding of the substrate S and release of the product P take place at the internal surface. The electrical potentials are ψ_i and ψ_o at the inner and outer bulk phases, respectively, and ψ_i' and ψ_o' at planes close to the inner and outer membrane surfaces that correspond, as indicated, to the positions of the binding sites. J_1 and J_2, the transport and reaction flows, refer to the processes shown, their conjugate forces being X_1 and X_2. (From Caplan.[31])

The Diagram Method

Directional Diagrams: Probabilities of States

For purposes of discussion, we consider the simple hypothetical five-state model of an ion pump treated by Caplan,[31] in which an intrinsic uncoupling or slip transition is not excluded (Fig. 2). This model is explicitly a "reduction" of a six-state model in which two states are considered to be in rapid equilibrium (but it may also be regarded as a reduction of a model comprising a much larger number of microstates). During one turnover of the enzyme, one molecule of substrate S is converted to product P and n ligand ions L^+ are transported from the inner phase to the

[31] S. R. Caplan, *Proc. Natl. Acad. Sci. U.S.A.* **78**, 4314 (1981).

outer phase. The voltage-dependent transition here is 11' (which has been assumed to be very fast); however, for present purposes it is not necessary to take this dependence into consideration, i.e., we assume $\Delta\psi = 0$. Note that we are dealing with a large number, N, of equivalent and independent units, each of which may exist in any one of the states enumerated. Using a graphic method, a complete kinetic description can be obtained in terms of the first-order or pseudo-first-order rate constants for each transition. The latter correspond to the ligand-binding steps, where it is assumed that the binding rate is determined by the activity of the ligand in the compartment from which it binds. Referring to Fig. 2, and replacing activities by concentrations, we have

$$\alpha_{12} = \alpha_{12}^* c_{L_f^+}^n, \qquad \alpha_{13} = \alpha_{13}^* c_{L_t^+}^n,$$
$$\alpha_{24} = \alpha_{24}^* c_S, \qquad \alpha_{35} = \alpha_{35}^* c_P \tag{23}$$

where α_{12}^* and α_{13}^* are $(n + 1)$th-order rate constants, and α_{24}^* and α_{35}^* are second-order rate constants. The successive steps in this procedure are outlined below.

Step 1. Draw the corresponding diagram. In Fig. 2 (lower left-hand side) the diagram is shown both before and after reduction to five states. The states are indicated by numbered nodes, and each transition by a line corresponding to two rate constants, one for the forward rate and one for the reverse. Arrows entering the diagram indicate binding (release is not indicated). The three possible cycles associated with the five-state diagram, a, b, and c, appear on the lower right-hand side of Fig. 2; we shall deal with these later.

Step 2. Construct a complete set of partial diagrams, each of which contains the maximum number of lines that can be included without forming *any* closed cycle. In our example there are eleven such partial diagrams. These can be seen by considering Fig. 3 with the arrowheads removed.

Step 3. Construct a set of directional diagrams by introducing arrows into the partial diagrams so as to make all connected paths flow toward a given state. One such subset is constructed for each state, making a complete set of 55 directional diagrams in the present example. The subset corresponding to state 2 of the model is illustrated in Fig. 3. By denoting the first-order rate constant for the transition $i \rightarrow j$ as α_{ij}, an algebraic value can be assigned to each directional diagram as indicated, i.e., the product of the rate constants along all "streams" leading to the given state. The steady-state probability of the ith state, p_i, is given by

$$p_i = (\text{sum of directional diagrams of state } i)/\Sigma \tag{24}$$

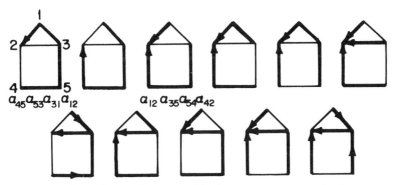

FIG. 3. Directional diagrams for state 2 of the five-state model shown in Fig. 2. Algebraic values are given for the first and third directional diagrams. (From Caplan and Essig,[4] after Hill[16]; reprinted by permission.)

where

$$\Sigma = \text{sum of directional diagrams of all states} \qquad (25)$$

If we denote the net mean steady-state transition flux between states i and j in the direction $i \rightarrow j$ as J_{ij}, we have

$$J_{ij} = N(\alpha_{ij} p_i - \alpha_{ji} p_j) \qquad (26)$$

Note that the number of independent, nonzero, steady-state transition fluxes for a diagram is given by one plus the number of lines minus the number of nodes. In the present model there are just two. These may be described as the operational fluxes: for example J_{12} and J_{24} represent such fluxes, viz. transport of ligand and rate of reaction, respectively.

Flux Diagrams: Probabilities of Cycles

Although Eq. (26) is, in principle, sufficient to characterize the kinetics of the system, a far more powerful algorithm, which in addition expresses the fundamental relationship between kinetics and nonequilibrium thermodynamics, is obtained from the following graphical procedure (the final step of the diagram method). Construct a complete set of flux diagrams by adding one line to each of the partial diagrams (a certain amount of replication occurs). These bear the same relation to cycles as directional diagrams bear to states. The set corresponding to the five-state model is shown in Fig. 4. A flux diagram may be defined as a diagram containing one cycle only, with associated "streams" (if present) flowing into it at one or more nodes. Thus the set of flux diagrams will include diagrams, or at

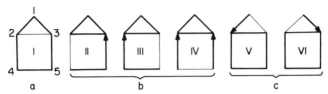

FIG. 4. Six flux diagrams for the five-state model of Fig. 2, classified according to cycle. (From Caplan and Essig,[4] after Hill[16]; reprinted by permission.)

least one diagram, corresponding to each of the component cycles of the system (cf. Fig. 3). For present purposes an algebraic value can be assigned to each flux diagram in the following way. We adopt the convention that the counterclockwise direction around a cycle is to be deemed positive, as indicated in Fig. 2. Then each flux diagram may be regarded as the difference between a pair of cyclic diagrams, one for the counterclockwise and one for the clockwise direction, each being the product of rate constants around the cycle and along the stream or streams leading to the cycle. Thus for flux diagram III of Fig. 4 we have

$$\text{III} = \text{III}_+ - \text{III}_- = (\alpha_{12}\alpha_{23}\alpha_{31} - \alpha_{13}\alpha_{32}\alpha_{21})\alpha_{54}\alpha_{42} \qquad (27)$$

The algebraic value of a given cycle is now defined as the sum of the flux diagrams associated with it. Referring again to Fig. 4,

$$\text{cycle } a = \text{I}, \qquad \text{cycle } b = \text{II} + \text{III} + \text{IV}, \qquad \text{cycle } c = \text{V} + \text{VI} \qquad (28)$$

The individual cycle fluxes are then given by

$$J_a = N(\text{cycle } a)/\Sigma, \qquad J_b = N(\text{cycle } b)/\Sigma \qquad (29)$$

etc.

It is convenient at this point to generalize Eq. (29) and to write it explicitly for the kth cycle flux, J_k:

$$J_k = N(\Pi_{k+} - \Pi_{k-})\Sigma_k/\Sigma \qquad (30)$$

where Π_{k+} and Π_{k-} are the products of rate constants round the cycle in the counterclockwise and clockwise directions, respectively, and Σ_k is the sum, over all flux diagrams belonging to the kth cycle, of the products of rate constants along the "appendages" feeding into the cycle. (Remember: each partial diagram generates a flux diagram on "closing" the kth cycle, which must be done with one line only. Two or more partial diagrams may generate the same flux diagram.) If *no* appendages feed into the cycle, $\Sigma_k \equiv 1$ (for instance, cycle a in the model—cf. flux diagram I of Fig. 4). Note that the number of independent cycle fluxes is equal to the number of cycles. The transition flux J_{ij} can be expressed as an algebraic sum of the cycle fluxes of the different cycles which contain the transition $i \rightarrow j$.[16] Equation (26) is thus replaced by the following expression:

$$J_{ij} = N(\text{sum of } i \rightarrow j \text{ cycles})/\Sigma = \text{sum of } i \rightarrow j \text{ cycle fluxes} \quad (31)$$

where an $i \rightarrow j$ cycle is a cycle that includes the transition $i \rightarrow j$ (if $i \rightarrow j$ occurs in the clockwise direction the cycle must be considered negative). As will be seen, the cycle fluxes represent a much more intrinsic aspect of the kinetic behavior than do the transition fluxes. For example, if two or more operational fluxes involve the same cycle flux when expressed according to Eq. (31), they must be thermodynamically coupled. It should be emphasized that in modeling ion pumps and similar energy-transducing mechanisms, it is fruitless to search for one or more crucial steps in the completely coupled enzymatic cycle at which free energy transfer between ligands supposedly occurs. Free energy transfer between the ligand molecules is an indivisible property of the entire cycle.[32]

It can be shown that, in conformity with one's intuition, each cycle flux can be expressed as the difference between two unidirectional cycle fluxes. Rewriting Eq. (30),

$$J_k = J_{k+} - J_{k-} = (N\Pi_{k+}\Sigma_k/\Sigma) - (N\Pi_{k-}\Sigma_k/\Sigma) \quad (32)$$

where J_{k+} and J_{k-}, which correspond to the first and second terms on the right-hand side, respectively, are both positive, while J_k may have either sign. [Note that unidirectional cycle fluxes are related to cyclic diagrams as cycle fluxes are to flux diagrams, cf. Eq. (27).] Now suppose that, over a long time period in the steady state, the fractions of completed unidirectional cycles of type a^+, a^-, b^+, b^-, etc., are p_{a+}, p_{a-}, p_{b+}, p_{b-}, etc. These "cycle probabilities" (which sum to unity) are given by[16]

$$p_{k+} = \Pi_{k+}\Sigma_k/(\Pi_{a+}\Sigma_a + \Pi_{a-}\Sigma_a + \Pi_{b+}\Sigma_b + \cdots) \quad (33)$$

i.e.,

$$p_{k+} = \frac{(\text{sum of cyclic diagrams belonging to } k^+)}{(\text{sum of all cyclic diagrams})} \quad (34)$$

Microscopic Reversibility: Thermodynamic Forces

It will now be seen that the rate constants of the model are not independent of each other. Interdependence of the rate constants is a consequence of microscopic reversibility (i.e., the principle of detailed balancing), a condition which must be satisfied.[16] Consider cycle b at equilibrium. From

---

[32] T. L. Hill and E. Eisenberg, *Q. Rev. Biophys.* **14**, 463 (1981).

microscopic reversibility,

$$\Pi_{b+} = \Pi_{b-} \quad \text{or} \quad \alpha_{12}\alpha_{23}\alpha_{31} = \alpha_{13}\alpha_{32}\alpha_{21} \qquad \text{(equilibrium)} \qquad (35)$$

However, since $c_{L_i^+} = c_{L_o^+}$ at equilibrium, we find

$$\alpha_{12}^*\alpha_{23}\alpha_{31}/\alpha_{13}^*\alpha_{32}\alpha_{21} = 1 \tag{36}$$

This is a condition between invariant rate constants which always holds. Similarly, for cycle c,

$$\Pi_{c+} = \Pi_{c-} \quad \text{or} \quad \alpha_{24}\alpha_{45}\alpha_{53}\alpha_{32} = \alpha_{35}\alpha_{54}\alpha_{42}\alpha_{23} \qquad \text{(equilibrium)} \quad (37)$$

and since $Kc_S = c_P$ at equilibrium, where K is the equilibrium constant of the reaction,

$$\alpha_{24}^*\alpha_{45}\alpha_{53}\alpha_{32}/\alpha_{35}^*\alpha_{54}\alpha_{42}\alpha_{23} = K \tag{38}$$

No new relation is added by cycle a, since there are only two independent operational flows and their conjugate forces.

For any steady state we can now write for cycles b and c, from Eqs. (36) and (38)

$$\Pi_{b+}/\Pi_{b-} \equiv J_{b+}/J_{b-} = c_{L_i^+}^n/c_{L_o^+}^n = e^{nX_1/RT} \tag{39}$$

$$\Pi_{c+}/\Pi_{c-} \equiv J_{c+}/J_{c-} = Kc_S/c_P = e^{X_2/RT} \tag{40}$$

Similarly, for cycle a, it is readily seen that

$$\Pi_{a+}/\Pi_{a-} \equiv J_{a+}/J_{a-} = Kc_S c_{L_i^+}^n/c_P c_{L_o^+}^n = e^{(nX_1+X_2)/RT} \tag{41}$$

Here X_1 is the chemical potential difference of the ligand $\Delta\mu_{L^+}$, and X_2 is the reaction affinity A. It is a straightforward procedure to relax the condition $\Delta\psi = 0$ (an example will be given below). When this is done, $X_1 = \Delta\tilde{\mu}_{L^+}$.

Steady-State Operational Fluxes: Coupling and Stoichiometry

We have already seen that for the model of Fig. 2 the transition fluxes J_{12} and J_{24} may be identified with the operational (or observable) fluxes, and we shall denote them by J_1/n (since we are interested in the rate of transport of single ligand molecules) and J_2, respectively. Then, from Eq. (31)

$$J_1/n = J_a + J_b, \qquad J_2 = J_a + J_c \tag{42}$$

Thus we have expressed the operational fluxes in terms of the cycle fluxes. Each operational flux is the sum of all the cycle fluxes in which it participates. It is useful now to use the abbreviated notation

$$a = N\Pi_a\text{-}\Sigma_a/\Sigma, \qquad b = N\Pi_b\text{-}\Sigma_b/\Sigma \tag{43}$$

etc. From Eqs. (30) and (39)–(41),

$$J_a = a(e^{(nX_1+X_2)/RT} - 1), \qquad J_b = b(e^{nX_1/RT} - 1)$$
$$J_c = c(e^{X_2/RT} - 1) \tag{44}$$

In Eqs. (44) the cycle fluxes are expressed in terms of the forces operating in each cycle. This is the most useful form in which the cycle fluxes can be written. It will be noted that cycle a, which is the source of the coupling between the operational fluxes, is acted on by a complex force [cf. Eq. (20)]. The kinetic coefficients a, b, and c are nothing other than the clockwise unidirectional cycle fluxes, i.e., $a = J_{a-}$, etc. [cf. Eq. (32)], and, of course, are functions of the external concentrations (and electrical potentials).

With the above expressions for the cycle fluxes the operational fluxes can be given in full. From Eqs. (42) and (44)

$$J_1 = na(e^{(nX_1+X_2)/RT} - 1) + nb(e^{nX_1/RT} - 1)$$
$$J_2 = a(e^{(nX_1+X_2)/RT} - 1) + c(e^{X_2/RT} - 1) \tag{45}$$

In the Onsager domain $nX_1 \ll RT$ and $X_2 \ll RT$. On expanding the exponentials for this condition we find

$$J_1 = n^2(a^{eq} + b^{eq})X_1/RT + na^{eq}X_2/RT$$
$$J_2 = na^{eq}X_1/RT + (a^{eq} + c^{eq})X_2/RT \tag{46}$$

where the concentrations (and $\Delta\psi$) in a^{eq}, b^{eq}, and c^{eq} are those at equilibrium. Hence we have recovered linear and symmetrical flux relations of the form of Eq. (9)—but only for the near-equilibrium domain. Under these conditions, and for models of this kind, we have for the degree of coupling and the phenomenological stoichiometry

$$q^{ons} = 1/\sqrt{(1 + b^{eq}/a^{eq})(1 + c^{eq}/a^{eq})} \tag{47}$$

$$Z^{ons} = n\sqrt{(1 + b^{eq}/a^{eq})/(1 + c^{eq}/a^{eq})} \tag{48}$$

These quantities are essentially constant within the Onsager domain, and may be regarded as fundamental kinetic properties of the system. From Eqs. (47) and (48) it can be shown that q^{ons}, Z^{ons}, and n are related by the condition

$$1/q^{ons} \geq Z^{ons}/n \geq q^{ons} \tag{49}$$

and thus at high degrees of coupling Z^{ons} approaches n in value.

Determination of q^{ons} and Z^{ons}. Referring to Eq. (16), it is seen that in general, for systems in the linear domain near equilibrium, the level flow ratio of flows is

$$(J_1/J_2)^{lf} = q^{ons} Z^{ons} \qquad (50)$$

It turns out that this relation is also obeyed exactly far from equilibrium.[31] Hence forward and reverse level flow measurements in kinetically reversible systems at high values of the forces should yield individual values of q^{ons} and Z^{ons}.

Multidimensional Inflection Points

Flux relations such as Eqs. (45) describe a sigmoid type of behavior, in other words each flow bears a sigmoid relationship to each force when the other force is held constant, and an inflection point will be observed between two regions of saturation. It was shown by Rothschild *et al.*[33] that providing a kinetic system satisfies certain conditions which, in our terms, are not overly restrictive, it should exhibit one or more multidimensional inflection points (MIPs). These are steady states in the vicinity of which linear behavior occurs over a considerable range for all flows, irrespective of which force is chosen for variation. If the MIP coincides with equilibrium, symmetry will be observed and in general the Onsager domain of linearity may be widened. If the MIP is situated far from equilibrium, symmetry does not normally occur; however, it was found by Caplan[31] that in the special case of complete coupling symmetry always occurs at the MIP.

Solutions for the MIPs of a system (if they exist) are found as follows.[31,33] Suppose c_1 and c_2 are the concentrations to be varied, i.e., c_1 may, in the above model, be either $c_{L_i^+}$ or c_{L_i}, and c_2 may be either c_S or c_P. Then Σ must have the following structure if an MIP exists when concentrations are varied:

$$\Sigma = f + f_1 c_1^n + f_2 c_2 + f_{12} c_1^n c_2 \qquad (51)$$

in which the f values are constants. Then, at the MIP,

$$c_1^n = \sqrt{ff_2/f_1 f_{12}}, \qquad c_2 = \sqrt{ff_1/f_2 f_{12}} \qquad (52)$$

We have not yet dealt explicitly with variations in $\Delta\psi$ (see next section). However, when $\Delta\psi$ is varied instead of c_1, Σ must have a structure completely analogous to Eq. (51) if an MIP exists, in which c_1^n is replaced by $\exp(nzF\Delta\psi/RT)$.

[33] K. J. Rothschild, S. A. Ellias, A. Essig, and H. E. Stanley, *Biophys. J.* **30,** 209 (1980).

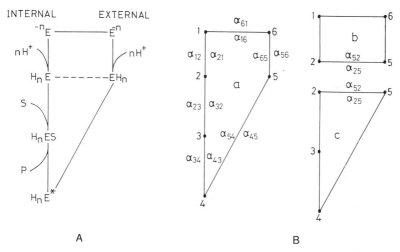

FIG. 5. Six-state model of a proton pump. (A) Complete diagram. E is a membrane-bound protein with n negatively charged proton-binding sites accessible either to the internal medium or to the external medium. Substrate(s) S and product(s) P are bound or released on the internal side only. The reaction $S \rightarrow P$ brings about a conformational change in E to E*, in which state it undergoes a transition which changes the accessibility of the bound protons from the internal to the external medium. The transition indicated by a dashed line introduces the possibility of slip (intrinsic uncoupling). (B) The cycles of the diagram. The states are assigned numbers and the α_{ij} are first-order or pseudo-first-order rate constants. Cycle a, completely coupled cycle; cycle b, proton slip cycle; cycle c, reaction slip cycle. [From Pietrobon and Caplan,[30] reprinted with permission. Copyright (1985) American Chemical Society.]

Force–Flow Relationships with and without Intrinsic Uncoupling

A Six-State Pump Model

The model we have used to study the pattern of flow–force relationships that a representative two-flow coupled process (such as a proton pump) can generate, and the properties of the regions of approximate linearity, are shown in Fig. 5A.[30] We consider this the minimal model that can be used as a simplified representation of, for example, the ATPase proton pump, where S represents ATP and P corresponds to both ADP and P_i. The transition indicated by the dashed line introduces the possibility of uncoupled cycles (slips). Both the completely coupled cycle a and the uncoupled cycles b and c are shown in Fig. 5B. In this model the substrate–product conversion is assumed to be very fast, as has been found experimentally for the mitochondrial ATPase,[34,35] but this assumption is not made for the voltage-dependent step, which corresponds to a conformational change translocating or otherwise changing the accessibility of the negatively charged empty proton binding sites from one side of the membrane to the other. The rate constants α_{16} and α_{61} are both dependent on the electrical potential gradient. Thermodynamic considerations[16,31] lead to the following relation:

$$\alpha_{16}/\alpha_{61} = [\alpha_{16}(0)/\alpha_{61}(0)]e^{-(nF/RT)\Delta\psi_m} \tag{53}$$

where $\Delta\psi_m$ ($=\psi_m^{in} - \psi_m^{ex}$) is the electrical potential difference between the planes in the membrane corresponding to the positions of the n empty binding sites in states 1 and 6, and $\alpha_{16}(0)$ and $\alpha_{61}(0)$ are the values of the rate constants when $\Delta\psi_m = 0$. If we assume that these planes coincide with the two surfaces of the membrane, $\Delta\psi_m$ is the electrical potential difference across the membrane, which is related to the bulk phase potential difference, $\Delta\psi$, through the phase boundary potentials $\Delta\psi_b^{in}$ ($=\psi_m^{in} - \psi^{in}$) and $\Delta\psi_b^{ex}$ ($=\psi_m^{ex} - \psi^{ex}$):

$$\Delta\psi = (\psi^{in} - \psi^{ex}) = \Delta\psi_m + (\Delta\psi_b^{ex} - \Delta\psi_b^{in}) \tag{54}$$

For simplicity, we consider a symmetrical Eyring barrier.[6,36,37] The individual rate constants are then given by

$$\alpha_{16} = \alpha_{16}(0)e^{-(nF/2RT)\Delta\psi_m}, \qquad \alpha_{61} = \alpha_{61}(0)e^{(nF/2RT)\Delta\psi_m} \tag{55}$$

In this model the binding of protons is considered to be a diffusion-controlled process. Hence we assign the effects of the phase boundary potentials $\Delta\psi_b^{in}$ and $\Delta\psi_b^{ex}$ to the rates of unbinding, i.e., to the rate constants α_{21} and α_{56}, respectively.[31] Note that the planes corresponding to the positions of the binding sites need not necessarily coincide with the inner and outer membrane surfaces. If not, part of the voltage drop would lie between the surfaces and the binding sites and the binding constants α_{12} and α_{65} would become $\Delta\psi$ dependent.[9]

If the number of protons translocated per cycle a is higher than 1, the transitions between states 1 and 2, and 5 and 6, are actually "reduced" transitions[16] involving a number of partially protonated intervening transient states. The corresponding pseudo-first-order rate constants are functions of the proton concentrations and the rate constants of the elementary transitions. If the reaction gives rise to more than one product (as in the case, for example, of ATP hydrolysis), the transitions between states 3 and

[34] C. Grubmeyer, R. L. Cross, and H. S. Penefsky, *J. Biol. Chem.* **257**, 12092 (1982).

[35] C. C. O'Neal and P. D. Boyer, *J. Biol. Chem.* **259**, 5761 (1984).

[36] P. Läuger and G. Stark, *Biochim. Biophys. Acta* **211**, 458 (1970).

[37] T. L. Hill and Y. Chen, *Proc. Natl. Acad. Sci. U.S.A.* **66**, 607 (1970).

4 are reduced transitions involving one transient intervening state, and the rate constants are functions of those of the elementary transitions and the product concentration. A detailed description of the reduction procedure used to eliminate explicit consideration of transient intermediates is beyond the scope of this review, but the general approach has been described by Hill[16] and Pietrobon and Caplan.[30] Usually the procedure determines upper limits for the values of the reduced rate constants.

Characteristics of Complete Coupling: Kinetically Reversible and Irreversible Pumps

Computer simulations of models such as that of Fig. 5 are readily carried out by means of the diagram method, using, for example, straightforward FORTRAN programming to evaluate the cycle fluxes. In this way one can generate families of force–flow relationships under various conditions. To commence with, consider only the completely coupled cycle *a*. The steady-state reaction flow J_r and proton flow J_H are given by

$$J_r = a(e^{(A + n\Delta\tilde{\mu}_H)/RT} - 1) \qquad [a = f_a(c_H^{in}, c_H^{ex}, c_S, c_P, \Delta\psi)] \qquad (56)$$

$$J_H = nJ_r \qquad (57)$$

where, as discussed earlier, a is a function of all the concentrations and the electrical potential difference, as well as all the rate constants of the diagram. Figures 6 and 7 show simulations of the flow–force relationships computed for Eqs. (56) and (57) for the two sets of rate constants specified in the legends.[30] In both figures, panel a depicts the variation of J_r with $\Delta\tilde{\mu}_H$ at different constant values of A, while panel b depicts the variation of J_r with A at different constant values of $\Delta\tilde{\mu}_H$. Figure 6 corresponds to the case of a kinetically reversible pump (pump A). The simulations are performed on the basis of the ATPase pump, using the equilibrium constant for hydrolysis of ATP and $n = 3H^+/ATP$. In this case A is the affinity of the ATP hydrolysis reaction and J_r is positive for the hydrolysis reaction (and negative for the ATP synthesis reaction), since the reaction term in the dissipation function $\Phi = J_r A + J_H \Delta\tilde{\mu}_H$ is regarded as the input term. On the other hand, Fig. 7 corresponds to the case of a kinetically irreversible (or practically irreversible) pump (pump B). Here the equilibrium constant of the electron transfer reaction from succinate to oxygen is used with $n = 4H^+/e^-$. Note that the sets of rate constants apart from satisfying microscopic reversibility and the constraints due to the reductions of the model are chosen so as to situate the flow-controlling ranges of forces and the values of the flows close to the experimentally determined ones.[11] From Figs. 6a and 7a it can be seen that the total range of control of the flows by $\Delta\tilde{\mu}_H$ is confined within very narrow limits, approximately 1 kcal/mol.

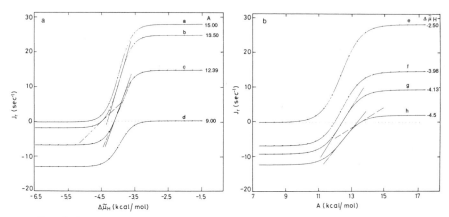

FIG. 6. Completely coupled pump (pump A). Dependence of J_r on $\Delta\tilde{\mu}_H$ for different constant values of A (a) and of J_r on A for different constant values of $\Delta\tilde{\mu}_H$ (b). The reaction is assumed to be ATP hydrolysis, so that positive values of J_r represent rates of ATP hydrolysis, whereas negative values of J_r represent rates of ATP synthesis. A is varied by varying c_{ATP}; $\Delta\tilde{\mu}_H$ is varied by varying $\Delta\psi$. The parameters of the simulation are as follows: $n = 3$, $K = 1.71 \times 10^6 M$, $c_{ADP} = 10^{-4} M$, $c_{P_i} = 10^{-2} M$, $pH_{in} = pH_{ex} = 7.4$, $\Delta\psi_b^{in} = 0$, $\Delta\psi_b^{ex} = -50$ mV, $\alpha_{12} = 4 \times 10^2$ sec^{-1}, $\alpha_{21} = 40$ sec^{-1}, $\alpha_{32} = 5 \times 10^3$ sec^{-1}, $\alpha_{34} = 10^2$ sec^{-1}, $\alpha_{43} = 10^2$ sec^{-1}, $\alpha_{45} = 10^2$ sec^{-1}, $\alpha_{54} = 10^2$ sec^{-1}, $\alpha_{56} = 10^3$ sec^{-1}, $\alpha_{65} = 10^3$ sec^{-1}, $\alpha_{25} = 5.85 \times 10^{-30}$ sec^{-1}, $\alpha_{52} = 1 \times 10^{-20}$ sec^{-1}, $\alpha_{23}^* = 5 \times 10^6 M^{-1}$ sec^{-1}, $\alpha_{16}(0) = 10^2$ sec^{-1}, $\alpha_{61}(0) = 4.98 \times 10^7$ sec^{-1}. [From Pietrobon and Caplan,[30] reprinted with permission. Copyright (1985) American Chemical Society.]

Sigmoidal flow–force relationships are a general characteristic of models of ion pumps or cotransport systems involving cyclic reaction schemes with one voltage-dependent step.[7,8] This characteristic is not conditional on the reaction sequence or on the position in this sequence of the step chosen to be voltage-dependent.[8] An important consequence of the sigmoidicity of the curves is illustrated in Fig. 6. In Fig. 6a, curve c shows the expected behavior for a reversible pump: when $|\Delta\tilde{\mu}_H| > |\Delta\tilde{\mu}_H^{eq}|$ (i.e., the value of $\Delta\tilde{\mu}_H$ at which $J_r = 0$, $|\Delta\tilde{\mu}_H^{eq}| = A/n$), ATP synthesis occurs, while when $|\Delta\tilde{\mu}_H| < |\Delta\tilde{\mu}_H^{eq}|$, hydrolysis occurs. However, for the same pump, curve a (at higher affinity) shows virtually zero synthesis at $|\Delta\tilde{\mu}_H| > |\Delta\tilde{\mu}_H^{eq}|$, while curve d (at low affinity) shows virtually zero hydrolysis at $|\Delta\tilde{\mu}_H| < |\Delta\tilde{\mu}_H^{eq}|$. The typical relationship between rate of ATP synthesis and $\Delta\tilde{\mu}_H$, characterized by a threshold value of $|\Delta\tilde{\mu}_H|$ below which no ATP synthesis occurs,[38–40] resembles curve d in Fig. 6a. It may therefore be, in principle, a simple consequence of the kinetic properties of the enzyme. Depending on the value of the affinity ("ΔG_p"), the generally kinetically reversible pump can behave as if it were kinetically irreversible. Hence it is quite likely that

[38] A. Baccarini Melandri, R. Casadio, and B. A. Melandri, *Eur. J. Biochem.* **78**, 389 (1977).
[39] M. Zoratti, D. Pietrobon, and G. F. Azzone, *Eur. J. Biochem.* **126**, 443 (1982).
[40] J. E. G. McCarthy and S. J. Ferguson, *Eur. J. Biochem.* **132**, 425 (1983).

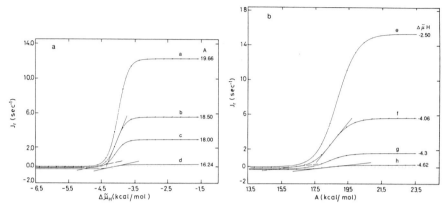

FIG. 7. Completely coupled pump (pump B). (a and b) As for Fig. 6, but the reaction is assumed to be electron transfer between succinate and oxygen so that J_r represents electron flow. The affinity A corresponds to the transfer of a single electron and is varied by varying $c_{\text{succinate}}$. $\Delta\tilde{\mu}_H$ is varied by varying $\Delta\psi$. The parameters of the simulation are as follows: $n = 4$, $K = 1.31 \times 10^{13} M$ (at pH = 7.4 and $P_{O_2} = 0.2$ atm), $c_{\text{fumarate}} = 5 \times 10^{-5} M$, $pH_{in} = pH_{ex} = 7.4$, $\psi_b^{in} = 0$, $\psi_b^{ex} = -50$ mV, $\alpha_{12} = 4 \times 10^2$ sec^{-1}, $\alpha_{21} = 5$ sec^{-1}, $\alpha_{32} = 8 \times 10^4$ sec^{-1}, $\alpha_{34} = 4 \times 10^2$ sec^{-1}, $\alpha_{43} = 4 \times 10^2$ sec^{-1}, $\alpha_{45} = 40$ sec^{-1}, $\alpha_{54} = 0.4$ sec^{-1}, $\alpha_{56} = 10^2$ sec^{-1}, $\alpha_{65} = 10^5$ sec^{-1}, $\alpha_{25} = 6.75 \times 10^{-33}$ sec^{-1}, $\alpha_{52} = 1 \times 10^{-20}$ sec^{-1}, $\alpha_{23}^* = 1 \times 10^6 M^{-1/2}$ sec^{-1}, $\alpha_{16}(0) = 1.3 \times 10^2$ sec^{-1}, $\alpha_{61}(0) = 10^{12}$ sec^{-1}. [From Pietrobon and Caplan,[30] reprinted with permission. Copyright (1985) American Chemical Society.]

at least some cases of kinetic irreversibility are attributable to sigmoidicity in an appropriate range of the flows and forces.[7] Gating effects, observed for example in the conductance of F_0, as well as in the behavior of many carriers, may be predicted from the regions of saturation. (This is readily seen on examination of the curve for $q^{\text{ons}} \approx 0$ in Fig. 9b below.)

In these simulations $\Delta\tilde{\mu}_H$ is varied by varying $\Delta\psi$. However, in general, kinetic equivalence of $\Delta\psi$ and ΔpH is not expected, and hence the flow–force relationships depend on which of the two parameters is varied. The presence or absence of kinetic equivalence of this kind is governed by a set of specific conditions limiting the relative magnitudes of the rate constants.[7,30]

Characteristics of Intrinsic Uncoupling

We now consider the entire model, including the slip cycles b and c. The equations become [cf. Eqs. (45)]

$$J_r = a(e^{(A+n\Delta\tilde{\mu}_H)/RT} - 1) + c(e^{A/RT} - 1)$$
$$[c = f_c(c_H^{in}, c_H^{ex}, c_S, c_P, \Delta\psi)] \qquad (58)$$

$$J_H = na(e^{(A+n\Delta\tilde{\mu}_H)/RT} - 1) + nb(e^{n\Delta\tilde{\mu}_H/RT} - 1)$$
$$[b = f_b(c_H^{in}, c_H^{ex}, c_S, c_P, \Delta\psi)] \qquad (59)$$

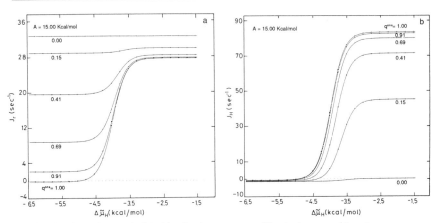

FIG. 8. Flow–force relationships in the presence of intrinsic uncoupling for pump A at high fixed affinity. Families of curves corresponding to increasing α_{52} and α_{25} (degree of coupling q^{ons} varies from 1 to 0). $\Delta\tilde{\mu}_H$ varied by varying $\Delta\psi$. (a) J_r versus $\Delta\tilde{\mu}_H$: reaction slip decreases as $|\Delta\tilde{\mu}_H|$ decreases. (b) J_H versus $\Delta\tilde{\mu}_H$: proton slip negligible, proton flux reduced due to competition between cycles c and a. [From Pietrobon and Caplan,[30] reprinted with permission. Copyright (1985) American Chemical Society.]

It is seen that the stoichiometric relationship between J_r and J_H no longer exists. Figures 8 and 9 show simulations of the flow–force relationships [Eqs. (58) and (59)] for the two sets of rate constants specified in the legends to Figs. 6 and 7, but allowing α_{52} and α_{25} to increase so as to give rise to families of curves corresponding to different degrees of coupling. The parameter used here is the degree of coupling q^{ons} characterizing the linear domain near equilibrium where Onsager symmetry holds. The limiting curve at $q^{\text{ons}} = 1$ in Fig. 8 is identical to curve a of Fig. 6a. This is the ATP-driven pump operating at high affinity in the proton-pumping direction. Decreasing q^{ons} results in an increase in the hydrolysis rate at all values of $\Delta\tilde{\mu}_H$ which primarily reflects the contribution of cycle c, the reaction slip. It is seen that this contribution increases as $|\Delta\tilde{\mu}_H|$ increases. Evidently, the higher the force against which the pump moves protons, the higher the probability of reaction slip. In fact, the probability of cycle c increases at the expense of that of cycle a. In a sense the two cycles may be regarded as in "competition" with each other. Thus the decrease in J_H at low $\Delta\tilde{\mu}_H$ in Fig. 8b as q^{ons} decreases reflects primarily a decreasing probability of cycle a as a consequence of such competition. There appears to be only a very small contribution of cycle b at high $|\Delta\tilde{\mu}_H|$. On the other hand, Fig. 9 shows that at low affinity, when the same pump works in the ATP synthesis direction, cycle b is the main uncoupled cycle while the redox slip is negligible at all values of $\Delta\tilde{\mu}_H$.

The lowest curve in Fig. 9b, for which $q^{\text{ons}} \approx 0$, gives the relationship

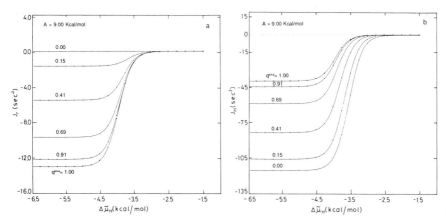

FIG. 9. Flow–force relationships in the presence of intrinsic uncoupling for pump A at low fixed affinity. Families of curves correspond to increasing α_{52} and α_{25} (degree of coupling q^{ons} varies from 1 to 0). $\Delta\tilde{\mu}_H$ varied by varying $\Delta\psi$. (a) J_r versus $\Delta\tilde{\mu}_H$: reaction slip negligible, ATP synthesis decreased due to competition between cycles b and a. (b) J_H versus $\Delta\tilde{\mu}_H$: proton slip occurring, with steep dependence of J_H on $\Delta\tilde{\mu}_H$. [From Pietrobon and Caplan,[30] reprinted with permission. Copyright (1985) American Chemical Society.]

between J_H and $\Delta\tilde{\mu}_H$ characteristic of completely uncoupled pumps, which under these conditions act as proton leaks. Note the very steep dependence of J_H on $\Delta\tilde{\mu}_H$ above a threshold value. Figure 9a shows the decrease in the rate of ATP synthesis as a result of uncoupling. This decrease, as before, is due to the decreased probability of cycle a, which is now in competition with cycle b. The bottom line is that both the relative contributions and the magnitude of the cycles depend on the magnitude of the forces. As the affinity of the reaction increases the reaction slip increases while the proton slip decreases. As $|\Delta\tilde{\mu}_H|$ increases both the proton and the reaction slip increase. To the extent that the contribution of one or other uncoupled cycle is increased the contribution of the fully coupled cycle is decreased due to the competition between them. This regulation of the slip by the forces may well have physiological significance (for the efficiency of oxidative phosphorylation, for example[22]) and is in accord with present data in the literature. Another possible level of regulation of the slip could be achieved by specific physiological effectors.

Figure 10 illustrates the dependence of the flow ratio $J_H/J_r n$ on the force ratio $n\Delta\tilde{\mu}_H/A$ for pump B at several different values of q^{ons}.[41] The behavior shown is instructive with regard to stoichiometry measurements both on flow and force ratios in systems which exhibit only intrinsic uncoupling. It is seen that as $\Delta\tilde{\mu}_H$ approaches zero J_H/J_r becomes constant, i.e., the condition $\Delta\tilde{\mu}_H = 0$ need not be attained to determine $(J_H/J_r)^{lf}$. At

[41] D. Pietrobon, M. Zoratti, G. F. Azzone, and S. R. Caplan, *Biochemistry* **25**, 767 (1986).

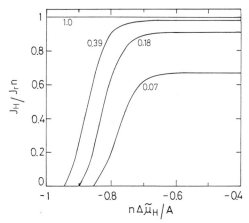

FIG. 10. Dependence of the flow ratio $J_H/J_r n$ on the force ratio $n\Delta\tilde{\mu}_H/A$ for pump B at various indicated values of q^{ons}. $A = 19.66$ kcal/mol; $\Delta\tilde{\mu}_H$ varied by varying $\Delta\psi$. [After Pietrobon *et al.*,[41] with permission. Copyright (1985) American Chemical Society.]

static head the ratio $(n\Delta\tilde{\mu}_H/A)^{\mathrm{sh}}$ remains close to unity even at quite low values of q^{ons}.

Linear Domain

In the completely coupled case near equilibrium, when both $|A| \ll RT$ and $|n\Delta\tilde{\mu}_H| \ll RT$ or when $|A + n\Delta\tilde{\mu}_H| \ll RT$, Eqs. (56) and (57) become, respectively,

$$J_r = (a^{\mathrm{eq}}/RT)(A + n\Delta\tilde{\mu}_H) \tag{60}$$

$$J_H = (na^{\mathrm{eq}}/RT)(A + n\Delta\tilde{\mu}_H) \tag{61}$$

where $a^{\mathrm{eq}} = f_a(c_H^{\mathrm{in(eq)}},\ c_H^{\mathrm{ex(eq)}},\ c_S^{\mathrm{(eq)}},\ c_P^{\mathrm{(eq)}},\ \Delta\psi^{\mathrm{(eq)}})$. The condition $|A + n\Delta\tilde{\mu}_H| \ll RT$ defines an extremely restricted linear region in which A may be varied around its equilibrium value A^{eq} while maintaining $|\Delta\tilde{\mu}_H|$ ($= A^{\mathrm{eq}}/n$) constant, or $\Delta\tilde{\mu}_H$ varied around its equilibrium value $\Delta\tilde{\mu}_H^{\mathrm{eq}}$ while maintaining A ($= |n\Delta\tilde{\mu}_H^{\mathrm{eq}}|$) constant. The extent of the variations over which Eqs. (60) and (61) hold will generally not exceed ± 0.01 kcal/mol.

The regions of approximate linearity around an inflection point arbitrarily far from equilibrium (which extend over a range of 0.5–1.5 kcal/mol) can be described by empirical equations of the type discussed earlier[26,28] [cf. Eqs. (22)]

$$J_r = L_r A + L_{rH}\Delta\tilde{\mu}_H + K_r \tag{62}$$

$$J_H = L_{Hr} A + L_H \Delta\tilde{\mu}_H + K_H \tag{63}$$

Since coupling is complete, $L_{Hr} = nL_r$, $L_H = nL_{rH}$, and $K_H = nK_r$. For purposes of discussion, consider curves b and h of Fig. 6. The slopes of the tangents (dashed lines) to the equilibrium point, defined for both curves by $A = 13.5$ kcal/mol and $\Delta\tilde{\mu}_H = -4.5$ kcal/mol, represent na^{eq} and a^{eq} respectively. The slopes of the tangents (solid lines) to the inflection points L_{rH} and L_r, respectively, are different. It is convenient to write[30]

$$L_{rH} = L_H/n = \alpha na^{eq}/RT \tag{64}$$

$$L_r = L_{Hr}/n = \beta a^{eq}/RT \tag{65}$$

Equations (62) and (63) for the extended region then become

$$J_r = (a^{eq}/RT)(\beta A + \alpha n\Delta\tilde{\mu}_H) + K_r \tag{66}$$

$$J_H = (na^{eq}/RT)(\beta A + \alpha n\Delta\tilde{\mu}_H) + nK_r \tag{67}$$

where $\alpha = \alpha(A)$ depends on the value of A held constant while $\Delta\tilde{\mu}_H$ is varied over an experimentally established linear range, while $\beta = \beta(\Delta\tilde{\mu}_H)$ depends on the value of $\Delta\tilde{\mu}_H$ held constant while A is varied over an experimentally established linear range. Similarly, K_r depends on the particular values of A and $\Delta\tilde{\mu}_H$ around which a region of linearity is found. These equations are similar to those suggested by van Dam et al.[29]:

$$J_r = L(A + \gamma n\Delta\tilde{\mu}_H) + K_r \tag{68}$$

$$J_H = nL(A + \gamma n\Delta\tilde{\mu}_H) + nK_r \tag{69}$$

where $L = \beta a^{eq}/RT$ and $\gamma = \alpha/\beta$.

Figures 6 and 7 clarify several aspects of the behavior of completely coupled pumps in terms of the parameters of Eqs. (66) and (67). First, in general, $\alpha \neq \beta$ (i.e., $L_{rH}/L_{Hr} = \alpha/\beta \neq 1$). This means that in general the two forces A and $\Delta\tilde{\mu}_H$ are kinetically inequivalent. Second, the symmetry-generating condition $\alpha = \beta$ (or $\gamma = 1$), considered by van Dam and co-workers to be associated with equilibrium,[29] occurs only at an MIP. For example, compare curves b and f in Fig. 7: here the forces characterizing the two curves combine to give an MIP far from the equilibrium point of both curves. The closer the MIP is to equilibrium, the closer α and β are to unity and the smaller is the value of the additive constant K_r. On the other hand, if the equilibrium state is not an MIP, then even if it is within an extended range of linearity when $\Delta\tilde{\mu}_H$ is varied ($\alpha = 1$), in general, $\beta \neq 1$ and hence $\gamma \neq 1$.

From Eqs. (62) and (63) one can formally write the parameters Z and q:

$$Z \equiv \sqrt{L_H/L_r}$$
$$q \equiv \sqrt{L_{Hr}L_{rH}/L_H L_r} \tag{70}$$

In the completely coupled case [cf. Eqs. (64) and (65)] $Z = n\sqrt{\alpha/\beta}$ and $q = 1$. Hence the phenomenological stoichiometry is a measure of the mechanistic stoichiometry only when $\alpha/\beta = 1$, i.e., at the MIP, even though the system is fully coupled. Away from the MIP Z has no meaning. To determine n in completely coupled systems the appropriate measure is not Z but L_H/L_{rH} [see Eq. (64)]. In the incompletely coupled case, meaningful parameters q and Z can be obtained from the slopes of the flow–force curves if the pump is operating in the vicinity of an MIP. However, these parameters are not simply related to q^{ons} and Z^{ons}.[30] It is found that in general q deviates very markedly from q^{ons} toward higher values, and may be quite close to unity even at low q^{ons}. This effect becomes larger as the MIP moves further away from the equilibrium point of the fully coupled cycle $(A + n\Delta\tilde{\mu}_H = 0)$. Furthermore, uncoupling introduces asymmetry: with decreasing q^{ons} the ratio L_{rH}/L_{Hr} falls from the value of unity at $q^{ons} = 1$. Z^{ons}/n increases rather rapidly with decreasing q^{ons}, but Z/n measured at the MIP decreases from the value of unity at $q^{ons} = 1$, although rather slowly. From the above observations it is clear that the phenomenological Eqs. (62) and (63) are of limited use in analyzing experiments in which both A and $\Delta\tilde{\mu}_H$ are varied.

On the other hand, many studies have been carried out in different systems in which only one force is varied, while the other (usually A) is kept constant (or assumed to be constant). In this case the question of kinetic equivalence is irrelevant. In the completely coupled case, instead of Eq. (65) one can write

$$L_r = L_{Hr}/n = \alpha a^{eq}/RT \qquad (A \text{ constant}) \tag{71}$$

giving, instead of Eqs. (66) and (67),

$$J_r = (\alpha a^{eq}/RT)(A + n\Delta\tilde{\mu}_H) + K'_r \qquad (A \text{ constant}) \tag{72}$$

$$J_H = (n\alpha a^{eq}/RT)(A + n\Delta\tilde{\mu}_H) + nK'_r \qquad (A \text{ constant}) \tag{73}$$

where K'_r is simply the intercept of the line J_r versus $(A + n\Delta\tilde{\mu}_H)$. The symmetry which appears in the above equations is entirely formal. However, if $\Delta\tilde{\mu}_H^{eq}$ is within the linear region (strictly, at the inflection point) of J_r versus $\Delta\tilde{\mu}_H$, $\alpha = 1$, and $K'_r = 0$. In general it is useful for experimental purposes to rewrite Eqs. (72) and (73) in the form

$$J_r = \left(\frac{\alpha a^{eq}}{RT} + \frac{K'_r}{A}\right)A + n\left(\frac{\alpha a^{eq}}{RT}\right)\Delta\tilde{\mu}_H = L_r^{eff}A + L_{rH}^{eff}\Delta\tilde{\mu}_H \tag{74}$$

$$J_H = n\left(\frac{\alpha a^{eq}}{RT} + \frac{K'_r}{A}\right)A + n^2\left(\frac{\alpha a^{eq}}{RT}\right)\Delta\tilde{\mu}_H = L_{Hr}^{eff}A + L_H^{eff}\Delta\tilde{\mu}_H \tag{75}$$

remembering that A is constant. These four effective phenomenological coefficients are readily determined by measuring the slopes and intercepts of the linear ranges of the curves J_r and J_H versus $\Delta\tilde{\mu}_H$.[22-24] From Eqs. (74) and (75),

$$\left(\frac{L_{rH}}{L_{Hr}}\right)^{\text{eff}} = \frac{1}{1+\delta}, \quad Z^{\text{eff}} = \frac{n}{\sqrt{1+\delta}}, \quad q_{rH}^{\text{eff}} = \frac{1}{\sqrt{1+\delta}}, \quad q_{Hr}^{\text{eff}} = \sqrt{1+\delta} \quad (76)$$

where

$$\delta = \frac{RTK_r'}{\alpha a^{\text{eq}} A} \quad (77)$$

and $q_{rH}^{\text{eff}} q_{Hr}^{\text{eff}} = 1$. If $\Delta\tilde{\mu}_H^{\text{eq}}$ is within the range of extended linearity around the inflection point, as in the case of curve c of Fig. 6a and curve d of Fig. 7a, $(L_{rH}/L_{Hr})^{\text{eff}} = 1$ and $Z^{\text{eff}} = n$, i.e., $\delta = 0$. With increasing distance of $\Delta\tilde{\mu}_H^{\text{eq}}$ from the region of approximate linearity (i.e., as we move from curve c to curve a in Fig. 6a or from curve d to curve a in Fig 7a) δ becomes increasingly negative (due to the increasingly negative value of K_r') and Z^{eff} increasingly overestimates n while $(L_{rH}/L_{Hr})^{\text{eff}}$, q_{rH}^{eff}, and q_{Hr}^{eff} increasingly deviate from 1 ($(L_{rH}/L_{Hr})^{\text{eff}} > 1$, $q_{rH}^{\text{eff}} > 1$, and $q_{Hr}^{\text{eff}} < 1$).[30]

An experimental determination of the symmetry parameter $(L_{rH}/L_{Hr})^{\text{eff}}$ gives an idea of how far from equilibrium the pumps are actually operating, since it becomes progressively larger than 1 as this distance increases. It should be noted that a finding that $(L_{rH}/L_{Hr})^{\text{eff}} \approx 1$[22,24] does not necessarily mean that the region of linearity considered is centered on equilibrium and that it coincides with the Onsager domain. Nevertheless, for completely coupled pumps one can safely assume in this case that the region of linearity is not far from equilibrium (δ is small) and hence Z^{eff} and q_{rH}^{eff} are not much greater than n and 1, respectively. This may justify, as a reasonable approximation, the use of proportional phenomenological equations for regions of extended linearity not containing the equilibrium point but sufficiently close to it.

In the case of incompletely coupled pumps, however, the analysis is complicated by the fact that the ratio $(L_{rH}/L_{Hr})^{\text{eff}}$ decreases with increasing uncoupling.[30] Hence increasing uncoupling can lead to apparent symmetry, by mutual compensation of the effects of the distance from equilibrium and $q^{\text{ons}} < 1$. This compensation also occurs in the behavior of q_{rH}^{eff} and q_{Hr}^{eff}, where an apparent complete coupling is found at a value of $q^{\text{ons}} < 1$. The parameter Z^{eff} as well as the ratio $(L_H/L_{rH})^{\text{eff}}$ ($=L_H/L_{rH}$), which always has the value n at $q^{\text{ons}} = 1$ and is the most informative parameter with respect to the mechanistic stoichiometry, always increase with increasing uncoupling.

Cautionary Note

The above considerations indicate that the application of the linear phenomenological equations of near-equilibrium nonequilibrium thermodynamics to regions of approximate linearity situated away from equilibrium is at best a simplification to be treated with great caution. The values of q and Z that have been reported for mitochondrial proton pumps are for the most part effective values.[3,22,24,42] They are not simply related to q^{ons} and Z^{ons}, and in the absence of independent information regarding the distance from equilibrium of the flow-controlling ranges of the forces (i.e., how far from equilibrium the pumps normally operate), they give very little information about the degree of coupling and the mechanistic stoichiometry. In fact, their values are essentially governed by the position of the extended region of approximate linearity relative to the equilibrium point. The useful relationships describing the static head force ratio $(X_1/X_2^{sh} = -q^{ons}/Z^{ons})$ and the maximum efficiency of energy conversion [cf. Eq. (19)] in the Onsager domain do not hold for the extended regions of linearity at physiological (i.e., high) values of the forces, even on replacing the Onsager parameters by q^{eff} and Z^{eff} or by $q(MIP)$ and $Z(MIP)$. Maximal efficiency far from equilibrium η_{max} is higher than η_{max}^{ons} [30]; its value depends on the magnitudes of the forces so that the more the flow-controlling ranges of the output force are removed from equilibrium, the lower the maximal efficiency. A low static head force ratio does not necessarily indicate a low degree of coupling of the pump but rather the presence of a leak in parallel with a pump operating far from equilibrium. In fact, in the presence of a leak, at a given degree of uncoupling, the extent to which the ratio of output to input force underestimates the mechanistic stoichiometry increases as the flow-controlling ranges of the output force move further from equilibrium.

Applications Especially in Oxidative Phosphorylation

Use of the Dissipation Function in Estimating Stoichiometry

In this section we take the mitochondrial system as an example and illustrate our argument using the method outlined by Walz.[43] Figure 11 shows the processes pertinent to the redox-driven proton pump in an experiment in which ferricyanide is used as an electron acceptor in a suspension of mitochondria. For simplicity the phosphate carrier and other

---

[42] J. J. Lemasters, R. Grunwald, and R. K. Emaus, *J. Biol. Chem.* **259**, 3058 (1984).

[43] D. Walz, *in* "Biological Structures and Coupled Flows" (A. Oplatka and M. Balaban, eds.), p. 45. Academic Press, New York, and Balaban Publ., Philadelphia, Pennsylvania, 1983.

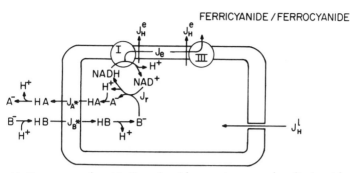

FIG. 11. Processes pertinent to the redox-driven proton pump in mitochondria. A^- and B^- represent acetoacetate and β-hydroxybutyrate, respectively. See the text for details. (After Walz.[43])

proton carriers are ignored. The substrate β-hydroxybutyrate (B^-) moves as a protonated species (HB) across the membrane, whereupon it is oxidized by NAD^+ to acetoacetate (A^-), which again permeates as a protonated species (HA). The resulting NADH injects electrons into the respiratory chain via NADH dehydrogenase (ubiquinone) (NADH-ubiquinone oxidoreductase) (complex I); these electrons are eventually transferred to the ferricyanide via the ubiquinol–cytochrome-c oxidoreductase (complex III). The substrate and product fluxes are denoted J_{B*} and J_{A*} respectively, where the asterisk indicates that the species concerned is protonated (we arbitrarily choose the direction in → out as positive). The flow of electrons through the respiratory chain, J_e, is coupled to the extrusion of protons, giving rise to the flux J_H^e. In addition, there is a leakage flux of protons, J_H^l. Considering the steady state of the membrane, the dissipation function is readily written down:

$$\Phi = J_e(\tilde{\mu}_e^{d(i)} - \tilde{\mu}_e^{a(o)}) + J_H^e \,\Delta\tilde{\mu}_H + J_H^l \,\Delta\tilde{\mu}_H$$
$$+ J_{B*} \,\Delta\tilde{\mu}_{B*} + J_{A*} \,\Delta\tilde{\mu}_{A*} + J_r A_r \qquad (78)$$

Here $\tilde{\mu}_e^{d(i)}$ and $\tilde{\mu}_e^{a(o)}$ represent, respectively, the electrochemical potential of the electron as defined by the inner electron-donating and the outer electron-accepting redox couple. The electrochemical potential of the electron in solution is given by[44]

$$\tilde{\mu}_e = (\tilde{\mu}_{red} - \tilde{\mu}_{ox})/\nu_e \qquad (79)$$

where the subscripts red and ox denote the reduced and oxidized forms of a redox couple which exchanges ν_e electrons. Thus for the ferricyanide/ferro-

[44] D. Walz, *Biochim. Biophys. Acta* **505**, 279 (1979).

cyanide couple

$$\tilde{\mu}_e^{a(o)} = \tilde{\mu}_{F2}^{(o)} - \tilde{\mu}_{F3}^{(o)} \tag{80}$$

where the subscripts F2 and F3 refer to bivalent and tervalent iron in the hexacyano complexes, and the superscript (o) denotes the outer medium. For the $NAD^+/NADH$ couple,

$$\tilde{\mu}_e^{d(i)} = [\tilde{\mu}_{NADH}^{(i)} - (\tilde{\mu}_H^{(i)} + \tilde{\mu}_{NAD^+}^{(i)})]/2 \tag{81}$$

where the dissociation equilibrium of the oxidized form is taken into account, and the superscript (i) denotes the inner medium. The oxidation of B^- to A^-, involving the passage of two electrons, may be written explicitly as

$$B^- + (NAD^+ + H^+) \rightarrow A^- + NADH + 2H^+ \tag{82}$$

The velocity and affinity of this reaction have been denoted J_r and A_r, respectively. Hence, since, in general, $A = -\Sigma_i \nu_i \tilde{\mu}_i$ [cf. Eq. (1)],

$$A_r = (\tilde{\mu}_B^{(i)} + \tilde{\mu}_{NAD^+}^{(i)}) - (\tilde{\mu}_A^{(i)} + \tilde{\mu}_{NADH}^{(i)} + \tilde{\mu}_H^{(i)}) \tag{83}$$

The donation of two electrons by NADH to complex I releases the proton bound in the reduction reaction. Thus each rotation of the cycle is associated with the generation of two protons in the inner medium. These are frequently referred to as the "scalar" protons.

It is convenient to complete this example on a "single-electron" basis. Consider the net redox reaction observable in the outer medium as a consequence of electron flow. Denoting the affinity of this reaction as $A_e^{(o)}$, we have

$$A_e^{(o)} = (\tilde{\mu}_B^{(o)} - \tilde{\mu}_A^{(o)})/2 - \tilde{\mu}_H^{(o)} - (\tilde{\mu}_{F2}^{(o)} - \tilde{\mu}_{F3}^{(o)}) \tag{84}$$

Combining this with Eqs. (80), (81), and (83) we find after some algebra that

$$A_e^{(o)} = (\tilde{\mu}_e^{d(i)} - \tilde{\mu}_e^{a(o)}) - (\Delta\tilde{\mu}_B - \Delta\tilde{\mu}_A - A_r)/2 + \Delta\tilde{\mu}_H \tag{85}$$

Since protonation reactions of small molecules in aqueous solution are very fast and may be assumed to be close to equilibrium,

$$\Delta\tilde{\mu}_A + \Delta\tilde{\mu}_H = \Delta\tilde{\mu}_{A^\bullet} \quad \text{and} \quad \Delta\tilde{\mu}_B + \Delta\tilde{\mu}_H = \Delta\tilde{\mu}_{B^\bullet} \tag{86}$$

Using Eqs. (85) and (86), further algebraic manipulation leads to the transformation of Eq. (78) into an alternative form:

$$\Phi = J_e A_e^{(o)} + (J_H^e + J_H^1 - J_e + J_{B^\bullet} + J_{A^\bullet}) \Delta\tilde{\mu}_H + (J_{B^\bullet} + J_e/2) \Delta\tilde{\mu}_B$$
$$+ (J_{A^\bullet} - J_e/2) \Delta\tilde{\mu}_A + (J_r - J_e/2) A_r \tag{87}$$

If we now consider the steady state of the mitochondrion as a whole, it is evident that the last three terms of this form of the dissipation function must be zero and that $J_{B^*} + J_{A^*} = 0$. Hence we arrive at the conclusion that

$$\Phi = J_e A_e^{(o)} + J'_H \Delta\tilde{\mu}_H \tag{88}$$

where $J'_H = J_H^e + J_H^1 - J_e$. This is an important result, in that it clarifies a commonly used method of evaluating the stoichiometry, assuming there are no slips or leaks. Adopting this assumption, we denote the mechanistic stoichiometry J_H^e/J_e by n_e. Then $J'_H/J_e = n_e - 1$ and we have

$$\Phi = J_e[A_e^{(o)} + (n_e - 1)\,\Delta\tilde{\mu}_H] \tag{89}$$

Consequently, at static head (when $\Phi = 0$),

$$A_e^{(o)}/-\Delta\tilde{\mu}_H = n_e - 1 \tag{90}$$

where $n_e - 1$ represents the stoichiometry based on the net proton efflux. The essential point to grasp here is that the ratio of the affinity of the redox reaction measured in the outer medium to the proton potential difference at static head gives the appropriate stoichiometry. It can be concluded that the stoichiometry of oxidative phosphorylation will likewise be given by the ratio of the affinity of the redox reaction to the "phosphate potential," both determined in the outer medium [cf. Eq. (11)]. Evidently, using the same assumption (no slips or leaks), a relationship analogous to Eq. (89) can be derived for the phosphorylation site:

$$\Phi = J_p(A_p^{(o)} + n_p\,\Delta\tilde{\mu}_H) \tag{91}$$

were n_p is the mechanistic stoichiometry J_H^p/J_p. Then, at static head,

$$A_p^{(o)}/-\Delta\tilde{\mu}_H = n_p \tag{92}$$

By combining Eqs. (90) and (92) we find for the static head of oxidative phosphorylation,

$$A_e^{(o)}/A_p^{(o)} = (n_e - 1)/n_p \tag{93}$$

(In a detailed analysis of the phosphorylation site, n_p would be replaced by $n_p + 1$ in the above equation, owing to the operation of the adenine nucleotide translocator.)

Forman and Wilson[45] used an experiment of this kind to determine the stoichiometry of oxidative phosphorylation from NADH to cytochrome c in rat liver mitochondria, obtaining a value of 2.0 ATP/2e^-. This result was criticized by Scholes and Hinkle[46] and by Lemasters et al.[42] on

[45] N. G. Forman and D. F. Wilson, J. Biol. Chem. 257, 12908 (1982).
[46] T. A. Scholes and P. C. Hinkle, Biochemistry 23, 3341 (1984).

grounds that the free energy change of the redox reaction $-\Delta G_e$ (which is equivalent to its affinity) was calculated on the basis of the internal rather than the external pH. Indeed, as pointed out by Lemasters et $al.$, since the internal pH was higher by about 0.9 pH units, the $-\Delta G_e$ values calculated with reference to internal pH were about 2.5 kcal/mol higher than values calculated with respect to bulk phase pH. Making this correction to the data analysis of Forman and Wilson, we find a stoichiometry of 1.8 ATP/2e^- (in agreement with the results of Lemasters et $al.$[42]).

Forman and Wilson argued that since the "midpoint" or half-reduction potential of the acetoacetate/β-hydroxybutyrate couple is pH dependent, and since the β-hydroxybutyrate dehydrogenase is located at the inner surface of the membrane, the appropriate midpoint potential to be used in estimating $-\Delta G_e$ is that which corresponds to the pH of the mitochondrial matrix. This argument is contradicted by the rigorous demonstration above that the dissipation function of this type of system involves the affinity measured in the outer medium. However, it is illuminating to consider more closely how midpoint potentials, which are associated with the measurement of redox potentials, enter into these considerations. With complete generality one may write for the acetoacetate/β-hydroxybutyrate couple in the outer medium [cf. Eq. (79)]

$$\tilde{\mu}_e^{d(o)} = [\tilde{\mu}_B^{(o)} - (2\tilde{\mu}_H^{(o)} + \tilde{\mu}_A^{(o)})]/2 \tag{94}$$

which together with Eqs. (80) and (84) gives

$$A_e^{(o)} = \tilde{\mu}_e^{d(o)} - \tilde{\mu}_e^{a(o)} \tag{95}$$

Now the electrochemical potential of the electron in solution as defined by a given redox couple may be expressed in terms of the redox potential of the couple E, which simply reflects the activity-dependent part of $\tilde{\mu}_e$, i.e., the chemical potential of the electron, μ_e:

$$\tilde{\mu}_e = \mu_e - F\psi = -F(E + \psi) \tag{96}$$

E is usually calculated from the ratio of the concentrations of the oxidized and reduced species, and tabulated values of the midpoint potential. The latter is sensitive not only to pH but to the distribution of liganded species, and is not simply related to the standard redox potential (a full discussion of this question has been given by Walz[44]). We define the redox potential differences

$$\Delta E^{(o)} = E^{a(o)} - E^{d(o)}, \qquad \Delta E^{(oi)} = E^{a(o)} - E^{d(i)} \tag{97}$$

Then, from Eqs. (95)–(97)

$$A_e^{(o)} = F\Delta E^{(o)} \tag{98}$$

which is the operational measurement of $A_e^{(o)}$.

Returning to the assumptions above, and introducing Eqs. (96) and (97) into Eq. (85), we find at static head, which of course corresponds to equilibrium ($\Delta\tilde{\mu}_B = \Delta\tilde{\mu}_A = A_r = 0$),

$$A_e^{(o)} = F(\Delta E^{(oi)} - \Delta\psi) + \Delta\tilde{\mu}_H = F(\Delta E^{(oi)} + \Delta\mu_H/F) \qquad (99)$$

where $\Delta\mu_H = -2.303RT\Delta pH$. Substituting Eq. (99) into Eq. (90) gives a further two equivalent static head relations:

$$\frac{\Delta E^{(oi)} + \Delta\mu_H/F}{-\Delta\tilde{\mu}_H/F} = n_e - 1 \qquad (100)$$

or

$$\frac{\Delta E^{(oi)} - \Delta\psi}{-\Delta\tilde{\mu}_H/F} = n_e \qquad (101)$$

Since Forman and Wilson calculate $-\Delta G_e$ values from $\Delta E^{(oi)}$ rather than $\Delta E^{(o)}$, their values clearly need correction before they can be used to estimate stoichiometry, either by using the pH difference as discussed above or by using the electrical potential difference with due attention to the meaning of the resulting flow ratio. [In this regard two interesting limiting cases of both Eqs. (100) and (101) should be noted: if $\Delta pH = 0$, $\Delta E^{(oi)}/(-\Delta\psi) = n_e - 1$; if $\Delta\psi = 0$, $\Delta E^{(oi)}/(-\Delta\mu_H/F) = n_e$. These cases have been extensively discussed by Scholes and Hinkle.[46]]

To sum up, it should be appreciated that the dissipation function can be a powerful tool in all energetic considerations of this nature.

Use of the Linearized Approach

The four effective phenomenological coefficients for oxidative phosphorylation[22] and for the redox pumps[24] have been measured in rat liver mitochondria from the slopes and intercepts of the linear relationships between the flows and the output force, with the input force kept constant (see, for example, Fig. 12). In both cases the ratio of the coupling coefficients was approximately 1 (less than 10% deviation from complete symmetry). The authors therefore considered the proportional phenomenological equations of the Onsager domain [cf. Eq. (12)] as a good description of the data, and derived the parameters q and Z from the phenomenological coefficients. Unfortunately, as the analysis developed in the previous section shows, in the absence of independent evidence a finding of approximately equal effective coupling coefficients does not necessarily mean that the region of linearity considered is centered on equilibrium and coincides with the Onsager domain (especially if the pumps are not completely coupled). The parameters q and Z calculated from the effective phenome-

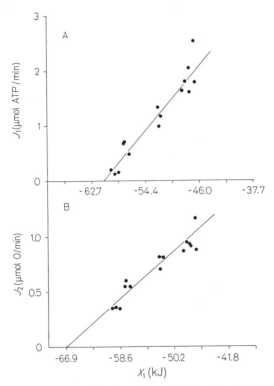

Fig. 12. Oxidative phosphorylation in rat liver mitochondria. (A) J_1 (net flow of ATP) versus X_1 (phosphate potential) at constant X_2 (redox potential difference) as determined from mitochondrial incubations. (B) J_2 (net oxygen consumption) versus X_1 determined similarly; X_2 was about 188 kJ/mol. (From Stucki.[22])

nological coefficients as well as by the method suggested by Rottenberg[3] and recently used by Lemasters[42,47] have a physical meaning only if the extended region of linearity for both forces are real extensions of the Onsager domain. To demonstrate proportionality and reciprocity the relationships between the flows and the input force (at a constant value of the output force) must also be measured (if possible), and the resulting effective coupling coefficients shown to be equal to those obtained by varying the output force. It is worthwhile noting that this result would in any case be condition dependent, i.e., it would depend on the value of the force kept constant (cf. previous section, *Linear Domain*). This prediction has been

[47] J. J. Lemasters, *J. Biol. Chem.* **259**, 13123 (1984).

experimentally verified by van Dam *et al.*,[48] who found different values of the factor γ [cf. Eqs. (68) and (69)] depending on the values of A and $\Delta\tilde{\mu}_H$ kept constant. Rottenberg,[26] in an experimental study of oxidative phosphorylation in which both the oxidation and phosphorylation affinities were varied, found symmetry but not proportionality. His results suggest that the mitochondria, under the conditions obtaining, were functioning close to an MIP, but it should be remembered that this is a system of greater complexity than the representative two-flow coupled process we have studied, since two pumps are involved. Complex systems may well give rise to new phenomena. Thus epithelial systems show remarkably broad ranges of linearity, sometimes extending over 4–5 kcal/mol, which generally include static head. Linear nonequilibrium thermodynamics has been extensively applied to these systems.[4,23] Assuming symmetrical and proportional phenomenological equations, the affinity of the driving reaction, as well as the degree of coupling, has been calculated from the slopes and intercepts of the linear relationships between the flows and $\Delta\tilde{\mu}_{Na}$. In these studies the driving affinity is inaccessible, and must be assumed constant while the imposed $\Delta\tilde{\mu}_{Na}$ across the tissue is varied. (It has been shown independently that the affinity does in fact remain constant under these conditions.[49]) The flows measured are the sodium flux and the respiration rate as reflected in O_2 consumption or CO_2 production. For this kind of application, if the sodium pump is highly coupled and the region of extended linearity is sufficiently close to equilibrium, the linearized treatment is a reasonable approximation, and the internal consistency of the results obtained indicates that this is the case. For example, in studies of frog skins exposed to levels of 2-deoxy-D-glucose (2DG) or ouabain sufficient to inhibit short-circuit current by about 50%, 2DG significantly lowered the measured reaction affinity without significantly altering the coupling coefficient, while ouabain did the reverse.[50] Since 2DG is not metabolized but interferes with cellular energy metabolism, it may be expected to promote depletion of ATP. Ouabain, on the other hand, is a potent inhibitor of the sodium pump. These results are therefore consistent with the interpretation of the data based on a linearized model.

Other applications of the linearized approach have been made, *inter*

[48] K. van Dam, H. V. Westerhoff, M. Rutgers, J. A. Bode, P. C. de Jonge, M. M. Bos, and G. van den Berg, *in* "Vectorial Reactions in Electron and Ion Transport in Mitochondria and Bacteria" (F. Palmieri, E. Quagliariello, N. Siliprandi, and E. C. Slater, eds.), p. 389. Elsevier, Amsterdam, 1981.

[49] M. M. Civan, K. Peterson-Yantorno, D. R. DiBona, D. F. Wilson, and M. Erecińska, *Am. J. Physiol.* **245**, F691 (1983).

[50] A. Owen, S. R. Caplan, and A. Essig, *Biochim. Biophys. Acta* **394**, 438 (1975).

alia, to the Na^+/aminoisobutyrate cotransport in Ehrlich cells[51,52] and to the light-driven proton pump bacteriorhodopsin from *Halobacterium halobium*.[53,54] A particularly interesting application to the characterization of the Na^+/H^+ exchange system in subbacterial particles prepared from *H. halobium* has been described by Cooper *et al.*[55] In this study, the need to measure the internal pH and pNa was avoided by using the "potential-jump" method. On suppression of $\Delta\psi$, the flow equations (assuming linearity and proportionality) are

$$J_H = -L_H \Delta pH + L_{HNa} \Delta pNa$$
$$J_{Na} = L_{NaH} \Delta pH - L_{Na} \Delta pNa \tag{102}$$

The experimental approach is to produce an abrupt change (at time $t = 0$) in either the pH or the pNa of the outer solution, and to determine J_H or J_{Na} before and after the occurrence of the potential jump. It should be emphasized that initial equilibrium conditions are not required. Consider the proton flux, and denote the conditions just before and just after a potential jump by I and II, respectively. (Conditions II are obtained by measurement of flow relaxation kinetics, extrapolating to $t = 0$.) Then, since the inner concentrations are not changed at $t = 0$, $pH^{iI} = pH^{iII}$ and $pNa^{iI} = pNa^{iII}$. Accordingly,

$$\Delta J_H = J_H^{II} - J_H^I = -L_H (pH^{oII} - pH^{oI}) + L_{HNa} (pNa^{oII} - pNa^{oI}) \tag{103}$$

By restricting the potential jump to either pH or pNa only, one can determine L_H or L_{HNa} from the slopes of the plots of J_H versus ($pH^{oII} - pH^{oI}$) or ($pNA^{oII} - pNa^{oI}$), respectively. These plots were found to be linear for sufficiently small values of the jumps. A valuable bonus of this method is that it gives a direct measure of the inner buffering capacity when flow relaxation follows a single exponential decay. This involves three steps:

1. Determination of the buffering capacity of the outer solution $[-(dpH^o/dn_H^o)^{-1}]$, where n_H^o represents the amount of H^+ in the outer medium per unit of cellular or vesicular protein.
2. Determination of L_H.
3. Determination of the slope, $-T_H$, of the linear plot of $\ln J_H$ versus time.

[51] P. Geck, E. Heinz, and B. Pfeiffer, *Biochim. Biophys. Acta* **288**, 486 (1972).
[52] E. Heinz and P. Geck, *Biochim. Biophys. Acta* **339**, 426 (1974).
[53] H. V. Westerhoff and K. van Dam, *Curr. Top. Bioenerg.* **9**, 1 (1979).
[54] J. C. Arents, K. J. Hellingwerf, K. van Dam, and H. V. Westerhoff, *J. Membr. Biol.* **60**, 95 (1981).
[55] S. Cooper, I. Michaeli, and S. R. Caplan, *Biochim. Biophys. Acta* **736**, 11 (1983).

Then the inverse of the internal buffering capacity is given by[55]

$$-\frac{d\mathrm{pH}^i}{dn_H^i} = \frac{T_H}{L_H} + \frac{d\mathrm{pH}^\circ}{dn_H^\circ} \qquad (104)$$

Leak versus Slip: Intrinsic Uncoupling of Mitochondrial Proton Pumps

In static head mitochondria extensive inhibition of the rate of electron transfer (or ATP hydrolysis) by specific inhibitors or by changes in the substrate concentration is accompanied by little change in $\Delta\tilde{\mu}_H$.[12,56,57] This is not expected on the basis of Eq. (105) below, derived from the stationary state condition of zero net proton flow under the assumptions of complete coupling of the proton pumps and ohmic conductance of the membrane:

$$n_e J_e^{sh} = L_H^1 \Delta\tilde{\mu}_H^{sh} \qquad (105)$$

(where n_e is the stoichiometry, H^+/e^-, of the redox pumps and L_H^1 is the proton conductance of the membrane). A nonohmic membrane conductance and/or an uncoupled reaction rate due to a certain amount of broken or damaged mitochondria cannot account for the experimental observation.[58] It has been proposed that the above-mentioned typical relationship between the rate of the driving reaction and $\Delta\tilde{\mu}_H$ at static head may well be a manifestation of a certain extent of intrinsic uncoupling (i.e., slip) in the mitochondrial proton pumps.[12] While it is clear that, if the redox pumps (for example) are completely coupled and the nonzero rate of electron flow at static head is due only to extrinsic "leaks," the relationship between J_e^{sh} and $\Delta\tilde{\mu}_H^{sh}$ simply reflects the relationship between passive proton flow and $\Delta\tilde{\mu}_H$, it is much less clear what such a relationship should be in the case of "slipping" pumps. Simulations of this relationship for different values of mixed intrinsic (slip) and extrinsic (leak) uncoupling can be performed with the model of Fig. 5 (with a leak in parallel) by introducing the reaction with an inhibitor and varying the inhibitor concentration, or by varying the substrate concentration. For each concentration of inhibitor and/or substrate $\Delta\tilde{\mu}_H^{sh}$ is calculated as the $\Delta\tilde{\mu}_H$ at which the total proton flow becomes zero, while J_e^{sh} is the value of J_e at $\Delta\tilde{\mu}_H^{sh}$. Figure 13 shows an example of simulated relationships between J_e^{sh} and $\Delta\tilde{\mu}_H^{sh}$ obtained with the kinetic parameters of pump B (cf. Fig. 7) and with a combination of leak and slip which accounts for the electron transport at static head—the combination being in the proportion indicated by direct measurement of the passive

[56] D. G. Nicholls, *Eur. J. Biochem.* **50**, 305 (1974).
[57] D. Pietrobon, M. Zoratti, and G. F. Azzone, *Biochim. Biophys. Acta* **723**, 317 (1983).
[58] M. Zoratti, M. Favaron, D. Pietrobon, and G. F. Azzone, *Biochemistry* **25**, 760 (1986).

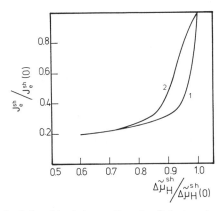

FIG. 13. Simulated relationships between the rate of electron transfer at static head J_e^{sh} and $\Delta\tilde{\mu}_H^{sh}$ on progressive inhibition of a slipping redox proton pump with a constant proton leak conductance in parallel. $J_e^{sh}(0)$ and $\Delta\tilde{\mu}_H^{sh}(0)$ refer to zero inhibitor concentration. Curves 1 and 2 correspond to two different types of inhibitor (see the text). [From Pietrobon *et al.*,[41] reprinted with permission. Copyright (1985) American Chemical Society.]

proton conductance of the mitochondrial membranes.[58] Curve 1 was obtained with an inhibitor which binds to all the states in the cycle with the same affinity, thus decreasing the population of the states and hence all the rate constants by a factor related to the inhibitor concentration (noncompetitive inhibition). Curve 2 was obtained with an inhibitor which reduces the rate constants α_{23}, α_{32} (of the substrate binding transitions) to practically zero in a fraction of the enzymes, while the remaining fraction is unaffected (competitive inhibition). The different simulated behaviors obtained with the two hypothetical inhibitors are strongly reminiscent of the different experimental results found in rat liver mitochondria with the two inhibitors antimycin and malonate.[12,58] The same simulated curve (2) is obtained by varying the substrate concentration in the model, and in rat liver mitochondria the same relationship between J_e^{sh} and $\Delta\tilde{\mu}_H^{sh}$ is also obtained either by varying the succinate concentration or by varying the malonate concentration. The different behavior with the two inhibitors in Fig. 13 arises from the fact that inhibitor 1 (curve 1) inhibits simultaneously to the same extent the coupled cycle a and the uncoupled cycles b and c, while inhibitor 2 (curve 2), or a decrease in the substrate concentration, inhibits the coupled reaction cycle and the redox slip (cycle c) but not the proton slip (which is actually relatively increased due to the released competition between cycles a, b, and c).

The recognition of a certain degree of "physiological" intrinsic uncoupling in the proton pumps lends a new perspective to the analysis of the unphysiological uncoupling induced by certain agents or conditions. The analysis of the effect of increasing intrinsic uncoupling (cf. Fig. 9) or extrinsic uncoupling (not shown) on the simulated flow–force relationships provides criteria enabling one to distinguish whether a certain uncoupling agent acts by increasing the "leak" through the membrane, or by increasing molecular slipping, or by completely uncoupling a certain number of pumps. Considering the redox proton pump, all kinds of uncoupling agents induce an increase in J_e^{sh} and a decrease in the flow ratio J_H/J_e measured at low $\Delta\tilde{\mu}_H$. Clearly if these effects are not accompanied by any increase in the passive conductance of the membrane as measured directly, the agent is an intrinsic uncoupler or slip inducer. If the uncoupling agent has a mixed behavior (both "leak" and "slip" inducer), J_e^{sh} increases more than the proton influx through the leak divided by n_e. An intrinsic uncoupler induces an increase in the apparent leak conductance, $n_e J_e^{sh}/\Delta\tilde{\mu}_H^{sh}$ [cf. Eq. (105)]. This increase is, however, progressively decreased in a titration with redox inhibitors. Moreover, the increase in J_e^{sh} in the presence of an intrinsic uncoupler is accompanied by a decrease in $\Delta\tilde{\mu}_H$ which may be negligible, and anyway is lower than that which accompanies the same increase in J_e^{sh} induced by an extrinsic uncoupler (leak inducer). Note that the flow–force analysis provides a criterion enabling one to decide whether or not a certain agent is an intrinsic uncoupler without any need to measure the passive proton influx or proton conductance of the membrane directly. The slope of the region of approximate linearity of the rate of electron transfer (or in general of the rate of the driving reaction) as a function of $\Delta\tilde{\mu}_H$ (or in general of the output force), L_{eH}, is lower in the presence of an intrinsic uncoupler (see Fig. 8) while it is unchanged in the presence of an uncoupler which only increases the extrinsic leak. Experimentally the check can be done by measuring J_e and $\Delta\tilde{\mu}_H$ at different concentrations of K^+ (+ valinomycin) in the presence and absence of the uncoupling agent.

Double Inhibitor Titrations (and Localized versus Delocalized Chemiosmosis)

In the chemiosmotic view energy transduction is accomplished by two linked proton pumps: a $\Delta\tilde{\mu}_H$-generating primary (i.e., redox) pump and a $\Delta\tilde{\mu}_H$-utilizing secondary (i.e., ATPase) pump. There is much supporting evidence for this "delocalized" chemiosmotic model of energy coupling, but there are some results, from different experimental approaches, which have been considered in conflict with it.[59,60] Among these are the results of

[59] H. V. Westerhoff, B. A. Melandri, G. Venturoli, G. F. Azzone, and D. B. Kell, *Biochim. Biophys. Acta* **768**, 257 (1984).

[60] S. J. Ferguson, *Biochim. Biophys. Acta* **811**, 47 (1985).

the well-known double inhibitor titrations.[61-63] It was found that the relative inhibition of the $\Delta\tilde{\mu}_H$-driven reaction (i.e., ATP synthesis) by a given concentration of an inhibitor of the primary (i.e., redox) pumps was unchanged (or even, in some cases, higher) when the secondary (i.e., ATPase) pumps were partially inhibited (and vice versa). This result has been considered as unequivocally incompatible with delocalized models of energy coupling, since (it has been argued) the extent to which one of the pumps controls the flux should be decreased when the other pump is inhibited. Therefore, the relative inhibition of the rate of ATP synthesis by a given concentration of redox inhibitor should be lower when the ATPases are partially inhibited than when they are not. The use of nonequilibrium thermodynamics allows a test as to whether or not this conclusion is general (as has been claimed[62]), independent of the flow–force relationships of the three basic elements of the chemiosmotic protonic circuit (primary pump, secondary pump, and leak). This example of application is particularly interesting since it shows the utility and at the same time the limitations of linear nonequilibrium thermodynamics, and how the linear and nonlinear approaches can be integrated to obtain a deeper understanding of a phenomenon.

The simplest possible nonequilibrium thermodynamic description of the proton flow through the redox (J_H^e) and ATPase (J_H^p) proton pumps is given by[29]

$$J_H^e = n_e J_e = n_e L_e (A_e + n_e \Delta\tilde{\mu}_H) \tag{106}$$

$$J_H^p = n_p J_p = n_p L_p (A_p + n_p \Delta\tilde{\mu}_H) \tag{107}$$

From Eqs. (106) and (107) (to which an equation for the proton flow through the leak can be added, for example $J_H^l = L_H^l \Delta\tilde{\mu}_H$) and the condition of zero net proton flow, the values of $\Delta\tilde{\mu}_H$ and J_p in the stationary state of phosphorylation (state 3) can easily be derived as functions of the conductances of the three elements of the chemiosmotic circuit and the stoichiometries and affinities of the reactions. The double-inhibitor titrations can then be simulated by multiplying the phenomenological coefficients L_p and L_e by the inhibition factors f_p and f_e, which vary from 1 at zero inhibitor concentration to 0 at a concentration of inhibitor which inhibits all the pumps. Alternatively, the double-inhibitor titrations can be

[61] H. Baum, G. S. Hall, J. Nalder, and R. B. Beechey, in "Energy Transduction in Respiration and Photosynthesis" (E. Quagliariello, S. Papa, and G. S. Rossi, eds.), p. 747. Adriatica Editrice, Bari, Italy, 1971.

[62] G. D. Hitchens and D. B. Kell, Biochem J. 206, 351 (1982).

[63] H. V. Westerhoff, A.-M. Colen, and K. van Dam, Biochem. Soc. Trans. 11, 81 (1983).

analyzed using Eqs. (106) and (107), the condition of zero net proton flow, and control theory.[63-66]

It turns out that in the presence of a negligible leak, the relative inhibition of the rate of ATP synthesis, by a given combination of inhibitors, is exclusively dependent on the ratio between the conductances of the two proton pumps in the chemiosmotic protonic circuit,[67] i.e., $n_p^2 L_p / n_e^2 L_e$. For the titration of the redox pump:

$$J_p^e / J_p(0) = \left(1 + f_p \frac{n_p^2 L_p}{n_e^2 L_e}\right) \Big/ \left(1 + \frac{f_p n_p^2 L_p}{f_e n_e^2 L_e}\right) \tag{108}$$

For the titration of the ATPase pump:

$$J_p^p / J_p(0) = f_p \left(1 + \frac{1}{f_e} \frac{n_p^2 L_p}{n_e^2 L_e}\right) \Big/ \left(1 + \frac{f_p n_p^2 L_p}{f_e n_e^2 L_e}\right) \tag{109}$$

The flux controls of the two proton pumps are also uniquely dependent on the same ratio.[64] As $n_p^2 L_p / n_e^2 L_e$ increases, or in other words as the "readiness" with which the ATPases change their rate in response to a change of $\Delta \tilde{\mu}_H$ increases with respect to that of the redox pumps, the flux control of the ATPases decreases while that of the redox pumps increases. Consequently an inhibitor of the ATPases becomes less effective in inhibiting the rate of ATP synthesis while an inhibitor of the redox pumps becomes more effective. With linear flow–force relationships governing a double-inhibitor titration, the relative inhibition of the rate of ATP synthesis by a given concentration of redox inhibitor will always be lower in the presence of a concentration of ATPase inhibitor which reduces the number of active ATPases. If this number is reduced by, say, 50%, $n_p^2 L_p / n_e^2 L_e$ is also reduced by 50% with a consequent decrease of the flux control of the redox pumps. With nonlinear flow–force relationships, the parameter equivalent to $n_p^2 L_p / n_e^2 L_e$ is the ratio $n_p(\partial J_p / \partial \Delta \tilde{\mu}_H) / n_e(\partial J_e / \partial \Delta \tilde{\mu}_H)$, which, in general, in contrast to the linear case, is different at different $\Delta \tilde{\mu}_H$ values. The analysis developed above for the linear case makes it clear that when nonlinear flow–force relationships govern a double-inhibitor titration, the behavior will be different from that found with the linear model whenever the changes in $\Delta \tilde{\mu}_H$ which accompany partial inhibition of one of the pumps, at any given concentration of an inhibitor of the other pump, give rise to

[64] H. V. Westerhoff, Ph.D. thesis. University of Amsterdam, Amsterdam, The Netherlands, 1983.
[65] H. Kacser and J. A. Burns, in "Rate Control of Biological Processes" (D. D. Davis, ed.), p. 65. Cambridge Univ. Press, London, 1973.
[66] R. Heinrich and T. A. Rapoport, Eur. J. Biochem. 42, 89 (1974).
[67] D. Pietrobon and S. R. Caplan, Biochemistry 25, 7690 (1986).

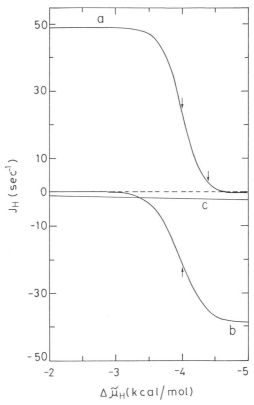

FIG. 14. The stationary state of oxidative phosphorylation (total proton flow zero). (Curve a) Dependence on $\Delta\tilde{\mu}_H$ of proton flow J_H (positive efflux) using the model in Fig. 5 as a representation of a redox pump at high constant affinity. (Curve b) Dependence on $\Delta\tilde{\mu}_H$ of proton flow (negative influx) when the model is used as a representation of an ATPase pump at low affinity of the hydrolysis reaction. (Curve c) Dependence on $\Delta\tilde{\mu}_H$ of proton leak (assumed ohmic). Opposing arrows, Stationary state in the absence of ATPase inhibitor; unopposed arrow, stationary state with complete inhibition of ATPases. (From Pietrobon and Caplan.[67])

different values of $n_p(\partial J_p/\partial\Delta\tilde{\mu}_H)/n_e(\partial J_e/\partial\Delta\tilde{\mu}_H)$. The relative inhibition of the rate of ATP synthesis by a given concentration of redox inhibitor can remain unchanged (or even become higher) in the presence of an ATPase inhibitor if $\partial J_p/\partial\Delta\tilde{\mu}_H$ increases and/or $\partial J_e/\partial\Delta\tilde{\mu}_H$ decreases as $|\Delta\tilde{\mu}_H|$ increases. This type of kinetics makes it possible that at the higher $|\Delta\tilde{\mu}_H|$ obtaining in the presence of an ATPase inhibitor, the ratio $n_p(\partial J_p/\partial\Delta\tilde{\mu}_H)/n_e(\partial J_e/\partial\Delta\tilde{\mu}_H)$ does not necessarily decrease (as it would always do in the linear case); indeed it may increase.

Curve a in Fig. 14 shows the dependence on $\Delta\tilde{\mu}_H$ of the proton flow

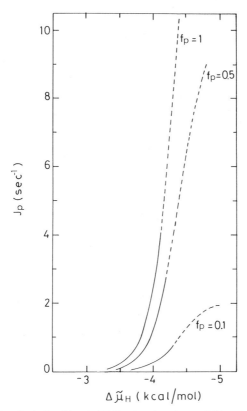

FIG. 15. Simulated relationships (solid lines), using the model in Fig. 5, between rate of ATP synthesis J_p and $\Delta\tilde{\mu}_H$. These relationships would be obtained on decreasing $\Delta\tilde{\mu}_H$ below its state 3 value with uncouplers or redox inhibitors. The factor f_p indicates the fraction of ATPases still active after addition of an ATPase inhibitor (in the absence of ATPase inhibitor $f_p = 1$). The dashed lines complete the flow–force curves. (From Pietrobon and Caplan.[67])

(positive efflux) generated by the model in Fig. 5 used as a representation of a $\Delta\tilde{\mu}_H$-generating redox proton pump, at high constant affinity, while curve b shows the dependence on $\Delta\tilde{\mu}_H$ of the proton flow (negative influx) when the model is used as a representation of an ATPase pump, at low affinity of the ATP hydrolysis reaction. The relative position of the two sigmoidal curves, in other words the range of $\Delta\tilde{\mu}_H$ controlling the influx through the ATPases as compared to that controlling the efflux through the redox pumps, determines the location of the stationary state of phosphorylation (the state at which the total proton flow is zero, i.e., $J_H^e + J_H^p + J_H^l = 0$), as indicated by the matched arrows on curves a and b. This location in turn determines the kind of change in $(\partial J_p/\partial\Delta\tilde{\mu}_H)/(\partial J_e/\partial\Delta\tilde{\mu}_H)$ at the $\Delta\tilde{\mu}_H$ obtaining with inhibitor relative to that at the $\Delta\tilde{\mu}_H$ obtaining without inhibitor. Figure 15 (continuous lines) shows an example of simulated relationships between rate of ATP synthesis and $\Delta\tilde{\mu}_H$ (as would be obtained experimentally by decreasing $\Delta\tilde{\mu}_H$ below its state 3 value with uncouplers or redox inhibitors) in the absence of ATPase inhibitor ($f_p = 1$) or in the presence of concentrations of inhibitor which decrease the number of ATPases by the factors $f_p = 0.5$ or 0.1. These kinds of flow–force relationships give rise to similar titration curves by a redox inhibitor both in the presence and absence of an ATPase inhibitor.[67] An inspection of the figure shows why. Upon inhibition of the ATPases state 3 shifts to higher $|\Delta\tilde{\mu}_H|$ (from -4.12 kcal/mol with $f_p = 1$ to -4.31 kcal/mol with $f_p = 0.1$), moving upward in the sigmoidal flow–force curve (from below the beginning of the region of approximate linearity until well into it). In each curve $(\partial J_p/\partial\Delta\tilde{\mu}_H)$ at the higher $|\Delta\tilde{\mu}_H|$ obtaining with inhibitor is higher than at the $|\Delta\tilde{\mu}_H|$ obtaining without inhibitor. Therefore, upon inhibition of the ATPases, $(\partial J_p/\partial\Delta\tilde{\mu}_H)$ decreases much less than by the factor f_p, and as a consequence there is only a small redistribution of flux control between the two pumps so that similar titration curves are obtained.

It can be concluded that the double-inhibitor titration approach by itself cannot unequivocally discriminate between "delocalized" and "localized" chemiosmotic mechanisms. To do this, it is necessary to examine carefully the relationships between rate of ATP synthesis and $\Delta\tilde{\mu}_H$ and between rate of electron transfer and $\Delta\tilde{\mu}_H$.

Acknowledgment

The authors are indebted to Dr. D. Walz for helpful discussions.

[21] Cell Isolation Techniques: Use of Enzymes and Chelators

By MIGUEL J. M. LEWIN and A. M. CHERET

Introduction

The use of cell suspensions in transport studies, although with obvious limitations (e.g., transport vectoriality is lost), does offer several advantages over tissue preparations. These include better reproducibility and quantification of samples, easier control of the extracellular medium due to the absence of connective tissue diffusion barrier, and the short-circuiting of paracellular pathways, which is of special interest in studies concerned with ion transport. In addition, isolated cell systems are particularly useful when dealing with complex tissues containing a mixture of different cell types. In this instance, further sorting of isolated cells may allow obtention of suspensions enriched in a specific cell type. Cell isolation can also be required in studies aimed at achieving monolayer cultures and cell cloning.

Ideally, one would expect a good isolation procedure not to alter any function of the cell *in vivo*. In practice, however, tissue dissociation requires a combination of enzymatic, chemical, and mechanical procedures, which are all damaging, to various degrees. Furthermore, isolated cell sorting by centrifugation results in structural alterations whose magnitude depends on the type and the speed of the rotor as well as on the chemical composition of the medium. In addition, because of differences in cell surfaces and intercellular matrix composition, procedures suitable for a given tissue of a given species may not be appropriate for another tissue or even for the same tissue in another species. For all these reasons no universal method exists and the investigator must select the most appropriate one for his particular problem. This study will describe some of the possible approaches relating to gastric mucosal cell isolation.

Dissociation of Gastric Mucosa into Individual Cells

Owing to the presence of a strong connective architecture, gastric epithelial cells cannot be easily separated from each other without the intervention of proteolytic enzymes. Collagenase and pronase are the most convenient for this purpose.

Collagenase from *Clostridium histolyticum* (clostridiopeptidase A; EC 3.4.24.3) cleaves collagen and other proline-containing peptides at glycine residues. Commercially available collagenases are of various types and degrees of purity. We found that crude type I or V collagenase from Sigma Chemical Co. (St. Louis, Missouri) or Worthington Biochemical Corp. (Freehold, New Jersey) gave excellent results (a more expensive high-purity type VII collagenase, largely devoid of other nonspecific protease activity, is also available from these firms). An important point to keep in mind when using collagenase for tissue dissociation is the fact that the activity of this enzyme is dependent on the presence of Ca^{2+} (which is activating) and is sensitive to other divalent cations, which can be either activators (e.g., Zn^{2+}) or inhibitors (e.g., Mg^{2+}). It is therefore clear that the ionic composition of the dissociation medium must be carefully controlled. In particular, calcium chelators such as EGTA should not be used simultaneously with collagenase.

Pronase is not a well-defined enzyme, but rather a mixture of several proteinases isolated from *Streptomyces griseus*. Because of this, it has a wide spectrum of proteolytic activity. In the presence of EDTA or EGTA, however, calcium-independent esterase activity is largely predominant.[1] Pronase has proved essentially efficient in the dissociation of gastric mucosal cells in amphibian as well as mammalian species.[2,3] In practice, the use of this enzyme is required to obtain individualized cells (not clusters) with a reasonable yield.[4,5] Good quality lyophilized pronase can be obtained from Boehringer (Mannheim, Federal Republic of Germany) or Merck (Darmstadt, Federal Republic of Germany). However, because of variations which may occur from one batch to another, it is recommended that the enzyme activity be tested first on a sample and if found suitable, that an amount large enough to allow for work over an extended period be ordered.

In addition to collagenase and pronase, EDTA and EGTA are extremely helpful to achieve dissociation of gastric epithelium. This may be explained by the important role played by divalent cations and especially by calcium in cell attachment. The use of these chelators in conjunction with vigorous shaking can be sufficient to separate the epithelial layer from the connective stroma and to produce suspensions of isolated tubular pits.[6] However, EDTA and EGTA are poorly effective in the dissolution of the intercellular junctions, which is required to obtain isolated cells. Therefore, protocols for gastric epithelial cell isolation involve either concomitant or successive use of proteolytic enzymes (i.e., collagenase and/or pronase) and calcium chelators.

Pronase EDTA Method

This method, originally developed for the dissociation of rat gastric epithelium,[3,7] is also suitable for dissociating epithelial cells from large pieces excised from the human stomach. Its particular feature is the preparation of everted pouches (epithelial layer outside), which are filled with a pronase dissociation mixture. This procedure allows improved cell viabil-

---

[1] Y. Narahashi and M. Yanagita, *J. Biochem. (Tokyo)* **62**, 633 (1967).

[2] A. L. Blum, G. T. Shah, V. D. Wiebelhaus, F. T. Brennan, H. F. Helander, R. Ceballos, and G. Sachs, *Gastroenterology* **62**, 189 (1971).

[3] M. J. M. Lewin, A. M. Cheret, A. Soumarmon, and J. Girodet, *Biol. Gastro-Enterol.* **7**, 139 (1974).

[4] S. Muallen and G. Sachs, *Biochim. Biophys. Acta* **805**, 181 (1984).

[5] K. F. Sewing, P. Harms, G. Schulz, and H. Hannemann, *Gut* **24**, 557 (1983).

[6] C. Gespach, D. Bataille, C. Dupont, G. Rosselin, E. Wünsch, and E. Jaeger, *Biochim. Biophys. Acta* **630**, 433 (1980).

[7] M. J. M. Lewin, A. M. Cheret, A. Soumarmon, J. Girodet, D. Ghesquier, F. Grelac, and S. Bonfils, *in* "Stimulation Secretion Coupling in the GI Tract" (R. M. Case and H. Goebell, eds.), p. 371. MTP, Lancaster, England, 1976.

ity due to more efficacious oxygenation, along with a larger bathing volume. Furthermore, because the proteolytic mixture is confined with the pouch interior, the epithelial cells are released in a pronase-free medium. This method is also relatively inexpensive because it requires a minimal volume of concentrated pronase.

The animals used are male Wistar rats, weighing from 100 to 340 g; they are usually not fasted before the experiment, and are killed by cervical dislocation and are then laparotomized. Two ligatures are made, one at the level of the cardioesophageal junction, the other at the level of the antropyloric junction. The stomach is removed and eversed using a glass rod passed through a small incision made in the rumen. A third ligature is performed at the rumen–fundus junction, and the rumen is discarded. At this stage one may perform a further ligature at the antrofundic junction and discard either the fundic or the antral part of the stomach for the specific preparation of antral or fundic epithelial cells, respectively. The everted pouch thus prepared is then cleansed of food debris using tap water and is gently wiped with a paper towel to remove surface mucus. It is filled with the proteolytic solution, using a syringe with a 26½-gauge needle, and placed in the incubation chamber. This chamber consists of a porous glass filtration funnel (No. 4 pore size) with capacity varying from 50 to 200 ml according to the number of everted pouches treated at the same time (usually three to five). Gassing of the medium is achieved through porous glass with 95% oxygen and 5% CO_2 under 3 mbar of pressure.

In experiments on the human stomach the fundic mucosa is first dissected from the muscularis layer with a pair of scissors, leaving a border of muscularis on the sides which are sewn together.

The three following solutions are used:

Medium A

NaH_2PO_4, 0.5 mM; Na_2HPO_4, 1 mM; $NaHCO_3$, 20 mM; NaCl, 80 mM; KCl, 5 mM; N-2-hydroxyethylpiperazine-N'-2-ethanesulfonic acid (HEPES), 50 mM; glucose, 11 mM; bovine serum albumin (BSA), 2%; EDTA, 2 mM; (pH 7.4).

Medium B

The same composition as medium A, but EDTA free, and supplemented with 1 mM Ca^{2+}, 1.5 mM Mg^{2+}, 1% BSA (pH 7.4).

Medium C

The same as medium B, but with only 0.1% BSA (pH 7.4).

The proteolytic mixture injected into the pouches consists of pronase diluted in medium A (20 mg/ml, Boehringer; 1000 U/ml, Merck). Tissue

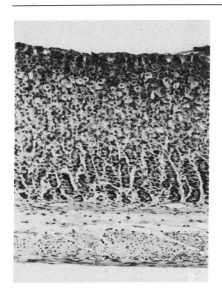 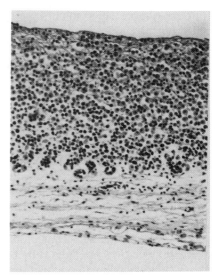

FIG. 1. Pronase method for gastric mucosal cell isolation. Appearance of rat fundic epithelium after 1 hr of incubation with a calcium-free pronase-containing medium at 37°, pH 7.4, as described in the text, as compared to control mucosa incubated under the same experimental conditions in standard Krebs medium. Note the apparent dissolution of the connective stroma and the dissociation of the tissue into individualized cells.

dissociation is allowed to proceed for 90 min at 37°, using medium A as the incubation medium. During this period, the incubation medium is changed every 30 min. The everted pouches are then transferred to a beaker containing medium B at room temperature, and cell dispersion is obtained by gentle magnetic stirring. The dispersion phase lasts 30 min. The cells are collected every 10 min, harvested by centrifugation (100 g, 5 min), and resuspended in medium C. This procedure is repeated twice. During all these phases of dissociation the pH of the incubation medium is carefully controlled and adjusted, if necessary, to pH 7.4 (Figs. 1 and 2).

Collagenase–Pronase–EDTA Method

This method is suitable for guinea pig gastric mucosa (too fragile to be everted) and for small pieces, including biopsies of the human stomach. It can also be used, with modifications, for other mammalian species such as rabbit and pig.[4,8,9] The strategy, here, is first to achieve a predissociation of

[8] S. Mardh, L. Norberg, M. Ljungström, L. Humble, T. Borg, and C. Carlsson, *Acta Physiol. Scand.* **122**, 607 (1984).

[9] H. R. Koelz, S. J. Hersey, G. Sachs, and C. S. Chew, *Am. J. Physiol.* **243**, G218 (1982).

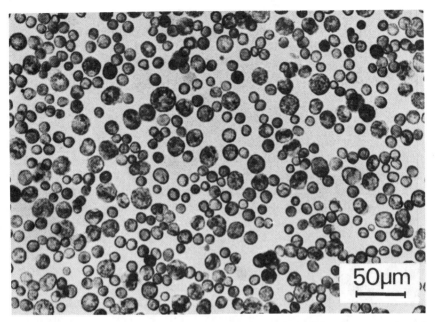

FIG. 2. Pronase method for gastric mucosal cell isolation. Light microscope view of isolated cell suspension as obtained from pronase-incubated mucosa (see Fig. 1) by gentle stirring in pronase-free calcium-supplemented Krebs medium. Cell size heterogeneity accounts for the presence of the different cell types naturally occurring in the fundic epithelium. Parietal cells (mean diameter 16.0 ± 0.8 μm) represents $30.3 \pm 0.8\%$ of the total population, with the remaining corresponding to chief and mucous cells (mean diameter 9.0 ± 0.5 μm). Currently, 8×10^7 to 1×10^8 cells are obtained per stomach.

the tissue using collagenase and EDTA, and then to disperse the tubules or clusters obtained into individualized cells with the aid of pronase.

The animals used are unfasted 300- to 600-g male guinea pigs from either the Hartley or the tricolor strain. They are killed and operated on as previously described for rats. The stomach is removed and opened along the greater curvature. The mucosa is carefully washed, wiped as described above, and gently scraped from the muscularis mucosa with the edge of a glass slide. For mucosal pieces from the human stomach, the epithelium is first dissected from the submucosa and coarsely minced with a pair of scissors.

Mucosal fragments (or biopsy pieces) are collected in a medium B (see above) containing 2% BSA and incubated in the presence of collagenase (0.4 mg per ml) for 45 min at $37°$, while maintaining a vigorous gassing of the medium by 95% O_2 and 5% CO_2. After this period, the fragments are transferred to medium A for 15 min at room temperature. The incubation medium is then discarded and replaced by a fresh solution containing 1 mg/ml pronase in medium A. The dissociation is completed after 10 min of continuous gassing and gentle magnetic stirring at room temperature. Cells are collected, washed twice (100 g, 5 min centrifugation) and resuspended in medium C containing 0.5% BSA.

Purification of Isolated Gastric Cells

The gastric mucosa is typically a heterogeneous epithelium: besides the mucus-secreting cells distributed all over the stomach, several types of endocrine cells mainly occurring in the lower part of the stomach (i.e., the antrum), as well as pepsin-secreting "chief" and acid-secreting "parietal" cells specifically located in the upper part (the fundus), are also found there. As pointed out above, it is possible to limit tissue dissociation to either the antral or the fundic part. Moreover, one can take advantage of the anatomical disposition of the epithelium layer: mucous cells are distributed from the surface down to the neck of the glandular pits, while parietal and chief cells are located deeper, in the middle and the bottom of these pits, respectively. Sequential collection of isolated cells during dissociation of gastric fundic mucosa could thus first provide suspensions enriched in mucous cells, and then suspensions enriched in parietal and chief cells. Although helpful, sequential dissociation does not, however, allow a high degree of purity and the problem of purifying one given cell type from a mixed population of isolated cells remains crucial. In an attempt to solve this problem several centrifugation methods have been developed. These are based on differences in cell size and/or cell density and can be classified into two groups. Analytical methods are in essence similar to those previously developed by De Duve and co-workers for subcellular fractions of tissues, i.e., they achieve quantitative and recuperative resolution of isolated cells into subpopulations characterized by specific morphological or biochemical markers. Such methods are of interest in studies aimed at characterizing the cellular origin of a particular component such as a receptor or an enzyme system: this is done by comparing the distribution of this component to that of the various cell types. Preparative methods have a different purpose. They are intended to provide substantial amounts of purified cells within a minimum time and with minimal handling.

In both cases, the markers used to monitor the purification may be morphological, biochemical, and/or pharmacological. As far as the fundic gastric mucosa is concerned, parietal cells can be distinguished from the other epithelial cells by their typical ultrastructure (namely the presence of intracellular canaliculi and a particular abundance of mitochondria) by their enrichment in mitochondrial enzymes, such as cytochrome c oxidase, by the presence of specific membrane enzymes such as cytochrome-b_5 reductase, H^+,K^+-ATPase, and histamine-stimulated adenylate cyclase, or by their ability to undergo morphological changes on stimulation. Similarly, chief and mucous cells can be morphologically as well as biochemically recognized by their characteristic secretory granules. However, not all the cellular types of the gastric mucosa can be readily purified by centrifugation. These methods are very effective in separating parietal from nonparietal cells, which are markedly smaller and denser, but they are much less efficient in separating chief cells from mucous cells, whose sedimentation characteristics are too similar. Below, three centrifugation methods in current use in our laboratory[10] to purify parietal cells from rat and guinea pig gastric mucosa are described. Another centrifugation method suitable for the sorting of gastric mucosal cells (i.e., elutriation) will be found in another chapter of this volume (Sanders and Soll [23]).

Density Gradient Centrifugation (Analytical)

In this method isolated cells are forced to sediment in a medium of increasing density up to their isopycnic equilibration, i.e., when cellular and external densities are the same. The medium consists of 50 mM HEPES buffer at pH 7.4 in which a linear density gradient from 1.035 to 1.400 g/ml is achieved by mixing two appropriate sucrose solutions. Depending upon the number of cells to be treated, centrifugation is performed either in a 600-ml zonal rotor (Ti 14, Beckman) or in 30-ml tubes of a swinging-bucket rotor (SW 25, Beckman). Cells are introduced on top of the gradient either with 50-ml syringes in a zonal rotor maintained at a constant speed of 4500 rpm or with a 2-ml pipet in swinging-bucket tubes. After 1 hr of centrifugation at 5000 rpm, fractions of 20 ml (zonal) or 2 ml (swinging bucket) are collected and their density is measured using a refractometer. This method is highly efficient in the separation of parietal from nonparietal cells, but it is much less so in the separation of chief from mucus cells, due to their comparable buoyant densities. It is appropriate for biochemical studies dealing with intracellular as well as membrane components. Because of the possibility of treating large samples (using the zonal rotor) and the use of sucrose density medium, the method is also particularly suitable for studies involving further subcellular fractionation of isolated cells. However, as far as the functional ability of the isolated cells is concerned, it should be kept in mind that the high osmotic pressure

[10] M. J. M. Lewin, A. M. Cheret, A. Soumarmon, F. Grelac, and G. Sachs, *in* "Cell Separation: Methods and Selected Applications" (T. G. Pretlow II and T. P. Pretlow, eds.), p. 223. Academic Press, New York, 1982.

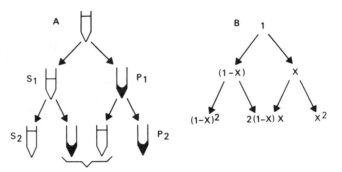

Fig. 3. Iterative centrifugation method for gastric cell purification. (A) Only two steps are shown (S_1, S_2 and P_1, P_2 would be the successive supernatants and pellets; see the text). (B) The corresponding mathematical model, with x representing the fractional amount of parietal cells.

of the sucrose density medium can alter the permeability and the transport properties of the cellular membrane.

Iterative Differential Centrifugation (Analytical)

This method is based on a series of centrifugations, as shown in Fig. 3. Cells are diluted in 15 ml of medium C (detailed above) and submitted to centrifugation (100 *g*) for 45 sec in a swinging-bucket-equipped rotor (No. 269, International Equipment Company). This allows incomplete sedimentation, with the pellet (P_1) being enriched in parietal cells and the supernatant (S_1) enriched in nonparietal cells. Pelleted cells are collected, diluted in 15 ml of fresh medium, and submitted to a second 100 *g* centrifugation. The pellet of this centrifugation (P_2) shows increased enrichment in parietal cells. It is collected and diluted in fresh medium, and the supernatant is used to dilute the pellet from the centrifugation of the first supernatant. A third centrifugation is then performed and the entire procedure is repeated. This method has the great advantage of not requiring the use of any density medium. Therefore, it allows the purification of cells under optimal survival conditions and is, for this reason, very suitable for preparing cells for functional studies. Since the method is iterative and based on sedimentation velocity, it is potentially capable of separating cells differing even slightly with respect to size. However, beyond a certain number of centrifugation runs (generally 10), the benefit of cell purification is counterbalanced by an excessive reduction in yield.

Percoll "Isoneutral" Centrifugation (Preparative)

Polyvinylpyrrolidone-coated colloidal silica particles (Percoll, Pharmacia Fine Chemicals, Uppsala, Sweden) are mixed with isolated cells in medium C to a density of 1.041 g/ml [29% (v/v) Percoll]. The mixture is then subjected to centrifugation (190 g) for 15 min in swinging-bucket adapted to a low-speed centrifuge. Parietal cells are collected on the surface while nonparietal cells are found pelleted on the bottom of the tubes (Fig. 4). This method is easy and rapid to perform: only one 15-min step is sufficient to obtain large amounts of highly purified cells. It is particularly suitable for preparing suspensions of parietal cells enriched up to 90%, but other cell types may be purified in the same way by modifying Percoll concentration. Another advantage of the method is the overall preservation of cell integrity. In contrast to sucrose, Percoll does not significantly change medium osmolarity nor does it penetrate the cell membrane. Furthermore, Percoll particles can easily be washed out of the cell suspension by centrifugation. For all these reasons we find Percoll purification the most suitable preparative method to date available in the preparation of gastric epithelial cell suspensions for the purpose of biochemical as well as functional studies.

Assessment of Cell Purification

One of the best means of estimating the degree of cell purification is direct observation of the sorted cell populations under electron microscope. For this purpose, the cells are quantitatively pelleted by a 10-min (100 g) centrifugation or are layered on a Millipore filter, fixed in 5% glutaraldehyde, and embedded in Epon. Ultrathin sections are then made and stained with uranyl acetate and osmium tetroxide. Chief and mucous cells are recognizable because of their small size and the appearance of their secretory granules (large and light in aspect in the case of chief cells and smaller as well as more electron-dense for mucous cells). Parietal cells are of a larger size and have a typical ultrastructure, previously described.

As far as the purification of parietal cells is concerned, one can rapidly estimate the degree of enrichment by counting under light microscope the proportion of "large" (i.e., parietal) and "small" cells in fresh, unfixed preparations (Fig. 5). However, light microscope observation does not allow the distinction of mucous from chief cells.

Alternatively, biochemical markers can be used to assess cell purification. Thus, separation of parietal from nonparietal cells can be monitored by the recovery of cytochrome c oxidase, K^+-PNPPase, and cytochrome b_5 as markers for parietal cells, on the one hand, and pepsinogen as well as RNA as markers for nonparietal cells, on the other[11] (Fig. 6). Receptors for histamine and peptide hormones are also useful for this purpose. Thus,

[11] M. J. M. Lewin, D. Ghesquier, A. Soumarmon, A. M. Cheret, F. Grelac, and J. Guesnon, *Acta Physiol. Scand. (Spec. Suppl.)*, p. 267 (1978).

FIG. 4. Electron microphotographs of isolated rat gastric epithelial cells. (A) Scanning microscopy of crude isolated cell population. (B) Transmission microscopy of Percoll-purified parietal cells.

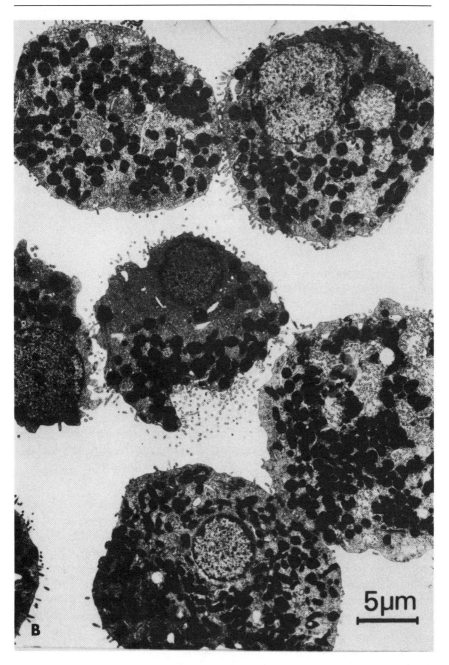

FIG. 4B.

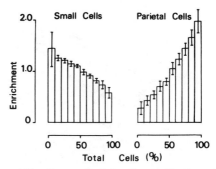

FIG. 5. Iterative centrifugation method for the purification of isolated gastric cells. Diagrammatic representation of the relative enrichment (ordinate axis) in parietal or small cells isolated from rat gastric mucosa by classes representing 10% of the total cells population after x centrifugation steps (abscissa) (mean values ± SEM).

gastrin and histamine H2 receptors are chiefly, if not exclusively, associated with parietal cells[12-14] (Fig. 7).

Assessment of Cell Viability

As in the case of other cell systems one cannot give any exhaustive definition of viability for isolated gastric cells. This notion refers rather to a series of criteria which are related to the intended use of the cells (i.e., for biochemical or pharmacophysiological studies). It should be pointed out that cell isolation per se can produce several artifacts resulting from protein redistribution around the pericellular membrane, whereas the necessary use of proteolytic enzymes and calcium chelators for this isolation inevitably results in a more or less severe alteration of cell integrity.

Morphological Criteria

In contrast to isolated chief and mucous cells, which show a smooth-surfaced or embossed appearance, isolated parietal cells are characterized by a profusion of microvilli over their entire surface. At the optical microscope level, ruthenium red staining may be used to assess the preservation of the glycoprotein coat on the plasma and intracellular canaliculi membrane of the parietal cell, whereas specific histological reactions can be used to assess the nondegranulation of mucous and chief cells, respectively (e.g.,

[12] A. Soumarmon, A. M. Cheret, and M. J. M. Lewin, *Gastroenterology* 73, 900 (1977).
[13] A. M. Cheret, F. Pignal, and M. J. M. Lewin, *Mol. Pharmacol.* 20, 326 (1982).
[14] A. M. Cheret, C. Scarpignato, M. J. M. Lewin, and G. Bertaccini, *Pharmacology* 28, 268 (1984).

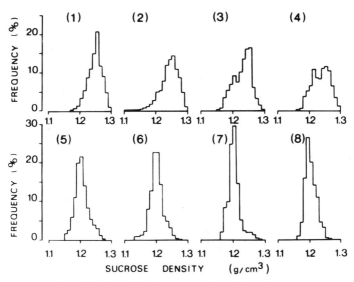

FIG. 6. Isopycnic equilibration of isolated rat gastric cells in a linear sucrose gradient. (1) Small cells, (2) pepsin activity, (3) RNA, (4) DNA, (5) parietal cells, (6) cytochrome c oxidase, (7) K$^+$-PNPPase, and (8) cytochrome b_5 distributions.

PAS and Bowie reactions). At the electron microscope level, isolated gastric cells show an ultrastructure closely resembling that observed in these cells *in situ*. As far as the inner architecture of the cells is concerned, one observes an apparent preservation of cell polarity. The apical abutment of intracellular canaliculi remains visible in the isolated parietal cell and the apical localization of secretory granules is retained in chief and mucous cells. Furthermore, freeze-fracture studies indicate preservation of pericellular and intracellular membranes, membrane particles being of the same appearance and distribution as for the cells *in situ*.

Biochemical Composition

The analysis of the biochemical content of isolated gastric cells is of critical importance because of the use of proteolytic enzyme to dissociate the epithelium. In addition, the assay of endocellular enzymes can yield important information concerning the preservation of cell membrane impermeability. As an example, Table I gives the mean biochemical composition of isolated rat fundic gastric cells as compared to the homogenate of (undissociated) fundic mucosa. The two preparations are seen to have a very close biochemical profile. However, acid phosphatase activity is lower in isolated cells than in homogenates, which may reflect a partial release of lysosomal enzymes. Similarly, the activity of ectocellular enzymes such as

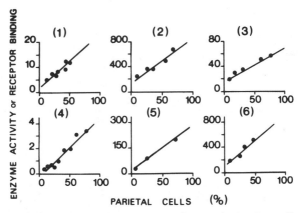

Fɪɢ. 7. Correlation between the percentage of rat parietal cells purified by iterative centrifugation (abscissa) and (1) O_2 consumption rate (nmol/min/10^6 cells), (2) [^{3}H]gastrin binding (cpm/10^6 cells), (3) carbonate dehydratase activity (μmol/min/10^6 cells), (4) cytochrome c oxidase activity (U/min/10^6 cells), (5) histamine (10^{-3} M) adenylate cyclase stimulation (%) in rat (Percoll-purified cells), and (6) in guinea pig (iterative centrifugation-purified cells).

5′-AMPase is 20–30% lower in isolated cells than in homogenates, which may indicate partial inactivation or the partial removal of such enzymes during the course of tissue dissociation. Providing good evidence for cell membrane impermeability, endocellular components such as K$^+$-stimulatable phosphatase (PNPPase) and cytochrome b_5 are 90–95% latent in

TABLE I

Bɪᴏᴄʜᴇᴍɪᴄᴀʟ Pʀᴏꜰɪʟᴇ ᴏꜰ Iꜱᴏʟᴀᴛᴇᴅ Rᴀᴛ Gᴀꜱᴛʀɪᴄ Cᴇʟʟꜱ ᴀꜱ
Cᴏᴍᴘᴀʀᴇᴅ ᴛᴏ Hᴏᴍᴏɢᴇɴᴀᴛᴇ Pʀᴇᴘᴀʀᴀᴛɪᴏɴꜱ ᴏꜰ Gᴀꜱᴛʀɪᴄ Mᴜᴄᴏꜱᴀ[a]

Components	Homogenate	Isolated cells
DNA (μg)	31.6 ± 5.7	43.1 ± 5.4
RNA (μg)	52.7 ± 3.4	56.9 ± 9.4
Cytochrome c oxidase (log OD/min)	8.6 ± 1.2	10.2 ± 2.5
Acid phosphatase (nmol/min)	12.0 ± 1.0	7.6 ± 1.4
Carbonate dehydratase (μmol/min)	725 ± 60	311 ± 74
Mg^{2+}-ATPase (nmol/min)	159 ± 30	144 ± 40
ATPase HCO$_3^-$ (nmol/min)	224 ± 24	185 ± 50
K$^+$-PNPPase (nmol/min)	18.5 ± 3.4	14.6 ± 2.0
Cytochrome b_5 (pmol)	26.6 ± 0.6	28.8 ± 2.3
Pepsin (μmol/min)	1.63 ± 0.09	1.61 ± 0.10

[a] The correspondence is based on the mean value ($\pm$SEM) of 0.178 mg ($\pm$0.002) protein per 10^6 cells. Units are per milligram of protein.

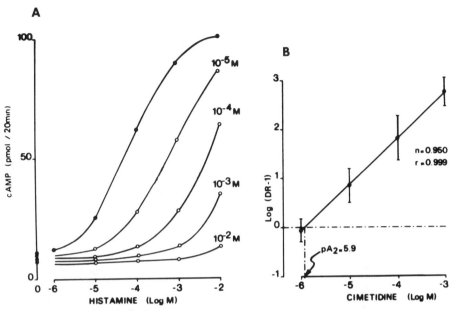

FIG. 8. Histamine stimulation of isolated guinea pig gastric mucosal cell adenylate cyclase. (A) Concentration effect curves obtained in the presence of histamine alone (control) (●) or together with increased concentration of the specific H2 receptor antagonist cimetidine (○). (B) Schild representation of the same experiments allowing extrapolation of pA_2 value for the antagonist.[13]

intact isolated cells, whereas they are fully recovered after cell sonication.[11] Isolated gastric cells are also characterized by a histamine-stimulated adenylate cyclase. This enzyme, located on the parietal cells (Fig. 7), is specifically inhibited by histamine H2 receptor antagonists such as cimetidine and ranitidine[13,14] (Fig. 8). Its activity is apparently higher in guinea pig cells than in rat and human cells (Fig. 9), which emphasizes on the suitability of guinea pig cells as experimental model in studies concerned with histamine H2 receptor.[15]

Cell Membrane Permeability

Plasma membrane permeability of isolated gastric cells can be tested by the classical Trypan blue method. As a rule, a dye exclusion percentage of more than 85% is observed, even after a 3-hr incubation of the cells in medium C at room temperature. Another useful test consists in measuring the ATP retention index calculated as the ratio of intracellular ATP over total ATP content of the cell suspension.[10] This index must remain above

[15] S. Batzri and J. D. Gardner, *Mol. Pharmacol.* **16**, 406 (1979).

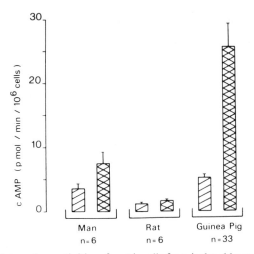

Fig. 9. Adenylate cyclase activities of gastric cells from isolated human, rat, and guinea pig fundic mucosa. These activities were measured under standard conditions, in the presence of histamine ($10^{-3}M$)(⬚) or without histamine (⬚). Note that the mean value of the basal adenylate cyclase of guinea pig gastric cells is 4 times higher than that of rat cells and 1.4 times higher than that of human cells, whereas in the presence of histamine, guinea pig cell adenylate cyclase activity is 15 times that of rat cells and 3.4 times that of human cells (n is the number of experiments with triplicate determinations; bars stand for ±1 SEM).

95% after up to a 3-hr incubation of the cells in medium C at room temperature.

Respiration

A typical mixed population of isolated rat gastric cells displays a mean oxygen consumption of 3.5 ± 0.1 nmol/min per 10^6 cells at 37°. Such a value is comparable to those for isolated mouse and dog gastric cells, i.e., 4.8 ± 1.1 and 2.3 ± 0.1 nmol/min per 10^6 cells, respectively.[16,17] As mentioned above, this rate is correlated to the enrichment of the suspension in parietal cells (Fig. 7) and it usually exceeds 10 nmol/min per 10^6 cells for 80–90% pure parietal cell suspensions. The cells also take in radioactive glucose and convert it to CO_2 at a rate of 0.66 ± 0.07 nmol CO_2/min per 10^6 cells (glucose, 1 mM).

Oxidative Phosphorylation

Oxygen consumption by isolated gastric cells is dramatically increased by uncouplers such as dinitrophenol, with a concomitant decrease in intracellular ATP content. It is blocked by mitochondrial inhibitors such as

[16] L. J. Romrell, R. M. Coppe, D. R. Munro, and S. Ito, *J. Cell Biol.* **65**, 428 (1975).
[17] A. H. Soll, *J. Clin. Invest.* **61**, 370 (1978).

rotenone and KCN. Addition of the electron acceptor menadione on rotenone-treated isolated cells restores the level of intracellular ATP to that of controls, which account for the cells' capacity to synthesize ATP in a respiration-coupled dependent process.[10]

Secretion and Transport Properties

Isolated rat parietal cells are able to maintain ionic gradients between the intracellular and extracellular space.[18,19] They respond to histamine by increased transport of Cl^- [20] and increased accumulation of amino-[^{14}C]pyrine (as an index of H^+ secretion).[21,22] The same effects can be obtained with dBcAMP, and can be blocked by histamine H2 receptor antagonists such as cimetidine and ranitidine. The cells also respond to β-adrenergic agonists and to acetylcholine, as well as other muscarinic cholinergic agonists by an increased accumulation of amino-[^{14}C]pyrine.[23-25]

The cholinergic response further includes changes in the fluxes of Ca^{2+} and K^+.[4] The histamine (or dibutyril cyclic AMP) stimulation of amino-[^{14}C]pyrine accumulation by isolated parietal cells could result from the activation of cAMP-dependent protein kinases,[26] however, that the stimulation is inhibited by tumor-promoting phorbol esters[27] suggests further involvement of protein kinase C. Histamine, cAMP, and cholinergic stimulations of amino-[^{14}C]pyrine accumulation are totally blocked by omeprazole, an inhibitor of the H^+,K^+-ATPase.[5] Isolated rat gastric cells are also capable of synthetizing proteins[28] and they can secrete pepsin (unpublished observations) as well as the intrinsic factor.[29]

[18] H. R. Koelz, J. A. Fischer, G. Sachs, and A. L. Blum, *Am. J. Physiol.* **235**, E16 (1978).

[19] A. Sonnenberg, T. Berglindh, M. J. M. Lewin, J. A. Fischer, G. Sachs, and A. L. Blum, *in* "Hormone Receptors in Digestion and Nutrition" (G. Rosselin, P. Fromageot, and S. Bonfils, eds.), p. 337. Elsevier, Amsterdam, 1979.

[20] A. Soumarmon, D. Ghesquier, and M. J. M. Lewin, *in* "Hormone Receptors in Digestion and Nutrition" (G. Rosselin, P. Fromageot, and S. Bonfils, eds.), p. 349. Elsevier, Amsterdam, 1979.

[21] W. Schepp and H. J. Ruoff, *Eur. J. Pharmacol.* **98**, 9 (1984).

[22] H. K. Heim and H. J. Ruoff, *Naunyn-Schmiedeberg's Arch. Pharmacol.* **330**, 147 (1985).

[23] G. Rosenfeld, *J. Pharmacol Exp. Ther.* **229**, 763 (1984).

[24] H. J. Ruoff, M. Wagner, C. Günther, and S. Maslinski, *Naunyn-Schmiedeberg's Arch. Pharmacol.* **320**, 175 (1982).

[25] R. Ecknauer, E. Dial, W. J. Thompson, L. R. Johnson, and G. C. Rosenfeld, *Life Sci.* **28**, 609 (1981).

[26] P. Mangeat, G. Marchis-Mouren, A. M. Cheret, and M. J. M. Lewin, *Biochim. Biophys. Acta* **629**, 604 (1980).

[27] N. G. Anderson and P. J. Hanson, *Biochem. Biophys. Res. Commun.* **121**, 566 (1984).

[28] C. N. A. Trotman, J. R. Greenwell, J. M. Carrington, A. K. MacGill, N. Taylor, and W. P. Tate, *Comp. Biochem. Physiol. B* **79**, 583 (1984).

[29] W. Schepp, H. J. Ruoff, and S. E. Miederer, *Biochim. Biophys. Acta* **763**, 426 (1983).

[22] Cell Separation by Gradient Centrifugation Methods

By THOMAS G. PRETLOW and THERESA P. PRETLOW

Introduction

Methods for the purification of individual kinds of cells have increased markedly in variety and sophistication during the past two decades.[1,2] In addition to gradient centrifugation, the separation of cells by velocity sedimentation can be accomplished in the absence of gradients by a procedure termed "counterstreaming centrifugation" by Lindahl,[3] its inventor, and termed "elutriation" by the commercial manufacturer of the apparatus required for its use in a very slightly modified form; the theory and applications of this procedure have been reviewed.[4-7] Similar separations are possible by velocity sedimentation in gradients without centrifugation at unit gravity as first described by Mel.[8-10] The most commonly employed version of sedimentation at unit gravity was described in 1967 by Peterson and Evans[11] and has been reviewed recently.[12] The same physical properties of cells, i.e., density and diameter, govern the sedimentation of cells in gradient centrifugation, sedimentation at unit gravity, and elutriation. The differences in these techniques caused by different apparatus and the associated different artifacts have been discussed in detail as relevant to sedimentation at unit gravity,[2,12] elutriation,[2,4] and gradient centrifugation.[2]

[1] K. Shortman, *Annu. Rev. Biophys. Bioeng.* **1**, 93 (1972).

[2] T. G. Pretlow, E. E. Weir, and J. G. Zettergren, *Int. Rev. Exp. Pathol.* **14**, 91 (1975).

[3] P. E. Lindahl, *Nature (London)* **161**, 648 (1948).

[4] P. E. Lindahl, *Biochim. Biophys. Acta* **21**, 411 (1956).

[5] T. G. Pretlow and T. P. Pretlow, *Cell Biophys.* **1**, 195 (1979).

[6] M. L. Meistrich, *in* "Cell Separation: Methods and Selected Applications" (T. G. Pretlow II and T. P. Pretlow, eds.), Vol. 2, p. 33. Academic Press, Orlando, Florida, 1983.

[7] M. J. Sanders and A. H. Soll, this volume [23].

[8] H. C. Mel, *J. Theor. Biol.* **6**, 159 (1964).

[9] H. C. Mel, *J. Theor. Biol.* **6**, 181 (1964).

[10] H. C. Mel, *J. Theor. Biol.* **6**, 307 (1964).

[11] E. A. Peterson and W. H. Evans, *Nature (London)* **214**, 824 (1967).

[12] W. C. Hymer and J. M. Hatfield, *in* "Cell Separation: Methods and Selected Applications" (T. G. Pretlow II and T. P. Pretlow, eds.), Vol. 3, p. 163. Academic Press, Orlando, Florida, 1984.

Theory

Since Lindahl's[3] early work with the sedimentation of cells without gradients, in what is now called an "elutriator," and Brakke's[13] subsequent introduction of gradient centrifugation, there have been many detailed reviews of the theory of sedimentation as applied to mammalian cells; three of these [1,2,4] are cited above. We shall present the theory briefly here.

In a centrifugal field, the sedimentation of cells is described by the equation

$$\frac{dr}{dt} = \frac{a^2(D_c - D_m)\omega^2 r}{k\eta}$$

In this equation, r is the distance of the cell from the center of revolution, a is the diameter or radius (depending on k) of the cell, D_c is the density of the cell, D_m is the density of the gradient at the location of the cell, ω is the angular velocity (speed of centrifugation), t is the duration of centrifugation, and η is the viscosity of the medium at the location of the cell.

From a superficial examination of this equation, it is apparent that the term that represents the size of the cell is squared and therefore relatively important. Diameter has the greatest influence on the sedimentation of the cell until the cell approaches its density in the gradient and $D_c - D_m$ becomes small. Despite the relative importance of diameter, we wish to emphasize the fact that cells are not, as is so often stated in the literature, separated by velocity sedimentation "according to diameter." It is not unusual for the densities of single kinds of mammalian cells to cover the entire range of densities between 1.060 and 1.110 g/ml. If we select a D_m that is low and consequently results in a large $D_c - D_m$, the density of the cell continues to be an important determinant of the velocity (dr/dt) of sedimentation. For example, if we assign a density of 1.010 g/ml (a density only slightly greater than that of tissue culture medium) to D_m and consider cells with identical diameters, the least dense and the most dense cells of a particular type—i.e., hepatocyte, pancreatic acinar cell, cardiac myocyte, or other type of cell—with the range of density described above will differ twofold in velocities of sedimentation *only* because of the differences in density exhibited among cells of the same type, i.e., the values 1.060 − 1.010 and 1.110 − 1.010 are different by a factor of two. As shown in an earlier publication,[2] cells that have the same diameter (13 μm) but differ by 0.010 g/ml in density will be separated by approximately 1 cm after velocity sedimentation in the isokinetic gradient described by our labora-

[13] M. K. Brakke, *J. Am Chem. Soc.* **73**, 1847 (1951).

tory.[14] Because of the large sizes of mammalian cells, diffusion, electrical charge, and other factors that influence the sedimentation of molecules are negligible in the consideration of the sedimentation of mammalian cells.

Densities of Cells

Most mammalian cells are characterized by densities between 1.055 and 1.110 g/ml. Most individual kinds of mammalian cells exhibit a broad range of densities and cover at least one-half and often more than three-fourths of the range of density given in the previous sentence. The densities of a large number of different kinds of mammalian cells as determined by isopycnic centrifugation were reviewed by us previously.[2] We would agree with Shortman's[15] statements that "The direct osmotic contribution of the gradient substance itself is reduced by choosing a material of high molecular weight. In the case of albumin (or say Ficoll) the direct osmotic pressure contribution is minimal ...." When the reported densities of the same kinds of cells were compared as determined in different laboratories by isopycnic centrifugation in gradients of albumin, Ficoll, or colloidal silica,[2] the values were found to be quite similar, i.e., the densities did not seem to be significantly influenced by the choice of one or another of these media. One would not expect that a gradient medium of lower molecular weight, such as metrizamide, could be used for the measurement of cell densities by isopycnic centrifugation without adjustment of the salt concentration in the gradient to compensate for the change in osmolarity that results from the change in the concentration of the gradient medium. The importance of pH as it affects the sedimentation of some kinds of cells has been discussed by Shortman.[15]

In 1970, Abeloff[16] found that the most highly purified blasts were among the least dense cells after isopycnic centrifugation in Ficoll gradients. While larger lymphoid and hemopoietic cells often exhibit lesser densities than the corresponding mature types of cells, we would not agree with Miller[17] that "cell density is roughly inversely proportional to cell size." Many of the very dense cells that we have studied have been much larger than the average cell. Examples of large cells with comparatively high densities that have been characterized include mast cells,[18] cardiac

[14] T. G. Pretlow, *Anal. Biochem.* **41**, 248 (1971).
[15] K. Shortman, this series, Vol. 108, p. 102.
[16] M. D. Abeloff, R. J. Mangi, T. G. Pretlow, and M. R. Mardiney, *J. Lab. Clin. Med.* **75**, 703 (1970).
[17] R. G. Miller, this series, Vol. 108, p. 64.
[18] T. G. Pretlow and I. M. Cassady, *Am. J. Pathol.* **62**, 323 (1970).

myocytes,[19] parotid acinar cells,[20] and pancreatic acinar cells.[21] In our experience, the densities, diameters, and/or rates of sedimentation under defined conditions must be tested experimentally. As reviewed previously,[2] if any two of these parameters is known, the third can be calculated.

Suspension of Cells

Numerous authors[1,2] have emphasized the fact that effective cell separations will be achieved only after cells are obtained as suspensions of single cells as free as possible of aggregates, debris, dead cells, and DNA gel. We mention this fact here because of its importance to cell separation by gradient centrifugation and refer the reader to other reviews[2,22,23] that describe the detailed methods for obtaining cells in suspension. In our experience, the aggregation of cells can be minimized by (1) working at sufficiently dilute concentrations of cells, (2) working at low temperatures (0–4°), and (3) introducing cells into density gradients in medium that contains protein. Specifically, we generally layer cells over gradients in 10% serum; others have used 3% albumin in their sample suspension of cells in order to minimize the aggregation of cells.

Type and Design of Gradients

There are two general types of gradient centrifugation; isopycnic centrifugation and velocity centrifugation. It is important to be aware that the separations achieved with these two kinds of gradient centrifugation depend on different physical properties of cells.

Isopycnic Centrifugation

In isopycnic centrifugation, cells are separated according to their different densities. This is accomplished simply by centrifuging a cell with sufficient centrifugal force and for a sufficient period of time for it to arrive at its isopycnic density in the gradient, i.e., the location in the gradient where the density of the cell is the same as the density in the gradient. This kind of gradient centrifugation is also called buoyant density sedimenta-

---

[19] T. G. Pretlow, M. R. Glick, and W. J. Reddy, *Am. J. Pathol.* **67**, 215 (1972).

[20] D. Aspray, A. Pitts, T. G. Pretlow, and E. E. Weir, *Anal. Biochem.* **66**, 353 (1975).

[21] J. Blackmon, A. Pitts, and T. G. Pretlow, *Am. J. Pathol.* **72**, 417 (1973).

[22] C. Waymouth, *in* "Cell Separation: Methods and Selected Applications" (T. G. Pretlow II and T. P. Pretlow, eds.), Vol. 1, p. 1. Academic Press, New York, 1982.

[23] T. G. Pretlow and T. P. Pretlow, *In Vitro Monogr.* **5**, 4 (1984).

tion. While isopycnic centrifugation is occasionally useful as an additional step used in conjunction with velocity sedimentation, review[2] of the very large number of cell separations accomplished by isopycnic centrifugation and by velocity sedimentation leads one inevitably to the conclusion that it is extremely rare that cells can be purified as effectively by isopycnic centrifugation as by velocity sedimentation. This fact results from the general, poorly understood fact that the densities of the different types of cells found in most tissues overlap broadly, while, in contrast, the diameters of different kinds of cells in the same tissues are most commonly different.

As discussed in detail by Shortman,[15] there are many kinds of gradient media that are suitable for use with the isopycnic centrifugation of cells. The most widely employed such media are albumin, Ficoll, and Percoll. Of these three media, the newest, Percoll, appears to offer some advantages. Since it does not have to be dissolved, it is more easily prepared than albumin. It offers the additional advantage over Ficoll that it is much less viscous than Ficoll. The advantages and disadvantages of Percoll have been reviewed recently by the investigator who invented it.[24] If Percoll is selected for isopycnic centrifugation, one should be aware of a recent, well-documented report[25] that the traditional method for the dilution of Percoll must be modified if one is to obtain an isosmotic density gradient.

Unlike the slopes of gradients for velocity sedimentation, the slope (g/ml/cm) of a gradient for isopycnic sedimentation is not of great importance. Since cells will closely approach their respective densities in gradients during isopycnic centrifugation, they will be separated from cells with different densities as long as sufficiently small fractions are collected from the gradients. In constructing gradients for isopycnic centrifugation, it is important that the most dense portion of the gradient be more dense than any cells to be separated by isopycnic centrifugation and that the least dense portion of the gradient be of as low density as possible. The latter condition is important because, when the transition from the starting sample suspension of cells to the gradient is too sharp, the cells may be pelleted against the sample–gradient interface. When this occurs, there may be aggregations of cells that will then sediment, not as single cells, but as aggregates. Aggregates will be sedimented isopycnically to densities that are not simple functions of the densities of the individual cells that compose the aggregates.

Like gradients for velocity sedimentation, cells may be introduced into gradients for isopycnic centrifugation by being suspended in a solution of

[24] H. Pertoft and T. C. Laurent, in "Cell Separation: Methods and Selected Applications" (T. G. Pretlow II and T. P. Pretlow, eds.), Vol. 1, p. 115. Academic Press, New York, 1982.
[25] R. Vincent and D. Nadeau, Anal. Biochem. 141, 322 (1984).

low density and being layered over the least dense portion of the density gradient; however, there are other methods of introducing cells into gradients for isopycnic centrifugation. Cells can be suspended in very dense media and layered under the most dense portion of the gradient. This procedure has been followed only rarely; however, it offers the advantage (see below) that, at least in theory, one might avoid the loss that results from the wall effect artifact.[2] In the case of Percoll, isopycnic centrifugation can be accomplished by mixing the cells with a homogeneous solution of Percoll and centrifuging with sufficient force to cause the Percoll to sediment to form a gradient. This kind of gradient has been called a "self-generated" gradient. While this procedure offers the advantage that cells can be introduced into the gradient at sufficiently low concentrations to minimize the aggregation of cells prior to sedimentation, it is associated with the disadvantage that the high centrifugal forces needed to generate a Percoll gradient are much higher than those needed for the isopycnic centrifugation of cells and are commonly injurious to cells. Percoll gradients can be preformed with a gradient maker or may be self-generated before cells are layered either under or over them.

Velocity Centrifugation

As mentioned above, we[2,26] have previously reviewed the several methods available for velocity sedimentation. Here, we shall discuss the velocity sedimentation of cells only as accomplished by gradient centrifugation.

The design of the density gradient for the separation of cells by velocity sedimentation is critical. It is possible to separate cells that have different densities but the same diameters by velocity sedimentation; however, in most cases, one uses velocity sedimentation to separate cells when one hopes to maximize the effect of diameter on the separation that is achieved.

For velocity sedimentation, it is important that the gradient medium have a defined viscosity that will be constant from batch to batch. It also seems advantageous that it contribute negligibly to the osmotic pressure of the gradient. If the gradient medium contributes importantly to the osmotic pressure in the gradient, large, rapidly sedimenting cells will encounter increased osmotic pressures as they sediment into locations in the gradient that contain higher concentrations of the gradient medium, i.e., Ficoll or albumin. As this occurs, the higher osmotic pressures will cause

---

[26] T. G. Pretlow and T. P. Pretlow, in "Cell Separation: Methods and Selected Applications" (T. G. Pretlow II and T. P. Pretlow, eds.), Vol. 1, p. 41. Academic Press, New York, 1982.

the cells to lose water, shrink, and sediment less rapidly. This will result in a diminished rate of sedimentation for the more rapidly sedimenting cells and a less effective separation of these cells from the more slowly sedimenting cells. It has therefore seemed advantageous to us to select a synthetic polymer of high molecular weight that will produce solutions of known viscosity. In our laboratory, Ficoll (polysucrose, average MW, 400,000, Pharmacia Fine Chemicals, Piscataway, New Jersey) has been useful for this purpose.

Previously, we[27] discussed the principles of the design of gradients that give high resolution in velocity sedimentation. If one is to maximize the separation that can be achieved by velocity sedimentation, (1) the gradient should have a very low concentration of gradient medium at the sample–gradient interface; and (2) the slope (g/ml/cm) of the gradient should be as small as is consistent with the stability of the gradient. The low concentration of gradient medium at the sample–gradient interface will minimize the tendency of cells to be packed against the sample–gradient interface. The small slope of the gradient will mean that the viscosity will increase as slowly as possible as the cell sediments through the gradient and the effective mass $(D_c - D_m)$ of the cell will decrease as slowly as possible as the cell sediments through the gradient. For the separation of cells as the result of differences in diameter, the conditions will be optimized by maximizing $D_c - D_m$. Here $D_c - D_m$ will be maximized when conditions 1 and 2 above are satisfied. In our experience, these conditions are closely approached by an isokinetic gradient[14] described previously from our laboratory. As discussed in detail previously,[2,14] this gradient offers the additional advantage that cells sediment in it at constant velocities that are linear functions of centrifugal force and duration of centrifugation. These conditions make it possible for one to predict the locations of cells and the conditions that will permit optimal purifications of cells without the integration of the Stokes equation.

Purpose of a Density Gradient for Velocity Sedimentation

It is possible to achieve a degree of separation of cells by the velocity sedimentation of cells in homogeneous solutions in the absence of a density gradient. Unfortunately, in the absence of a density gradient, a homogeneous column of fluid is unstable and mixing occurs both while cells are sedimenting and while fractions are being collected. The stability of the column of fluid is particularly important in gradient centrifugation, as

[27] T. G. Pretlow, J. I. Kreisberg, W. D. Fine, G. A. Zieman, M. G. Brattain, and T. P. Pretlow, *Biochem. J.* **174**, 303 (1978).

opposed to sedimentation at unit gravity, since both Coriolis effects[28] and "swirling" of the gradient during acceleration and particularly deceleration of the gradient will cause mixing of successive layers in the gradients if the column of fluid is unstable. The stability of a gradient during centrifugation is related both to centrifugal force and to its slope. Swirling, in this context, refers to what de Duve[29] has termed the "rotational movements of the fluid" in the centrifuge tube. During early acceleration and late deceleration of swinging buckets, the tangential velocities of fluid at the same level in the centrifuge tube are different at the centrifugal and centripetal walls of the centrifuge tube because the arc described by the centrifugal edge of the tube is greater than the arc described at the centripetal edge of the tube. This difference in force and velocity on the centrifugal and centripetal edges of the tube causes the gradient solution to rotate. The slope (g/ml/cm) of the density gradient must be sufficiently great to cause the gradient to be stable and rotate within the tube in a radially symmetrical fashion. Particularly during deceleration, one hopes to avoid the alternative, i.e., mixing of cells that have sedimented to different levels in the gradient.

Band or Gradient Capacity

Pragmatically, band capacity or gradient capacity is one of the most important and commonly ignored considerations in planning the separation of cells by gradient centrifugation. The theoretical treatment of "band capacity" as applied to gradient centrifugation has been presented by Brakke.[30] Brakke has pointed out that there is often a very substantial discrepancy between the theoretical and observed band capacity. Practically, it is important for the investigator to be aware that each gradient is characterized by an upper limit with respect to the number of a particular kind of cell that will sediment ideally, i.e., without artifacts. Experimentally, when the capacity of the gradient is exceeded, the changes that are seen in the sedimentation of cells are similar to those illustrated for the sedimentation of viruses by Brakke[31] in 1965. The modal population of particles (cells or viruses) becomes progressively broader as the load of particles is increased. When one plots the distribution of particles as a function of fraction number, a previously symmetrical or almost symmetrical modal population becomes asymmetrical such that the fractions that contain the most rapidly sedimenting particles will be disproportionately

[28] A. S. Berman, *Natl. Cancer Inst. Monogr.* **21**, 41 (1966).
[29] C. de Duve, J. Berthet, and H. Beaufay, *Prog. Biophys. Biophys. Chem.* **9**, 325 (1959).
[30] M. K. Brakke, *Adv. Virus Res.* **7**, 193 (1960).
[31] M. K. Brakke, *Science* **148**, 387 (1965).

increased relative to the fractions that contain the less rapidly sedimenting cells. When gradients are severely overloaded, entire modal populations of cells may be shifted into fractions that would contain more rapidly sedimenting cells in gradients that are not overloaded. When gradients are markedly overloaded, modal populations of cells may be physically quite remote from their appropriate locations in the absence of this artifact. The artifacts that are associated with exceeding the gradient capacity can result in the generation of specious subpopulations of cells that differ in their rates of sedimentation. Reports of such subpopulations, identified only by their rates of sedimentation, can be found in the literature and serve to emphasize the importance that should be attached to the quantitative and qualitative analysis of purified cells identified by characteristics independent of their rates of sedimentation.

Whether the theoretical concept of "band capacity" adequately explains the artifact that is observed in the laboratory is uncertain; however, in the laboratory, the kinds of artifacts that are observed when gradients are overloaded are characteristic and easily recognized. Before loading large numbers of cells on a gradient, the investigator should investigate the sedimentation of a small number of cells. The actual number of cells will be determined by the system used. For our isokinetic gradient,[14] we know that the band capacity has never been exceeded with fewer than 20 million cells. For the SZ-14 and TZ-28 reorienting zonal rotors (rotors that have identical properties for our purposes), the gradient capacity was exceeded with 213 million very large ascites tumor cells[32]; was exceeded by 1 billion liver cells[33]; and failed to be exceeded by 692 million human tonsillar cells,[34] i.e., all we could get from the largest tonsil examined. Sample graphs of the kinds of changes that occur when one loads gradients progressively more heavily until their capacities are exceeded have been published both for graded doses of hepatocytes[33] and for the ascitic form of a mast cell tumor.[32] The gradient capacity is generally much larger for small cells such as lymphocytes than for large cells such as hepatocytes or very large cells such as the ascitic tumor cell mentioned above. The capacity of the gradient can be increased by increasing the slope (g/ml/cm) of the gradient; however, small increases in the slope of the gradient will often result in markedly decreased resolution.

[32] C. L. Green, T. P. Pretlow, K. A. Tucker, E. L. Bradley, W. J. Cook, A. M. Pitts, and T. G. Pretlow, *Cancer Res.* **40**, 1791 (1980).

[33] S. B. Miller, T. P. Pretlow, J. A. Scott, and T. G. Pretlow, *J. Natl. Cancer Inst.* **68**, 851 (1982).

[34] D. F. Daugherty, J. A. Scott, and T. G. Pretlow, *Clin. Exp. Immunol.* **42**, 370 (1980).

Continuous and Discontinuous Gradients

There are few adequate reasons for the choice of a discontinuous gradient. While discontinuous gradients become less prone to cause artifacts as one decreases the differences in solute concentration (density and viscosity) between contiguous layers of the gradient, discontinuous gradients have inherent in their design several disadvantages as compared with continuous gradients. The presence of interfaces in discontinuous gradients results in the accumulation of cells at high concentrations at these interfaces, and the result is often the incorporation of different kinds of cells in the same aggregates. Obviously, these heterogeneous aggregates will prevent the cells that are contained in them from being as highly purified as they might have been if they had not become adherent to cells of another kind. It seems absurd that an investigator would deliberately compress together cells that have different densities or rates of sedimentation by the construction of gradients in which cells can be found only at a limited number of interfaces. If one wishes deliberately to combine cells that have been separated according to their different physical properties, cells from fractions from continuous gradients can be combined without sacrificing the superior purity that is available when one avoids the artifacts that are inherent in the use of discontinuous gradients. Discontinuous gradients do not permit cells to be distributed according to their natural densities or rates of sedimentation.

Because of the widespread use of discontinuous gradients, it should be pointed out that we are not aware of any published examples of comparisons of continuous and discontinuous gradients that have shown the separations achieved with discontinuous gradients to be anything other than inferior to those achieved in continuous gradients. Several authorities on gradient centrifugation have expressed similar views. Shortman[15] stated: "A continuous density gradient ... will always provide the maximum information, the maximum resolution, and will have the maximum cell load [gradient capacity]." In speaking of the common practice of collecting cells from the interfaces of discontinuous gradients, Shortman continued: "It should be remembered that these bands of cells are quite arbitrary cuts, essentially artifacts of the technique, rather than natural subpopulations. In addition, since it is these sharp narrow zones of density change which carry almost all the cells and constitute the separation zones, the cell capacity of discontinuous gradients for effective separation is much less than for continuous gradients." de Duve[35] stated: "The discontinuous gradient is essentially a device for generating artificial bands ... it is also a

[35] C. de Duve, *J. Cell Biol.* **50,** 20d (1971).

very dangerous procedure, in that it creates the illusion of clear-cut separation." In describing the continuous gradient, he continued by stating: "It lends itself to an entirely objective assessment of the frequency-distribution curves of certain physical properties, such as density or sedimentation coefficient, from which in turn other characteristics of a population, including its size distribution, can be derived ... " The use of discontinuous gradients for the purification of cells from the lymphoid system would seem particularly hazardous since it is ever more apparent that small subpopulations of specialized lymphoid cells can alter the functions of other lymphoid cells dramatically. If small numbers of unwanted cells adulterate the "purified" lymphoid cells because an inferior technique was used for their purification, the biological functions measured after purification might be and often have been very misleading.

Choice of Centrifuges, Rotors, and Centrifuge Tubes

For the separation and/or characterization of large numbers of cells by gradient centrifugation, there is only one currently commercially available zonal rotor that has had important applications for the separation of large numbers of viable cells by gradient centrifugation, the Sorvall TZ-28 rotor. Work with this rotor has been reviewed.[36]

Most separation of cells by gradient centrifugation has been carried out in centrifuge tubes. It has been demonstrated that the resolution of cells and other particles with different rates of sedimentation improves as the density gradient is moved closer to the center of revolution, i.e., the center of the centrifuge[27,37]; however, despite the improved resolution that is achieved, moving the gradient close to the center of revolution (1) markedly decreases the recovery of cells because of the wall effect artifact and (2) increases swirling artifacts (*vide supra*). In well-designed gradients, such as an isokinetic, 2.7–5.5% (w/v), 13-cm Ficoll gradient,[14] the resolution that can be achieved even when the gradient is a great distance from the center of revolution is much in excess of that which is required for the separation of cells that can be separated by velocity sedimentation. Practically, it is best to use a centrifuge that allows one to place the gradient as far as possible from the center of revolution in order to minimize the loss of cells that results from the wall effect artifact.

In the selection of a centrifuge and rotor that will permit one to move the gradient away from the center of revolution, it is convenient to use a centrifuge that has a heavy head that will decelerate slowly because of a

---

[36] T. P. Pretlow and T. G. Pretlow, *in* "Cell Separation: Methods and Selected Applications" (T. G. Pretlow II and T. P. Pretlow, eds.), Vol. 2, p. 221. Academic Press, New York, 1983.
[37] T. G. Pretlow and C. W. Boone, *Science* **161**, 911 (1968).

high angular momentum. When light heads are used, we would recommend that all swinging buckets be filled with the largest available tubes filled with water. In our experience, with few exceptions, low-speed, refrigerated centrifuges made by most manufacturers are quite satisfactory. In our opinion, the MSE LR-6 centrifuge with the six-place, swing-out head that takes 1-liter buckets has the slight advantage over other centrifuges that it (1) allows one to place the gradient as far as any commercially available centrifuge from the center of revolution and (2) has one of the heaviest heads available. With this machine and rotor, the sample-gradient interface of the previously described,[2] 13-cm isokinetic gradient is 13.7 cm from the center of revolution when the gradient rests on a 5.5-ml cushion.

Empirically, we have tested several kinds of centrifuge tubes. We use the International Equipment Company (I.E.C.) polycarbonate tube #2806. This tube can be autoclaved. Additional positive features of this tube include the fact that it holds a 13-cm, 85-ml gradient and has a diameter of 32 mm. While we doubt that the use of a tube with a smaller diameter would be a disadvantage, we have tested tubes with larger diameters under identical experimental conditions and have found that shallow density gradients are less stable in tubes of larger diameter. One possible explanation for this observation is that tubes of larger diameter fail to dampen the swirling artifact. A small tube offers an additional advantage that results from the fact that the band capacity in a centrifuge tube is related to the circumference of the tube (see previous discussion[2] of the wall effect artifact). The circumference/volume ratio is larger for tubes of small diameter.

Choice of Methods for Unloading Gradients

There have been many studies of methods for the collection of fractions from gradients. Unquestionably, the most artifact-prone method that is commonly used for the collection of fractions is the aspiration of cells from gradients (often from interfaces) with Pasteur pipets. Many immunologists use this method routinely and emphasize their confidence in this method. Both of the present authors spent a year in different, very well-known immunology laboratories recently; and we both finished our time in those laboratories with the strong conviction that this method of collecting fractions is less than quantitative, very imprecise, and the cause of much experimental inconsistency.

Most investigators who have critically examined several systems for the collection of gradient fractions prefer the upward displacement of gradients through a conical, watertight tapping cap that is inserted in the top of the tube. This upward displacement is accomplished by the introduction into the bottom of the centrifuge tube of a material of greater density than the most dense portion of the density gradient. Generally, we use 55% sucrose as the material to be introduced into the bottom of the centrifuge tube. Suitable tapping caps have been illustrated diagrammatically by Shortman[15] and photographically by us.[2] The important features of a tapping cap to be used for this purpose is that it (1) fit the centrifuge tube to give a watertight seal and (2) has an angle at its apex of less than 37°. The tapping cap that we use with the I.E.C. tube #2806 is available commercially (Halpro, Inc., Rockville, Maryland).

The sizes of fractions to be collected can be varied according to individual needs; however, we would caution against using fractions that are so large than an entire modal population of cells is found in a single fraction. Fractions should be small enough to give several data points that characterize the leading and trailing edges of each modal population.

Preparation of Ficoll

For velocity sedimentation, solutions of Ficoll in tissue culture medium can be prepared by stirring Ficoll with tissue culture medium in an ice bath with a stirring rod. For velocity sedimentation, most commonly, we have used 5.5% (w/w) Ficoll as our most concentrated gradient solution. The highest concentration of Ficoll that we have ever used for velocity sedimentation is 18.5% (w/w)[19]; this concentration of Ficoll can be sterilized by filtration through a 0.2-μm filter. For velocity sedimentation, we prefer to use Ficoll because it is a synthetic polymer of defined viscosity and, to our knowledge, has no documented toxicity for any kind of cell when used in the concentration (5.5%) employed for most of our experiments.

For both velocity and isopycnic sedimentation, Ficoll can be divided into aliquots of the desired size and stored at −20° for at least three months with no change in physical properties. We generally make a 3-month supply, sterilize it by filtration, and keep it frozen. Unfrozen, at pH 6.8–7.8, Ficoll is stable for approximately 3 days. Gradients may be used on the day when they are made, 1 day later, or 2 days later. After 2–3 days, one begins to encounter changes in the viscosity of Ficoll stored as a liquid at approximately neutral pH presumably secondary to a slow depolymerization. This problem is not encountered for at least 3 months when solutions are kept frozen.

For isopycnic centrifugation, Ficoll is satisfactory; however, Percoll offers some advantages over Ficoll as detailed above. We have generally used solutions of approximately 41% Ficoll for cushions below our density gradients. Ficoll has the advantage over many media used for cushions in that its high molecular weight, low molar concentration, and high viscosity retard diffusion. Similar Ficoll solutions can be used as the heavy medium for the construction of gradients for isopycnic centrifugation.

Because of difficulty that others have described and that we encountered initially in dissolving Ficoll at high concentrations, we (see p. 160 in Ref. 2) have described in detail a convenient method for obtaining Ficoll in solution at high concentrations. We place 500 g of Ficoll in a 1-liter Erlenmeyer flask in an ice bath. Stirring the material, add 700 ml of distilled deionized water to the Ficoll, approximately 100 ml every 5–10 min. When using a 1-liter flask, the undissolved lumps of Ficoll will float on top of the solution and will be forced into the narrow neck of the flask. These lumps can be crushed with a rod and mixed with the solution below. After approximately 1–2 hr, the flask should contain a thick, opaque, white paste. If this is stirred approximately every 15 min, the Ficoll will form a transparent solution over 3–6 hr. We generally hasten this process after the first 2 hr by autoclaving the flask for 12 min at 121°. The autoclaving of Ficoll is desirable both to hasten its being dissolved and because dry Ficoll often contains small numbers of yeast, as it comes as a powder from the manufacturer. If it is autoclaved, it is very important that no more than 16 min should elapse between the beginning of the sterilization process and the time when the Ficoll is removed from the autoclave. If the Ficoll is left in the autoclave longer, it may carmelize and have markedly altered viscosity. Sterile water (10 ml) is layered over the top of the Ficoll solution in the Erlenmeyer flask, and it is allowed to cool slowly at room temperature.

Because Ficoll contributes less than 25 mOsm/liter to a solution of 40% Ficoll, it is necessary to dissolve it in a salt solution if it is to provide an isotonic solution for cells. When it is to be mixed with commercially available powdered tissue culture medium, the medium should be washed into a 1-liter volumetric flask with 15 ml of sterile water. Only after the bubbles have all popped, the concentrated Ficoll can be added to the volumetric flask. If the Ficoll is added to the medium before the bubbles have popped, it may take days for the bubbles to pop. After the volume is made up to a liter, the wet tissue culture ingredients and Ficoll are mixed by pouring the material back and forth between sterile containers several times.

Procedure for Making Density Gradients

The procedure given below is that which we use for making isokinetic gradients. For isopycnic gradients, the procedure would be modified only by substituting for 5.5% (w/w) Ficoll a solution of Ficoll, Percoll, albumin, or whatever medium is preferred at a concentration that gives a density higher than the most dense cell to be separated by isopycnic centrifugation.

Materials and Reagents

85 ml of tissue culture medium that is selected based on the kind of cell studied.

80 ml 5.5% (w/w) Ficoll in culture medium (refractive index = 5.5 sucrose equivalents) + (refractive index of salt solution in sucrose equivalents) = 7.4 – 7.7 sucrose equivalents for 5.5% Ficoll solutions in most culture media in our laboratory. Many refractometers are calibrated in sucrose equivalents.

5.5 ml 40% or > 40% (w/w) Ficoll in tissue culture medium.

Gradient generator (your own or commercially available as illustrated previously[2] from Lido Glass, Stirling, New Jersey 07980).

Centrifuge tubes, polycarbonate (I.E.C. #2806), or others as preferred.

Graduated cylinder.

Polystaltic pump (gravity can be used instead, with less convenience).

Wide-bore and narrow-bore 10-ml pipets.

Procedure

1. Place polycarbonate centrifuge tubes (H; see Fig. 1) in clamps on ringstand such that they deviate slightly (approximately 15°) from the vertical.

2. Connect gradient makers to tubing (F) (Tygon, $\frac{1}{16}$-in. internal diameter) that conducts the gradient solutions to the tops of the centrifuge tubes. The rate of flow is most conveniently controlled by passing the tubing through a pump (G); however, the rate of flow can be controlled by raising and lowering the centrifuge tube on a ringstand.

3. With a wide-bore pipet, introduce 5.5 ml of 40% or > 40% Ficoll in tissue culture medium into the bottom of the centrifuge tube without wetting the sides of the tube.

4. Insert bubblers (D) made of 9-in. Pasteur pipets into gradient makers such that the tips of the Pasteur pipets are $\frac{1}{8}$ – $\frac{1}{4}$ in. from the bottoms of the gradient maker (cylinder C).

5. Put 30 – 40 ml of tissue culture medium in cylinder A and tilt the gradient maker until the glass connector B is filled with medium; close stopcock (E).

6. Fill A with culture medium to the 80-ml graduation point.

7. With tubing F clamped closed, introduce 80 ml of 5.5% (w/w) Ficoll into cylinder C.

8. Bubble Ficoll solution in cylinder C with 5 or 10% CO_2 (as recommended for the kind of culture medium used as solvent) until the color indicator in the medium indicates a pH between 7.0 and 7.4.

9. Open stopcock E to allow communication between chambers A and C; there should be a very small backflow from C to A.

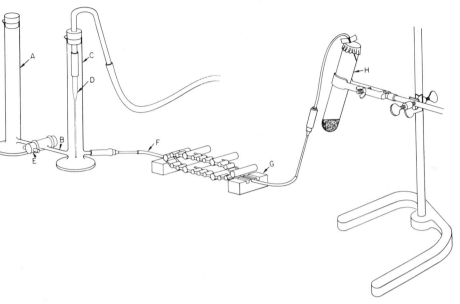

Fig. 1. Apparatus used for making sterile Ficoll gradients. (A and C) Graduated cylinders, (B) a connecting segment of glass tubing, (D) a Pasteur pipet used to mix gradient solution by bubbling during the generation of the gradient, (E) a Teflon stopcock, (F) tubing connected to the gradient maker, (G) a proportioning pump, (H) a sterile polycarbonate (Autoclear) centrifuge tube. The gradient maker is sterilized by autoclaving. (From Pretlow *et al.*[2])

10. Release the clamp on tubing F and allow the gradient to flow into the centrifuge tube (H) at a rate not to exceed 0.5 ml/min.

11. Allow gradient to form until the gradient is 2 cm from the top of the tube when the tube is vertical; the tube should be made vertical toward the end of this procedure.

12. After the gradient is formed, remove the next drop of gradient solution from the end of tubing F; it should have a refractive index of 4.1–4.2 sucrose equivalents (the refractive index of 2.5% Ficoll + the refractive index of the culture medium).

13. Place rubber stoppers in gradient tubes (above gradients), and place the gradients in the refrigerator until they are used on the day they are made, 1 day later, or 2 days later. Discard the gradients if they are not used by the end of the second day after they are made.

Procedure for Centrifuging Gradients

1. Turn on centrifuge to allow cooling with rotor turning at 200–300 rpm 2 hr before needed. Adjust temperature to 0–4°.

2. If a stroboscope is to be used (this is desirable but not essential), turn it on at least 30 min before checking its calibration.

3. Place gradients in swinging buckets.

4. Layer a known number of cells over gradients in 5–7 ml of medium with 10% serum or 3% bovine serum albumin. Save some cells for permanent slides.

5. Balance buckets. Always balance a gradient with a gradient; never balance a gradient in one swinging bucket with water in the opposite bucket.

6. Centrifuge the gradient as soon as possible after it is loaded. Accelerate up to a centrifuge speed of up to 200 rpm (no more) over the first 15 sec; 500 rpm, over the next 15 sec; and final speed for velocity sedimentation (usually a maximum of 99 g; 800 rpm on the M.S.E. LR-6 centrifuge) over the next 15 sec. For isopycnic centrifugation, accelerations should start slowly similarly to velocity sedimentation; however, the time course to arrive at final speed is not important since the distance sedimented is not affected.

7. Time centrifugation from the time when final centrifuge speed is reached until the switch has been turned off and deceleration begins. Select a centrifuge speed that is consistent with centrifugation for 10–20 min; under these circumstances, the sedimentation that occurs during acceleration and deceleration is negligible.

Procedure for Sterile Gradients

If one wishes to perform experiments under sterile conditions, this can be accomplished easily since all of the equipment is small and can be wrapped with paper or aluminum foil and autoclaved. It is important to include a gas filter in the gas line for the bubblers (D).

Procedure for Processing Gradient Fractions

1. Carefully remove gradient from centrifuge and place it in a clamp on a ringstand.

2. Insert the watertight tapping cap (available commercially from Halpro, Inc., Rockville, Maryland) into the tube.

3. Introduce 55% (w/w) sucrose into the bottom of the tube to float the gradient out through the tapping cap. Fractions can be collected at speeds up to 2 ml/min without causing mixing of the gradient.

4. Perform cell counts on every fraction.

5. Take the refractive index of each fraction to confirm the linearity of the gradient.

6. Make permanent preparations of the cells in each fraction. We generally use a Wright–Giemsa or hematoxylin, eosin, and azure stain.

Centrifugal Force and Duration of Centrifugation

Isopycnic centrifugation can be accomplished in continuous Ficoll gradients with a centrifugal force of 1200 g for 1 hr. Szenberg and Shortman[38] found that isopycnic centrifugation in albumin gradients could be accomplished with 1300 g for 90 min. With less viscous gradient media such as Percoll, less centrifugal force and time are necessary. The most common error in the selection of appropriate centrifugal forces and times for centrifugation is the use of severalfold too much force. Centrifugal force is not innocuous for cells, and it appears that high centrifugal forces are much more dangerous for cells after they have ceased sedimentation, i.e., reached their isopycnic densities. With 1200 g for 3 hr at 4°, we have seen HeLa cell nuclei centrifuged right through the cell wall. In that experiment, nuclei could be seen indenting the cell walls, breaking the cell walls, and after they had left the cells. In practice, there are very few situations in which one can obtain cells in as high purity by isopycnic centrifugation as by velocity centrifugation; however, on those occasions when one wishes to purify cells by isopycnic centrifugation or when one simply wants to measure the densities of cells by isopycnic centrifugation, excessive centrifugal force should not be used. Raidt et al.[39] found that some lymphoid cell functions remained after they separated cells with 20,000 g for 30 min in albumin gradients; however, most cells will not survive these conditions. One never knows if functions that survive these conditions are representative of the functions that might have been present if one had used a more appropriate, lower centrifugal force.

For velocity sedimentation, the appropriate speed and duration of centrifugation can be determined by a variety of methods depending on the kinds of gradient media employed. If experiments are to be repeatable, it is essential that the gradient medium have the same relationship between viscosity and concentration from batch to batch, unlike some biological materials that have been used as media for the construction of gradients. In our hands, Ficoll has been very satisfactory for this purpose. If one uses Ficoll, the fact that its viscosity is an exponential function of its concentration[40] greatly simplifies the integration of the Stokes equation in order to allow one to predict the locations of cells with known diameters and densities in Ficoll gradients of known concentrations and dimensions. A computer program in BASIC language for this purpose has been published.[2] If the isokinetic gradient[2,14] is used, a computer program is not necessary since (1) cells sediment in the isokinetic gradient with velocities that vary by less than 3% throughout the entire length of the gradient, (2)

[38] A. Szenberg and K. Shortman, *Ann. N.Y. Acad. Sci.* **129**, 310 (1966).
[39] D. J. Raidt, R. I. Mishell, and R. W. Dutton, *J. Exp. Med.* **128**, 681 (1968).
[40] T. G. Pretlow, C. W. Boone, R. I. Shrager, and G. H. Weiss, *Anal. Biochem.* **29**, 230 (1969).

the degree of separation of cells with different rates of sedimentation in this gradient is a constant function of this distance sedimented, and (3) the distance sedimented is directly proportional to the centrifugal force and to the duration of centrifugation.

As explained in detail,[2] when one takes 4-ml fractions from the isokinetic gradient, cells are maximally separated in the isokinetic gradient by sedimenting them to approximately fraction 16–19 of the 21–22 fraction gradient. At that location, they are maximally separated from the less rapidly sedimenting cells and not yet at the gradient–cushion interface. Empirically, one can determine how this is accomplished by centrifuging with a low centrifugal force for a short time (for example, 30 g for 10 min). After fractions are collected and cells counted, one can determine the velocity of sedimentation of the cells that one wishes to purify. If the desired cells traveled 2 cm and one wants to purify them further by having them travel 10 cm through the 13-cm gradient (fivefold farther), one simply quintuples the centrifugal force or the duration of centrifugation. Use of this isokinetic gradient obviates the need for the computer integration of the Stokes equation and results in a resolution that approaches the limit of resolution possible by velocity centrifugation.

In altering the centrifugal force for velocity sedimentation, one has to recall that centrifugal force is a function of the square of the speed of centrifugation, i.e., centrifugal force = (radius in cm)(1.11 × 10^{-5})(revolutions per minute).[2]

Analysis of Experimental Cell Separations

There are several kinds of data that should be analyzed routinely after any kind of separation of cells. Many of these data are often omitted; without these data, one cannot interpret experimental cell separations.

Probably the most widely omitted calculation is the calculation of the recovery of each cell type layered over the gradient or introduced into any other procedure for the separation of cells. Often, investigators report that they obtain a particular kind of cell in 99% purity; however, they omit data that would allow us to know how representative the cells recovered were of those that were introduced into the procedure for the separation of cells. It is important to be aware that, if only one cell were recovered, it would be 100% pure. We doubt that any procedure for the separation of cells allows one a quantitative recovery of all cells introduced into the procedure. While recoveries from the reorienting zonal rotor reviewed by us[36] are often in the range of 90%, after most kinds of purifications, the recovery of cells ranges between 40 and 90% of cells introduced into the system for purification. This has been true for cells separated by electrophoresis,[41] elutriation,[42] sedimentation at unit gravity,[12] and isokinetic sedimentation.[43,44]

It would seem unnecessary to emphasize the importance of quantifying the results of all cell separation procedures; however, reports in the literature continue to describe purified cells as "essentially free" of other kinds of cells or "highly purified." Sophisticated biochemical and immunological studies of the function of "purifed" cells are often published in journals without any numerical, quantitative description of the degree of purity. Such biochemical or immunological studies are no more valuable than the degree to which the object of the study is identified. One wonders if these same journals would seriously consider publication of kinetic studies of enzymes described as "essentially pure" as judged by unspecified criteria. de Duve[35] has emphasized the difficulty with conclusions based on poorly purified materials: "The history of tissue fractionation is replate with examples of properties that were erroneously attributed to the object of incomplete purification." Incompletely purified cells will probably always be with us to some degree; it is important to remember that uncharacterized, incompletely purified cells are even less valuable.

We should mention explicitly the fact that the gross appearance of "bands" of cells in a gradient is misleading. Some journals continue to publish occasional articles in which photographs of different "bands" of cells in gradients are taken as evidence of the purification of cells. In fact, in our experience, cells in gradients that scatter light sufficiently to be photographed scatter light because they are aggregated. Light scattering is a function of the size of particles. We have seen several situations in which aggregates of cells formed visible bands while, in the same gradients, monomeric cells at higher concentrations failed to scatter light sufficiently to be detected photographically. We have also refereed papers purporting to separate organelles in which protein and enzyme concentrations showed maxima at locations other than those of the photographable bands. In our opinion, the photography of bands was an acceptable first approach in the early days of centrifugation when the state of technology made it difficult to quantify cells or organelles in gradients adequately. Today, much more adequate means to quantify purified populations of cells and organelles are available.

[41] T. G. Pretlow and T. P. Pretlow, *Int. Rev. Cytol.* **61**, 85 (1979).
[42] T. G. Pretlow and T. P. Pretlow, *Cell Biophys.* **1**, 195 (1979).
[43] T. G. Pretlow and T. P. Pretlow, *in* "Cell Separation: Methods and Selected Applications" (T. G. Pretlow II and T. P. Pretlow, eds.), Vol. 5, p. 281. Academic Press, Orlando, Florida, 1987.
[44] T. G. Pretlow and T. P. Pretlow, *Nature (London)* **333**, 97 (1988).

Finally, in the analysis of cell separations, we cannot emphasize too strongly the value of graphs that show the number of purified cells of each type plotted against the fraction numbers. It is often only in making such distribution graphs that one realizes that one is dealing with more than one population of a "particular kind" of cell. In making graphs, one also becomes aware of the location of the majority of cells of a particular type. Without such a graph, one may not become aware that the fractions that contain highly purified cells may not be the fractions that contain the majority of cells of a particular type. In addition, such graphs often make one sufficiently aware of the relative locations of different modal populations of cells in gradients to improve the experimental design for cell separations.

Acknowledgment

Supported by USPHS Grant CA-36467 and American Cancer Society Grant BC-437B.

[23] Cell Separation by Elutriation: Major and Minor Cell Types from Complex Tissues

By Martin J. Sanders and Andrew H. Soll

Introduction

The ability to disperse and enrich subpopulations of cells from a complex tissue is essential for allowing investigation of the functional interrelationships among component cells. Centrifugal elutriation has been a very useful technique for cell separation, with certain unique advantages and applications.[1-3] The definition of elutriation is separation or removal by washing. Centrifugal elutriation of cells is accomplished in a rotor that allows a counterflow of medium through a separation chamber opposing the centrifugal force; the balance of the opposing counterflow of medium and centrifugal force underlies the separation. The design of a set of rotating seals that allows perfusion of the chamber in the spinning rotor was a critical step in the development of the elutriator rotor. Elutriator

[1] R. J. Sanderson, *in* "Cell Separation: Methods and Selected Applications" (T. G. Pretlow II and T. P. Pretlow, eds.), Vol. 1. Academic Press, New York, 1982.

[2] T. G. Pretlow and T. P. Pretlow, *Cell Biophys.* **1**, 195 (1979).

[3] M. L. Meistrich, *in* "Cell Separation: Methods and Selected Applications" (T. G. Pretlow II and T. P. Pretlow, eds.), Vol. 2. Academic Press, New York, 1983.

rotors are manufactured by Beckman Instruments (Palo Alto, California). Over the past 8 years our laboratory has utilized elutriation as the primary method for enriching dispersed populations of canine fundic mucosal cells.

As pointed out by Meistrich,[3] the advantages of elutriation are that large numbers of cells can be rapidly processed. The technique is reproducible and allows use of any medium, rather than the expensive and possibly toxic media necessary for density gradient separations. Cell viability in general is maintained, although there are some cells that are sensitive to the shear forces encountered. Another major advantage of elutriation is that bacteria, with their small size, are flushed out during the initial wash; therefore, when sterile medium is used and the rotor is properly cleansed, sterile fractions result which are suitable for cell culture.

Factors Affecting Elutriation

Elutriation separates cells by velocity sedimentation. Centrifugal or centripetal movement of cells in the elutriation chamber results from the balance of the opposing centrifugal force and centripetal flow of medium. The design of the elutriator chamber is critical[4,5] in that the walls of the elutriation chamber are cone shaped, with the apex of the cone directed centrifugally (Fig. 1). This design results in an increasing cross-sectional area (A) and therefore a decreasing medium flow velocity (V_f) toward the center of the rotor. After loading, cells are pushed centrifugally against the increasing flow rate until they reach a point where their sedimentation velocity equals V_f. These factors allow the formation of a gradient of cells in the rotor as a function of their sedimentation velocity, V_s. Under these conditions, Stokes law relates these velocities to cell size and density:

$$V_s = V_f = F/A = \frac{d^2(\rho_p - \rho_m)(\omega^2 r)}{18\eta} \tag{1}$$

F is the counterflow rate (ml/min), A is the cross-sectional area (cm^2), d is the particle diameter (μm), ρ_p and ρ_m are the densities of cell and medium respectively, η is the viscosity of the medium, and $\omega^2 r$ is the angular velocity of the rotor.

Cells are eluted from the chamber when the counterflow velocity exceeds the sedimentation velocity at a given $\omega^2 r$. Solving Eq. (1) for F and assuming constant viscosity, medium, and particle densities (terms combined as C including chamber dimensions), one can determine the combi-

[4] P. E. Lindahl, *Nature (London)* **161**, 648 (1948).
[5] R. J. Sanderson, K. E. Bird, N. F. Palmer, and J. Brenman, *Anal. Biochem.* **71**, 615 (1976).

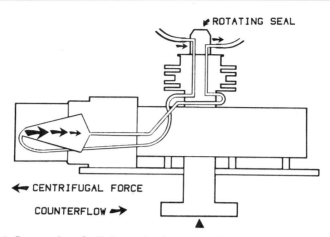

Fig. 1. Cross section of a Beckman elutriator rotor (JE-10X) illustrating the path of flow of medium through the cone-shaped chamber, in opposition to the centrifugal force. The rotating seal is contained within the top headpiece of the rotor.

nation of flow rate and rotor speed to elute a cell of a given diameter (D):

$$F = CD^2 \ (\text{rpm}/1000)^2 \tag{2}$$

In order to optimally utilize an elutriator rotor, precise control of both revolutions per minute (rpm) and flow rate is essential. We currently use both the JE-10X rotor in a J-6 centrifuge and the standard JE-6 rotor in a J-21 centrifuge. For both centrifuges we use a three-pole, two-position switch that permits bypass of the normal speed control, with a voltage divider consisting of a 10-turn potentiometer in series with a fixed resistor so that the rotor speed can be precisely regulated in the low-speed range for elutriator operations, and yet leaving the normal speed range available for other uses. For the J-21 centrifuge we used a 500-Ω potentiometer (Bournes 3400S-1-501) and a 2500-Ω resistor. For the J-6 centrifuge we used a 2000-Ω potentiometer (Bournes 3540S-1-202) and a 1000-Ω resistor. Flow of medium through the rotor is controlled by a Cole–Parmer Masterflex peristaltic pump, with the speed control modified by substituting a 10-turn potentiometer in place of the factory-installed single-turn speed control. For the standard elutriator rotor we use a 1- to 100-rpm pump (R-7553-10) and for the 10X rotor a 6- to 600-rpm pump (R-7553-00). Both motors are connected to 7015 pump heads; silicon tubing is used throughout the system.

Most laboratories manipulate rotor speed and/or flow rate to achieve separation of cells of differing sedimentation velocities. For cells that are similar in diameter, Figdor and co-workers[6] have evaluated increasing the density of the medium flowing through the chamber by adding Percoll and increasing the viscosity of the medium by adding polyethylene oxide.

Dispersion of Fundic Mucosal Cells

Optimal enrichment of cells from complex tissue requires a single cell suspension as the starting point. To prepare cells for elutriation, mucosal tissue is bluntly dissected from the remaining muscularis mucosa and submucosa. The tissue is diced into small pieces with scalpels and placed into 250-ml plastic Erlenmeyer flasks (Corning). For canine tissue, we have found that crude collagenase (Sigma Type I) works well in the range of 0.2 to 0.35 mg% for intestinal and fundic tissue. The collagenase is dissolved in Basal Medium Eagle's (BME, Gibco) supplemented with 0.1% BSA (solution 1). Following a 15-min treatment in this solution, the supernatant is discarded and replaced with Ca^{2+}- and Mg^{2+}-free Earle's salt solution containing 2 mM EDTA. After a 15-min incubation, the supernatant is again discarded and the tissue is further digested for 45 min in solution 1. To decrease the adhesiveness of the freshly dispersed cells, the cell suspension is washed twice by centrifugation at 1000 rpm in Hanks' salt solution and filtered through coarse (62 mesh) and fine (205 mesh) nylon mesh. The cell suspension is electronically counted (Celloscope, Particle Data) and diluted to about 5×10^6 cells/ml in Hanks' salt solution containing penicillin (100 μg/ml), streptomycin (100 μg/ml), and 10 mM HEPES, pH 7.4. Cells are stored at room temperature prior to elutriation, and are fine filtered again immediately before loading into the chamber.

Other Dispersion Media and Supplements

In addition to collagenase, other enzymes have been used, either alone or in combination, to produce isolated suspensions of cells; pronase has been used for fundic mucosa,[7] elastase for lung, and trypsin for pituitary and islets of Langerhans.[8-12] The enzyme and conditions must be opti-

[6] C. G. Figdor, J. M. M. Leemans, W. S. Bont, and J. E. De Vries, *Cell Biophys.* 5, 105 (1983).
[7] S. Mardh, L. Norberg, M. Ljungström, L. Humble, T. Borg, and C. Carlsson, *Acta Physiol. Scand.* 122, 607 (1984).
[8] H. Chamras, J. M. Hershman, and T. M. Stanley, *Endocrinology* 115, 1406 (1984).
[9] B. J. Undem, F. Green, T. Warner, C. K. Buckner, and F. M. Graziano, *J. Immunol. Methods* 81, 187 (1985).
[10] R. D. Greenleaf, R. J. Mason, and M. C. Williams, *In Vitro* 15, 673 (1979).
[11] S. E. S. Brown, B. E. Goodman, and E. D. Crandall, *Lung* 162, 217 (1984).
[12] D. G. Pipeleers and Pipeleers-Marichal, *Diabetologia* 20, 654 (1981).

mized for each tissue, depending upon cell yield, viability, functional responsiveness, and tendency to clump during manipulations.

Clumpiness can result when cells are damaged by the isolation procedure. For fundic or antral cells, release of mucus by mucous cells can be a major obstacle to successful elutriation. If cell clumping occurs and is not reduced by washing, Cleland's reagent (dithiothreitol, 0.01–0.05%) is added to the cell suspension prior to elutriation. In most cases this treatment significantly reduces the severity of the problem. Stickiness of cells can also be due to DNA released from damaged nuclei; addition of DNase (0.3–1 mg/liter) to the incubation solution has been reported to decrease clumping.[13,14]

In some tissues, the behavior of cells during subsequent elutriation may be greatly influenced by the choice of enzyme: Meistrich[3] noted that both a trypsin–DNase combination and collagenase dispersed testis into single cells, but the collagenase-dispersed cells displayed a strong tendency to clump. In contrast, we have not found that collagenase-dispersed fundic cells tend to clump excessively, compared to cells dispersed with other treatments. It is obvious that methodology for cells from different tissues and different species must be individualized.

Practical Considerations of Elutriation

Rotor Preparation

The rotor is reassembled each day and the pieces are inspected for wear and damage. The O rings and bearing assembly need special attention. We have found it essential to check the fit of the rotating seal with its O ring on the rotor spindle. The rotating seal is pushed into place by a spring; if the fit with the O ring is too tight, or the spring tension too weak, the rotating seal will not be in the proper position and there will be mixing of the incoming and outgoing streams of medium passing through the elutriation chamber. Following assembly, we sterilize the rotor by circulating 70% ethanol through the rotor for 15 min. Ethanol is flushed with distilled water and the rotor brought up to speed to calibrate flow rates.

Composition of Media

For elutriation we use Hanks' salt solution containing 0.1% BSA, 10 mM HEPES, penicillin (100 μg/ml), and streptomycin (100 μg/ml). We

[13] W. Beeken, M. Mieremet-Ooms, L. A. Ginsel, P. C. J. Leijh, and H. Verspaget, *J. Immunol. Methods* **73**, 189 (1984).

[14] M. G. Irving, J. F. Roll, S. Huang, and D. M. Bissell, *Gastroenterology* **87**, 1233 (1984).

have found no differences in cell function when cells are elutriated in a more complete medium such as culture medium. BSA can be replaced by serum (1%) without a compromise in cell function, but gelatin is a poor substitute for serum proteins for some cell types.

Cell Load and Collection

In our system, we load into the chamber via a three-way stop valve placed in between the reservoir of medium and pump. Cells thus pass through the pump to get to the rotor; we have not found this procedure to compromise cell viability and it saves considerable time. Others note a similar experience.[3] We elute fundic mucosal cells at 25°; greater than 90% viability is maintained for 4 hr under these conditions. At this temperature membrane fluidity is greater than at 4° and therefore cells may withstand stress with less damage. Others have reported that cell viability is poorly maintained at ambient temperature, compared to 4°.[3] Temperature may also induce the clumping of some cell types and is another important variable that requires careful evaluation for each cell type.

Temperature variations may also influence the reproducibility of the elutriator separation. Small changes in viscosity will influence sedimentation velocity [Eq. (1)], thereby shifting the elutriation profile.[3] For those working at 4°, temperature fluctuation of the medium reservoir, the silicone tubing, and rotor itself may be considerable, especially with the rapid transfer of heat across silicone tubing and the several hours required for temperature equilibration of the rotor itself.[15]

We have found essentially comparable results with cell loads between 50×10^6 and 200×10^6 cells; we usually load 200×10^6 cells into the JE-6 elutriation chamber. Although the JE-10X rotor physically has 10 times the capacity of the standard JE-6 rotor, we find that delivery of 200×10^7 cells results in suboptimal yields due to cell clumping. We have found for canine fundic mucosal cells that a seven times greater load generally provides no compromise in cell recovery. We collect a fraction size of 100 ml for the JE-6 rotor and 500 ml for the JE-10X rotor. Larger fraction volumes do not significantly increase cell recovery. Sterile 50-ml plastic centrifuge tubes are used to collect cells from the JE-6 rotor, and the 500-ml glass media bottles are used to collect cells from the JE-10X rotor.

Cell Identification

Equally important to the methodology used for cell separation is the ability to identify specific cell types. For routine histological and immuno-

---

[15] M. L. Meistrich and N. M. Hunter, *Cell Biophys.* 3, 127 (1981).

cytochemical screening we prepare cells for staining on slides using a cytocentrifuge (Shandon Southern, Sewickley, Pennsylvania). The periodic acid–Schiff stain allows identification of parietal cells, chief cells, and mucous cells. Mast cells are identified using toluidine blue. We have also used immunochemical staining for pepsinogen to identify chief cells (antibody provided by Dr. I. M. Samloff), H^+,K^+-ATPase to mark for somatostatin in parietal cells (antibody provided by Dr. A. Smolka), and glucagon to identify endocrine cells. The antibodies directed against these cell products are revealed using second antibodies coupled to fluorescein isothiocyanate (FITC) or to the chromophore diaminobenzidine (DAB) reacting with a peroxidase–antiperoxidase complex. Using radioimmunoassay the peptide contents of cells can also be followed as an index of enrichment for populations of cells.

Separation Protocols

Cell size can be used as a first approximation in designing a separation protocol, with flow rate and centrifuge speed derived from Eq. (2). However since a given cell population is likely to be heterogeneous in size, a broader fraction will probably have to be collected to optimally enrich the desired cell type. Thus, in addition to cell size, it is necessary to have some means of cell identification to follow cell enrichment. The separation protocols from cells isolated from the fundic mucosa of both canine and rabbit are given in Table I. The protocol for the small rotor is unusual in

TABLE I
COMPARATIVE FLOW RATES AND RPM FOR SMALL AND LARGE ROTORS

Elutriator fraction number	Small rotor			Large rotor		
	rpm	Flow rate (ml/min)	Apparent cell diameter (μm)	rpm	Flow rate (ml/min)	Apparent cell diameter (μm)
1	2800	21.5	8.2	1200	30	8.0
2	2300	21.5	10.1	1200	55	10.0
3	2100	21.5	11.6	1200	75	11.6
4	2000	24.0	12.6	1200	87.5	12.6
5	2000	32.0	14.7	1200	100	13.6
6	2000	38.0	15.8	1200	142.5	16.1
7	2000	44.5	17.3	1200	185	18.5
8	2000	61	19.5	1100	192.5	20.5
9	2000	104	25.5	1100	200	23.0
Rest	0	104	>26	0	200	>23

that both the flow rate and rotor speed are altered during a separation. In addition to achieving separation of the major cell types found in the mucosa, this protocol is also designed to minimize the time cells are in the elutriator chamber and reduce the chance for cells clumping. Instead of maintaining a rotor speed of 2000 and using very slow flow rates for the small cell fractions, we use an intermediate flow rate (20 ml/min for the JE-6 rotor) and a higher rotor speed to elute these smaller cells. The use of high rotor speeds and flow rates has to be tempered by the higher shear forces encountered by cells crossing the rotating seal and by turbulence within the chamber. Because of the broad range of cells that we separate, we decrease the rotor speed to avoid excessive flow rates otherwise necessary for the large cell fractions.

Monitoring the size distribution of the cells being separated can be very useful; we have used the computerized size-analysis functions of the Particle Data Elzone for this purpose. A typical size distribution for the unfractionated canine fundic cell suspension is illustrated; large numbers of red blood cells and other debris comprise the peak in the 5-μm region (Fig. 2). The peak at 14 μm is composed predominantly of chief cells and mucous cells. A smaller parietal cell peak is evident at 19 μm. Material larger than 19 μm generally represents partially digested gastric gland fragments or capillary fragments.

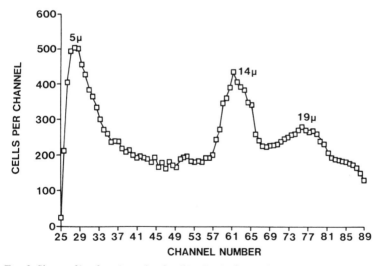

FIG. 2. Size profile of crude canine fundic cells obtained using computerized size analysis (Celloscope, Particle Data). The channel numbers on the ordinate correlate with cell size; the ordinate depicts the number of cells per channel. The instrument was calibrated with latex beads.

In our studies with canine fundic cells we have found that the elutriation profile for a variety of cell types is surprisingly constant (Fig. 3). We generally discard fraction 1 because of the high content of red blood cells, debris, and bacteria. The small cell fractions (fractions 2 and 3, Fig. 3) are maximally enriched in a large number of different cell types. In addition to mast cells and somatostatin-containing cells (Fig. 3A), we find glucagon- and serotonin-containing cells, macrophages, lymphocytes, and single endothelial cells, plus a few of the three cell types predominantly eluting in the larger cell fractions. Mucous cells are most abundant in fractions 4 and 5, chief cells in fractions 6 and 7, and parietal cells in fractions 8 and 9 (Fig. 3B). We routinely collect these paired fractions to save time and reduce cell clumping.

Density Gradient Separation

While elutriation enriches selected cell populations in respective fractions, the degree of enrichment is relative. Depending upon the questions being asked, additional cell separation techniques may be necessary. For example, a high degree of cell purity is needed in experiments to determine receptor populations on a specific type. Canine mucous, chief, and parietal cells are each populations with a relatively wide range of cell size and density. This heterogeneity is evident as a limit on the maximal elutriator enrichment of parietal cells of between 50 and 65% in fractions 8 and 9, with the chief cell the major contaminant (Fig. 3B). Parietal and chief cells differ in density, and we have used an albumin–Ficoll medium for step density gradient separation; in the light and dense fractions, respectively, we generally achieve greater than 85% enrichment of parietal cells and of chief cells.[16] The sequential use of elutriation and density separation removes small cells, such as mast and endocrine cells, that would otherwise be present in the step density fractions. Most mucous cells are also removed from fractions 7–9 by elutriation, thus reducing the chance of clumping during density gradient separation. Using 50-ml polypropylene centrifuge tubes, we load 45–50 million cells from fractions 7–9 per tube. Several other laboratories have used somewhat different protocols and media to enrich parietal and chief cells, including Nicodenz,[17] metrizamide,[18] and Percoll.[7,19] Most of these latter studies have not utilized se-

[16] A. H. Soll, D. A. Amirian, L. P. Thomas, T. J. Reedy, and J. D. Elashoff, *J. Clin. Invest.* **73**, 1434 (1984).

[17] C. S. Chew and M. R. Brown, *Biochim. Biophys. Acta* **888**, 116 (1986).

[18] W. S. McDougual and J. J. Decosse, *Exp. Cell Res.* **61**, 203 (1970).

[19] A. Sonnenberg, T. Berglindh, M. J. M. Lewin, J. A. Fischer, G. Sachs, and A. Blum, *in* "Hormone Receptors in Digestion and Nutrition" (G. Rosselin, P. Fromageot, and S. Bonfils, eds.), p. 337. Elsevier/North-Holland, Amsterdam, 1979.

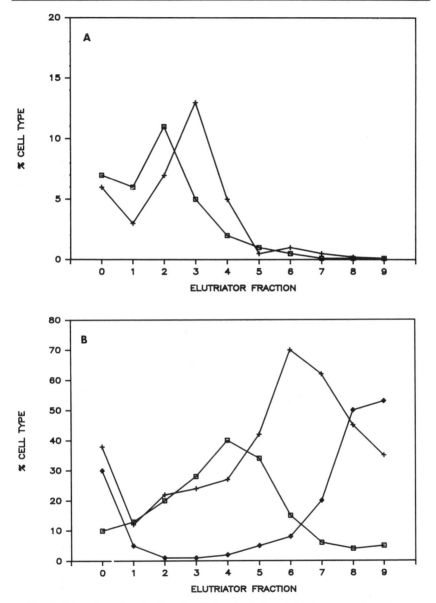

FIG. 3. Separation of canine fundic cells by elutriation. (A) The separated cell fractions were stained with toluidine blue to reveal mast cells (+); cells containing somatostatin-like immunoreactivity (□) were detected by immunohistochemistry. (B) Cytocentrifuge slides of the separated cell fractions were stained with periodic acid–Schiff reagent to reveal the characteristic mucopolysaccharides present in mucous cells (□) and to allow identification of chief (+) and parietal (◆) cells. (Data from Ref. 20.)

quential size and density separation to remove small cells, nor have they marked for these minor populations.

Unit Gravity Sedimentation

As an alternative to centrifugal methods, cells can be separated by sedimentation through a Ficoll gradient at unit gravity. We have evaluated this method utilizing the Wescor CelSep (Logan, Utah) as a means to enrich parietal and chief cells. Cells were first elutriated to remove debris and 100×10^6 cells from elutriator fractions 2–8 were resuspended in 1% Ficoll in Hanks' balanced salt solution containing 1% fetal calf serum. This suspension was allowed to settle through a linear gradient of 2–4% Ficoll with a 20% Ficoll cushion at the bottom. After 90 min, 40-ml aliquots were collected and cells were pelleted and stained for analysis. The results of a typical separation are illustrated in Fig. 4. Chief and parietal cells can be separated when compared to the starting mixture, but the number of cells recovered per fraction is small, on the order of $2–4 \times 10^6$ cells/ml. Given the time required and the overall yield obtained, unit gravity sedimentation is not a practical method for enriching cells from large quantities of dispersed tissue.

Enrichment by Cell Culture

The above methods utilize physical characteristics of cells to achieve enrichment of selected cell populations. Another approach to achieve enrichment is to utilize short-term culture and take advantage of selective adherence of some cell type to culture substrates. For example, in studying the small cell fractions (2–3) from canine fundic mucosa, somatostatin cells can be separated from other endocrine cells and mast cells by virtue of somatostatin cells preferentially adhering to collagen-coated surfaces in short-term culture.[20] These culture techniques can result in a considerable enrichment of somatostatin cells over the starting fraction; however, variable contamination with small chief cells can compromise this enrichment.

As noted above, chief cells also adhere to collagen substrate and we have also used cell culture methods to produce highly enriched chief cell monolayers for studies of pepsinogen secretion[21] and electrophysiological investigations.[22,23] Both canine fundic mucous and parietal cells do not

[20] A. H. Soll, T. Yamada, J. Park, and L. P. Thomas, *Am. J. Physiol.* **247**, G567 (1984).

[21] M. J. Sanders, D. A. Amirian, A. Ayalon, and A. H. Soll, *Am. J. Physiol.* **245**, G641 (1983).

[22] A. Ayalon, M. J. Sanders, L. P. Thomas, D. A. Amirian, and A. H. Soll, *Proc. Natl. Acad. Sci. U.S.A.* **79**, 7009 (1982).

[23] M. J. Sanders, A. Ayalon, M. Roll, and A. H. Soll, *Nature (London)* **313**, 52 (1985).

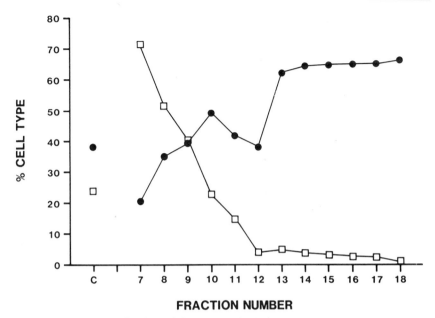

FIG. 4. Separation of canine chief (●) and parietal (□) cells by unit gravity sedimentation in a Wescor CelSep. The starting material (fraction C) consisted of elutriator fractions 2–8.

readily adhere to culture substrates and are generally excluded from the monolayers prepared from medium-sized canine fundic cells. In contrast to fundic cells, however, antral mucous cells do adhere to culture surfaces. Species differences are an important variable; in a recent report, mucous cells from the guinea pig fundus, but not chief cells, adhered to tissue culture plastic.[24]

We have not found conditions that allow parietal cells and mast cells to adhere to culture substrates. However, both cell types survive in short-term suspension culture. During a 2- to 18-hr culture period, marked improvement in both parietal cell and mast cell function occurs, greatly enhancing usefulness of these cells for functional studies (M. C. Y. Chen, D. Amirian, M. Toomey, M. Sanders, and A. H. Soll, manuscript in preparation). For these studies, flasks designed to hold a suspended magnetic stir bar are used (Spinner flask #1967, Bellco Glass).

Because of the wide availability of culture media of different compositions, the decision as to which combination of culture ingredients to use must be empirically determined for each cell type. For parietal cells we have found that 5% canine serum in Ham's F12:DMEM (Dulbecco's

[24] M. Rutten, D. Rattner, and W. Silen, Am. J. Physiol. 249, C503 (1985).

minimum essential medium) (1 : 1) supplemented with insulin (8 μg/ml) and hydrocortisone (10 nM) will maintain parietal cell function in overnight culture.

Elutriation of Cells from Other Complex Tissues

Intestine

Recently Dr. Diane Barber in our laboratory has applied elutriator separation to canine intestinal tissue to produce enriched populations of endocrine cells, including cholecystokinin cells from the jejunum and neurotensin cells from the ileum.[25,26] Intestinal tissue contains diverse populations of endocrine cells, but because the cells are widely dispersed throughout the tissue, they constitute a small proportion of the total mucosal cell population. Using a sensitive radioimmunoassay to follow the population of CCK cells through the separation protocol, it was found that two sequential elutriator separations were necessary. Following tissue dispersion with collagenase and EDTA, jejunal cells were first elutriated at 30 ml/min and 1800 rpm to remove most of the non-CCK-containing cells. This fraction was then reloaded into the elutriator at 25 ml/min and 2300 rpm and a fraction collected at 40 ml/min at 1800 rpm. This last fraction was enriched 20-fold in cholecystokinin-like immunoreactivity over the initial dispersed starting material.

Neurotensin cells were enriched from the ileum using a single pass through the elutriator. Four fractions were collected, at flow rates of 25, 30, 55, and 100 ml/min, with the centrifuge speed at 2500 rpm for fraction 1, 2200 rpm for fraction 2, and 2000 rpm for fractions 3 and 4. Fraction 3 was the most highly enriched in neurotensin cells, with about a fivefold enrichment in neurotensin-like immunoreactivity over the starting material. NTLI cells accounted for 12.6% of the total cells in fraction 3. It was found that freshly isolated cells used in functional studies contained enzyme activity that degraded most of the NTLI released by the cells. While the source of this protease activity was not determined, it disappeared when cells were placed in 48-hr culture. NT cells selectively adhere to collagen-coated substrates and are thus further enriched to 40% of the total cells present.

In a recent study Beeken *et al.*[13] have obtained a macrophage-enriched population from human intestinal lamina propria. Adaptation of elutria-

---

[25] D. L. Barber, J. H. Walsh, and A. H. Soll, *Gastroenterology* **91**, 627 (1986).
[26] D. L. Barber, A. M. J. Buchan, J. H. Walsh, and A. H. Soll, *Am. J. Physiol.* **250**, G385 (1986).

tion methodology for this cell type was more effective in producing enriched macrophages than that obtained from Ficoll–Hypaque density gradients. Clumping of cells in the rotor was noted and reduced by adding DNase (1 mg/liter), using a larger load volume, and lower initial rotor speed.

Lung Cell Separation and Enrichment

The lung is a heterogeneous organ containing a large number of different cell types. Three cell types that have been enriched from lung are the type II alveolar epithelial cell,[10,11] the Clara cell,[27] and the mast cell.[9] Isolation of type II pneumocytes by conventional density gradient methods are usually first preceded by methods that reduce contaminating pulmonary macrophages. The macrophages in the cell suspension are allowed to either adhere to a solid surface or phagocytose a fluorocarbon suspension and can be removed by density centrifugation. In contrast, elutriation of a suspension of lung cells can produce an 85% enriched population in type II cells without prior manipulation of the cells. Type II pneumocytes are identified by staining with phosphene 3R, a dye that visualizes the characteristic lamellar bodies contained within the cell. Elutriation produced a 30% enrichment in Clara cells. Further enrichment of this population of cells was achieved using a two-polymer aqueous phase system that consisted of 3.8% polyethylene glycol 6000 and 5% dextran T-500 prepared in 0.15 M phosphate buffer.[28] Unit gravity separation of cells added to this system resulted in a 70% enrichment of Clara cells in the upper polyethylene glycol phase. Mast cells isolated from guinea pig lung were first enriched by elutriation to about 23% of the cell population and then this fraction was further purified in a Percoll density gradient to 37%.

Liver Cell Separation and Enrichment

Isolation of liver cells is a several-step procedure that first separates hepatocytes (parenchymal cells) from nonparenchymal cells. The latter are endothelial cells, Kupffer cells, fat-storing cells, and pit cells.[29] A variety of digestion protocols have been developed with crude collagenase to disperse hepatocytes and additional pronase treatment to further disperse the nonparenchymal cells.[14,29] Since hepatocytes are large and dense cells, they can be separated from nonparenchymal cells by any number of methods,

---

[27] T. R. Devereux and J. R. Fouts, *In Vitro* **16**, 958 (1980).

[28] H. Walter, *in* "Methods of Cell Separation" (N. Catsimpoolas, ed.), p. 307. Plenum, New York, 1977.

[29] R. N. Zahlten, H. K. Hagler, M. E. Nejtek, and C. J. Day, *Gastroenterology* **75**, 80 (1978).

including density gradient centrifugation and elutriation. A sequential elutriation procedure has been developed by Irving and co-workers.[14] With this protocol, dispersed cells are loaded into the elutriator chamber at a low rotor speed (1100 rpm) and a flow rate varied between 23 and 34 ml/min; under these conditions, large debris and undigested cells are left in the rotor and then discarded. The elutriated suspension is then reintroduced into the rotor at 2400 rpm and a flow rate varied between 27 and 53 ml/min to capture single hepatocytes in the chamber and elutriate nonparenchymal cells. The latter eluant is then subjected to further pronase treatment and reintroduced into the elutriator rotor at 2400 rpm with flow rate varied between 15 and 62 ml/min. The elutriated fractions from this last pass through the rotor contain a large distribution of mostly endothelial cells that overlap with Kupffer cells (40% of the total) in the largest fractions. Better separation of Kupffer cells is possible if the liver is first perfused with heat-treated red blood cells so that Kupffer cells are heavier due to phagocytosis.

Separation of Cells That Vary in Their Mitotic or Developmental State

In dividing populations of cells, it is possible to use elutriation to separate cells in different phases of the cell cycle,[30] taking advantage of the fact that cell size shows an exponential increase during the cell cycle.[31] The separated cells can be used in studies in which synchronized cells are required. From a starting fraction of mouse L-P59 cells, in which 36% of the cells are in G_1, 54% are in S, and 10% are in G_2M, peak fractions can result in a $G_1:S:G_2M$ composition of 90:10:0% and a $G_1:S:G_2M$ composition of 3:16:81%.

Cells in different stages of development present in a complex tissue can also be resolved by elutriation. Beaven et al.[32] separated rat peritoneal mast cells by elutriation and found an inverse relationship between cellular histamine content and histamine synthetic activity. Smaller (younger) cells contained high activities of histidine decarboxylase and low levels of histamine, whereas larger (older) cells had lost the histidine decarboxylase activity but contained more histamine.

Testicular spermatogenic cells have also been separated on the basis of cell developmental stage.[33] Elutriation allowed enrichment of three types

[30] M. L. Meistrich, R. E. Meyn, and B. Barlogie, *Exp. Cell Res.* **105**, 169 (1977).

[31] E. C. Anderson, G. I. Bell, D. F. Petersen, and R. A. Tobey, *Biophys. J.* **9**, 246 (1969).

[32] M. A. Beaven, D. L. Aiken, E. Woldemussie, and A. H. Soll, *J. Pharmacol. Exp. Ther.* **224**, 620 (1983).

[33] R. J. Grabske, S. Lake, B. L. Gledhill, and M. L. Meistrich, *J. Cell. Physiol.* **86**, 177 (1975).

of cells: spermatozoa (100%), round spermatids (68%), and pachytene spermatocytes (58%). Although the degree of enrichment by elutriation was slightly lower than that obtained with unit gravity separations, much less time was required.

Acknowledgment

Supported by NIAMDD Grants AM 19984, AM 17328, AM 30444 and by the Medical Research Service of the Veterans Administration. A.H.S. is the recipient of Medical Investigatorship, VA Wadsworth Hospital Center.

[24] Cell Separation Using Velocity Sedimentation at Unit Gravity and Buoyant Density Centrifugation

By JOHN WELLS

Introduction

Reliable methods to purify subpopulations of cells from heterogeneous mixtures are needed by an increasing variety of laboratories. No single separation technology is optimal for all types of tissue, but some procedures do offer advantages for certain studies. For example, we have found unit gravity separations, where a spectrum of fractions of increasing size is obtained, very effective for cell cycle studies. To be useful for most laboratories, cell separation methods should be reproducible, relatively simple to perform, and maintain the viability of cells. It is advantageous if the procedures allow sterility to be maintained.

In this chapter I describe procedures that are used in our laboratory on a routine basis by students and technicians. A system using velocity sedimentation is presented and two methods of buoyant density centrifugation are discussed. These descriptions focus on the details of performing the separations, improving the quality of the separations, and some of the limitations of each methodology.

Velocity Sedimentation at Unit Gravity

Recent advances in unit gravity separations evidence its usefulness for applications in separations of various cell types and cellular components

METHODS IN ENZYMOLOGY, VOL. 171

such as blood and bone marrow cells,[1-7] glandular cells,[8-12] tumor cells,[13-15] testicular cells,[16,17] kidney and liver cells,[18-21] gastrointestinal cells,[22] and chromosomes.[23] Refinement of systems for velocity sedimentation has increased the range of applications to a larger variety of tissues.

Theory

Velocity sedimentation-based separations typically allow the enrichment of subpopulations of cells that are differentiated from other subpopulations of cells by a size difference greater than 0.5 μm or by a very large density difference. Since both size and density differences are important in velocity sedimentation separations the additive effects of the two parameters must be considered.

The theory of particle separation in a fluid at unit gravity has been reviewed elsewhere.[24,25] A particle falling in a fluid reaches a velocity

[1] H. E. Broxmeyer, N. Jacobsen, and M. A. S. Moore, *J. Natl. Cancer Inst.* **60**, 513 (1978).

[2] W. S. Bont and J. E. de Vries, *in* "Cell Populations" (E. Reid, ed.), p. 53. Horwood, Chichester, England, 1979.

[3] J. R. Wells, *in* "Cell Separation: Methods and Selected Applications" (T. G. Pretlow II and T. P. Pretlow, eds.), Vol. 1, p. 169. Academic Press, New York, 1982.

[4] J. R. Wells, *in* "Methods in Hematology: Hematopoiesis" (D. Golde, ed.), p. 333. Churchill Livingstone, New York, 1984.

[5] H. R. MacDonald and R. G. Miller, *Biophys. J.* **10**, 834 (1970).

[6] K. Zeiller and E. Hansen, *Cell. Immunol.* **44**, 381 (1979).

[7] F. L. Harrison, T. M. Beswick, and C. J. Chesterton, *Biochem. J.* **194**, 789 (1981).

[8] J. F. Crivello and G. N. Gill, *Endocrinology* **113**, 1 (1983).

[9] C. Henriksson and A. Soome, *Soc. Int. Chir.* **September**, 166 (1979).

[10] W. C. Hymer and J. M. Hatfield, this series, Vol. 103, p. 247.

[11] A. M. Mulder, J. M. Durdik, P. Toth, and E. S. Golub, *Cell. Immunol.* **40**, 326 (1978).

[12] J. F. Tait, S. A. S. Tait, R. P. Gould, and M. S. R. Mee, *Proc. R. Soc. London, Ser. B* **185**, 375 (1974).

[13] W. S. Bont and J. M. Hilgers, *Prep. Biochem.* 7, 45 (1977).

[14] S. Becker and S. Haskill, *J. Natl. Cancer Inst.* **65**, 469 (1980).

[15] K. Moore and W. H. McBride, *Int. J. Cancer* **26**, 609 (1980).

[16] M. Loir and M. Lanneau, *Methods Cell Biol.* **15**, 55 (1979).

[17] M. L. Meistrich, *Methods Cell Biol.* **15**, 15 (1979).

[18] J. Deschenes, L. LaFleur, and N. Marceau, *Exp. Cell Res.* **103**, 183 (1976).

[19] E. Parthenias, A. Soots, A. Nemlander, E. von Willebrand, and P. Häyry, *Cell. Immunol.* **57**, 92 (1981).

[20] R. G. Price, *in* "Cell Populations" (E. Reid, ed.), p. 105. Horwood, Chichester, England, 1979.

[21] G. Steinheider, I. Bertoncello, and H. J. Seidel, *Exp. Hematol.* **8**, 779 (1980).

[22] C. N. A. Trotman, *in* "Cell Populations" (E. Reid, ed.), p. 111. Horwood, Chichester, England, 1979.

[23] J. G. Collard, E. Phillippus, A. Tulp, R. V. Lebo, and J. W. Gray, *Cytometry* **5**, 9 (1984).

[24] H. C. Mel, *J. Theor. Biol.* **6**, 307 (1964).

[25] R. G. Miller, *in* "New Techniques in Biophysics and Cell Biology" (R. H. Pain and B. J. Smith, eds.), Vol. 1. Wiley, New York, 1973.

dependent on its size and density and the density and viscosity of the suspending medium. A constant velocity is reached when the force of gravity is balanced by the resistance to movement. The gravitational force is defined by the equation $F = 4/3\pi r^3 (\rho_p - \rho_m)g$, where ρ_p is the density of the particle, ρ_m is the density of the medium, and r is the radius of the particle. The resistance to movement at equilibrium, μ, is given by Stokes law, $F = 6\pi\eta r\mu$, where η denotes the absolute viscosity of the medium. The simplified equation when these forces are in balance yields the general sedimentation law:

$$\mu = \frac{2(\rho_p - \rho_m)gr^2}{9\eta}$$

where μ is in millimeters per hour, ρ_p and ρ_m are in grams per milliliter, and r is in micrometers.

A condition that is often overlooked when a separation is evaluated is the starting zone (sample volume) size. Since sedimentation velocities for many mammalian cell types in heterogeneous mixtures are not widely different, the depth of the starting zone can effect the resolution of the separation. Figure 1 illustrates conditions where a broad starting zone (A) results in overlap between B and C, whereas a thinner starting zone (D) gives separation of bands E and F after a given period of sedimentation. The path length required to separate two cell types with different sedimentation velocities is proportional to the starting zone width. Large samples can reduce resolution when the sedimentation time or the path length is limited.

Since the sample zone must be less dense than the initial portion of the gradient, there is usually an increase in viscosity as the cells move from the

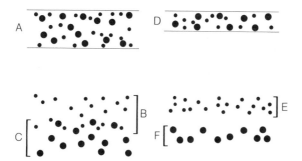

Fig. 1. Comparison of resolution resulting from sedimentation of particles starting from a wide zone (A) or a narrow zone (D). The path length required to separate particles with different sedimentation velocities is proportional to the starting zone width. The wider starting zone (A) will not result in the separation of zones B and C in the same path length that will allow resolution of E and F starting from zone D.

starting zone into the gradients. If this change is large, the starting zone will be compressed and can lead to streaming or nonideal sedimentation. This condition is known as exceeding the streaming limit and in some cases causes clumps (of cells) to form, and these clumps sediment rapidly. The streaming limit for mammalian cells is usually around 1×10^6 cells/ml but can vary due to the size or surface characteristics of the cells.

The CelSep System

For unit gravity separations we use the CelSep system (Wescor Inc., Logan, Utah 84321) because it gives excellent resolution. The apparatus is not complex and the critical components can be autoclaved for sterile work. Another important consideration is that the separations are highly reproducible, which allows the location of enriched subpopulations to be located by fraction number. This allows a rapid transition to the next phase of the experiment.

The CelSep system is composed of a sedimentation chamber, a base for tilting the chamber, and a gradient maker (Fig. 2). The sedimentation chamber consists of upper and lower clear plastic plates that are clamped to

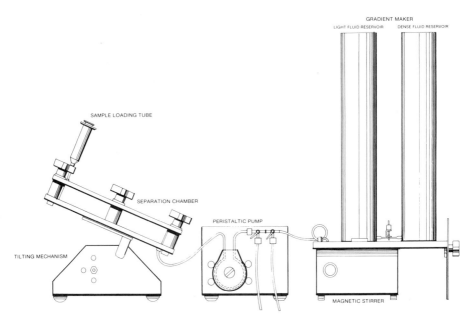

FIG. 2. Components used for unit gravity cell separation. CelSep apparatus includes sedimentation chamber and tilting base. Gradient maker includes bubble trap and requires stirrer motor for operation. Peristaltic pump is used to load gradient, cushion, and cell sample. Pump is also used to unload separated cells.

a clear plastic cylindrical section. The autoclavable chamber has an aluminum, Teflon-coated center section and glass end plates. The chamber holds approximately 1000 ml. A peristaltic pump and magnetic stirrer are also required for operation.

The tilting assembly is motor driven and moves the chamber through an angle of 30° from the filling position to the horizontal or separation position. This orientation takes 5 min and the unit automatically stops when it reaches either the loading or separation positions.

The design of the separation chamber provides unique baffles arranged around the inlet ports that allow rapid liquid loading and unloading without disturbing the gradient, interfaces, or the separated cell populations. The design of the chamber provides maximum utilization of the gradient volume for sedimentation without wall effects. Figure 3 shows the sequence of loading a gradient, cushion, cell sample, and overlay, followed by the separation phase and the unloading procedure. A continuous gradient is pumped into the chamber light end first through the lower inlet (1). The chamber is held in the tilted position, creating a funnel around the lower inlet. The gradient is followed by a dense cushion with higher viscosity to prevent rapidly sedimenting cells and clumps from contacting the bottom of the chamber. The chamber is completely filled with the gradient and cushion. A cell suspension is then loaded through the top inlet by reversing the pump and removing a portion of the cushion (2). An overlay is then loaded to create a sample zone of constant depth. The separation chamber is slowly oriented to the horizontal position and sedimentation is allowed until separation of subpopulations is achieved (3 and 4). After the separation is completed, the chamber is reoriented (5) to the tilted position and the separated fractions are unloaded through the bottom port (6).

Sterile operation of CelSep is performed using the aluminum and glass sedimentation chamber that can be autoclaved. The gradient maker is also autoclavable and the gradient solutions are filter sterilized. The entire procedure is performed in a tissue culture hood to decrease the opportunity for contamination during loading and unloading.

Temperature control is an important factor for reproducible results. Most laboratories provide sufficient temperature control, but air conditioning and heating vents should be avoided. Operation in a cold room or refrigerator is also feasible.

Standard Procedure for Separating Mammalian Cells

1. Preparing the solutions, gradient, and cell suspensions: A shallow density gradient is necessary to provide stability during loading and un-

1 Chamber in Loading Position

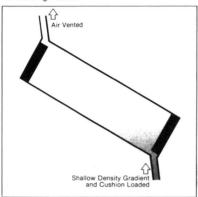

Air Vented

Shallow Density Gradient and Cushion Loaded

2 Cell Suspension and Overlay Loaded

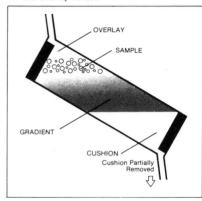

OVERLAY

SAMPLE

GRADIENT

CUSHION

Cushion Partially Removed

3 Chamber Oriented to Separation Position

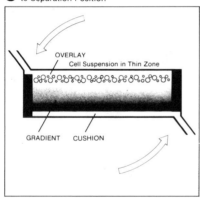

OVERLAY
Cell Suspension in Thin Zone

GRADIENT   CUSHION

4 Separation Unit Gravity

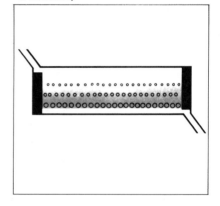

5 Chamber Reoriented to Unloading Position

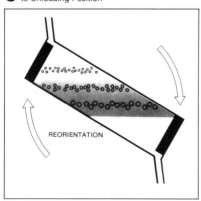

REORIENTATION

6 Separated Subpopulations

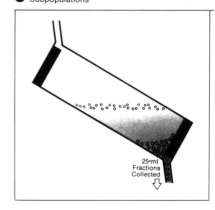

25-ml Fractions Collected

TABLE I
PHYSICAL PROPERTIES OF FICOLL SOLUTIONS

Property	Light 2% (w/w) Ficoll	Heavy 4% (w/w) Ficoll
Density (24°)	1.009 g/mg ± 0.0014	1.015 g/ml ± 0.00114
Refractive index (24°)	1.3370 ± 0.0003	1.3400 ± 0.0003
Osmolality	275 mOsm ± 3	272 mOsm ± 4

loading and to prevent mixing from thermal effects. The increased viscosity of the gradient solutions causes a reduction in the sedimentation rate and therefore the slope of the gradient should be kept as shallow as possible. Gradients of 1–2% (w/w) Ficoll have been used, but for most separations a 2–4% (w/w) Ficoll linear gradient allows a more concentrated cell suspension to be used. The cell sample is suspended in 1% (w/w) Ficoll and the overlay is saline or culture media. The gradient and starting sample must provide uniform osmolality so that cell size is not altered during sedimentation.

The 2–4% (w/w) Ficoll gradient with an osmolality suitable for human cells (and many other mammalian cells) using Hanks' balanced salt solution and Ficoll 400 (Pharmacia Fine Chemicals, Piscataway, New Jersey) is generated from two solutions.

2% (w/w) Ficoll. Ficoll (2%, w/w) is prepared in a blender by combining 14 g of Ficoll with 686 ml of Hanks' 0.97× dilution. The solution is adjusted to pH 7.2 with 7.5% sodium bicarbonate and is filtered through a 0.45-μm filter when it is stored for any length of time.

4% (w/w) Ficoll. Ficoll (4%, w/w) is prepared by combining 28 g of Ficoll in 672 ml of Hanks' 0.95× dilution, followed by pH adjustment and filtration.

Both solutions are sampled for refractive index (for density determination) and osmolarity. Table I gives the measured physical properties for these solutions. The cells are suspended in 25–150 ml 1% (w/w) Ficoll in Hanks' 1.0x. A solution of 20% (w/w) Ficoll in Hanks' 1.0× dilution is used as an underlayer for the gradient.

The cushion prevents rapidly sedimenting cells or clumps from contacting the bottom of the chamber. The volume of the cushion should be 50 ml or greater if clumps are present. Most of the cushion can be reused if

FIG. 3. Diagrams of the sequence of operation of CelSep reorienting chamber cell separation device. (1) Loading of continuous density gradient and cushions. (2) The cell suspension (sample) and overlay are loaded by removing a portion of the cushion. (3) The chamber is oriented to the horizontal position and (4) the separation takes place usually in 90–180 min. (5) The chamber is slowly reoriented to the unloading position and (6) the gradient is unloaded by collecting fractions.

care is taken during the loading and unloading. For conditions where maximum resolution is desired the sample volume should be 25–40 ml. Larger sample suspensions of up to 150 ml can be used if the cells to be separated have sufficiently different sedimentation rates. During a series of separations the volume of the sample, overlay, and gradient should be carefully measured so that fraction peaks are reproduced. About 200 ml of cushion should be available for each separation. A variable-speed peristaltic pump capable of 10–50 ml/min is required for normal operation. Silicone tubing is used to connect the gradient maker, pump, and chamber since it is flexible and autoclavable.

The loading procedure including gradient and sample is completed with the chamber in the tilted position. The gradient is loaded at 45–55 ml/min followed by the cushion, completely filling the chamber. The cell suspension is loaded on top of the gradient (through the upper port) by reversing the pump and removing some of the cushion. The overlay is loaded in the same manner. The loading rate for these steps is 15–20 ml/min. The chamber is then reoriented to the horizontal position and the separation period is started. Most mammalian cell separations at unit gravity are accomplished in 1.5–3 hr.

Unloading is accomplished with the separation chamber in the tilted position at a flow rate of 30–40 ml/min. The volume of fractions can range from 25–50 ml. The number and proximity of separated subpopulations of cells will determine the number and volumes of fractions necessary to obtain maximum resolution. Fraction volumes below 20 ml are usually not necessary. Collection of fractions must be precise to ensure reproducible results.

After a standard separation procedure is determined, it is not necessary to collect the entire gradient as fractions. The unloading tubing can be directed into a graduated vessel at first and only the portion of the gradient containing the cells of interest is collected as fractions.

Buoyant Density Cell Separations

Cell separation by differences in buoyant density or isopycnic centrifugation is the most widely used physical method. In our laboratory this method usually provides the greatest increases in purity. This technique was tedious for many years due to the lack of a readily available acceptable centrifugation medium. Procedures using BSA, sucrose, and other media required careful preparations or subjected the cells to hypertonic conditions that altered their density.

New gradient materials are now commercially available that allow rapid gradient construction and isopycnic separations under isotonic conditions. I will limit this discussion to two of these materials, Percoll and Nycodenz.

Nycodenz

Nycodenz is a new gradient medium developed by Nyegaard that replaced metrizamide (also developed by Nyegaard). Several articles have been published describing Nycodenz and its use for the separation of cells[26,27] and other particles.[28] A number of publications detailing separation of specific subpopulations of cells have also recently been published.[29-36] These articles and our own tests indicate that Nycodenz is probably the current best choice for buoyant density separations of mammalian cells and many other particles. Nycodenz is a substituted triiodobenzene ring compound with the chemical name N,N'-bis(2,3-dihydroxypropyl)-5-[N-(2,3-dihydroxypropyl)acetamido]-2,4,6-triiodoisophthalamide). It has a molecular weight of 821 and a density of 2.1 g/ml.

Nycodenz is available in powder form and is readily dissolved in water to over 80% (w/v). This solution can be autoclaved and stored for long periods. Nycodenz is not toxic and is not metabolized by mammalian cells.[37,38]

At concentrations useful for most cell separation techniques (5–15%), Nycodenz has a low osmolality and low viscosity. Density measurements can be made by refractometry. Table II gives some of the physical properties of solutions of Nycodenz.

Although it is possible to generate gradients by centrifugation with Nycodenz, the high g forces and time required make this procedure unsuitable for most laboratories interested in cell separation.

[26] D. Rickwood, T. Ford, and J. Graham, *Anal. Biochem.* **123**, 23 (1982).
[27] T. C. Ford and D. Rickwood, *Anal. Biochem.* **124**, 293 (1982).
[28] T. Ford, D. Rickwood, and J. Graham, *Anal. Biochem.* **128**, 232 (1983).
[29] A. Bøyum, *Scand. J. Immunol.* **17**, 429 (1983).
[30] A. Bøyum, this series, Vol. 108, p. 88.
[31] R. Blomhoff, W. Eskild, and T. Berg, *Biochem. J.* **218**, 81 (1984).
[32] M. O. Symes, B. Fermor, H. C. Umpleby, C. R. Tribe, and R. C. N. Williamson, *Br. J. Cancer* **50**, 423 (1984).
[33] H. C. Umpleby, *Br. J. Surg.* **71**, 659 (1984).
[34] T. Berg, T. Ford, G. Kindberg, R. Blomhoff, and C. Drevon, *Exp. Cell Res.* **156**, 570 (1985).
[35] F. Hartl, W. W. Just, A. Köster, and H. Schimassek, *Arch. Biochem. Biophys.* **237**, 124 (1985).
[36] E. Rodahl, *Scand. J. Immunol.* **21**, 93 (1985).
[37] S. Salvesen, *Acta Radiol. Suppl.* **362**, 73 (1980).
[38] W. Mutzel and U. Speck, *Acta Radiol. Suppl.* **362**, 87 (1980).

TABLE II
PHYSICAL PROPERTIES OF NYCODENZ SOLUTIONS[a]

Concentration [%(w/v)]	Refractive index (n)	Density (g/ml)	Osmolality (mOsm)
0	1.3330	0.999	0
10	1.3494	1.052	112
20	1.3659	1.105	211

[a] At 20°.

Solutions for preparing gradients using solid Nycodenz that are isotonic for most mammalian cells have been described by Ford and Rickwood.[27] Alternatively, a 27.6% (w/v) Nycodenz solution can be purchased from Nyegaard (Diagnostic Division, Torshov, Oslo 4, Norway). This solution has a density of 1.15 g/ml and can be diluted using tissue culture medium.

Percoll

A second material in widespread use for buoyant density cell separations is Percoll. We have discontinued using this material because it can produce artifacts if used improperly and it requires a second procedure to remove the Percoll following separation.

Some of the problems associated with gradients using colloidal silica such as Ludox or Percoll are a result of the particulate nature of the suspension. These problems include particle–particle association, adherence of particles to cells, and multiple bands composed of similar subpopulations.

The artificial band formation is particularly disturbing since it can be misinterpreted as a true density separation. Colloidal silica consists of particles 10 to 20 nm in diameter and g forces above a critical $\omega^2 t$ will cause changes in gradient shape and stability. Colloidal silica gradients analyzed with optics in an analytical centrifuge demonstrated that these gradients become microscopically discontinuous in a few minutes at 14,000 g. These tests showed a homogeneous population of red blood cells (RBCs) produced as many as 20–40 bands when centrifuged in a gradient that was spun at 12,000 g for 12 min.[39] To avoid these problems the centrifugation should be kept well below the critical threshold for gradient instability.

Most separations of mammalian cells run under 12 min at less than 3000 g (or an equivalent $\omega^2 t$) will avoid gradient instability while provid-

[39] C. M. West and D. McMahon, *Anal. Biochem.* **76**, 589 (1976).

ing sufficient g minutes to band cells isopycnically.[39,40] Problems can also result if gradients using Percoll are formed by centrifugation at high speed.

Methods for Buoyant Density Cell Separation

Preformed gradients can be continuous or can have density steps. We use both continuous and step gradients in our laboratory. Continuous gradients can be formed by layering various density solutions into tubes and allowing diffusion to occur. Reorienting the tube to a horizontal position will reduce the time required to form linear gradients. A continuous gradient using Nycodenz in a 15-ml disposable tube can be formed in 1 hr using four to five density steps when the tube is closed and slowly rotated to the horizontal position. Continuous gradients can also be made using a standard two-chamber gradient maker.

When we begin working with a new mixture of cell types we construct a continuous gradient with a wide density range. Usually this range is approximately 1.045 to 1.095 g/ml. Cells are then banded in this gradient, the gradient is collected, and the peak density of each subpopulation is determined. If the cells of interest have pelleted or if they remain on the top of the gradient the density range of the gradient is expanded appropriately. The density profile for each subpopulation of cells we can identify is plotted. These profiles are then used to design a step gradient that will give either maximum purity of the desired subpopulations or maximum yield. Some trial and error may be necessary to adjust the step gradient, but if careful attention is given to density determinations a step gradient will provide rapid and reproducible separations that can enrich some subpopulations more than 50-fold.

Preparative step gradient cell separations in our laboratory are performed in 15- and 50-ml clear disposable tubes. The 15-ml tubes provide higher resolution conditions since problems with mixing during centrifugation (acceleration and deceleration) and unloading are reduced. These tubes allow sterile conditions to be maintained when loading and unloading is completed in a hood.

A long thin glass tube is used to load in the cell suspension and then the gradient steps from the bottom of the tube. The gradient steps are of equal volume. As each is loaded the tube is marked to locate the position of each interface. A peristaltic pump is used for this purpose and a very small bubble is allowed to enter the inlet tubing after each density solution. When this bubble reaches the bottom of the centrifuge tube during the loading of each density step, the location of the interface is marked on the

---

[40] A. Butriago, E. Gylfe, C. Henriksson, and H. Pertoft, *Biochem. Biophys. Res. Commun.* **79**, 823 (1977).

centrifuge tube. This marking system will correspond to the location of isolated subpopulations after centrifugation. The interface zones and the intermediate zones are collected with a sterile polypropylene pipet. Often, bands of cells can be visualized by holding the centrifuge tube against a black background and using a microscope light or other source.

We routinely use five or six steps in 15-ml tubes and maintain interfaces with density differences of 0.0015 g/ml. These steps are 1.5–2 ml and the sample volume is 2–5 ml with $1-8 \times 10^7$ cells. For separations in 50 ml tubes we use 5-ml steps with density differences of 0.003 g/ml or greater.

For shallow continuous gradients and step gradients, slow acceleration and deceleration of the rotor are critical. Most modern centrifuges have a rate control feature. It is also advantageous to use a large swinging-bucket rotor with all of the carriers filled to further extend the deceleration.

Maintaining the proper osmolality of gradients is as critical as reproducing density. In the past the lack of reproducible isoosmotic gradients was the biggest drawback of buoyant density separations. Materials such as Nycodenz have greatly facilitated constructing isotonic gradients. Even with the availability of pure compounds such as Nycodenz, the osmolality of each new stock solution should be measured when it is prepared.

Density Gradient Fractionation

Collection of the separated subpopulations from continuous or step gradients in tubes can be done using several methods. For continuous gradient separations the most precise method is to drip the gradient out the bottom of the tube. This will require tubes that can be punctured, such as disposable ultracentrifuge tubes. Newer versions of these tubes, such as the new Crimp seal (DuPont Co., Wilmington, Delaware 19898) tubes, can be used when the sizes can be matched to the available low-speed holders. These Crimp seal tubes can be adapted to sterile procedures if gas sterilization or UV sterilization is available. The tubes are run with the plug seal and metal cap in place but are not crimped so they can be easily opened. The tubes are punctured using a fractionation device available from Hoeffer (Hoeffer Sci., San Francisco, California 94107). Fractions as small as five to eight drops to 0.5 ml are collected. We generally divide each gradient into 25–35 fractions.

Collection of bands from step gradients can be accomplished by sequentially removing aliquots from the top of the tube. We usually collect the volume above a band and then collect zones where a density interface was located using polypropylene disposable pipets. Often these bands are not visible and only by using the marked location on the side of the tube will the correct volume be collected.

Improving the Quality of Separations

Velocity Sedimentation

Sedimentation rates of different cell types in heterogeneous mixtures are usually not very different; therefore, it is helpful to use a narrow starting zone to improve resolution. The starting zone width is decreased when the cells sediment into the higher viscosity of the gradient, but this effect must be limited or aggregation and streaming will occur.

Streaming begins when the cell concentration exceeds a "streaming limit" and clumps of cells form and sediment rapidly. The streaming limit is different for each type of cell. Gross streaming can be observed visually but more often it is indicated by the presence of small cells in the more rapidly sedimenting zones containing the larger cells.

We perform useful separation above the streaming limit when only the slower portion of the gradient containing the smaller cells is used. Streaming often begins due to the presence of aggregates or clumps in the starting sample. A low-speed spin or simply letting the aggregates settle out for a few minutes can be quite effective in removing most of them from the starting sample.

For most mammalian cells below 10 μm, a cell concentration of 1×10^6/ml will provide excellent sedimentation conditions without clumping and streaming. Separations in the CelSep with a sample volume larger than 75 ml give acceptable separations when the differences in sedimentation velocity are sufficient to allow separation in the path length available in the separation chamber.

Analysis of the separation can include cell counts, volume measurement, specific stains, and an activity measurement. The conditions for these assays must be standardized before starting the cell separation experiments. A portion of the starting suspension is set aside and assayed with the separated fractions so that yields may be determined. Microscopic examination to determine if clumps are present is important, since certain doublets or triplets can sediment as discrete zones and are later disrupted during collection or washing, giving false peaks in cell counts. Cytocentrifuge preparations of separated fractions for staining and morphological identification are very useful. Once the pattern of separation is known the fractionation will be reproducible for most cell mixtures, and subpopulations can be located by cell counts around peak fractions.

After the cell counts are made in critical fractions some pooling is usually helpful to avoid an excessive number of assays. In many cases a second separation technique can be performed, such as cell sorting by flow cytometry.

Density Determinations

Once the gradient is fractionated, the location of cells must be determined along with other parameters. With Nycodenz the density of fractions can be determined by measuring the refractive index. The relationship between refractive index η and density for solutions of Nycodenz is given by the formula

$$\text{Density} \quad (g/ml) = 3.242\eta - 3.323$$

In practice we construct density steps from two stock solutions (low density and high density). We determine the density of these using a 5-ml pycnometer and measure the refractive index. The individual steps are also measured and a graph is made. The readings for fractions after centrifugation are then read from this graph.

Osmolality Determinations

As osmometer is used to determine the effects of solute concentration in gradient solutions. These measurements are based on the changes in the free energy of solvent molecules that occur when a solute (or several) is dissolved in the solvent. Two types of osmometers can be employed to measure gradient solutions, freezing point and vapor pressure osmometers. These instruments determine the total concentration or osmolality of the solutions. Each mole of particles dissolved in 1 kg water will depress the freezing point by 1.86° and the vapor pressure by 0.3 mm Hg (at 20°), relative to pure water.

Even though the two types of instruments measure unique parameters, the most important differences for cell separation work are the volume differences, the time required, and the ability to obtain accurate readings with solutions containing particulate matter.

Freezing point osmometers require sample volumes from 0.050 to 2 ml and each reading can require several minutes. The vapor pressure systems require about 10 μl and are very rapid.

Volume Measurements

The frequency/volume profiles that are generated from the Coulter Counter–Channelyzer system are very useful for identifying cell populations after the separation and for rapidly quantitating the resolution of the separation. Separations of mammalian cells are particularly well suited to this analysis, since density differences are usually not a significant factor and the location of cells in the gradient are usually a direct result of size differences.

The volume profiles will quickly locate cell subpopulations by comparing profiles to data from previous separations. These profiles can usually provide accurate enrichment calculations and allow pooling of fractions of like cells.

Figure 4 shows an example of profiles of 6 fractions from a separation of human monocytes and lymphocytes from a 2-hr CelSep run. Fraction volume was 25 ml. The profiles allow a rapid assessment of the quality of the separation and information about which fractions contain the desired subpopulations.

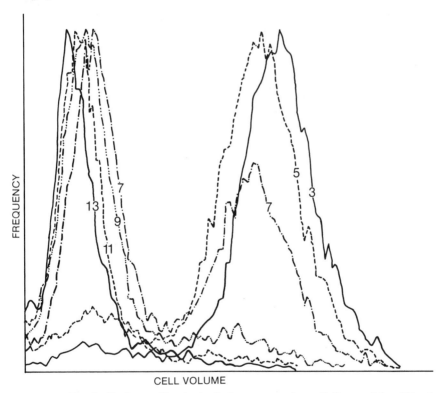

FIG. 4. Size distribution profiles obtained when mononuclear cells from peripheral blood were separated for 2 hr in a 2–4% (w/w) Ficoll gradient at room temperature. Fractions 3–5 contain monocytes. Fraction 7 contains both monocytes and large lymphocytes. Fractions 9–13 contain lymphocyte subpopulations of decreasing size.

The profiles can also indicate when poor resolution is achieved and provide information on possible causes. The presence of small cells in the large cell fractions indicates streaming or aggregation occurred during the separation. A possible solution is to use a less concentrated sample. The presence of large and small cells throughout the gradient can indicate improper gradient formation, unstable (too rapid) unloading, or a large variation in cell density.

Cytology

One of the most valuable methods of determining the degree of enrichment (or contamination) is to use a cytocentrifuge to make a cytological analysis of the cells in each fraction. We use a Cytospin (Shandon Inc., Pittsburgh, Pennsylvania 15275) for this purpose, but modifications of the standard procedure are necessary for good recovery of cells. The Cytospin uses a filter paper to remove the fluid from the cell suspension. In order to prevent excessive cell loss into the filter we prewet the filters with two drops of Hanks' balanced salt solution (BSS) or phosphate-buffered saline (PBS) and spin the rotor for a brief period before we load the cell suspension. Even using this step over 50% of the cells are lost and, unfortunately, small cells are preferentially lost. Certain cell types can be lost due to breakage or by not sticking to the slide. The addition of the one drop of 22% BSA solution (Sigma, St. Louis, Missouri 63178) can improve this, as can longer spin times at lower rpm. In spite of these problems the cytological information that can be gained from the procedure will greatly assist in charting improvements in resolution.

Examination of a wet preparation immediately after the collection of fractions can also be quite useful, since small clumps can often contaminate the larger fractions. If a fluorescence microscope is available a wet preparation stained with one of several dyes such as acridine orange or euchrysin can often help to distinguish different cell types.

Growth Assays

Isolating cells using sterile techniques is feasible with the procedures described in this chapter. However, the longer the cells remain in a hostile environment and the greater the number of manipulations the cells are put through, the more diminished is their cloning efficiency. Several things can be done to help keep the cells viable. The cells should never be suspended in solutions that do not contain at least a small amount of protein. We put 1% FBS into the gradient and wash solutions. Excessive spinning of the cells during washing or allowing the cells to stand for long periods at a high concentration should be avoided. It is also important to avoid large changes in pH and osmolality when transferring cells between solutions.

[25] Separation of Cell Populations by Free-Flow Electrophoresis

By Hans-G. Heidrich and Kurt Hannig

There are various methods for obtaining homogeneous cell populations from cell mixtures, the majority of which exploit physical or chemical differences between the cells. The physical differences, which can be used for separation, include density, used in gradient centrifugation, and size, used in 1 *g* sedimentation. The chemical differences are generally related to the properties of cell surface molecules and include, for example, the presence of different antigens, recognized in immunoadsorbent and fluorescent labeling techniques. Free-flow electrophoresis makes use of differences in the electrical charge of the cell surface molecules to separate cell populations. Characteristic changes in cell surface charge are sometimes associated with various physiological processes such as differentiation or stimulation. Free-flow electrophoresis can be used not only to separate different types of cells, but also, in many cases, to separate similar cells on the basis of their functional properties.

The Method in General

There are several difficulties involved in the electrophoretic separation of cells or other bioparticles (for background reading, see Refs. 1–4). First, the cells to be separated must be present as a single-cell suspension. Assemblies of cells or cell aggregates cannot be separated. Thus cells from the blood, the lymph, or other body fluids can be prepared fairly easily,[2] whereas cells from tissues must first be isolated by mechanical disintegration. Many procedures for isolating cells result in cell damage or disruption, or release of nuclei or DNA, giving a preparation which is not suitable for electrophoresis. The amount of disruption can be reduced by treating tissues with various enzymes in combination with weak mechanical forces, such as gentle stirring or suction through a wide-mouth pipet. This procedure is the only way to obtain a single-cell suspension from, for example, tumor tissue. Enzymes which have proved to be suitable for obtaining

---

[1] K. Hannig and H.-G. Heidrich, this series, Vol. 31, p. 746.

[2] K. Hannig, H. Wirth, B. H. Meyer, and K. Zeiller, *Hoppe-Seyler's Z. Physiol. Chem.* **356**, 1209 (1975).

[3] K. Zeiller, R. Löser, G. Pascher, and K. Hannig, *Hoppe-Seyler's Z. Physiol. Chem.* **356**, 1225 (1975).

[4] K. Hannig, *Electrophoresis* 3, 235 (1982).

METHODS IN ENZYMOLOGY, VOL. 171

single-cell suspensions from various tissues are trypsin, chymotrypsin, collagenase, hyaluronidase, and dispase. Such enzymes, however, often have the disadvantage of cleaving off the specific surface groups and charged molecules which are a prerequisite for electrophoretic separation. This is one of the main reasons why relatively few examples of successful electrophoretic separations of tissue cells can be found in the literature. A different approach, useful for cells such as those from the nephron, which are connected to each other via Ca^{2+}-dependent junctional complexes, is to treat the tissue with weak Ca^{2+} binders such as citrate, EDTA, or EGTA in combination with slight stirring or shaking. Generally any nuclei set free during the isolation procedure can be removed by filtering the suspension over glass wool loosely packed into a Pasteur pipet. Free DNA, which may appear if nuclei are broken, must also be removed since it produces aggregation of cells in the electric field. A short treatment with ultrapure DNase ($20-30$ μg/ml) at room temperature for $5-10$ min is generally sufficient.

Another problem is the choice of the preparation and separation medium. Most of the commercially available free-flow electrophoresis instruments now have powerful cooling devices so that it is possible to raise the ionic strength of the separation medium to $8-10 \times 10^2$ μmho/cm. This is still one power below physiological conditions, however, so that cells are often not viable after separation. In order to mitigate the problem the cells should be collected in more physiological solutions, such as culture media, after they have passed through the separation chamber. The most suitable composition for the separation buffer must be found for each cell type. Triethanolamine, together with the counterion acetate, has proven to be better than phosphate, Tris, or any other buffer tested so far (Table I). Ions such as Mg^{2+}, Ca^{2+}, or Mn^{2+} can be added to the medium, together with physiological reducing agents such as glutathione. The pH should be in the physiological range, between 7.0 and 7.5. The osmolarity should also be close to the physiological value, i.e., around 300 mOsm. It can be adjusted with sugars, such as sucrose, ribose, or mannose, or with glycine. The latter is not suitable for cells, such as renal tubule cells, which tend to take up molecules very actively via pinocytotic processes.

The temperature at which the cells are prepared, isolated, and separated must also be chosen carefully. Cells grown in culture at 37°, for example, cannot be treated at ice temperature because they might suffer from thermal shock and become altered. In this case room temperature should be used for the preparation, and temperatures between 10 and 15° for the electrophoretic separation. Theoretically cells prepared at temperatures below 10° can be separated at 4°. Because of the phase transition of membrane lipids, however, the temperature in the electrophoresis apparatus should be kept above 6°.

TABLE I
MEDIA FOR CELL SEPARATION USING
FREE-FLOW ELECTROPHORESIS

Cell type and medium composition	Separation medium		Electrode medium	
	g/liter	mM	g/liter	mM
Lymphocytes[a]				
Triethanolamine	2.2	15	11.2	75
Potassium acetate	0.4	4	2.0	20
Glycine	18.0	240	—	—
Glucose	2.0	11	—	—
Malaria parasites, kidney cells[b]				
Triethanolamine	1.64	11	16.4	110
Acetic acid (100%)	0.66	11	6.6	110
Glucose	1.0	5	—	—
$CaCl_2 \cdot 2H_2O$	0.0074	0.05	—	—
$MgCl_2 \cdot 6H_2O$	0.1	0.5	—	—
Sucrose	92.5	270.0	—	—

[a] Adjust to pH 7.2 with acetic acid. Adjust electrophoresis–separation medium with sucrose to 310 mOsm. Conductivity, 900 μS.
[b] Adjust to pH 7.5 with either 2 N KOH or 2 N NaOH. Electrophoresis–separation medium has an osmolarity of 335 mOsm and a conductivity of 650 μS.

The cell concentration in the suspension to be separated is important. Too high a concentration of cells causes aggregation, making a good separation impossible. Too low a concentration results in not finding sufficient cells in the individual separated fractions. The optimum concentration lies between 20 and 60 × 10⁶ cells per ml.

Finally, the running conditions in the electrophoresis apparatus itself play a key role in achieving a good separation. The apparatus should always be run at maximum power, which is often limited by the cooling capacity. The proportion of buffer flow to sample injection should be 200–100 : 1–2; in other words, when 200–100 ml of separation buffer are run through the chamber per hour, only 1–2 ml of cell suspension can be injected. It is advisable to coat the glass walls of the separation chamber with albumin prior to each separation in order to bring the ζ potential of the glass as close as possible to the ζ potential of the bioparticles (in the range of −27 to −45 mA) and avoid disturbances through electroosmosis. A procedure for coating is described in Ref. 2.

The strategy to be used in approaching the complete electrophoretic separation can then be summarized as follows. A single-cell suspension is prepared and freed from aggregates and any contaminating nuclei or DNA. The cells are washed gradually into separation media with different pH or ionic composition and are then observed for about 2 hr to check their stability and maintenance of properties. The best buffer is selected and the cells are gradually transferred into this buffer to give a suitable concentration. The separation conditions are chosen such that the fastest moving cells are deflected toward the anode into about fraction 15 of the electrophoresis. If cells aggregate, the preparation procedure is not optimal and must be improved. The cells are collected in culture medium or a medium as close to physiological as possible. Cell counts may be made in the individual collection tubes during separation in order to produce a separation profile, on the basis of which analysis of the various fractions can be started immediately after the injection is finished. If no separation is achieved the pH and the buffer composition should be altered. If it is still not possible to achieve separation it is likely that the surface charge differences between the cell types is not large enough to give a clear separation. In general the chances for a successful separation are high when there are clear morphological differences between the various cell types. Such differences are apparent, for example, between renal proximal and renal distal tubule cells. The former have long and numerous microvilli and are less negatively charged than the latter, which have shorter and fewer microvilli. In this case the basis for the difference in electrophoretic mobility is that the proximal cells have more microvillous membrane, which is less negatively charged and therefore has a lower electrophoretic mobility, and the distal cells have more basal membrane, which is more negatively charged.[5,6]

In the following discussion we describe the separation of four fundamentally different types of cells, immunocompetent cells, malaria parasites from culture and from *Anopheles,* kidney cells, and tumor cells. The results obtained with these cells, and the different methods developed to deal with the specific problems and queries give a good idea of the possibilities and limitations of the free-flow electrophoresis method. For the latest and most updated reference list on cell separations using the free-flow electrophoresis system, see a review entitled "Free-flow particle electrophoresis in cell biology and biochemistry—An updated review" by H.-G. Heidrich.[6a]

[5] H.-G. Heidrich, R. Kinne, E. Kinne-Saffran, and K. Hannig, *J. Cell Biol.* **54,** 232 (1972).

[6] H.-G. Heidrich and R. Geiger, *Kidney Int.* **18,** 77 (1980).

[6a] H.-G. Heidrich, *in* "Electrophoresis '88, Proceedings of the Sixth Meeting of the International Electrophoresis Society," July 1988, Copenhagen (C. Schafer-Nielsen, ed.), pp. 187–220. VCH Verlagsgesellschaft mbH, Weinheim, 1988.

Immunocompetent Cells

Investigations of cells from the immune system have excited considerable interest in recent years and it is not surprising that free-flow electrophoresis has gained particular importance in this area, not least because cells tend to be separated on the basis of functional differences (for reviews, see Refs. 4 and 7–10). The electrophoretic separation of lymphocytes was described in detail in an earlier volume in this series.[10] In this chapter we shall restrict ourselves to a summary of the most important points together with a description of more recent applications and results.

Separation of Murine Lymphocytes

Bimodal distributions are observed following free-flow electrophoresis of murine peripheral lymphoid cells from spleen, lymph nodes, Peyer's patches, blood, or lymph.[11,12] T cells are found in the region of high, and B cells in the region of low, electrophoretic mobility. Light microscopy shows that in all preparations the fast fraction tends to consist entirely of small lymphocytes, whereas the slow fraction also contains a small number of larger lymphocytes, blast cells, and plasma cells.[13] Eighty-two hours after immunization of rats with sheep erythrocytes, the B cell region of the distribution profile from lymph node cells contains two fractions of hemolysin-producing cells (Fig. 1). This separation runs parallel with that of plasma and blast cells.[13,14]

The clear separation of T and B lymphocytes from mice and rats has been shown by analysis of a number of parameters in the separated cell fractions. These include analysis of the cell surface proteins, Fc receptors, complement receptors, rosette formation with sheep erythrocytes, graft-versus-host reactions, mixed lymphocyte cultures, stimulation with specific mitogens and antigens, cytotoxicity, cell cooperation, and immune responses[8,12,14–17] (Table II). Monoclonal antibodies have also been used to

[7] P. Häyry and L. C. Andersson, *Scand. J. Immunol.* **5**, 31 (1976).

[8] K. Shortman, *Methods Cell Sep.* **1**, 229 (1977).

[9] E. Hansen, in "Cell Electrophoresis" (W. Schütt and H. Klinkmann, eds.), p. 287. de Gruyter, Berlin, Federal Republic of Germany, 1985.

[10] E. Hansen and K. Hannig, this series, Vol. 108, p. 180.

[11] K. Zeiller and K. Hannig, *Hoppe-Seyler's Z. Physiol. Chem.* **352**, 1162 (1971).

[12] L. C. Andersson, S. Nordling, and P. Häyry, *Cell. Immunol.* **8**, 235 (1973).

[13] K. Hannig and K. Zeiller, *Hoppe-Seyler's Z. Physiol. Chem.* **350**, 467 (1969).

[14] P. Häyry, L. C. Andersson, C. Cahmberg, P. Roberts, A. Ranki, and S. Nordling, *J. Med. Sci. Israel* **11**, 1299 (1975).

[15] G. V. Sherbet, "The Biophysical Characterization of the Cell Surface." Academic Press, New York, 1978.

[16] H. Boehmer, K. Shortman, and G. J. V. Nossal, *J. Cell. Physiol.* **83**, 231 (1974).

[17] T. G. Pretlow and T. P. Pretlow, *Int. Cytol.* **61**, 85 (1979).

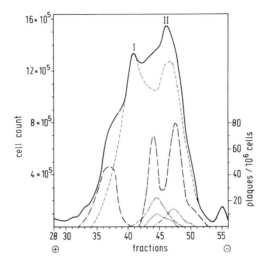

FIG. 1. Electrophoretic separation of lymph node cells. Cell isolation and separation was carried out 82 hr after injecting Wistar rat with sheep red blood cells.[13] (——) Cell counts in the various fractions; (– – –) hemolysin-producing cells; (-------) lymphocytes (I is the T cell region; II is the B cell region); (– · – · –) erythrocytes; (– · · · – · · ·) plasma cells; (– · · · – · · ·) blast cells; (· · · · · · · · · ·) plasmocytes.

TABLE II

ANALYSIS OF MURINE B AND T LYMPHOCYTES SEPARATED BY
FREE-FLOW ELECTROPHORESIS

Lymphocytes	B	T	Reference
Surface charge density	Low	High	11–14
Surface glycoproteins			7, 17, 21
Receptors	Ig, C′-rec		14, 16, 17
Surface antigens			
Mouse	MBLA	Thy-1, MTLA	3, 12, 13, 21
Rat	RBLA	RHLA, RTLA	24, 25
	Ia	Lyt-1, Lyt-2	19
Histocompatibility antigen density	Higher	Lower	19, 20
EM morphology, volume			11, 24
Thymus dependency	−	+	9, 10, 24
Mitogenic response	LPS	PHA, Con A	8, 12
Immune functions	PFC,B memory		11, 21, 25
		GvH	11, 12, 14
		CTL	12, 14
		MCL	7, 14, 16
		Help/suppress	23–25

characterize B and T cells, with Thy-1.2 and Lyt-2 antigens defining mouse T cells.[18] Beside these, quantitative differences have been observed between lymphocytes of different electrophoretic mobility, for example, in the amount of histocompatibility antigens in both mouse[19] and rat cells.[20] A graft-versus-host reaction was only observed when spleen cells of high, but not of low, electrophoretic mobility were injected into irradiated recipients,[21] showing these cells to be T cells.

After bone marrow cells have been freed from nonlymphoid cells by 1 g sedimentation, immature lymphoid precursor cells and mature B and T lymphocytes can be separated electrophoretically.[14,22-24] This offers the possibility of preparing stem cells free of T cells (killer cells) for use in bone marrow transplantation. Such experiments have been successfully performed in mice.[25]

These and other results have shown the high separation potential of free-flow electrophoresis for murine lymphocyte subpopulations with different characteristics, functions and stage of maturation or activation (reviewed in Refs. 2, 8, 14, 17, and 24). Electrophoretic separation has also been demonstrated with lymphocytes from other species, including chicken, guinea pig, cat, and nonhuman primates (for references, see Refs. 2 and 25).

Separation of Human Lymphocytes

In contrast to the murine system human peripheral blood lymphocytes do not have a bimodal electrophoretic distribution.[14,26-30] There is only a small shoulder on the main peak. A number of authors have questioned whether there are differences in the electrophoretic mobility of human T

[18] F. Dumont and R. C. Habbersett, *J. Immunol. Methods* **53,** 233 (1982).
[19] F. Dumont, R. C. Habberstettand, and A. Ahmed, *Electrophoresis* **81,** 813 (1981).
[20] E. Hansen, *Blut* **44,** 141 (1982).
[21] K. Zeiller, K. Hannig, and G. Pascher, *Hoppe-Seyler's Z. Physiol. Chem.* **352,** 1168 (1971).
[22] D. G. Osmond, R. G. Miller, and H. Boehmer, *J. Immunol.* **114,** 1230 (1975).
[23] K. Zeiller, E. Hansen, D. Leihener, G. Pascher, and H. Wirth, *Hoppe-Seyler's Z. Physiol. Chem.* **357,** 1309 (1976).
[24] K. Zeiller and E. Hansen, *Histochem. Cytochem.* **26,** 369 (1978).
[25] K. Zeiller, *Behring Inst. Mitt.* **52,** 11 (1972).
[26] G. Stein, *Biomedizine* **23,** 5 (1975).
[27] P. Chollet, P. Herve, J. Chassagne, M. Masse, R. Plagne, and A. Peters, *Biomedicine* **28,** 119 (1978).
[28] E. M. Levy, S. Silverman, K. Schmid, and S. R. Cooperband, *Cell. Immunol.* **40,** 222 (1978).
[29] E. Hansen and K. Hannig, *J. Immunol. Methods* **51,** 197 (1982).
[30] K. A. Ault, A. L. Griffith, C. D. Platsoucas, and N. Catsimpoulas, *J. Immunol.* **117,** 1406 (1976).

and B cells (reviewed in Ref. 17). In preparative cell electrophoresis the nature of any observed electrophoretic distribution of cells can be analyzed directly in the fractions. The shoulder in the electrophoretic profile of human blood mononuclear cells is caused by monocytes, which have a lower surface charge density than the lymphocytes.[26,28,29,31] The B and T cells do not have a measurably different mobility. The observed results can sometimes be misleading, however, since chlorine dioxide, which can be formed in small quantities at the electrodes during electrophoresis, causes a nonspecific slowing effect in a portion of the cells (K. Hannig, unpublished observations).

Despite the similarity in their electrophoretic mobility, human B and T lymphocytes can be separated electrophoretically by selectively altering the surface charge with specific cell markers.[10,29] This technique is called ASECS, for antigen-specific electrophoretic cell separation. The method is based on the fact that immunoglobulin has an eight times lower negative charge than the cell membrane. Cells loaded with antibody thus have a lowered electrophoretic mobility. In general the receptor density is too low for a measurable effect to be obtained with a simple antibody reaction. The effect can be intensified, however, by using a second antibody directed against the first. In special cases a third antibody can be used. The masking of the negative cell surface charge is then generally sufficient for the subpopulation of interest to be separated. Figure 2 shows a flow diagram of the procedure (details are given in Refs. 10 and 29). This method has been used successfully to isolate human B cells with 90% purity by applying an antibody against human IgM. T lymphocyte subpopulations can be separated using monoclonal antibodies. The helper/inducer T cell subset can be separated from the cytotoxic/supressor subset using OKT 4 or OKT 8 antibodies. The helper/inducer subset is OKT 4⁻ (cells in the low-electrophoretic-mobility region following ASECS with OKT 4 antibody) and OKT 8⁻ (cells in the high-electrophoretic-mobility region following ASECS with OKT 8 antibody), whereas the cytotoxic supressor subset is OKT 4⁻ and OKT 8⁺.[9,10,29] There is a mitogenic response to the T cell mitogens phytohemagglutinin (PHA) and concanavalin A (Con A), of human lymphocyte subpopulations isolated by ASECS using OKT 8 monoclonal antibody.[9] The T cell subpopulations differ in their response to PHA. The bound antibodies seem not to interfere with this polyclonal activation. This assay can be used to study the differential effects of drugs on human lymphocyte subsets. The helper and suppressor activities of human lymphocyte subpopulations separated by ASECS can be investigated in an *in*

[31] J. Bauer and K. Hannig, *Electrophoresis* **5**, 155 (1984).

Fresh human blood, defibrinized

| Centrifugation according to Böyum in Ficol ($d = 1.07$ kg/liter), 30 min, 400 g, 4°. Wash
| cells twice in Puck G Medium at 4°.
↓
Lymphocytes

| First incubation: centrifuge antiserum 30 min, 18,000 g. Resuspend 50×10^6 lympho-
| cytes in a 1:4 dilution of the antiserum in Puck G Medium. Keep 30 min at 4°.
| Centrifuge 10 min, 100 g. Discard supernatant and wash cells twice in Puck G Medium
| at 4°.
↓
Lymphocytes, loaded with antibody I

| Second incubation: resuspend washed cells in 1 ml of a 1:2 dilution of TRITC-conju-
| gated goat antirabbit IgG immunoglobulin. Keep 30 min at 4°. Centrifuge and wash cells
| as above and resuspend cells in 2 ml of electrophoresis separation medium.
↓
Lymphocytes, loaded with antibody I and antibody II

| Filter cells through glass wool loosely packed in a Pasteur pipet. Electrophoretic separa-
| tion at 200 mA and 100 V/cm at 4°. Collect cells in tubes containing 1 ml Puck G
| Medium supplemented with 1% BSA.
↓
Separated cell populations

FIG. 2. Procedure for ASECS.[10,29]

vitro assay of humoral immune response.[9] Strong supression of B cells is observed after incubation of isolated B cells with *Staphylococcus aureus.*

These results demonstrate the functional integrity of lymphocytes after ASECS. It should be possible to use the same method to further separate murine lymphocyte subpopulations. Future research is likely to use charge-modified antibodies and the application of molecules other than antibodies, to specifically modify the surface charge density of cells. The ASECS method allows separation at a very high throughput, more than 100,000 cells/sec, much higher than that in comparable fluorescence-activated cell-sorting methods.

Malaria Parasites

In the past few years research into *Plasmodium falciparum,* the parasite which causes the lethal malaria tropica, has become very active. One of the main aims is the development of a vaccine against the parasite. As a preliminary it was clearly desirable to have homogeneous preparations of parasites free from erythrocyte components. Equally, it was necessary to develop large-scale isolation techniques for parasite stages such as mero-zoites and sporozoites. Free-flow electrophoresis has proved to be useful for isolating such parasites as well as for separating infected and noninfected intact host erythrocytes, and membranes from infected and noninfected host cells.[32]

After the merozoite has invaded the host cell, intraerythrocytic stages of the malaria parasite develop differentiating within the erythrocyte from early trophozoites (often called "ring" stages) to late trophozoites, schizonts, segmenters, and merozoites. The latter is the invasive stage of the parasite and the only stage that, for a short period of time, is free in the bloodstream. The development of culture systems[33] was a major advance, considerably facilitating research into the malaria parasite. These cultures contain all the intraerythrocytic stages normally present in the living host, thus providing suitable material for the isolation of different parasite stages. Before isolating the parasites they must be liberated from the host erythrocytes, a difficult task as they are rather fragile. An excellent technique was developed using quasi-cross-linking of the erythrocytes with lectins, such as lectin from *Phaseolus vulgaris* (hemagglutinin), followed by gentle breakage of the aggregates by running them over a series of nylon sieves with mesh sizes decreasing from 100 to 1 μm. The erythrocyte membrane fractures and the parasites are set free.[34,35] This procedure has been used for various species of rodent *Plasmodium* as well as for the human strain *P. falciparum*. The mixture of unbroken erythrocytes, erythrocyte vesicles, and free parasites of the various stages can then be separated by free-flow electrophoresis. During this separation several sharp bands, all deflected toward the anode, can be observed in the separation chamber. Examination of the various fractions (Fig. 3) by Giemsa-stained smears, electron microscopy, gel electrophoresis, and enzyme tests (i.e., acetylcholinesterase for erythrocyte constituents; glutamate dehydrogenase for parasites) proved that erythrocyte material was only present in fractions 28–37, whereas all the other fractions contained free parasites without erythrocyte contamination. A clear separation was observed in the erythrocyte-containing fractions between noninfected erythrocytes and their membranes, and infected erythrocytes and their (knobby) membranes. Infected and noninfected erythrocytes were also separated with good purity when intact, unbroken erythrocytes were used for the electrophoresis.[34,36] However, intact infected and noninfected erythrocytes can be separated

[32] H.-G. Heidrich, *in* "Electrophoresis '81" (R. Allen and P. Arnaud, eds), p. 859. de Gruyter, Berlin, Federal Republic of Germany, 1981.

[33] W. Trager and J. B. Jenssen, *Science* **193**, 673 (1976).

[34] H. G. Heidrich, L. Rüssmann, B. Bayer, and A. Jung, *Z. Parasitenkd.* **58**, 151 (1979).

[35] H.-G. Heidrich, J. E. K. Mrema, D. L. Vander Jagt, P. Reyes, and K. H. Rieckmann, *J. Parasitol.* **68**, 443 (1982).

[36] M. Suzuki, I. Sawasaki, S. Waki, H. Iwaoka, T. Asada, and H. Nakajima, *Bull. W.H.O.* **57**, 129 (1979).

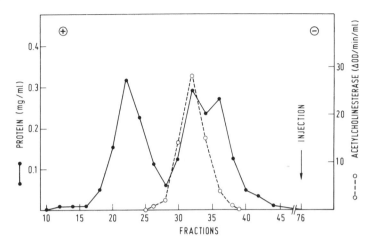

Fig. 3. Free-flow electrophoresis isolation of intracellular parasites from *Plasmodium falciparum* cultures. Parasites were isolated from their host cells with the lectin technique.[34,35] (●) Protein distribution after electrophoresis. Fractions 12–27 and fractions 33–45 contained free parasite stages. Intact infected and uninfected erythrocytes and erythrocyte membrane vesicles were found in fractions 28–36. (○) Acetylcholinesterase activity as a marker for erythrocyte membranes.

much better, and in higher yields, using the Plasma gel procedure.[37] In the electrophoresis separations in which parasites had been liberated with the lectin technique, young trophozoites were present in fractions 12–20 and more mature ones in fractions 20–27. Schizonts were found in fractions 35–38 and merozoites, the least negatively charged stage of the parasites, in fractions 39–44. The latter possessed a distinct surface coat, a possible marker for invasive ability. Thus a separation of the parasites according to stages was achieved. The molecular basis for this separation has not yet been elucidated.

Various tests and experiments have been carried out with these electrophoretically purified parasite stages. *Plasmodium falciparum* merozoites were lyophilized and, after emulsifying with Freund's adjuvant, injected intramuscularly into *Aotus trivirgatus* monkeys (3 times 5 mg of merozoite protein at intervals of 3 weeks). Control animals only received Freund's adjuvant. The first injection was with complete adjuvant, the second and third with incomplete adjuvant. Three weeks after the last injection all animals were challenged with 10^5 virulent parasites from a malaria infected monkey. Control animals developed severe parasitemias after 6–9 days and had to be treated and cured by chemotherapy. In contrast some of the

[37] G. Pasvol, R. J. Wilson, M. E. Smalley, and J. Brown, *Ann. Trop. Med. Parasitol.* **72,** 87 (1978).

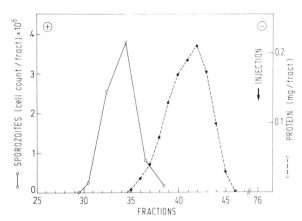

FIG. 4. Free-flow electrophoresis isolation of whole body *Plasmodium berghei* sporozoites. Details of the procedure are described in Ref. 39. (●) Protein distribution after
electrophoresis. Fractions 37–45 contained mosquito debris, mitochondria, mitoplasts, and
unidentified contamination. (○) Sporozoite counts. Fractions 31–35 contained clean sporozoites.

animals which had received merozoites, were partially protected and developed only very low parasitemias, which were reduced to zero following a
second challenge 6 months later. However, these animals developed ane-
mias.[38] Thus it appears that the electrophoretically purified merozoites
contain the immunogens necessary for a protective immunization against
malaria and that these immunogens are not severely altered by the isolation and separation procedure.

Sporozoites, the stages of the malaria parasite which are injected into
the host by the infected female *Anopheles* during the blood meal, and
which then enter the liver thus transmitting the disease, can also be isolated
by free-flow electrophoresis in excellent purity and high yield.[39] About 400
Plasmodium berghei-infected female mosquitoes from an insectory were
killed with chloroform, ground and layered on a two-step (40 and 50%,
v/v) Hypaque gradient.[40] After spinning (12 min, 16,300 *g,* 4°) the inter-
phase layer was diluted to give about $6–9 \times 10^6$ sporozoites per/ml and
then separated in a free-flow electrophoresis apparatus. There was only one
diffuse band in the electrophoresis chamber. Light and electron microscopic examination, protein assay in the individual fraction tubes, PAGE
analysis, and bacterial counts on agar plates showed that this main protein
peak (Fig. 4), fractions 37–45, consisted of contamination, i.e., mitochon-

[38] C. Espinal, K. H. Rieckmann, and H.-G. Heidrich, unpublished observations.
[39] H.-G. Heidrich, H. D. Danforth, J. L. Leef, and R. L. Beaudoin, *J. Parasitol.* **69,** 360
(1983).
[40] N. P. Paccheo, C. P. A. Strome, F. Mitchell, M. P. Bawden, and R. L. Beaudoin, *J.
Parasitol.* **65,** 414 (1978).

dria, mitoplasts, membrane vesicles, and debris. Virtually all the sporozoites and no contamination were present in fractions 31–35. On average, $6-8 \times 10^6$ sporozoites were isolated from 400 mosquitos. The surface appeared not to be altered. The serological status was tested by indirect immunofluorescence using serum from mice immunized with *P. berghei* sporozoites. A combination of surface iodination, SDS–PAGE, and autoradiography showed a protein with an apparent molecular weight of about 44,000 to be the predominant constituent of the sporozoite membrane. The sporozoites were infective even after electrophoresis lasting for about 2 hr. Mice were inoculated with a rather high inoculum, 50,000 sporozoites per animal, but only 20% of those were potentially infective salivary gland forms (the remaining 80% being midgut sporozoites of varying degrees of maturity). All the inoculated mice were infected.

An important step for obtaining clean sporozoite preparations is the removal of bacterial contamination which stems from the insectory. In electrophoresis the contaminating bacteria of the mosquitos from two different insectories showed different electrophoretic mobilities. However, in both cases bacteria were removed from the sporozoites, either to the more anodic or the more cathodic fractions. This was shown by colony counts on agar plates at 24 and 48 hr at a serial dilution of 1 : 1,000 or 1 : 10,000. Sporozoites were virtually free from bacterial contamination.

Cells from the Renal Proximal and Distal Tubules

In recent years most of the cell types in the various segments of the renal cortical nephron have been identified using a combination of microdissection, biochemical marker techniques,[41] and ultrastructural analysis.[42] However, it has proved more difficult to isolate and culture these cell populations for further physiological experiments. One recently developed method is to use microdissected segments as a basis for cell culture.[43-45] This method shares a problem common to all cell culture methods that the highly specialized cells from such nephron segments tend to dedifferentiate in long-term culture. It has a further major disadvantage, however. As well as nephron cells the microdissected fragments contain vascular and connective tissue, and cells from these lines tend to overgrow the actual cells of interest. Thus it would still be an advantage to obtain pure cell populations for long-term culturing.

[41] F. Morel, *Am. J. Physiol.* **240,** F159 (1981).

[42] B. Kaissling and W. Kriz, *in* "Structural Analysis of the Rabbit Kidney" (A. Brodal, ed.), p. 3. Springer-Verlag, Berlin, Federal Republic of Germany, 1978.

[43] M. F. Horster, J. Fabritius, and M. Schmolke, *Pfluegers Arch.* **405,** 158 (1985).

[44] N. Green, A. Algren, J. Hoyer, T. Triche, and M. Burg, *Am. J. Physiol.* **249,** C97 (1985).

[45] M. F. Horster and M. Stopp, *Kidney Int.* **29,** 46 (1986).

Two kidneys from an anesthetized New Zealand rabbit

Work at room temperature. Perfuse each kidney with 50 ml Earle's Balanced Salt Solution supplemented per 1000 ml with 1.0 g glucose, 17.11 g sucrose, 2.2 g sodium bicarbonate, 13 ml of a 200 mM glutamine solution, and 10 mg BSA (pH 7.4; 330 mOsm) for 5 min. Remove cortex and carefully press through a fine tissue press onto a 40-mesh nylon screen. Stir the suspension through the sieve while rinsing with a 1:1 dilution of the Earle's solution and electrophoresis separation medium.

Cell filtrate

Run the combined cell filtrates from both kidneys (about 200 ml) through a 20-mesh nylon screen. From now on work at 4°. Centrifuge 5 min, 150 g. Discard supernatant and resuspend pellet in 40 ml electrophoresis separation medium. Spin again as above, discard supernatant and resuspend pellet in 20 ml electrophoresis separation medium.

Cell suspension

Rotate cell suspension (about 20 ml) carefully for about 5 min in order to disperse aggregates. Filter suspension through glass wool loosely packed in a Pasteur pipet. Centrifuge suspension 5 min, 150 g. Resuspend pellet gently in 1 ml electrophoresis separation medium.

Single cell suspension

There will be about 25–50 × 10⁶ cells/ml. Dilute suspension with electrophoresis separation medium to give a cell concentration of 25 × 10⁶ cells/ml. Electrophoretic separation at 170 mA, 130 V/cm, 6°, sample injection 1.85 ml/fraction/hr. Collect cells in tubes containing 1 ml of the above described Earle's solution.

Separated cell populations

FIG. 5. Procedure for isolating cell populations from rabbit kidney cortex.[46-48]

Most of the methods used for separating kidney cells have been based on density differences.[46-56] Free-flow electrophoresis, however, can be successfully used to separate at least four different cell populations on the basis of surface charge, two from the proximal tubule, one from the distal tubule, and a group from the collecting duct.

[46] H.-G. Heidrich and M. Dew, *J. Cell Biol.* **74**, 780 (1977).
[47] A. Vandewalle, B. Köpfer-Hobelsberger, and H.-G. Heidrich, *J. Cell Biol.* **92**, 505 (1982).
[48] H.-G. Heidrich, in "Cell Function and Differentiation" (J. Glozgatsos, G. Palaiologos, A. Trakatellis, and C. P. Tsiganos, eds.), Part A, p. 115. Liss, New York, 1982.
[49] B. Thimmappayya, R. Ramachandra-Reddy, and P. M. Bhargava, *Exp. Cell Res.* **63**, 333 (1970).
[50] J. I. Kreisberg, R. L. Hoover, and M. J. Karnovsky, *Kidney Int.* **14**, 21 (1978).
[51] D. W. Scholer and J. S. Edelman, *Am. J. Physiol.* **237**, F350 (1979).
[52] J. I. Kreisberg, G. Sachs, T. G. Pretlow II, and R. A. McGuire, *J. Cell Physiol.* **93**, 169 (1977).
[53] J. Eveloff, W. Haase, and R. Kinne, *J. Cell Biol.* **87**, 673 (1980).
[54] J. I. Kreisberg, A. M. Pitts, and T. G. Pretlow II, *Am. J. Pathol.* **86**, 591 (1977).
[55] J. Bosackova, *Physiol. Biochem.* **11**, 39 (1962).
[56] D. P. Jones, G.-B. Sundby, K. Ormstad, and S. Orrenius, *Biochem. Pharmacol.* **28**, 929 (1979).

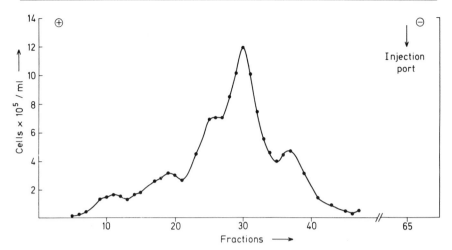

FIG. 6. Electrophoretic separation of cells from the renal cortex. Detailed procedure for producing the cell suspension used in the electrophoretic separation is described in Fig. 5 and in Refs. 46 and 47. All the cells were deflected toward the anode (see the text).

We have not found it possible to separate cell suspensions isolated from rabbit kidney tissue by means of enzyme digestion with collagenase, trypsin, dispase, or hyaluronidase[49-67] using electrophoresis. Apparently surface molecules specific for the various cell populations, and necessary for their electrophoretic separation, are cleaved off by the enzymes. Cells from the renal nephron, however, are connected to each other via junctional complexes which are labile in the presence of Ca^{2+} chelators,[49] and can thus be obtained as a single-cell suspension without using enzymes.[46-48] The method we used is shown in Fig. 5. Cells prepared in this way have intact surface groups, i.e., the groups responsible for the different surface charge of the various cell types were kept intact.[46] These cell suspensions were separated by free-flow electrophoresis into a main cell peak, with various shoulders and minor peaks (Fig. 6). Red blood cells were found in fractions 36–40. The cells in fractions 25–33 possessed nuclei and numer-

[57] L. Pine, G. Brandley, and D. Miller, *Cytobios* **4**, 347 (1969).
[58] H. J. Phillips, *Can. J. Biochem.* **45**, 1495 (1976).
[59] H. Montes De Oca and T. I. Malinin, *J. Clin. Microbiol.* **2**, 243 (1975).
[60] D. Cade-Treyer and S. Tsuji, *Cell Tissue Res.* **163**, 15 (1975).
[61] V. C. Shah, *Experientia* **18**, 239 (1962).
[62] R. K. Studer and A. B. Brole, *J. Membr. Biol.* **48**, 325 (1979).
[63] P. J. Baker, B. P. Croken, and S. G. Osofsky, *Kidney Int.* **20**, 437 (1981).
[64] H. J. Lyons and P. C. Churchill, *Am. J. Physiol.* **228**, 1835 (1975).
[65] H. Bojar, K. Balzer, F. Boeminghams, and W. Staib, *Z. Klin. Chem. Klin. Biochem.* **13**, 31 (1975).
[66] D. L. Dworzack and J. J. Grantham, *Kidney Int.* **8**, 191 (1975).
[67] W.-T. H. Chao and L. R. Forte, *Endocrinology* **112**, 745 (1983).

ous long microvilli. The cells in fractions 10–20 were also nucleated but had less, and much shorter, microvilli. The cells appeared not to be altered by the separation procedure. They had an excellent morphological appearance in the electron microscope and about 90% of their ATP could only be assayed after treatment with digitonin. The behavior of the cells in culture provided additional evidence of their excellent condition. Cells were plated out, stained with eosin, the protein content per cell assayed, and [14C]uridine incorporation measured. Cell breakdown was not observed until after 3 days in culture.[48] The good quality of the isolated cells was shown more conclusively by their ability to produce cAMP in culture, without addition of ATP, for up to 48 hr. cAMP is formed from the internal precursor ATP, which is only present in cultured cells if they are intact, i.e., not leaky.[48]

A similar separation has been tried with rat kidney cortex using the citrate perfusion technique. However, the cells were not as well preserved as those from rabbit (B. Köpfer-Hobelsberger and H.-G. Heidrich, unpublished observations). Kreisberg et al.[52] have obtained good quality cells from rat kidney cortex using a different technique for preparing the cell suspensions, i.e., repeated digestion of the minced tissue with 0.25% trypsin. The isolated cells were separated by free-flow electrophoresis into proximal tubule cells and other cells which have not been characterized.

The rabbit kidney cortex, which was used in the electrophoresis experiments described above, consists mainly of proximal segments of the nephron, with a lesser amount of distal segments and collecting ducts. Thus we expected that the main fraction of the separated cells would be cells from the proximal segment. Cells from this segment are characterized by long microvilli and specific enzyme activities such as alkaline phosphatase. Cells from fractions 25–33 of the separation, the main peak, had these characteristics (Fig. 7) and were therefore thought to be proximal tubule cells. The analysis of other enzyme markers provided further proof. The cells had high values of phosphoenolpyruvate carboxykinase (PEPCK), a marker for proximal tubule cells, and low values of hexokinase, as described by Vandewalle et al.[68] The analysis of the various hormone-stimulated adenylate cyclases, as described by Morel,[41] also supported the identification of these cells as proximal tubule cells. According to Morel[41] the nephron is finely segmented into tubule portions containing a series of adenylate cyclase sites sensitive to stimulation by different hormones or adrenergic agonists. In the proximal part of the nephron, the enzyme, which appears at high concentrations, can only be stimulated by the parathyroid hormone (PTH); in the distal part the enzyme can be

---

[68] A. Vandewalle, G. Wirthensohn, H.-G. Heidrich, and W. G. Guder, Am. J. Physiol. **240,** F492 (1981).

FIG. 7. Scanning electron microscopy of proximal tubule cells. Cells from fractions 25–33 (Fig. 6) possessed long microvilli and other characteristics typical for proximal tubule cells (see the text). Bar, 5 μm.

stimulated by both PTH and isoproterenol (ISO); and in the collecting duct the enzyme can only be stimulated by arginine-vasopressin (AVP). The results obtained with the isolated cells demonstrate that the cells in fractions 25–33 (Fig. 5) were proximal tubule cells.[47] The enzymatic characteristics of these cells were analyzed further to discover whether there were any differences within the fractions 25–33. Both γ-glutamyl-transferase, a marker for the straight portion of the proximal tubule,[69] and pyruvate kinase, more active in the straight portion of the proximal tubule,[69] were higher in fractions 30–33 than in fractions 25–28. In contrast, alkaline phosphatase activity, which is higher in the proximal convoluted tubule than in any other segment,[69] was higher in fractions 25–28. The cells in fractions 25–28 were larger in diameter, and possessed more densely arranged microvilli, than the cells in fractions 30–33. The mitochondria in cells in fractions 25–28 were rod-shaped and those from fractions 30–33 had more rounded profiles. All these results point to a separation of cells from the convoluted (fractions 25–28) and straight (fractions 30–33) portions of the proximal tubule.[42]

[69] B. D. Ross and W. G. Guder, *in* "Metabolic Compartmentation" (H. Sies, ed.), p. 363. Academic Press, New York, 1982.

The electrophoretically fast-moving cells in fractions 15–25 had only a few short microvilli and were characterized by low values of alkaline phosphatase and high values of hexokinase, a distal convoluted and connecting tubule marker enzyme.[68] In addition these cells possessed adenylate cyclase activity which could be stimulated by both PTH and ISO, as well as by AVP. Thus the fast-moving cells appear to be a mixture of mainly distal cells with some cells from the collecting duct.

An interesting contribution to kidney cell separation by free-flow electrophoresis was made in space under microgravity conditions. Morrison *et al.*[70] separated a cell suspension from cultured primary human embryonic kidney cells. Various cell populations were obtained, a few of them able to produce high activities of urokinase (plasminogen activator) in culture. Whether this technique can be used for isolating large quantities of this cell population for the production of urokinase remains to be shown.

Tumor Cells

The high hopes raised by early experiments with tumor cells have not been fulfilled in practice. Originally it appeared that malignant neoplastic cells had a higher electrophoretic mobility than their healthy counterparts, but more recent research has shown that this is not a general characteristic of tumor cells.[17] Treatment of tumor cells with sialidase, however, reduces their electrophoretic mobility drastically.[71] However, some progress has been made and the development of modern techniques for preparative electrophoresis has enabled unequivocal identification and characterization of separated cells.

Ascites tumor cells have been isolated intact using free-flow electrophoresis.[72] Figure 8 shows the results of a typical experiment using the Walker ascites carcinoma of the rat. Over 90% of the cells in fraction 2 are carcinoma cells, as shown by light microscopy of stained smears. Mayhew[73] reported similar results with Ehrlich ascites tumor cells, and Pretlow *et al.* with tumor ascites mast cells.[74] Neoplastic cells were enriched 65% with 95% viable cells. Other studies have attempted to find a basis for cell separation. Brossmer *et al.*[75] investigated changes in cell surface charge

[70] D. R. Morrison, G. H. Barlow, C. Cleveland, M. A. Farrington, R. Grindeland, J. M. Hatfield, W. C. Hymer, J. W. Landham, M. L. Lewis, D. S. Nachtweg, P. Todd, and W. Wilfinger, *Adv. Space Res.* **4**, 67 (1984).

[71] F. Hartveit, D. B. Cater, and J. N. Mehrishi, *J. Exp. Pathol.* **49**, 643 (1968).

[72] K. Hannig and H. Wrba, *Z. Naturforsch. B* **19**, 860 (1964).

[73] E. Mayhew, *Cancer Res.* **28**, 1590 (1968).

[74] T. P. Pretlow, G. Sachs, T. G. Pretlow, and A. M. Pittis, *Br. J. Cancer* **43**, 537 (1981).

[75] R. Brossmer, C. T. Nebe, B. Bohn, W. Tilgen, V. Schirrmacher, R. Schwarz, J. Sonka, M. Volm, B. Doerken, M. Kaufmann, and D. Fritze, *J. Cancer Res. Clin. Oncol.* **105**, A4 (1983).

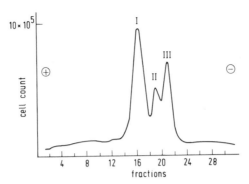

FIG. 8. Electrophoretic separation of cells from the Walker ascites tumor.[72] (I) Erythrocytes; (II) tumor cells; (III) leucocytes.

related to stage of malignity, differentiation, and viral transformation. They showed, for example, that highly metastatic mouse T lymphoma cells have a lower electrophoretic mobility than their nonmetastatic parent cell line.

Not all investigators have discovered differences in cell surface charge. Schubert et al.[76,77] used free-flow electrophoresis to show that bone marrow cells from patients with chronic myeloid leukemia, chronic lymphotic leukemia, typical and atypical plasma cell myeloma, and Waldenstrom's disease, with the exception of myeloma cells, all have the same electrophoretic mobility as their normal counterparts. Only the myeloma cells showed a marked increase in electrophoretic mobility in comparison with normal plasma cells.

Retention of normal properties, however, can also open up possibilities for separation and/or diagnosis of tumor cells. The electrophoretic separation profile of normal human peripheral blood lymphocytes shows two subpopulations with only slightly different electrophoretic mobilities. They are characterized by IgD and IgM immunoglobulin, the IgD cells being the faster fraction. The lymphocytes from four patients with chronic, IgD-positive lymphatic leukemia had the same electrophoretic characteristics as the normal IgD cells. The same appears to be true for IgM leukemia cells.

In summary, it can be said that we still have a very limited knowledge of the electrophoretic characteristics of tumor cells, and most of this is confined to cells which naturally occur in suspension. Until now the difficulties associated with the preparation of single-cell suspensions from tumor tissue have prevented much progress in the investigation of cells from solid tumors.

[76] J. C. F. Schubert, F. Walther, E. Holzberg, G. Pascher, and K. Zeiller, *Klin. Wochenschr.* **51,** 327 (1973).
[77] J. C. F. Schubert, K. Schopow, and F. Walther, *Blut* **35,** 135 (1977).

[26] Separation of Cells and Cell Organelles by Partition in Aqueous Polymer Two-Phase Systems

By PER-ÅKE ALBERTSSON

Cells, cell organelles, and membrane vesicles can be separated by partition between two immiscible phases composed of aqueous polymer solutions.[1] By this method cell particles are separated according to their surface properties and therefore this method complements centrifugation methods.

A large number of different cells and cell organelles have been studied by phase partition. For reviews, see the Refs. 1–3.

Cell particles are distributed between one of the bulk phases and the interface. The distribution is usually characterized as the percentage of the material present in the upper phase. If the distribution is very different for two particles, they can be separated by one or a few partition steps. If they are closer to each other in their behavior, one has to use a multistage procedure such as countercurrent distribution.[1]

Phase Systems

The phase systems mainly used are composed of dextran, polyethylene glycol (PEG), and water. Other polymer–polymer two-phase systems and their general properties have been described by Albertsson.[1]

The phase diagram of a dextran–PEG system is shown in Fig. 1. The phase diagram depends on the temperature and on the molecular weight of the polymers. This means that, if different commercial batches of a polymer vary in molecular weight, the phase diagram will also vary.

The most prominent feature of these phase systems is that both phases are rich in water. The upper PEG phase contains 90–95% water and the lower dextran phase contains 80–90% water (see Table I). Sucrose, sorbitol, salts, and low-molecular-weight compounds distribute almost equally between the phases and can be included in the phase system (without gross

[1] P.-Å. Albertsson, "Partition of Cell Particles and Macromolecules," 3rd Ed. Wiley, New York, 1986.
[2] P.-Å. Albertsson, B. Andersson, C. Larsson, and H.-E. Åkerlund, *Methods Biochem. Anal.* **28**, 115 (1982).
[3] H. Walter and D. Fisher, *in* "Partitioning in Aqueous Two-Phase Systems" (H. Walter, D. E. Brooks, and D. Fisher, eds.), pp. 378–414. Academic Press, Orlando, Florida, 1985.

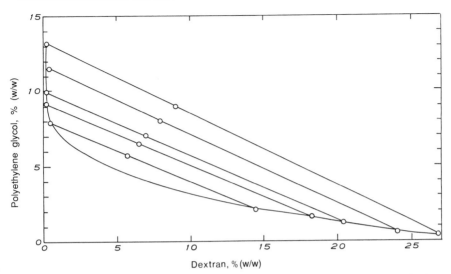

FIG. 1. Phase diagram of the Dextran 500–PEG 4000–water system at 0°. (From Ref. 1.)

change in the phase diagram) to give an optimal environment for the stability of cells and cell organelles. The contribution of the polymers alone to the tonicity is about 30 mOsm.[4] The surface tension of the interface is very small, 0.0001–0.1 dyn/cm.[1]

The polymer two-phase systems have a relatively long settling time (from 3 min to hours) due to the high viscosity of the phases and the small differences in density. The settling time may be reduced to less than a minute by low-speed centrifugation, provided the partitioned particles do not sediment.

Generally, the aqueous polymer two-phase systems are mild toward biological material. This is due to the high water content, low interfacial tension between the phases, and the presence of the polymers which protect enzymes and cell membranes against denaturation.

Factors Determining Partition of Cells and Cell Organelles

Partition of cell particles depends on several factors[1]: the surface properties of the cell particles, including surface charges, hydrophobicity of the surface, and specific receptors on the membrane surface; and the ionic

---

[4] S. Bamberger, D. Brooks, K. A. Sharp, J. Van Alstine, and T. J. Webber, in "Partitioning in Aqueous Two-Phase Systems" (H. Walter, D. E. Brooks, and D. Fisher, eds.), pp. 378–414. Academic Press, Orlando, Florida, 1985.

TABLE I
Composition of the Phases in Dextran 500–PEG 4000–H$_2$O Systems[a]

System	Total system			Bottom phase			Top phase		
	Dextran (%, w/w)	Polyethylene glycol (%, w/w)	H$_2$O (%, w/w)	Dextran (%, w/w)	Polyethylene glycol (%, w/w)	H$_2$O (%, w/w)	Dextran (%, w/w)	Polyethylene glycol (%, w/w)	H$_2$O (%, w/w)
A	5.70	5.70	88.00	14.38	2.57	83.05	0.40	7.92	91.68
B	6.50	6.50	87.00	18.21	2.05	79.74	0.18	9.13	90.69
C	7.00	7.00	86.00	20.31	1.81	77.88	0.20	9.96	89.84
D	8.00	8.00	84.00	23.96	1.49	74.55	0.35	11.53	88.12
E	9.00	9.00	82.00	26.80	1.26	71.94	0.26	13.20	86.54

[a] Described in Fig. 1.

composition of the phase system, the type and molecular weight of the polymers, and their concentration.

The different factors which determine partition can be explored separately or in combination to achieve an effective separation. We can also enhance some of the factors so that they will dominate the partition behavior. With respect to the partitioned substance the following types of partitions can be distinguished:

1. Size-dependent partition: molecular size or the surface area of the particles is the dominating factor.

2. Electrochemical: electrical potential between the phases is used to separate molecules or particles according to their charge.

3. Hydrophobic affinity: hydrophobic properties of a phase system are used for separation according to the hydrophobicity of molecules or particles.

4. Biospecific affinity: the affinity between sites on the molecules or particles and ligands attached to the phase polymers is used for separation.

5. Conformation dependent: conformation of the molecules and particles is the determining factor.

6. Chiral: enantiomeric forms are separated.

Formally, we can split the logarithm of the partition coefficient into several terms:

$$\ln K = \ln K^0 + \ln K_{el} + \ln K_{hfob} + \ln K_{biosp} + \ln K_{size} + \ln K_{conf}$$

where the subscripts el, hfob, biosp, size, and conf stand for electrochemical, hydrophobic, biospecific, size, and conformational contributions to the partition coefficient, and $\ln K^0$ includes other factors.

The different factors are more or less independent of each other, though none is probably completely independent of the others. For example, when we introduce a hydrophobic group on one of the phase-forming polymers, we may slightly influence both the distribution of ions and the electrical potential. When the molecular weight of a macromolecule is increased its net charge may also increase. When the conformation of a macromolecule is changed new groups of atoms are exposed to the surrounding phase.

Ionic Composition of the Phase System

Different ions have different chemical affinity for the two phases. An electrical potential is therefore created between the two phases when a salt is included in the phase system.[1] For cells and cell organelles which usually carry a net negative charge, the positively charged ions cause the cell particles to favor the upper phase in the following order: $Cs^+ \sim K^+ <$

$Na^+ < Li^+$. The negative ions have the same effect in the following order: $Cl^- \cong H_2PO_4 < HPO_4^{2-} \cong SO_4^{2-} \cong$ citrate. To get a maximal distribution into the upper phase one should therefore use Li_2HPO_4 or Li_2SO_4 and to get a maximal distribution into the lower phase one should use NaCl or KCl. Figure 2 demonstrates the effect of salt composition on the partition.

Polymer Concentration

An increase in polymer concentration of the dextran – PEG system will favor the collection of particles at the interface (Fig. 3).

Molecular Weight of Polymer

If for a given concentration of polymers, one of them is replaced by the same type of polymer having a smaller molecular weight, particles will favor the phase containing the polymer with decreased molecular weight. For example, cells and cell organelles will favor the upper phase more, if the molecular weight of PEG is decreased, and the bottom phase more, if the molecular weight of dextran is decreased. It seems as if the cell particles are relatively attracted by smaller polymer molecules and repelled by larger ones.

Choice of Phase System

Most applications have used the dextran – PEG phase system because the polymers are available in fairly well-defined fractions and the settling time of the phase system is relatively short.

Stock Solutions

Dextran; 20% (w/w). Dextran T-500 (Pharmacia, Uppsala, Sweden) is dissolved directly in water. It is desirable to heat the solution up to boiling to minimize subsequent microbiological contamination. The moisture content of the dextran powder can be supposed to be 10%. The exact concentration of dextran in the solution is determined polarimetrically as follows: 10 g of the solution is diluted to 25 ml; the optical rotation is determined with a tube 20 cm long, and the concentration is calculated using an $|\alpha|_D^{25}$ value of $+199°$.

PEG 6000 or PEG 4000; 40% (w/w) Carbowax 6000 (renamed Carbowax 8000) or Carbowax 4000 (renamed Carbowax 3350) (Union Carbide, New York) does not contain moisture and may be dissolved directly in water.

Trimethylaminopolyethylene glycol (TMA-PEG) and polyethylene

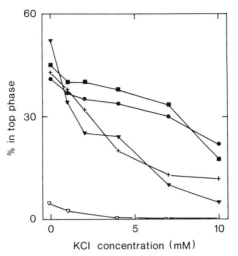

FIG. 2. Effect of salt composition on the partition of rat liver microsomes in a dextran–PEG phase system; 6% (w/w) Dextran 500, 6% (w/w) PEG 4000, 15 mM Tris–sulfate, pH 7.8. Marker enzymes in top phase as function of KCl. The curves represent the particles as follows (marker enzyme in parentheses): plasma membrane (5′-nucleotidase), ■; lysosomes (acid phosphate), ●; Golgi (N-acetylglucosamine galactosyltransferase), ▼; endoplasmic reticulum (NADPH–cytochrome c_2 reductase), ▽; protein, +.[18]

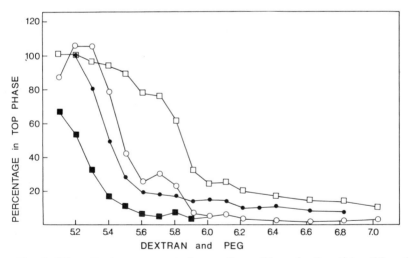

FIG. 3. Effect of polymer concentration on the partition of cell particles. When the concentration of the two polymers, dextran and polyethylene glycol (PEG), are each increased, the phase system is more and more removed from the critical point and the tie line of the phase diagram is increased. The particles favor more the interface, and the bottom phase and concentration in the top phase decreases. Symbols: ○, chloroplasts (class II); □, press-disintegrated chloroplast vesicles; ●, small stroma thylakoid vesicles (100K fraction); ■, inside-out thyalkoid vesicles.

TABLE II
PROTOCOL FOR PARTITION OF CELL ORGANELLES OR MEMBRANE VESICLES IN
PHASE SYSTEMS DIFFERING IN POLYMER CONCENTRATION AT
CONSTANT IONIC COMPOSITION[a]

Tube no.	Dextran 500 (20%)	PEG 4000 (40%)	Sucrose (1 M)	Sodium phosphate (0.1 M, pH 8)	H_2O (ml)	Sample (ml)
1	1.38	0.69	1.0	0.5	0.93	0.5
2	1.43	0.72	1.0	0.5	0.86	0.5
3	1.5	0.75	1.0	0.5	0.75	0.5
4	1.63	0.81	1.0	0.5	0.56	0.5
5	1.75	0.88	1.0	0.5	0.37	0.5

[a] 10 nM sodium phosphate, pH 8. All quantities in grams except the sample.

glycol sulfonate (S-PEG) are synthesized as described in Ref. 5. They are supplied by Aqueous Affinity (Johansson & Co. HB, Nygatan 9, 23200 Arlöv, Sweden), which also supplies other derivatives of PEG such as esters with fatty acids or derivatives with specific ligands.

Salt and buffer concentrations. Stock solutions of these are made in concentrations 10 times higher than that of the final concentration in the phase systems.

To find a system with a suitable composition for separation it is desirable to have a fractional distribution of the mixture between the upper phase and the interface plus lower phase, i.e., all material should not be entirely in the upper or in the lower phase. Since cells and cell organelles may have different requirements regarding the ionic composition for their stability, the different protocols given in Tables II–V are suitable for testing different particles.

Countercurrent Distribution

The principle and application of countercurrent distribution to cells and cell organelles is described in several reviews (see Ref. 1). A number of examples which illustrate the versatility and the selectivity of the method are given below.

[5] G. Johansson, A. Hartman, and P.-Å. Albertsson, *Eur. J. Biochem.* **33,** 379 (1973).

TABLE III
Protocol for Partition of Cell Organelles or Membrane Vesicles in a Phase System of 6% Dextran 500 and 6% PEG 4000 with Different Ionic Compositions[a]

Tube no.	Dextran 500 (20%)	PEG 4000 (40%)	Sucrose (1 M)	NaPB (0.1 M, pH 8)	LiPB (0.1 M, pH 8)	LiCl (0.1 M)	NaCl (0.1 M)	KCl (0.1 M)	H₂O (ml)	Sample (ml)
1	1.5	0.75	1.0	0.5	—	—	—	—	0.75	0.5
2	1.5	0.75	1.0	—	0.5	—	—	—	0.75	0.5
3	1.5	0.75	1.0	—	—	0.5	—	—	0.75	0.5
4	1.5	0.75	1.0	—	—	—	0.5	—	0.75	0.5
5	1.5	0.75	1.0	—	—	—	—	0.5	0.75	0.5

[a] PB, Phosphate buffer. All quantities in grams except sample.

TABLE IV
PROTOCOL FOR PARTITION OF CELLS IN PHASE SYSTEMS DIFFERING IN
POLYMER CONCENTRATION[a]

Tube no.	Dextran 500 (20%)	PEG 4000 (40%)	Sodium phosphate (0.5 M, pH 7)	H$_2$O (ml)	Sample (ml)
1	1.38	0.69	1.0	0.93	1.0
2	1.43	0.71	1.0	0.86	1.0
3	1.5	0.75	1.0	0.75	1.0
4	1.63	0.81	1.0	0.56	1.0
5	1.75	0.88	1.0	0.37	1.0

[a] At 100 mM sodium phosphate, pH 7. All quantities in grams except sample.

Applications

Phase partition can be used for several purposes in studying cells, cell organelles, and membrane properties, such as to (1) gain information of the surface properties of cells and cell organelles, (2) gain information on the heterogeneity of cell and cell organelle populations, (3) study changes in surface properties as a function of the cell's stage in its growth cycle, (4) separate subpopulations of cells and cell organelles, and (5) purify cell organelles and membrane vesicles.

Microorganisms

Several bacteria, algae, yeast cells, and spores have been studied by partition in the dextran–PEG phase system and by countercurrent distri-

TABLE V
PROTOCOL FOR PARTITION OF CELLS (ERYTHROCYTES) IN PHASE SYSTEMS WITH
CONSTANT POLYMER CONCENTRATION BUT WITH DIFFERENT IONIC COMPOSITIONS[a]

Tube no.	Dextran 500 (20%)	PEG 6000 (20%)	Sodium phosphate (0.4 M, pH 6.8)	NaCl (0.4 M)	H$_2$O (ml)	Packed erythrocytes (ml)
1	2	1.6	2.2	—	2.2	0.1
2	2	1.6	1.8	0.6	2.0	0.1
3	2	1.6	1.2	1.5	1.7	0.1
4	2	1.6	0.6	2.4	1.4	0.1
5	2	1.6	0.2	3.0	1.2	0.1

[a] All quantities in grams except samples.

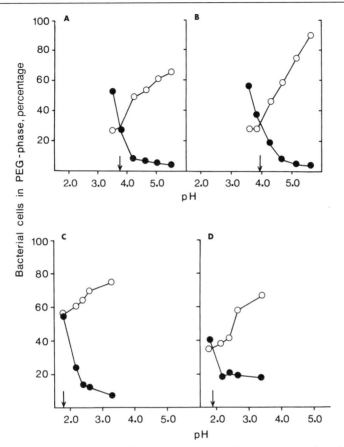

FIG. 4. The isoelectric point of cells and membrane vesicles can be determined by cross partition. (A–D) Bacterial species were partitioned in two different phase systems [(O) TMA–PEG and (●) S–PEG], having different interfacial potentials, at different pH. The pH curves for the two systems cross at the isoelectric point.[10] (A) Group A streptococcus, strain AR1; (B) group A streptococcus, strain HA1; (C) *Staphylococcus aureus,* strain Cowan I; and (D) *S. aureus,* strain HS1.

bution.[6] Closely related strains may be separated.[7] Mutants can be isolated.[8,9] The surface properties, notably charged groups and the hydropho-

[6] K. E. Magnusson and O. Stendahl, *in* "Partitioning in Aqueous Two-Phase Systems" (H. Walter, D. E. Brooks, and D. Fisher, eds.), pp. 416–452. Academic Press, Orlando, Florida, 1985.

[7] P.-Å. Albertsson and G. D. Baird, *Exp. Cell Res.* **28,** 296 (1962).

[8] O. Stendahl, C. Tagesson, and M. Edebo, *Infect. Immun.* **8,** 36 (1973).

[9] L. G. Wayne and H. Walter, *Antimicrob. Agents Chemother.* **5,** 203 (1974).

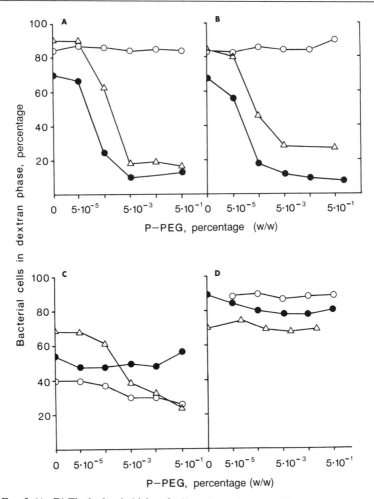

Fig. 5. (A–D) The hydrophobicity of cells and membrane vesicles can be determined by hydrophobic affinity partition, i.e., partition in a dextran–PEG phase system with and without replacement of some of the PEG by palmitoyl–PEG (P–PEG). The P–PEG in the top phase binds to hydrophobic domains on the cell surface and removes the cells from the lower dextran phase. (From Ref. 10.) (A) Group A streptococci, (B) group C streptococci, (C) group G streptococci, and (D) staphylococci.

bicity of the cell surface, have been studied by phase partition.[10] Different stages in the growth cycles of cells can be separated by countercurrent distribution.[11,12]

[10] H. Miörner, P.-Å. Albertsson, and G. Kronvall, *Infect. Immun.* **36**, 277 (1982).
[11] H. Walter, G. Ericksson, Ö. Taube, and P.-Å. Albertsson, *Exp. Cell Res.* **64**, 486 (1971).
[12] G. Pinaev, B. Hoorn, and P.-Å. Albertsson, *Exp. Cell Res.* **98**, 127 (1976).

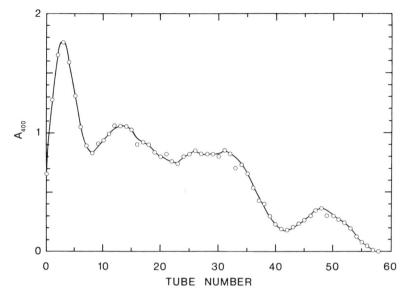

Fɪɢ. 6. Countercurrent distribution of bakers' yeast (59 transfers). Phase system: 8% (w/w) dextran 500, 8% PEG 4000, and with 1.1% (w/w) trimethylamino–PEG. (From Ref. 13.)

Figure 4 shows how the isoelectric point of different bacteria has been determined by cross partition. The two *Streptococcus* strains have an isoelectric point of 3.7–3.9 while that of the *Staphylococcus* strains is 1.9. The hydrophobicity of different strains is shown in Fig. 5.

Countercurrent distribution can resolve a population of cells into different subpopulations (Fig. 6).[13]

Animal Cells

Several animal cells have been studied by phase partition with respect to their surface properties. By comparing cells in phase systems with different electrical interfacial potential, conclusions can be drawn as to the surface charge of the cells.[3] The hydrophobicity of the cell surface has been studied by using PEG esterified with fatty acids.[14] Figure 7 shows the distribution of different erythrocytes as a function of different concentrations and types of PEG esters. Very small additions of PEG esters (0.01 – 0.1% of total PEG) is needed to transfer cells selectively from the lower dextran phase to the upper PEG phase. The effectiveness in transferring the

[13] G. Johansson, *Mol. Cell. Biochem.* **4**, 169 (1974).
[14] E. Eriksson, P.-Å. Albertsson, and G. Johansson, *Mol. Cell. Biochem.* **10**, 123 (1976).

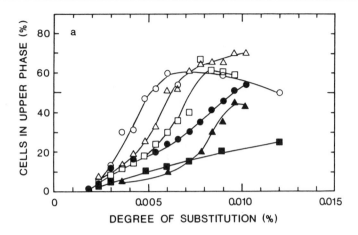

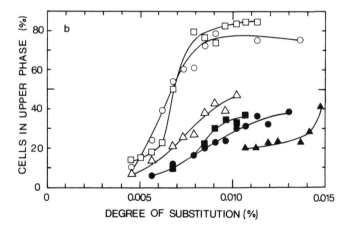

FIG. 7. Hydrophobic affinity partition of erythrocytes. The percentage of erythrocytes in the upper phase is a function of the degree of substitution of PEG with esterified hydrophobic groups (percentage PEG replaced by esterified PEG). The erythrocytes are from O, dog; △, rat; □, guinea pig; ●, sheep; ■, rabbit; and ▲, human. PEG was esterified with (a) palmitic acid, (b) oleic acid, (c) linoleic acid, (d) linolenic acid, and (e) deoxycholic acid.[14]

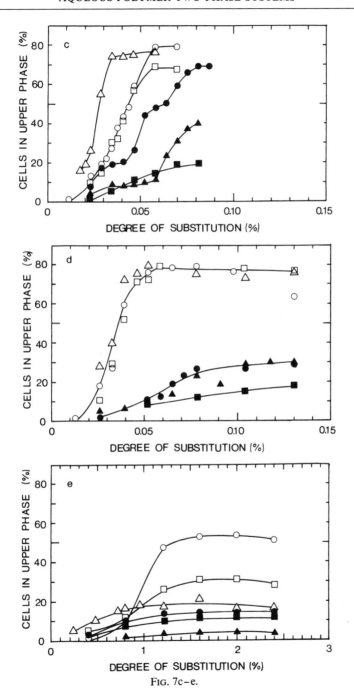

FIG. 7c–e.

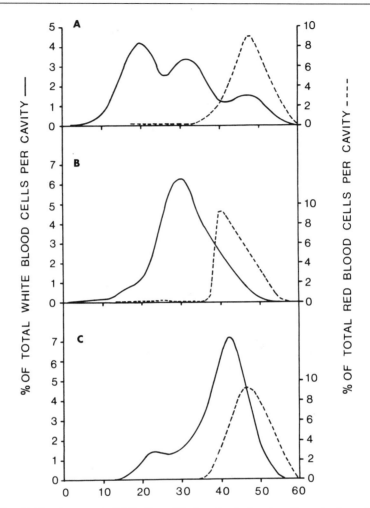

FIG. 8. Countercurrent distribution of leukocytes (---) and erythrocytes (——) from different organs [(A) spleen, (B) lymph node, and (C) thoracic duct] from the rat. (From Ref. 15.)

cells into the upper phase decreases with increasing unsaturation. PEG palmitate is about 1.8 times more efficient than PEG oleate and 6 times more efficient than PEG linolate or PEG linolenate. The erythrocytes fall into two groups independent of the fatty acid esterified with PEG — one with a relatively higher affinity for the PEG esters (rat, dog, and guinea pig) and one with relatively lower affinity (rabbit, sheep, and human). It is of interest that the latter group has a higher sphingomyelin content of the

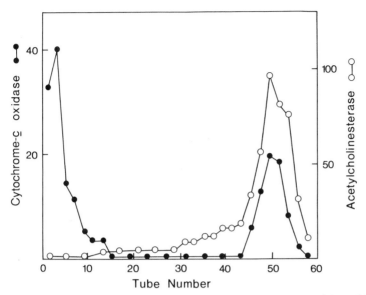

FIG. 9. Separation of free mitochondria (left peak) from synaptosomes (right peak) of rat brain by countercurrent distribution. (From Ref. 17.)

plasma membrane than the other group. Thus it seems to be the outer layer of the plasma membrane bilayer that is an important factor in determining partition.

Several different animal cells have been analyzed by countercurrent distribution.[3] Figure 8 shows a countercurrent distribution experiment with leukocytes.[15] The high selectivity of phase partition was demonstrated by countercurrent distribution of erythrocytes from two different individuals.[16] Erythrocytes from different individuals have small but significant differences in partition which reflect variations in cell surface properties.[16] Any pair of erythrocyte samples from two individuals, except those from identical twins or triplets, displayed these differences, indicating that the differences are probably genetic.

Cell Organelles and Membrane Vesicles

Organelles from both animal and plant cells have been purified by phase partition and their homogeneity has been analyzed by countercurrent distribution. The applications include, among others, mitochondria,[17]

[15] P. Malmström, K. Nelson, A. Jönsson, H.-O. Sjögren, H. Walter, and P.-Å. Albertsson, *Cell. Immunol.* **37,** 409 (1978).
[16] H. Walter and E. J. Krob, *Biochem. Biophys. Res. Commun.* **120,** 250 (1984).
[17] M. J. Lopez-Perez, G. Paris, and C. Larsson, *Biochim. Biophys. Acta* **635,** 359 (1981).

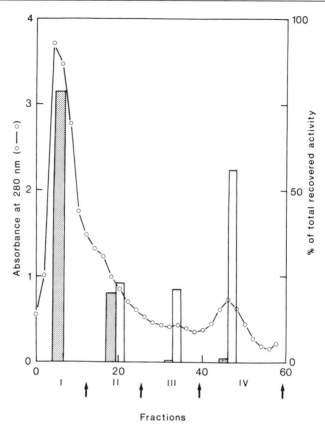

FIG. 10. Countercurrent distribution of a microsomal fraction from oat root. The peak to the right represents plasma membrane (glucan synthase II is a marker for plant plasma membrane). (□) Glucan synthase II and (■) cytochrome-*c* oxidase. (From Ref. 24.)

microsomes,[18] chloroplasts,[19] inside-out thylakoid vesicles,[20,21] chloroplast fragments,[22,23] and plasma membrane.[24-26]

[18] P. Gierow, B. Jergil, and C. Larsson, *Biochem. J.* **235,** 686 (1986).

[19] C. Larsson, C. Collin, and P.-Å. Albertsson, *Biochim. Biophys. Acta* **245,** 425 (1971).

[20] B. Andersson, H.-E. Åkerlund, and P.-Å. Albertsson, *Biochim. Biophys. Acta* **423,** 122 (1976).

[21] H.-E. Åkerlund, B. Andersson, and P.-Å. Albertsson, *Biochim. Biophys. Acta* **449,** 525 (1976).

[22] P.-Å. Albertsson, *Physiol. Veg.* **23,** 731 (1985).

[23] A. Melis, P. Svensson, and P.-Å. Albertsson, *Biochim. Biophys. Acta* **850,** 402 (1986).

[24] S. Widell, T. Lundberg, and C. Larsson, *Plant Physiol.* **70,** 1429 (1982).

[25] P. Kjellbom and C. Larsson, *Physiol. Plant.* **62,** 501 (1984).

[26] C. Larsson, *in* "Modern Methods of Plant Analysis" (H. F. Linskens and J. F. Jackson, eds.), Vol. 1, pp. 85–104. Springer-Verlag, Berlin and New York, 1985.

As with cells, phase partition can be used both for characterizing the surface properties or organelles and for separating different organelles from each other.

Cross partition has been used for determining the isoelectric point of mitochondria,[27] peroxisomes,[28] chloroplasts,[29] and chloroplast membrane fragments.[30] Figure 9 shows the separation of synaptosomes from mitochondria by countercurrent distribution of a centrifugal fraction from rat brain. Figure 10 shows the purification of plasma membranes from a microsomal fraction obtained from plant cells. Plasma membranes, both of plant[25] and animal origin,[18] can be effectively purified by a few batch partition steps.

[27] I. Ericson, *Biochim. Biophys. Acta* **356**, 100 (1974).
[28] S. Horie, H. Ishii, H. Nakazawa, T. Suga, and H. Orii, *Biochim. Biophys. Acta* **585**, 435 (1979).
[29] H. Westrin, V. P. Shanbhag, and P.-Å. Albertsson, *Biochim. Biophys. Acta* **732**, 83 (1983).
[30] H.-E. Åkerlund, B. Andersson, A. Persson, and P.-Å. Albertsson, *Biochim. Biophys. Acta* **522**, 238 (1979).

[27] Flow Cytometry: Rapid Isolation and Analysis of Single Cells

By BRUCE D. JENSEN and PAUL KARL HORAN

Introduction

Although more than 10 years have passed since the introduction of the commercial flow cytometer and sorter, these instruments continue to be highly complex and require a considerable amount of experience for their proper operation. The complex nature of the flow cytometer and sorter makes it impossible to provide complete instructions in this chapter for the operation of such an instrument. The general principles of operation are the same for all flow cytometers and sorters, and it is these general operating principles which we will describe. However, one should be aware that the ways in which the operating principles are satisfied are often specific to each instrument. For these reasons, we will restrict ourselves to a discussion of the basics with only occasional forays to describe interesting applications and features of specific instruments.

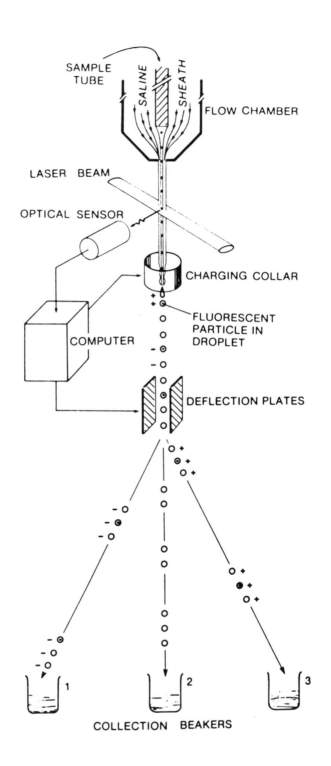

SAMPLE TUBE

SALINE

SHEATH

FLOW CHAMBER

LASER BEAM

OPTICAL SENSOR

COMPUTER

CHARGING COLLAR

FLUORESCENT PARTICLE IN DROPLET

DEFLECTION PLATES

COLLECTION BEAKERS

Following our discussion of the instrumentation, we will present an in-depth discussion of a number of commonly used sample preparation and staining techniques, which are applicable to all cell sorters.

What is a Flow Cytometer and Sorter?

Sample Analysis

A flow cytometer is an instrument which provides for the serial delivery of individual cells in flow for analysis of parameters, such as fluorescence intensity and light scatter. If the flow cytometer is also a sorter, provision is made for isolating individual cells from the flow stream based on analysis parameters. The basic components of such an instrument include a sample handling system, light source, optical sensors, analog measurement circuits, analog-to-digital convertors, a computer, and the sorting apparatus. These components and their spatial relationships are illustrated in Fig. 1.

Sample Handling. Environment control. Most commercial instruments make available the capability to control the environment of the sample. Environmental control usually takes the form of temperature control and sample mixing. Temperature control is very important when working with samples in which changes in metabolism can adversely affect the parameter being measured. Such parameters include membrane potentials, cellular pH, calcium levels, pyridine nucleotide levels, etc. Temperature control is also important when specific cell types capable of capping and shedding are labeled with immunofluorescent markers. Controlling the temperature usually consists of pumping temperature-controlled water through an aluminum sample holder. Some, but not all instruments, provide sample mixing as well as temperature control. Mixing is usually accomplished by agitating the sample rather than stirring, since stirring can damage delicate cells and can also result in nonrandom sampling. Mixing is important to environment control, since cells which settle out become anoxic almost immediately.

Sample transport and delivery. Many commercially available flow cytometers provide a sample transport system whose motive force is gas pressure. Commonly used gases include compressed air and nitrogen. Where high sample throughput is important, a syringe drive sample transport is used instead of a gas pressure-driven transport. However, this type

---

FIG. 1. Generalized schematic of a flow cytometer/cell sorter. Cells are injected into the flow stream and carried past the laser in single file. Quantitative measurements are made on which sort decisions are based. Cells satisfying the sort criteria are charged at the charging collar and deflected by an electric field into the appropriate collection beaker.

of system does not easily allow for temperature control and mixing. In all cases, the sample is driven through tubing into a flow chamber and introduced into a rapidly moving saline sheath where there is no mixing of the cell stream and sheath (laminar flow). The laminar flow of the sheath fluid through a typical 50- to 100-μm orifice causes the sample stream to be hydrodynamically focused, resulting in the cells flowing out of the orifice in single file and constrained primarily to the center of the stream. The stream with the entrained cells flows past a light source. The principle of hydrodynamic focusing guarantees that each cell will experience the same intensity and duration of excitation light. Therefore, the fluorescence intensity of each cell is proportional to the amount of the fluorochrome associated with the cell. In some instruments, the sample stream is enclosed in a quartz cuvet until it passes the light source, while others present the stream in air to the light source. Either system works well, but if the instrument is capable of sorting, it must have a "stream in air" configuration at some point.

Microsampling. With the advent of robotics, some manufacturers have developed the capability to automatically access small samples (100 μl), either singly or in multiwell plates. This type of apparatus is very convenient for routine analysis of clinical samples where the machine settings do not have to be changed during the run. It is also very useful in situations where sample size is problematic.

Excitation. The sample transport system reproducibly delivers the sample to a point in space where it intersects an excitation light beam. The excitation light is either elastically scattered or excites fluorescence from the cell. The requirements for an exciting light to be used in flow cytometry are stability, uniformity, wavelength availability, and intensity. There are two things which affect the quality and quantity of exciting light delivered to the sample, the light source, and the optics which focus the light.

Light sources. Two different types of light sources have been used in flow cytometry, lasers, and mercury arc lamps. Typical laser light sources include argon ion lasers, krypton ion lasers, helium–cadmium lasers, and helium–neon lasers. A list of these commonly used light sources along with the wavelengths available from each is presented in Table I. Most helium–cadmium and helium–neon lasers used are low powered, providing only a few milliwatts output power. This necessitates their use on systems with high-efficiency collection optics.

Optics. The beam-focusing/shaping optics are at least as important to a uniform intense excitation source as is the light-generating component. The two types of beam-shaping optics which have frequently been used include spherical and crossed cylindrical lenses. The spherical lens configuration focuses the beam to a single point, with the focal point set off axis.

TABLE I
COMMON LIGHT SOURCES USED FOR FLOW CYTOMETRY

Light source[a]	Wavelengths available (nm)
Mercury arc lamp	313–334, 365, 405, 436, 546, 578
Argon ion laser	351–363, 457, 466, 473, 476, 488, 496, 502, 514
Krypton ion laser	350–356, 406–415, 531, 568, 647, 676, 752
Helium–cadmium laser	325, 441
Helium–neon laser	632

[a] For the high-powered argon and krypton ion lasers, only those wavelengths which are available at 100-mW output power or greater are listed.

The result is a high-intensity beam path whose length depends on the focal length of the lens and how far the focal point is from the stream. The advantage of this setup is that it provides the maximum intensity available from the light source. The disadvantage of the spherical lens configuration is that the light intensity falls off very rapidly in a direction orthogonal to the stream–laser intersection point. Small fluctuations in the position of cells as they traverse the beam result in significant variations in the amount of exciting light that each cell experiences. The result can be reasonably large sample coefficients of variation (CVs), especially when sample aggregates disturb the laminar flow properties of the stream.

The most commonly used optical setup today is the crossed cylindrical lens configuration. This optical configuration shapes the light beam so that it has both long and short axes. Its long axis is orthogonal to the stream–laser intersection point. The advantage of this configuration is that the light intensity along the long axis of the beam is quite uniform; therefore, differences in excitation intensity due to positional variations between cells within the sample stream are minimized. The main disadvantage of this optical orientation is that a large part of the available excitation intensity is thrown away for the sake of uniform sample illumination. Inefficient use of the available excitation light is often a minor consideration, particularly if the light source is a high-intensity laser where there is more light available than needed.

Signal Types and Their Detection. Signal detection begins with the light collection optics, which is made up of lenses and a pinhole/aperture (see Fig. 2). The purpose of the first lens in the detection system is to focus the fluorescent light from the laser–stream intersection point on the aperture (aperture not shown in figure). This aperture minimizes interfering signals, such as stray room light or the Raman scattered light due to the laser interacting directly with the stream. After the pinhole, the light is divided

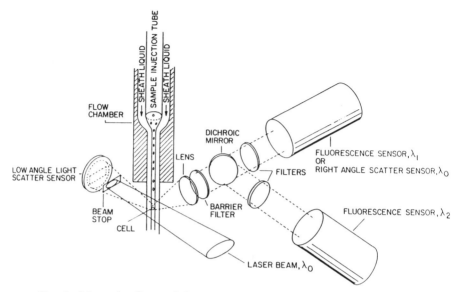

FIG. 2. Schematic of general detector arrangement. Forward-angle scattered light is measured at a photodiode. Fluorescent and 90° scattered light is collected at 90° to the laser–stream intersection point. First, a lens focuses the light on an aperture (not shown). The signal is then split by dichroic mirrors and filtered into appropriate wavelength ranges before encountering the photomultiplier tube.

and filtered by a series of dichroic mirrors and optical filters. The light from specific wavelength regions is reflected or transmitted by specific filter arrangements into individual photomultiplier tubes. Figure 2 illustrates the detector configuration for a simple flow cytometer with one light scatter sensor and two photomultiplier tubes.

There are three types of detectors used on various flow cytometers. The one which is common to virtually all flow cytometers is the photomultiplier tube (PMT). PMTs are used to quantitate fluorescent light or 90° angle scattered light, where low light levels are encountered and high sensitivity is required. Other types of sensors include photodiodes and impedance measurement electronics. Each of these is used for specific applications and will be discussed below in conjunction with those applications.

Multiple color fluorescence. Fluorescence is by far the most widely measured parameter on the flow cytometer. With the development of fluorochrome-conjugated monoclonal antibodies, fluorescent biochemical probes, and nucleic acid stains an almost unlimited number of biophysical and biochemical problems can be addressed in a quantitative fashion. Fluorescent light is collected at right angles to the laser–stream intersection point, and is fractionated by dichroic mirrors and barrier filters into specific wavelength regions, as was previously described.

Because the flow cytometer possesses more than one PMT, it is possible to quantitate multiple fluorescent signals, simultaneously. Quantitation of multiple probes is dependent on being able to split the total signal such that each PMT sees the signal from only one fluorochrome. The best possible situation is where all of the fluorochromes can be excited by the same excitation source, simultaneously, and the emission maxima of all probes are far enough apart that simple optical filtering is sufficient to separate the signals. Often this is not the case. Many investigators have resorted to using multiple excitation sources and special timing circuitry to simultaneously quantitate signals whose excitation peaks do not overlap sufficiently or whose emission peaks overlap too much.[1,2]

An alternative to using multiple lasers and special timing circuitry is color compensation.[3] In this procedure, the contaminating overlap signal in a secondary emission window is subtracted from the total signal in that window based on the value of the overlapping fluorochrome in its primary emission window. The color compensation procedure is analogous to the correction procedure used in dual-label liquid scintillation counting.

Light scatter at a 90° angle. This parameter is reasonably nonspecific and gives a measure of the granularity or surface ruffling of the particle in the beam. It is in fact a measure of the refractility of the particle and generally shows up even slight differences or changes in the internal or external organization between cells. Like fluorescence, this parameter is collected at right angles to the laser–stream intersection point and a PMT is the quantitation device. A dichroic mirror is used to separate this signal from fluorescent light, before the fluorescence is quantitated, and a narrow bandpass filter is generally used to further clean up the signal.

Low-angle light scatter. This parameter is also called forward-angle light scatter, or FALS. It is collected over a narrow angle by a photodiode positioned on either side of the beam dump, which absorbs the transmitted laser light. The low-angle light scatter signal is so intense that the relatively insensitive photodiode is sufficient to give good quantitation (see Fig. 2 for the optical configuration). The configuration shown is for a flow cytometer whose laser–stream intersection point is in air. This configuration results in a very large amount of light scattered from the stream itself. The light scattered from the stream intersects the photodiode as a narrow band perpendicular to both the laser and stream. It is very intense and could

[1] M. Stohr, "Pulse Cytophotometry," 2nd Int. Symp., pp. 39–45. European Press, Ghent, Belgium.

[2] H. A. Crissman and J. A. Steinkamp, *Cytometry* **3**, 84 (1982).

[3] M. R. Loken, D. R. Parks, and L. A. Herzenberg, *J. Histochem. Cytochem.* **25**, 899 (1977).

damage the photodiode, so a beam stop is placed in front of the photodiode.

Light scattered from the sample is collected over an angle which varies with different flow cytometers, but is generally between 2 and 12°. This parameter has been found to vary monotonically with the volume of the cell; narrower collection angles give better correlations with volume.

Absorbance and axial light loss. These two parameters are often confused. Although attempts have been made to use absorbance, it is little used today because of the considerable interference from scattered light.[4] The axial light loss parameter is available on a number of instruments. This measurement is made with a narrow beam and light is collected over a very narrow angle axial with the beam. Therefore most scattered light is eliminated from the analysis and the resulting signal is related to the cross-sectional area of the cell.[5]

Electronic volume. Some instruments offer the feature of electronic volume sizing.[6] These instruments measure the impedance of the cell as it traverses an orifice upstream of the laser intersection point and before it enters the jet stream for sorting. While this is not an optical signal, it can be correlated with optical measurements made downstream.

Pulse shape. The information obtainable by analyzing the shape of the pulse as the cell traverses the laser beam is dependent on the relative sizes of the cell and laser. If the beam is large with respect to the cells of interest, both the pulse height and area will reflect the total fluorescence intensity of the cell. However, if the beam is small with respect to the cell diameter, only the pulse area will reflect the fluorescence intensity of the cell. The pulse height indicates the relative dye distribution or the fluorescence topography of the sample. This parameter can be very useful for scanning chromosomes in flow, or if one is interested in the relative distribution of the dye between the nucleus and the cytoplasm.[7,8]

Combinations. The great strength of the flow cytometer lies in its ability to serially examine single cells, instead of measuring bulk population fluorescence. Most flow cytometers provide a minimum of four to six data channels. This makes it possible to accumulate combinations of the above-mentioned parameters, simultaneously, resulting in multiparameter correlations which can be used to identify subpopulations in a cell suspension. These multiparameter correlations can then be used as the basis for cell sorting.

[4] L. A. Kamentsky, M. R. Melamed, and H. Derman, *Science* **150,** 630 (1965).

[5] L. A. Kamentsky, *Adv. Biol. Med. Phys.* **14,** 93 (1973).

[6] W. H. Coulter, *Proc. Natl. Elect. Conf.* **12,** 1034 (1956).

[7] T. K. Sharpless and M. R. Melamed, *J. Histochem. Cytochem.* **24,** 257 (1976).

[8] J. A. Steinkamp, *Rev. Sci. Instrum.* **55,** 1375 (1984).

Signal Processing. Each of the signal types mentioned above is detected by either a PMT, a photodiode, or in the case of volume estimates, impedance measurement circuitry. In all cases, the result is an analog voltage signal proportional to the parameter of interest. This analog voltage signal is then further processed by an analog-to-digital convertor (ADC) before being input to a computer. The computer then compiles the results from each of the measurements in a correlated fashion. These results are then available to the investigator and can also be used as the basis for a sort decision.

Electrostatic Cell Sorting

Electrostatic cell sorting has become the accepted method of achieving cell separation in flow. There are four components to this process; the sort decision, sort timing and droplet formation, droplet charging, and droplet deflection/collection.

The Sort Decision. The sort decision is based on the sample analysis process, previously described. This decision can be based on any of the parameters or combinations of the parameters which have been discussed. The flow cytometrist must first analyze a small fraction of the cell suspension for the parameters of interest. He or she must then determine which cells are to be selected, based on the level of expression of each of the parameters of interest. This selection process consists of setting software gates in the flow cytometer's computer. Therefore, any cell which falls within the gated regions for each of the parameters of interest will be selected.

Sort Timing and Droplet Formation. Any flow cytometer which is capable of electrostatic sorting has a jet in air configuration after the analysis point. The natural tendency of flowing water in the stream is to break up into nonuniform droplets. To make the droplet formation more uniform and regular in time, most cell sorters have a piezoelectric crystal oscillator coupled to the flow tip. The oscillating crystal causes the formation of a standing wave in the stream with uniformly spaced nodes from which uniform droplets are formed at the crystal drive frequency. The droplet volume is a function of the crystal frequency and flow rate. The droplet break-off point is determined by the power applied to the crystal and therefore the amplitude of the oscillation. The uniformity of the droplet is important in sorting so that the deflection force will result in equal deflections for all droplets. Given that the cell concentration flowing through the system is low enough, very few droplets will contain more than one cell. In fact, under standard sorting conditions as few as 1 in 60,000 drops may contain a cell. This way, the probability of any drop containing more than one cell (coincident events) is minimized.

Sort Timing and Droplet Charging. Once a cell of interest has been analyzed and a positive sort decision is made, a voltage pulse is applied to the stream before the droplet containing the cell breaks away from the stream. This pulse causes migration of charge toward or away from the forming droplet, based upon the polarity of the applied voltage. At the point of separation, a net charge is left on the droplet. Correctly timing the charging pulse is critical so that multiple drops do not receive partial charges, resulting in a range of stream deflection angles when sorting. The phase relationship between the crystal drive and the nodes prior to droplet formation is arbitrary, but constant. A time delay corresponding to this phase relationship must be adjusted so that each droplet receives the same charge. The phase delay is set correctly when a single deflected sort stream is present. There are two methods of applying a voltage pulse to the stream. One involves direct electrical contact with the stream and the other uses a charging collar (see Fig. 1) to apply a voltage pulse by capacitive coupling through an electric field.[9]

The sample stream is moving at a speed between 1 and 10 m per second. Consequently, the timing of the voltage pulse is critical in sorting, if the correct droplet is to be charged. In setting up an instrument, the time delay should be optimized by sorting microspheres onto a microscope slide using various time delay settings. The spheres can then be counted and the setting which gives the highest number of spheres sorted is the optimal setting. While this time delay is optimal for most spheres, slight variations in the flow rate cause this delay to be suboptimal for a fraction of the spheres. One can improve the sorting efficiency by sorting more than one droplet. Three-drop sorting is often performed to compensate for slight flow rate variations.

After sorting is complete, the efficiency of the sort should be assessed. If the sorted samples are large enough, a portion of each should be reanalyzed on the flow cytometer to determine the purity of the sample. This purity is the fraction of the sorted population which falls within the computer's sort gates on reanalysis. If the sorted populations are too small to waste, microsphere sorting may again be used to determine whether the instrument has remained stable during the sort.

Droplet Deflection and Sample Collection. After the droplet has been charged, it travels past a pair of deflection plates (see Fig. 1) to which a high voltage has been applied. The high electric field between the plates causes the charged droplets to be deflected toward one or the other of the plates, depending on the polarity of the droplet's charge. The trajectory of the

---

[9] P. K. Horan, K. A. Muirhead, and S. E. Slezak, *in* "Flow Cytometry and Sorting" (M. Melamed and M. Mendelsohn, eds.), 2nd Ed. Liss, New York, 1989 (in press).

droplets is adjusted by varying the voltage on the deflection plates until the droplets fall into the collection vessel.

Biological Parameters for Analysis and Sorting

Immunofluorescence

The development of fluorophore-conjugated monoclonal antibodies for the identification of leukocyte surface differentiation antigens has caused immunofluorescence labeling to become the most widely used staining technique applied to flow cytometry, both in clinical and research laboratories. Monoclonal antibodies to leukocyte differentiation antigens have been made at an ever-increasing rate and have caused the development of a large number of different nomenclatures. Recently, binding data for many of these antibodies have been compiled and have lead to the identification of 15 clusters of differentiation antigens. A common nomenclature has been proposed and published based on these 15 differentiation clusters.[10]

The great success in the identification of different leukocyte subsets has stimulated many laboratories throughout the world to produce antibodies to both surface and intracellular markers in an ever-expanding variety of cells. This has lead to the use of monoclonal antibodies for the identification of many cell types of both normal and neoplastic origin.[11-14]

The great specificity of monoclonal antibodies has also lead to their use by the biochemist. The stoichiometric nature of the binding reaction allows the biochemist to detect the level of a specific intracellular or extracellular enzyme (see last entry of Table II). Even small differences between cell types or small changes induced in a single cell type can be accurately quantitated.

One of the strong advantages in using immunofluorescence to detect cellular enzyme levels lies in the fact that it is conceptually easier to develop a monoclonal antibody to an enzyme than to develop a fluoro-

[10] J. Immunol. **134**, 659 (1985).
[11] G. Kohler and C. Milstein, Nature (London) **256**, 495 (1974).
[12] R. H. Kennett, Z. L. Jonak, and K. B. Bechtol, in "Monoclonal Antibodies. Hybridomas: A New Dimension in Biological Analyses" (R. H. Kennett, T. J. McKearn, and K. B. Bechtol, eds.), pp. 155–170. Plenum, New York, 1980.
[13] D. Colcher, J. Esteban, and F. Mornex, this series, Vol. 121, p. 802.
[14] D. W. Mason, R. J. Brideau, W. R. McMaster, M. Webb, R. A. White, and A. F. Williams, in "Monoclonal Antibodies. Hybridomas: A New Dimension in Biological Analyses" (R. H. Kennett, T. J. McKearn, and K. B. Bechtol, eds.), pp. 251–274. Plenum, New York, 1980.

TABLE II

COMMON FLUORESCENT SUBSTRATES USED IN FLOW CYTOMETRIC
ENZYME ANALYSIS

Target enzyme	Substrate[a]
Nonspecific esterase	Fluorescein diacetate
	BHNA acetate
	α-Naphthyl acetate + fast blue BB
Myeloperoxidase	H_2O_2 + 4-chloro-1-naphthol
β-Glucuronidase	BHNA -β-D-glucuronide
Acid phosphatase	Methylumbelliferone phosphate
	3-O-Methylfluorescein phosphate
	BHNA phosphate
	Hydroxyflavone phosphate
Alkaline phosphatase	α-Naphthol phosphate
	BHNA phosphate + CDND
	BHNA phosphate
Cathepsin B	CBZ-Ala-Arg-Arg-MNA + NSA
Dihydrofolate reductase	Methotrexate – fluorescein
Dipeptidyl-aminopeptidase I	Pro-Arg-MNA + NSA
Dipeptidyl-aminopeptidase II	Lys-Ala-MNA + NSA
γ-Glutamyltransferase	γ-Glutamy-4-methoxy-2-naphthylamide + NSA
Lactate dehydrogenase	Sodium lactate + NAD
Terminal deoxynucleotidyl- transferase	Fluorescein-conjugated antibody

[a] BHNA, 7-Bromo-3-hydroxy-2-naphtho-o-anisidide; CDND, 4-chloroto-luene-1,5'-diazonium naphthalene disulfonate; MNA, 4-methoxy-2-naph-thylamide; NSA, 5-nitrosalicylaldehyde; and fast blue BB, 4-benzoylamino-2,5-diethoxybenzene diazonium chloride.

genic substrate, given that one can isolate sufficient quantities of the pure enzyme for monoclonal antibody production and screening. Further, one can produce antibodies to enzymes whose purpose is not yet known or for which there are no good histochemical or biochemical methods for detection. An advantage of the immunofluorescence approach to enzymology is that antibodies can be generated to substrates, products, or the enzyme itself, making it possible to detect the presence of the enzyme and its functional state.

Enzyme Activity and Concentration

There are two basic techniques used to detect enzymes in cells, immunofluorescence and histochemical methods. Table II lists a number of enzymes and the methods used to detect them.[15,16] The histochemical

[15] H. S. Kruth, *Anal. Biochem.* **125,** 225 (1982).
[16] F. Dolbeare, *in* "Oncodevelopmental Markers: Biologic, Diagnostic, and Monitoring Aspects" (W. H. Fishman, ed.), pp. 207–217. Academic Press, New York, 1983.

methods which have been used employ either absorbance or fluorescence for detection. Absorbance-type probes have not been of great utility because they do not provide quantitative results, but only qualitative information. The reason for this has been previously discussed. The greater sensitivity of fluorogenic probes and the quantitative nature of the results provided make this type of histochemical technique more desirable. However, there are a number of problems with the existing techniques which have caused them to be of limited use. Many of the probes have limited permeability to the cell membrane and can generate artifacts based on this permeability. In these cases, fixation of the cells is necessary. Fixation can also adversely affect the analysis if the fixative damages the enzymes of interest. In a large number of cases, the fluorescent substrate formed in the enzyme reaction rapidly diffuses out of the cells. In these cases it has been necessary to look for ways of coupling the reaction product to compounds which trap the indicator in the cell. Diazo salts have been useful for this purpose.

Metabolic Activity

The basis for many biochemical enzyme assays is the UV-excited fluorescence of reduced pyridine nucleotides [NAD(P)H]. The "autofluorescence" of this natural substrate has been recognized for a number of years and has been exploited in bulk metabolic studies. These measurements have been extended to the single-cell level using a flow cytometer with the intention of monitoring the metabolic state of the cell under completely noninvasive conditions. Most recently, autofluorescence measurements have been carried out by exciting fluorescence in two different wavelength regions, simultaneously. UV excitation was used for the detection of reduced pyridine nucleotides as described above. By employing a slight time delay, it was possible to excite fluorescence from oxidized flavin nucleotides and flavoproteins at the same time.[17]

DNA and/or RNA Content

The first widespread use of flow cytometric fluorescence analysis for biological purposes was in the measurement of DNA content in whole cells and nuclei. This remains one of the most widely used application of flow cytometry today.[8,15,18] A number of the probes used to measure DNA content have specificity for RNA as well, making them useful for measuring either kind of nucleic acid, depending on how the cells are prepared

[17] B. Thorell, *Cytometry* **4**, 61 (1983).
[18] H. M. Shapiro, *Cytometry* **3**, 976 (1983).

TABLE III
DNA- AND RNA-SPECIFIC DYES USEFUL FOR
FLOW CYTOMETRIC ANALYSIS

Dye	Target
Acriflavine	DNA
Auromine O	DNA
Hoechst 33258	DNA
Hoechst 33342	DNA
DAPI	DNA
DIPI	DNA
Mithramycin	DNA
Chromomycin A$_3$	DNA
Olivomycin	DNA
Ethidium bromide	DNA/RNA
Propidium iodide	DNA/RNA
Acridine orange	DNA/RNA
Pyronin Y	RNA
Di-O-C$_1$ (3)	RNA
Thioflavin T	RNA

prior to staining. Table III is a list of probes which have been used for the measurement of DNA and/or RNA content. Each of these probes binds stoichiometrically to the listed target and the nature of the binding is different in many cases. For example, some bind covalently (acriflavine and auromine O), while others bind noncovalently (ethidium bromide, Hoechst 33258, mithramycin). The three which show specificity for both DNA and RNA (propidium iodide, ethidium bromide, and acridine orange) are intercalators and are actually specific for double-stranded nucleotides. Measurement of DNA with these probes requires elimination of double-stranded RNA. Acridine orange is an exception to this rule. It is a metachromatic dye showing green fluorescence when intercalated into double-stranded nucleotides and red fluorescence when bound to single-stranded nucleotides.[19] Groove binding dyes show specificity for (A–T)-rich DNA regions (Hoechst 33258/33342, DAPI, DIPI) or (G–C)-rich regions (mithramycin, chromomycin A$_3$, olivomycin). Because of the differences in specificity between these probes one can expect that the measurement of DNA content will depend on the probe one chooses for the measurement. Variations in A–T and G–C contents can significantly affect the measured total DNA content. If this is a problem, multiple probes should be used and the results compared. Another problem is that

[19] Z. Darzynkiewicz, F. Traganos, T. Sharpless, and M. Melamed, *Exp. Cell Res.* **95,** 143 (1975).

chromatin proteins may interfere in the binding of some probes (acridine orange), making it necessary to eliminate these proteins before staining of the cells.

The incorporation of bromodeoxyuridine (BrdUrd) into cells in the DNA synthetic phase of growth has been useful in identifying and quantitating S phase cells. The most widely used technique is based on the development of a monoclonal antibody which binds to BrdUrd.[20] The binding of this antibody to DNA requires that the histone proteins are removed and the DNA be denatured. Obviously, the cells must be fixed for this to occur; therefore, viable sorting based on this parameter is not possible. However, sorting for further morphological or biochemical analysis can be achieved.

DNA stains are also useful for flow karyotyping and chromosome sorting. However, the very small size and low fluorescence intensity of individual chromosomes make optical alignment and the degree of hydrodynamic focusing very important. To date it has proved possible to resolve 18 out of 24 human chromosomes.[21] Because of this, a collaborative effort has been arranged between the Lawrence Livermore Laboratory and the Los Alamos National Laboratory for the purpose of establishing a human chromosome bank for use in human gene mapping.

Table III also lists a number of dyes which have been used for RNA analysis. Correlation of both RNA and DNA content simultaneously has been used to better define the cycle and the events which take place in the individual compartments of the cycle.[22]

Protein Content

Protein content can be measured by either covalent or noncovalent dyes, much like DNA and RNA. In fact, total cellular RNA content over the cell cycle seems to correlate very well with the total cellular protein content. Table IV lists different dyes which have been used to measure protein content.[23-25]

[20] F. Dolbeare, H. Gratzner, M. G. Pallavicini, and J. W. Gray, *Proc. Natl. Acad. Sci. U.S.A.* **80**, 5573 (1983).

[21] A. V. Carrano, M. A. Van Dilla, and J. W. Gray, *in* "Flow Cytometry and Cell Sorting" (M. R. Melamed, P. L. Mullaney, and M. L. Mendelsohn, eds.), pp. 421–451. Wiley, New York, 1979.

[22] Z. Darzynkiewicz, F. Traganos, and M. Melamed, *Cytometry* **1**, 98 (1980).

[23] D. A. Freeman and H. A. Crissman, *Stain Technol.* **50**, 279 (1975).

[24] M. Stohr, M. Vogt-Schaden, M. Knoblock, R. Vogel, and G. Futterman, *Stain Technol.* **53**, 205 (1978).

[25] H. M. Shapiro, *J. Histochem. Cytochem.* **25**, 976 (1977).

4-Acetamido-4′-isothiocyanatostilbene-2,2′-disulfonic acid
Brilliant sulfaflavine
5-Dimethylamino-1-naphthalene sulfonylchloride
Fluorescein isothiocyanate
Rhodamine 101 isothiocyanate (Texas Red)
Sulforhodamine
Tetramethylrhodamine isothiocyanate

Cell Volume Distribution

Some instruments are available which have both electronic volume measurement capability and sorting capability. These instruments have a volume measurement orifice upstream of the laser and a jet-in-air configuration below the laser. In these particular instruments, it is possible to sort cells based on volume.[6]

Light Scatter

Another parameter which gives information related to volume is forward-angle light scatter. As was previously mentioned, if scattered light is collected over a very narrow angle in front of the beam, the signal is monotonically related to the cellular volume. Although it is not linear, this signal can also be used to achieve cell separation based on volume. This measurement capability is much more common than the electronic volume measurement capability.

Sample Requirements in Flow Cytometry

Single-Cell Suspension

What Does "Single-Cell Suspension" Really Mean? The most fundamental requirement of all flow cytometers is a single-cell suspension. It is important to realize that what the microscopist calls a single-cell suspension may not be a single-cell suspension to the flow cytometrist. Oftentimes the cell culturist may remove the monolayer culture from a flask by treating it with a trypsin–EDTA solution, followed by rapidly pipeting the cell suspension to break up cell clumps. If we put an aliquot of this suspension under the microscope, we will see primarily single cells. It is likely, however, that we will find an appreciable number of cell doublets, triplets, and higher order aggregates. This is not a single-cell suspension to the flow cytometrist for a number of reasons, which are both physical and analytical in nature.

It is unlikely that one can completely eliminate cell aggregates from the sample without damaging a significant fraction of the population. Therefore, it is necessary to establish what the acceptable level of aggregates is for a particular application. There is no acceptable level for large aggregates capable of plugging the orifice. Acceptable levels for small aggregates (doublets and triplets) depend on the ability to detect these aggregates and exclude them from analysis and sorting. Let us look at an example where the doublets cannot be discriminated from the rest of the population. If you are trying to analyze or sort cells from a very rare population (1 in 10^5 cells) whose fluorescence intensity equals that of the doublets from the major population, then the acceptable level for doublets must be set far below that of the rare population of interest. In more common cases, 0.1 to 1.0% doublets in the sample may be acceptable. Various methods of eliminating aggregates or detecting and dealing with them are described below.

Problems Caused by Cell Aggregates. Plugging. The first and perhaps most irritating problem caused by cell aggregates is plugging of the orifice. As was previously mentioned, the orifice is typically $50-100$ μm in diameter. Depending on the cell size, an aggregate containing very few cells can completely plug the orifice, making it necessary to remove it for cleaning. This necessitates realigning the optics and restandardizing.

Although cleaning a totally plugged orifice is time consuming and inconvenient, this type of plug is an easily recognized problem. One that is not so easy to recognize is a partial plug at the orifice which causes turbulence in the flow, but does not stop the flow. This is problematic, because it can go unrecognized for long periods of time during a sort, leading to broadened CVs and inefficient sorting. A partial plug is the most common problem encountered with the flow cytometer. Detection of a partial plug is much easier if the flow cytometer has a jet-in-air configuration immediately after the sample injection point. This is because the plug causes turbulence in the stream just before it exits the flow tip, usually resulting in deflection of the sample stream. If the flow cytometer forms the jet in air past the analysis point, then deflection of the sample stream may be less pronounced and a partial plug less obvious. One is then dependent on stability of the flow rate and fluorescence intensity to detect a plug.

There are two common methods for the elimination of cell aggregates. The first employs nylon mesh (typically 37 to 44 μm mesh) to remove aggregates by filtration. This is only good for larger aggregates, not doublets or triplets. In the second method, the cell suspension is passed through a syringe with a fine-bore needle to more thoroughly break up the aggregates. It is important to be aware that neither of these techniques is universally applicable. However, one or the other of these techniques can usually be employed with success.

Misclassification of aggregates. Under most circumstances, the flow cytometer is used to measure the integrated fluorescence intensity of a particle passing through the light beam. The integrated fluorescence intensity is the summed total of the fluorescent light emitted as the particle traverses the full diameter of the light beam. When examining this parameter, the flow cytometer cannot tell the difference between a doublet of two small cells and a single large cell, which has twice the fluorescence intensity of a small cell in the same suspension. If the number of aggregates is appreciable, the result is analytical confusion and a skewed flow cytometric histogram. It can also lead one to mistakenly conclude that subpopulations exist in the cell suspension.

Electronic discrimination methods for doublet detection. The most common way to detect aggregates is to examine the forward- and right-angle light scatter patterns in a correlated fashion. This works well for eliminating large aggregates, but in situations where better aggregate discrimination is needed electronic means have been developed. There are two basic electronic discrimination techniques which have been used to detect aggregates in a suspension. We are interested in the shape or size of the pulse with each of these techniques. Consequently, the excitation light beam must be narrow with respect to the size of the cells being analyzed. In the following discussion, we will define a small single cell to have a volume of $1 \times$ and a large single cell to have a volume of $2 \times$.

The first technique employs high-resolution slit scanning and pulse-shape analysis to detect aggregates in the cell suspension. One can use this technique to detect multiple peaks in the sample, the presence of multiple peaks being indicative of an aggregate.[8,26] The only problem here is in the situation where multiple peaks exist in a single cell, such as a multinucleated cell stained for DNA. Alternatively, a nonlinear clump of cells may not show multiple peaks.

The second method employs a narrow light beam to analyze the pulse width as the particle traverses the light beam.[27] Understandably, large particles have larger pulse widths than do small particles and, as long as the particles are spherical, the width of the pulse is directly related to the particle volume. Because there is hydrodynamic focusing within the sample stream, the vast majority of the doublets will line up along their long axis such that the cells in the doublet pass into the light beam in single file.

[26] T. K. Sharpless, F. Tranganos, and Z. Darzynkiewicz, *Acta Cytol.* **19**, 577 (1975).
[27] L. L. Wheeless, *in* "Flow Cytometry and Cell Sorting" (M. R. Melamed, P. L. Mullaney, and M. L. Mendelsohn, eds.) pp. 125–136. Wiley, New York, 1979.

The pulse width for a large cell (volume $2\times$) is shorter than that of a small cell doublet of the same volume, since the axial diameter of the doublet is larger than the diameter of the large cell. Therefore, a time gate can be set based on the pulse width of the largest single cell in the suspension. Any particle with a larger pulse width is an aggregate.

Nonrandom Cell Losses. Another sample requirement in flow cytometry is concerned with the maintenance of a random population of cells in the suspension. If the sample preparation or handling techniques results in nonrandom cell losses, the histogram representing a particular parameter will be skewed and will not represent the true population average.

Problems which can lead to nonrandom losses of cells include aggregate filtration, differential adherence, and sample sedimentation. If the disaggregation technique employed for cultured cells or tissue samples generates a nonrandom distribution of cells in the aggregates, filtration will cause nonrandom cell losses. Another problem results from differential unit gravity sedimentation of cells. Many flow cytometers do not provide a means of stirring the sample. Because of this the cells settle out over time. If the run time for a particular sample is long, the cells will distribute themselves according to size and will be selected according to the depth of the sample tube in the suspension. This issue is most important when the run time is long, such as during rare event analysis or during a long sort. Some cell types, such as lymphocytes and monocytes, adhere differently to various substrates (i.e., glass, plastics). If steps are not taken to prevent this, nonrandom cell losses can be very significant.

Cell Preparation Procedures for Flow Cytometry

Trypsin–EDTA for Monolayer Cultures. The most common cell preparation procedure used with adherent cell cultures employs enzymes and divalent cation chelators to release the adherent cells from the culture dish.[28] Trypsin (0.5 g/liter) and EDTA (0.2 g/liter) in Hanks' balanced salt solution is commonly used for this purpose. The monolayer culture is rinsed with nonprotein-containing media and enough of the trypsin–EDTA solution is added to just coat the bottom of the dish. The culture is incubated at 37° until the cells are released and then serum-containing media is added to stop the protease activity. Cell clumps are dissociated by rapidly pipeting the suspension. Aggregated cells are then treated as previously described. However, the cell preparation technique cannot alter or destroy the parameter of interest, if this parameter is to be used as the basis for sorting of cells. Oftentimes, monoclonal antibodies conjugated with fluorescent molecules are used to identify specific cell types. If the antigen

---

[28] M. J. M. Lewin and A. M. Cheret, this volume [21].

to which the monoclonal antibody is directed is a protein, it may be trypsin labile, in which case a different means of preparing the sample must be sought.

Tissue Disaggregation. Whole tissues and tumors are often disaggregated so that cell subtypes can be identified and isolated. Both enzymatic and nonenzymatic techniques have been used for this purpose. The nonenzymatic techniques generally employ chelators to induce the disaggregation of tissues. Simple mechanical disruption is another nonenzymatic disaggregation technique which has been used. A wide variety of proteolytic enzymes have also been used, depending on the tissue being treated and the measurements to be performed after disaggregation. The same considerations are important here as were previously mentioned for the trypsin–EDTA technique. The enzyme used to prepare the cell suspension must not alter or destroy the parameter to be measured. A thorough treatment of this subject is presented both in this volume and elsewhere.[29,30]

Recently, a variation of the tissue disaggregation technique has been employed to prepare nuclei for DNA analysis from archival paraffin-embedded tissues. The reader is referred to the work of Hedley and co-workers for a description of this procedure.[31]

Cooperative Use of Other Cell Separation Methods. The major difference between flow cytometric cell sorting and the other bulk cell separation techniques, such as gradient separation or elutriation, is that cell sorting is a serial technique, while the other techniques are basically parallel. A serial technique is one which is performed on one cell at a time, parallel techniques being applied to the entire population, simultaneously. The fundamental difference in the nature of cell sorting causes this technique to be a very precise but low-capacity cell separation method. Bulk cell separation methods tend to have high capacity, but less precision. The relative low capacity of flow cytometric cell sorting is a significant problem when one is interested in isolating a minor subpopulation from a cell suspension. A common case in point can be used to illustrate this problem: sorting T helper cells from a sample of whole blood. In whole blood, about 1 out of every 1000 cells is a white blood cell, and approximately 25% of these are T helper cells. To isolate 10^6 T helper cells, it is necessary to examine 4×10^9 cells. A medium purity sort (93–95% pure) can be performed at a rate of approximately 50,000 cells per min (see Table V). Therefore, isolating 10^6 T helper cells will take about 55.5 days. It is obvious that sorting subpopu-

[29] C. Waymouth, *in* "Cell Separation: Methods and Selected Applications" (T. G. Pretlow II and T. P. Pretlow, eds.), Vol. 1, pp. 1–29. Academic Press, New York, 1982.

[30] M. G. Brattain, *in* "Flow Cytometry and Cell Sorting" (M. R. Melamed, P. L. Mullaney, and M. L. Mendelsohn, eds.), pp. 193–206. Wiley, New York, 1979.

[31] D. W. Hedley, M. L. Friedlander, I. W. Taylor, C. A. Rugg, and E. A. Musgrove, *J. Histochem. Cytochem.* **31**, 1333 (1983).

TABLE V
MAXIMUM EXPECTED PURITY OF
SORTED SAMPLES[a]

Analysis rate (cells/min)	Fraction of cells of interest in the population				
	90%	50%	10%	1%	0.1%
300,000	96.8	84.0	71.3	68.4	68.1
100,000	98.6	93.2	87.8	86.6	86.5
50,000	99.3	96.3	93.4	92.8	92.7
25,000	99.6	98.1	96.6	96.3	96.2

[a] No coincidence detection.

lations of lymphocytes out of whole blood is not feasible. The answer to this problem has been the development of a symbiotic relationship between cell sorting and the other bulk (parallel) cell separation methods. To understand the utility of partially purifying the sample before sorting, consider another example. If most of the red cells have been eliminated from our sample of whole blood, the resultant suspension should be about 80% white blood cells. One out of every five cells is now a T helper cell. Therefore, the sort time is reduced by a factor of almost 800, giving a much improved time of 1.7 hr. The examples described above were calculated for the case where coincidence detection is not being used. The time differential is not as bad if coincidence detection is used, but the principle is the same.

A wide variety of bulk separation techniques have been used in a prepurification step before sorting. These techniques include velocity sedimentation, gradient centrifugation, centrifugal elutriation, and affinity columns. Thorough discussions of each of these techniques may be found throughout this volume and elsewhere.[32-36]

Important Considerations in Cell Sorting

Sterility

Until recently, the flow cytometer/cell sorter has been the only instrument capable of isolating viable and essentially pure populations of cells.

[32] T. G. Pretlow and T. P. Pretlow (eds.), "Cell Separation: Methods and Selected Applications," Vol. 1. Academic Press, New York, 1982.
[33] T. G. Pretlow and T. P. Pretlow, this volume [22].
[34] M. J. Sanders and A. H. Soll, this volume [23].
[35] J. Wells, this volume [24].
[36] H.-G. Heidrich and K. Hannig, this volume [25].

This unique ability has lead to the development of sterile sorting techniques. The large size and high price of most cell sorters has made it impractical to confine these instruments to tissue culture hoods or to restrict their use to sterile applications. Fortunately, the sheath and sample compartments, as well as the flow system on all cell sorters, are enclosed up to the flow tip, where the jet stream forms. Beyond this, the jet stream and collection vials are also enclosed in a compartment where contamination from room air is minimized.

Sterilizing the Sorter. Sterilizing the flow system consists of first running sterile saline through the system to dislodge any loose contamination. All accessible sample tubing is changed for new clean tubing. The sheath tank and tubing leading to the flow chamber is then removed and 10% chlorine bleach in sterile distilled water is driven into the flow and sample chambers by syringe. The flow system is exposed to bleach for 30 min to overnight. Sterile distilled water is then driven through the system by syringe to flush out the bleach. A clean sheath tank and tubing, which has been autoclaved, is attached to the flow system through a new 0.22 μm filter. The sheath tank should be filled with sterile saline in a tissue culture hood prior to being attached to the system. The system is then pressurized and flushed with the sterile saline for at least 30 min before sorting. Contamination can also be minimized by including antibiotics in all solutions which the cells will contact.

This procedure is sufficient to decontaminate most cell sorters. However, if the system has been exposed to bacteria or yeast at any time, the chlorine bleach solution may not be a strong enough treatment to eliminate all contamination. In this circumstance, a formalin solution may be substituted for bleach to sterilize the flow system.

Viability

It was previously pointed out that the jet stream is moving at 1 to 10 m/sec as it exists the flow tip. The deceleration force on the cell when it lands in the collection vial must, therefore, be quite high. It is not surprising that some cell types are more susceptible to the impact damage caused by high flow rates than other cell types. In reality, viabilities may range from 10% or less to nearly 100% in particularly sturdy cell lines. A number of things can be done to increase the viability of a sample. Often the viability of a delicate sample can be increased by sorting into culture media containing an elevated level of serum proteins or into pure serum. This provides the most favorable environment immediately after the cell has been sorted. If this doesn't work, the pressure settings on the cell sorter (sample and sheath pressures) may be changed so that the flow rate and impact forces are minimized.

Sorting Speed Relative to Purity and Percentage Recovery

The sorting speed directly depends on the desired purity and recovery of the sample. Purity and recovery are in many cases inversely related to one another. For example, if high purity (99%) is an absolute requirement, coincidence detection can be used to maximize purity at the expense of throwing away some of the cells of interest, which appeared as coincident events. On the other hand, if recovery is more important than absolute purity, the sort can be performed without coincidence detection, so that all cells of interest are kept. The cost here is decreased purity due to coincident events. The operator-adjustable parameters which determine the sort speed, purity, and recovery are (1) the sample analysis rate, (2) the droplet formation frequency, and (3) the number of drops per sort pulse. The initial sample purity is also important if coincidence detection is not used.

Table V lists the maximum expected sorting purity as a function of analysis rate and initial fractional representation of cells of interest in the population. Table V corresponds to the case where sorting is being performed without coincidence detection. The values are based on a Poisson probability distribution for a typical three-drop sort and a droplet formation frequency of 32,000/sec.

If purity is the principle consideration, then coincidence detection should be used. This fixes the purity at a high level (e.g., 99%), but causes the percent recovery to vary with the analysis rate, the droplet formation frequency, and the number of drops per sort pulse. The recovery can also be calculated based on a Poisson probability distribution according to the equation

$$\text{Percentage recovery} = 100 \times \exp(-A_r n/F)$$

where A_r, n, and F are the analysis rate, number of drops per sort pulse, and droplet formation frequency, respectively.

Standardization

Instrument Standards. For purposes of analysis or sorting, it is important that the performance of the instrument be the same from day to day. Instrument performance can be monitored using standard particles that have a stable well-defined fluorescence intensity and size. The coefficients of variation for these parameters should be small for the chosen standard. Commonly used instrument standards include fluorescent microspheres or glutaraldehyde-fixed chicken erythrocytes. It is also important that the standard chosen be of approximately the same size as the cells to be analyzed or sorted, since size can affect the flow characteristics of the cell.

Aside from monitoring the day-to-day performance of the instrument, standards can also be used to monitor the sample-to-sample variations due to instrument instabilities. If microspheres can be distinguished from the cells of the sample by differences in volume or light scatter, then changes in the fluorescence intensity of the sphere and cell populations can be examined in a mixed sample. Therefore, in cell samples to which microspheres have been added, changes in the average intensity of the microsphere peak can be used to correct the fluorescence intensity of the cell population for minor variations in the instrument response.

A third use of instrument standards is in determining the plating efficiency of cells which have been sorted. The use of microspheres for maximizing the sorting efficiency has already been described. If we now want to sort 100 cells into a petri dish for a colony-forming assay, the instrument will count 100 events. But the actual number of cells plated is a function of the sorting efficiency. A knowledge of this efficiency will allow us to determine the actual colony-forming efficiency for a sorted sample.[37]

Stain Standards. So far we have only considered errors which can be introduced by variations in machine performance. We have not yet considered errors which can be introduced by variations in the performance of the experimentor or the condition of the samples. The best example of this is in DNA analysis. For a stoichiometrically binding DNA stain, the fluorescence intensity of the cell is a function of the amount of dye bound to the DNA. Assuming the dye is at saturating concentrations, the amount of dye bound to the DNA is a function of the relative amounts of DNA and the accessibility of the dye binding site to the fluorochrome. So variations in fluorescence intensity from sample to sample reflect differences in DNA content only if the configuration of the DNA is the same between samples. To test for this, chick erythrocyte nuclei are included in the sample to be analyzed. The DNA content of these nuclei is about 25% of the normal diploid human DNA content and, therefore, the chick erythrocyte peak will not overlap the DNA distribution of the sample. Changes in the fluorescence intensity of the chick peak reflect staining variations, which can be used to correct the sample DNA distribution.[37] Any differences between samples after correction for instrument and staining errors reflect true differences in DNA content. (See peak 1 of Fig. 4.)

Selected Cell Preparation and Labeling Protocols for Flow Cytometric Analysis and Sorting

For reasons previously discussed, immunofluorescence probes and DNA stains are the two most commonly used probe types in flow cytom-

[37] P. K. Horan and M. R. Loken, *in* "Flow Cytometry: Instrumentation and Data Analysis" (M. A. Van Dilla, O. D. Laerum, M. R. Melamed, and P. N. Dean, eds.), pp. 259–280. Academic Press, Orlando, Florida, 1985.

etry. We will, therefore, present protocols below for general immunofluorescence staining (direct and indirect) as well as a newly developed procedure for the staining and simultaneous analysis of five lymphocyte subsets using only one laser and two fluorochromes. We will also present procedures for cell preparation and staining using various DNA probes.

Immunofluorescence Staining

The procedures described below are applicable to virtually any cell type and antibody combination. Solutions which are common to all of the immunofluorescence methods presented here are listed below. All solutions which are to be added to live cells, with the exception of fixatives, should be isosmotic.

   PBS — Phosphate-Buffered Saline (pH 7.2)
   Dulbecco's phosphate-buffered saline or any equivalent should work
   for this purpose.
   PBA — Phosphate-Buffered Saline (pH 7.2), Supplemented
   The supplement is 1.0% bovine serum albumin (Fraction V, Sigma
   Chemical Co., St. Louis, Missouri) and 0.05% sodium azide (Fisher
   Chemical Co., King of Prussia, Pennsylvania).
   FBS — Fetal Bovine Serum
   For some applications the serum (Gibco Laboratories, Grand Island,
   New York) should be heat inactivated at 56° to destroy complement activity.
   4% Paraformaldehyde Fixative
   Four grams of paraformaldehyde (Sigma Chemical Co.) should be
   added to 100 ml of PBS and stirred while heating until the solution
   boils. If the solution is not clear, 10 N NaOH should be added
   dropwise until clarity is achieved. The pH should be readjusted to
   7.2. For use in the flow cytometer, the solution should be filtered
   through a 0.22-μm filter.
   Propidium Iodide for Live/Dead Discrimination
   Dissolve propidium iodide powder in PBS (50 μg/ml pH 7.2). Solution
   must be protected from light and may be stored for months if kept
   refrigerated.

Direct Immunofluorescence. Direct immunofluorescence has the advantage that the antibody specific for the property of interest is directly labeled with a fluorochrome, so a single binding reaction is required and multiple immunofluorescence probes can be used simultaneously. The disadvantage of this technique lies in the fact that you cannot choose the fluorochrome bound to the antibody unless the manufacturer conjugates the antibody with various fluorochromes.

The solution volumes and cell concentrations to be discussed are appropriate for small (5-ml) test tubes. However, for the procedures herein described the cells may be washed and suspended in any convenient container. The solution volumes and cell concentrations should be adjusted to maintain a total of 10^6 cells/sample. If the cells to be stained tend to stick to plastic, the tubes may be treated with a siliconizing solution, such as Prosil-28 (Specialty Chemicals, Gainsville, Florida). If a large number of samples are to be stained, Costar 96-well V-bottom microtiter plates (Bellco Glass, Vineland, New Jersey) are convenient. The cells should be isolated as a single-cell suspension and prepurified if necessary. Procedures for isolation and prepurification are described elsewhere in this volume.[28,33-36] From this point, all procedures should be carried out at 4°.

Aliquots (1 ml) of a cell suspension at 10^6 cells/ml should be added to each tube. The cells should be washed in PBA with centrifugation at 400 g for 10 min to pellet the cells. Resuspend the cells in 50 μl of PBA to a final concentration of 2×10^7 cells/ml. A saturating amount of the fluorochrome-labeled antibody should be added to each aliquot of cells. Typically, 5 to 20 μl of a commercially prepared reagent is added to 10^6 cells. However, volumes vary with different manufacturers and should be carefully checked for each antibody. Incubate the cells with the antibodies for 30 min to allow for complete binding. After the incubation, wash the cells three times in cold PBA and then resuspend in 1 ml of PBA to give a final concentration of 10^6 cells/ml. Longer incubations (up to 2 hr) may be required with particulate fluorochromes, where steric hindrance is problematic. Such particulate fluorochromes include antibody-conjugated fluorescent microspheres or liposomes.

An alternate rapid washing procedure may be used in which the stained cells are separated from unbound antibody by a fetal bovine serum step gradient. If this procedure is to be used, then all previous steps should be carried out in conical bottom tubes or plates. After the antibody incubation, dilute each sample with PBS to a volume of 1 ml. Add 3 drops of FBS directly to the center of each tube so that the FBS falls to the bottom and forms an obvious pool. Centrifuge the tubes at 400 g for 10 min and aspirate the supernatant. The cell pellet will be free of unbound antibody and should not require further washing. Resuspend the pellet in 1 ml of cold PBA to a final concentration of 10^6 cells/ml. The cells are now ready to run on the flow cytometer.

If quantitative estimates of specific binding are to be made, then controls must be done to estimate the amount of cellular autofluorescence (unstained control) as well as the amount of antibody bound nonspecifically to the cells. This nonspecific binding can be estimated by labeling an aliquot of cells with an irrelevant antibody of the same isotype. Care must be taken that the specific activity of the fluorochrome on the specific and nonspecific antibodies is the same.

If flow cytometric measurements are to be made in viable cells, then it is possible to discriminate between live and dead cells by propidium iodide dye exclusion. This works much like the trypan blue exclusion test, except the nuclei of the dead cells now become fluorescent. If live/dead gating is to be done, add 0.1 ml of the stock propidium iodide solution to each 1-ml sample and mix thoroughly (5 μg/ml final concentration). Those cells which fluoresce brightly in the red (> 590 nm) are dead and can be excluded from analysis and sorting.

If the viability of the sample is high ($\geq$ 95%) it may be fixed in 2% paraformaldehyde for later analysis without appreciably affecting the analysis. If this is to be done, add 1 volume of 4% paraformaldehyde stock solution to 1 volume of cell suspension and incubate at 4° overnight.

Indirect Immunofluorescence. This method of staining requires the use of two antibodies and two separate binding reactions. A nonfluorochrome-conjugated (nonfluorescent) primary antibody specific for the parameter of interest is first bound to the target cells. After washing out the unbound primary antibody, a fluorochrome-conjugated secondary antibody specific for the primary antibody type is added to the cells and allowed to bind. After a second washing step, the resultant cells have been indirectly labeled for the parameter of interest. The advantage of this technique is that the fluorochrome can be chosen to quantitate the binding reaction and is not usually limited to a specific fluorochrome, as with direct immunofluorescence. Also, if more than one secondary antibody binds to each primary antibody, the cells will ultimately be labeled brighter by the indirect method. The major disadvantage of this technique arises when it is necessary to simultaneously measure more than one parameter by immunofluorescence staining. Common secondary antibodies are specific for immunoglobulin (G or M) from the species of animal in which the primary antibody was produced. Two primary antibodies which were produced in the same species of animal and are both immunoglobulins of the same class, cannot be used simultaneously in an indirect immunofluorescence assay. Any chosen secondary antibody will recognize both primary antibodies.

To perform immunofluorescence labeling, follow the procedure for direct immunofluorescence labeling, substituting a nonfluorescent primary antibody for the direct antibody. After the cells have been washed to remove the unbound primary antibody, once again resuspend the cells in 50 μl of cold PBA. Next add an appropriate amount of fluorochrome-conjugated secondary antibody and incubate for 30 min to allow for complete binding. Afterward, wash the cells three times in cold PBA and resuspend in 1 ml PBA to a final concentration of 10^6 cells/ml for analysis.

As with the direct staining method, quantitative estimates of binding require the measurement of cellular autofluorescence (unstained control) and nonspecific binding of the fluorochrome-conjugated secondary antibody. In this case, nonspecific binding of the fluorochrome-conjugated antibody is estimated by first binding an irrelevant primary antibody of the same isotype as the original primary antibody. The secondary antibody is then bound afterward. This way nonspecific bindings due to both the primary and secondary antibodies are estimated, simultaneously.

If it is desired, both unbound primary and secondary antibodies may be eliminated using the rapid wash procedure previously described. Live/dead gating as well as paraformaldehyde fixation may be performed as for the direct staining method.

Simultaneous Analysis of Five Leukocyte Subsets. The simultaneous five-part subset analysis applied here to leukocytes is applicable to any cell type where multiple subsets exist and antibodies to each subset are available.[38] An appropriate antibody cocktail must be prepared which will separate the subsets of interest in two dimensions.

Flow cytometric analysis of human peripheral blood leukocyte subsets has typically been achieved by staining multiple aliquots of the same sample with different immunofluorescent reagents specific for the subsets of interest and serially analyzing the aliquots. Fluorochromes which are spectrally discrete have been developed and conjugated to various antibodies for applications where the simultaneous identification of multiple subsets is accomplished by the use of multiple reagents per aliquot. Extending the analysis to more than two reagents per sample has required the development and use of large dual-laser flow cytometers possessing complex instrumentation and analytical capabilities. Recently, a technique was developed as an alternative to the dual-laser approach to multiple subset analysis. The technique described here requires only one laser, two fluorochromes, and commercially available reagents.

Under normal circumstances, one would not be able to simultaneously analyze two different subsets which are labeled with the same fluorochrome, since their profiles would either overlap or be completely indistinguishable. In fact, the present technique uses dilution of commercial fluorochrome-conjugated reagents with competing unlabeled reagents to adjust the relative fluorescence intensities of different subsets, which are labeled with the same fluorochrome. The fluorochrome-labeled antibody is diluted with unlabeled antibody so that binding saturation is maintained under conditions where fluorescent antibody binding is decreased. This concept can be extended for use in two dimensions, allowing the quantitation of at least five subsets with only two fluorochromes.

[38] P. K. Horan, S. E. Slezak, and G. Poste, *Proc. Natl. Acad. Sci. U.S.A.* **83,** 8361 (1986).

The reagents to be used in this assay are the same as those used for direct immunofluorescence. The staining procedure may be carried out in any convenient container, but will be described for a 96-well V-bottom microtiter plate. A list of the immunofluorescence probes used here is given below along with their specificities and manufacturers.

Coulter Immunology (Hialeah, Florida)

FITC-B1   (fluorescein-conjugated B lymphocyte marker)
FITC-Mo2   (fluorescein-conjugated monocyte marker)

Becton–Dickinson Monoclonal Center (Mountain View, California)

PE-Leu 2a   (phycoerythrin-conjugated suppressor T lymphocyte marker)
Leu 3a   (unconjugated helper T lymphocyte marker)
PE-Leu 3a   (phycoerythrin-conjugated helper T lymphocyte marker)
Leu 4   (unconjugated T lymphocyte marker)
FITC-Leu 4 (fluorescein-conjugated T lymphocyte marker)
PE-Leu 11c (phycoerythrin-conjugated natural killer cell marker)
PE-M3   (phycoerythrin-conjugated monocyte marker)

The unconjugated PE-Leu 3a and FITC-Leu 4 antibodies are to be used to dilute the fluorochrome-conjugated antibodies with the same specificity. For the distribution shown in Fig. 3, the Leu 4 reagent was diluted by mixing 3 μl of FITC-Leu 4 with 1 μl of unconjugated antibody. Leu 3a was diluted by mixing 13 μl of PE-Leu 3a with 1.5 μl of unconjugated Leu 3a. The reader should note that the specific activity of each antibody may vary between lots. Consequently, it may be necessary to adjust the dilution factor to optimize the separation with different lot numbers of the same antibody.

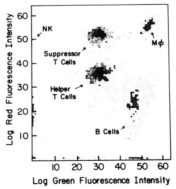

FIG. 3. Log red versus log green fluorescence scatter pattern for an antibody cocktail specific for five leukocyte subsets.

The components of the multiple subset cocktail are listed below.

5 μl FITC-B1
5 μl FITC-Mo2
20 μl PE-M3
20 μl PE-Leu 11c
20 μl PE-Leu 2a
15 μl diluted PE-Leu 3a
5 μl diluted FITC-Leu 4

Before staining, the mononuclear cells must be isolated from whole blood. To do this, 8 ml of whole blood is layered on 5 ml of Lymphoprep sodium metrizoate/Ficoll separation medium (Nyegaard & Co., Oslo, Norway) and centrifuged at 400 g for 40 min at 20°. The interface layer contains the lymphocytes and monocytes and should be aspirated and washed twice in PBA solution. The resultant pellet is resuspended in cold PBA to a final concentration of 2×10^7 cells/ml and 50-μl aliquots (10^6 cells) are placed in individual wells of the microtiter plate. All subsequent steps should be carried out at 4°.

The cells should be pelleted at 400 g for 10 min and the supernatant aspirated. Then 90 μl of the antibody cocktail is added to the wells to be stained and the plate is incubated for 30 min. Control wells should be prepared with appropriate individual antibodies or antibody combinations and PBA solution added to a final volume of 90 μl. After the incubation is complete, 50 μl of PBS is added to each well followed by 1 drop (20 μl) of FBS. The plates are centrifuged at 400 g for 10 min and the supernatants are aspirated. The cells in each well should be resuspended in 200 μl of PBA solution for flow cytometric analysis.

The instrument setup described here is for an EPICS 753 flow cytometer (Coulter Electronics, Hialeah, Florida), but may be used directly or adapted for use on any flow cytometer capable of measuring forward-angle light scatter, right-angle light scatter, and at least two colors of fluorescence, simultaneously. On some instruments, electronic volume sizing may be substituted for FALS. The laser (argon ion) is set to an exciting wavelength of 488 nm and 500-mW power. A 488-nm dichroic mirror and 488-nm bandpass filter are used to measure the RALS signal. A 515-nm interference filter and a 515-nm long pass absorbance filter are employed as laser-blocking filters. The FITC and PE fluorescent light is split by a 560-nm short pass dichroic mirror; 575- and 525-nm bandpass filters are used in front of the phycoerythrin and fluorescein PMTs, respectively. Finally, a 1.5 OD neutral density filter is used in front of the FALS detector to attenuate the direct laser radiation. Software gates should be set on the forward- and right-angle light scatter parameters selecting the mononuclear cells and excluding cell clumps and debris, as discussed by Muirhead *et al.*[39] Color compensation should be used if possible to correct for spectral crossover between the PMTs.[3]

The fluorescence intensity of the lymphocyte and monocyte subpopulations can then be determined in a correlated two parameter analysis. Figure 3 is a typical two parameter histogram showing the five labeled subsets (monocytes, natural killer cells, T helper lymphocytes, T suppressor lymphocytes, and B lymphocytes).

DNA Staining

Vital DNA Staining. Vital DNA staining has been accomplished using a class of bisbenzimidazole dyes.[40] Hoechst 33342 and Hoechst 33258 have both been useful in staining viable cells for cell cycle analysis and sorting. Hoechst 33342 has become the more accepted probe, since cell membranes are more permeable to it, thus less time is required for maximal staining.

A 250 μM stock solution of Hoechst 33342 is made up in phosphate-buffered saline. This solution should be protected from light. Cells to be stained are isolated as a single-cell suspension and washed in PBS with centrifugation at 400 g for 10 min at 25°. The cell pellet is resuspended in PBS at 10^6 cells/ml and Hoechst 33342 is added from the stock solution to a final concentration of 5 to 10 μM. This concentration may vary with the cell line being stained. The cells are incubated in a 37° water bath for 60 to 90 min, followed by flow cytometric analysis. The cells should not be washed out of the dye before analyzing. Using this staining technique, the G_1 peak CV should be in the range of 6%.

For this experiment, the argon ion laser should be adjusted for ultraviolet output (351–363 nm) with an output power of 200 mW. The laser-blocking filter is a 418-nm interference filter with a 470-nm bandpass filter in front of the photomultiplier tube. A 0.5 OD neutral density filter is used to attenuate direct laser radiation at the light scatter sensor. Software gates should be set based on light scatter to exclude dead cells, cell clumps, and debris.

If sorted cells are to be put back into culture, it is advisable to wash the cells in media after analysis. Washing the cells a second time, after incubation in fresh media for 2 hr, helps to ensure that all of the stain has been removed from the cells.

Nonvital DNA Staining. DNA staining of fixed cells can be performed with a number of the dyes listed in Table III, with good results. The

[39] K. A. Muirhead, P. K. Horan, and G. Poste, *Biotechnology* **3**, 337 (1985).
[40] D. J. Arndt-Jovin and T. M. Jovin, *J. Histochem. Cytochem.* **25**, 585 (1979).

procedure described here is for propidium iodide, but works just as well for ethidium bromide.

Cells to be stained are isolated as a single-cell suspension and are washed in PBS with centrifugation at 400 g for 10 min at 4°. The cell pellet is resuspended in ice-cold (0–5°) PBS at 10^6 cells/ml. This sample is fixed by injection (22-gauge needle) into ice-cold methanol to a final methanol concentration of 70% and is incubated overnight. After fixation the cells are spun at 400 g for 10 min at 25° and the methanol is aspirated.

Before further treatment, chick erythrocyte nuclei are added as stain standards. The same amount of the chick nuclei should be added to each sample to a concentration equivalent to 1–10% of the cell concentration. The cellular RNA is then digested by resuspending the cells in 1 mg/ml RNase A (Sigma Chemical Co., St. Louis, Missouri) and incubating at 37° for 1 hr. The cell/standard samples are then spun at 400 g for 10 min at 25° and the supernatant is aspirated. The cells are washed twice in PBS and finally resuspended in a 50 μg/ml stock solution of propidium iodine in PBS. The samples are then ready for flow cytometric analysis.

The instrument setup described here is for an EPICS 753 flow cytometer (Coulter Electronics, Hialeah, FL), but a similar analysis may be performed on any commercially available flow cytometer. The laser (argon ion) is set to an exciting wavelength of 488 nm and 500 mW power. A 488-nm dichroic mirror and 488-nm band pass filter are used to measure the RALS signal. A 515-nm interference filter and a 515-nm long pass

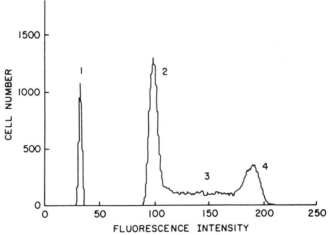

Fig. 4. DNA histogram of propidium iodide–stained HL60 cells (human promyelocytic leukemia). (1) Chicken erythrocyte nuclei, (2) G_0/G_1 phase cells, (3) S phase cells, (4) G_2/M phase cells.

absorbance filter are employed as laser-blocking filters. The propidium iodide signal is passed through a 595-nm long pass interference filter followed by a 610-nm long pass absorbance filter. Finally, a 1.5 OD neutral density filter is used in front of the FALS detector to attenuate the direct laser radiation. Once again, software gating is used to exclude cell clumps and debris based on the forward- and right-angle light scatter distributions.

Figure 4 is a typical one-parameter DNA histogram showing the cell cycle distribution of exponentially growing HL60 cells (human promyelocytic leukemia).

[28] Use of Cell-Specific Monoclonal Antibodies to Isolate Renal Epithelia

By ARLYN GARCIA-PEREZ, WILLIAM S. SPIELMAN,
WILLIAM K. SONNENBURG, and WILLIAM L. SMITH

Cultured cells are gaining popularity as tools to study cellular and biochemical aspects of renal function. Although one would like to have available highly differentiated cell lines derived from the glomerulus and from each segment of the renal tubule, this ideal has yet to be realized. Instead, investigators have resorted either to culturing microdissected tubules,[1-3] cloning specialized lines expressing one or two functions of interest,[4] or using established Madin–Darby canine kidney (MDCK)[5-7] or LLC-PK$_1$[8,9] cell lines. Unfortunately, microdissection is laborious, requires considerable expertise, and yields amounts of tissue which, even following culture ($\sim 10^5$ cells), are inadequate for most biochemical measurements. The various specialized lines which are now available retain some, but not all, of the properties of the parent cells.

[1] R. J. Anderson, P. D. Wilson, M. A. Dillingham, R. Breckon, U. Schwerschlaf, and J. A. Garcia-Sainz, *Am. Soc. Nephrol. Meet., 17th,* 154A (1984).
[2] M. G. Currie, B. R. Cole, K. Deschryver-Kecskemeti, S. Holmberg, and P. Needleman, *Am. J. Physiol.* **244,** F724 (1983).
[3] N. Green, A. Algren, J. Hoyer, T. Triche, and M. Burg, *Am. J. Physiol.* **249,** C97 (1985).
[4] S. Sariban-Sohraby, M. B. Burg, and R. J. Turner, *Am. J. Physiol.* **245,** C167 (1983).
[5] D. S. Misfeldt, S. T. Hamamoto, and D. R. Pitelka, *Proc. Natl. Acad. Sci. U.S.A.* **73,** 1212 (1976).
[6] M. J. Rindler, L. M. Churman, L. Shaffer, and M. H. Saier, Jr., *J. Cell Biol.* **81,** 635 (1979).
[7] M. H. Saier, Jr., *Am. J. Physiol.* **240,** C106 (1981).
[8] D. S. Misfeldt and M. J. Sanders, *Am. J. Physiol.* **240,** C92 (1981).
[9] J. M. Mullin, C.-J. M. Cha, and A. Kleinzeller, *Am. J. Physiol.* **242,** C41 (1982).

We have developed immunoadsorption procedures for isolating large, homogeneous populations of both canine and rabbit collecting tubule cells.[10-12] Cultures of these cells exhibit a number of the differentiated properties of the parent cells even after several passages. In principle, immunoadsorption should be applicable to isolating any cell type against which a specific antibody is available.[11] We describe here the procedures for isolating canine and rabbit cortical collecting tubule cells, CCCT[10] cells and RCCT[12] cells, respectively.

Preparation of Monoclonal Antibodies

Unless one already has available antibodies specific for a cell surface protein on the cell of interest, it is necessary to prepare monoclonal antibodies. For immunodissection, the nature of the antigen is of secondary importance, but it must be an ectoantigen present in relative abundance and only on the cell of interest.[11] It is also important that the antibody be of the immunoglobulin G class because IgGs are more stable and easier to purify.

Immunization

Cell lines may be used as immunogens if they have or are expected to have some of the same cell surface proteins as the cells to be isolated. For example, to prepare monoclonal antibodies against the canine collecting tubule, we immunized rats with MDCK cells[10]; MDCK cells have morphological characteristics similar to those of canine collecting tubule epithelia. More commonly, protease dispersions of the tissue containing the cell of interest may be used for immunizations.[11-13] Methods for preparing single-cell dispersions from rabbit and canine renal cortex are detailed below.

Reagents

10^7 cultured cells or cells from kidney tissue dispersed by treatment with a protease resuspended in 0.20 ml Krebs buffer (composition, millimolar: 118 NaCl, 25 NaHCO$_3$, 14 glucose, 4.7 KCl, 2.5 CaCl$_2$, 1.8 MgSO$_4$, and 1.8 KH$_2$PO$_4$), pH 7.3, per immunization
2-week-old outbred rats or 4-week-old outbred mice
Method. Each animal is immunized using a tuberculin syringe equipped with a 22-gauge needle on days 1 and 15, first intravenously then

[10] A. Garcia-Perez and W. L. Smith, *Am. J. Physiol.* **244,** C211 (1983).
[11] W. L. Smith and A. Garcia-Perez, *Am. J. Physiol.* **248,** F1 (1985).
[12] W. S. Spielman, W. K. Sonnenburg, M. L. Allen, K. Gerozissis, and W. L. Smith, *Am. J. Physiol.* **251,** F348 (1986).
[13] M. L. Allen, A. Garcia-Perez, and W. L. Smith, *Fed. Proc.* **41,** 1693 (1982).

intraperitoneally, or three times intraperitoneally on days 1, 15, and 29; 10^7 cells in 0.2 ml Krebs buffer, pH 7.3 are used per immunization.

Fusion

There are a number of published procedures for B lymphocyte–myeloma cell fusions and nutritional selection of resulting hybridoma cells.[11,14-16] We perform cell fusions on the third day after the last injection, when the number of fusion-active lymphocytes is at a maximum.[11,16] We have routinely fused with the SP2/0-Ag14 myeloma line,[17] which does not secrete antibody fragments and forms stable hybrids with either rat or mouse B lymphocytes.

Screening of Hybridoma Cultures for Monoclonal Antibodies

Indirect immunofluorescence is the most efficient technique to test for antibodies to specific renal cell types. In this procedure, cryotome sections of kidney are overlayed first with media from each hybridoma-containing well. If antibody against a renal cell antigen is present in the culture medium, the antibody will bind the antigen in the tissue section of this step. The sections are washed to remove unbound antibody and then overlayed with fluorescein isothiocyanate (FITC)-labeled antimouse (or rate) IgG. The second antibody binds and labels any monoclonal antibody bound to the tissue section.[11] In most cases, identification of fluorescently stained cell types can be made by visual examination under the light microscope. Identification is confirmed, as necessary, by histochemistry of serial sections using combinations of tubule-specific stains.[10,11,18]

Reagents

Sections (10 μm), cut on a cryotome, from quick-frozen blocks (about 1 cm^3) of unfixed kidney and mounted on glass cover slips

Phosphate-buffered saline (PBS; composition, millimolar: 151 NaCl, 45 KH$_2$PO$_4$, and 2.5 NaOH) pH 7.2–7.4

100 μl of media from each tissue culture well containing hybridomas

Fluorescein isothiocyanate-labeled antimouse (or rat) IgG (Miles Laboratories)

[14] G. Galfre, S. C. Howe, C. Milstein, G. W. Butcher, and J. C. Howard, *Nature (London)* **256**, 495 (1977).

[15] J. J. Langone and H. Van Vunakis (eds.), this series, Vol. 92.

[16] B. B. Mishell and S. M. Shiigi, "Selected Methods in Cellular Immunology." Freeman, San Francisco, California, 1980.

[17] M. Shulman, C. D. Wilde, and G. Kohler, *Nature (London)* **276**, 269 (1978).

[18] R. B. Nagle, R. E. Bulger, R. E. Cutler, H. R. Jervis, and E. P. Benditt, *Lab. Invest.* **28**, 456 (1973).

Glycerol

Glass slides

Method. Details for preparing cryotome sections of unfixed kidney tissue are presented elsewhere in this series.[19] Each section is overlayed with 0.3 ml of a 1:3 dilution of hybridoma culture medium in PBS, pH 7.3, and incubated at 24° for 30 min. The sections are rinsed three times with 0.3 ml of PBS, pH 7.3, to remove unbound antibody and are then overlayed with 0.3 ml of a 1:20 dilution of FITC-labeled rabbit antimouse (or rat) IgG in PBS, pH 7.3. After a 30-min incubation at 24° each section is rinsed as before with PBS, pH 7.3. The cover clip is mounted on a glass slide with a drop of glycerol and examined for specific fluorescent staining by fluorescence microscopy.

Once the medium in a culture well has been identified as containing an antibody of interest, one clones the hybridoma cells present in the tissue culture well as soon as possible. Cloning techniques for hybridoma cells are detailed elsewhere.[11,16,20] Freezing and storage of the cloned line for future use is performed using common tissue culture techniques.[11,21] The cloned hybridoma is routinely grown in serum-free or IgG-free media to obtain the monoclonal antibody it produces.[22]

As discussed below, it is necessary to purify the antibody before its use in immunodissection. IgG molecules are easily purified from media containing serum free of bovine IgG[22] or from commercially available serum-free culture media by column chromatography on protein A–Sepharose (Pharmacia).[10–12,22] Protocols for purification of mouse IgG$_1$ and mouse IgG$_{2b}$ are presented elsewhere in this series.[22] Rat IgG$_{2c}$[10] and mouse IgG$_3$[12] may be purified using the same protocols; these immunoglobulins are eluted from protein A–Sepharose at pH 4.5

Immunodissection

In theory, a variety of immunochemical separation techniques can be used to isolate specific cells from fresh tissue once an appropriate antibody has been prepared. We describe here only the use of antibody-coated polystyrene culture dishes (Fig. 1). This is the most convenient and least expensive procedure, and we utilize it to isolate CCCT cells[10] and RCCT cells.[12] All procedures are done under sterile tissue culture conditions.

[19] W. L. Smith and T. E. Rollins, this series, Vol. 86, p. 213.

[20] L. A. Herzenberg, L. A. Herzenberg, and C. Milstein, *Handb. Exp. Immunol. (3rd Ed.)* **2**, 25.1 (1979).

[21] L. L. Coriell, this series, Vol. 58, p. 29.

[22] D. L. DeWitt, J. S. Day, J. A. Gauger, and W. L. Smith, this series, Vol. 86, p. 229.

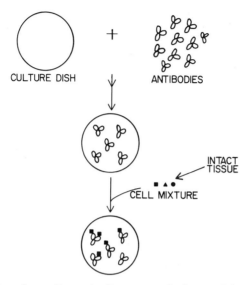

FIG. 1. Adsorption of a specific renal cell type on antibody-coated tissue culture dishes.

Coating of Culture Dishes with Antibody

Monoclonal antibodies bind to polystyrene and other unsubstituted polymers [e.g., polypropylene, poly(vinyl chloride), and polyethylene] irreversibly.[23,24] Because proteins other than antibodies can bind to these polymers[24] and because only a limited amount of protein can bind, highly purified antibody should be used to coat the culture dishes. This maximizes the number of antigen-combining sites available on the dish. Saturation by mouse IgG_3 of the protein binding sites of a Corning 100-mm polystyrene culture dish (~ 40 μg of protein) is obtained by incubation with $200 - 400$ μg of antibody in 9 ml of PBS, pH 7.3, for 3 – 4 hr at 24° or 16 hr at 4°. Conditions (i.e., exposure time, temperature, and pH) for optimal binding of antibody to culture plates should be explored with each antibody.[23] It is also important to note that there are major differences in the extent of cell immunoadsorption with tissue culture dishes available from different suppliers. In our hands, antibody-coated Corning dishes provide two to five times the yields of antibody-coated Costar dishes, while Falcon dishes are not effective.

[23] K. Catt and G. W. Tregear, *Science* **158,** 1570 (1967).
[24] R. I. Leininger, C. W. Cooper, M. M. Epstein, R. D. Falb, and G. A. Grode, *Science* **152,** 1625 (1966).

Reagents (per dish to be coated)

200–400 μg purified monoclonal antibody
9 ml of PBS, pH 7.3
0.45-μm sterilization filters (Nalgene)
100-mm tissue culture dishes (Corning)
15–20 ml of 1% bovine serum albumin (w/v) in PBS, pH 7.3

Method. The antibody is diluted in the PBS, pH 7.3, to a final volume of 9 ml per dish (e.g., 45 ml for five dishes). This solution is sterilized by filtration through a 0.45-μm Nalgene filter unit. Aliquots of 9 ml are pipetted into each dish and the dishes incubated at 24° for 3 hr or overnight at 4°. Most immunoglobulins are sufficiently stable so that the dishes may be stored for up to 7 days at 4° if they are not to be used immediately. Just prior to using the antibody-coated dishes for cell isolation, the IgG solution is aspirated and the dishes washed three to four times with 3–5 ml of 1% BSA (w/v) in PBS to coat any free protein-binding sites. This step is important to minimize nonspecific binding of cells.

Preparation of Cell Dispersions

Successful immunodissection requires the preparation of dispersions containing primarily single cells or short strings of 2–10 tubule cells. Sufficient yields of dispersed cells are required so that $10^7–10^8$ cells can be applied to each of 5–10 antibody-coated dishes. Generally, collagenase is effective in dispersing renal tissue. Red blood cells, filamentous material, and tissue fragments must be removed from cell dispersions for efficient immunodissection. We present here the protocol used for selective adsorption of CCCT and RCCT cells.

Reagents

Fresh canine or rabbit kidneys
PBS, pH 7.3
24 ml of 0.1% collagenase (Worthington, CLS II) (w/v) in Krebs buffer, pH 7.3 (for CCCT cells)
24 ml of 0.1% collagenase (GIBCO) (w/v) in Krebs buffer, pH 7.3 (for RCCT cells)
10 ml of 0.2% NaCl (w/v) in distilled water
10 ml of 1.6% NaCl (w/v) in distilled water
10–12 stainless-steel meshes (Gelman Sciences, syringe filter holder, 0.25-mm pore size), sterilized by autoclaving
10 ml of 10% BSA (w/v) in PBS, pH 7.4

The preparation of cell dispersions from canine and rabbit renal cortex is done under sterile conditions. The kidney is removed immediately after sacrifice and rinsed with PBS, pH 7.3. This eliminates some excess blood. The clear capsule surrounding the kidney is peeled away. The kidney is then placed in a 100-mm plastic culture dish. The organ is cut sagittally to obtain slices about 0.5 cm thick. A thin region about 0.2 cm thick is removed from the circumference of the slice using a sterile blade and is discarded. Next, the outer cortical region is dissected. The accumulated tissue (~ 5 g from two rabbit kidneys; ~ 5 g from one dog kidney) is diced thoroughly with a sterile razor blade for 2–10 min in a 100-mm plastic culture dish until a very fine mash about the consistency of thick tomato soup is obtained. It is very important to obtain a fine tissue mash to maximize cell yields. The minced tissue is then incubated with 24 ml of 0.1% collagenase in Krebs buffer, pH 7.3, for 40–60 min at 37° under a water-saturated 5–10% CO_2 atmosphere. At intervals during the incubation period the tissue suspension is drawn up and down 5–10 times in a large-bore 10-ml pipet to further disperse the tissue fragments. In the case of canine renal cortex, we incubate the tissue with Worthington CLS II collagenase for 40 min, disperse the tissue with a 10-ml pipet, and then incubate the sample for another 20 min. With rabbit renal cortex we incubate the minced tissue with GIBCO collagenase for a total of 40 min, dispersing the sample every 10 min with a 10-ml pipet. After treatment with collagenase, the cell suspension is centrifuged at 800 g for 10 min and the supernatant is discarded. Red blood cells in the pellet are removed by hypotonic lysis by resuspending the pellet in 10 ml of 0.2% saline for 30 sec followed by the addition of 10 ml of 1.6% saline and incubation for 30 sec. Finally, 20 ml of PBS, pH 7.3, is added. This suspension is drawn into a 10-ml pipet using an automatic pipettor and expelled through a sterile Gelman stainless-steel mesh (0.25-mm pore size), which filters clumps of tissue and some filamentous material. It is usually necessary to use three or four meshes, as they become clogged with chunks of tissue. The filtrate containing single cells and short strings of cells is centrifuged at 800 g for 10 min at ambient temperature and the supernatant discarded. The cell pellet is resuspended in 10% bovine serum albumin in PBS, pH 7.3, and again centrifuged at 800 g for 10 min at ambient temperature. The resulting cell pellet is suspended in 1–5 ml of Krebs buffer, pH 7.3, or PBS, pH 7.3, and the cells are counted using a hemacytometer. We typically obtain 10^7–10^8 live renal cells per gram of renal cortex.

The conditions for collagenase treatment (e.g., time, collagenase concentration, type of collagenase) must be determined experimentally for each different tissue to maximize the yield of viable single cells or short strings of cells. The number of live cells is determined by counting total and stained (dead) cells in a hemocytometer following staining with a vital dye.[25]

Immunoadsorption

Dispersed cells are added in a small volume to the antibody-coated dishes to maximize collisions between the stationary antibody molecules and the floating cells. Exposure time is short (≤5 min) to minimize nonspecific binding of cells (especially fibroblasts).

Reagents

Pellet of dispersed cells
PBS, pH 7.3

Method. For isolation of CCCT cells, the pellet ($\sim 10^9$ cells) is suspended in 10 ml of PBS, pH 7.4. For isolation of RCCT cells, the pellet is smaller ($1-3 \times 10^8$ cells) and is suspended in 5 ml of PBS, pH 7.4. Aliquots of suspended cells are added to antibody-coated dishes, 1 ml per dish, and the samples allowed to stand for 3–5 min in 24°. The dishes are promptly and vigorously washed three to five times with 5 ml of PBS, pH 7.4, and are expelled from a 10-ml pipet to remove unbound or loosely bound cells.

Specific binding is considered to be achieved when the number of cells bound to the antibody-coated culture dish is 100 times the number of cells bound to a control BSA-coated dish. With a $10\times$ eyepiece and a $10\times$ objective, one should see 5–10 cells per field on the antibody-coated dish. More accurate counts can be obtained by detaching the cells from the dish immediately after the isolation with 5 ml of 0.1% trypsin (w/v) and 0.05% EDTA (w/v) in PBS, pH 7.3, at 37° for 45–60 min. Typically, we obtain about 10^6 CCCT or RCCT cells per dish. This compares with 10^3 cells bound to control dishes coated with bovine serum albumin.

The isolated cells may either be grown on the antibody-coated dishes or transferred immediately to other types of culture dishes (e.g., 24-well Costar dishes on which experiments can be performed directly).[10,12] If the isolated cells are to be grown in medium containing serum, serum that has been decomplemented (incubated at 56° for 15–20 min) should be used initially to avoid complement-mediated cell lysis.

Characterization of Isolated Cells

Cells isolated by immunodissection need to be well characterized to confirm their site of origin and their degree of homogeneity. Morphologi-

[25] M. K. Patterson, Jr., this series, Vol. 58, p. 141.

cal criteria include such features as the shape of the cells in culture and the specific subcellular characteristics of the cells as seen by transmission electron microscopy.[10-12,26] In our specific instances, where the isolated cells should behave like transporting renal epithelia, the ability of the cells to form domes, exhibit membrane asymmetry and tight junctions, and develop transepithelial potential differences of the correct sign are used as renal segment-specific criteria.[10,11,26]

Isolated cells should also be stained histochemically for marker enzymes[11,18,27] and the staining compared with that observed in cryotome sections of kidney stained under similar conditions. Homogeneous cell populations yield homogeneous staining when examined histochemically. Binding of antibodies or lectins which are tissue specific may also be used. Peanut agglutinin has been shown to specifically bind the intercalated cells of the rabbit collecting tubule,[28] and we have used this lectin to characterize RCCT cells.[12]

A key factor in identifying isolated cells is their response to cell-specific hormones.[10-12,29] This is important not only for characterization but can also provide a quantitative indication of the degree of differentiation. For example, CCCT cells form cAMP in response to arginine-vasopressin and isoproterenol, but not to parathyroid hormone. CCCT cells also synthesize prostaglandins. These properties are unique to collecting tubules and provided evidence that cultured CCCT cells are an appropriate model for studying arginine-vasopressin – prostaglandin interactions in the collecting tubule.[10,26]

Isolation of CCCT cells and RCCT cells provide the first two examples of the use of immunodissection in nephrology. It is likely that this technique will be useful in preparing homogeneous populations of other hormonally responsive renal cells and possibly cells from other tissues containing significant numbers of several different types of cells.

Acknowledgments

This work was supported in part by National Institutes of Health Grants AM 22042 and AM 36485 and was done under the tenure of an Established Investigatorship from the American Heart Association (W.L.S.), a NIH predoctoral traineeship HL 07404 (A.G.P. and W.K.S.), and a NIH Career Development Award HL 01010 (W.S.S.).

[26] A. Garcia-Perez and W. L. Smith, *J. Clin. Invest.* **74**, 63 (1984).
[27] J. D. Bancroft, "Histochemical Techniques." Butterworths, London, 1975.
[28] M. Lehir, B. Kaissling, B. Koeppen, and J. B. Wade, *Am. J. Physiol.* **242**, C117 (1982).
[29] F. Morel, *Am. J. Physiol.* **240**, F159 (1981).

[29] Pancreatic Acini as Second Messenger Models in Exocrine Secretion

By JERRY D. GARDNER and ROBERT T. JENSEN

Exocytosis refers to the process whereby secretory granules release their contents from a cell. The membrane surrounding the secretory granule fuses with the plasma membrane, which ruptures, and the contents of the secretory granule are discharged from the cell. In pancreatic acinar cells, digestive enzymes are packaged in secretory granules called "zymogen granules," and these zymogen granules release their contents by exocytosis in response to appropriate secretagogues.[1] Pancreatic acini, which are clusters of pancreatic acinar cells,[2] constitute a particularly useful preparation for examination of the cellular messengers that mediate the stimulation of enzyme secretion caused by various secretagogues.

Previously[3] we have described in detail the procedures for preparing dispersed pancreatic acinar cells and dispersed pancreatic acini. The procedure for preparing dispersed pancreatic acinar cells is a minor modification of the technique developed by Amsterdam and Jamieson.[4-6] This preparation contains at least 96% acinar cells and these cells retain their responsiveness to pancreatic secretagogues for up to 5 hr in vitro. These cells have been used to measure binding of secretagogues to their cell surface receptors, changes in cellular cyclic nucleotides such as cyclic AMP and cyclic GMP, changes in the transport of cations such as calcium, sodium, and potassium, and stimulation of secretion of pancreatic enzymes. Dispersed acinar cells have been prepared using pancreas from mouse, guinea pig, or rat. The procedure for preparing dispersed pancreatic acini is a minor modification of the technique reported by Peikin et al.[2] Pancreatic acini can be used for at least 5 hr in vitro and have been used to measure the same functions that have been measured using dispersed acinar cells. The major advantage of pancreatic acini is that they secrete much more enzyme in response to pancreatic secretagogues than do dispersed acinar

[1] J. D. Jamieson, in "Secretin, Cholecystokinin, Pancreozymin and Gastrin" (J. E. Jorpes and V. Mutt, eds.), pp. 195–217. Springer-Verlag, New York, 1973.
[2] S. R. Peikin, A. J. Rottman, S. Batzri, and J. D. Gardner, Am. J. Physiol. 235, E743 (1978).
[3] J. D. Gardner and R. T. Jensen, this series, Vol. 109, p. 288.
[4] A. Amsterdam and J. D. Jamieson, Proc. Natl. Acad. Sci. U.S.A. 69, 3028 (1972).
[5] A. Amsterdam and J. D. Jamieson, J. Cell Biol. 63, 1037 (1974).
[6] A. Amsterdam and J. D. Jamieson, J. Cell Biol. 63, 1057 (1974).

cells. Dispersed acini have been prepared using pancreas from mouse, guinea pig, rat, rabbit, dog, or human.

The various secretagogues that act on pancreatic acini stimulate enzyme secretion by activating one of two functionally distinct processes.[7] One process involves binding of the secretagogue to its receptors, increased turnover of phosphatidylinositol, mobilization of cellular calcium, activation of phospholipid-dependent, calcium-sensitive protein kinase (protein kinase C), modification of the phosphorylation of one or more cellular proteins, and, after a series of presently undefined steps, stimulation of enzyme secretion. Receptors which when occupied cause activation of the calcium pathway are those for muscarinic cholinergic agents, cholecystokinin (CCK)-related peptides, bombesin-related peptides, or substance P-related peptides.[8] The other process involves binding of the secretagogue to its receptors, activation of adenylate cyclase, increased cellular cyclic AMP, activation of cyclic AMP-dependent protein kinase (protein kinase A), modification of the phosphorylation of one or more cellular proteins, and, after a series of presently undefined steps, stimulation of enzyme secretion. Receptors which when occupied cause activation of the cyclic AMP pathway are those for vasoactive intestinal peptide (VIP)-related peptides, calcitonin gene-related peptide, or cholera toxin.[8-12] The initial steps in the calcium and cyclic AMP pathways are functionally distinct. Secretagogues that cause mobilization of cellular calcium do not increase cellular cyclic AMP, and secretagogues that increase cyclic AMP do not cause mobilization of cellular calcium. At some unidentified later step these two pathways are functionally coupled. The result of this coupling is potentiation of enzyme secretion when both processes are activated simultaneously by combining a secretagogue that mobilizes cellular calcium with a secretagogue that increases cellular cyclic AMP.

There are a number of different results that have established that cyclic AMP, the classical second messenger, mediates the action of VIP-related peptides, calcitonin gene-related peptide, and cholera toxin on enzyme secretion from pancreatic acini. Each of these secretagogues increases cel-

[7] J. D. Gardner and R. T. Jensen, in "Physiology of the Gastrointestinal Tract" (L. R. Johnson, ed.), pp. 831–871. Raven, New York, 1981.

[8] J. D. Gardner and R. T. Jensen, Annu. Rev. Physiol. 48, 103 (1986).

[9] J. D. Gardner and R. T. Jensen, in "The Exocrine Pancreas" (V. L. W. Go, F. P. Brooks, E. P. Dimagno, J. D. Gardner, E. Lebenthal, and G. A. Scheele, eds.), pp. 109–122. Raven, New York, 1986.

[10] J. D. Gardner and A. J. Rottman, Biochim. Biophys. Acta 585, 250 (1979).

[11] H. Seifert, J. C. Sawchenko, J. Rivier, W. Vale, and S. J. Pandol, Am. J. Physiol. 249, G147 (1985).

[12] Z.-C. Zhou, M. L. Villaneuva, M. Noguchi, S. W. Jones, J. D. Gardner, and R. T. Jensen, Am. J. Physiol. 251, G391 (1986).

lular cyclic AMP in pancreatic acini.[8-12] VIP activates adenylate cyclase in broken pancreatic acinar cells.[13] Exogenous derivatives of cyclic AMP stimulate enzyme secretion from pancreatic acini.[7,14] Inhibitors of cyclic nucleotide phosphodiesterase, the enzyme that degrades cyclic AMP, augment the action of secretagogues on cellular cyclic AMP and on enzyme secretion.[15] Finally, the dose–response curve for occupation of receptors by a particular secretagogue is the same as the dose–response curve for the secretagogue-induced increase in cellular cyclic AMP.[9] One exception to the above statements relates to the action of high concentrations of secretin, a VIP-related peptide, on enzyme secretion from rat pancreatic acini.[16] Although cyclic AMP appears to mediate the actions of secretin in most systems studied, in rat pancreatic acini high concentrations of secretin stimulate enzyme secretion by some presently unknown process that does not involve cyclic AMP as a mediator.[16]

For a number of years calcium was considered to be the second messenger mediating the action on enzyme secretion of those secretagogues that mobilize cellular calcium in pancreatic acini. As it became clear that calcium was being mobilized from intracellular stores,[17] investigators realized that a mechanism was required to transmit a signal from the cell surface receptors for secretagogues to the intracellular calcium stores. An elegant experiment from Schulz's laboratory showed that inositol 1,4,5-trisphosphate (IP$_3$), which was formed in response to receptor occupation,[18,19] could mobilize cellular calcium from pancreatic acinar cells[20] and that this calcium was mobilized from the endoplasmic reticulum.[20]

During the past several years a reasonably clear picture has begun to emerge concerning the second messengers that may mediate the actions on pancreatic enzyme secretion of those secretagogues that mobilize cellular calcium. The following working hypothesis can be set forth to account for the actions of muscarinic cholinergic agents, of CCK-related peptides, of

---

[13] H. L. Klaeveman, T. P. Conlon, and J. D. Gardner, in "Gastrointestinal Hormones" (J. C. Thompson, ed.), pp. 321–344. Univ. of Texas Press, Austin, 1975.

[14] S. R. Peikin, C. L. Costenbader, and J. D. Gardner, J. Biol. Chem. 254, 5321 (1979).

[15] J. D. Gardner, L. Korman, M. Walker, and V. E. Sutliff, Am. J. Physiol. 242, G547 (1982).

[16] B. M. Bissonnette, M. J. Collen, H. Adachi, R. T. Jensen, and J. D. Gardner, Am. J. Physiol. 246, G710 (1984).

[17] J. A. Williams and S. R. Hootman, in "The Exocrine Pancreas" (V. L. W. Go, F. P. Brooks, E. P. Dimagno, J. D. Gardner, E. Lebenthal, and G. A. Scheele, eds.), pp. 123–139. Raven, New York, 1986.

[18] J. W. Putney, G. M. Burgess, S. P. Halenda, J. S. McKinney, and R. P. Rubin, Biochem. J. 212, 483 (1983).

[19] H. Streb, J. P. Heslop, R. F. Irvine, I. Schultz, and M. J. Berridge, J. Biol. Chem. 260, 7309 (1985).

[20] H. Streb, R. F. Irvine, M. J. Berridge, and I. Schultz, Nature (London) 306, 67 (1983).

bombesin-related peptides, and of substance P-related peptides. For those secretagogues receptor occupation is followed by activation of a phospholipase C in the plasma membrane of pancreatic acinar cells.[18,19] Activation of this phospholipase C catalyzes the breakdown of phosphatidylinositol 4,5-bisphosphate to form IP_3 and diacylglycerol.[18,19,21] IP_3 mobilizes cellular calcium, which is involved in the activation of protein kinase C as well as calmodulin-dependent protein kinase. Diacylglycerol activates protein kinase C. Thus, for those secretagogues that mobilize cellular calcium the second messengers appear to be IP_3 and diacylglycerol.

The hypothesis set forth in the preceding paragraph has not been established for all classes of secretagogues that mobilize cellular calcium. This is important because each class of secretagogues has a different dose-response curve for stimulation of amylase secretion[7] and these differences may reflect differences in second messenger function. Furthermore, the relation between occupation of receptors by a given secretagogue and the increase by IP_3 or diacylglycerol has not been established. Moreover, the relation between second messenger generation and stimulation of enzyme secretion has only been examined in one study.[21] In this study CCK and carbachol each stimulated formation of diacylglycerol; however, the concentrations of secretagogue that stimulated diacylglycerol formation were supramaximal for stimulating enzyme secretion. Bombesin, which stimulates enzyme secretion almost as effectively as CCK and carbachol, caused little or no increase in diacylglycerol.

In summary, pancreatic acini constitute a particularly advantageous preparation to use to examine second messenger function in exocytosis. For some secretagogues it is unequivocally established that their ability to stimulate enzyme secretion is mediated by cyclic AMP. For other secretagogues (those that mobilize cellular calcium), IP_3 and diacylglycerol are probably involved as second messengers; however, additional studies are necessary to establish the quantitative relationships between receptor occupation, messenger generation, and stimulation of enzyme secretion.

[21] S. J. Pandol and M. S. Schoenfield, *J. Biol. Chem.* **261**, 4438 (1986).

[30] Ascites Cell Preparation: Strains, Caveats

By ROSE M. JOHNSTONE and PHILIP C. LARIS

Prior to the now prevalent use of cultured mammalian cell lines, investigators depended on slices or minces of tissues for experimental work. Although these preparations enabled investigators to make many basic discoveries about the behavior and properties of mammalian cells, there were uncertainties about these properties since none of these preparations represented a uniform population of a single type of cell. Hence the mammalian red cell has become a prototype for many studies of mammalian cell function, due in considerable degree to its ready availability, simplicity, ease of preparation, as well as its unique function in mammalian physiology. Yet, without denying the importance of red cell work to an assessment of the physiology of mammalian cells, the absence of nuclear events and the high degree of specialization of red cells leave large areas of cell physiology unaddressed. The ability to grow cell lines in ascitic form provides an opportunity to obtain large amounts of a relatively pure population of cells. While ascitic cell lines do not provide the investigator with the same type of control over conditions of growth or absolute freedom from control from contaminating cells that are provided with *in vitro* methods, the ascites cell lines generally produce large numbers of cells which are easy to harvest and to maintain. A single mouse, for example, can yield 10^9 or more Ehrlich ascites tumor cells after a culture period of 7 days. At the cost of less than $2.00 per mouse, the Ehrlich cell is less expensive to produce than most *in vitro* culture lines.

As shown in Table I,[1-44] a number of malignant cell lines have been developed which can be grown in the peritoneal cavities of laboratory animals. Of these, the most frequently studied for over 30 years has been

[1] A. P. Sherblom, J. W. Huggins, R. W. Chesnut, R. L. Buck, C. L. Ownby, G. B. Dermer, and K. L. Carraway, *Exp. Cell Res.* **126,** 417 (1980).

[2] S. Matzku, D. Komitowski, M. Mildenberger, and M. Zoeller, *Invasion Metastasis* **3,** 109 (1983).

[3] E. C. Cadman, D. E. Dix, and R. E. Handschumacher, *Cancer Res.* **38,** 682 (1978).

[4] K. Nakaya, T. Shimizu, Y. Segawa, S. Nakajo, and N. Nakamura, *Cancer Res.* **43,** 3778 (1983).

[5] T. Yoshida, *Methods Cancer Res.* **6,** 97 (1971).

[6] D. F. Smith, E. F. Walborg, Jr., and J. P. Chang, *Cancer Res.* **30,** 2306 (1970).

[7] P. Lando, J. Gabriel, K. Berzins, and P. Perlmann, *Scand. J. Immunol.* **14,** 3 (1982).

[8] A. B. Okey, G. P. Bondy, M. E. Mason, D. W. Nebert, C. J. Forster-Gibson, J. Muncan, and M. J. Dufresne, *J. Biol. Chem.* **255,** 11415 (1980).

[9] A. B. Novikoff, *Cancer Res.* **17,** 1010 (1957).

[10] F. Nato, M. Aubery, and R. Bourrillon, *Biochim. Biophys. Acta* **718,** 11 (1982).

[11] A. Tax and L. A. Manson, *Cancer Res.* **39,** 1739 (1979).

[12] C. Miyaji, Y. Ogawa, Y. Imajo, K. Imanaka, and S. Kimura, *Oncology* **40,** 115 (1983).

[13] A. M. Hughes and M. Calvin, *Cancer Lett.* **6,** 15 (1979).

[14] G. Klein, *Cancer* **3,** 1052 (1950).

[15] P. Ehrlich and H. Apolant, *Z. Naturforsch.* **42,** 871 (1905).

[16] H. Loewenthal and G. John, *Z. Krebsforsch.* **37,** 439 (1932).

[17] G. Klein and E. Klein, *Cancer Res.* **11,** 468 (1951).

[18] A. W. Thomson, D. J. Tweedie, R. G. P. Pugh-Humphreys, and M. Arthur, *Br. J. Cancer* **38,** 106 (1978).

the Ehrlich ascites tumor cell. This cell was derived from a spontaneous mammary carcinoma in the mouse and was then grown in the peritoneal cavity in ascites form after inoculation with a minced soma suspension. It was first made popular by Klein[45] in 1950. The cell line established by Klein appears to be the source of the material for subsequent work with Ehrlich cells in much of North America and Europe (Christensen, Jacquez, Hempling, Levinson, Johnstone, Maizels, Heinz). The simple original protocol reported by Klein seems to be adequate to produce an ascitic tumor from a solid, i.e., lump-type tumor, precursor.

[19] G. S. Tarnowski, I. M. Mountain, C. C. Stock, P. G. Munder, H. V. Weltzien, and O. Westphal, *Cancer Res.* **38**, 339 (1978).

[20] H. Taper, G. W. Woolley, M. N. Teller, and M. P. Lardis, *Cancer Res.* **26**, 143 (1966).

[21] J. L. Portis, *J. Natl. Cancer Inst.* **62**, 611 (1979).

[22] J. Hilgers, J. Haverman, R. Nusse, W. J. van Blittersivijk, F. J. Cleton, P. C. Hageman, R. van Nil, and J. Calafat, *J. Natl. Cancer Inst.* **54**, 1323 (1975).

[23] C. W. Jackson, R. A. Krance, C. C. Edwards, M. A. Whidden, and P. A. Gauthier, *Cancer Res.* **40**, 667 (1980).

[24] R. T. Mulcahy, M. N. Gould, E. Hidvergi, C. E. Elson, and M. B. Yatvin, *Int. J. Radiat. Biol. Relat. Stud. Phys. Chem. Med.* **39**, 95 (1981).

[25] W.-T. Liu, M. J. Rogers, L. L. Law, and K. S. S. Chang, *J. Natl. Cancer Inst.* **58**, 1661 (1977).

[26] E. Luevano, V. Kumar, and M. Bennett, *Scand. J. Immunol.* **13**, 563 (1981).

[27] W. W. Young, Jr., Y. Tamura, H. S. Johnson, and D. A. Miller, *J. Exp. Med.* **157**, 24 (1983).

[28] A. Ray, R. Barzilai, G. Spira, and M. Inbar, *Cancer Res.* **38**, 2480 (1978).

[29] J. M. Davis, A. K. Chan, and E. A. Thompson, Jr., *J. Natl. Cancer Inst.* **64**, 55 (1980).

[30] G. Mercier and J. Harel, *Eur. J. Biochem.* **123**, 407 (1982).

[31] M. C. Cohen, J. P. Brozna, and P. A. Ward, *Am. J. Pathol.* **94**, 603 (1979).

[32] C. Reyero, D. E. Burger, S. Ritacco, and J. M. Varga, *Cell. Immunol.* **78**, 290 (1982).

[33] A. W. Hamburger, F. E. Dunn, and K. L. Tencer, *J. Natl. Cancer Inst.* **70**, 157 (1983).

[34] M. Morimoto, D. Green, A. Rahman, A. Goldin, and P. S. Schein, *Cancer Res.* **40**, 2762 (1980).

[35] A. Zachowski and A. Parof, *Biochem. Biophys. Res. Commun.* **57**, 787 (1974).

[36] H. C. Morse III, M. E. Neiders, R. Lieberman, A. R. Lawton III, and R. Asofsky, *J. Immunol.* **118**, 1682 (1977).

[37] A. Liepens and E. de Harven, *Scanning Electron Microsc.* **11**, 37 (1978).

[38] R. H. Mathews, D. Dewar, and T. E. Webb, *Physiol. Chem. Phys.* **8**, 151 (1976).

[39] S. H. Lee and A. C. Sartorelli, *Cancer Res.* **41**, 1086 (1981).

[40] M. J. Marinello and A. Levan, *Hereditas* **96**, 42 (1982).

[41] J. T. Meyer, P. M. Thompson, R. Behringer, R. C. Steiner, W. M. Saxton, and S. B. Oppenheimer, *Exp. Cell Res.* **143**, 63 (1983).

[42] J. Fujimoto, *J. Natl. Cancer Inst.* **50**, 79 (1979).

[43] T. C. Hamilton, R. C. Young, W. M. McKay, K. R. Grotzinger, J. A. Green, E. W. Chu, J. Whang-Peng, A. M. Rogan, W. R. Green, and R. F. Ozols, *Cancer Res.* **43**, 5379 (1983).

[44] T. C. Hamilton, R. C. Young, K. G. Lowe, B. C. Behrens, W. M. McKay, K. R. Grotzinger, and R. F. Ozols, *Cancer Res.* **44**, 5286 (1984).

[45] G. Klein, *Cancer* **3**, 1052 (1950).

TABLE I

Sample of Ascites Cell Lines Derived from a Literature Search (1976–1983)[a]

Type	Source	Designation	Comments
Rat			
Adenocarcinoma	Mammary gland	13672[1]	Sublines can grow in mice
Adenocarcinoma	Pancreas	BSP-73[2]	—
Adenocarcinoma (carcinosarcoma)		W-256,[3] Walker	—
Hepatoma	Liver	AH-66[4]	Maintained in Donryu rats
Hepatoma	Liver	AH-130,[5] Yoshida	Induced by feeding 4-dimethyl-aminoazobenzene
Hepatoma	Liver	AS-30D,[6] Chang	Induced by 3'-methyl-4-dimethyl-aminobenzene
Hepatoma	Liver	D-23 asc.[7]	
Hepatoma	Liver	HTC[8]	Derived from the "minimal duration" hepatoma 7288C
Hepatoma	Liver	NHAC,[9] Novikoff	Induced by feeding 4-dimethylami-noazobenzene
Hepatoma	Liver	Zajdela[10]	—
Mouse			
Adenocarcinoma	Mammary gland	A-10[11]	Spontaneous tumor which arose in a female A/He mouse in 1968 and subsequently transformed into an ascites form
Adenocarcinoma	Mammary gland	MM-46[12]	Spontaneous tumor in a C3H/He mouse
Adenocarcinoma		TA-3[13]	
Carcinoma	Mammary gland	Ehrlich[14]	Most commonly used ascites cells; solid tumor originally described in 1905[15] transformed into ascites in 1932[16]*
Carcinoma		Krebs II[17]	Second most commonly used ascites cell originated spontaneously as a solid carcinoma in the inguinal region of a hybrid male mouse
Carcinoma	Mammary gland	LAC,[18] Landschuetz	Derived from the Ehrlich cell
Fibrosarcoma	—	Meth A[19]	—
Hepatoma	Liver	Taper[20]	Poorly differentiated primary liver tumor of spontaneous origin in the Swiss mouse
Leukemia	—	FBL-3[21]	Induced by Friend virus
Leukemia	—	GRSL14[22]	Spontaneous origin, grown in GR mice
Leukemia	—	L-1210[23]	*
Leukemia	—	P-388[24]	*
Leukemia	—	RBL-5[25]	Virus induced
Lymphoma	—	EL-4[26]	—
Lymphoma	—	L-5178Y[27]	—
Lymphoma	—	YAC[28]	Virus induced

TABLE I *(continued)*

Type	Source	Designation	Comments
Lymphosarcoma	—	P1798[29]	
Lymphosarcoma	—	FLS[30]	Initiated in suckling Swiss mice which had been injected ip with nucleic acids from lymphoid tissue
Mastocytoma	—	P-815[31]	—
Myeloma	—	MOPC-315[32]	*
Myeloma	—	MPC-11[33]	*
Plasmacytoma	—	LPC-1,[34] Lieberman	—
Plasmacytoma	—	MOPC-173[35]	Line MF_2F^+ grows in mice as ascites
Plasmacytoma	—	SAMM 368[36]	Arose as an ascitic tumor in a pathogen-free BALB/c mouse after three ip injections of peritane (2,6,10,14-tetramethylpentadecane)
Sarcoma (reticulum)	—	J774[37]	—
Sarcoma	—	S-37[38]	—
Sarcoma	—	S-180[39]	*
Sarcoma	Bone	SEWA[40]	Induced in 1959 by inoculation of poliovirus. The primary tumor was an osteosarcoma. After it had been placed in serial *in vivo* transfer, it dedifferentiated into an anaplastic sarcoma at passage 10 and has been carried since as an ascites in ASW mice, the strain of origin
Teratocarcinoma	—	OTT6050-2568[41]	Ref.: Meyer *et al.* (1983); an embryoid body (now single-cell) form used
—	Embryo cells	Fujimoto[42]	Ref.: Fujimoto (1979); derived from C3H/He mouse embryo cells treated with 4-nitroquinoline 1-oxide and maintained by passage through strain of origin
Human Adenocarcinoma	Ovary	NIH:OVCAR-3[43,44]	From malignant ascites of a patient with common epithelial ovarian cancer. Cell line subjected to selection procedures for subpopulations which will grow in heterologous hosts. Adapted to subcutaneous injection in nude athymic mice (BALB/c) and then as ascites

[a] Asterisk indicates cell line is available from American Type Culture Collection, 12301 Parklawn Drive, Rockville, Maryland 20852.

Procedure for Establishing and Maintaining an Ascites Cell Line

After excision of the solid tumor, any necrotic areas are removed and the residual tissue minced with scissors in the presence of a small quantity of isotonic saline. Sterile saline is added to the finely chopped tissues to bring the tissue concentration to approximately 10% (by volume) and the suspension is aspirated into the barrel of a sterile syringe. The suspension is then forced through a large-gauge hypodermic needle several times to obtain a fine tissue suspension, which is filtered through gauze. The filtered homogenate is used as a source of cells for intraperitoneal (ip) injection.

An inoculation of 10^7 Ehrlich cells is adequate to produce a substantial (5×10^8 cells) harvest in 6–7 days. A smaller inoculum (10^6 cells) will also produce an ascitic tumor. It has been reported[46] that the TD_{50} (the number of cells which produces a fatal take in 50% of the mice) for an Ehrlich ascites tumor strain is close to 250 cells and that as few as 10 FBL-3 ascites cells produced ascites and ultimately the death of the recipient.[21]

Some investigators have passed a primary tumor subcutaneously prior to cultivation of the cell line in the peritoneal cavity. This precaution may remove some unwanted cells from the primary transplant.

Once established, the ascitic cell lines can be maintained by passage of the ascitic cells to a new host. For Ehrlich cells, the optimum time for transplantation is 6–8 days after inoculation. With other cell lines the culture time for optimal growth may be as long as 120 days.[47] The cells may be transferred directly in the ascitic fluid or can be isolated under sterile conditions, washed, resuspended in isotonic saline, and inoculated at a known concentration of cells. Alternatively, the cell number in the ascitic fluid may be determined prior to inoculation so that a known number of cells is transplanted. The first method is simpler and quicker. Although the other procedures permit standardization of the number of cells in the inoculum with the Ehrlich cell and similar types of cells (e.g., S_{37} sarcoma), little advantage is obtained for predicting the yield of cells after 6–8 days of cultivation.

Maintenance of the Line

In our hands it has not been possible to maintain the cell line indefinitely by continuous passage of an ascitic culture. Eventually the line will die out if handled in this fashion. Diminished viability becomes evident

[46] P. K. Lala, *in* "Growth Kinetics and Biochemical Regulation of Normal and Malignant Cells" (B. Drewinks and R. M. Humphrey, eds.), p. 509. Williams & Wilkins, Baltimore, Maryland, 1977.

[47] Y. Ueyama, Y. Kondo, N. Tamaoki, and N. Ohsawa, *J. Natl. Cancer Inst.* **57**, 965 (1976).

earlier by the appearance of solid tumor nodules generally adhering to the intestine in addition to the decreased formation of a free cell suspension. Two procedures have been used to maintain the cell lines: (1) storing a cell suspension at $-70°$ in liquid nitrogen or (2) maintenance of a solid tumor in the animal. For freezing, either a 10% solution of glycerol or dimethyl sulfoxide in isotonic saline can be used. For solid tumor transfer, a subcutaneous injection of $\sim 10^7$ Ehrlich cells will yield a significant solid tumor in approximately 3 weeks. The tumor can then be prepared for ip transfer as described for the primary tumor. It has been our experience with Ehrlich cells that the first passage to ascitic form yields a highly concentrated cell population, which is somewhat more difficult to harvest than that obtained with subsequent passages. The cell yield is also somewhat less on average (30–50%). However, the variability in cell yield is large on first and subsequent passages. Because the cells in the first ascitic passage are generally different from those of subsequent passages, we have generally used cells from the second and subsequent passages for routine experiments. A cycle of approximately 10 weeks for solid to ascitic transfer is recommended.

Preparation of Ascitic Cells for Experimental Use

Ascitic cells are readily harvested by aspirating the fluid from the peritoneal cavity. After the animal is sacrificed, a small cut is made in the abdominal wall. It is preferable to cut through the surface layer, exposing the abdominal wall before cutting through it. The fluid in the peritoneal cavity can be excised with a syringe or Pasteur pipet through a small incision in the wall. The peritoneal cavity can be rinsed with saline to help harvest most of the cells.

Ehrlich cells and many other tumor lines can be readily separated from contaminating blood cells by differential centrifugation. The ascitic cells form a loose pellet after centrifugation for 20–30 sec at 1400 g, whereas blood cells usually remain in suspension. Two or three such centrifugations will yield a suspension almost totally free from red cells. Many investigators choose to reject cells highly contaminated with erythrocytes. A preparation of approximately 10^8 cells can generally be obtained within 10 min of sacrificing the animal.

Cells kept on ice retain the capacity to transport materials and to synthesize proteins for 2–3 hr. It is common practice to isolate and wash cells with cold buffered saline. While this procedure probably maintains cell viability, it leads to changes in intracellular ion composition which may introduce unwarranted conclusions to subsequent experiments unless the cells are first incubated at warmer temperatures so as to restore normal ionic balance.

Host Specificity

While the commonly used Ehrlich cell and the Krebs II ascites cell appear to grow in most strains of mice, other tumor lines show varying degrees of host preference. In some cases the cells are typically grown in the mouse strain of their origin, e.g., Fujimoto cells.[42] The ability of the Ehrlich cell to grow in most strains of mice is limited to the ascitic form of growth. When grown as solid tumors, the tumorigenic potential may vary from high to low, depending on the ascites cell and the host.[48] The number of cells required to produce a solid tumor, a fatal take, subcutaneously is generally much greater than that necessary via the intraperitoneal route, e.g., according to Lala,[46] the TD_{50} for the Ehrlich cell in CF_1 strain mice is approximately 250 cells intraperitoneally and 10^6 cells subcutaneously. The difference may be due to environmental limitations for specific growth requirements or the greater susceptibility of the subcutaneous transplants to immune mechanisms. Subcutaneous transplants which grew poorly in some male mice grew well in hosts that had been irradiated.[49,50] Furthermore, mice carrying Ehrlich ascites do not exhibit splenomegaly whereas marked splenomegaly is seen in mice bearing the same tumor in solid form.[51] Splenomegaly is usually accompanied by a parallel decrease in thymus weight. These phenomena are expressions of the host's response to a tumor.

Some tumors will also grow in heterologous hosts. Two of five sublines, MAT-C and MAT-C1, of the 13762 rat mammary adenocarcinoma, for example, can be transplanted into mice.[1] Although these two of the five sublines have the greatest amounts of total, trypsin-releasable, and neuraminidase-releasable sialic acid, the transplantation differences between these sublines cannot be explained solely by the presence of a major sialoglycoprotein at the cell surface.[1]

Since heterologous cell lines can grow in immunologically compromised hosts such as nude mice, it should be possible to propagate human cell lines as ascites cells in small laboratory animals. The growth of ascites cells derived from a human stomach in nude mice has been reported.[21]

Variability in Rate of Growth and Yield of Cells

Transplanted ascites tumors are a very special class of cell populations adapted to fast growth as free cells suspended in a serous cavity.[46] Most

[48] P. K. Lala, *Methods Cancer Res.* **6,** 3 (1971).
[49] S. Thunold, *Acta Pathol. Microbiol. Scand.* **68,** 37 (1966).
[50] S. Thunold, *Acta Pathol. Microbiol. Scand.* **70,** 259 (1967).
[51] A. Rivenson, V. Schnelle, H. Moroson, R. Madden, and A. Herp, *Experientia* **37,** 195 (1981).

ascites cells are highly transformed cells and show similar traits probably because they have been selected to grow in the peritoneal cavity. During the initial growth phase, the exponential phase, the doubling time of the cell depends on the transplant size; it increases with an increase in the initial inoculation size.[52] The same tumor mass at the plateau growth phase is, however, obtained with transplants of different sizes.

Most noncycling cells in ascites tumors can resume dividing on transplantation to another host or upon the aspiration of some cells from the original host. Cell division in most of these cells appears to be restricted by nutritional deprivation or toxic factors and these cells have a limited life span. Others may be held back by specific feedback inhibitors or by immune inhibition.

Use of Ascites Tumors as an Experimental Tool to Measure the Response to Drugs and Antitumor Agents

From the time that Ehrlich cells were introduced, it was proposed that ascitic forms held great promise as a means of testing the antitumor efficacy of experiment drugs in a host line. The reality, however, has been less than the promise. The reasons are probably manyfold. Among the prime problems are (1) the variability of the cells produced per unit inocculum and (2) instability of the karyotype and changes in cell properties with the number of cell passages or relatively small differences in the culture conditions. The latter point is illustrated by a report[21] of a large difference in the virulence of murine leukemia FBL-3 cells, depending on the time they were grown as ascites cells. Cells grown as ascites for 7 days prior to subcutaneous injection produced tumors which were rejected 20–40 days after injection. In contrast, this line caused 50–70% mortality in the host when injected subcutaneously after 14 days of culture as ascites. This difference in virulence appeared to be due to a change in the cells, for day 7 ascites cells (nonlethal) became lethal when grown as ascites for 14 days and day 14 ascites cells (lethal) became nonlethal when passed as ascites for only 7 days. Hence, even a single culture of ascites cells may vary significantly during the period of cultivation, in this case 7 days differing from 14.

Experimental Work and the Problem of Multiple Cell Lines

The rapid rate of growth of ascites cells and the ease with which they can be cultured and handled have made ascites cells, especially the Ehrlich

---

[52] P. K. Lala, "Effects of Radiation on Cellular Proliferation and Differentiation," p. 463. Int. At. Energy Ag., Vienna, 1968.

ascites tumor cell and the Krebs II ascites cell (these cells were used in more than 70% of the papers on ascites cells found in our search), favorite research materials.

Because of their general utility, Ehrlich cells have been grown in many laboratories for over 30 years. Given the earlier discussion of changes in these cells during their culture, one should expect that differences must be present in these lines, although they apparently were derived from a common source. According to Lala et al.[53] "possibly no two lines are alike." Yet relatively few sublines have been described and most investigators simply refer to the Ehrlich ascites tumor cell. The differences between two sublines may be extensive. For example, the glycogen-free and glycogen-containing sublines[54] have been shown to differ also in colchicine resistance, lipid synthesis, cell surface properties such as sialic acid content, electrophoretic mobility, enzyme activity, lipid composition, and membrane fluidity. When reports from different laboratories are not in agreement, investigators should consider the possibility that the differences result from variants in the cells as well as differences in experimental design. Recently, for example, the membranes of Ehrlich cells were reported to exhibit different properties when incubated with glucose. Using a voltage-sensitive dye to monitor the membrane potential, Heinz et al.[55] recorded a hyperpolarization with glucose metabolism which they attributed to proton pump. Employing the same dye, Laris and Henius[56] observed no change in membrane potential but instead recorded cell shrinkage resulting from an apparently electrically silent loss of K^+Cl^-. When the two cell lines were tested in the same (Laris) laboratory, the original observations of each laboratory were repeated. There may be many other such apparent discrepancies in the literature.

There is evidence that variants of the Ehrlich cell may arise spontaneously by the fusion of these cells and host cells in vivo.[53] Host-specific T_6 chromosome markers appeared in Ehrlich cells propagated in CBA/HT$_6$ mice. Tumor–host cell hybrids with nearly complete sets of parental chromosomes had a limited life span. Survival value was increased with the elimination of some unidentified chromosomes, resulting in karyotypes with a chromosome composition similar to the original line. The frequency of the hybrid rose to 18% of the population and remained close to this figure after 4 years of weekly passage. It is not known whether the observed host–tumor cell fusion and the consequent alterations of the properties of

[53] P. K. Lala, V. Santer, and K. S. Rahil, Eur. J. Cancer 16, 487 (1980).
[54] E. W. Haeffner, B. Heck, and K. Kolbe, Biochim. Biophys. Acta 693, 280 (1982).
[55] A. Heinz, J. W. Jackson, B. E. Richey, G. Sachs, and J. A. Schafer, J. Membr. Biol. 62, 149 (1981).
[56] P. C. Laris and G. V. Henius, Am. J. Physiol. 242, C326 (1982).

the tumor represent a natural phenomenon or a laboratory artifact created by repeated transplantations of tumor cells. Neither is it known if this mechanism may lead to changes in tumor cell phenotype during the development of a spontaneous tumor.[53]

Section IV

Polar Cell Systems

[31] Direct Current Electrical Measurement in Epithelia: Steady-State and Transient Analysis

By WARREN S. REHM, M. SCHWARTZ, and G. CARRASQUER

Introduction

The purpose of this chapter is to describe the direct current method for determining the resistance of biological tissues and to provide examples how resistance measurements advance our understanding of the mechanisms of biological transport. For our purpose we will use the findings on the *in vitro* frog gastric mucosa. Step current applied to this tissue produces a small initial step increase in the transmucosal potential difference (PD), which is followed by a rapid exponential-like rise which seems to level off in the millisecond range. However, the leveling off in the millisecond range is only apparent because the PD continues to increase, but at a much slower rate, and reaches a final level in about 2 min. The continued increase in PD after about 100 msec is referred to as the long-time constant transient (LTCT). Some epithelial tissues do not possess a LTCT and some other tissues do. The response of the PD over the whole time domain has not received much attention for many epithelial tissues. However, it has been investigated, under a variety of conditions, for the *in vitro* frog fundus. Hence the frog fundus is the appropriate tissue for our review. Most of the work has been performed on *Rana pipiens* but substantial amounts of work have also been performed on other species such as, *Rana esculenta, Rana temporaria,* and *Rana catesbeiana.* In general, the overall picture of the transport mechanisms is not dependent on the species of frog. Most of the results reported herein were obtained from *R. pipiens.*

Response of PD to Step Currents

In determining the resistance of an ideal resistor, the effect of an applied current *(I)* on the PD is obtained and the resistance is calculated as $\Delta PD/I$. For an ideal resistor, application of a step current results in a step increase in PD which is maintained during current flow, and following the break of the current the PD decreases stepwise to the original level. With an ideal resistor, the PD, after the make of the current, is not a function of time and there is no problem in determining the resistance.

METHODS IN ENZYMOLOGY, VOL. 171

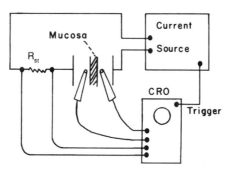

Fig. 1. Diagram of method of measuring response of PD across frog gastric mucosa to single pulses of direct current of constant magnitude. Current source consists of two Tektronix pulse generators that trigger the horizontal sweep of a dual-beam cathode-ray oscilloscope (Tektronix 502A) at a fixed time before sending a direct current across the gastric mucosa. The PD across a standard resistor R_{st} was recorded on one beam of the oscilloscope ($I = PD/R_{st}$), while the PD across the mucosa was recorded on the second beam. The responses were then photographed by a Polaroid camera. (Reproduced from Ref. 1.)

In contrast, with living tissue, a step current produces, in general, a small increase in PD, which is followed by an exponential-like increase of the PD which appears to level off in the millisecond range. In order to accurately record the PD versus time, a cathode ray oscilloscope is used. Figure 1 shows the circuit—a current source with a rise time of a few microseconds is connected to the tissue via a standard resistor. The PD across the resistor and across the tissue is measured by means of a dual-beam oscilloscope.[1] Figure 2 shows a typical experiment where the PD did not level off before the break of the current.[1] Figure 3 shows an experiment in which the PD was recorded over a much longer period.[2] After the application of a step current the PD appears to level off in the range between 50 and 100 msec. However, this leveling off is only apparent—the PD continues to increase to a final level in about 2 min. The rate of increase in PD in the millisecond range is very high compared to the rate of increase after the apparent leveling off in the millisecond range. In order to present the total ΔPD we have used five linear time scales, each one being a 10-fold change from the previous one.[2]

The question arises as to how to determine the transmucosal resistance, R_t. As a first approach we use a model (as shown in Fig. 3) with five resistors in series, with four of the resistors shunted by capacitors (C). Before analyzing these findings on the basis of this model we will show that the gastric mucosa is a linear–bilateral system, at least in the millisecond range, i.e., if the ΔPD/I versus time is independent of the direction of I, it is

[1] W. S. Rehm, *Fed. Proc. Am. Soc. Exp. Biol.* **26,** 1303 (1967).
[2] D. H. Noyes and W. S. Rehm, *Am. J. Physiol.* **219,** 184 (1970).

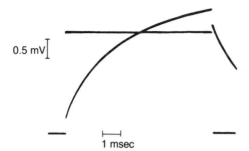

0.5 mV⌶

⊢—⊣
1 msec

FIG. 2. Typical response of the PD across the mucosa to a pulse of direct current (10 μA/1.3 cm²) of about an 8-msec duration. Mucosa bathed in standard Cl⁻ bathing media, nonsecretory frog mucosa, inhibition by thiocyanate (SCN). The stepwise increase and decrease of the one beam is due to the change in PD across the standard resistor (see Fig. 1). (Reproduced from Ref. 1.)

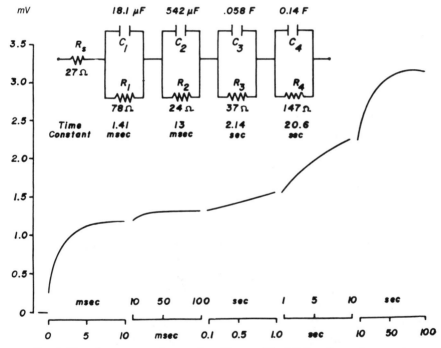

FIG. 3. Plot of a representative transient. Current (10 μA/1.3 cm²) sent from nutrient to secretory side at $t = 0$. Values on ordinate are changes in PD (PD during current minus PD before current) following step current and were obtained from oscilloscope records and oscilloscope photographs. Abscissa consists of five linear time scales, each one being a 10-fold reduction of the previous one. Numerical values given for the equivalent circuit are those for this transient for an area of 1.3 cm². (Reproduced from Ref. 2.)

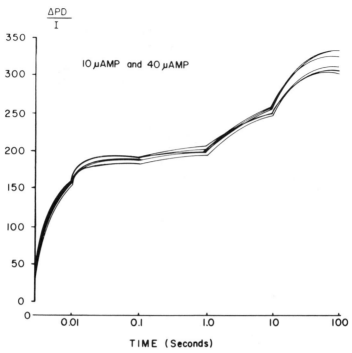

FIG. 4. A plot of $|V/I|$ versus time for eight transients from one mucosa. First, 10 μA/1.3 cm² sent nutrient to secretory (two transients, one on and one off), then from secretory to nutrient (two transients), then 40 μA/1.3 cm² sent nutrient to secretory, and then secretory to nutrient. Abscissa consists of five linear time scales 0–0.01, 0.01–0.1, 0.1–1.0, 1.0–10, and 10–100, but gaps in the time scale as in Fig. 3 are omitted. Data obtained from oscilloscope and oscillograph records were then plotted for each transient on separate graphs. All eight transients were then traced on the same graph without separate designations. Total duration of experiment was about 20 min. (Reproduced from Ref. 2.)

a linear–bilateral system.[2] Figure 4 shows the superposition of eight transients in which ΔPD/I is plotted versus time.[2] Currents of 10 μA and then 40 μA were sent first in one direction and then in the opposite direction. The ΔPD/I response is shown for charging and discharging for each direction of current flow and for the two current densities (i.e., four responses of 10 μA and four for 40 μA). It can be seen that the curves in the millisecond range are superimposable, but there are deviations from linear–bilateral behaviors in the range from 1 to 100 sec.

With the response of the PD, as shown in Fig. 3, the question arises as how to determine R_t. Is R_t a function of all of the R values or only some of them? The values of the capacitances and resistances were initially obtained by the classical "stripping techniques" (but vide infra).

Method of Analysis for Obtaining the R and C
Values of the Circuit in Fig. 3

Since the PD is flat between 50 and 100 msec, the transient can be separated for analysis into two parts, i.e., periods from about 0.05 to 100 msec and from 100 msec to 100 sec. We assume that the total transient can be accurately represented by the type of circuit illustrated in Fig. 3, in which each of the elements is assumed to be linear. It should be pointed out that this circuit is an equivalent circuit and does not necessarily represent the actual circuit. A PD response (V) to a step current (I) would, on the basis of circuit analysis (see Fig. 3 for definition of symbols), be given by

$$V = R_s I + \sum_{i=1}^{M} R_i I [1 - \exp(-t/T_i)] \tag{1}$$

where the time constant, T_i, is given by

$$T_i = C_i R_i \tag{2}$$

and the C and R values are those shown in the circuit in Fig. 3; V represents the response of the PD to the current and equals the PD during current flow minus the PD before the application of current. At infinite time, i.e., the time when the PD has reached a plateau, the voltage would be given by

$$V_\infty = R_s I + \sum_{i=1}^{M} R_i I \tag{3}$$

and subtracting Eq. (1) from Eq. (3) yields

$$\Delta V = V_\infty - V = \sum_{i=1}^{M} R_i I \exp(-t/T_i) \tag{4}$$

It follows from circuit analysis that for the circuit in Fig. 3, the discharge voltage $V_d = \Delta V$ of Eq. (4), provided that during the charging process V reaches a plateau. The expression V_d represents, after the step decrease following the break of the current, the PD during the discharge minus the PD at the end of the discharge transient. A plot of log ΔV versus time (or log V_d versus time for a discharge transient) for the part of the transient shown in Fig. 3 between 0.1 and 100 sec is shown in Fig. 5. It can be seen that the plot is linear from about 7 sec on, and hence this part of the transient can be accurately represented by the RC circuit with the longest time constant $T_4 = C_4 R_4$, i.e.,

$$\Delta V = R_4 I \exp(-t/T_4) \tag{5}$$

Taking the log of both sides of Eq. (5) yields

$$\log \Delta V = \log R_4 I - t/2.3 R_4 C_4 \qquad (6)$$

The linear portion of the curve is extrapolated to $t = 0$ and equals $R_4 I$. Since I is known, R_4 is obtained. The value of C_4 is then obtained from the value of R_4 and the slope $(-1/2.3 C_4 R_4)$. The values from the extrapolated line are subtracted from the actual values and the log of these values are plotted versus time; it can be seen in Fig. 5 that the plot is linear. From the slope of this line and its intercept the values of C_3 and R_3 are obtained.

As pointed out above, the analysis can be applied in the region from 0.05 to 100 msec since the PD reaches a relatively flat level between 50 and 100 msec. Hence the values for C_1, R_1, R_2, and C_2 are obtained. Figure 6 shows plots for the millisecond region similar to those in Fig. 5 and in similar fashion $C_1 R_1$ and $C_2 R_2$ are obtained. The total PD response can be represented by values of C and R. The value for R_s (see Fig. 3) equals the initial step ΔV divided by I after subtracting the resistance of the bathing fluids (from the probes to the tissue).

An objection to the stripping technique is that the error in drawing the initial straight line is carried over to the rest of the data and the other data have no input in determining the initial straight line. A more adequate method is that in which an exponential regression analysis is used with the aid of a computer. With this analysis the computer is asked for a given

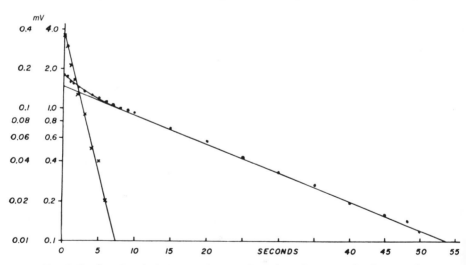

FIG. 5. Semilog plot of ΔV ($V_{100} - V$) versus time of transient shown in Fig. 3, starting at 0.1 sec. Dots are original values for ΔV. Crosses are values for ΔV after subtracting values from longest time constant circuit. Right-hand ordinate scale is for dots and left-hand one for crosses. (Reproduced from Ref. 2.)

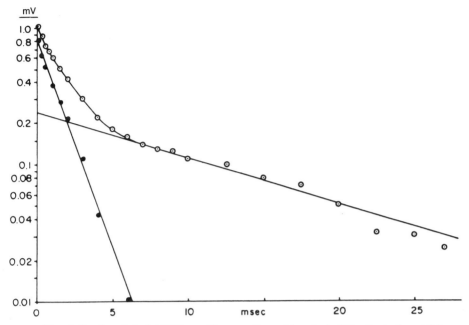

FIG. 6. Semilog plot of ΔV ($V_{100} - V$) versus time of transient of Fig. 3, starting at 0.1 msec. Dots surrounded by circles are original values for ΔV, and solid dots are values after subtracting the $R_2 C_2$ subcircuit values. (Reproduced from Ref. 2.)

transient to give the best fit for a given number of exponentials. Figure 7 shows a typical experiment in which the computer gave the best fit for two exponentials for the time domain of 0.1 to 100 msec.[3] The values for the time constants, the resistances and capacitances, and the "error" are printed in the upper right-hand portion of the figure. The error equals the sum of the squares of the vertical distances (in millivolts) between the data points and the theoretical curve. To obtain the variance, the error is divided by the degrees of freedom (equal to the number of data points minus twice the number of exponentials). For a given transient the computer was asked to give the best fit for one, two, or three exponentials. To determine if three exponentials give a better fit than two, and two a better fit than one, the mean of the ratio of the variances was obtained and the F test (tables in standard references) was applied. From common sense and from our statistical test it was clear that in every case two exponentials gave a significantly better fit than a single exponential. In general, a single

[3] W. S. Rehm, S. S. Saunders, M. G. Tant, F. M. Hoffman, and J. T. Tarvin, *in* "Gastric Hydrogen Ion Secretion" (D. Kasbekar, G. Sachs, and W. S. Rehm, eds.), pp. 29–53. Dekker, New York, 1976.

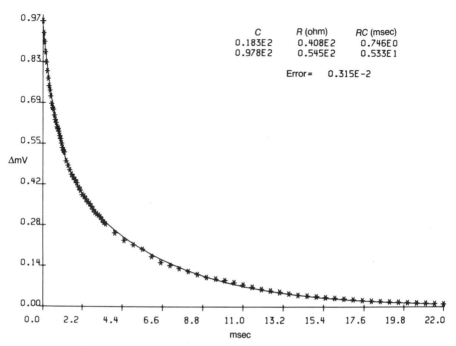

	C	R (ohm)	RC (msec)
	0.183E2	0.408E2	0.746E0
	0.978E2	0.545E2	0.533E1
	Error =	0.315E-2	

FIG. 7. Computer printout of voltage response (ΔmV) to a step current change of 10 μA/1.3 cm^2. The continuous line is the best fit (least-squares method) for a model with two exponentials. The time constants RC (msec), the resistance R (ohms), the capacitances C (μF), and the "error" are printed in upper right. (Reproduced from Ref. 3.)

exponential gave a very poor fit. On the other hand, two exponentials gave what looked like on the basis of a simple inspection a very good fit. In some transients, on the basis of the F test, three exponentials gave a significantly better fit than two and in other cases three was no better than two.[3]

Using the model of Fig. 3, the average values of the capacitances for 1 cm^2 gross area of the four RC subcircuits for the secreting fundus are 27, 480, 72 $\times$ 10^3, and 190 $\times$ 10^3 μF^2. It is well accepted that the capacitance of biological and also of black lipid membranes is about 1 μF/cm^2 of actual membrane area.[4] However, the gastric mucosa is not a flat sheet — there are numerous microvilli and tonguelike projections from the limiting membranes of the cells of the secreting mucosa (e.g., Refs. 5 and 6). So the

[4] K. S. Cole, "Membranes, Ions and Impulses." Univ. of Calif. Press, Berkeley, 1968.
[5] J. G. Forte, J. A. Black, T. M. Forte, T. E. Machen, and J. M. Wolosin, *Am. J. Physiol.* **241,** G349 (1982).
[6] H. F. Helander, S. S. Saunders, L. L. Shanbour, and W. S. Rehm, *Acta Physiol Scand.* **95,** 353 (1975).

question arises: Is the actual area of the membranes of the mucosa large enough to account for the apparent capacitances? The use of morphometric techniques on electron micrographs of the secreting frog mucosa revealed, for the tubular cells, that the actual area of the secretory (lumen-facing) membrane was about 200 cm^2 and the nutrient membrane about 100 cm^2 for 1 cm^2 of gross area (chamber area).[6-9] Due to the complexity of the tissue and the limitations of the morphometric technique, the above values can only be considered an approximation of the actual areas.

In order to check on the adequacy of our morphometric measurements we have used what we refer to as a "common sense" technique. We assume that the histological techniques are adequate to reveal all the projections that exist and that the diameter of these projections in the fixed tissue are approximately those in the unfixed tissue. We assume that the actual area can be approximated by calculating the area of an array of cylinders with the dimensions of the cylinders and the density of the cylinders determined by the actual histological findings. We examined a large number of electron micrographs of the secreting mucosa and found that (1) there are about nine microvilli in the secretory membrane of the tubular cells per micrometer, (2) that the average diameter of the microvillous projections was 0.06 μm (radius = 0.03 μm), and (3) the height (L) of the projections is 2 μm.[5-9] With the above values, the number of cylinders per square centimeter is 81 $\times$ 10^8 and the area of a given cylinder is 38 $\times$ 10^{-10} cm^2 (i.e., 2 rL = 38 $\times$ 10^{-10} cm^2). The total area is the product of the number of cylinders per square centimeter and the area of a cylinder which equals 31 cm^2. There is another factor which results in an increase in area; it is the invagination of the pits and tubules and we found this to be about a factor of 13 in the dog's stomach, and similar considerations for the frog stomach indicate that this factor is about sevenfold.[10] Hence, the total area of the secretory membrane of the frog stomach would be equal to 7 $\times$ 31, or 217, cm^2. It is possible that the values of L, r, and the density of microvilli (cylinders) per unit length are too low and the area is actually larger than 217 cm^2. In an attempt to find an upper limit of the actual area, our observations indicate that L is probably not larger than 3 μm, and that the number of projections per unit length is not greater than 13. With 13 per micrometer, the maximum diameter of the cylinders has to be less than 0.077 μm (i.e., 13 $\times$ 0.077 = 1 μm) otherwise the cylinders would overlap.

[7] H. F. Helander, S. S. Sanders, W. S. Rehm, and B. I. Hirschowitz, in "Gastric Secretion" (G. Sachs, E. Heinz, and K. L. Ullrich, eds.), pp. 69–88. Academic Press, New York, 1972.

[8] C. D. Logsdon and T. E. Machen, Am. J. Physiol. 242, G388 (1982).

[9] S. Ito, Handb. Physiol. 2, 705 (1967).

[10] C. Canosa and W. S. Rehm, Biophys. J. 10, 215 (1958).

But assuming that the cylinders touch each other, which is unlikely, the area per cylinder would be 73×10^{-10} cm^2 and the number of cylinders would now be 169×10^8 and the total area would be 861 cm^2. We believe this is an upper limit of the actual area.

It should be noted that recent work (vide infra) has shown that the resistance of the lumen–tubular cell pathway of the secreting fundus accounts for the low value of R_t and that the parallel pathways—the surface cell and the transintercellular (TIC) pathways—have a much higher resistance. In other words the high-resistance parallel pathways can be ignored in relating the transient results to the capacitance and resistance of the two opposing membranes of the tubular cells.

The question arises as to the relation between the area and the capacitance. The capacitance is given by

$$C = \epsilon\epsilon_0 A/d \qquad (7)$$

where ϵ is the dimensionless dielectric coefficient ϵ_0, the permittivity of free space, which equals 8.85×10^{-14} F/cm, d is the thickness of the dielectric, and A is the area. With $\epsilon = 6$ and $d = 50 \times 10^{-8}$ cm, C is approximately 1 μF for 1 cm^2. In relating the capacitance to the area, the assumption is made that the resistance of the fluid surrounding the cylinders and also the resistance of the fluid inside the cylinders is much less than the actual resistance of the membranes. The capacitance would be less if the above assumption is not approximately correct. Within the framework of the above analysis the capacitance of the secretory membrane could be around 500 μF for 1 cm^2 per gross area. In other words, the 480 μF/cm^2 seems like a reasonable value for the secretory membrane of the tubular cells.

The area of the basolateral membrane of the tubular cells is substantially larger than the gross area and the 27 μF/cm^2 could be attributed to this membrane. It becomes obvious that the 72×10^3 and $190 \times 10^3 \mu$F/cm^2 cannot be explained on the basis of dielectric capacitors—there must be other mechanisms to account for these apparently high capacitances.

There are three possible mechanisms which could explain the exponential-like increase in PD due to the application of step currents.[2] They are (1) actual RC circuits, (2) nonlinear resistors, and (3) polarization of electromotive forces (EMFs). It is clear that the actual areas of the limiting plasma membranes are much too low to account for the 72×10^3 and $190 \times 10^3 \mu$F, the mechanisms for which we refer to as the long-time constant transients (LTCT). The nonlinear resistor mechanism is ruled out by the fact that the gastric mucosa is to a good first approximation a linear–bilateral system for the range from 0 to 100 msec (vide infra). In other words, the long-time constant transients must be due to polarization of EMFs.

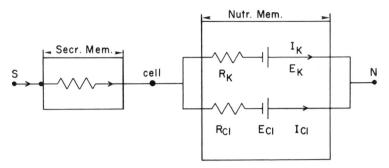

FIG. 8. Equivalent circuit for long-time constant transient. See the text.

Kidder and Rehm[11] provided a satisfactory explanation of the long-time constant transient on the basis of polarization of EMFs. Figure 8 represents the Kidder–Rehm model. It is postulated that upon the application of current across the mucosa, about 100% of the applied current across the secretory membrane is due to Cl^- transport while initially across the nutrient membrane it is due to the sum of K^+ and Cl^- transport. The net transmucosal transport of K^+ is very low,[12] but transport across the nutrient membrane is relatively high so upon application of current the K^+ concentration in the cell changes, and as it changes the diffusion EMF of the K^+ changes, so that eventually there will be no current flow in the K^+ limb and all of the current across the nutrient membrane, in the steady state, will be due to Cl^-. We developed equations for this model which are qualitatively identical to those for an RC circuit, i.e., where K and T are constants. So the capacitative-like type of response for the long-time constant transients can be explained on the basis of polarization of EMFs.

Noyes and Rehm[13] developed equations on the basis of a model with unstirred layers and in which the transference number for K^+ across the membrane is assumed to be unity and the equations, again for low current densities, are essentially the same as for RC circuits, i.e., the same as Eq. (1) but the K and T parameters have meanings different than those for the Kidder–Rehm model. It is reasonable to believe that both the Noyes–Rehm and Kidder–Rehm models are operative for the long-time constant transients.

On the basis of the above discussion it is clear that the long-time constant transients cannot be due to charging of dielectric capacitances and are due to changes in EMFs. Therefore, the PD response after the response in the millisecond range should not be included in the determination of the resistance of the tissue. The two shorter time constant transients are un-

[11] G. W. Kidder III and W. S. Rehm, *Biophys. J.* **10**, 215 (1970).
[12] W. W. Reenstra, J. Bellencourt, and J. G. Forte, *Am. J. Physiol.* **250**, G455 (1986).
[13] D. H. Noyes and W. S. Rehm, *J. Theor. Biol.* **32**, 25 (1971).

doubtedly due to charging of the dielectric capacitors and hence the resistance of the tissue R_t is equal to $R_s + R_1 + R_2$. On the basis of the areas, R_1 is identified with the secretory membrane and R_2 with the nutrient membrane of the tubular (the acid-secreting) cells.

For those readers not too conversant with some of the aspects of the foregoing analysis, a simple experimental approach demonstrates the validity of our explanations for the long-time constant transients. On the basis of the above models, it should be apparent that a K^+ conductance blocker would eliminate the long-time constant transients. Barium blocks K^+ conductance in the nutrient membrane of the frog gastric mucosa (vide infra). Barium produces a marked increase in R_t, and, in its presence, as shown in Fig. 9, the LTCT, as would be predicted, is not present.[2,14,15] This is convincing evidence that the two long-time constant transients are not due to capacitative reactances. Hence R_t can be measured by sending a step current and determining the change in PD after the dielectric capacitors are charged and before the long-time constant transients have an appreciable influence on the PD. So when the PD for the frog mucosa is measured about 0.5 sec after the application of the current, we have an accurate measurement of the resistance of the tissue. The rate of change of V due to the long-time constant transients ($C_3 R_3$ and $C_4 R_4$) is relatively small so, using the ΔPD even at 1 sec, the accuracy is still excellent. However, it is obvious that using the ΔPD after 10 or 100 sec yields a substantial overestimate of R_t.

With this background we are in a position to review some of the accomplishments of the measurements of R_t in relation to the physiological and biophysical characteristics of the gastric mucosa. However, before presenting this aspect of our work we will comment on important papers by Clausen and colleagues.

Use of Distributed Parameter Models

In the past we used lumped parameter models rather than distributed parameter models for electrical events.[1-3] However, for other purposes, such as the diffusion of ions and water via the tubular lumina, we have used distributed parameter models.[16] The complexity of epithelial tissues is such that one realizes that lumped parameter models may yield only

[14] M. Schwartz, A. D. Pacifico, T. N. MacKrell, A. Jacobson, and W. S. Rehm, *Proc. Soc. Exp. Biol. Med.* **127**, 223 (1968).
[15] A. D. Pacifico, M. Schwartz, T. N. MacKrell, S. G. Spangler, S. S. Sanders, and W. S. Rehm, *Am. J. Physiol.* **216**, 536 (1969).
[16] W. S. Rehm, T. C. Chu, M. Schwartz, and G. Carrasquer, *Am. J. Physiol.* **245**, G143 (1983).

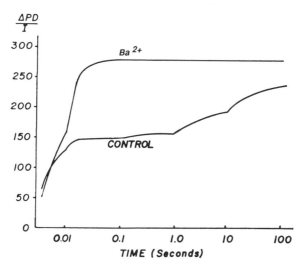

FIG. 9. A plot of $\Delta PD/I$ versus time for a mucosa before and after addition of Ba^{2+} (to final concentration of 1.0 mM) to nutrient fluid. Five linear time scales are used as in Fig. 3. Current sent nutrient to secretory for both transients. (Reproduced from Ref. 2.)

approximate evaluations—and that distributed parameter models would be the models of choice if it could be shown that they were viable models.

Clausen *et al.*[17,18] determined with an ac technique the impedance of the gastric mucosa. They plotted the resistive impedances and the phase angles versus the log of frequency. They used three models: one model was a lumped one-cell model, the same as our model for the millisecond domain. The second model was a model consisting of four resistors in series, with three resistors shunted by capacitors. They point out that this model can approximate a model with RC circuits for two cells in parallel. Their third model was a one-cell distributed parameter model. They found with the ac technique that the theoretical curves for the resistive impedance fit the data relatively well for all three models. However, the phase angle curves gave a better fit for the distributed parameter model. Apart from the particular model used, the transient method has the advantage that it is determined in less than 1 sec whereas it takes 12 min[17,18] to obtain data for a run with the ac model. Clausen and colleagues argue that the effects of higher frequencies are underestimated by the transient method and hence the ac method is superior. It is implied that in using the transient method the data are taken at uniform time intervals for the analysis. However, one can vary the time intervals for obtaining data for various parts of the

[17] C. Clausen, T. E. Machen, and J. M. Diamond, *Science* **217**, 448 (1982).
[18] C. Clausen, T. E. Machen, and J. M. Diamond, *Biophys. J.* **41**, 167 (1983).

transients so that the high-frequency end of the spectrum can even be overestimated. Figure 7 shows an analysis in a resting frog mucosa of a transient to a step current in which the time interval is abruptly changed.[3] In other words, one can emphasize various domains of the transient response. Clausen and colleagues[17,18] find for the secreting mucosa larger values for the capacitance of the secretory membrane than we obtain by the lumped model (C_2 of our model). However, some of their findings are difficult to reconcile with the characteristics of the gastric mucosa. For example, in three consecutive runs on a given mucosa (runs 2-1, 2-2, and 2-3 of Table 2 in Ref. 18) the H^+ secretory rate was about the same but the capacitances increased in consecutive runs from 1900 to 8500 and finally to 20,000 $\mu F/cm^2$. The mechanism for this increase in the capacitance is difficult for us to understand. Furthermore, in a plot of the capacitance of the secretory membrane (their C_a) of the tubular cells versus the H^+ rate, the capacitance increases about 10-fold when the H^+ rate increases by only a small amount. They point out that the abrupt increase in the capacitance of the secretory membrane as the H^+ rate increases by a small amount does not reflect a true increase in area (i.e., capacitance). They suggest other explanations for this abrupt increase in this capacitance. The increase in capacitance from lower to higher rates before the abrupt change is also puzzling.

Regardless of the exact values of the capacitance of the secretory membrane of the gastric mucosa, it is clear that during secretion the area of the secretory membrane of the tubular cells is several hundred times the gross area[5-9] and this membrane will possess a capacitance many times the gross area.[2,17,18]

An important contribution of the work of Clausen *et al.* is that they showed that the capacitance of the secretory membrane of the tubular cells increases dramatically during the transitions from the resting to secretory state.

Clausen *et al.*[18] suggest that in the transition between the secreting and the resting states the change in R_t may be due entirely to a change in the area of the secretory membrane and that·there is no change in the specific ionic conductances of the secretory membrane. It will be shown below that there is a change in the specific ion conductances of this membrane. They further suggest that the characteristics of the nutrient membrane of the tubular cells are not changed in transition between the two states (but see below). They find that C_b, the capacitance of the basolateral membrane of the tubular cells, is 99 $\mu F/cm^2$, while our value is 27 $\mu F/cm^2$. Our morphometric value[6] of 89 $\mu F/cm^2$ agrees quite well with their value for C_b and not well with our value. However, examination of their results[18] reveals that the results were obtained on six mucosa with a total of 17 runs. There were

11 runs on four mucosas (mucosas 1 through 4) with an average C_b of 135 (SD $\pm$ 45) μF/cm^2, while in six runs on the last two experiments (mucosas 5 and 6) the average value of C_b is 31.7 (SD $\pm$ 20) μF/cm^2. The latter value is very close to our value of 27 μF/cm^2.

Clausen *et al.* obtained values for R_s, the series resistance of the tissue, that are puzzling.[18] This value should be independent of the method, i.e., either the transient or the ac method. They give an average value of R_s of 130 Ωcm^2 and after subtraction of their blank value for the resistance in the absence of the tissue of 40 Ωcm^2, R_s is 90 Ωcm^2. They attribute this high value of R_s to "the tissue must restrict diffusion more than does an equivalent thickness of free solution." In contrast to their value for R_s our value is 22 (SD $\pm$ 7.5) Ωcm^2. The value of R_t varies considerably, but in many mucosa that secrete at high H$^+$ rates, R_t is often less than 100 Ωcm^2. It has been shown, as indicated above, that the low resistance of R_t is due to a low resistance of the lumen–tubular cell pathway and that all the parallel pathways have much higher resistances.[16,19] In other words, their value of R_s is difficult to accept.

Use of Ba^{2+} in Determining the Mechanisms of Ion Transport

The usefulness of Ba^{2+} was discovered accidentally. We studied the effects of removing Ca^{2+} from bathing media on the frog fundus. Sedar and Forte[20] reported that removing Ca^{2+} from the bathing media produced a marked decrease in R_t and suppressed the H$^+$ secretion. In our studies[21] we found that the removal of Ca^{2+} initially produced an increase in R_t (first phase), which was followed by a decrease to very low levels (the second phase), confirming Sedar and Forte. We measured R_t at frequent intervals while Forte and Sedar measured R_t at infrequent intervals and so did not see the first phase, i.e., the phase in which R_t increased. It appears that the first effect of removal of Ca^{2+} is to decrease the K$^+$ conductance of the nutrient membrane; subsequently a lack of Ca^{2+} opens up the tight junctions and results in the second phase, i.e., the marked decrease in R_t. We attempted to determine if other divalent cations could reverse the effect of the removal of Ca^{2+}. We found that during the second phase Ba^{2+} had no effect on R_t. We then attempted to see if Ba^{2+} would reverse the first phase. During the first phase, addition of Ca^{2+} to the nutrient side brings the resistance down to the control levels. We found that Ba^{2+}, instead of acting like Ca^{2+} and bringing the resistance of the first phase down to control levels, markedly increased the R_t. We then found that with the normal

[19] W. S. Rehm, M. Schwartz, and G. Carrasquer, *Am. J. Physiol.* **250**, G639 (1986).
[20] A. W. Sedar and J. G. Forte, *J. Cell Biol.* **22**, 173 (1964).
[21] A. Jacobson, M. Schwartz, and W. S. Rehm, *Am. J. Physiol.* **209**, 134 (1965).

Ca^{2+} in the nutrient fluid Ba^{2+} produced large increases in R_t.[14] While this was, as far as we know, the first use of Ba^{2+} in epithelial tissues, previous work on excitable tissue showed that Ba^{2+} blocked K^+ channels.[22] It was soon apparent that this was so for the frog fundus.[14,15] The very high value for R_t in the presence of Ba^{2+} was reversed in a matter of seconds by the increase of K^+ concentration in the nutrient from 4 mM to larger values (e.g., 80 mM), while Ba^{2+} was still maintained in the nutrient solution.[15] The rapid return of R_t to control levels as the result of elevation of K^+ indicated that the effect of Ba^{2+} must be on the nutrient membranes of the mucosa. Subsequent work with microelectrodes, placed in the surface cells of the frog mucosa, showed that this large increase in resistance due to Ba^{2+} was due to its acting on the nutrient membrane of the tubular cells.[23]

Use of Ba^{2+} in Determining the Mechanism of HCO_3^- and Cl^- Transport across the Nutrient Membrane

The marked increase in R_t due to Ba^{2+} is not, in general, accompanied by much change in the H^+ rate, i.e., in the presence of the very large value of R_t due to Ba^{2+}, and H^+ rate is maintained at a high level. In our first experiment, we only maintained the Ba^{2+} in the nutrient membrane for a short time and found that, after removal of Ba^{2+}, R_t would return to the control conditions. Up to this time no one had actually looked at the possible mechanisms for Cl^- entry and HCO_3^- exit from the tubular cells. It had been shown, because of electroneutrality, that the rate of HCO_3^- exit from the cell must match H^+ secretory rate—CO_2 per se exiting would violate electroneutrality.[24] After the first experiment we attempted to account for the movement of these ions through separate channels in the nutrient membrane. In attempting to account for the movement of these ions through conductive channels we leaned over backward, so to speak, and assumed that the total conductance of the nutrient membrane equals the sum of the Cl^- and HCO_3^- conductances and that the conductances for K^+ and other ions were zero. In the first experiment, R_t increased by about 700 Ωcm^2 and we made the reasonable assumption that this represented the minimum resistance of this membrane. Figure 10 represents the model we used in these calculations. It consists of two limbs, a Cl^- and an HCO_3^- limb, and with the resistance between N and the cell being 700 Ωcm^2, it is obvious that the minimum resistance around the loop would be 4×700 (2800 Ω) for 1 cm^2 and this would occur only when R_{Cl^-} equaled $R_{HCO_3^-}$.

[22] N. Sperekalis, M. F. Schneider, and E. J. Harris, *J. Gen. Physiol.* **50**, 1565 (1967).

[23] J. O'Callaghan, S. S. Sanders, R. L. Shoemaakerr, and W. S. Rehm, *Spring Meet. Assoc. Res. Vision Ophthalmol.*, 67 (1954).

[24] R. E. Davies, *Biochem. J.* **42**, 609 (1948).

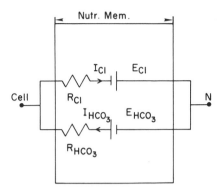

FIG. 10. Equivalent circuit for Cl⁻ and HCO₃⁻ exchange across nutrient membrane. See the text.

The driving force for the movement of Cl⁻ into and HCO₃⁻ out of the cell would be the sum of the diffusion potentials for Cl⁻ and HCO₃⁻. With 25 mM HCO₃⁻ in the bathing media the maximum concentration of HCO₃⁻ in the cell could not be much above 50 mM. Hence, the value of the EMF for HCO₃⁻ would be about 18 mV. In the steady state the Cl⁻ moving in and the HCO₃⁻ moving out must equal the rate of H⁺ secretion, and this, translated to current, for typical H⁺ rates, equals 0.11 mA, so the sum of E_{Cl^-} and $E_{HCO_3^-}$ would be about 308 mV, [i.e. (2800 × 0.11 = 308)]. Therefore the magnitude of E_{Cl^-} would be about 290 mV and with Cl⁻ concentration in the nutrient fluid of 80 mM the concentration of Cl⁻ in the cell would be about 1.3×10^{-3} mM, an absurdly low value.

However, in the first experiment, Ba²⁺ was only present for a relatively short period of time and we considered the possibility that HCO₃⁻ could pile up in the cell and Cl⁻ be depleted from the cell so that in this short time the H⁺ rate may not change. Calculations revealed that this state of affairs could only last for about 10 min. At this stage we did not know what subsequent experiments would reveal. However, they did reveal that with Ba²⁺, R_t and the H⁺ rate were maintained at high levels indefinitely. The conclusion became obvious that the entrance of Cl⁻ and the exit of HCO₃⁻, the amounts that match the H⁺ rate, cannot be via conductance channels, they must be via neutral mechanisms.

What makes the evidence more compelling is that, at a moderate to high H⁺ rate, the HCO₃⁻ conductance of the nutrient membrane is essentially zero.[25] Parenthetically, it was found by Flemstrom and Sachs[26] that

[25] S. S. Sanders, J. O'Callaghan, C. F. Butler, and W. S. Rehm, *Am. J. Physiol.* **222**, 1348 (1972).
[26] G. Flemstrom and G. Sachs, *Am. J. Physiol.* **228**, 1188 (1975).

the HCO_3^- conductance of the nutrient membrane of the fundus during inhibition of acid secretion is substantial — a finding which we confirm.[27]

Mechanisms for Change in R_t in Transition between the Resting and Secretory State

It is well accepted that in the transition from the resting to the secretory state, R_t, in general, decreases and the area of the secretory membrane of the tubular cells markedly increases. As indicated above, Clausen *et al.*[18] suggest that the mechanism for proton secretion is neutral in the intact tissue and that the decrease in R_t is due solely to the increase in area, i.e., no change in the specific resistance of the secretory membrane. However, this suggestion turns out not to be the case. Carlisle *et al.*[28] found that histamine in the presence of SCN produces the typical increase in the area of the secretory membrane without the establishment of H^+ secretion. On the basis of the suggestion by Clausen *et al.*, it would be predicted that the resistance would show the typical decrease. Contrary to this prediction, in the presence of SCN, histamine did not produce a significant decrease in R_t.[16] However, upon the subsequent removal of SCN the typical decrease in R_t occurred and H^+ secretion was established, i.e., R_t showed the typical decrease when the area was not changing. The decrease in R_t is obviously related to the activation of the HCl secretory mechanism and indicates a decrease in the resistance of ion transport mechanisms. This is an important finding in attempts to evaluate the validity of various models, electrogenic or neutral, for HCl production in intact tissue.[16]

Resistance of the Lumen–Tubular Cell Pathway and That of the Parallel Pathways: The Surface Cell and Transintercellular (Paracellular) Pathways

The frog gastric mucosa has a very low R_t and, in general, it is assumed that epithelial tissues with low resistances have leaky transintercellular (TIC) pathways. However, many lines of evidence indicate that the TIC pathways of the frog fundus have very high resistances and that the lumen–tubular cell pathway is the low-resistance pathway which accounts for the low resistance of the gastric mucosa. It is beyond the scope of this chapter to present all of the evidence in support of the above. However, we will present interesting evidence from recent work.

It was found with an isotonic nutrient solution that changing from a hypertonic to a hypotonic secretory produced only a small increase in R_t in

[27] M. Schwartz, G. Carrasquer, and W. S. Rehm, *Biochim. Biophys. Acta* **819,** 187 (1985).
[28] K. S. Carlisle, C. S. Chew, and S. S. Hersey, *J. Cell Biol.* **76,** 31 (1978).

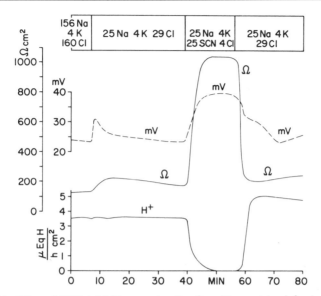

FIG. 11. Effect of SCN inhibition on *in vitro* frog *(Rana pipiens)* fundus. The PD (reference to secretory side), resistance, and H⁺ rate are plotted versus time. Bar shows composition of secretory (mucosal) fluid. Standard nutrient fluid (see *Methods*) on nutrient side.

the stimulated fundus.[16] However, inhibition with a hypotonic secretory solution, with an isotonic nutrient solution, produced a huge increase in R_t. This is illustrated in Fig. 11. It can be seen that changing from a hypertonic secretory solution to a hypotonic one during secretion produced only a small increase in R_t. However, without materially changing the osmotic pressure of the secretory solution, substituting 25 mM SCN for Cl⁻, the H⁺ rate went to zero and R_t rapidly increased to about 1000 Ωcm². It should be noted that SCN is a well-known inhibitor of acid secretion and is effective when added to either side. It is also seen that changing back to the SCN-free hypotonic solution resulted in a rapid return to the control levels.

The conductive pathways for the fundus are illustrated in Fig. 12. It is to be noted that the subscripts in Fig. 12 do not have the same meaning as in Fig. 3. The pathways are the surface cell pathway, R_{sur} $(=R_1 + R_2)$, the TIC pathway between the surface cells, the TIC′ pathways between the tubular cells, the tubular cell pathway R_{TUB} or $(=R_3 + R_4)$, and R_L, the lumen pathway. The lumen–tubular cell pathway is denoted by R_{LC} $(=R_L + R_{TUB})$. It has been shown that for the secreting fundus, the surface cell and TIC pathways have high resistances and that the huge increase in R_t seen in Fig. 11 is due to an increase in R_{LC}, which is due to a large

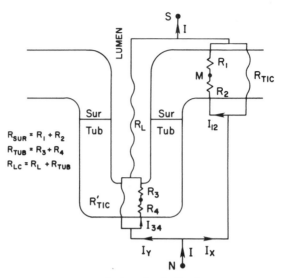

FIG. 12. Equivalent circuit. See the text.

increase in resistance of the lumen, R_L. The increase in R_L has been shown to be due to a large decrease in the area of the lumen.[16,29] A significant part of the evidence for the above is illustrated in Fig. 13. With a hypotonic SCN secretory solution similar to that in Fig. 11, mannitol is added to the secretory solution (the low ion concentration is maintained). It can be seen that the addition of mannitol results in a marked decrease in R_t and a marked increase in PD.[16,19,29] It has been shown that the surface cells are practically impermeable to H_2O in the net transport sense,[30] so the mannitol should have relatively little effect on the resistance of the surface cells. The secretory solution with mannitol is now hypertonic and hence the patency of the lumina is restored and as would be expected, R_t decreases. Hence, the low resistance of the secretory fundus is due to the low resistance of the lumen–cell pathway, and the resistance of the surface cell and the TIC pathways are very large. In the inhibited fundus with the hypotonic secretory solution the characteristics of the surface cells can be studied with relatively little interference from the tubular cells. With this technique it has been shown, using other inhibitors (e.g., cimetidine or omeprazole), that SCN has only a small effect on the surface cell resistance.[19] It produces a small but significant increase in resistance rather than a decrease.

[29] W. S. Rehm, G. Carrasquer, and M. Schwartz, *Am. J. Physiol.* **250,** G511 (1986).
[30] W. S. Rehm, H. Schlesinger, and W. Dennis, *Am. J. Physiol.* **175,** 473 (1953).

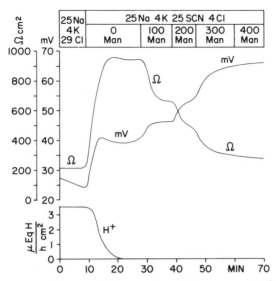

FIG. 13. Protocol of the first part of this experiment is the same as in Fig. 11. After H^+ went to zero, the PD increased by about 17 mV and R_t increased to about 950 Ωcm^2. Mannitol (Man) was added stepwise to a final concentration of 400 mM. [Reproduced from W. S. Rehm, M. Schwartz, G. Carrasquer, and M. Dinno, in "Membrane Biophysics III. Biological Transport" (M. Dinno, ed.), pp. 1–22. Liss, New York, 1988.]

It was also shown with the hypotonic technique that Ba^{2+} has only a small effect on the surface cells and TIC pathways, i.e., a small increase in resistance.[19] This confirms the work with the microelectrodes in which it was shown that the large increase due to Ba^{2+} is due to an increase in the resistance of the nutrient membrane of the tubular cells.[23]

The foregoing discussion presents some of the evidence showing that in the secreting fundus, R_{LC} is a low-resistance pathway and that the parallel pathways (surface cell and TIC pathways) have a high resistance.

Summary

A method has been presented for the determination of resistance of biological tissues in which the PD response to step currents is determined. The ΔPD after the dielectric capacitors are charged, divided by the current, gives the resistance, provided the current density is low enough so that the tissue behaves as a linear–bilateral system. In the gastric mucosa the PD continues to increase after the dielectric capacitors are charged and it is shown that this part of the ΔPD is due to polarization of EMFs and should not be used in determining the resistance. It has been shown that (1) resistance measurements have enabled us to demonstrate that during acid secretion there is a neutral mechanism(s) for the movement of HCO_3^- out of and the entrance of Cl^- into the oxyntic cells, (2) the transmucosal resistance varies inversely with the rate of acid secretion, and (3) the low resistance of the secreting frog fundus is due to the low resistance of the lumen–tubular cell pathway—the parallel pathways (the TIC or paracellular and surface cell pathways) have high resistances. The results of both the resistance and PD measurements have recently been analyzed[16,19,29] with respect to the problem of whether the proton pump is neutral or electrogenic in the intact tissue.

[32] Impedance Analysis in Tight Epithelia

By CHRIS CLAUSEN

Introduction

When investigating ionic transport mechanisms in tight epithelia, one is faced with the problem of resolving transepithelial measurements of membrane conductance into the respective conductances of the different membranes, namely the apical and basolateral membranes, and the paracellular pathway formed by the tight junctions and lateral spaces. In addition, since epithelia are generally highly distensible tissues, it is important to develop techniques which are capable of yielding estimates of exposed membrane areas. This is important since it allows one to determine specific conductances (ionic conductance per unit membrane area), which are indirect measurements of ionic permeability. Finally, over the past decade, investigators have found that regulation of ionic transport often involves endo- and exocytotic fusion processes resulting in alterations in the exposed area of the apical and/or basolateral membranes. Here, membrane area measurements are necessary since these processes result in transport-related changes in membrane conductance which may not result from changes in ionic permeability.

In many cases, the quantitative analysis of transepithelial impedance is capable of yielding reliable estimates of the different membrane ionic conductances. In tight epithelia, where it is known *a priori* that the paracellular ionic pathway is negligible, analysis of impedance yields direct estimates of the apical and basolateral membrane conductances. In moderately tight epithelia, analysis of impedance, coupled with either microelectrode measurements of the apical-to-basolateral membrane resistance ratio or independent estimates of the paracellular conductance, also yields estimates of the different membrane conductances. These analyses also yield estimates of the different membrane electrical capacitances. Since membrane-specific capacitance has been shown to be remarkably constant at 1 $\mu F/cm^2$,[1] these techniques provide an indirect measure of exposed membrane areas. Finally, in some cases, impedance analysis techniques allow one to monitor transport-related changes in cell geometry (notably, the geometry of the lateral spaces).

In this chapter the specific details involved in the measurement of transepithelial impedance are discussed, as well as the subsequent quantitative analysis using equivalent epithelial circuits. The techniques described have been used to investigate ionic transport mechanisms in a number of different preparations, notably sodium transport in amphibian and mammalian urinary bladder,[2,3] gastric acid secretion in amphibian gastric mucosa,[4,5] sodium and potassium transport in mammalian descending colon,[6] chloride secretion in amphibian cornea,[7] and proton secretion in renal epithelia.[8,9] Specific methodology applicable to similar studies in leaky epithelia can be found elsewhere in this volume ([33]).

What Is Impedance?

Transepithelial impedance is simply the ratio of the transepithelial steady-state voltage $v(t)$ resulting from an applied sinusoidal current $i(t)$ at a given angular frequency $\omega (= 2\pi f; f$ in Hz). At each frequency of interest, one actually measures two quantities. The impedance magnitude is simply the ratio of the amplitudes of $v(t)$ and $i(t)$, and the phase angle is the normalized time delay between $v(t)$ and $i(t)$. Measurements of impedance are generally performed at a range of different frequencies, typically between 0.001 and 10 kHz, logarithmically spaced in frequency.[2] Since the subsequent analyses involve the extraction of a number of membrane parameters (at minimum five) using curve-fitting techniques, the number of frequencies must be sufficient in order to obtain statistical significance. Typically, 100 frequencies encompassing the complete range are sufficient.[7]

[1] K. S. Cole, "Membranes, Ions, and Impulses," p. 12. Univ. of California Press, Berkeley, 1972.
[2] C. Clausen, S. Lewis, and J. M. Diamond, *Biophys. J.* **26**, 291 (1979).
[3] N. K. Wills, S. A. Lewis, and C. Clausen, *J. Membr. Biol.*, submitted (1989).
[4] C. Clausen, T. E. Machen, and J. M. Diamond, *Biophys. J.* **41**, 167 (1983).
[5] J. M. Diamond and T. E. Machen, *J. Membr. Biol.* **72**, 17 (1983).
[6] N. K. Wills and C. Clausen, *J. Membr. Biol.* **95**, 21 (1987).
[7] C. Clausen, P. S. Reinach, and D. C. Marcus, *J. Membr. Biol.* **91**, 213 (1986).
[8] C. Clausen and T. E. Dixon, *J. Membr. Biol.* **92**, 9 (1986).
[9] T. E. Dixon, C. Clausen, D. Coachman, and B. Lane, *J. Membr. Biol.* **94**, 233 (1986).

Two important points must be stressed regarding the measurement of impedance. Since impedance describes the *steady-state* voltage response to an applied sinusoidal current, a period of time must elapse between changing the applied current and measuring the resulting voltage, otherwise the resulting voltage will be contaminated by the transient response to the change in current. Also, since the analysis of impedance is done using an equivalent circuit composed of linear circuit elements, one must ensure that the voltage response is independent of applied-current amplitude. Since voltage-dependent (gated) channels appear as inductances[10] which alter phase angle and magnitude measurements, it is therefore essential to verify linearity of the epithelium by demonstrating that measurements made at different current amplitudes yield indistinguishable estimates of impedance.

Measuring Transepithelial Impedance

Conceptually, the simplest method for measuring impedance involves using a sine wave generator to generate the current signal, and then measuring the amplitude and phase angle of the voltage response using a phase-lock amplifier and/or phase meter and root mean square (rms) voltmeter.[2,11] One of the problems with this method is that each frequency must be applied sequentially. Because of the time required to attain a steady-state response after changing frequency—approximately 15 min is required to measure impedance at 25 different frequencies spanning the range of 0.001 to 10 kHz—the method is cumbersome. It can only be utilized to characterize epithelial impedance when the preparation is stable over long periods.

In order to measure impedance rapidly over a wide range of frequencies, one can exploit the fact that impedance describes the sinusoidal current–voltage response of a circuit composed of linear circuit elements. The composite voltage response resulting from the simultaneous application of several sinusoidal currents is equal to the sum of the individual responses at each frequency. Techniques have been developed to measure the impedance using so-called wide-band current signals composed of the sum of a number of different frequency components. After the application of a wide-band current, one needs to wait only a period of time sufficient to attain a steady-state response at the lowest frequency component; by that time, all the simultaneously applied higher frequencies will also be at steady state. These techniques all involve standard Fourier analysis tech-

[10] H. M. Fishman, D. Poussart, and L. E. Moore, *J. Membr. Biol.* **50,** 43 (1979).
[11] R. Valdiosera, C. Clausen, and R. S. Eisenberg, *J. Gen. Physiol.* **63,** 460 (1974).

niques[12] implemented on a laboratory computer, to separate the composite voltage response into the responses at each of the simultaneously applied frequencies. They typically allow one to measure impedance in the range of 0.001 to 10 kHz in less than 5 sec, with a resolution of several hundred different frequencies.

Basically two classes of wide-band signals have been used. Random (white) noise signals are suitable,[13] but their analysis requires sophisticated recording instrumentation (simultaneous recording of the applied current and resulting voltage), and a significant amount of numerical processing, which in turn, requires expensive special-purpose laboratory instrumentation (e.g., dual-channel Fourier processors). We prefer instead to utilize pseudo-random noise signals which are deterministic (not random) and periodic. Compared to the random signals, the pseudo-random signals are particularly advantageous for several reasons: total data acquisition time (i.e., recording of the epithelial voltage response) is minimized; so-called spectral leakage is avoided; the number of numerical computations required in computing the impedance is reduced and computations can be readily performed using a general-purpose laboratory microcomputer; data acquisition hardware and ancillary electronics (amplifiers, filters) are simplified; and signal amplitude characteristics can be better controlled (for a complete discussion, see Ref. 14).

Pseudo-Random Binary Signals

Clausen and Fernandez[14] describe in detail the use of signals based on pseudo-random binary sequences (PRBS signals) to measure impedance in biological systems in general; the use of this method for epithelial impedance analysis is described by Clausen et al.[7]

PRBS signals are square-wave-like waveforms which exhibit random periods between the transitions from high to low states. They are wide band signals with similar characteristics as white noise[15] in that each sinusoidal (frequency) component is present with nearly equal power (i.e., a "flat" power spectral density). The signal is produced using a digital circuit based around a clocked shift register. The clock rate determines the maximum bandwidth, and the length of the shift register determines the

[12] J. S. Bendat and A. G. Piersol, "Random Data: Analysis and Measurement Procedures." Wiley (Interscience), New York, 1971.

[13] R. T. Mathias, *in* "Membranes, Channels and Noise" (R. S. Eisenberg, M. Frank, and C. F. Stevens, eds.), p. 49. Plenun, New York, 1984.

[14] C. Clausen and J. M. Fernandez, *Pfleugers Arch.* **390**, 290 (1981).

[15] P. Z. Marmarelis and V. Z. Marmarelis, "Analysis of Physiological Systems." Plenum, New York, 1978.

number of frequency components. Several commercial PRBS signal generators are available, but these instruments are not readily adaptable to biological impedance analysis because they do not provide a convenient means to synchronize acquisition of the voltage response with generation of the signal.

Clausen and Fernandez[14] provide complete details, including a complete schematic, of a digital circuit for generating the PRBS current signal, coupled with timing circuitry to control the acquisition of the voltage response. The circuit is constructed using readily available low-cost components, which can easily be assembled in a basic electronics shop facility. The circuit uses a 10-bit shift register, clocked at three different bandwidths (0.2 to 100 Hz, 2 to 1000 Hz, and 20 Hz to 10 kHz), comprising approximately 500 linearly spaced discrete frequencies for each bandwidth.

Instrumentation

A block diagram of the recording configuration is shown in Fig. 1. Measurement of transepithelial impedance is performed using a four-electrode configuration, where a pair of electrodes mounted at opposite ends of the transepithelial chamber are used to pass current, and a second pair of electrodes mounted as close as possible to the epithelial surfaces are used to measure transepithelial potential. Besides the PRBS signal generator described above, the following instrumentation is also required: a circuit for converting the voltage output of the PRBS signal generator to a constant current; a differential amplifier for measuring transepithelial impedance; an antialiasing filter; and sample-and-hold amplifier; and analog-to-digital (A-to-D) converters. The characteristics of each of these items is described below.

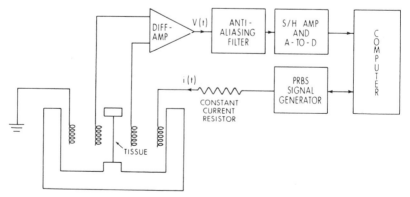

FIG. 1. Block diagram showing the configuration of the instrumentation required to measure transepithelial impedance.

Constant Current Generation

The simplest way to generate a constant transepithelial current is simply to utilize a high-value carbon resistor in series with the PRBS signal generator and current electrodes. An automatic current-clamp circuit, which generates constant current using an active feedback circuit, tends to produce phase artifacts at high frequencies (> 1 kHz) resulting from the finite speed of the clamp amplifier. Since it is difficult to devise appropriate correction procedures for these artifacts, these circuits should not be used.

The choice of series resistor value depends on the epithelial impedance, the resistance of the current electrodes, and the desired current amplitude. At minimum, the resistor value must be at least 100 times greater than the sum of the electrode resistance and the dc transepithelial resistance. Ag–AgCl wires and/or pellets provide the lowest resistance current electrodes (~ 1 kΩ), but can only be used when mucosal and serosal bathing solutions contain normal extracellular levels of chloride, and when the epithelial preparation is insensitive to low levels of Ag^+ arising from the finite solubility of AgCl. The resistance of salt bridges is significantly higher than that of AgCl electrodes. In order to avoid the use of an extremely high-value series resistor, which limits the transepithelial current amplitude, bridges should be kept as short as possible, they should be constructed from wide-diameter tubing, and they should be filled with agar containing high ionic strength.

Transepithelial Voltage Amplifier

The transepithelial differential voltage amplifier should have the following characteristics (apart from suitable speed and input impedance): low-input capacitance (~ 15 pF), AC coupled inputs, and switchable gain to values as high as 1000. A commercially available amplifier that meets these specifications is the PARC-113 battery-powered amplifier (Princeton Applied Research, Princeton, New Jersey), but similar performance can also be achieved by the use of a unity-gain field effect transistor (FET) preamplifier (i.e., "standard" transepithelial voltage amplifier) with a subsequent switchable-gain instrumentation amplifier.

Low-input capacitance and low-voltage electrode resistance are both of primary importance since high values limit the speed of the amplifier. The resistance of the voltage electrodes coupled with the input capacitance results in a one-pole RC filter which attenuates the voltage at high frequencies. The resulting phase "error" is given by $\tan^{-1}(\omega RC)$, and at 10 kHz equals $0.05°$ when using 1-kΩ AgCl electrodes, but increases to over $5°$ if the voltage electrodes are 100-kΩ bridges (not an unusual value for narrow bridges several centimeters in length). Again, when using bridges, their resistance must be minimized by keeping them as short as possible. Under no circumstances should shielded wiring be used to connect the electrodes to the amplifier; shielded wire typically adds 30 pF input capacitance per foot of length.

High gain is required to amplify the millivolt-level signals to a level suitable for analog-to-digital conversion (most A-to-D converters have fixed dynamic range of ~ 10 V). Therefore, the amplifier must be ac coupled in order to remove the dc transepithelial potential which, at high gains, would result in amplifier saturation.

Antialiasing Filter

Prior to analog-to-digital conversion, the voltage signals must be low-pass filtered at the maximum bandwidth (set by the PRBS signal generator), in order to avoid so-called aliasing artifacts (a complete discussion of aliasing artifacts can be found in Refs. 12 and 13). The filters should be of the elliptic design and exhibit extremely high roll-off ($\geq$ 100 dB/octave). A recommended commercially available filter is the model LP-120 (Unigon, Mt. Vernon, New York), which has the added feature of being computer programmable.

Sample-and-Hold Amplifier and A-to-D Converters

Most commercially available high-speed data acquisition interfaces (e.g., Data Translation, Marlborough, Massachusetts) can be used to acquire the signals. The system should have 12-bit (or better) precision in order to resolve the high-frequency voltage response; impedance magnitude decreases with increasing frequency, resulting in small high-frequency voltage amplitudes. The sample-and-hold amplifier must be capable of rapidly acquiring and holding a sample ($<$ 1 μsec), and the A-to-D converter and computer must be capable of digitizing and acquiring data at a rate of better than 20 kilowords/sec (50 μsec per point). The system must be capable of utilizing an external clock source, provided by the PRBS signal generator, which synchronizes the acquisition of each point of the voltage response with the generation of the PRBS current signal. Finally, the interface must possess suitable hardware (programmable external digital output, or trigger input) in order to synchronize the data acquisition with the beginning of the period of the PRBS signal (see Ref. 14).

Computational Procedure Involving Fourier Analysis

The specific computational procedure for measuring impedance is described in Clausen and Fernandez[14] and Clausen et al.[7] A detailed discussion of the Fourier analyses involved can be found in Bendat and Piersol.[12]

Prior to beginning an experiment, the PRBS current signal $i(t)$ for each PRBS bandwidth should be recorded and stored in the memory of the computer; since the signal is constant, owing to the large-value series resistance, this procedure need only be performed once. The procedure involves measuring the voltage drop across a calibrated carbon resistor which has resistance comparable to the dc transepithelial resistance. Metal film and wire-wound resistors should be avoided since they inhibit small but finite reactive properties (i.e., capacitance and inductance). Once $i(t)$ is determined from the voltage drop across the calibrated resistor, it should be Fourier transformed, producing $I(f)$. For details regarding the fast Fourier transform (FFT) algorithm, the reader is directed to Brigham,[16] which describes the properties of the FFT, and to Bergland and Dolan,[17] which provides efficient computer programs implementing the FFT algorithm.

The epithelial impedance is computed by recording the transepithelial voltage response $v(t)$, Fourier transforming the response and thereby producing $V(f)$, and then computing the cross-spectral density of the voltage and the current divided by the power-spectral density of the current, namely,

$$Z(f) = I(f)^*V(f)/I(f)^*I(f)$$

(the asterisk denotes complex conjugate). In order to improve the signal-to-noise ratio, signal averaging can be performed in the time domain by averaging successive blocks of the voltage response $v(t)$, or in the frequency domain by averaging successive blocks of the cross-spectral density $I(f)^*V(f)$. Note that if frequency domain averaging is performed, one can also compute the associated coherence function, a useful quantity for assessing the degree of extraneous noise present in the measurements, and for assessing the linearity of the epithelial voltage response. The properties of the coherence function, as well as formulas used in its computation, are discussed by Bendat and Piersol.[12]

Curve-Fitting Procedure

Determination of the epithelial circuit parameters (membrane resistances and capacitances) is done by fitting the measured impedance by that predicted for a morphologically based equivalent-circuit model of the

[16] E. O. Brigham, "The Fast Fourier Transform." Prentice-Hall, Englewood Cliffs, New Jersey, 1973.
[17] G. D. Bergland and M. T. Dolan, in "Programs for Digital Signal Processing" (A. C. Schell et al., eds.), pp. 1.2-1. IEEE Press, New York, 1979.

epithelium (see below). The curve-fitting procedure involves adjusting the circuit elements in a systematic manner such that the sum of the squared deviations between the data and the model (the sum-square error) is minimized. Any derivative-free nonlinear curve-fitting algorithm can be used for this purpose (e.g., subroutine ZXSSQ, a derivative-free implementation of the Levenberg–Marquardt algorithm, available from IMSL, Inc., Houston, Texas); for a practical discussion of curve-fitting algorithms, see Press et al.[18]

The transepithelial impedance is generally measured using two PRBS signal maximum bandwidths, typically 1 and 10 kHz, thereby providing for good low-frequency and high-frequency resolution (2 and 20 Hz, respectively). When using 120-dB/octave antialiasing filters, the upper 17% of the frequency range for each bandwidth is contaminated by aliasing artifacts and therefore must be discarded (less steep filters require the deletion of a larger percentage; see Ref. 12). The final effective bandwidths are therefore approximately 0.830 and 8.3 kHz, respectively, comprising impedance measurements at 850 total frequencies. The data are subsequently reduced to a manageable number by selecting 50 points from each bandwidth (100 frequencies total) based on a logarithmic distribution of frequency (i.e., each decade in frequency should have approximately an equal number of frequencies). The data are then converted to Bode plots by computing the phase angle and log impedance magnitude at each frequency.

The Bode representation of the impedance (phase angle and log impedance magnitude plotted against log frequency) was found empirically to be the most useful representation for the equivalent circuit analysis.[2] Plotting the data against linear frequency tends to place undue weight on the high-frequency data, simply due to the high relative number of high-frequency data points. The Nyquist (or impedance locus) representation unduly weights the low-frequency data due to high relative impedance magnitude at these frequencies. Finally, phase angle is extremely sensitive to small changes in circuit parameters,[11] notably small series resistances (see below).

Finally, the phase angles and impedance magnitudes must each be normalized to unity, by dividing each series of points by its respective maximum value. This is required since phase measurements usually range in magnitude from 0 to $-90°$, whereas log impedance magnitude typically ranges between 1 (10 Ω) and 4 (10 kΩ). The resulting 200 data points (100 normalized phase and magnitude estimates) are then fitted by the equiva-

[18] W. H. Press, B. P. Flannery, S. A. Teukolsky, and W. T. Vetterling, "Numerical Recipes." Cambridge Univ. Press, New York, 1987.

lent-circuit model using the nonlinear curve-fitting algorithm. In evaluating the model, double-precision arithmetic should be used.[7] Suitable factors should be also used to scale the circuit parameters (being determined by the curve-fitting algorithm) to near unity.

Statistical Analyses

Three statistical tests are used to evaluate the quality of the curve-fitting procedure, the validity of the model, and the accuracy of the circuit parameters determined by the procedure. In addition, a test is needed to verify the independence of the individual parameters. It should be noted that these tests are essential to any analysis, not just equivalent-circuit analysis of transepithelial impedance, where multiple parameters are being determined by curve fitting a mathematical model to experimental data.

Quality of Fit

The Hamilton R factor[19] can be used as an objective measure of fit. The square of the R factor is simply the residual sum-square error at the best fit divided by the sum of the squares of the data, and is therefore interpreted as the fractional (or percentage) misfit between the model-predicted impedance and the data. The R factor cannot be used to validate the model being used. However, the R ratio, the ratio of two R factors derived from two different models, can be used to determine whether the quality of the data warrants a more complex model, which invariably results in a better fit to the data owing to additional circuit parameters. This test is a modified F test (variance ratio test), and is described by Hamilton.[19]

Accuracy of Circuit Parameters

The curve-fitting program internally computes the so-called Jacobian matrix — the matrix of first derivatives of the residual function (model minus data) with respect to each of the parameters. From this, one can estimate a parameter covariance matrix (see Refs. 18 and 19). The parameter covariance matrix is important for two reasons. First, the square root of the diagonal elements of the matrix provides an estimate of the standard deviations of the determined circuit parameters. These values *cannot* be used to compute confidence intervals, since, in reality, they are the standard deviations of a linear approximation to the nonlinear function describing the impedance.[11] However, the values are useful since large relative parameter standard deviations (e.g., 10% of the parameter value)

[19] W. C. Hamilton, "Statistics in Physical Science." Ronald Press, New York, 1964.

indicate that the circuit parameters cannot be accurately determined by the available data. Stated a different way, a large relative standard deviation indicates that large changes in the parameter itself produces negligible changes in the impedance, thereby indicating that the parameter is poorly determined.

The second use of the covariance matrix is in the computation of the parameter correlation matrix (see Ref. 19). If two parameters are found to be highly correlated (i.e., entries in the correlation matrix near ± 1), then the effects of the two parameters cannot be separately determined, for example, the effect of increasing one parameter can be perfectly compensated by simultaneously decreasing a different parameter. Again, this means that the circuit parameters are poorly determined from the available impedance data. For further discussion of this point regarding analysis of impedance data, the reader should consult Valdiosera et al.[11]

Equivalent Circuit Models

An appropriate equivalent-circuit model must conform closely to the morphological structure of the epithelium, as determined by observations of micrographs. Its impedance properties must also accurately fit the measured impedance producing an R factor near 1%. Moreover, deviations between the model-predicted and measured impedance should be explainable by the observed variance in the data, and not by systematic deviations. Finally, and most importantly, one must verify the model both quantitatively and qualitatively. For example, membrane capacitances must be consistent with observed membrane area amplification resulting from microvilli, canaliculi, tortuous lateral spaces, and the like; in aldosterone-sensitive sodium-reabsorbing epithelia, the application of amiloride must result in a impedance-predicted decrease in the apical membrane conductance; the application of agents known to result in changes in membrane area (e.g., histamine stimulation of gastric acid secretion in gastric mucosa) must result in concomitant changes in the appropriate membrane capacitances.

The simplest equivalent-circuit representation of a tight epithelium is termed the lumped model, and is shown in Fig. 2. The apical and basolateral membranes are represented as parallel resistor–capacitor (RC) circuits where G_a and G_b represent the respective ionic conductances and are proportional to ionic permeabilities, and C_a and C_b represent the respective membrane capacitances and are proportional to exposed membrane area (recall that biological membranes exhibit a specific capacitance of approximately 1 $\mu F/cm^2$). The paracellular ionic pathway via the tight junctions is represented by the conductance of G_j; junctional capacitance can be safely

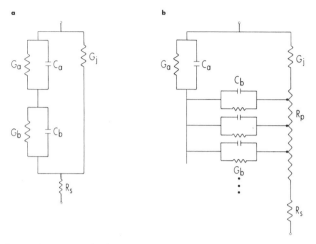

FIG. 2. Lumped (a) and distributed (b) equivalent-circuit models describing epithelial impedance. Consult the text for definition of the symbols.

assumed to be negligible owing to the small relative cross-sectional area of the junctions compared to the cell membranes. A small series resistance R_s is included in the model definition to account for the resistance between the voltage electrodes and the epithelial surfaces. The impedance of the lumped model is given by

$$Z(j\omega) = \frac{Y_a + Y_b}{Y_a Y_b + G_j(Y_a + Y_b)}$$

where $Y_a = G_a + j\omega C_a$ (apical membrane admittance), $Y_b = G_b + j\omega C_b$ (basolateral membrane admittance), $\omega = 2\pi f$ (f in Hz), and $j = \sqrt{-1}$.

In general practice, the lumped model is inadequate for accurately describing transepithelial impedance. The basolateral membrane lines the lateral spaces which, in many tight epithelia, are extremely narrow (~ 10 nm). The small dimensions of the fluid-filled lateral spaces thereby results in a significant paracellular path resistance (R_p) distributed along the lateral spaces from the tight junctions to the serosa. Since R_p is comparable to the basolateral impedance at even midrange frequencies, the basolateral membrane combined with the lateral spaces must be treated as a distributed impedance. The complete derivation of the resulting distributed model, shown in Fig. 2, can be found in Clausen et al.[2,7] The resulting impedance is given by

$$Z(j\omega) = \frac{Y_a + Y'_b + G_j[Y_a Y'_b/(Y_b/R_p) + 2(1 - \text{sech}\sqrt{Y_b/G_b})]}{Y_a Y'_b + G_j(Y_a + Y'_b)}$$

where $Y_b' = \sqrt{Y_b/R_p} \tanh \sqrt{Y_b R_p}$. Note that R_p provides an indirect measure of the geometry of the lateral spaces, since it is directly proportional to the length of the spaces and inversely proportional to the cross-sectional area of the spaces. In cases where the spaces become dilated (e.g., due to a high degree of mounting stretch or cell shrinkage), R_p decreases and the mathematical description of the distributed model simplifies to that of the lumped model.

The distributed model has been used successfully in the quantitative analysis of epithelial impedance from a number of different preparations, notably, rabbit urinary bladder,[2] toad bladder,[3] frog corneal epithelium,[7] and turtle urinary bladder.[8,9] In some cases, modifications of the basic model were required to account for serosal cell layers,[3,7] and apical membrane-penetrating tubular structures and glandular structures.[5-7]

Paracellular Pathway

In cases where G_j is known *a priori* to be negligible, then the parameters of the distributed model can be determined from analysis of transepithelial impedance alone. However, when G_j is comparable to the other membrane conductances, one needs additional information to determine the model parameters. This is because the impedance of the equivalent-circuit model can be represented mathematically by a function composed of *less* independent parameters than circuit elements (electrical circuits of this type are termed nonminimum circuits[20]). In the case of the lumped model, for example, the impedance function (above) is actually described by only five independent parameters, even though it contains six circuit elements.

There are two ways to handle this problem. The most obvious method involves measuring G_j using an independent experimental procedure. Subsequently, the other membrane parameters can be determined by fitting the impedance holding G_j constant at its measured value. Unfortunately, techniques for measuring G_j in tight epithelia are not straightforward. In sodium-transporting epithelia, one can use the method of Yonath and Civan[21] (see also Refs. 22 and 23), which involves the use of amiloride to increase selectively apical membrane resistance. Plotting transepithelial resistance versus potential yields a linear relationship where the ordinate

[20] H. Ruston and J. Bordogna, "Electric Networks: Functions, Filters, Analysis." McGraw-Hill, New York, 1966.
[21] J. Yonath and M. M. Civan, *J. Membr. Biol.* **5**, 366 (1971).
[22] S. A. Lewis, D. C. Eaton, C. Clausen, and J. M. Diamond, *J. Gen. Physiol.* **70**, 427 (1977).
[23] N. K. Wills, S. A. Lewis, and D. C. Eaton, *J. Membr. Biol.* **45**, 81 (1979).

intercept provides an estimate of $R_j = 1/G_j$. In other epithelia, one can use an analogous method where channel-forming antibiotics are used to increase selectively the apical membrane conductance.[22] Needless to say, these methods are specific to the epithelium of interest, and they rely on specific assumptions regarding the action of transport-altering agents. In addition, the agents used are often irreversible, permitting only a single measurement of G_j at the end of an experiment, and thereby not allowing investigation of transport-dependent changes in G_j.

Clausen and Wills[24] developed a second method which facilitates determination of all the model parameters. One first determines the membrane resistance ratio (α) by measuring the fractional voltage drop across the apical or basolateral membrane resulting from a pulse of transepithelial current. Subsequently, the transepithelial impedance is fitted by allowing one of the membrane conductances to vary, and by calculating the other membrane conductance using the previously measure α. Fitting the impedance subject to the constraint that the membrane conductances are consistent with α allows the determination of all the other circuit-element values, including G_j.

It should be noted that α is *not* simply equal to R_a/R_b ($R_a = 1/G_a$ and $R_b = 1/G_b$), since the voltage drop across the basolateral membrane results from the current flow across the basolateral membrane which is in series with a portion of the lateral-space resistance (R_p). Moreover, the series resistance (R_s) also affects the interpretation of α, although R_s can generally be minimized by careful placement of the transepithelial voltage electrodes close to the epithelial surfaces. Nevertheless, assuming that R_s is equally distributed on either side of the epithelium, then from the distributed model one can show that

$$\alpha = (R_a + R_s/2)/(R_b' + R_s/2)$$

where

$$R_b' = \sqrt{R_p R_b} \coth \sqrt{R_p/R_p}$$

Note that when $R_p/R_b < 0.15$, this expression simplifies to

$$R_b' \simeq R_b + R_p/3$$

and this latter expression is accurate to $> 98.5\%$ when $R_p/R_b < 1$.

When fitting the impedance data, one generally rearranges the above expression, thereby solving for R_a as a function of the variable parameters R_p and R_s and the constant independently measured parameter α.

[24] C. Clausen and N. K. Wills, *Soc. Gen. Physiol. Ser.* **36**, 79 (1981).

Conclusions

The purpose of this chapter was to summarize the methods involved in the quantitative analysis of transepithelial impedance. These methods, when properly implemented, comprise a powerful tool for the study of transport processes in tight (or moderately tight) epithelia. The method's major forte results from the ability to ascertain the conductive properties of the different epithelial membranes, as well as providing a means for determining the membrane areas. This allows the investigator to distinguish between conductance alterations resulting from changes in ionic permeability or from changes in membrane area.

It should be emphasized that other complementary methods exist for measuring and analyzing impedance, but it is beyond the scope of this chapter to discuss their relative merits and deficiencies. Needless to say, the methods presented here are straightforward and have been used extensively for performing impedance analyses in a number of different epithelial preparations.

[33] Electrical Impedance Analysis of Leaky Epithelia: Theory, Techniques, and Leak Artifact Problems

By L. G. M. GORDON, G. KOTTRA, and E. FRÖMTER

Introduction

Determination of the resistance to direct current (dc resistance) alone does not provide much information concerning the resistive and capacitive properties of the network of membrane barriers and ion-transporting systems in epithelia. More insight can be obtained from the frequency-dependent impedance to alternating current. For example, in the case of frog skin bathed in sodium sulfate solutions a Nyquist plot of the real component versus the imaginary component of the transepithelial impedance reveals two semicircles[1] from which one may infer that the epithelium consists of a series arrangement of two major capacitive and resistive membrane barriers and from which quantitative estimates of the resistances and capacitances of these membrane barriers can be obtained.

Such a straightforward analysis, however, is restricted to the case of membranes arranged in series as found in tight epithelia (e.g., toad

---

[1] P. G. Smith, *Acta Physiol. Scand.* **81,** 355 (1971).

bladder[2-4] or under certain conditions frog skin[5]), but it cannot be applied to cases in which membrane elements or current pathways are arranged in parallel as found in leaky epithelia (e.g., gallbladder[6] or small intestine[7]). In the latter case, to identify parallel elements and to quantify their relative importance, one has at best two options:

1. If one of two parallel pathways can be specifically blocked by an inhibitor the other pathway can be readily analyzed by transepithelial measurements. This option, therefore, depends on the *a priori* knowledge about the membrane and on the existence of specific blockers. Moreover, even if an inhibitor exists which binds exclusively to one membrane element and blocks all transport across it, this blockage may cause changes in ion activities or electrical gradients which may affect the other element or pathway indirectly.

2. The second option is to use microelectrodes which allow individual cells or lateral spaces between cells to be punctured or allow the surface of epithelia to be scanned for current inhomogeneities. While the first approach gains additional information from a hopefully well-controlled modification of the membrane, the latter uses measuring probes of high spatial resolution to detect transport inhomogeneities that may be associated with the inhomogeneous structure of the membrane. By adequately analyzing and comparing macroscopic and microscopic observations it may then become possible to determine the properties of each individual membrane element in the network.

Recently, we have applied the latter principle to impedance studies on *Necturus* gallbladder epithelium[8,9] and have noticed that under favorable conditions such studies allow the resistances of cell membranes, tight junctions, and lateral intercellular spaces to be determined and possible changes to be monitored with a time resolution of a few seconds. In the present chapter we shall describe the principles of this technique, define some of its limitations, review crucial questions of instrumentation and data analysis, and discuss some pitfalls which may occur from various leak artifacts that have never been reported before. Details about the technical

[2] E. Frömter and J. M. Diamond, *Nature (London), New Biol.* **235,** 9 (1972).

[3] E. Frömter and B. Gebler, *Pfluegers Arch.* **371,** 99 (1977).

[4] S. A. Lewis, D. C. Eaton, and J. M. Diamond, *J. Membr. Biol.* **28,** 41 (1976).

[5] H. H. Ussing, *in* "Functional Regulation at the Cellular and Molecular Levels" (R. A. Corradino, ed.), p. 285. Elsevier/North-Holland, Amsterdam, 1982.

[6] E. Frömter, *J. Membr. Biol.* **8,** 259 (1972).

[7] R. A. Frizzell and S. G. Schultz, *J. Gen. Physiol.* **59,** 318 (1972).

[8] G. Kottra and E. Frömter, *Pfluegers Arch.* **402,** 409 (1984).

[9] G. Kottra and E. Frömter, *Pfluegers Arch.* **402,** 421 (1984).

setup in our laboratory have been described in the above-mentioned publications.[8,9]

General Principles of the Impedance Technique

Cell membranes are essentially lipid bilayers with incorporated transport proteins. In the frequency range of 10 Hz to 100 kHz they exhibit capacitive and conductive responses to transmembrane current flow and can be represented in good approximation by an equivalent circuit of a capacitor and a resistor in parallel. Based on the appearance of a typical unilayered resorptive epithelium (such as gallbladder or renal proximal tubule) in the electron microscope, a single epithelial cell and the portion of the paracellular shunt path associated with that cell may be represented in good approximation by the equivalent circuit of Fig. 1a or more appro-

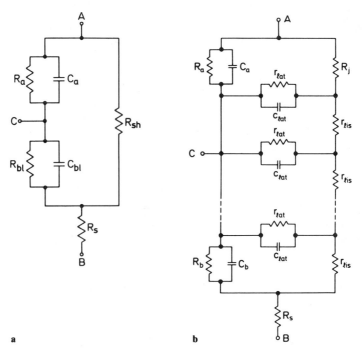

Fig. 1. (a and b) Electrical equivalent circuits of a single-layered epithelium, consisting of only one cell type. Points A, B, and C represent the mucosal, serosal, and intracellular fluid compartment. R_s is the series resistance of the serosal connective tissue. In the lumped model circuit (a) the longitudinal resistance of the lateral intercellular space is neglected. R_a, C_a, R_{bl}, and C_{bl} represent the resistances and capacitances of the apical and basolateral cell membranes, and the shunt resistor, R_{sh}, represents the resistance of the tight junction, R_j. In the distributed model circuit (b) the resistance and capacitance of the lateral cell membrane, r_{lat} and c_{lat}, and the resistance of the lateral intercellular space, r_{lis}, are distributed and form a cablelike structure which terminates in R_j.

priately by that of Fig. 1b. In these circuits we assume that (1) the intracellular compartment (represented by node C) is isopotential, (2) that a capacitive component of the tight junction may be neglected because of the negligible lateral extension (area) of the terminal bars in comparison with the cell membranes, and (3) that the longitudinal resistance of the lateral intercellular space can either be neglected so that the tight junction may be represented by a simple shunt resistor (lumped model, Fig. 1a) or that the resistances of the lateral intercellular space and of the lateral cell membrane form a cablelike structure which terminates in the tight junction resistance (distributed model, Fig. 1b).

The circuits of Fig. 1 can be analyzed by passing alternating sine wave currents of different frequencies, $I_T(f)$, from A to B and measuring the transepithelial, $V_T(f)$, and intracellular, $V_b(f)$, voltage responses from which the complex transepithelial impedance, Z_T, and a complex apparent basal membrane impedance, Z_b^{app}, can be calculated.[10] Using derivative-free least-square-fitting routines, appropriate equations for Z_T and Z_b^{app} are then fitted to the measured impedance data to obtain best-fit estimates of the apical and basolateral membrane resistances and capacitances (R_a, R_{bl}, C_a, C_{bl}), the resistance of the tight junction (R_j), and the resistance of the lateral intercellular space (R_{lis}).

The fact that intracellular measurements are required and must be compared with transepithelial measurements imposes some restrictions on the choice of the epithelia and on the development of the analysis system:

1. The epithelium must be uniform, consisting only of one cell type. In heterocellular epithelia equivalent circuits more complicated than those of Fig. 1 must be considered and measurements from at least each different cell type must be obtained. In addition, current density must be uniform in that there should not be any edge leaks in the chambers nor leaks in the cell membrane around the impaling microelectrode (see also below).

2. Intracellular and transepithelial measurements must be collected simultaneously to avoid changes of tissue function between measurements of Z_T and Z_b^{app}. In addition, while the frequency spectrum is being measured, the tissue must remain in a steady state or the duration of data collection must be short in comparison with the period in which changes in tissue conditions or experimental parameters occur (preferably in the order of ≈ 1 second).

---

[10] Note that Z_b^{app} does not represent the true basal membrane impedance, which would be calculated by dividing V_b by the current flow across the basal cell membrane, I_b, if this were known. However, since I_b is *not* known, we divide V_b by the total transepithelial current, I_T, to obtain the operational parameter Z_b^{app}, which also can be used for data fitting, provided appropriate equations are derived. [See Ref. 9. Note, however, that Eq. (A-13) of Ref. 9 contains two printing errors: Y_T should read V_T and the element a_{41} of the matrix should read -1 instead of 1.]

3. The frequency range investigated should cover the entire so-called α dispersion[11] of the cell membrane but should exclude other relaxation processes such as structural relaxation or ion concentration changes in unstirred layers. Furthermore, all measurements must be performed in the linear range of the system (small signal impedance), which means that minimal current densities must be applied. This may entail resolution problems and may require the use of averaging techniques to increase the signal-to-noise ratio.

These preconditions lead to the choice of the current waveform and instrumentation.

Choice of Current Waveform

In principle, a series of single sine wave currents of various frequencies could be used to determine Z_T and Z_b^{app}. This, however, would require at least several minutes time, since the voltage responses must be examined over the entire frequency band from a few hertz to 10 kHz or more with four to five frequencies per decade. One problem is that each frequency must be applied over at least two to five sine wave periods before the actual measurements can be taken, because time is required for the tissue response to build up. This considerably extends the duration of the analysis, particularly for low frequencies. It is preferable therefore to use current wave forms which allow multiple frequencies to be analyzed at a time with the help of Fourier analysis techniques.[12] Figure 2 shows a comparison of the amplitude and frequency content of different repetitive current wave forms which might be used. The square wave pulse (SQW) cannot be recommended because of the low power in the high-frequency end. Commercially available noise generators which produce a so-called pseudo-random binary sequence (PRBS) noise cover a broad frequency spectrum equally but must be synchronized with the Fourier analysis system because they repeat with a sequence of $2^n - 1$ clock pulses while fast Fourier transform calculations require block lengths of 2^n data.[13,14] In addition, they exhibit considerable attenuation of the single-frequency amplitude as compared to the peak-to-peak amplitude of the time domain signal, although with PRBS the attenuation is less pronounced than with Dirac pulses (D). The latter is also true for Fourier-synthesized signals as de-

[11] H. P. Schwan, *in* "Advances in Biological and Medical Physics" (J. J. Lawrence and C. A. Tobias, eds.), p. 147. Academic Press, New York, 1957.

[12] J. S. Bendat and A. G. Piersol, "Random Data: Analysis and Measurement Procedures." Wiley, New York, 1971.

[13] D. Poussart and U. S. Ganguly, *Proc. IEEE* **65**, 741 (1977).

[14] C. Clausen and J. M. Fernandez, *Pfluegers Arch.* **390**, 290 (1981).

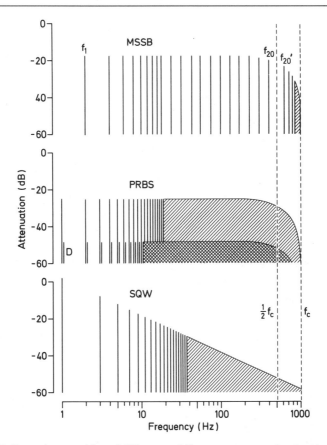

FIG. 2. Spectral composition of different multifrequency current signals. Abscissa; Frequency in hertz, on logarithmic scale; ordinate, attenuation of peak-to-peak amplitudes of individual frequencies relative to the maximum peak-to-peak amplitude of the composite signal, in decibels. All waveforms are activated with a clock frequency (f_c) of 1024 Hz. MSSB, Multiple superimposed sine wave bursts, consisting of 20 frequencies, f_1 to f_{20} (see Ref. 3). Since its 512 time points are addressed twice per second, the lowest frequency is 2 Hz. PRBS, Pseudorandom binary sequence noise ($2^{10} - 1$ time points, synchronized); D, quasi-Dirac pulse (duration, 1 clock pulse; interval, 1023 clock pulses). SQW, Square wave signal (pulse duration = interval = 512 clock pulses). Note that the single frequency intensity of the MSSB is 7.4 dB greater, or 2.34 times higher, than the PRBS at equal peak-to-peak amplitude and 30.6 dB greater, or 33.7 times higher, than the quasi-Dirac pulse. Here f'_{20} is the first harmonic of the MSSB. Without using antialiasing filters it would fold back in the Fourier transform onto f_{20}. The shaded areas indicate regions of dense frequency packing, in which individual frequencies cannot be resolved in the picture.

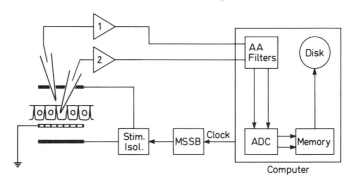

FIG. 3. Schematic diagram of experimental design for impedance measurements used in the authors' laboratory.[8] Whereas the voltage-measuring transepithelial and intraepithelial microelectrodes are connected to amplifiers 1 and 2, the current-passing Pt electrodes are connected to the stimulus isolator, which is driven from the MSSB signal (see Fig. 2 and text). The Ag/AgCl grid electrode grounds the serosal compartment. In addition, the major components of the analysis system are depicted (AA filters, antialiasing filters; ADC, analog-to-digital converter). The computer is a DEC PDP11/34 operated with Genrads Time-Series-Language TSALF under Fortran.

scribed by Nakamura et al.[15] and Fishman et al.[16] but is less pronounced in the Fourier-synthesized current signal which we have designed (MSSB).[8] This current wave form differs from the others by its nearly equidistant frequency spacing on a logarithmic rather than a linear frequency axis.

Instrumentation

The measurements can be performed in a modified Ussing chamber where the tissue is mounted, epithelial surface up, between an upper and lower compartment filled with Ringer's solution. The chamber must allow individual cells to be punctured with microelectrodes under microscopic inspection (Fig. 3). In our chamber current is applied from a stimulus isolation unit via a pair of ground-free Pt electrodes located in the fluids of the upper and lower compartments. The lower compartment is grounded and the cell potential (V_b) and the transepithelial potential difference (V_T) are measured with reference to ground with two microelectrodes. One is inserted into a cell and the other one is placed nearby in the mucosal fluid. Alternatively, one of the current electrodes could be grounded and the potential measurements could be performed with differential amplifiers.

[15] H. Nakamura, Y. Husimi, and A. Wada, *Jpn. J. Appl. Physiol.* **16,** 2301 (1977).
[16] H. M. Fishman, L. E. Moore, and D. Poussart, *in* "Biophysical Approach to Excitable Systems" (W. D. Adelman and D. E. Goldman, eds.), p. 65. Plenum, New York, 1981.

To avoid capacitive coupling from the bath and from nearby conductors into the microelectrodes and hence to ensure that the electrodes register only the potential that prevails at the very tip, the electrodes have to be completely shielded from the shaft up to approximately 30 μm from the tip. We use a convenient method that easily produces low-capacitance, high-resistance shields.[17] We usually break the tip of the extracellular microelectrode to a width of approximately 10 μm and fill it with Ringer's solution, rather than with 2.7 M KCl, in order to prevent contamination from outflowing KCl. Then V_T and V_b are recorded with feedback amplifiers similar to those described in Ref. 18, which use the electrode shield as feedback capacitor. In its present configuration our electrode system allows distortion-free measurements (phase shift less than 2°) of up to 10 kHz to be obtained with microelectrodes of ≈ 20 MΩ resistance (P. Langer and E. Frömter, unpublished observations).[19] In order to save dynamic range of the A/D conversion during the impedance measurements, the dc potential of the intracellular microelectrode is suppressed toward zero with an appropriate sample-and-hold circuit.

Data Sampling and Analysis

Since both V_T and V_b are to be sampled simultaneously, at minimum, a computer-controlled two-channel data acquisition unit is required. If a third channel is available, the current can be recorded concurrently with the voltages. However, if we use a constant repetitive current wave form, the current can be recorded before the experiment and recalled from the memory for impedance calculations, whenever required. Since even the Fourier-synthesized current signal may contain some high-frequency harmonics, it is necessary to use steep antialiasing filters (120 dB/octave) to prevent errors from back-folding frequencies in the Fourier calculations. Using the fast Fourier transforms of the voltage signals and of the current wave form, cross power spectra are calculated for the transepithelial and intracellular data, which are then divided by the autopower spectrum of the current signal to obtain the impedances Z_T and Z_b^{app}.

An example of an impedance measurement obtained from *Necturus* gallbladder is shown in Fig. 4. The different symbols at the low- and

---

[17] G. Kottra and E. Frömter, *Pfluegers Arch.* **395,** 156 (1982).

[18] K. Suzuki, V. Rohlicek, and E. Frömter, *Pfluegers Arch.* **378,** 141 (1978).

[19] To correct for the remaining distortion of the microelectrodes and amplifiers, we have adopted a technique of measuring the transfer function of the intracellular electrode while the electrode is inside a cell and to correct each measurement for the electrode transfer function by complex division.

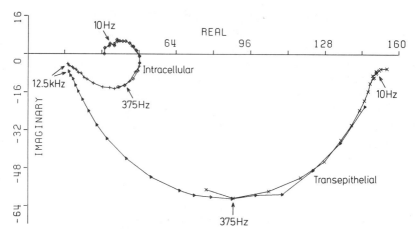

FIG. 4. Transepithelial and intracellular impedances of *Necturus* gallbladder epithelium (Nyquist plot). Abscissa and ordinate, Real and imaginary impedance components in Ω cm². As indicated by different symbols the data were collected in two sequential frequency runs with somewhat overlapping frequencies. (From Ref. 8.)

high-frequency ends of the impedance loci indicate that the measurement was divided into two consecutive operations: The MSSB signal was applied first with a low clock frequency to cover the spectrum from 2.5 to 500 Hz and then with a higher clock frequency to cover the spectrum from 62.5 Hz to 12.5 kHz. The time required to obtain the measurements and calculate both impedance loci over the entire frequency range was approximately 5 sec, with the actual measuring time amounting only to 1 to 2 sec.

Data Fitting

Fitting of an equivalent circuit to impedance data such as those plotted in Fig. 4 presents a number of problems:

1. The structure of the impedance loci is rather condensed and does not show conspicuous details.

2. Transepithelial and intracellular measurements of different absolute magnitude are being combined which poses the question of weighting factors.

3. The number of variables to be fitted is rather high. For example, in the case of the equivalent circuit of Fig. 1b, at least six parameters must be fitted.[20]

[20] Two parameters, the series resistance R_s and the ratio of basal to lateral cell membrane surface (β), can be directly estimated from the high-frequency end of the impedance plots or, respectively, from electron micrographs of the epithelium.

Using conventional least-square-fitting routines, it is therefore quite possible that the search is trapped in some side minima and does not find the proper minimum in reasonable time. Moreover, we must not forget that equivalent circuits such as those of Fig. 1 are only approximate representations of the true impedance characteristics of an epithelium. Our experience with fitting tells us that it is of advantage to approach the problem in two steps: (1) to perform prefits with a selected number of data points (to decrease computation time) starting from selectible parameter estimates and (2) use the best prefit results as starting parameters for the final fit. This procedure results in reasonably good fits with small mean deviations between 1 and 2% (defined as average deviation of a single data point from its theoretical location in percentage of the transepithelial dc resistance of the same tissue). Applying these fit procedures to 46 impedance measurements on 10 gallbladders under control conditions, we have been able to clearly demonstrate that the distributed equivalent circuit model of Fig. 1b, which represents the lateral space as a cablelike structure and the tight junction as a simple resistor, describes the impedance properties of gallbladder well, while the more simple lumped model of Fig. 1a, which consists only of two membrane equivalents and a resistor for the paracellular shunt, does not.[9]

Looking at best-fit estimates of individual circuit parameters, however, we find that they are determined with different degrees of reliability, as can be assessed from calculations of the marginal standard deviation according to Hamilton[21] and from the skewed distribution in histograms.[9,22] Under control conditions the greatest difficulties arise with determination of R_a, while C_a, C_{bl}, R_j, and R_{lis} are reasonably well defined. We hope that by combining impedance analysis with a modified 2D-cable analysis[6] the remaining uncertainty can be resolved. As long as this is not achieved, we are forced to increase the number of observations if we want to estimate change in R_a. This problem with R_a and to a lesser extent with R_{bl}, however, does not invalidate the reliability of the other parameter estimates. Thus, we have been able to demonstrate that R_{lis} changes appropriately if the lateral spaces are collapsed, that R_j changes on blockage of the tight junctions, and that estimates of C_a and C_{bl} compare well with the surface amplification of the respective membranes as observed in the electron microscope.[9,23,24]

[21] W. C. Hamilton, "Statistics in Physical Science: Estimation, Hypothesis Testing and Least Squares." Ronald Press, New York, 1964.
[22] P. Weskamp, *Pfluegers Arch.* **402**, R10 (1982).
[23] G. Kottra and E. Frömter, *Pfluegers Arch.* **400**, R26 (1984).
[24] H. Ewald, P. Langer, and E. Frömter, *Pfluegers Arch.* **405**, R26 (1985).

Inhomogeneity Problems in Impedance Measurements

As mentioned above, an essential precondition for the proper analysis of epithelial equivalent circuits (using impedance techniques or other techniques) is the macroscopic uniformity of the preparation and the necessity that the impaled cell properly represents the properties of all other cells of the tissue. There are essentially three ways in which this precondition may be violated.

1. The properties of the exposed tissue may vary from one end to the other (e.g., longitudinal or circumferential variation of transport properties in different parts of the gastrointestinal tract or different thickness of supporting layers if those have partially been dissected off to provide free access of the microelectrode to the epithelial cells).

2. The tissue may not be tightly sealed inside the chamber (e.g., edge leak or edge damage).

3. Some membrane elements may have been injured during the impalement with the microelectrode (impalement artifact).

Recognition of such problems is usually difficult and requires special techniques such as voltage scanning[6,25] or vibrating electrode measurements.[26] In addition, a number of techniques have been described to minimize edge leak and edge damage.[25,27] However, since even the best precautions can never guarantee the absence of leaks, it is important to discuss the problems of inhomogeneities in greater detail. This discussion is based on three lines of evidence: (1) a short theoretical analysis of the edge leak problem in epithelial impedance studies, (2) a series of impedance measurements obtained on a discrete resistor and capacitor model (dummy circuit of the epithelium), and (3) a series of measurements on intentionally injured *Necturus* gallbladders.

Simplified Theoretical Treatment of Edge Leaks

Figure 5 provides the most simple equivalent circuit of a membrane preparation in a chamber in the presence of edge leak. It depicts the frequency-dependent transepithelial impedance $Z_T(f)$ in parallel with an ohmic resistor R_k which represents the edge leak. In addition, we have inserted an ohmic resistor in series with the epithelial impedance (R_s), which represents the serosal subepithelial tissue.[28] The impedance of this

[25] J. T. Higgins, L. Cesaro, B. Gebler, and E. Frömter, *Pfluegers Arch.* **358**, 41 (1975).
[26] L. F. Jaffe and R. Nucitelli, *J. Cell Biol.* **63**, 614 (1974).
[27] S. I. Helman and D. A. Miller, *Science* **173**, 146 (1971).
[28] Alternatively, it is also possible that the leak only lies in parallel with Z_T. As $Z_T(0) \gg R_s$, the result will be essentially identical to the circuit of Fig. 5.

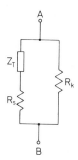

FIG. 5. Simplified equivalent circuit of an epithelial preparation with edge leak. Z_T is the frequency-dependent impedance of the epithelium, R_s is the resistance of the serosal connective tissue layer, and R_k is the resistance of the edge leak path.

circuit is

$$Z^*(f) = \frac{V_T(f)}{I(f)} = \frac{R_k[Z_T(f) + R_s]}{R_k + Z_T(f) + R_s} \tag{1}$$

At $f = 0$ we obtain

$$Z^*(0) = \frac{R_k[Z_T(0) + R_s]}{R_k + Z_T(0) + R_s} \tag{2}$$

which is a simple parallel combination of the ohmic resistances of both limbs.

At $f = \infty$, we obtain $Z_T(\infty) = 0$ and

$$Z^*(\infty) = \frac{R_k R_s}{R_k + R_s} \approx R_s \tag{3}$$

If $R_s \ll R_k$, $Z^*(\infty)$ approaches R_s. This means that the leak resistance R_k has little effect on the measurements at high frequencies but becomes increasingly influential toward the low-frequency end of the impedance locus. In other words, at high frequency, the capacitive elements of the epithelium are effectively short-circuited so that the current distributes uniformly over the exposed surface area, while at low frequency or under dc conditions the resistance of the epithelium is high and a considerable part of the current flows through the edge leak. This implies that, at low frequencies, the current density is low in the center of the exposed tissue, whereas at high frequencies it is uniformly high, and this phenomenon leads to a number of surprising observations (see below).

Measurements on an Electrical Analog of Necturus Gallbladder

Since an exact mathematical treatment of the edge leak problem in Ussing chambers is extremely involved we have assembled an electrical analog (dummy circuit) on which various types of leaks or inhomogeneities in Ussing chambers could be simulated so that their effects on the ac impedance measurements could be investigated. The analog was assembled from 17 similar sets of components (resistors and capacitors; see Fig. 6) arranged in a parallel array. Each set, consisting of 10 components, was connected via a pair of concentric silver/silver chloride electrodes to two buckets filled with dilute salt solution (3 mmol/liter KCl) simulating the resistive properties of the mucosal and serosal compartments (including the series resistance of the subepithelial tissue). The electrolyte concentration was deliberately decreased to such a low value (in comparison with Ringer's solution) to compensate for the exceedingly greater volume of the buckets as compared to the volume of the serosal (and mucosal) compartments in the Ussing chamber. The first set represented the central cell of the epithelium with associated tight junction and lateral intercellular space.

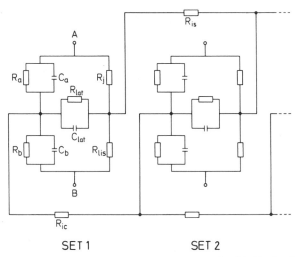

FIG. 6. Circuit diagram of gallbladder analog (dummy circuit). Depicted are 2 (from a total of 17) sets which represent a cell in the center of the preparation and the ring of cells surrounding it. The values of the resistors (R) and capacitors (C) of set 1 were $R_a = 10^9 \Omega$, $R_b = 2.2 \times 10^8 \Omega$, $R_{lat} = 5.1 \times 10^7 \Omega$, $R_j = 3.5 \times 10^7 \Omega$, $R_{lis} = 4.7 \times 10^6 \Omega$, $R_{ic} = 5.7 \times 10^5 \Omega$, $R_{ip} = 8.3 \times 10^4 \Omega$, $C_a = 3.3 \times 10^{-11}$ F, $C_b = 3.3 \times 10^{-11}$ F, and $C_{lat} = 1.1 \times 10^{-10}$ F. The indices are a, apical; b, basal; lat, lateral; j, tight junction; lis, lateral intercellular space; ic, intercellular (cell-to-cell coupling); and is, interspatial (horizontal space-to-space coupling). Related to 1 cm^2 of epithelium, the above numbers correspond to $R_a = 3890 \Omega$ cm^2, $R_{bl} = (g_b + g_{lat})^{-1} = 161 \Omega$ cm^2, $R_j = 136 \Omega$ cm^2, $R_{lis} = 18 \Omega$ cm^2, $C_a = 8.48 \mu$F/cm^2, and $C_{bl} = C_b + C_{lat} = 36.77 \mu$F/cm^2. The parameter values of sets 2–17 were calculated according to set 1 with appropriate correction for the change in geometry (see the text). At nodes A and B, the sets were connected via circular Ag/AgCl electrodes to buckets filled with dilute salt solution representing the mucosal and serosal fluid compartments and thus providing an equivalent of the subepithelial tissue resistance R_s.

The second set stood for the hexagonal ring of cells which surrounds the center cell, and the next seven sets (sets 3 to 9) stood for the following seven cell rings. Sets 10 to 16 represented more distant rings of cells of different width (3, 10, 40, 40, 40, 10, and 3 cells) and set 17 represented the outermost cell ring. In total the analog represented $\approx 72,000$ cells, which is approximately the number of cells contained in 0.28 cm^2 of *Necturus* gallbladder epithelium (the exposed surface area in our chamber).

As shown in Fig. 6 each set was composed of three parallel RC elements for the apical, basal, and lateral cell membranes plus four resistors representing the tight junction, the transverse and radial resistance of the lateral intercellular space, and the radial cell to cell connection. The magnitude of the individual resistors and capacitors of the first set was calculated from impedance data of *Necturus* gallbladder, multiplying or respectively dividing those by 72,000. The values for the other sets were determined by dividing or respectively multiplying with the number of cells in a set. Representation of the lateral space and estimation of the vertical and radial resistance to current flow in the spaces was more difficult because, for practical purposes, the analog had to be as simple as possible. On the other hand, it was also less critical, because the major scope of the investigation concerned questions of transepithelial rather than radial current flow. The final choice (see legend to Fig. 6) was a compromise, based on estimates of lateral space width and resistance, as determined previously.[9] The cell-to-cell coupling resistor was calculated from 2D-cable analysis data[6] under the assumption that the intracellular resistance was negligible. In sets 10 to 16 the values of the radial components were estimated from the theory of radial diffusion inside a continuous membrane, which indicates that the resistance (R) obeys the relation

$$R = A + B \ln r \tag{4}$$

where r is the radius and A and B are constants related to the parameters of the system.

Another problematic point was the coupling of the individual sets to the volume conductors representing the serosal and mucosal fluid bath. This coupling was achieved by concentric silver-plated electrodes, the area of which increased in proportion with the tissue area represented by each set. Impalement leaks were simulated by inserting appropriate leak resistors into set 1. Edge leaks were simulated by connecting the outermost pair of silver rings in the buckets (set 17) by appropriate leak resistors.

The impedance locus of the dummy circuit in its control configuration is shown (among other measurements) in Fig. 7. As can be seen, it compares well with the loci observed on *Necturus* gallbladders under control conditions.

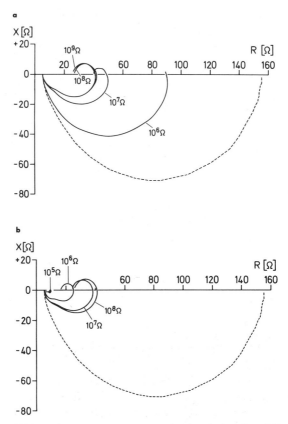

FIG. 7. (a and b) Impedance measurements on the electrical analog of *Necturus* gallbladder. Simulation of cell membrane leaks. Nyquist plot as in Fig. 4. For sake of clarity, instead of depicting individual data points the loci were traced by hand. The dashed lines indicate the transepithelial impedance and the solid lines indicate the intracellular impedance. (a) Variation of R_a in set 1 (see Fig. 6) from 10^9 to 10^6 Ω, simulating a leak in the apical cell membrane. (b) Corresponding variation of R_b from 10^8 to 10^5 Ω, simulating a leak in the basal cell membrane of the impaled cell. In the former case, R_b was constant at 2.2×10^8 Ω, in the latter case, R_a was constant at 10^9 Ω. Note adequate representation of gallbladder impedance by the electrical analog under control conditions (compare Figs. 4 and 7) and deformation of intracellular impedance locus with increasing leak conductance.

Production of Leaks in the Necturus Gallbladder Preparation

In gallbladder preparations three types of leaks were produced. (1) Leaks around the impaling microelectrode may be generated by repetitive forceful impalements or occur spontaneously and may be recognized by low, unstable cell membrane potentials. Their influence can be studied

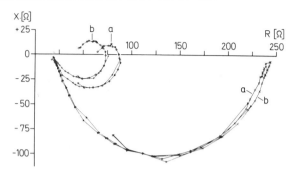

FIG. 8. Sequential impedance measurements on a *Necturus* gallbladder during supposedly increasing seal around the intracellular microelectrode. Loci a and b were obtained 3 and 7 min, respectively, after impalement during a phase of increasing cell membrane potential. Note similarity to Fig. 7a.

easily during spontaneous resealing of the cell membrane, when the membrane potential gradually increases. (2) Leaks at the edge were produced by cutting around the chamber edge with a scalpel. (3) In subepithelial punctures leaks were again produced by repeatedly and forcefully pushing the electrode across the epithelium into the upper part of the subepithelial connective tissue layer.

Effect of Cell Membrane Leaks

Figure 7 shows Nyquist plots from impedance measurements made on the dummy circuit during variation of R_a in set 1 (from its control value of 10^9 to 10^6 Ω) and of R_b in set 1 (from its control value of 10^8 to 10^5 Ω). These variations are equivalent to leaks introduced into the respective cell membranes of the impaled cell. It can be seen that a leak in the apical cell membrane expands the "intracellular" impedance locus while reducing simultaneously the magnitude of the apparent inductive semicircle. As these changes affect almost exclusively the impaled cell, the transepithelial impedance, which is determined by the total of 72,000 cells and their tight junctions, remains unchanged.

Exactly the same observations can be made in the gallbladder if one follows the change of intracellular impedance during sealing of a microelectrode in the apical cell membrane, as assessed from a gradual rise of the cell membrane potential (Fig. 8). Corresponding observations were also made earlier in a study of gallbladder impedance with square wave current pulses,[29] where a correlation was observed between the membrane poten-

[29] K. Suzuki, G. Kottra, L. Kampmann, and E. Frömter, *Pfluegers Arch.* **394**, 302 (1982).

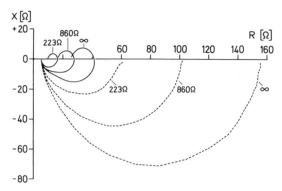

FIG. 9. Simulation of edge leak on gallbladder analog circuit. Details as in Fig. 7, except that R_a and R_b were 10^9 and 2.2×10^8 Ω throughout, and an edge leak resistor, R_k, was introduced in parallel to set 17. R_k bridged both fluid compartments through a pair of Ag/AgCl electrodes. R_k was varied from ∞ (control) to 860 and 223 Ω. Note progressive collapse of transepithelial semicircle and tendency of the intracellular locus to round up.

tial and the appearance of an overshoot phenomenon in the potential response of the apical or basolateral cell membrane, respectively.

In contrast, inserting a leak into the basal cell membrane squeezes the intracellular impedance locus. The latter simulation would be relevant for leaky punctures from the basolateral surface.

Effect of Edge Leaks

Figure 9 shows the effect of edge leaks on the impedance loci of the dummy circuit. As was to be expected in the presence of a leak path, the transepithelial locus shrinks from its control value (at $R_k = ∞$ Ω) to approximately $\frac{1}{3}$ (at $R_k = 223$ Ω) and a similar squeezing is observed with the intracellular impedance locus. In addition, however, a deformation appears in the intracellular impedance locus in that the radius of the apparent capacitive semicircle shrinks more than that of the apparent inductive semicircle.

Similar observations were made on *Necturus* gallbladders. Figure 10 depicts impedance measurements on two preparations, one of which may have been leaky spontaneously, while the other had been cut along the edge. In both cases the cell membrane potential measurements were stable at values of -85 mV so that leaks around the microelectrode may be excluded. The transepithelial impedance loci appear squeezed as expected, and the low-frequency end of the intracellular impedance locus is shifted toward the left with the size of the apparent inductive semicircle nearly equaling, or, in the other case, even exceeding that of the apparent capaci-

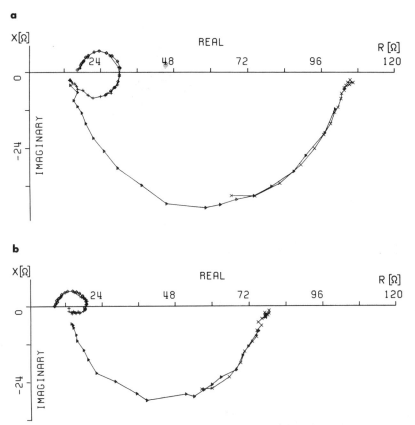

FIG. 10. (a and b) Impedance measurements on *Necturus* gallbladder in the presence of edge leaks. The cell potential was stable at −85 mV in both cases. Details as in Fig. 4. (a) The preparation was presumably spontaneously leaky at the edge; (b) an edge leak had been introduced deliberately. Note progressive squeezing of the transepithelial locus and creeping of the low-frequency intracellular data toward left, beyond the high-frequency points in b. For explanation see the text.

tive semicircle. The latter distortion is more pronounced than in the measurements on the dummy circuit, because the polarization of the electrodes in the dummy circuit at set 17 probably limited the leak current. In principle, however, the observations are similar.

The surprising fact that, in the presence of edge leaks, the real component of the intracellular impedance is definitely smaller at low frequencies than at high frequencies can well be explained by the simplified theory of the edge leak given above. At low frequencies a great deal of the current passes across the edge leak. As a consequence, in the center of the preparation, the current density decreases so that the voltage drop across connective tissue and basal cell membrane is now smaller than the voltage drop measured when the current density is uniform throughout (as in the case of high-frequency currents).

Supporting evidence for this interpretation can also be obtained from the analysis of impedance data generated by an electrode that has been advanced across the epithelium into the underlying subepithelial tissue layer to record the voltage signal developing across this tissue alone. In Fig. 11 are shown the real and imaginary components of the impedance obtained during such a puncture on the same gallbladder (with edge leak) used in the experiment from which Fig. 10b was derived. It can be seen that the high-frequency end of the transepithelial and subepithelial impedance loci extrapolates approximately to the same point on the real axis. With decreasing frequency, however, the subepithelial locus passes in counterclockwise direction through a semicircle which returns to the real axis far below the high-frequency intersection. Using the dummy circuit this behavior could be fully confirmed as shown for two cases of increasing edge leak in Fig. 12. Taken together, these observations confirm that in the presence of edge leaks the current density in the center of the preparation depends on the frequency of the current. At high frequency the current density is uniform throughout the chamber, while at low frequency it decreases to smaller values in the center region of the preparation.

Effect of Leaks in Subepithelial Punctures

Subepithelial punctures are often used to determine the resistance of the connective tissue layer for correcting dc voltage divider measurements.

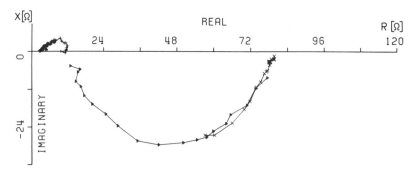

FIG. 11. Subepithelial puncture in a gallbladder preparation with edge leak. Same experiment as Fig. 10b, except that the microelectrode was advanced from the cell into the subepithelial tissue. Note that the subepithelial impedance locus appears as a somewhat deformed semicircle in the inductive plane with its low-frequency end nearer to the origin.

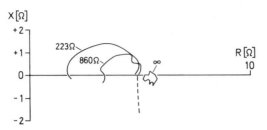

FIG. 12. Simulation of a subepithelial puncture in a preparation with edge leak in the analog circuit. Redrawn subepithelial impedance loci in the presence of increasing edge leak (decreasing R_k). The dashed line depicts the high-frequency end of the transepithelial impedance locus. At $R_k = \infty$ the subepithelial locus exhibits a simple dot (depicted by its contours), whereas at $R_k = 860\ \Omega$ and $R_k = 223\ \Omega$, semicircles appear in the inductive plane.

Our own experience with this approach, however, has not been fully satisfactory, because the data showed too much scatter. With the help of the dummy model circuit for *Necturus* gallbladder we have now recognized that the scatter most probably reflected leaks between the apical and basal fluid compartments that developed around the impaling microelectrode. Figure 13 shows a series of impedance measurements on the dummy

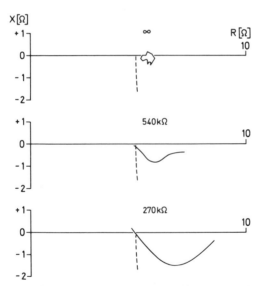

FIG. 13. Simulation of impalement leaks in subepithelial punctures in the analog circuit. Details as in Fig. 12, except that no edge leak was present but nodes A and B of set 1 (see Fig. 6) were connected with different resistors to simulate a transepithelial leak around the impaling microelectrode. Note appearance of a capacitive semicircle with decreasing shunt resistance instead of the originally expected single dot, which only shows up at $R = \infty$.

circuit pertaining to this problem. In the absence of any edge leak or microelectrode-induced leak, the subepithelial electrode records a single dot in the Nyquist plot as expected. If we insert now a leak resistor which connects the apical and basal fluid compartments at set 1, we obtain an apparent capacitive semicircle, the size of which increases with increasing leak conductance. This indicates that the current density in the center of the preparation increases with decreasing frequency so that the voltage response exceeds the response at high frequency (observed under uniform current distribution). If, in addition, one provides an edge leak in the preparation, which affects the current density in the opposite sense, one may observe all kinds of impedance loci, depending on which leak is quantitatively the most important. It is also possible that both effects may cancel each other and that we observe a single dot in the proper position, although both the edge and the center are leaky. Examples of this are shown in Fig. 14.

Conclusions

In the present chapter we have surveyed the principle technical concepts of the combined transepithelial and intraepithelial impedance analysis, which offers a new dimension to the exploration of the intraepithelial organization of ion transport barriers. In addition, we have summarized a series of experiments in which the influence of various leaks on impedance measurements was systematically investigated. The observations from the latter experiments, which were often surprising, could all be explained on

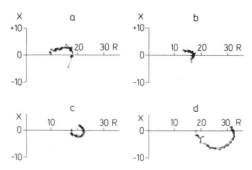

FIG. 14. (a–d) Influence of increasing leaks around a subepithelial puncture in a *Necturus* gallbladder preparation with moderate edge leak. Only subepithelial impedance loci are depicted. The first record (a) was taken immediately after gently advancing the electrode across the epithelium. The other records were taken after further forceful forward and backward movements with the microelectrode. Note that the high-frequency end of the impedance locus remains unchanged at about 18 Ω while the low-frequency end moves from the lower value measured in the presence of only an edge leak (a) to a higher value measured in the presence of an additional leak at the micropuncture site (d).

the basis of local inhomogeneities in current density which occur in the low-frequency range and disappear in the high-frequency range when the capacitive membrane elements effectively short-circuit the epithelia.

The reason for presenting the latter data at relatively great length was twofold: (1) the data can help the experimenter in recognizing leak artifacts in measurements and finding ways to avoid those artifacts, and (2) these observations may serve as a starting point for the development of new techniques of leak identification in a more general sense.

In fact, leak artifacts may not only disturb the presently described extended impedance measurements, but are equally important in practically all other approaches to determine cellular or transepithelial transport properties and in particular in all attempts to quantify equivalent circuit parameters of epithelia irrespective of the method used (measurements of dc potential and resistance and of voltage divider ratios, or measurements of ion activities and unidirectional fluxes). In the absence of simple detection techniques these problems have often enough been neglected in the past. However, the difference in low- and high-frequency current densities, in leaky preparations described here, promises to provide the basis for the development of simple edge leak detection methods applicable to all epithelial studies in Ussing chambers. Thus, by high-resolution measurements of the current density during high- or low-frequency alternating current flow in the center of a membrane preparation, it should be possible not only to detect edge leaks but also to quantify their magnitude. Experiments to develop such techniques are presently being performed in our laboratories.

Acknowledgment

This work was supported by the Deutsche Forschungsgemeinschaft Fr 233/5-3 and -4 and by the Medical Research Council of New Zealand.

[34] Patch-Clamp Experiments in Epithelia: Activation by Hormones or Neurotransmitters

By Ole H. Petersen

Introduction

Exocrine glands secrete fluid, electrolytes, and organic materials onto the outer or inner surfaces of the body. These glands secrete very little or not at all in the absence of stimulation, but when excited they can produce large quantities of fluid. Fluid secretion from sweat, lacrimal, and salivary glands, in particular, is finely controlled and can be switched on and off repeatedly and with great precision. It has been known for many years that exocrine secretion is controlled by chemical messengers secreted from nerve endings (neurotransmitters; for example, acetylcholine and noradrenaline) or from endocrine glands (hormones; for example, gastrin and cholecystokinin). These chemical messengers interact with specific sites (receptors) on the gland cell membrane and thereby induce transport events leading to fluid formation.[1]

From the extensive intracellular microelectrode investigations in mammalian exocrine gland cells, it is clear that hormones and neurotransmitters evoking secretion can only activate ion conductance pathways by binding to specific receptor sites on the outside, but not the inside, of the plasma membrane.[2]

Since the membrane conductance increase evoked by neurotransmitters and hormones occurs after a relatively long latency of about 0.5 sec, even in cases where very close iontophoretic agonist application is used, and where in the same experiments iontophoretic amino acid application results in immediate membrane depolarization (due to operation of Na^+–amino acid cotransport),[2,3] it has long been assumed that there is an involvement of an intracellular second messenger.[2] The first direct evidence for messenger-mediated activation of ion conductance pathways came from studies in which intracellular Ca^{2+} injections were shown to mimic the effects evoked by extracellular neurotransmitter applications.[2,4,5]

Investigations using classical intracellular microelectrode recording cannot, however, give detailed molecular information about the transduction mechanisms involved, and it has only been through the application of patch-clamp methods[6] that further progress has been made possible.

This chapter deals with patch-clamp single-channel and whole-cell current recording experiments designed to give precise and detailed information about the transduction mechanisms involved in stimulus–secretion coupling.

The Patch-Clamp Method

Electrical currents in cells are mediated by a single class of proteins in the plasma membrane called ion channels or pores. The first direct obser-

[1] O. H. Petersen, *Sci. Prog. (Oxford)* **70**, 251 (1986).

[2] O. H. Petersen, "The Electrophysiology of Gland Cells." Academic Press, New York, 1980.

[3] N. Iwatsuki and O. H. Petersen, *Nature (London)* **283**, 492 (1980).

[4] N. Iwatsuki and O. H. Petersen, *Nature (London)* **268**, 147 (1977).

[5] N. Iwatsuki and O. H. Petersen, *Pfluegers Arch.* **377**, 185 (1978).

[6] O. P. Hamill, A. Marty, E. Neher, B. Sakmann, and F. J. Sigworth, *Pfluegers Arch.* **391**, 85 (1981).

vation of current flowing through single-ion channels was a result of studies in bimolecular lipid membranes in the presence of antibiotics. In intact cells the problem is to detect single-channel currents in the presence of background electrical noise. Conventional intracellular microelectrode methods for current measurement are associated with a background noise of at least 100 pA, whereas the current flowing when a single channel opens is only a small fraction of this background noise.

Neher and Sakmann[7] solved this problem by the patch-clamp method. Instead of inserting a microelectrode into a cell, they pressed a microelectrode tip onto the surface of a cell, effectively isolating a patch of membrane. The intrinsic noise increases with the area of the membrane under study, and when a small area $(1 - 10 \ \mu m^2)$ is isolated, the extraneous noise levels are so low that the picoampere currents flowing through single channels can be measured directly.

The seal between the tip of the microelectrode and the outer surface of the cell membrane has, under suitable conditions (fire-polished and clean micropipet tip and clean membrane surface), a high electrical resistance [in the order of gigaohms $(10^9 \ \Omega)$] and is mechanically surprisingly stable. The discovery in 1980 of this high-resistance seal by Neher, Sakmann, and co-workers as well as by Horn and Patlak turned out to be very important, as it made entirely new types of experiments possible (Fig. 1).

The electrically isolated membrane patch can be pulled off the cell (excised) in such a way that the inside of the plasma membrane faces the bath solution (inside-out) or alternatively so that the inside faces the solution in the micropipet (outside-out). By breaking the patch membrane in the cell-attached recording conformation, the solution in the pipet interior gains direct access to the cell interior and cell dialysis is carried out under conditions in which the currents across the whole-cell membrane can be measured. Equilibration of the cell interior with the bath solution can be done while single-channel currents are recorded by making holes in the plasma membrane outside the isolated patch area with the help of detergents such as saponin.[8]

Single-Channel Currents Recorded in Excised Membrane Patches

Many of the ionic currents so far characterized in mammalian exocrine gland cells at the single-channel level are activated by intracellular Ca^{2+}. Ca^{2+}- and voltage-activated K^+ selective channels with a high unit conductance $(200 - 300 \ pS$ in symmetrical K^+-rich solutions) are present in

---

[7] E. Neher and B. Sakmann, *Nature (London)* **260,** 799 (1976).

[8] Y. Maruyama and O. H. Petersen, *J. Membr. Biol.* **81,** 83 (1984).

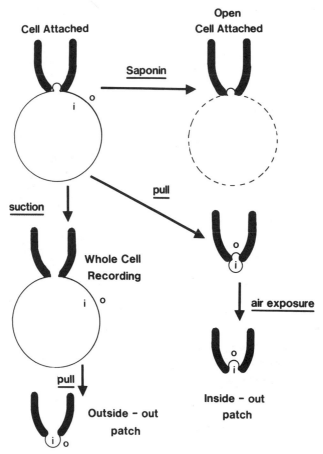

Fig. 1. Schematic representation of different patch-clamp recording configurations and procedures used to establish them. A patch-clamp experiment starts in the cell-attached configuration (top left). The tip of the fire-polished recording pipet is gently brought into contact with the cell surface, and slight suction is applied to the pipet, resulting in a high-resistance seal between tip of the glass pipet and the cell membrane. The detergent saponin can be introduced briefly into the bath to permeabilize the cell membrane (outside the isolated patch area), in this way allowing equilibration between cell interior and exterior (open cell-attached configuration). Starting from the original cell-attached situation, the patch membrane can be mechanically pulled off, leading to formation of a closed membrane vesicle in the pipet tip. The outer surface of this vesicle can be disrupted by passing the pipet tip briefly through the air–water interface of the bath, leading to the excised inside-out membrane configuration. If a short pulse of suction is applied to the pipet in the cell-attached mode, the membrane patch can be broken and direct continuity between pipet and cell interior established. Currents flowing across the entire cell membrane can now be recorded (whole-cell recording). If the pipet is thereafter pulled away from the cell, excised membrane fragments will reseal so that an excised outside-out membrane patch is obtained. (Adapted from Petersen and Findlay.[10])

many different epithelial cell types[9-11] and were first demonstrated in mouse and rat salivary acinar cells.[12] The Ca^{2+}-activated nonselective cation channels with a unit conductance of about 30 pS are present in some epithelial cells[9,10] and were first found in mouse and rat pancreatic acinar cells.[13] The Ca^{2+}-activated Cl^- channels with a very small unit conductance (1 – 2 pS) have been mainly studied in mouse and rat lacrimal acinar cells.[14-16] The basic characteristics of these various Ca^{2+}-activated ion channels are not discussed here, as they have been reviewed several times in recent years.[9-11,17]

Use of Patch-Clamp Single-Channel Current Recording Experiments to Demonstrate Messenger-Mediated Channel Opening

The first experiments in which hormonal activation of single-channel currents via an internal (second) messenger were demonstrated were carried out on mouse pancreatic acinar cells in which the Ca^{2+}-activated nonselective cation channel is dominant.[18] In such experiments, cholecystokinin octapeptide (CCK_8) is applied by leakage from an extracellularly placed micropipet close to the cell from which unit recording was made (Fig. 2). When the tip of the CCK-delivery pipet is far away from the acinar unit under investigation, the patch is always silent. When the pipet is brought close to the recording site, unit activity begins to appear, and after less than 1 min, vigorous activity can be observed (Fig. 3). The cell-attached configuration is very stable, and currents can be recorded at many different pipet potentials. The $I - V$ relationship obtained in these experiments is linear, with a single-channel conductance in the range 30 – 35 pS. Reversal of current polarity occurs at a pipet potential of about -25 mV. After the $I - V$ relationship has been established *in situ*, the patch can be excised and the membrane inside exposed to a solution containing a high Ca^{2+} concentration (for example, 10 μM) (Fig. 3). Single-channel currents can be recorded again, and a new $I - V$ relationship (linear) obtained. This again represents a single-channel conductance of 30 – 35 pS, but this time

[9] O. H. Petersen and Y. Maruyama, *Nature (London)* **307**, 693 (1984).

[10] O. H. Petersen and I. Findlay, *Physiol. Rev.* **67**, 1054 (1987).

[11] O. H. Petersen and D. V. Gallacher, *Annu. Rev. Physiol* **50**, 65 (1988).

[12] Y. Maruyama, D. V. Gallacher, and O. H. Petersen, *Nature (London)* **302**, 827 (1983).

[13] Y. Maruyama and O. H. Petersen, *Nature (London)* **299**, 159 (1982).

[14] A. Marty, Y. P. Tan, and A. Trautmann, *J. Physiol. (London)* **357**, 293 (1984).

[15] I. Findlay and O. H. Petersen, *Pfluegers Arch.* **403**, 328 (1985).

[16] M. G. Evans and A. Marty, *J. Physiol. (London)* **378**, 437 (1986).

[17] A. Marty, *Trends NeuroSci. (Pers. Ed.)* **10**, 373 (1987).

[18] Y. Maruyama and O. H. Petersen, *Nature (London)* **300**, 61 (1982).

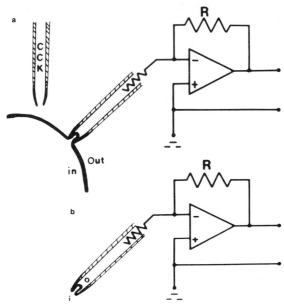

Fig. 2. Activation of single-channel current in mouse acinar cell by local CCK application. Schematic outline of the experimental system. (a) Recording from *in situ* membrane patch. In these experiments the recording pipet was always filled with extracellular bath solution. No agonist was present in the pipet solution. The pipet labeled CCK contained the octapeptide in a concentration of 5 μM. The spontaneous diffusion of CCK from the tip of the micropipet was used as the means of stimulation. (b) After the *in situ* recording, the patch-clamp pipet was withdrawn from the acinus to obtain an excised inside-out membrane patch. In these experiments single-channel recordings were made in symmetrical saline solutions. In other experiments the bath fluid was changed to one having an intracellular composition. (Adapted from Maruyama and Petersen.[18])

reversal of current polarity occurs at 0 mV pipet potential (same extracellular solution in pipet and bath). It would therefore appear that a membrane potential of about -25 mV is present in the acinar cell during CCK stimulation.

The effect of CCK has also been investigated in the presence of the CCK antagonist dibutyryl cyclic guanosine monophosphate.[19] In this case CCK always fails to activate the unitary inward currents, but after excision of the patch and exposure of the inner aspect of the membrane to the high Ca^{2+} concentration of the bath, single-channel currents representative of the nonselective cation channel invariably appear.[18]

These experimental results show that specific CCK receptor interaction can activate the 30- to 35-pS cation channel localized in the basolateral

[19] H. G. Philpott and O. H. Petersen, *Pfluegers Arch.* **382**, 263 (1979).

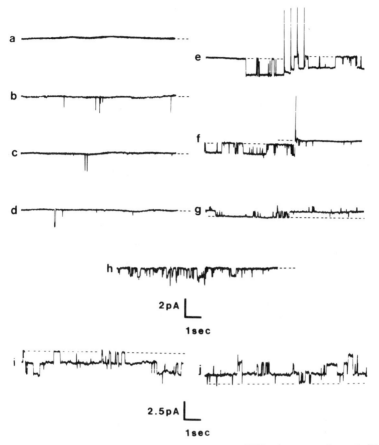

FIG. 3. Activation of nonselective cation channel by CCK using setup shown in Fig. 2. Consecutive traces from same membrane patch. (a) Recording situation is "cell attached" and the CCK pipet is far away from the acinus under investigation. Pipet potential, +40 mV. (b–d) Three continuous traces obtained 20, 30, and 40 sec, respectively, after the tip of the CCK pipet had been brought close to the patch pipet. Pipet potential, +40 mV. (e) Fifty seconds after start of CCK stimulation. The artifacts indicate that the potential of the patch pipet was changed (in steps of 5 mV) from +40 to +20 mV. (f) Pipet potential, +20 mV. At the artifact the potential was changed to zero. (g) Pipet potential, −40 mV. (h) Currents from the same patch after excision (inside-out). Symmetrical extracellular saline solutions. Pipet potential, +50 mV. (i and j) Currents recorded from another excised (inside-out) membrane patch with intracellular solution (high K^+, low Na^+) in the bath and the normal extracellular solution (high Na^+, low K^+) in the pipet. Pipette potential, +60 mV in i and −60 mV in j. Dashed lines represent the current level when all channels are closed. Downward deflections represent the current from extracellular to intracellular side of membrane. (Adapted from Maruyama and Petersen.[18])

plasma membrane by a mechanism that would appear to involve an internal messenger, since channels localized in an area to which CCK has no access (the patch area) are controlled by the agonist.

In a separate series of experiments with a similar type of protocol, the action of acetylcholine (ACh) has been investigated. ACh acts in a manner indistinguishable from that of CCK. Thus, it appears that different agonists interacting with two different receptor types control the activation of the same channels.[18,20]

The simplest interpretation of the experiments of the type represented by Fig. 3 would be that binding of hormones or neurotransmitters to specific receptor sites evokes an increase in intracellular free Ca^{2+} concentration, $[Ca^{2+}]_i$, and that Ca^{2+}, by interacting directly with the channel protein or a closely associated regulatory subunit, causes a conformational change in the channel, resulting in ion flux.

This simple interpretation is, however, too simple. In experiments on pig pancreatic acinar cells, which are dominated electrically by the high-conductance Ca^{2+}- and voltage-activated K^+ channel,[21] it became clear that there must be involvement of at least one other messenger in addition to Ca^{2+}.[22] The experimental approach shown in Fig. 2 was used again, but this time experiments were carried out both in the presence and absence of Ca^{2+} in the recording patch-clamp pipet. The experiments showed that sustained activation of the K^+ channels could only be evoked by excitation of CCK receptors if Ca^{2+} was present in the solution within the recording pipet.[22]

The large Ca^{2+}- and voltage-activated K^+ channel has been extensively characterized in excised membrane patches from a variety of cell types.[9-11] Channel opening is controlled exclusively by the membrane potential and $[Ca^{2+}]_i$, whereas changes in $[Ca^{2+}]_o$ have no effect. The external Ca^{2+} requirement in the isolated membrane patch area in the experiments demonstrating indirect CCK activation of K^+ channels cannot, therefore, be explained by the action of Ca^{2+} on the external site of the channel, but must be due to an influence on $[Ca^{2+}]_i$. Hormones and transmitters seem to mobilize intracellular Ca^{2+} via the intracellular messenger inositol 1,4,5-trisphosphate (IP_3), and this substance acts on the membrane of Ca^{2+}-storing organelles by opening a Ca^{2+} pathway.[23,24] However, many

[20] O. H. Petersen and Y. Maruyama, in "Single-Channel Recording" (B. Sakmann and E. Neher, eds.), p. 425. Plenum, New York, 1983.

[21] Y. Maruyama, O. H. Petersen, P. Flanagan, and G. T. Pearson, Nature (London) 305, 228 (1983).

[22] K. Suzuki, C. C. H. Petersen, and O. H. Petersen, FEBS Lett. 192, 307 (1985).

[23] M. J. Berridge and R. F. Irvine, Nature (London) 312, 315 (1984).

[24] M. J. Berridge, Annu. Rev. Biochem. 56, 159 (1987).

hormone- and transmitter-evoked responses in exocrine gland cells are dependent on extracellular Ca^{2+}, as documented for the stimulant-evoked hyperpolarization and increase in outward K^+ current in pig pancreatic acinar cells,[21,25] and pancreatic secretagogues increase unidirectional Ca^{2+} flux into acinar cells.[26] The experimental procedure in patch-clamp experiments of the type carried out on cell-attached pig pancreatic acinar cells[22] may emphasize the external Ca^{2+} dependence by physically moving the isolated membrane patch away from intracellular Ca^{2+} stores, and Ca^{2+} released in the pancreatic cell by the messenger following stimulation may not easily penetrate into the patch area due to restrictions of intracellular Ca^{2+} diffusion.[27] IP_3 or another messenger may diffuse more easily than Ca^{2+} and could perhaps mediate the CCK-evoked Ca^{2+}-controlled channel opening by allowing Ca^{2+} influx, raising $[Ca^{2+}]_i$ locally.[22]

In other words, the patch-clamp experiments have demonstrated an external Ca^{2+} requirement for sustained K^+ channel opening evoked by receptor (cholecystokinin) activation of pancreatic acinar cells. This Ca^{2+} requirement can be localized at a site separated from the hormone receptor site, which indicates messenger-mediated rather than direct receptor-operated Ca^{2+} entry.

Use of Patch-Clamp Whole-Cell Current Recording Experiments to Demonstrate Messenger-Mediated Channel Opening

The tight-seal whole-cell current recording configuration has been used extensively to characterize K^+ currents mediated by Ca^{2+}-activated K^+-selective channels in exocrine acinar cells.[9-11]

The whole-cell voltage-clamp current can be regarded as the sum of many small elementary currents flowing through individual high-conductance K^+ channels of the type characterized in single-channel studies. There are a number of strong arguments which support this conclusion. (1) The whole-cell K^+ current is Ca^{2+} activated, and the Ca^{2+} sensitivity is similar to that described for the single channels.[28] (2) The activation of the whole-cell currents occurs within the same voltage range as described for single-channel currents.[21,28] (3) Ca^{2+}-activated K^+ currents can be blocked by low concentrations (~ 2 mM) of tetraethylammonium acting specifically from the outside of the plasma membrane.[29] (4) Ensemble fluctuation analysis of

[25] G. T. Pearson, P. M. Flanagan, and O. H. Petersen, *Am. J. Physiol.* **247**, G520 (1984).
[26] S. Kondo and I. Schulz, *Biochim. Biophys. Acta* **419**, 76 (1976).
[27] P. F. Baker, *Ann. N.Y. Acad. Sci.* **307**, 250 (1978).
[28] Y. Maruyama and O. H. Petersen, *J. Membr. Biol.* **79**, 293 (1984).
[29] N. Iwatsuki and O. H. Petersen, *J. Membr. Biol.* **86**, 139 (1985).

the whole-cell currents indicates that the outward current can be entirely accounted for by a relatively small number (50–150) of high-conductance channels.[30,31]

The action of CCK-8 has been tested in the whole-cell recording configuration. In experiments of this type, the intracellular ionic composition is determined by the solution used to fill the recording pipet. When the pipet, and therefore the cell, is filled with a solution containing a high concentration (10 mM) of Ca^{2+} chelator [ethyleneglycol bis(aminoethyl ether)tetraacetic acid, EGTA], and no Ca^{2+} is added, CCK has no effect on the whole-cell currents. In this situation $[Ca^{2+}]_i$ is clamped at very low levels (probably $< 10^{-9}$ M) and this result is, therefore, in complete agreement with the hypothesis that the CCK-evoked conductance increase under physiologic conditions is mediated via an increase in $[Ca^{2+}]_i$. If, however, a low EGTA concentration is used in the pipet, CCK is able to increase the K^+ current. The effect of CCK is reversible on removal of the agonist, but can also be reversed by the removal of Ca^{2+} from the bathing solution in contact with the outside of the isolated cell. The effect of CCK added to the outside solution on the whole-cell currents is very similar to the action of internal Ca^{2+}, and taken together, these two types of experiments provide very strong evidence that the secretagogue-evoked increase in K^+ conductance is mediated by internal Ca^{2+} activating voltage-gated K^+-selective channels.[22,28]

Sustained stimulant-evoked activation of the K^+ conductance pathway requires the continued presence of extracellular Ca^{2+}, and this result suggests that stimulation, in addition to causing intracellular Ca^{2+} release, also promotes Ca^{2+} uptake from the extracellular fluid.[28]

Internal Cell Perfusion

Internal cell perfusion with exchange of solutions during individual patch-clamp experiments can be used to demonstrate the importance of inositol polyphosphates. The tight-seal whole-cell current recording configuration[6] has been very useful, but it has the important limitation that only one intracellular solution can be used in an individual experiment. By combining patch-clamp whole-cell current recording with internal perfusion of the patch pipet (Fig. 4) it is possible to carry out experiments in

[30] A. Trautmann and A. Marty, *Proc. Natl. Acad. Sci. U.S.A.* **81,** 611 (1984).
[31] Y. Maruyama, A. Nishiyama, T. Izumi, N. Hoshimiya, and O. H. Petersen, *Pfluegers Arch.* **406,** 69 (1986).

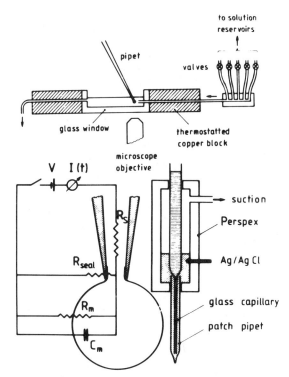

FIG. 4. The experimental arrangements for whole-cell current recording with internal perfusion of the patch pipet. (Upper part) Experimental setup for whole-cell recordings. The Teflon tube connecting the solution reservoirs to the measuring chamber was enclosed in the thermostatted copper block on a length of about 20 cm in order to ensure temperature equilibration. Solutions of different composition were separated by a small air bubble, which was injected into the manifold. (Lower right part) Internal perfusion of the patch pipet (not drawn to scale). The glass capillary had a length of about 5 cm and an internal diameter of 20–40 μm. The distance between the tip of the capillary and the tip of the patch pipet was 100–200 μm. (Lower left part) Circuit parameters of a whole-cell recording experiment; R_s, series resistance of the pipet; R_{seal}, pipet–membrane seal resistance; R_m and C_m, resistance and capacitance of the cell membrane. (Adapted from Jauch et al.[32])

which many different intracellular solutions can be tested in individual experiments.[32-34]

In the lacrimal acinar cells used for such a study, ACh evokes sustained membrane hyperpolarization due to a marked and sustained increase in

[32] P. Jauch, O. H. Petersen, and P. Läuger, J. Membr. Biol. 94, 99 (1986).
[33] J. Kimura, A. Noma, and H. Irisawa, Nature (London) 319, 596 (1986).
[34] J.-Y. Lapointe and G. Szabo, Pfluegers Arch. 410, 212 (1987).

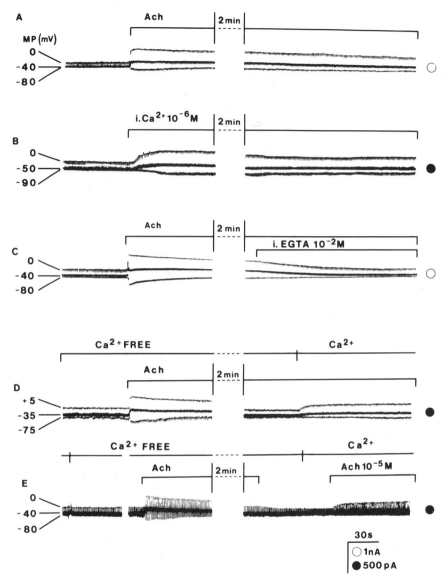

FIG. 5. Traces of whole-cell current from single acinar cells of mouse exorbital lacrimal glands. The acinar cells are voltage clamped at a holding potential very close (within 3 mV) to the zero current potential, namely, the resting membrane potential of the internally perfused cells. The internal solution contains 0.5 mM EGTA and no added calcium. From this holding potential a series of depolarizing and hyperpolarizing voltage jumps are superimposed (100–200 msec in duration) evoking outward K$^+$ (upward deflections) and inward Cl$^-$ (downward deflections) currents, respectively. Membrane potential (MP) (millivolts) indicates the holding potential and the membrane potential during the superimposed voltage jumps. (A) Currents in the unstimulated cell and changes on exposing the cell to ACh, 10^{-5} M. ACh stimulation is associated with a sustained increase in the amplitude of the outward current

outward K^+ current, mediated by high-conductance Ca^{2+} and voltage-activated K^+ channels. The activation of K^+ current by ACh is only sustained in the presence of extracellular Ca^{2+}. Figure 5 shows recordings of whole-cell currents from single isolated mouse lacrimal acinar cells. At the holding potential of -40 mV (in some experiments -35 or -50 mV), which is close to the normal resting potential, ACh evokes an outward current response which is substantially enhanced at a membrane potential of 0 mV. At more negative potentials (-75 to -90 mV), an inward current response is observed. These currents have been well characterized: the ACh-evoked outward current is carried by K^+ through high-conductance (200–300 pS) voltage-activated and Ca^{2+}-activated K^+ channels,[14,35] whereas the inward current is carried by Cl^- through Ca^{2+}-dependent Cl^- channels with a low unit conductance (1–2 pS).[14-16] Figure 5 shows that the effect of continued ACh stimulation is reversible and can be blocked either by introducing a high concentration (10 mM) of EGTA into the cell or by removing external Ca^{2+}. Readmission of extracellular Ca^{2+} during sustained ACh exposure brings back the full response. On the other hand, removal and readmission of external Ca^{2+} in the absence of ACh stimulation has no effect on the magnitude of the transmembrane currents. The effect of ACh can also be reversibly mimicked by introducing a solution with a high Ca^{2+} concentration (about 1 μM) into the cell. The results shown in Fig. 1 are very similar to those discussed in the previous section dealing with cholecystokinin-evoked Ca^{2+}-dependent K^+ currents in pig pancreatic acinar cells, but are more direct and convincing as a result of being able to change the intracellular ion composition during individual experiments.

---

and a largely transient increase in the inward currents. (B) At the point indicated, the internal perfusion fluid has been changed to a solution with a free calcium concentration of 10^{-6} M (i, intracellular). The rise in intracellular free calcium is associated with a sustained increase in the amplitude of the outward current. (C) Effects of extracellular application of ACh (10^{-5} M), as in A. At the point indicated, the control intracellular solution is replaced by one containing 10 mM EGTA and no added calcium. Dialysis of the cell with this calcium chelator reduces the ACh-activated currents to prestimulus levels. (D) The acinar cell is bathed in a calcium-free extracellular solution and at the point indicated is exposed to 10^{-5} M ACh. An initial increase in both outward and inward currents is not sustained in the absence of extracellular calcium. Reintroduction of extracellular calcium in the continued presence of the agonist (ACh) again gives a sustained increase in outward current amplitude. (E) Removal of extracellular calcium has no effect on whole-cell currents in the unstimulated lacrimal acinar cell. Stimulation with 10^{-5} M ACh is again associated with a transient increase in current amplitudes, and ACh removal after 4 min gives only a small further decrease in whole-cell currents. In the absence of ACh, the reintroduction of extracellular calcium produces no change in current amplitudes, but the cells respond upon readmission of ACh. (Adapted from Morris et al.[36])

[35] I. Findlay, *J. Physiol. (London)* **350**, 179 (1984).

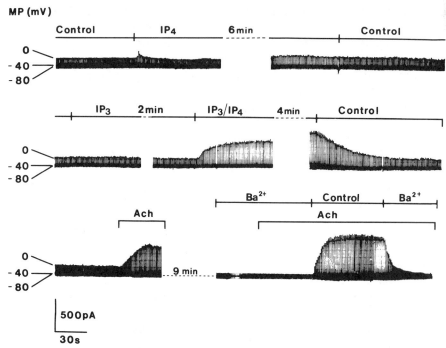

FIG. 6. The effects of IP_3 and IP_4 on K^+ currents. Traces are sections of one continuous recording from the same lacrimal acinar cell. The intracellular solution contained 0.5 mM EGTA with no added calcium. As for Fig. 5, the holding potential (-40 mV) is very close to the zero current potential and depolarizing and hyperpolarizing voltage jumps are superimposed (100 msec in duration). (Upper trace) The effect of changing the intracellular solution from the control to one containing 10^{-5} M IP_4. IP_4 is present for 8 min before return to control intracellular solution. (Middle trace) From left to right shows the effect of 10^{-5} M IP_3 in the internal solution, which after 4 min is replaced with a solution containing both IP_3 and IP_4. Only in the presence of the combined IP_3 and IP_4 is there any marked or sustained change in the amplitude of the outward currents. The increased amplitude of the outward K^+ current is sustained until IP_3/IP_4 is removed and the control intracellular solution reintroduced. (Bottom trace) Effect of ACh (10^{-5} M) in the extracellular solution (left), showing an increase in amplitude of the outward currents; effect of extracellular Ba^{2+} on whole-cell currents is shown to the right, when a marked reduction in the currents in the unstimulated cell is seen and the ACh-induced increase in outward current is abolished. (Adapted from Morris *et al.*[36])

Using the technique of internal patch pipet perfusion it is now possible to test the effects of various putative messenger molecules.[36] Figure 6 shows that neither IP_3 nor IP_4 alone evoked any clear current response in the

[36] A. P. Morris, D. V. Gallacher, R. F. Irvine, and O. H. Petersen, *Nature (London)* **330**, 653 (1987).

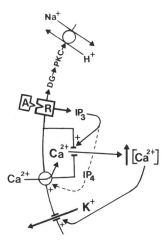

Fig. 7. Summary of stimulus–excitation coupling processes occurring in and around the basolateral plasma membrane. Agonist–receptor interaction causes breakdown of phosphatidylinositol bisphosphate, generating inositol 1,4,5-trisphosphate (IP$_3$) and diacylglycerol (DG). DG activates protein kinase C (PKC), stimulating Na$^+$–H$^+$ exchange. IP$_3$ opens Ca^{2+} channels in an intracellular store and thereby evoked Ca^{2+} release. IP$_3$ is also phosphorylated to IP$_4$ (inositol 1,3,4,5-tetrakisphosphate), and this agent controls Ca^{2+} entry into a pool connected to the intracellular Ca^{2+} store. The resulting elevation of [Ca^{2+}]$_i$ opens the Ca^{2+}-activated K$^+$ channels. (Adapted from Petersen and Gallacher.[11])

same lacrimal acinar cell, but when IP$_4$ was added on top of IP$_3$, a clear and sustained increase in the outward current at 0 mV was obtained. Wash-out of the two inositol phosphates reduced the currents toward the control level. ACh could thereafter evoke a response of similar magnitude to that induced by combination of the two inositol phosphates. The ACh-evoked outward current was subsequently reversibly abolished by the K$^+$ channel blocker, Ba^{2+}.[37]

In a further series of experiments it was shown that the IP$_3$/IP$_4$-evoked sustained outward current response was acutely dependent on the presence of extracellular Ca^{2+}.[36]

These results provide evidence that the messengers regulating Ca^{2+} entry in mammalian exocrine acinar cells, and probably also in other cell types, are IP$_3$ and IP$_4$ acting together, as previously proposed on the basis of injection studies on sea urchin egg cells.[38] There is evidence for the rapid generation of first IP$_3$ and then IP$_4$ in rat pancreatic acinar cells, following activation of muscarinic ACh receptors as well as cholecystokinin receptors, with a maximal increase in IP$_4$ concentration occurring within 15–30

[37] N. Iwatsuki and O. H. Petersen, *Biochim. Biophys. Acta* **819**, 249 (1985).
[38] R. F. Irvine and R. M. Moor, *Biochem. J.* **240**, 917 (1986).

sec.[39] Ca^{2+} probably enters the cell through an IP_3-sensitive Ca^{2+} store, as previously discussed,[11] and IP_4 modulates Ca^{2+} entry into that Ca^{2+} store (Fig. 7). The data do not, however, distinguish between the contributions made individually by the two inositol phosphates, and the only certain conclusion is that both are necessary to control Ca^{2+} entry in the acinar cells. Several criteria for a second messenger function for IP_4 have now been fulfilled and it would appear that we may now have the key elements to understand the link between Ca^{2+} entry and the increased turnover of phospholipids after stimulation, as originally proposed by Michell.[40]

Conclusion

Figure 7 shows a simplified model that may account for the results discussed in this short review. The precise nature of the IP_4-controlled Ca^{2+} uptake mechanism is obscure, and more work on this topic is urgently needed. The activation of $Na^+ - H^+$ exchange mediated by diacylglycerol[11] may be of physiological relevance, since an increase in internal pH may promote IP_3-evoked Ca^{2+} release.[41]

[39] E. R. Trimble, R. Bruzzone, C. J. Meehan, and T. J. Biden, *Biochem. J.* **242**, 289 (1987).
[40] R. H. Michell, *Biochim. Biophys. Acta* **415**, 81 (1975).
[41] W. Siffert and J. W. Akkerman, *Nature (London)* **325**, 456 (1987).

[35] Ionic Permeation Mechanisms in Epithelia: Biionic Potentials, Dilution Potentials, Conductances, and Streaming Potentials

By PETER H. BARRY

This chapter covers the techniques and approach that can be applied to the mechanism of passive (diffusive) ion permeation through paracellular ("tight junction") and cellular (apical and basolateral membrane) pathways in epithelia. It also touches very briefly on an associated electrical method of measuring nonelectrolyte permeation in epithelia. A discussion of the other transport mechanisms, such as cotransport and exchange transport, is given by Stein in another chapter in this volume.[1]

[1] W. D. Stein, this volume [3].

Electrical Measurements: Experimental Considerations

Introduction

Permeation properties of epithelia can be investigated at three levels: (1) transepithelial measurements, which, in the case of leaky epithelia, give permeation information predominantly about the paracellular tight junction pathway; (2) cellular transmembrane measurements to derive overall permeation parameters for the individual apical and basolateral membranes; and (3) at the single-channel level to derive information about the individual channels in each membrane.

Transepithelial Measurements. Two types of approach can be used for electrical measurements: a sac preparation, for organs such as the gallbladder, in which the organ can be everted so that the apical (mucosal) side, which has the smaller unstirred layer, faces outward. The epithelium can be cannulated with a small polyethylene cannula and tied off with silk thread (Fig. 1). The mucosal solution can be stirred by means of bubbling O_2 into the beaker, whereas the (now internal) serosal solution can be left unstirred. Solution changes are usually made by simply lifting up the organ and exchanging solution beakers. Electrical measurements are made by

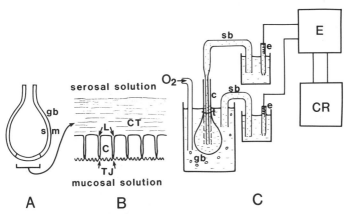

FIG. 1. Schematic diagram of the everted sac preparation, as used to measure transepithelial potentials (diffusion and streaming potentials) across epithelia such as the gallbladder. Here, A and B are similar to their counterparts in Fig. 1 of P. H. Barry and J. M. Diamond, *J. Membr. Biol.* **4**, 331 (1971). gb, Gallbladder (everted); s, serosal (basolateral) side; CT, connective tissue layer with large unstirred layer; m, mucosal (apical) side with the smaller unstirred layer; c, layer of cells; L, lateral intercellular spaces; TJ, tight junctions. (C) Experimental setup for measuring the transepithelial potentials. t, Silk thread; c, polyethylene cannula; sb, salt bridges; e, Ag/AgCl electrodes (or calomel half-cells); E, high-impedance electrometer (or preamplifier); CR, chart recorder. The mucosal (apical) solution is stirred by bubbling water-saturated O_2 through it.

virtue of small tubular polyethylene salt bridges filled with electrolyte, the one which is to be placed in the cannula having been pulled out to a fine diameter by having been drawn down gently over a low flame.

In another approach, the tissue can be held clamped as a planar sheet in a split Ussing-type of Lucite (Perspex, Plexiglas) chamber (Fig. 2A). This setup is appropriate for rapid measurements of transepithelial potentials, conductances, and streaming potentials in different solutions, as well as for tracer flux measurements. Often the tissue is held flat across an aperture in the chamber by a series of pins (e.g., ends of syringe needles) embedded in one of the chamber sections. One of the potential problems with this type of setup is that of edge damage, in which the cells around the perimeter of the hole have been damaged by the compression of the two chambers onto the epithelium, the relative area of damage for a damaged-tissue width, δ, being $2\delta/r$ and decreasing as the aperture radius, r, increases. Edge damage can be reduced by including a slight cut-away section around the perimeter of the hole on the mucosal half-chamber and by very careful lining up and clamping together of the chambers. The contribution of such edge damage has been estimated in electrophysiological measurements.[2,3] Additional precautions to reduce edge effects include gluing the epithelium to a Lucite ring and sealing the combination to the chambers with soft silicon rubber rings.[4] Lewis and Diamond[5] (Fig. 2B) have also designed a special chamber to minimize edge-damage effects and reported a virtual elimination of such effects. Typical aperture sizes range from about 2.6 mm^2 for choroid plexus to about 2 cm^2 for urinary bladders. Each chamber can then be stirred by either magnetic stirring bars or by a stream of water-saturated O_2 bubbles injected through syringes. In some circumstances, the connective tissue on the basolateral surface of the epithelial layer can be reduced in thickness by dissection[5] in order to reduce the electrical or diffusional resistance of the connective tissue. This is particularly important when the conductance measurements of the cell layer of nontight epithelia, such as the gallbladder, are being measured. Salt bridges, close to the epithelium to reduce solution series resistance, and current-passing electrodes could be introduced into both chambers, as shown in Fig. 2A.

Apical and Basolateral Membrane Measurements. The advent of microelectrode measurements in epithelia made the determination of the separate permeation parameters of the apical and basolateral membranes possible. It also allowed an investigation of the equivalent electrical circuit of the epithelium and provided a direct method of determining the pro-

[2] J. W. Prather and E. M. Wright, *J. Membr. Biol.* **2,** 150 (1970).
[3] A. Smulders and E. M. Wright, *J. Membr. Biol.* **5,** 297 (1971).
[4] H. Gogelein and W. Van Driessche, *J. Membr. Biol.* **60,** 187 (1981).
[5] S. A. Lewis and J. M. Diamond, *J. Membr. Biol.* **28,** 1 (1976).

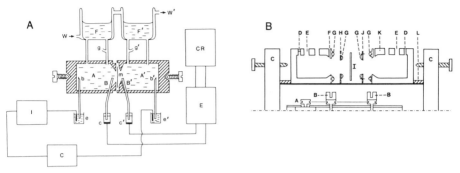

FIG. 2. Schematic diagrams of two planar-sheet Ussing-type chambers used for relatively fast measurements of transepithelial potentials, conductances, and streaming potentials. (A) Cross section of one set of chambers with two separate reservoirs in the context of the whole measuring circuit [based on Fig. 1 of A. Smulders and E. M. Wright, *J. Membr. Biol.* **5**, 297 (1971) and Fig. 1 of A. Lasansky and F. W. de Fisch, *J. Gen. Physiol.* **49**, 913 (1966)]. m, Epithelial membrane; A and A', two Lucite half-chambers; F and F', fluid reservoirs (for solutions); g and g', gas lifts for solution mixing; w and w', inlet and outlet of two glass water jackets for temperature control; e and e', current Ag/AgCl plate electrodes; b and b', 3 M KCl salt bridges; I, electrometer for current; C, potentiometer circuit for applied voltage; B and B', salt bridges (placed close to the epithelium and similar in composition to bathing solutions) for transepithelial potential; c and c', calomel half-cells or Ag/AgCl electrodes; E, electrometer for potential; CR, chart recorder. Note slight conical recess around the rim of the right half-chamber for reducing edge damage to epithelium. (B) Another modification of the Ussing chamber, made of Lucite, specially designed to reduce edge damage (redrawn from part of Fig. 1 of Lewis and Diamond[5]). Temperature-control water jackets are not shown. A, Electric motor; B, external magnets (Teflon-coated magnetic spin bars in solution chambers, not shown); C, vise ends for clamping chambers together; D, current electrode ports; E, gassing solution ports; F, voltage electrode ports; G, layers of silicone sealant; H and J, two plastic rings, with 20 holes or pins, respectively, around their circumferences; I, epithelium; K, chambers; L, grooved platform into which a ridge on the bottom of the chambers inserts to prevent lateral movement. The area of exposed tissue was 2 cm^2 and chamber volume was about 18 ml.

portions of permeation via the paracellular and cellular routes. Again the epithelium is held as a planar sheet, but now the apparatus is arranged so as to allow access to microelectrodes impaling the epithelium. For details of a typical setup see Gordon *et al.*[6] in this volume and Fig. 4 of Frömter.[7] An example of a smaller setup for use with continuous perfusion under a water-immersion microscope lens is given in Fig. 3.

These measurements are best done in conjunction with ion-selective electrodes[8,9] to monitor both the cell potential and ionic activities within

[6] L. G. M. Gordon, G. Kottra, and E. Frömter, this volume [33].

[7] E. Frömter, *J. Membr. Biol.* **8**, 259 (1972).

[8] T. Zeuthen, *in* "The Application of Ion-Selective Microelectrodes" (T. Zeuthen, ed.), p. 27. Elsevier/North-Holland, Amsterdam, 1981.

[9] J. De Long and M. M. Civan, *J. Membr. Biol.* **72**, 183 (1983).

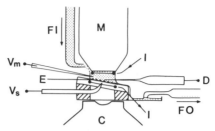

FIG. 3. Schematic diagram of small chamber used for intracellular recording in epithelia and especially designed for continuous perfusion and easy microscopic observation (redrawn from Fig. 1 of Zeuthen[24]). V_m and V_s, Agar salt bridges for monitoring the potentials in the mucosal and serosal solutions, respectively; FI and FO, fluid inflow (via peristaltic pump) and fluid outflow (via a wick into another chamber where it can be removed by suction); I, two ring-shaped silver wires for passing current; D, double-barrelled microelectrode (one barrel with ion-exchange resin for monitoring intracellular ion activities and the other reference barrel used for recording the potential of the intracellular solution); M, microscope water-immersion lens (40×) used with Nomarski optics for visual recording and TV monitoring; C, Nomarski condenser; E, the epithelium, apical side up.

the cell. The internal composition of the cell can then be altered by the use of ionophores and transport blockers. Ionophores can either be irreversible (e.g., gramicidin) or reversible (e.g., nystatin or amphotericin B). For example, the application of ionophores to the mucosal solution will allow the intracellular cation concentrations to approach the mucosal solution values. Blockers of the Na^+,K^+-active pump (e.g., ouabain) will also result in an increase in internal Na^+ and decrease in internal K^+.

Another approach, which is especially useful for evaluating the relative contributions of different pathways to permeation across the epithelium, uses the technique of impedance analysis and will be briefly mentioned later (for experimental details see Clausen[10] and Gordon et al.[6] in this volume).

A different approach that can also be used to investigate the electrical properties of the apical or basolateral membranes makes use of membrane vesicles produced from them. Ion transport can then be measured using atomic absorption spectroscopy and radioactive tracer techniques.[11] More recently, membrane potential-sensitive optical cyanine dyes have been used with these vesicles to measure their membrane potentials.[12] Another new technique using lipophilic cations (e.g., tetraphenylphosphonium; TPP^+) tagged with ^{14}C was also used for measuring intestinal cation and anion permeabilities.[13] In these measurements the influx of TPP^+ was

[10] C. Clausen, this volume [32].
[11] R. D. Gunther and E. M. Wright, J. Membr. Biol. 74, 85 (1983).
[12] R. D. Gunther, R. E. Schell, and E. M. Wright, J. Membr. Biol. 78, 119 (1984).
[13] G. A. Kimmich, J. Randles, D. Restrepo, and M. Montrose, Am. J. Physiol. 248, C399 (1985).

taken to be a measure of the membrane potential. Both methods used calibration procedures involving K^+ and valinomycin.

Single-Channel Properties. A third approach is to measure the permeation properties of single ionic channels. This can now be done in one of two different ways. One way uses the indirect fluctuation analysis (noise analysis; see Van Driessche and Lindemann[14]) to *indirectly* measure the *average* single-channel properties of a population of channels.

The other uses the *direct* patch-clamp technique to measure the currents flowing through *single* channels in small portions of cell membrane electrically isolated by a patch pipet.[15,16] This is the most direct method of investigating the properties of a particular channel and some of the principles will be outlined below (for additional experimental details, see Petersen[17] in this volume).

A special heat-polished glass patch pipet is constructed so that it has a smooth, thickened tip that will not pierce the cell membrane, but rather will allow an omega-shaped patch of membrane to be sucked up into it, resulting in a very high-resistance seal between membrane and pipet of the order of gigaohms ($> 10^9$ Ω). The setup is illustrated in Fig. 4, and provided that the pipet resistance is very small in comparison with the membrane–pipet seal resistance, virtually all of the current flowing through an open channel will flow through the patch pipet and will be measured by an appropriate patch-clamp amplifier. Provided that the background electrical noise in the system is low ($\lesssim 0.2$ pA), the channel conductance is not too small, and that the number of channels opening within the patch is small (but not zero), then clear resolution of single-channel currents will be achieved. The patch-clamp technique can be used to provide a number of different patch configurations—intact, whole cell, and excised (outside or inside membrane surface out)—as indicated in Fig. 4B.

In the case of the intact patch, there is only control over the composition of the solution within the pipet—none over the intracellular composition. In order to determine the voltage across the patch, an intracellular microelectrode must also be inserted into the cell to measure the membrane potential (ϵ_m). The voltage (V_m) across the patch (see Fig. 4A) is given by

$$V_m = \epsilon_m - V_P \tag{1}$$

where V_p is the potential applied to the patch amplifier. In addition, various junction potential corrections need to be applied to correct V_m (see

[14] W. Van Driessche and B. Lindemann, *Rev. Sci. Instrum.* **49**, 52 (1978).
[15] E. Neher, B. Sakmann, and J. H. Steinbach, *Pfluegers Arch.* **375**, 219 (1978).
[16] O. P. Hamill, A. Marty, E. Neher, B. Sakmann, and F. J. Sigworth, *Pfluegers Arch.* **391**, 85 (1981).
[17] O. H. Petersen, this volume [34].

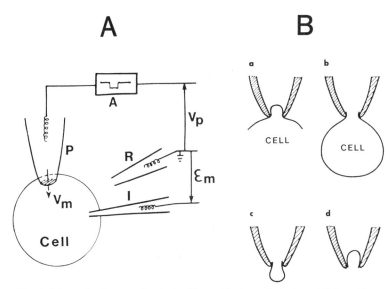

FIG. 4. Schematic diagrams illustrating the patch-clamp technique. (A) Overall setup for isolating single ionic channels in a small (intact) patch of cell membrane. P, Patch pipet; R and I, reference and intracellular microelectrodes; V_p, applied patch potential; ϵ_m, membrane potential; V_m, potential across patch; A, patch-clamp amplifier. (B) Four different recording setups (after Fig. 9 of Hamill et al.[16]). (a) Intact patch, (b) whole cell recording, (c) outside-out patch, and (d) inside-out patch.

later) for both the patch pipet zeroing and the membrane potential electrodes. The obvious advantage of the excised patch setup is the fact that, in such a case, complete control is available of both solutions bathing the dissociated membrane patch adhering to the pipet, and, apart from an additional reference electrode, no microelectrodes are necessary. A simple technique has recently been reported for changing solutions around an excised membrane patch by means of a crimped polyethylene sleeve that can slide along the shaft of the patch pipet to allow transfer of the patch between solutions of different composition or temperature.[17a,17b]

Two other points need to be made about the patch-clamp measurements. One concerns the nature of the cell membrane, which should be clean for successful gigaohm seals. This can be achieved either by cleaning the membranes enzymatically (e.g., using collagenase, trypsin, or protease), in the case of adult cells to remove any adhering connective tissue, etc., from the membrane surface or else by using tissue-cultured cells, which by their very nature tend to have fairly clean cell membranes. The other point

[17a] N. Quartararo and P. H. Barry, *Pfluegers Arch.* **410**, 677 (1987).
[17b] J. W. Lynch, P. H. Barry, and N. Quartararo, *Pfluegers Arch.* **412**, 322 (1988).

concerns the requirement for activating a small number of channels only. In the case of acetylcholine-sensitive channels, Ca^{2+}-activated K^+ channels, or voltage-sensitive channels, this is managed by controlling the amount of ACh, Ca^{2+} in the patch pipet or the pipet voltage (V_p), and endeavoring to block chemically any abundant channels of ions not under investigation.

Reference Electrodes and Microelectrodes

Although calomel electrodes (Hg/HgCl) have been used for certain electrophysiological measurements in the past, the most widely used electrode is the Ag/AgCl electrode. It is easy to make from sterling silver wire (or can even be obtained commercially as Ag/AgCl pellets, e.g., from W-P Instruments, Inc., New Haven, Connecticut); it is generally very stable, reversible, and, provided that there is some Cl^- in the solution, well defined. It can be used also for passing currents, though in most situations it is advisable to use separate current and voltage electrodes in the same experiment.

In order to produce extremely accurate Ag/AgCl electrodes, rather lengthy procedures need to be followed.[18] However, for most purposes, a very satisfactory method for producing Ag/AgCl electrodes with electrode potentials within 0.1 mV is as follows: Use sterling silver wire, physically cleaned with fine wet-and-dry sandpaper and preferably handled throughout with rubber gloves. The surface area of the wire may be increased by coiling it around a small former. After washing in distilled water and drying, plate the wires (up to about three at a time) with chloride for about 30 min in 0.1 M HCl with a 1-mA $\cdot$ cm^{-2} (or slightly greater) current. For potential-measuring electrodes, make the required electrode the anode all the time. The electrodes should look a uniform mauve–brown when fully chloride plated. The Ag/AgCl wires should be placed in suitable glass tubes and it is generally advisable to tape them to the outside of the tube to hold them in place during the next process of filling and to secure them during later use. Then aspirate (e.g., using a polyethylene tube over the end of the pipet) a molten dissolved solution of 3–4% agar (by weight) added to a 3 M KCl solution. Make a batch of about 10–20 electrodes at a time and ensure that only the Ag/AgCl-coated part of the wire is in contact with the KCl–agar solution. When the solutions have gelled and cooled, it is important to seal off the ends of the pipets with Sylgard silicon rubber encapsulating resin (e.g., #184 from Dow Corning, Minnesota) since deterioration of the electrodes is fairly rapid at the AgCl–KCl–air interface. The electrodes can then be checked electrically with a high-input imped-

[18] D. J. G. Ives and G. J. Janz, "Reference Electrodes." Academic Press, New York, 1961.

ance (e.g., $> 10^8 \, \Omega$) digital voltmeter and electrode pairs selected which have a potential difference (pd) less than about 0.2 mV. With care, many may be less than 0.1 mV. Store in a 3 M KCl solution in a sealed container to minimize evaporation of 3 M KCl, having covered exposed silver wires and connectors with Parafilm, for example, to reduce corrosion. For long periods, the electrodes should be stored in a refrigerator. Provided the Cl⁻ activity remains constant close to the electrodes and the Ag/AgCl coating does not deteriorate, the electrodes can be kept and used for at least a year.

The potential difference (E; electrode with respect to solution) of the Ag/AgCl simple junction[19] is given by

$$E = E_0^{Ag} - (RT/F) \ln([Cl^-]) \qquad (2)$$

where E_0^{Ag} is a constant standard state potential term; E is, of course, extremely sensitive to chloride activity, but with matched pairs of electrodes (both embedded in 3 M KCl–agar) the difference is small and constant.

As already mentioned, commercially available Ag/AgCl pellets can similarly be used in contact with the 3 M KCl solution. If a Ag/AgCl wire is used in glass microelectrodes or patch pipets in contact with 3 M KCl solution by itself, chloridizing may be required before each day's experiment. The same is generally true for current-passing electrodes.

Single-barrelled conventional microelectrodes can be prepared from fiber-filled borosilicate glass tubing (e.g., 1 mm o.d., 0.58 mm i.d., obtained from W-P Instruments). These are easily back-filled by capillarity to the shank and then topped up with electrolyte solution from the top with a very fine syringe needle (e.g., #30). The microelectrodes can be pulled on a two-stage microelectrode puller (e.g., Model PD-5, Narishige Scientific Instruments, Tokyo, Japan, or Brown-Flaming, Model P-77, Sutter Instruments, San Francisco, California), and it can be helpful to bevel the tips to minimize impalement damage (e.g., with Model 1300 Beveler, W-P Instruments).[20] Potentials must be measured with high-impedance preamplifiers or electrometers (e.g., Model M-4, W-P Instruments), and good-quality micromanipulators should be used to position the microelectrodes. In addition, a stepping motor can be used for easier and more controlled insertion of the microelectrodes or an inexpensive piezoelectric device can be made up to do a similar job.[21] The electrolyte solution to be used should be filtered with at least a 0.2-μm Millipore filter in order to reduce the possibility of tip blockage by debris, so that a free-flowing junction might

[19] D. A. MacInnes, "The Principles of Electrochemistry." Dover, New York, 1961.
[20] W. B. Guggino, E. L. Boulpaep, and G. Giebisch, *J. Membr. Biol.* **65,** 185 (1982).
[21] W. J. Lederer, *Pfluegers Arch.* **399,** 83 (1983).

be assured. However, it is also important with intracellular measurements that contamination of intracellular composition is not caused by electrolyte leakage, and for this reason it is sometimes advisable to use a smaller electrolyte concentration[22] (e.g., 1 M KCl rather than 3 M KCl). Normally, two internal microelectrodes should be used to measure both current and voltage concurrently, where current has to be passed between an intracellular electrode and a bathing solution (rather than transepithelially) and the membrane potential monitored. However, an alternative arrangement is to use a single microelectrode clamp (e.g., Axoclamp-2A, Axon Instruments, Burlingame, California) in which the same electrode switches between current passing and voltage recording.[23] For particularly accurate measurements, however, it is better to use two electrodes, perhaps combined for convenience, as a double-barrelled microelectrode (silicanized to minimize shunt resistance through the electrode walls; for further details of such microelectrodes, see Zeuthen[24]), since then minimal current is passed across the Ag/AgCl junction of the potential-recording electrodes.

It is often also necessary to monitor the internal ionic composition of the cells with an ion-sensitive electrode, the potential of which (V_i, with respect to a reference electrode in the same solution) is related to the activity a_i of the ion to be measured by

$$V_i = E_0 + S \log(a_i + K_{ji} a_j) \tag{3}$$

where E_0 is the standard electrode potential, S is the sensitivity of the electrode (ideally RT/zF), a_j is the activity of another ion j to which the electrode may have some selectivity, and K_{ji} is the equilibrium selectivity constant of ion j with respect to ion i. Thus a measurement of this potential with respect to a reference electrode in the same environment enables the ion activity to be calculated provided that correction can be made for the activity of any other selected ions and for the junction potential of the reference electrode. Again, such an electrode can be combined with a conventional reference electrode as a double-barrelled combination.[24] For details concerning the preparation of, and measurements using, ion-sensitive electrodes, refer to the comprehensive monographs of Thomas[25] and Ammann[25a] and also to Zeuthen.[26]

[22] M. Fromm and S. G. Schultz, *J. Membr. Biol.* **62**, 239 (1981).
[23] A. S. Finkel and S. Redman, *J. Neurosci. Methods* **11**, 101 (1984).
[24] T. Zeuthen, *J. Membr. Biol.* **66**, 109 (1982).
[25] R. C. Thomas, "Ion Sensitive Intracellular Microelectrodes." Academic Press, New York, 1978.
[25a] D. Ammann, "Ion-Selective Microelectrodes." Springer-Verlag, New York, 1986.
[26] T. Zeuthen, *J. Membr. Biol.* **76**, 113 (1983).

Further details on microelectrode techniques and ion-selective electrode measurements are discussed elsewhere.[9,20,26,27]

Salt Bridges and Junction Potentials

Salt bridges, or microelectrodes, are necessary for almost all electrophysiological measurement, and a correct estimate of the magnitudes and sign of all appropriate junction potential corrections is essential for accurate estimates of derived permeation parameters. Further detailed discussion of the use of salt bridges in electrophysiological measurements is given elsewhere.[19,28-30] In general, two types of salt bridges are often used, in each case the electrolyte being gelled with 3–4% agar to prevent bulk flow out of the bridge. In one type, 1, 2, or 3 M KCl is used to provide a low-resistance salt bridge, with a minimal junction potential correction between the recording electrodes and the bathing solutions. The ideal salt bridge is one in which the junction potential is small, well defined, and independent of the solution into which the bridge is placed. Such a condition is somewhat approximated by a freely flowing 1–3 M KCl junction, because the high KCl concentration dominates the junction potential, and yet, because $u_K/u_{Cl} \approx 0.96$, the junction potential is small. Were the mobility ratio to be closer to 1.0 (as in the case of RbCl or CsCl), although the magnitude of the potential would be reduced, it would become more dependent on the composition of the solution into which the bridge is placed. Provided that there is either a free-flowing KCl electrode or at least a good, fresh, uncontaminated end to the salt bridge (e.g., cut off a few millimeters of the end of the bridge after each solution change), such a KCl bridge provides a reasonable compromise and gives a fairly small junction potential, somewhat independent of the bathing solution composition (see Table II). The magnitude of such a junction potential can be most easily estimated using the generalized Henderson liquid junction potential equation.[31] For N monovalent ions the potential ϵ of the solution (S) with respect to the

---

[27] J. Narvarte and A. L. Finn, *J. Membr. Biol.* **84**, 1 (1985).

[28] P. C. Caldwell, *Int. Rev. Cytol.* **24**, 345 (1968).

[29] P. H. Barry and J. M. Diamond, *J. Membr. Biol.* **3**, 93 (1970).

[30] K. R. Page, *Comp. Biochem. Physiol. A* **67**, 637 (1980).

[31] Two different approaches have been used to calculate liquid junction potentials: the Henderson assumption, that the solutions within the region of mixing should be a series of mixtures of the two bulk solutions, leads to straightforward solutions; the Planck assumption, that the boundary region is well defined and that within it electroneutrality is obeyed, generally requires an interative solution. See MacInnes[19] for further details.

bridge (B) is given by

$$\epsilon^S - \epsilon^B = (RT/F)S_F \ln\left[\sum_{i=1}^{N} a_i^B u_i \Big/ \sum_{i=1}^{N} a_i^S u_i\right] \tag{4}$$

where

$$S_F = \sum_{i=1}^{N}\left[(u_i/z_i)(a_i^S - a_i^B)\right] \Big/ \sum_{i=1}^{N}\left[u_i(a_i^S - a_i^B)\right] \tag{5}$$

where u, a, and z represent the mobility, activity, and valency (including sign) of each ion species. Equation (4)[19] has been modified[29] to use activities rather than concentrations, with assumption that the ions obey the Guggenheim condition (i.e., that individual ionic activity coefficients are equal, so that $\gamma_1 = \gamma_2 = \gamma_3 = \cdots = \gamma$ everywhere in the solution).

A list of the more common ionic mobilities is given in Table I. The values for many other ions may be found in Robinson and Stokes.[32] Alternatively, if reported values of mobility and activity coefficients are not available, they will have to be directly measured. This is particularly likely to be true of many of the organic ions, and involves measuring their conductances with a conductance bridge and their activity coefficients by either freezing point or vapor pressure measurements. Estimates of activity

TABLE I
RELATIVE MOBILITIES OF SOME COMMON CATIONS AND ANIONS[a]

Cation	Mobility			Anion	Mobility		
	15°	20°	25°		15°	20°	25°
H^+	5.04	4.88	4.76	OH^-	—	2.68	2.70
Li^+	0.507	0.518	0.525	F^-	—	0.745	0.754
Na^+	0.666	0.674	0.682	Cl^-	1.030	1.035	1.039
K^+	1.000	1.000	1.000	Br^-	1.059	1.064	1.063
Rb^+	1.064	1.046	1.059	I^-	1.042	1.042	1.045
Cs^+	1.059	1.049	1.050	NO_3^-	—	0.974	0.972
NH_4^+	—	1.000	1.000				
Ca^{2+}	0.393	0.399	0.405				

[a] Calculated relative to K^+ from values of limiting equivalent conductivity ($\Lambda°$) of ions in water (obtained from Appendix 6.2 of Robinson and Stokes[32]), using the relationship $[u = \Lambda°/(|z|F^2)]$. The values of $\Lambda°$ at 20° were interpolated from values at 18 and 25°.

[32] R. A. Robinson and R. H. Stokes, "Electrolyte Solutions," 2nd Ed. Butterworths, London, 1965.

coefficients can be made with an ion-selective electrode selective for a counterion with known properties and a free-flowing 3 M KCl reference electrode, but the junction potential correction required for these measurements requires an estimate of the mobility and activity of the test ion, and so the method is rather circular and requires successive approximations to obtain the most consistent fit to the measurements.

However, it needs to be stressed that if the KCl bridge is not fresh, it becomes contaminated with any previous solutions into which it was dipped and the junction becomes totally unreliable, being both time and history dependent (see Fig. 4 of Barry and Diamond[29]). Even with good free-flowing junctions, the greatest deviation between calculated and measured junction potentials seems to occur with KCl salt bridges.

The other alternative approach in the minimization and correction of junction potentials is to use a salt bridge containing a solution very similar in composition to one of the bathing solutions.[29] For example, if an epithelium were to be bathed with both 150 mM NaCl and 150 mM CsCl solutions, then either a 150 mM NaCl or a 150 mM CsCl salt bridge could be used. Although such a junction potential is of larger magnitude (and the resistance of the salt bridge significantly higher) than for a 3 M KCl bridge, it is very much better defined, exhibits minimal salt bridge contamination (any contamination is in any case a mixture of the two end solutions), and hence it is time independent and relatively easy to estimate. Any junction potentials arising from the ends of the salt bridges (e.g., NaCl) in contact with the 3 M KCl solutions, in which the electrodes are placed, are equal and opposite and hence cancel out. For such a bi-ionic situation (e.g., NaCl:CsCl) or for a dilution situation [e.g., NaCl(150):NaCl(75)] the junction potential equations radically simplify. For example, for the dilution potential[29] both the Henderson and Planck approaches give an identical solution, which is

$$\epsilon^S - \epsilon^B = (RT/F) \frac{(u_+ - u_-)}{(u_+ + u_-)} \ln(a^B/a^S) \qquad (6)$$

where u_+ and u_- are the relative mobilities of the cations and anions, which are both assumed to have the same activity coefficient ($\gamma_+ = \gamma_-$; the Guggenheim assumption), whereas for the biionic junction it becomes

$$\epsilon^S - \epsilon^B = (RT/F) \frac{a^S(u_+^S - u_-^S) - a^B(u_+^B - u_-^B)}{a^S(u_+^S + u_-^S) - a^B(u_+^B + u_-^B)} \ln\left[\frac{a^B(u_+^B + u_-^B)}{a^S(u_+^S + u_-^S)}\right] \qquad (7)$$

if $a^B = a^S$ and the anion is common, these equations simplify even more radically, again reducing to the equivalent Planck equation, to give

$$\epsilon^S - \epsilon^B = (RT/F) \ln\left[\frac{(u_+^B + u_-^B)}{(u_+^S + u_-^S)}\right] \tag{8}$$

Estimates of junction potentials can most easily be achieved by simple calculations using one or more of the above equations. The only other alternative is experimental measurement, which is not without complications, because, first, it is impossible to measure a single electrode potential by itself, and, second, even when an ion-selective electrode is involved, the potential has to be measured in conjunction with a counter-ion and assumptions made about the activity coefficients of the cations and anions. Using the Guggenheim activity coefficient assumption (i.e., $\gamma_{Cl} = \gamma_X = \gamma_{XCl}$), γ_{Cl} may either be obtained from tables (see, e.g., Robinson and Stokes[32]) or even measured directly and hence the junction potential correction can be evaluated. For further examples and details, see Barry and Diamond.[29] A comparison and list of both experimental and calculated junction potentials for some commonly used solutions is given in Table II.

TABLE II
THEORETICAL AND EXPERIMENTAL SALT BRIDGE JUNCTION POTENTIALS

				$\epsilon_S - \epsilon_B$ (mV)	
Solution (S)	Concentration	Salt bridge (B)	Concentration	Theoretical	Experimental
LiCl	75[a]	LiCl	150[a]	−6.0	−6.0
NaCl	75[a]	NaCl	150[a]	−3.7	−3.7
KCl	75[a]	KCl	150[a]	−0.3	−0.4
CsCl	75[a]	CsCl	150[a]	+0.1	+0.5
LiCl	150[a]	NaCl	150[a]	+2.9	+2.9
KCl	150[a]	NaCl	150[a]	−4.8	−5.7
CsCl	150[a]	NaCl	150[a]	−5.5	−5.7
LiCl	150[a]	KCl	3 m^b	−0.2	—
NaCl	150[a]	KCl	3 m^b	−0.6	—
KCl	150[a]	KCl	3 m^b	−1.3	—
CsCl	150[a]	KCl	3 m^b	−1.5	—

[a] Taken from Table 2 of Barry and Diamond,[29] where the concentrations are millimolal. Calculated RbCl values (not shown) were in all cases within 0.2 mV of CsCl values. The experimental values were measured at 23° using Ag/AgCl electrodes and corrected using the Guggenheim activity coefficient assumption. Experimental errors were considered to be within 0.2 mV.

[b] Calculated using Eqs. (4) and (5) with activity coefficients from Robinson and Stokes,[32] together with mobilities taken from Table I at 25°.

Having either calculated or measured the appropriate junction potential corrections, it is critically important that they be applied in the right direction. Procedures for ensuring this are illustrated in Fig. 5. For standard transmembrane measurements, with microelectrodes in either solution, if the directions of the potentials are defined as in Fig. 5A (neglecting the two Ag/AgCl potentials, which will either cancel out or be backed off with some additional offset voltage when the setup is zeroed), then defining the experimentally measured potential as V_e and the membrane potential as V_m and using Kirchoff's law (in the direction indicated), the sum of the electromotive forces (EMFs) around the circuit is zero, and, as indicated, $V_e + \epsilon_L^i - \epsilon_m - \epsilon_L^o = 0$ so that

$$\epsilon_m = V_e - (\epsilon_L^o - \epsilon_L^i) \qquad (9)$$

where ϵ_L^o and ϵ_L^i are the two junction potentials measured as solution potential with respect to microelectrode. For example, for a typical

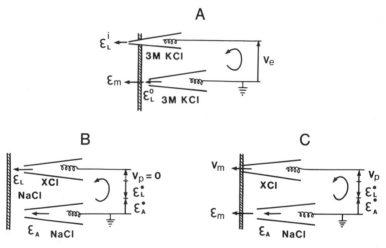

Fig. 5. The application of junction potential corrections to the measurement of membrane potentials and patch-clamp potentials. (A) Measurement of the membrane potential ϵ_m with normal $3\,M$ KCl–Ag/AgCl microelectrodes balanced (or backed off) before electrode insertion. V_e, Experimentally measured potential; ϵ_L^i and ϵ_L^o, microelectrode junction potentials. (B) Zeroing of the patch-clamp pipet before a seal is established. Bathing solution is NaCl; reference electrode solution is NaCl; patch pipet solution is XCl. Two sources of unbalanced potential need backing off by patch-clamp amplifier: ϵ_A, small potential difference between two Ag/AgCl electrodes; ϵ_L, junction potential between NaCl and XCl solutions; ϵ_L^* and ϵ_A^*, backed off voltage components in amplifier. With no applied potential, $\epsilon_L^* = \epsilon_L$ and $\epsilon_A^* = \epsilon_A$. (C) Patch-clamp pipet sealed against the membrane. ϵ_L is now eliminated, but ϵ_L^* is still present. For a voltage V_p, applied by the patch-clamp amplifier, V_m is given by Eq. (10).

NaCl–Ringer's solution and high K^+ and low Cl^- intracellular composition and $3\ M$ KCl electrodes, $\epsilon_L^o - \epsilon_L^i \approx 2.0$ mV.

Equation (9) also applies to the use of salt bridges (replacing microelectrodes) for transepithelial measurements if the superscripts o and i refer to the two bathing solutions.

The second situation (Fig. 5B and C) refers to the patch-clamp measurement itself (intact patch), in which an additional junction potential component arises by virtue of the necessary zeroing of the patch-clamp amplifier, whenever the solution in the patch-clamp pipet differs from that in the bathing solution. Before the patch pipet touches the membrane (Fig. 5B), there will be two potential components that need to be backed off: one is the junction potential (ϵ_L) arising because of the different solution in the patch pipet compared to the bathing solution, whereas the other is due to any asymmetry in the two Ag/AgCl electrode potentials (ϵ_A). These will need to be backed off by a voltage with two components ϵ_L^* and ϵ_A^*. Provided that $V_P = 0$, it may similarly be shown that $\epsilon_L = \epsilon_L^*$ and $\epsilon_A = \epsilon_A^*$. When the pipet is finally sealed against the membrane (Fig. 5C), the liquid junction potential correction, ϵ_L, is eliminated, but the backed-off component, ϵ_L^*, is still present. Thus

$$V_m = (\epsilon_M - V_P) + \epsilon_L \tag{10}$$

where V_P is the additional potential applied by the patch-clamp amplifier, ϵ_m is obtained from Eq. (9), and ϵ_L is the potential of the external bathing solution (with respect to the patch pipet). The same equations apply for the excised patch, together with the condition that $\epsilon_m = 0$.

Minimization and Definition of Unstirred Layers

For most epithelia the serosal connective tissue layer provides the greatest contribution to unstirred-layer effects, in magnitude typically equivalent to about 800 μm (see Ref. 33 for further discussion). Apart from physically dissecting away part of this[5] or treating it enzymatically, its magnitude cannot be reduced. However, many measurements involve solution changes at the apical surface, where, as also in the case of the additional bathing solution region beyond the serosal tissue layer, unstirred layers can be typically reduced to about 100 μm or so by stirring either with a magnetic stirrer or with a stream of water-saturated O_2 bubbles. In some rare cases they have even been further reduced to below 50 μm.

It is generally of value to determine the effective magnitude of the unstirred layers adjacent to the permeation barrier. This may be done by

[33] P. H. Barry and J. M. Diamond, *Physiol. Rev.* **64**, 763 (1984).

using the simple relationship, suggested by Diamond,[34] relating the half-time $t_{1/2}$ for the build-up of the solute concentration at the membrane–solution interface for solute diffusing from a well-stirred region separated from a planar membrane by an unstirred layer of thickness δ. In general,

$$t_{1/2} = a_v \delta^2 / D \tag{11}$$

where D is the diffusion coefficient in the unstirred layer and a_v is a constant of proportionality, which in the absence of any volume flow is ≈ 0.38. In the presence of a volume flow, however, a_v can be considerably reduced below 0.38,[33,35] and so the method should not be used where the volume flow is significant. It can be used with the measurements of dilution potentials (but not streaming potentials), provided $t_{1/2}$ is taken to be the time for the concentration at the membrane to reach half the concentration in the bulk solution. For a simple dilution potential equation, the fraction of the final membrane potential change at which this occurs is given by

$$\frac{\Delta\epsilon(\Delta C_m/2)}{\Delta\epsilon(\Delta C_m)} = \frac{\log[(f+1)/2f]}{\log[1/f]} \tag{12}$$

where f represents the fractional dilution of the salt concentration and is given by the ratio of the diluted salt concentration to its initial value. For example, for a 2:1 dilution, $f = 2$, this occurs at 0.415 of the final membrane potential change.

An alternate variation of the method[36] has been used to measure mucosal unstirred-layer thicknesses from the half-times for the abolition of short-circuit current in toad urinary bladder after exposure to $10^{-4}\,M$ amiloride in the mucosal solution, and based on an estimate of $6.35 \times 10^{-6}\,cm^2\,s^{-1}$ for the amiloride diffusion coefficient.

Electrical Measurements: Analysis and Interpretation of Data

Permeation Mechanisms

In general, passive permeation across membranes involves either a channel or carrier mechanism, and both mechanisms exhibit some sort of ion selectivity due to the presence of either charged or neutral (polar) sites lining the aqueous interior of the carrier or channel. Charged refers to the

[34] J. M. Diamond, *J. Physiol. (London)* **183**, 83 (1966).

[35] P. H. Barry, *Alfred Benzon Symp.* **15**, 132 (1981).

[36] D. Wolff and A. Essig, *J. Membr. Biol.* **55**, 53 (1980).

sites bearing net charge, whereas neutral or polar refers to the fact that although the sites may be the negative or positive ends of the dipoles lining the interior walls of the channel, the net charge of each dipole is zero. Channels may appear to behave as long or short (for further discussion, see Barry et al.[37]), depending on whether the Debye length, within the channel, a measure of the distance over which electroneutrality is violated, is much smaller or longer than the channel length. One of the uncertainties within aqueous channels in membranes is the magnitude of the electrolyte concentration there, which is dependent on the number of sites per channel. Since this helps to determine the effective Debye length there, its value is not known.

Channels can be distinguished from mobile carriers in three main ways[38]: known carrier transport rates are at least two orders of magnitude less than conductances observed for ionic channels; carrier transport $I-V$ curves should be sigmoidal, the current saturating at high voltages, in contrast with the generally linear $I-V$ curves expected for channels in symmetrical solutions; and the presence of significant electrokinetic coupling between ions and water in aqueous channels contrasts with its absence in the case of carrier transport. If permeation is via a mobile carrier mechanism, then only the neutral carrier (e.g., valinomycin) can contribute to any electrical current. In contrast, the charged carrier will be electrically silent when transporting an ion, and will behave as an exchange diffusion carrier only contributing to equal and opposite flux components. Similarly, cotransport (symport) and countertransport (antiport) systems (not included here, but in another chapter[1]), can either contribute electrical current if they are electrogenic or at least modify ion concentrations if they are electroneutral.

If the channels are "long," then neutral (polar) and charged site channels can be readily distinguished by the following opposite predictions. The long charged site channels will tend to have the sites balanced by counterions so that the number of ions within the channel and hence the channel conductance will tend to be independent of the external bathing solution concentration of the ions. In contrast, the long neutral site channels, even though they provide a favorable electrical environment for the counterions, do not require electrical balancing by them and so their concentration and hence their conductance will tend to increase linearly with the bathing solution concentration. If, however, the channels are short, then it would be expected that even the charged site channel would exhibit a linear dependence of conductance with external ion

[37] P. H. Barry, P. W. Gage, and D. F. Van Helden, *J. Membr. Biol.* **45,** 245 (1979).
[38] P. H. Barry and P. W. Gage, *Curr. Top. Membr. Transp.* **21,** 1 (1984).

concentration and so the distinction between neutral and charged site channels would be lost for such measurements.

It is also possible, especially when the membrane is very thick, as in the tight junction shunt regions between epithelial cells, that the channel may be some heterogeneous composite of long and short neutral site and charged site regions.[39] Similarly, channels may also open cooperatively in parallel, and this may explain, for example, some of the extremely high single-channel conductances observed for some anion-selective channels in alveolar epithelial cells.[40] For certain ion channels, structural evidence suggests that the channel opens out into a vestibule region at the end of the channel near the solution interface. It has been suggested that such a vestibular region has a significant effect on ion permeation (see, e.g., Dani[40a] and Dani and Eisenman[40b]).

Ion Permeation Models

Ion permeation across a membrane can be investigated by measurements of diffusion potentials and conductances in the presence of different salt solutions on either side of a membrane. In all cases, a precise evaluation of the permeability parameters requires fitting of the most appropriate membrane potential and current–voltage equations to the data, and this is determined by choosing the most relevant ion permeation model. For some measurements the choice of model does not make much difference, whereas for others it is critical and the data distinguish between models. For this reason, a number of different permeation models are listed in Table III.

Where *ions of one sign only* permeate across a membrane (or, in certain cases, as in amphibian skeletal muscle, where the only oppositely charged permeant ions are normally distributed in equilibrium), the zero-current diffusion potential, often in the case of individual channels referred to as the "null potential" or "reversal potential," reduces to a fairly simple expression similar in form to the Goldman–Hodgkin–Katz equation,[41,42] but lacking its restrictive assumptions. This equation, which has been referred to as the "generalized null potential equation," [38] under these conditions is applicable to a large number of different electrodiffusion

[39] J. H. Moreno and J. M. Diamond, *in* "Membranes: A Series of Advances" (G. Eisenman, ed.), Vol. 3, p. 383. Dekker, New York, 1975.

[40] G. T. Schneider, D. I. Cook, P. W. Gage, and J. A. Young, *Pfluegers Arch.* **474**, 354 (1985).

[40a] J. A. Dani, *Biophys. J.* **49**, 607 (1986).

[40b] J. A. Dani and G. Eisenman, *J. Gen. Physiol.* **89**, 959 (1987).

[41] D. Goldman, *J. Gen. Physiol.* **27**, 37 (1943).

[42] A. L. Hodgkin and B. Katz, *J. Physiol. (London)* **108**, 37 (1949).

TABLE III
ION PERMEATION MODELS

Model number	Model	Notes	Reference(s)[a]
1	Takeuchi equations	b	1
2	Goldman–Hodgkin–Katz (GHK) equation	c	2–4
3	Goldman equation with mixed valencies	d	5
3	Modified Goldman with active transport	e	6, 7
4	Electrodiffusion with cooperative sites	f	8, 9
5	Long channels with neutral sites	g–i	4, 8, 10
6	Long channels with charged sites	h,i	4, 8–10
7	Channels in parallel with shunt	g,h	8, 10–12
8	Short cation and anion channels	h,i	8
9	Rate-theory model (1 site)	i	4, 13, 14, 17
10	Rate-theory model (3 sites)	i	4, 15
11	Mobile carrier model		16

[a] References: (1) A. Takeuchi and N. Takeuchi, *J. Physiol. (London)* **154**, 52 (1960); (2) D. Goldman, *J. Gen. Physiol.* **27**, 37 (1943); (3) A. L. Hodgkin and B. Katz, *J. Physiol. (London)* **108**, 37 (1949); (4) P. H. Barry and P. W. Gage, *Curr. Top. Membr. Transp.* **21**, 1 (1984); (5) N. L. Lassignal and A. R. Martin, *J. Gen. Physiol.* **70**, 23 (1977); (6) T. Zeuthen, *in* "The Application of Ion-Selective Microelectrodes" (T. Zeuthen, ed.), p. 27. Elsevier/North-Holland, Amsterdam, 1981; (7) T. Zeuthen and E. M. Wright, *J. Membr. Biol.* **60**, 105 (1981); (8) P. H. Barry and J. M. Diamond, *J. Membr. Biol.* **4**, 295 (1971); (9) F. Conti and G. Eisenman, *Biophys. J.* **5**, 511 (1965); (10) P. H. Barry, P. W. Gage, and D. F. Van Helden, *J. Membr. Biol.* **45**, 245 (1979); (11) E. M. Wright, P. H. Barry, and J. M. Diamond, *J. Membr. Biol.* **4**, 331 (1971); (12) P. H. Barry, J. M. Diamond, and E. M. Wright, *J. Membr. Biol.* **4**, 358 (1971); (13) P. Lauger, *Biochim. Biophys. Acta* **311**, 423 (1973); (14) T. Begenisich and M. Cahalan, *in* "Membrane Transport Processes" (C. F. Stevens and R. W. Tsien, eds.), Vol. 3, p. 113. Raven, New York, 1979; (15) B. Hille, *J. Gen. Physiol.* **66**, 535 (1975); (16) S. Ciani, G. Eisenman, and G. Szabo, *J. Membr. Biol.* **1**, 1 (1969); (17) J. A. Dani, *Biophys. J.* **49**, 607 (1986).
[b] Didactically useful; inappropriate for more than one ion species per channel.
[c] Widely used, assumes constant electric field and flux independence.
[d] Same assumptions as model 2, but can handle mixed-magnitude valencies.
[e] Useful for modeling electrogenic transport across cell membranes.
[f] For $n \neq 1$, (K_a) replaced by $(K_a)^{1/n}$ and RT by nRT in Eq. (13) for permeant cations alone.
[g] Potentially useful for ion permeation through paracellular pathways.
[h] Potentially useful for ion permeation across cell membranes.
[i] Potentially useful for ion permeation through single channels.

(models 5 and 6 in Table III) and rate theory models (models 9 and 10), with very different underlying assumptions. It is of the form

$$\epsilon'' - \epsilon' = (RT/F) \ln \left[\frac{(P_1 a_1' + P_2 a_2' + P_4 a_4')}{(P_1 a_1'' + P_2 a_2'' + P_4 a_4'')} \right] \tag{13}$$

where the subscripts 1, 2, and 4 refer to ions 1, 2, and 4 of the same sign (ion 3 being reserved for a permeant ion of the opposite sign), P and a refer to the ions' relative permeabilities and activities, and the primes and double primes refer to solutions 1 and 2, respectively. The relative permeabilities are proportional to two underlying permeation parameters: the equilibrium constant K (or relative partition coefficient) for ions interacting with charged or polar sites lining the permeation channel and the ionic mobility u (or transition rate constant k) for ions moving between intrachannel sites. In other words

$$P_i/P_j = (u_i/u_j)K_{ij} = (u_i/u_j)(K_i/K_j) \tag{14}$$

where in rate theory modeling each u is replaced by k. This condition, of permeant ions of only one sign, is often met in ion permeation through single channels of one type (e.g., ACh channels, Ca^{2+}-activated K^+ channels) and would also be applicable to most patch-clamp measurements of single channels in epithelial cell membranes.

For a thick membrane (with long channels; models 5 and 6), with separate or combined cation and anion channels and only one permeant species of cation and anion, the null potential, ϵ_0, in the dilution potential situation is given by

$$\epsilon_0 = (RT/F)\left[\frac{(P_1 - P_3)}{(P_1 + P_3)}\right]\ln(a'/a'') \tag{15}$$

for the charged site model, whereas for the neutral site model P_1 and P_3 are replaced by u_1 and u_3.

However, for transepithelial permeation or whole-cell diffusion potential measurements, not only is there generally significant permeation of oppositely charged ions, but in addition to the presence of regular ion channels, there can also be the presence of a free-solution shunt pathway (model 7, particularly if K^+ or phosphate is omitted from the bathing solution[39]), which may increase with time following the dissection of the epithelium. Whenever ions of both signs are allowed to permeate significantly through a channel (or combination of channels), the null potential expression greatly increases in complexity and becomes very model dependent, since solution of the equation depends on balancing the different heavily model-dependent expressions for the cation and anion components of current across the membrane.

Unstirred-Layer Effects

Unstirred layers can affect electrophysiological measurements in two main ways (for a review and for further detail of mechanisms, see Barry and Diamond[33]). These are as follows. Water flow across a membrane in the presence of an electrolyte solution on either side of a membrane will cause a perturbation of the salt concentration in the unstirred layers, of thickness δ' and δ'' on each side of a membrane, as illustrated in Fig. 6A. This will result in a time-dependent change in the diffusion potential across the membrane which is proportional to the volume flow (normally substantially water). This pseudo-streaming potential is in the same direction as that expected for a true steaming potential, although its time course should be very much slower.

The second way in which unstirred layers can affect electrophysiological measurements occurs because of the transport number effect (for further details, see Ref. 33). This is illustrated in Fig. 6B for a cation-selective membrane separating two KCl solutions. The transport number (t_i) of an ion i reflects the fraction of current carried by that ion and is given by

$$t_i = z_i u_i C_i / \left[\sum_j z_j u_j C_j \right] \tag{16}$$

where z, u, and C refer to the magnitude of the valency, and mobility, and concentration of the appropriate ions and Σ refers to a summation over all the ions in the solution. For a KCl solution, $t_K \approx t_{Cl} \approx 0.5$. Thus, when one Faraday of charge is passed across the membrane, 0.5 mol of K^+ arrives at

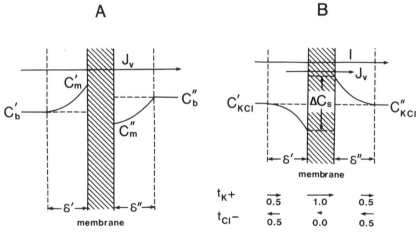

FIG. 6. Unstirred-layer effects near membranes (or epithelia), with adjacent unstirred-layer thicknesses, δ' and δ'', in the two solutions denoted by superscripts prime and double prime (similar to Figs. 10B and 23 of Barry and Diamond[33]). (A) The sweeping away and enhancement of salt concentration profiles by a volume flow J_v in the unstirred layers and balanced by diffusion down the concentration gradient, so created. C_m and C_b, Concentrations of salt adjacent to membrane and in bulk region (assumed to be perfectly stirred). (B) The transport number effect for a perfectly cation-selective membrane (or epithelium) separating two KCl solutions with bulk concentrations, C'_{KCl} and C''_{KCl}. t_{K^+} and t_{Cl^-}, Transport numbers of K^+ and Cl^- (direction and approximate magnitude, as indicated); J_v, resultant local osmotic flow due to local concentration change (ΔC_s) across the membrane.

the left-hand side of the membrane but 1.0 mol goes across it. Similarly, no Cl⁻ crosses the membrane but 0.5 mol of Cl⁻ leaves the solution–membrane interface to move to the left. These effects therefore result in a loss of 0.5 mol of KCl from that solution–membrane interface and a similar enhancement of KCl at the right solution–membrane interface, the rate of depletion or enhancement of KCl at each interface being proportional to the transport number difference (between membrane and solution) and the current. These concentration profiles, which will take some time to build up, will result in both a local osmotic water flow and also an increase in the transmembrane pd, which will mimic a slow capacitative transient on top of the normal voltage response, generally of much slower time course than the real capacitative transient. The magnitude of such transport number effects is greatest when the concentration of the major permeant ion (e.g., K^+ in many situations) is low. In conductance measurements only passing transmembrane currents for as short a time as possible, provided that it is still significantly greater than the real capacitative time constant, will tend to minimize contamination of real membrane properties by these unstirred-layer transport number effects.

Membrane Surface Charge Effects

This effect of surface charge is both pH and ionic strength dependent. It arises because most membranes have negatively charged groups on both their exterior and interior membrane surfaces. Typical surface charge densities are of the order of 10^{-5} C cm^{-2} and it is considered that the internal (cellular) surface normally has a smaller charge density than the outer one.[43] Such negative surface charges give rise to a negative surface potential, which increases the concentration of cations (and decreases that of anions) adjacent to the membrane interface. This surface potential contributes to the overall potential by linearly adding to the potential just across the membrane. However, with only permeant monovalent cations present, the effect of increased membrane interface concentration and surface potential exactly cancel out as far as the current–voltage relationships are concerned.

Surface charge effects can be modulated by ionic strength, pH, and divalent ions in one of two ways—screening or binding. In the case of screening there is a double-layer effect, which affects the surface potential, reducing its magnitude as the ionic concentrations are increased (further details are given elsewhere[38,40a,44,45]). Such screening, exponentially depen-

[43] B. Hille, A. M. Woodhull, and B. I. Shapiro, *Philos. Trans. R. Soc. London B* **270**, 301 (1975).

[44] D. C. Grahame, *Chem. Rev.* **41**, 441 (1947).

[45] K. Takeda, P. W. Gage, and P. H. Barry, *Proc. R. Soc. London B* **216**, 225 (1982).

dent on the valency of the ions, can help to explain the effect of divalent ions on some ion permeation properties. In addition to *screening* (in which ions remain hydrated), ions can also *bind* to these surface charges, the ion losing at least part of its hydration shell and coming into direct apposition with the negative charge groups on the surface of the membrane. Binding directly reduces the surface charge density and hence is very much more potent at decreasing the surface potential. Whereas *screening* is essentially independent of ion species of the same valency, *binding* is very ion specific, being dependent on the nature of the particular cation (its radius and hydration energy) and site radius and spacing between sites for divalent ions (see discussion in Barry and Gage[38]). Binding is also generally effective at much lower concentrations than screening. The binding of protons in low-pH solutions (protonation) reduces the surface charge density and hence the contribution of surface potentials to permeation. Similar effects have been observed for apical K^+ permeability. Ca^{2+}, Mg^{2+}, and Ni^{2+} added to the mucosal solution result in a decrease in P_K, suggestive of screening, whereas Sr^{2+} and Ba^{2+} are more potent, implicating a binding mechanism.[46]

In order to allow for surface charge effects, the surface potentials can be calculated for a particular value of surface charge and this potential in turn used to obtain values of membrane interface concentrations and to correct the measured values of membrane potential.[38,40a,45]

Specific Measurements

In order to characterize the mechanisms of passive ion permeation in epithelia, it is necessary to appreciate the various pathways that are involved. A diagram of some of the major pathways and the equivalent electrical circuit of one typical type of transporting epithelium (e.g., gallbladder, proximal convoluted tubule, intestine) is given in Fig. 7. In various other epithelia (cf. Fig. 1 of Baerentsen *et al.*[47] and Figs. 6 and 9 of Young *et al.*[48]) the situation is not as simple and various more complex cotransport (symport) and countertransport (antiport) systems may also be involved (see Stein[1]). For example, $2Cl^-$, K^+, Na^+-cotransporters are postulated for some basolateral (secretory epithelia[48]) and some apical (thick ascending limb of the loop of Henle[48a]) membranes. Evidence suggests that Na cotransport of organic solutes is electrogenic and contributes to electri-

[46] L. Reuss, L. Y. Cheung, and T. P. Grady, *J. Membr. Biol.* **59**, 211 (1981).

[47] H. Baerentsen, F. Giraldez, and T. Zeuthen, *J. Membr. Biol.* **75**, 205 (1983).

[48] J. A. Young, D. I. Cook, E. W. Van Lennep, and M. L. Roberts, *in* "Physiology of the Gastrointestinal Tract" (L. R. Johnson, J. Christensen, M. Jackson, E. Jacobson, and E. Walsh, eds.), 2nd Ed., Vol. 2, p. 773. Raven, New York, 1987.

[48a] R. Greger, *Physiol. Rev.* **65**, 760 (1985).

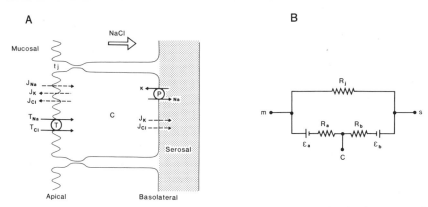

FIG. 7. A simplified schematic model and equivalent electrical circuit of one type of transporting epithelium (e.g., gallbladder, proximal convoluted tubule, intestine; cf. Fig. 2 of Zeuthen[8] and Fig. 1 of Baerentsen et al.[47]). (A) Some of the pathways and fluxes for K^+, Na^+, and Cl^-. Such epithelia have active Na^+,K^+-ATPase pumps (with probable ratio of 3Na/2K) on the basolateral membrane (as shown above), whereas the choroid plexus has the pump on the apical side. T, An electrically silent equivalent neutral cotransport system (symport), which in reality may be a combination of two countertransport (antiport) systems (Na^+/H^+ and Cl^-/HCO_3^-); J_K, J_{Na}, and J_{Cl}, passive electrodiffusive fluxes. Other epithelia have more complex coupled transport systems such as $2Cl^-,Na^+,K^+$-cotransporters (see the text). (B) A simplified electrical circuit of an epithelium. R_a, R_b, and R_j, Resistances across the apical, basolateral, and tight junction (tj) pathways, respectively; ϵ_a and ϵ_b, potential differences across the apical and basolateral membranes; m, s, and C, mucosal, serosal, and cell interior, respectively. The possible presence of a small potential difference across the tight junctions is not shown.

cal current across cell membranes.[49] The presence of electrogenic transport systems in operation would make the analysis of passive ion permeation very difficult. However, even electroneutral coupled cotransport or countertransport systems would result in fluxes that would complicate electrical measurements if they were allowed to change the internal cell composition and would possibly also cause local concentration changes in the unstirred layers adjacent to these membrane barriers. In addition, they could also significantly contribute to tracer flux measurements. Bearing such possible effects in mind and taking appropriate steps to reduce their contribution with specific blockers (see Stein[1]), the following outline will give a suggested approach to characterizing ion permeation across a particular membrane and extracting the appropriate permeation parameters. Most of the approach has been used already to investigate the mechanism of ion permeation through the paracellular tight junction pathway of a very leaky

[49] E. M. Wright, Am. J. Physiol. 246, F363 (1984).

epithelium (the gallbladder),[39,50-52] where such a route dominates trans-epithelial ion movement. Then, methods for estimating the quantitative proportion of permeation through the paracellular pathway will be outlined and, finally, the specific application of such an approach and the isolation of each pathway in order to determine the mechanism of ion permeation across the three pathways—paracellular, apical membrane, and basolateral membrane—will be discussed.

For the analysis of any data, the following choice of permeation models is suggested (see Table III): paracellular pathway, model 5 or 7, even though the actual pathway may be a more complicated composite[39]; large area of apical or basolateral membrane (in absence of electrogenic transport), mosaic form of models 5, 6, or 8; single channels, models 5, 6, and 8–10.

In situations where only ions of one valency are permeant, the zero-current potential would be given by the generalized null potential equation so that the relative permeabilities would be model independent. For other situations, the data should be used to select which model is most appropriate. Further details and discussion about characterizing ion permeation are also given elsewhere.[37-39,50-52]

For convenience, in the following examples, it is assumed that the membranes are predominantly cation selective. However, for anion-selective membranes, simply transpose anion and cation where appropriate in the discussion and correct for valency in the equations.

Approach Outline for Characterizing Ion and Nonelectrolyte Permeation

Potential Measurements. These can involve either dilution or biionic measurements. The former involve diluting the salt concentration in one solution (e.g., for transepithelial measurements, 150 mM NaCl to 75 mM NaCl) at constant osmolality (by adding mannitol or sucrose), while keeping the other constant (e.g., 150 mM NaCl) and measuring the resulting diffusion potential with salt bridges (e.g., 150 mM NaCl) as discussed earlier. Neither cation nor anion should be impermeant and ideally it would be preferable to have only one of each species in both solutions. However, it is generally necessary to include a low concentration of certain ions in the solution, e.g., both 3–5 mM K$^+$ and 2.5 mM phosphate in order to keep the preparation viable and to minimize the development of a free-solution shunt.[39]

[50] P. H. Barry and J. M. Diamond, *J. Membr. Biol.* **4**, 295 (1971).
[51] E. M. Wright, P. H. Barry, and J. M. Diamond, *J. Membr. Biol.* **4**, 331 (1971).
[52] P. H. Barry, J. M. Diamond, and E. M. Wright, *J. Membr. Biol.* **4**, 358 (1971).

In biionic potential measurements for a cation-selective membrane, the two solutions bathing the membrane should have different cations and a common anion at the same concentration (e.g., 150 mM NaCl:150 mM CsCl). Ideally, for such a membrane, the most appropriate choice of anion would be one with zero permeability, in which case, the relative cation permeabilities extracted would be essentially model independent and given by the generalized null potential equation [Eq. (13)].

Preliminary Dilution Potential Measurements. Initially, a number of different cations and anions may be screened in order to estimate the anion–cation selectivity of the membrane and to locate any cations or anions that are impermeant. If the situation could be approximated as having only one major permeant anion or cation, then an equation such as Eq. (15) could be used. Alternatively, other equations, or more general ones allowing for permeation of more than one anion and cation, would have to be used (see references in Table III).

Biionic Potentials. As a first approximation the site cooperativity factor n may be taken as 1. However, if there does appear to be any evidence of cooperativity (e.g., cation permeability ratios dependent on solution composition) then the value of n may be determined as outlined in Barry et al.[52]

Using $n = 1$ (or an experimentally determined value of n, as above) the cation or anion permeability sequences could be investigated for the alkali cations (Li^+, Na^+, K^+, Rb^+, Cs^+) or halide anions (F^-, Cl^-, Br^-, I^-) using biionic potential measurements. This could then be followed by determination of relative permeability sequences of a large number of different organic cations or anions to determine the minimum structural dimensions of the channels (for further details, see also Moreno and Diamond[39] and Adams et al.[53]).

Dilution Potential Measurements. Cation permeability sequences can be investigated by remeasuring and analyzing dilution potentials (null potentials) for different cations with a common anion, provided that the membrane is not too selective so that the permeability ratio P_+/P_- is neither zero nor infinite. Ionic strength and osmolality should be kept constant in all these measurements.

Dilution potentials can also be measured for different total salt concentrations on both sides of the epithelium or membrane. Provided that the sites in none of the channels were completely saturated, either a mosaic of charged site cation and anion channels or one (or more) neutral site channels alone should result in the permeability ratio P_+/P_- remaining fairly constant. In contrast, with the same proviso, for a mosaic of long

[53] D. J. Adams, W. Nonner, T. M. Dwyer, and B. Hille, *J. Gen. Physiol.* **78**, 593 (1981).

charged site cation and neutral site (or shunt) anion channels, the permeability ratio should no longer be constant.

Ion permeabilities can also be measured as a function of ion gradient (e.g., 150 mM:75 mM; 150 mM:37.5 mM), preferably at constant ionic strength, balancing the ionic strength ($\Sigma z_i^2 C_i$) with *impermeant monovalent* ions or with *impermeant polyvalent* ions, being careful that only ions that screen surface charge are used. Such measurements are useful for distinguishing between different models.[29]

Ion permeability ratios could be measured as a function of pH and surface charge. For example, changing the pH can alter the selectivity sequence for ion permeation by directly affecting electronegative charged sites within membrane channels[39,46,54] (for a detailed discussion, see Moreno and Diamond[39]). It would be expected that decreasing the pH would result in protons binding to weakly acidic negative groups and reducing the number (or at least the time-average occupancy) of available sites for cation permeation. Thus measuring conductance as a function of pH should give information about the pK_a of the permeation sites within the membrane channels. Similarly, at low pH, basic groups (e.g., amines) can be protonated to give rise to positive sites which can increase the anion conductance contribution.[55] Epithelia can only tolerate low-pH solutions for very short periods of time and this makes it very difficult to make many measurements at low pH. However, many polyvalent ions (e.g., Ca^{2+}, Ba^{2+}, La^{3+}, and Th^{4+}) also seem able to shift membrane permeability from being cation selective to anion selective. In particular, adding $Th(NO_3)_4$ to both epithelial bathing solutions had the effect of lowering the pH in a less damaging way than did directly changing the pH itself, thus turning the epithelium into a stable and viable anion-selective barrier for very much longer times.[56]

Measuring P_+/P_- and K_+/K_- as a function of ionic strength, the changes being made with impermeant ions, will give useful information about surface charge effects. Similarly, K_+/K_- measurements as pH is varied (from about 10 to 3, if possible) may enable the pK_a of the ion permeation sites to be estimated.

Conductance Measurements in Symmetrical Solutions. Conductance measurements should preferably be done with essentially single-salt symmetrical solutions (no junction potential corrections needed) and minimum extra ions in them, beyond what is needed to retain preparation viability. In each case, measurements should be made after capacitative

---

[54] L. Reuss, *in* "Ion Transport by Epithelia" (S. Schultz, ed.), p. 109. Raven, New York, 1981.

[55] E. M. Wright and J. M. Diamond, *Biochim. Biophys. Acta* **163**, 57 (1968).

[56] T. Machen and J. M. Diamond, *J. Membr. Biol.* **8**, 63 (1972).

transients are over but before transport number effects develop.[33,51] The conductance, usually measured as the chord conductance, G, is defined by

$$G = I/(V - \epsilon_0) \qquad (17)$$

where I is the current, V is the applied voltage, and ϵ_0 is null potential. Any solution series resistance component should be subtracted, if significant.[51]

Conductance measurements as a *function of concentration* will distinguish between *long charged site* and *long neutral site channels,* since the conductance of *long charged site channels* will tend to remain *constant* until site saturation is exceeded, whereas, it should *increase linearly with permeant salt concentration* for *neutral site channels* or *short charged site channels.*[37,51]

Conductance measurements *in different* (essentially single-salt) *salt solutions* will primarily reflect relative mobilities (u) of the individual ions. For *long charged site channels,* it primarily depends on u alone, whereas, for *neutral site channels* it also depends on K. A comparison of diffusion potentials and such measurements should enable extraction of relative values of u and K.[51]

Conductance should be *independent of voltage* (in symmetrical solutions) in the absence of dielectric breakdown or a mobile carrier system, otherwise an *asymmetrical energy profile* for ion permeation is implicated.

Conductance–Voltage Curves in Asymmetrical Solutions. For a *long channel,* even in asymmetrical single-salt solutions, the predicted *current–voltage curve* is *linear* and hence the conductance is independent of voltage. For a *short channel,* however, the *current–voltage curve* will in general be *nonlinear* unless the anion–cation permeability ratio equals 1.0. Hence measurement of current–voltage curves in the single-salt dilution situation serves as a test of whether the channel is long or short.[50,51]

In the biionic case, electrodiffusion models (and multisite rate theory models with uniform symmetrical barriers) predict that channel conductance measurements across a membrane (two solutions with different cations A and B and a common anion, X, e.g., $A^+X^-:B^+X^-$, ideally with X^- impermeant) as a function of potential should in general result in a sigmoidal conductance–voltage curve. The relative conductance values at extreme positive and negative potentials should reflect the relative mobilities of the A and B within the channel (further details are given elsewhere[37,38]).

Hence, a measurement of zero-current null potentials for a pair of cations A and B, together with the fitting (using nonlinear least-squares numerical analysis techniques) of either conductance–voltage or current–voltage curves in such a biionic situation, will enable the evaluation of both relative permeabilities (e.g., P_A/P_B) and ionic mobilities (u_A/u_B) or k_A/k_B)

and hence equilibrium constants ($K_{A/B}$ or K_A/K_B) to be extracted for these ions within such membrane channels.

The equilibrium constant sequence immediately gives information about the field strength of the sites. Similarly, channel mobility sequences and the relationship between mobilities and equilibrium constants can further help to characterize the mechanism of permeation. Further details are given elsewhere.[38,57-59]

Conductance–voltage curves measured as a *function of temperature* should enable determination of the activation energy for permeation and hence of the magnitudes of the energy barriers involved.[60] In addition, a plot of conductance as a function of $1/T$ will uncover any phase transitions, such as those that lipids might experience in the vicinity of ion channel proteins, which will manifest themselves in the form of a sudden change in slope of that curve.

Tracer Flux Studies. Such studies should also give information about individual ion permeabilities. There are two main sources of error, however, that can affect the interpretation of such measurements: (1) unstirred-layer effects which result in the real permeability being underestimated if the unstirred layer has a sufficient diffusional resistance (see Ref. 33 for more details) and (2) single-file effects which result in the tracer flux underestimating the true flux by ($N + 1$), where N is the number of sites in the channel.[61] Since they also include components of active transport and electroneutral transport, tracer measurements do not provide very reliable and accurate estimates of passive diffusive permeabilities. However, when used in conjunction with electrical measurements they can give useful information about these other nondiffusive and nonpassive transport systems.

Electrokinetic Measurements. The measurement of significant electrokinetic coupling in membranes is good evidence for aqueous ionic channels in which water and ions can interact, and its magnitude should give information about the density of electronegative or electropositive sites lining those channels. Electrokinetic coefficients can either be measured as an electroosmotic coupling coefficient, β (the volume flow produced by a current, with $\Delta P = 0$), or as a streaming potential $\Delta \epsilon$ developed by a hydrostatic pressure ΔP (with $I = 0$), to which it may be shown to be equal.

[57] J. M. Diamond and E. M. Wright, *Annu. Rev. Physiol.* **31**, 581 (1969).
[58] E. M. Wright and J. M. Diamond, *Physiol. Rev.* **57**, 109 (1977).
[59] G. Eisenman and R. Horn, *J. Membr. Biol.* **76**, 197 (1983).
[60] B. Hille, "Ionic Channels of Excitable Membranes." Sinauer, Sunderland, Massachusetts, 1984.
[61] H. Davson, "A Textbook of General Physiology," 4th Ed., Vol. 1. Churchill, London, 1970.

In contrast to channels, ions carried by mobile carriers, such as antibiotic ionophores, do not have any significant ion–water coupling.

Various techniques need to be employed in order to distinguish true electrokinetic coupling coefficients, determined by the properties of the channels, from the unstirred-layer effects that can give rise to both pseudo-electroosmosis and pseudo-streaming potentials discussed earlier (for a general discussion of these effects, see Ref. 33). Probably, the most useful involve (1) using ionophores to estimate unstirred-layer effects; (2) measuring the fast "instantaneous" component of current-induced volume flows to estimate the true electroosmotic component[62]; (3) increasing the concentration of the dominant permeant ion to reduce the contribution of any pseudo-streaming potential component to the total streaming potential; and (4) endeavoring to directly calculate the unstirred-layer contribution.

The last approach is probably the most satisfactory one that can be used for estimating true streaming potential components for epithelia and most biological membranes. Unfortunately, the time lag required for osmotic probe solutes to diffuse across unstirred layers to the membrane generally precludes the second technique. Estimates generally indicate that the predominant contribution in epithelia to "streaming potentials" is due to unstirred-layer effects.[33]

Another important point that should be considered, for measurements involving both osmotic and hydrostatic pressures, is that considerable tissue distortion involving lateral intercellular spaces and tight junctions can sometimes occur, which can give rise to a transient peak in the streaming potentials when an osmotic probe (even an impermeant one) is placed in the mucosal solution.[63]

Nonelectrolyte Permeation. There is another way in which both real and pseudo-electrokinetic components can be used to derive information about permeation, this time about permeant nonelectrolytes. Both real and pseudo-streaming potentials can be used to measure their relative permeabilities. The procedure makes use of the fact that, in both cases, the measured potential, $\Delta\epsilon$, is proportional to the volume flow, J_v. Hence, a measurement of the ratio of the total streaming potential, $\Delta\epsilon_s$, caused by a permeant solute to that caused by an impermeant one, $\Delta\epsilon_i$, at the same concentration ($\Delta C_s = \Delta C_i$), enables an estimation of the reflection coefficient, σ_s, and hence permeability, P_s, of the permeant solute, given by

$$P_s \propto (1 - \sigma_s) = [1 - (\Delta\epsilon_s/\Delta\epsilon_i)] \qquad (18)$$

Thus, the relative permeabilities of a number of different solutes may be

[62] P. H. Barry and A. B. Hope, *Biophys. J.* **9**, 729 (1969).
[63] E. M. Wright, A. P. Smulders, and J. M. Tormey, *J. Membr. Biol.* **7**, 198 (1972).

readily and rapidly estimated by measuring the ratio $\Delta\epsilon_s/\Delta\epsilon_i$ from the above equation.[64] $\Delta\epsilon_s$ is measured as the maximum change in pd for a permeant solute, the measurement being bracketed before and after by measurements of the maximum change in pd for an impermeant solute, which are then averaged as $\Delta\epsilon_i$. Details of the application of the technique are given by Smulders and Wright[3] (further discussion is given elsewhere[33,64]). The method is especially suitable for ranking permeability coefficients of solutes, although a more quantitative comparison of the relative permeabilities of different solutes can be affected by unstirred-layer effects.[33]

Application to Individual Epithelial Pathways

Before specific membrane measurements in epithelia are made it is necessary to know what fraction of permeation is through the paracellular and transcellular routes and the relative resistances of the tight junction, apical, and basolateral membranes.

Fraction of Ion Permeation through Different Pathways. For an epithelium whose overall permeation properties have not been well established, the fraction of ion permeation through the paracellular pathway can be determined qualitatively by some simple experiments, designed to establish whether it behaves as a two-compartment system (a very leaky epithelium with a predominant paracellular pathway so that $R_j \ll R_a + R_b$), or as a three-compartment system (a very tight epithelium with a predominant transcellular pathway so that $R_j \gg R_a + R_b$). The basic approach was designed to determine whether the dilution potentials were dependent on the composition of the intracellular solution or of the opposite bathing solution. The basic principles are as follows (for details, see Barry and Diamond[29]). For a two-compartment system (paracellular route predominant), dilution potentials resulting from mucosal dilutions should be dependent on serosal composition and independent of intracellular composition, modified by osmotically increasing the concentration of both bathing solutions. In contrast, a three-compartment solution should have opposite predictions.

Such measurements in the gallbladder suggested a paracellular ion permeation route,[29] and this was later confirmed by impedance analysis methods,[7] which provide a more quantitative determination of the relative permeation pathways (further details on standard cable and impedance analysis methods are given elsewhere[6,10]).

In addition, various other quantitative methods have also been used by Lewis and co-workers to determine relative values of R_a, R_b, and R_j (see

[64] E. M. Wright and J. M. Diamond, *Proc. R. Soc. London B* **172**, 203 (1969).

discussion in Lewis and Wills[65]). One of these involved perturbing the apical membrane resistance by the addition of an ionophore (e.g., nystatin or gramicidin D) to the mucosal solution. This has the effect of radically decreasing only R_a so that $R_a/R_b \ll 1$ and the total resistance is predominantly just dependent on R_b and R_j. This value together with the original value of R_a/R_b and the original total resistance enables calculation of each resistance component.[66,67] Another alternative approach using intracellular recording made use of the fact that inhibiting and stimulating the electrogenic Na^+,K^+ pump in the basolateral membrane resulted in a current flowing through the circuit. From these resulting relative voltage changes across the apical and basolateral membranes ($\Delta V_a/\Delta V_b$), together with intracellular measurements of R_a/R_b and the total epithelial resistance, the individual resistive components could again be evaluated.[65]

A comparison of the measurements made using these three different methods is given by Lewis and Wills.[65] For a list of individual resistance values in various epithelia and compounds affecting the resistances of the different pathways refer, for example, to Powell.[68]

General Principles. Having determined the relative resistances of the three pathways, the ion permeation properties of each can be investigated as follows. In all cases it would be advantageous to work at lower temperatures than *in vivo* in order to reduce any Na^+,K^+-ATPase activity and rates of coupled transport systems, which should have a higher activation energy, or Q_{10}, than the passive diffusive systems, provided that this did not reduce the viability of the preparation. Generally, the Na^+,K^+-ATPase pump should be blocked by the addition of 10^{-4} M ouabain to the serosal solution, so as to reduce any electrogenic potential components and any transepithelial solute and water transport which could perturb solute concentrations in unstirred layers adjacent to each membrane. However, this should also be done in conjunction with the blocking of coupled NaCl transport across the apical membrane, or else continued entry of NaCl and subsequent cell swelling will occur.[69] Ericson and Spring[69] showed that such coupled transport block could be readily achieved, for example, by using relatively impermeant anions in the mucosal solution, a procedure that would also facilitate the evaluation of the relative cation permeabilities.

Measurements of electrokinetic coupling and the use of real and pseudo-streaming potentials to rank nonelectrolyte reflection coefficients

[65] S. A. Lewis and N. K. Wills, *in* "Ion Transport by Epithelia" (S. G. Schultz, ed.), p. 93. Raven, New York, 1981.
[66] S. A. Lewis, D. C. Eaton, C. Clausen, and J. M. Diamond, *J. Gen. Physiol.* **70**, 427 (1977).
[67] N. K. Wills, *Fed. Proc., Fed. Am. Soc. Exp. Biol.* **40**, 2202 (1981).
[68] D. W. Powell, *Am. J. Physiol.* **241**, G275 (1981).
[69] A.-C. Ericson and K. R. Spring, *Am. J. Physiol.* **243**, C140 (1982).

and permeabilities ideally require steady-state volume flow conditions, which could be approximated with transepithelial (or even transmembrane patch-clamp) flows. Fast volume flow measurements into or out of individual cells could be made using computerized video-recording techniques.[70] In principle, if fast enough volume changes or transmembrane potential measurements could be made and solutions changed rapidly enough, such measurements would enable true electrokinetic data to be obtained for epithelial cells and membrane vesicles. For very leaky epithelia ($R_j \ll R_a + R_b$), true electroosmotic coefficients would predominantly be determined by ion–water interactions within the paracellular shunt pathway, whereas the total streaming potential values would include nonelectrolyte permeation components from both paracellular and transcellular pathways. In very tight epithelia ($R_j \gg R_a + R_b$), such transcellular streaming potential values would be mainly determined by nonelectrolyte permeabilities in both apical and basolateral membranes. In order to isolate the properties of the individual membranes, either patch-clamp or fast vesicle measurements would be required along the lines suggested earlier for ion permeation.

Paracellular Pathway. If the epithelium is very leaky, then the application of the permeation approach outline should be straightforward and the ion permeation measurements could be carried out as suggested.

If, however, the epithelium is only slightly leaky ($R_j \sim R_a + R_b$), then, at least in principle, it can be transformed into a more leaky one by increasing the resistance of the cellular membranes, by blocking the apical Na^+ channels without affecting the shunt pathway. Amiloride ($10 \ \mu M$), when added to the mucosal solution, effectively blocks these channels. Provided it does not significantly block the transcellular pathway, unlike TAP^+ (2,4,6-triaminopyrimidinium), which blocks both pathways, it would be a suitable blocker. The permeation investigation could then proceed as above.

For very tight epithelia, where the above procedure is inadequate, the properties of the paracellular pathway are not easily investigated and in any case are probably not of much interest.

Apical Membrane Pathway. Various approaches can be made to selectively investigate the permeation properties of apical membranes. The most straightforward, although technically exacting, approach is to use conventional intracellular microelectrodes (either separate current and voltage electrodes, e.g., double barrelled, or a single-electrode clamp system, as discussed earlier[23]) to monitor potential and pass current across the apical membrane, while changing the composition of the mucosal solution. In general, it would also be helpful to measure the internal ionic composi-

---

[70] K. R. Spring and A. Hope, *J. Gen. Physiol.* **73**, 287 (1979).

tion of the cells, or at least adjacent cells, with an ion-sensitive electrode (e.g., especially K^+). These procedures would allow virtually all of the measurements in the permeation approach outline to be completed. The only difficulty would arise with measurements of membrane conductance for a leaky epithelium, since these measurements would include both R_a in parallel with $R_b + R_j$. For a tight epithelium, R_a could be readily isolated. One approach with a leaky one would be to find a selective blocker for the shunt pathway which did not affect the permeability of the apical membrane. In addition, the ionic composition of the intracellular solution could also be changed.[27]

Another technique requires the preparation of vesicles from apical brush border membranes, which can be preloaded with appropriately chosen solutions.[12] These are too small for the insertion of microelectrodes, but alternative techniques have been developed and used to monitor membrane potentials. One of these uses the fluorescence of a voltage-sensitive cyanine dye. The system is first calibrated with K^+-valinomycin and Gunther et al.[12] have measured the relative permeabilities of a number of cations and anions in such vesicles and have shown that they can quite precisely determine those values with biionic potential measurements. Interestingly, such a measurement is one of the very few that are independent of junction potential corrections, being only dependent on the calculation of K^+ equilibrium potentials. A similar technique has also been used for determining relative permeabilities in isolated whole epithelial cells that could easily be also used for vesicle measurements. This used a lipophilic cation (tetraphenylphosphonium), which can be tagged with ^{14}C and whose rate of influx could be calibrated to the membrane potential (again with K^+–valinomycin[13]). Although current–voltage relations cannot be measured with this technique, it is a valuable additional and independent approach to measuring diffusional permeabilities across apical membranes. It can also be combined with tracer flux studies to investigate both diffusive and coupled components of ion transport.[12]

A very different approach would be to use either noise analysis or single-channel recording with the patch-clamp technique to endeavor to focus in more specifically on the appropriate apical membrane channels. In both cases, this would require some means of modulating the number of channels that were opening. For example, Na^+ channels could be blocked with amiloride (or other blockers[71]), Ca^{2+}-activated K^+ channels[71a,71b] with Ba^{2+} or TEA (tetraethylammonium), and Cl^- ions by various other

[71] O. Christensen and N. Bindslev, J. Membr. Biol. **65**, 19 (1982).

[71a] N. Iwatsuki and O. H. Petersen, Biochim. Biophys. Acta **819**, 249 (1985).

[71b] C. Miller, R. Latorre, and I. Reisin, J. Gen. Physiol. **90**, 427 (1987).

blockers.[71c] Using the indirect noise analysis technique,[14] measurements of apical K^+ channels in frog skin have been made by Zeiske and Van Driessche[72] and of amiloride-sensitive Na^+ channels by Lewis et al.[73] These measurements require the use of intracellular microelectrodes to determine membrane potential and current across the apical membrane. Alternatively, the patch-clamp technique (e.g., Hunter et al.[74] and Guggino et al.[75]; see also Petersen,[17] in this volume) has been used to directly measure single-channel currents and conductances across apical membranes. Excised patches have the advantage that full control of the ionic composition of the two membrane bathing solutions can be achieved and solution composition on one side of a patch readily altered.[17a,17b] Practically, in order to obtain suitable preparations for patch clamping, it may be necessary to either use tissue-cultured or dissociated epithelial cells. For intact patch-clamp measurements it is vitally important to monitor membrane potentials with an intracellular electrode and perhaps also to check on intracellular composition with an ion-selective one. Single-channel currents and null potentials can then be measured and the permeation approach outline followed.

Another variant of the patch-clamp approach would be to use the whole-cell recording mode with a single-electrode clamp; this would have to be done on isolated cells. This technique would also enable an exchange of cell composition with whatever was in the patch pipet. However, such measurements would involve the composite properties of both apical and basolateral membranes in parallel.

Basolateral Membrane Pathway. This basolateral membrane is generally much more inaccessible than the apical one because of the presence of the serosal connective tissue. Nevertheless, many of the same techniques suggested for that latter membrane can also be applied to it. Conventional intracellular recording can be used with changes made to the salt composition of the serosal solution. In order to increase accessibility of the membrane to such solution changes and to minimize unstirred-layer effects, as much as possible of the serosal connective tissue should be dissected away or removed with an appropriate enzyme treatment. A suitable alternative would be to grow a layer of tissue-cultured or dissociated epithelial cells on

[71c] P. Wangemann, M. Wittner, A. Di Stefano, H. C. Englert, H. J. Lang, E. Schlatter, and R. Greger, *Pfluegers Arch.* **407** (Suppl. 2), S128 (1986).

[72] W. Zeiske and W. Van Driessche, *J. Membr. Biol.* **76,** 57 (1983).

[73] S. A. Lewis, M. S. Ifshin, D. D. F. Loo, and J. M. Diamond, *J. Membr. Biol.* **80,** 135 (1984).

[74] M. Hunter, A. G. Lopes, E. Boulpaep, and G. Giebisch, *Am. J. Physiol.* **251,** F725 (1986).

[75] S. E. Guggino, B. A. Suarez-isla, W. B. Guggino, and B. Sacktor, *Am. J. Physiol.* **249,** F448 (1985).

an appropriate grid that would allow ready diffusional access from the serosal side. For these membranes, amiloride and bumetanide (10^{-4} M) could be added to the mucosal solution, amiloride to increase apical resistance and bumetanide to stop NaCl and water transport into the cells after blocking the pump[69] with ouabain. Again, two internal microelectrodes (or single-electrode clamp) should be used for current–voltage measurements and ion-selective electrodes for checking intracellular concentrations.

An alternative approach, which would be suitable for measuring basolateral membrane properties, would be to use ionophores (e.g., nystatin or gramicidin D) in the mucosal solution to selectively lower the apical membrane resistance.[66,76] Nystatin generally allows exchange of both cations and anions between mucosal solution and cell interior and, unless impermeant anions such as sulfate are used, cell swelling occurs. Gramicidin D, which is cation selective, allows cation exchange between the two solutions without any cell swelling. Under such conditions, the transepithelial resistance of the membrane now becomes $R_b R_j/(R_b + R_j)$, due to only R_b and R_j in parallel. For a very tight epithelium ($R_j \gg R_b$), the total resistance would approximate that of the basolateral membrane alone. For a leaky epithelium the same conditions may be induced by blocking the tight junction pathway with a large organic cation such as TAP$^+$, which is known to block it in the gallbladder.[39] In both these cases, measurement of transepithelial potentials should be equivalent to measurements of potentials across the basolateral membrane by itself, and current–voltage relations and diffusion potential measurements could be done without microelectrodes. However, for accurate estimates of permeation parameters, such measurements should ideally be done in conjunction with one intracellular microelectrode to accurately monitor the basolateral membrane potential.

In principle, membrane vesicles could also be made from basolateral membrane fragments, and voltage-sensitive dyes could be used, as in the apical membrane vesicle measurements, to determine ion permeation properties across this membrane.

Noise analysis techniques would be essentially the same as in the apical membrane case with intracellular electrodes, except that accessibility of blocking agents is more restricted on the basolateral side. Patch-clamp measurements have already been made on dissociated basolateral cells (e.g., urinary bladder, by Lewis and Hanrahan[77]) and on the basolateral membrane of segments of proximal tubules by manually or enzymatically

[76] S. A. Lewis and N. K. Wills, *J. Membr. Biol.* **67**, 45 (1982).
[77] S. A. Lewis and J. W. Hanrahan, *Pfluegers Arch.* **405** (Suppl. 1), S83 (1985).

dissecting away the connective tissue (e.g., Hunter *et al.*[78]), or by the pipet being brought into contact with the lateral membrane of the noncannulated end of the tubule (e.g., Gogelein and Greger[79]), and similar procedures could be followed with other epithelia.

Acknowledgments

I would like to express my appreciation to Elaine Bonnet, Alex Kalaiziovski, and the Department of Medical Illustration, University of New South Wales, for their help in the preparation of the figures for this chapter, and to Professors J. M. Diamond and E. M. Wright and Drs. D. Cook and N. Quartararo for their critical reading of an earlier, longer version of the manuscript, and for helpful comments. The support of research grants from the National Health and Medical Research Council and the Australian Research Grants Scheme during the course of this work is also much appreciated.

[78] M. Hunter, K. Kawahara, and G. Giebisch, *Fed. Proc., Fed. Am. Soc. Exp. Biol.* **45**, 2723 (1986).
[79] H. Gogelein and R. Greger, *Pfleugers Arch.* **410**, 288 (1987).

[36] Use of Ionophores in Epithelia: Characterizing Membrane Properties

By Simon A. Lewis and Nancy K. Wills

Introduction

The ability of epithelia to perform directed and selective transport of solute and solvents is a consequence of its structure. However, it is the structure of the epithelium which has impeded the understanding of the mechanisms, at the individual membrane level, of such transport. At the microscopic level we find that an epithelium is made up of a sheet of cells joined at their apical–lateral membrane interfaces by a continuous bond of material called the tight junction. As a consequence, there are two parallel pathways for solute and solvent movement across the epithelium. One is between the cells, i.e., through the tight junction. In the absence of electrical and chemical gradients this route cannot perform net movement of solutes and solvents, but it can restrict the movement of compounds according to parameters such as size and charge. The other route occurs through the cells. Thus, mechanisms must be available for selected compounds to traverse the apical membrane into the cell from the lumen and, once in the cell, they must be moved across the basolateral membrane into the interstitial fluid, one or both steps requiring direct or indirect expenditure of metabolic energy. It is this series arrangement of cell membranes in parallel with the tight junction (paracellular) pathway that makes a rigorous study of the properties of these structures more complex.

Although one might initially consider the tight junctions impermeable to solutes and solvents, this is not the case. Frömter and Diamond[1] showed that the large range in electrical resistance among epithelia (from 5 Ω cm^2 to > 100,000 Ω cm^2) was due to the large range in resistance of the tight junctions and not the parallel cellular pathway. (The tight junctional pathway has a resistance range of 5 to > 200,000 Ω cm^2.)

The challenge for epithelial transport physiologists in recent years has been to unravel the mechanisms involved in the regulation of apical and basolateral membrane transport processes: specifically, which processes allow the cellular milieu to maintain relatively constant ion activities in the face of even large transcellular movement of water and ions. To this end, epithelial physiologists have utilized two strategies. The first is to remove the epithelium from the animal and mount it between Lucite chambers. Such an *in vitro* system eliminates uncertainties caused by circulating hormones and neurotransmitters and allows a number of advantages such as rapid modification and control of the composition of both luminal and blood side solutions; a well-defined and measured surface area; easy estimates of epithelial resistances; the use of conventional and ion-specific microelectrodes to calculate membrane and junctional resistances, ionic selectivities, and cellular ionic activities; and the determination using radioactive isotopes of net solute and solvent movements. This first strategy is the major focus of this paper. The second strategy that has been developed is to fracture the epithelia cells under conditions where the apical membrane and basolateral membrane reform into spherical vesicles of submicrometer diameter. Such a single membrane system offers a number of advantages over the whole tissue approach, the most obvious being a quantification of the individual membrane properties uncontaminated by the transport properties of the series membrane and/or tight junction.

Both of the above strategies have distinct advantages and limitations. The remainder of this chapter is devoted to a description of the possible use of ionophores to circumvent some of the limitations (listed in Table I) offered by either experimental approach.

---

[1] E. Frömter and J. A. Diamond, *Nature (London)* **235**, 9 (1972).

TABLE I
LIMITATIONS TO MEMBRANE TRANSPORT PROCESSES

System	Limitations
Intact epithelium	Difficulty in resolving the resistances of the apical and basolateral membranes and tight junctions
	Inability to selectively voltage clamp a single membrane
	Inability to regulate quantitatively or to alter intracellular ion activities
	Inability to measure isotopic fluxes across a single membrane (apical or basolateral) or to evaluate neutral exchange processes or active transport process
Vesicles	Inability to monitor, change, or clamp the vesicle membrane potential
	Inability to maintain or collapse ionic gradients

Some Commonly Used Ionophores

As the name implies, an ionophore facilitates ion movements across lipid bilayers and in this sense can be considered as a transport-inducing agent. Before describing how ionophores can be used to overcome various limitations in experimental systems, we first list some commonly used ionophores and their transport properties.

Pore Formers

Amphotericin B. This polyene antibiotic, molecular weight 924.11, is isolated from *Streptomyces nodosus,* culture M 4575. It is soluble in methanol, ethanol, and dimethyl sulfoxide (DMSO), and requires at least four molecules to make a pore in the membrane (barrel staves). The membrane must contain sterols. When added to only one side of the membrane it is weakly cation selective (about three times K^+ over Cl^-). Amphotericin B allows the movement of small monovalent cations and to a certain extent divalent cations, e.g., Ca^{2+} and Ba^{2+} and small monovalent anions. It excludes molecules with a Stokes–Einstein radius greater than 7 Å.

Nystatin. This polyene antibiotic in commercially available form is a mixture of types A_1 and A_2. It is isolated from *Streptomyces noursei* and is soluble in many organic solvents. Its mechanism of pore formation is similar to that of amphotericin B, as is its ion selectivity. It is applied in biological amounts of units/milliliter, not concentration (biological activity/milligram is batch dependent; see U.S. Pharmacopoeia for determination of biological activity).

Filipin. This polyene antibiotic in commercially available form is a mixture containing at least eight pentaene compounds. It is isolated from

Streptomyces filipenensis. Its action as a pore former is similar to that of nystatin and amphotericin, i.e., barrel stave configuration with an absolute requirement for membrane sterols. In addition, it forms micelles within the lipid bilayers, which can be observed with a scanning electron microscope or using freeze-fracture techniques, and, as such, can be used to determine qualitatively the presence of sterols in membranes.

Gramicidin(s). The commercially available preparation is composed of 87.5% gramicidin A, 7.1% gramicidin B, 5.1% gramicidin C, and 0.3% gramicidin D; all are isolated from *Bacillus brevis.* Gramicidins are a pentedecapolypeptide of helical structure with an internal diameter of 3 to 4 Å. Two gramicidin monomers are required to form a transmembrane channel. This channel is highly selective for small monovalent cations over anions whether in a black lipid membrane or biological membrane. Gramicidins are soluble in the lower alcohols, but are practically insoluble in water.

Carriers

Valinomycin. One of the best studied of the carrier systems valinomycin has a molecular weight of 1111.36. It is a cyclododecadepsipeptide produced by *Streptomyces fulvissimus* and is highly selective for K^+. Although some ions, e.g., Rb^+ and Cs^+, can be readily carried by this ionophore, others such as Na^+ are some 10,000 times less favored. As with most other ionophores the absolute conductance offered by a fixed concentration of ionophore and salt is very dependent on the lipid composition of the membrane, which then reflects both the surface charge properties as well as the membrane fluidity.

FCCP. Carbonyl cyanide *p*-trifluoromethoxyphenylhydrazone (FCCP) is one of the proton ionophores (MW 256.00). Its mode of action involves a dimer form, $HA \cdot A^-$. The major action of this and other proton ionophores is the dissipation of proton gradients.

Neutral Exchangers

Monensin. This compound has a molecular weight of 670.90 and is isolated from *Streptomyces cinnamonensis.* It is a neutral ionophore that exchanges Na^+ for H^+, but it can also exchange K^+ for H^+. It is soluble in hydrocarbons and very soluble in organic solvents; however, it is only slightly soluble in water.

Nigericin. This compound has a molecular weight of 724.98 and is isolated from *Streptomyces hygroscopicus* E-749. It is a neutral exchanger that exchanges K^+ for H^+, but, as for the case of monensin, it can also exchange Na^+ for H^+ (but to a lesser extent). It is soluble in the same solvents as monensin.

A23187. Known as calimycin or calcium ionophore (MW 523.6), this compound is isolated from *Streptomyces chartreusis.* Two molecules bind to Ca^{2+} to transfer it across the membrane. H^+ ions are carried by the ionophore on its return across the membrane. It is soluble in lower alcohols and only slightly soluble in H_2O. Divalent selectivity is $Mn^{2+} \gg Ca^{2+} \sim Mg^{2+} > Sr^{2+} > Ba^{2+}$.

Although other Ca^{2+} ionophores have been used to a limited extent (e.g., antibiotic X-537A or lasalocid A), it is the consensus in the literature that A23187 is favored because it causes less perturbations of the system under study.

In the remainder of this chapter we will demonstrate how one or more of the above ionophores can be used to overcome the limitations of membrane transport process (see Table I); in addition we will place strong emphasis on the assumptions used and possible pitfalls of data interpretation.

Determination of Membrane Resistances: Some Methods

As mentioned previously, an important parameter which should be determined for almost all epithelia studied is the electrical resistance of the apical and basolateral membranes and the parallel tight junction (Fig. 1). There are a number of methods available for separating these three resistors into their individual components. The initial method used was cable analysis (see Frömter[2]). In this procedure one injects current from a microelectrode into a cell and measures the decrement of this current (with a second microelectrode) as a function of the distance from the current source. This yields the length constant of the epithelium, which is then a function of the radial resistance (cytoplasm plus gap junction) within the plane of the epithelium, as well as the resistance of the apical and basolateral membranes. In addition to the length constant, one also needs to measure the transepithelial resistance as well as the ratio of apical-to-basolateral membrane resistance. This procedure is both difficult and time consuming, but is valid for most planar epithelia and, with modification, to kidney tubules.

The second method relies on the perturbation of the resistance of only one of the cell membranes, without altering the tight junctional resistance or the opposing membrane resistance. The applicability of this method then rests on being able to selectively change a single membrane resistance such that it is large enough to cause a significant change in the total

[2] E. Frömter, *J. Membr. Biol.* **8**, 259 (1972).

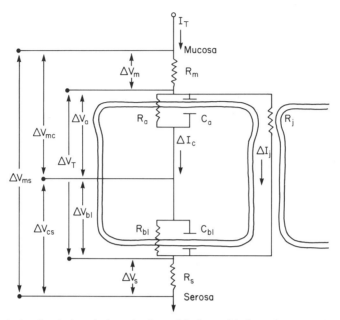

Fig. 1. An electrical equivalent-circuit model of an epithelium plus mucosal and serosal solution resistances. ΔV_X is the voltage response to a square current pulse (ΔI_T or ΔI_c), occurring either within a single compartment or between two compartments as denoted by the subscripts m (mucosa), c (cell), and s (serosa), or across the apical (a) and basolateral (bl) membranes, or the transepithelial response (T).

Connecting the macroscopic voltage-sensing electrodes and intracellular microelectrode to the appropriate electrometers, we measure ΔV_{ms}, ΔV_{cs}, and V_{mc}. Consequently, to calculate the voltage deflection across the epithelium and individual cell membranes, we must correct the measured voltages for the voltage deflection occurring in the mucosal and serosal solutions. To correct transepithelial response at time $\to \infty$ (steady state), $\Delta V_{ms}^{\infty} = \Delta I_T R_m + \Delta I_T R_T + \Delta I_T R_s$, where R_x is the resistance of the mucosal (m) and serosal (s) solutions and epithelium (T), respectively. At time $= 0$ plus (instantaneous voltage response to ΔI_T), $\Delta V_{ms}^0 = \Delta I_T(R_m + R_s)$. The impedance of the epithelium (i.e., frequency-dependent resistance) is zero after the current pulse starts because the apical and basolateral membrane capacitors (C_a and C_{bl}, respectively) act as short-circuits in parallel with the appropriate membrane resistor. Thus $\Delta V_T = \Delta V_{ms}^{\infty} - \Delta V_{ms}^0 = \Delta I_T R_T$.

To correct apical and basolateral voltage responses for solution resistance, place a microelectrode as close to the apical membrane as possible and measure the voltage response (ΔV_m) between the mucosal voltage electrode and microelectrode to a square current pulse. Mucosal resistance $\Delta V_m = \Delta I_T R_m$ and serosal series resistance $\Delta V_s = \Delta V_{ms}^0 - \Delta V_m = \Delta I_T R_s$. Thus $\Delta V_{bl} = \Delta V_{cs} - \Delta V_s = \Delta I_c R_{bl}$ and $\Delta V_a = \Delta V_{ms}^{\infty} - \Delta V_{cs} - \Delta V_m = \Delta I_c R_a$.

transepithelial resistance, as well as a significant change in the ratio of apical-to-basolateral membrane resistance. For certain amiloride-sensitive Na^+-transporting epithelia, the above criterion seems to be met. Thus amiloride specifically increases the apical membrane resistance by blocking, in a reversible manner, conductive Na^+ channels.

What if the epithelium of interest does not possess a readily blockable channel? Must one resort to cable analysis? With a few reservations the answer is *no*. If the epithelium does not provide a conductive pathway to block, or open up, we can use pore-forming ionophores (most of which are antibiotics) to provide the epithelium with a conductance or channel.

From Fig. 1, we generate the basic equations needed to calculate membrane resistance. The equations assumed that the ionophore perturbs only one of the membrane resistances and not the junction or the opposing membrane resistances.

As noted by Wills *et al.*,[3] since the macroscopic voltage-measuring electrodes are not immediately up against the apical and basolateral membranes, one must correct transepithelial, apical, and basolateral voltage response for the voltage drop that occurs in the solution on either side of the epithelium (for a detailed description of one method for measurement and correction of series resistance artifacts for both transepithelial and microelectrode measurements, see pp. 83–84 in Wills *et al.*[3]) A brief description of the method is supplied in Fig. 1.

Before addition of the ionophore, the total epithelial conductance (G_T) is calculated by measuring the voltage response across the epithelium (ΔV_T) to a square current pulse (ΔI_T) as

$$\Delta I_T/\Delta V_T = G_T = G_j + G_a G_{bl}/(G_a + G_{bl}) \tag{1}$$

and the resistance ratio (or conductance ratio) is given by

$$\alpha = R_a/R_{bl} = G_{bl}/G_a \tag{2}$$

and is calculated from the following measurement. After placing a microelectrode into the cell, a square current pulse is passed across the epithelium. A certain portion of the current (at long times, i.e., after membrane capacitors are fully charged) goes through the tight junction (ΔI_j) while the remainder (ΔI_c) goes through the apical membrane and then the basolateral membrane. This current flow through the individual membrane resistances results in a voltage deflection across both membranes (ΔV_a for the apical membrane and ΔV_{bl} for the basolateral membrane). The ratio of the apical-to-basolateral membrane voltage deflection is then equal to the ratio

[3] N. K. Wills, S. A. Lewis, and D. C. Eaton, *J. Membr. Biol.* **45**, 81 (1979).

of the apical resistance multiplied by the current flowing through this membrane, to the basolateral resistance multiplied by the current flowing through this membrane. Since the current which flows through the apical membrane is equal to that which flows through the basolateral membrane, the ratio of voltage deflections equals the ratio of resistances. Thus

$$\frac{\Delta V_a}{\Delta V_{bl}} = \frac{R_a \, \Delta I_c}{R_{bl} \, \Delta I_c} = \frac{R_a}{R_{bl}} \tag{3}$$

So far, we have two equations but three unknowns (R_a, R_{bl}, and R_j). Thus we cannot uniquely solve for these unknowns. However, if only one of the membrane resistors is changed to a new value (the apical membrane, R_a'), there are four equations with four unknowns, i.e., R_a, R_a', R_{bl}, and R_j. The equations are then

$$\alpha = \frac{R_a}{R_{bl}}, \qquad \alpha' = \frac{R_a'}{R_{bl}}, \qquad G_T = G_j + \frac{G_a G_{bl}}{G_a + G_{bl}}, \qquad G_T' = G_j + \frac{G_a' G_{bl}}{G_a' + G_{bl}}$$

Performing the appropriate substitutions into the above equations we achieve two equations with two unknowns.

$$G_T = G_j + G_{bl}/(1 + \alpha) \tag{4}$$

$$G_T' = G_j + G_{bl}/(1 + \alpha') \tag{5}$$

Thus we solve for G_j and G_{bl}. This then allows a calculation of G_a and G_a' using G_{bl}' and the appropriate α or α'. A more ideal system is to measure a series of G_T and α values, during perturbation of the apical membrane resistance. If only R_a changes then a plot of G_t versus $(1 + \alpha)^{-1}$ will be linear with a slope of G_{bl} and an intercept equal to G_j. If one perturbs only the basolateral membrane one can derive a comparable equation, which is

$$G_T = G_j + G_a \alpha/(\alpha + 1) \tag{6}$$

As above, a plot of G_t versus $\alpha/(1 + \alpha)$ will have a slope equal to G_a and an intercept equal to G_j.

It is important to note that deviation of data points from linearity suggests that one has violated the assumption of a constant R_j and/or a constant opposing series membrane resistance. Unfortunately, one cannot tell from the direction of the deviation whether it is a membrane or junctional resistance that has changed, nor the direction of resistance change.

An alternate approach to the determination of membrane resistances involves a pertubation of either a membrane voltage or current source. As in the above method one must alter a single membrane electromotive force (EMF or current source), while not altering the EMF or current source of the opposing membrane or tight junction. The following parameters must be measured; R_T, α, V_t (transepithelial potential), and V_{bl} (basolateral membrane potential) before and after changing either the apical or basolateral membrane EMF.

Altering the apical membrane resistance or EMF, one can demonstrate (see Frömter and Gebler[4]) that the ratio of the change in basolateral membrane potential (ΔV_{bl}) to transepithelial potential (ΔV_t) is equal to the ratio of basolateral resistance to junctional resistance.

$$\Delta V_{bl}/\Delta V_T = R_{bl}/R_j \equiv B \tag{7}$$

Since we also measure α (R_a/R_{bl}) and R_t before and after (α' and R_t') the apical change, we can calculate the individual resistances as

$$R_a = \alpha(1 + \alpha + B)/(1 + \alpha)B \tag{8}$$

$$R_b = R_a/\alpha \tag{9}$$

$$R_j = (1 + \alpha + B)R_T/(1 + \alpha) \tag{10}$$

Similarly for a basolateral perturbation one finds[5]

$$\Delta V_{bl}/\Delta V_T = 1 + R_a/R_j = 1 + \gamma \tag{11}$$

Thus we calculate

$$R_j = R_T \left[1 + \frac{\alpha}{\gamma(1 + \alpha)} \right] \tag{12}$$

$$R_a = \gamma R_j \tag{13}$$

$$R_b = R_a/\alpha \tag{14}$$

Again, one can test the assumption of a constant junction and opposing series membrane by plotting ΔV_{bl} versus ΔV_t; nonlinearity suggests that the assumptions have been violated.

The last method for determining membrane resistances involves the use of impedance analysis, which is discussed in detail in this volume [33].

All of the above methods (with the exception of cable analysis) work well on epithelia where the cellular resistance and parallel junctional resistances are within an order of magnitude of each other; if, however, the junctional resistance is an order of magnitude lower than the cell resistance, significant errors can occur when calculating apical and basolateral resistances.

[4] E. Frömter and B. Gebler, *Pfluegers Arch.* **371,** 99 (1977).
[5] S. A. Lewis and N. K. Wills, *J. Membr. Biol.* **67,** 45 (1982).

Ideally one should use both apical and basolateral perturbations to test the validity of the assumptions.

Use of Ionophores in Determining Membrane Resistance

From the basic principles described above it is easy to visualize how an ionophore (either a pore former or charge carrier) can be used to decrease the resistance of either apical or basolateral membranes. Thus one simply measures the change in V_t, R_t, and V_{bl} as a function of time after ionophore addition, applies the appropriate equations (which are picked determined by the membrane treated), and then calculates the individual membrane and junctional resistances, assuming that the only membrane resistance altered is the one to which the ionophore is added.

Certain precautions must be invoked to minimize the possibility of violation of the above assumptions. These precautions are dependent to a certain extent on the ionophore used. Listed below are some general rules that one might follow to achieve greatest accuracy in the determination of membrane resistances using ionophores.

1. Add the ionophore to the solution which bathes the higher resistance membrane, i.e., if $\alpha > 1$, add the ionophore to the apical bathing solution; if $\alpha < 1$, add the ionophore to the basolateral bathing solution.

2. Know possible solvent effects. This can be minimized by making the ionophore stock at the highest ionophore concentration possible; this reduces the increased osmolarity of the bathing solution when the ionophore stock is added.

3. Know the lipid requirements, permeability properties, and voltage sensitivities of the ionophores to be used.

4. The concentrations of the ions which pass through the ionophore must equal the cellular ion concentrations. This then reduces ionic and osmotic (i.e., volume) shifts.

5. Use the lowest concentration of ionophore which will yield an adequate transepithelial conductance change, this being dependent on the ionophore as well as the epithelium to be studied.

6. Determine whether you need a reversible or irreversible ionophore.

7. If possible, measure the permeability properties of the ionophore when it is in the membrane, and adjust the bathing solution composition as required.

8. Double check resistance values, e.g., use an ionophore on the apical membrane, then after full ionophore action, perturb the basolateral membrane resistance.

9. Ionophore-induced resistance decreases and voltage increases should be monophasic and reach steady-state plateau values. Secondary changes

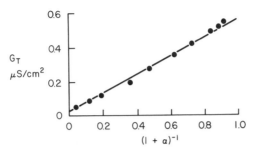

FIG. 2. An example of the relationship between transepithelial conductance (G_T) and $(1 + \alpha)^{-1}$ during nystatin action on the apical membrane. The intercept at zero $(1 + \alpha)^{-1}$ is equal to the tight junction conductance and the slope is equal to the basolateral membrane conductance. Linearity of the plot indicates that both junctional and basolateral membrane conductances are not significantly effected by nystatin. (Reproduced from *The Journal of General Physiology*,[6] 1977, Vol. **70**, p. 427 by copyright permission of The Rockefeller University Press.)

in resistance or voltage might suggest changes in one or more of the epithelial resistors, produced by a primary or secondary effect of the ionophore.

The most popular ionophores which have been used for epithelia are the polyene antibiotics, nystatin and amphotericin B. Typically they are applied (1.3 μl/ml) to the apical membrane bathing solution (as in most epithelia, $R_a > R_{bl}$) using a stock of 5 mg nystatin/ml of ethanol (a saturated solution). This is comparable to a solution of 24 mM ethanol, i.e., the bathing solution is hyperosmotic by 24 mOsm. In the mammalian urinary bladder, the dose–response curve (i.e., apical conductance versus nystatin activity) is steep (a decade change in nystatin causing a $> 10^4$ change in conductance), suggesting four to five individual polyene molecules form (in concert with membrane sterols) a transmembrane channel. Since nystatin (or amphotericin B) does not work instantaneously, but rather over a reasonable time domain (2 to 3 min), one can record a series of paired values for α and G_T, and from the shape of the line [plotting G_T versus $(\alpha + 1)^{-1}$] determine whether the rules listed above and assumptions implicit in such a plot have been violated. Figure 2 is an example of the effect of nystatin on the mammalian urinary bladder.[6] Note that the effect on G_T and α is complete within 120 sec. Extensive washing of the mucosal solution with a nystatin-free Ringer's solution results in a slow wash out of the nystatin from the apical membrane. The extent of the reversibility of nystatin is dependent on the length of time nystatin is left in the mucosal solution. The shorter the time of exposure, the more complete and rapid is the reversal. When nystatin is added to a mucosal solution containing high

[6] S. A. Lewis, D. C. Eaton, C. Clausen, and J. M. Diamond, *J. Gen. Physiol.* **70**, 427 (1977).

NaCl, there is a rapid increase in G_T and decrease in α, which resulted in a linear relationship between G_T and $(1 + \alpha)^{-1}$ (at least in the first 120 sec). However, after approximately 5 min, G_T continued to increase, as did α. This secondary effect was a consequence of an increase in basolateral membrane conductance induced by a continued influx of Na^+ and Cl^-, the end result being cell swelling and ultimately cell lysis. Such secondary effects can be overcome by designing the mucosal solution such that cell K^+ and mucosal K^+ are near equivalent and a nonpermeant anion replaces mucosal Cl^- (e.g., sulfate, methyl sulfate, gluconate). In this type of solution, secondary shifts in G_T and α are dramatically reduced and the probability of cell lysis is almost completely eliminated. Other refinements in the mucosal solution include a reduction in Ca^{2+} concentration (from a millimolar to a micromolar range), slight acidification of the solution (pH $\sim$ 7.0 from 7.4), addition of a small amount of Na^+ ($\sim$ 7 mM) and Cl^- (12 mM), and a reduction in HCO_3^- to 10 mM from 25 mM.

The time course for the serosal addition of nystatin is significantly longer ($\sim$ 30–60 min) than the apical effect, perhaps reflecting the sequestering of nystatin by membranes in series with the basolateral membrane of the cells of interest. Such membranous sinks might be capillary beds, muscle layers, connective tissue, and underlying epithelial cells of multicell layered epithelia. To date the addition of nystatin has been predominantly to the apical membrane of epithelial cells.

A second ionophore used has been gramicidin D. The advantages of this ionophore over nystatin are (1) its nonpermeability to Cl^- and (2) its lack of ready reversibility. This then simplifies the design of the solution to be used. Determination of the membrane and junctional resistances is the same as that described for nystatin or amphotericin B. Further characteristics of this antibiotic have been described by Lewis and Wills.[5]

Parameter Reliability

So far we have been concerned with possible secondary effects of mucosal antibiotic addition, i.e., whether the junctions and basolateral membrane are also influenced as a function of time after antibiotic challenge. In this section we will address the question of whether the resistances for the tight junction and basolateral measured using antibiotics truly reflect those for the intact epithelium (i.e., before changing the mucosal solution and adding the antibiotic). For the estimate of tight junction resistance (assessed using antibiotics) to equal the true junctional resistance, the following assumptions must be true.

1. Replacing the mucosal NaCl–Ringer's solution with the K^+ plus impermeant anion cannot alter junctional resistance. Thus the junctional

permeability for NaCl must be equal to the permeability of the replacement ions.

2. In a homogeneous cell population, the resistance ratio measured must represent the average ratio of all cells.

3. In a heterogeneous cell population, the resistance ratio measured must represent the average ratio of all cells, but more important, during antibiotic action the apical membrane of all cell populations must change in a similar time-dependent manner.

For the estimate of the basolateral membrane resistance (assessed using antibiotics) to equal the control basolateral membrane resistance, the following assumptions must be true. (1) The basolateral membrane voltage should remain nearly constant or hyperpolarize when the mucosal solution is changed from NaCl–Ringer's solution to K^+ plus impermeant anion, and (2) assumptions 2 and 3 for tight junctional resistance must be met.

Pseudo-Single Membrane Preparations

Given that the pore-forming antibiotics can increase the apical membrane conductance of the intact epithelia, it stands to reason that a saturating dose of the antibiotic might reduce the apical membrane resistance to a value not much different than zero. As such, this has been referred to as a single membrane preparation where the tight junctions and basolateral membrane are in parallel. This particular configuration can then be used to assess properties of the basolateral membrane, which are not possible in the presence of a high-resistance apical membrane. Using electrical techniques one can measure the conductive ionic selective permeability, as well as electrogenic active transport processes of this membrane. Isotopic techniques can be used to measure both electrogenic and electroneutral transport processes. Since the apical membrane is permeable to selected ions (depending on the ionophore used), one can study the effect of varying the cell ion composition on basolateral membrane conductive permeability, as well as electrogenic and electroneutral processes. In the section below we shall describe this experimental approach, as well as pitfalls involved in determining both ionic permeabilities and electrogenic properties of the basolateral membrane of epithelia.

Ionic Selectivity

For any single membrane preparation, there are two methods one can use to electrically measure the (conductive) selective permeability of the membrane. Both rely on the assumption that the permeabilities to be measured conform to the constant field equation. The first method relies on measuring the membrane potentials during partial and complete replacement of Na^+ for K^+, in, first, the presence of an impermeant anion, and, again, in the presence of the normal bathing solution anion, Cl^-.

First we take the voltages measured in the presence of an impermeant anion and calculate the Na^+-to-K^+ permeability ratio (P_{Cl}/P_K). This assumes of course that Na^+ and K^+ are the only permeant ions. Next we take the voltages measured while replacing Na^+ with K^+ in the presence of Cl^-, and using the previous value for Na^+-to-K^+ permeability, calculate the Cl^--to-K^+ permeability ratio (P_{Cl}/P_K), using the constant field equation shown below:

$$V_T = \frac{RT}{F} \ln\left[\frac{K_i + Na_i(P_{Na}/P_K) + Cl_o(P_{Cl}/P_K)}{K_o + Na_o(P_{Na}/P_K) + Cl_i(P_{Cl}/P_K)}\right] \tag{15}$$

where V_T is transepithelial voltage referenced to the serosal side, and subscripts i and o refer to inside (serosa) and outside (mucosa), respectively. Since the apical membrane has been permeabilized, the outside also refers to the cell cytoplasm. The permeability ratios as calculated from the above equation represent the combined permeability of the basolateral membrane plus parallel tight junctions. Depending on the selective permeability of the tight junctions, the basolateral selective permeability will be either overestimated or underestimated.

To determine the selective permeability of only the basolateral membrane, we must have a prior measurement of the junctional selective permeability. Figure 3 illustrates the electrical equivalent circuit for the single-membrane preparation in terms of resistors and batteries. The equation, which describes the transepithelial potential for this circuit, is

$$V_T = \left(\frac{E_{bl}}{R_{bl}} + \frac{E_j}{R_j}\right)\frac{R_{bl}R_j}{R_{bl} + R_j} \tag{16}$$

To calculate the basolateral membrane selective permeability we must be able to determine E_{bl} at each change in Na^+/K^+ concentration. A simple rearrangement of Eq. (16) demonstrates the need for additional independently measured circuit parameters:

$$E_{bl} = \frac{V_T(R_{bl} + R_j) - E_j R_{bl}}{R_j} \tag{17}$$

As is obvious from this equation, both E_j (junctional battery) and R_j (junctional resistance) must be accurately measured at each Na^+/K^+ concentration used. One approach is to force the apical membrane resistance to extremely high values (pharmacologically) before addition of the antibiotic and simple measure V_T and R_T (which will be equal to E_j and R_j, respectively) at each ion replacement. These values can then be used in the

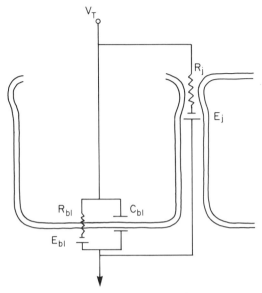

FIG. 3. An equivalent-circuit model of an epithelium after the apical membrane has been permeabilized with an ionophore. Since the mucosal solution and cell cytoplasm have (near) identical ionic composition for the permeable ions, the apical membrane possesses no electromotive force (EMF) nor significant resistance. Of importance is that the spontaneous potential (V_T) is a function of the basolateral membrane EMF (E_{bl}) and resistance (R_{bl}) and the junctional EMF (E_j) and resistance (R_j). Solution resistance (not shown) has no effect on V_T because there is no net transepithelial current. C_{bl}, Basolateral membrane capacitance.

above equation to calculate E_{bl}. The selective permeability can then be calculated for both the basolateral membrane and tight junction using the constant field equation. In many instances $R_j \gg R_{bl}$ and the correction becomes negligible.

The second approach is to perform a current–voltage relationship (I–V) on the single-membrane preparation. One applies a series of voltage-clamp steps across the epithelium and measures the corresponding transepithelial current flow. This I–V relationship is then fit (using a computer) by the constant field current equation (shown below) to yield estimates of P_{Na}/P_K and P_{Cl}/P_K.

$$I = (F^2/RT)VP_K\{[K_o + Na_o(P_{Na}/P_K) + Cl_i(P_{Na}/P_K)]$$
$$- [K_i + Na_i(P_{Na}/P_K) + Cl_o(P_{Cl}/P_{Na})]e^{-VF/RT}\}/(1 - e^{-VF/RT}) \quad (18)$$

Before performing such a curve-fitting routine, the I–V relationship has to be corrected for solution resistance and the parallel junctional I–V relationship. Figure 4 is an equivalent circuit similar to Fig. 3, but includes

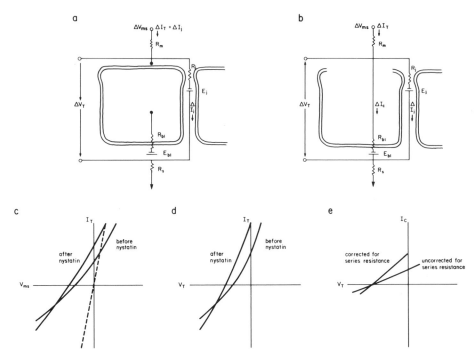

FIG. 4. Stepwise procedure for determining the current–voltage (I–V) relationship of the tight junctions and basolateral membrane of an epithelium: (a) This equivalent circuit describes the necessary conditions to measure the I–V of the tight junctions. As is obvious, the apical membrane resistance must be infinitely large compared to the junctional resistance. As such the total current flow (I_T) produced by a voltage-clamp step, ΔV_{ms}, is restricted to moving through the junctions, i.e., $\Delta I_T = \Delta I_j$. (b) This circuit describes the conditions after mucosal antibiotic administration; thus the apical membrane resistance is close to zero when compared to the basolateral membrane and tight junction. The total current flow to a voltage step (ΔV_{ms}) is the sum of the current through the two parallel paths, a cellular (ΔI_c) and junctional (ΔI_j). (c) I–V relationship uncorrected for the solution I–V series resistance, which is shown, for simplicity, as a linear dashed line. Curves labeled before and after nystatin refer to the circuits A and B, respectively. (d) The same as in a except after correcting for the solution series resistance. At each I_T the series solution voltage drop [$= I_T(R_s + R_m)$] was subtracted from the corresponding V_{ms} to yield V_T, i.e., the voltage drop across just the epithelium. Labels are as in c. (e) These curves represent the basolateral membrane I–V relationship; at each V_T, the difference in the currents after and before nystatin equals the current flowing across the basolateral membrane (I_c). As shown, if the solution series resistance is not corrected for, the shape of the I–V can be dramatically different when compared to one adequately corrected. Thus inadequate correction will yield incorrect values for the permeability properties of the basolateral membrane and tight junctions (compare shapes of curves before nystatin in c and d).

the bathing solution resistance present between the voltage-measuring electrodes and apical and basolateral membrane surfaces.

Since the experiments are performed under voltage-clamp conditions, we must take the current at each voltage, multiply it by the solution resistance, and subtract this voltage from the clamp voltage. This difference in voltage then represents the voltage drop across the single-membrane preparation (parallel combination of basolateral membrane and tight junction). As in the previous method, the values obtained for the selective permeabilities is the combination of the two parallel pathways.

In order to determine the $I-V$ relationship of the basolateral membrane we must first correct for the $I-V$ of the tight junctional complex. As in the previous section this can be performed (before nystatin addition) by forcing the apical membrane resistance (pharmacologically) to infinitely high values, and then measuring the current-voltage relationship of the epithelium. Since the cellular pathway has a nearly infinite resistance, the $I-V$ relationship then represents the junctional pathway (after correcting for the solution resistance). The $I-V$ relationship for the basolateral membrane is then the difference between the current that flows across the nystatin-treated preparation and the current that flows solely through the tight junctions, for matched (i.e., equal) values of voltage. This difference $I-V$ can then be fitted by the above equation using a computer-based curve-fitting routine. An example of the above procedure is demonstrated in Wills et al.[7]

Electrogenic Transport Systems

One of the ubiquitous features of the basolateral membrane of epithelia is the Na^+,K^+-ATPase, which is responsible for actively transporting Na^+ out of the cell (against a large net electrochemical gradient) into the plasma, in exchange for K^+, which is moved from plasma into the cell against a smaller net electrochemical gradient. Because of "cross-talk" between apical and basolateral membrane transport processes, a study of this pump is difficult in the intact epithelium. However, by artificially increasing apical membrane permeability, we can now study, both electrically and using radioisotopes, the kinetic properties of electrogenic transport systems as well as the influence of cell ionic composition on the kinetic properties of this system.

The experimental approach is similar to that used to determine basolateral membrane selectivity, or membrane resistances. First, the mucosal solution (normally, NaCl) is replaced with K^+ plus an impermeant anion

[7] N. K. Wills, D. C. Eaton, S. A. Lewis, and M. S. Ifshin, *Biochim. Biophys. Acta* **555**, 519 (1979).

(e.g., SO_4^{2-} or gluconate); the serosal solution is NaCl–Ringer's. Next, nystatin or amphotericin B is added to the mucosal solution, while monitoring transepithelial as well as microelectrode parameters.

Studies of Na^+ activation of the pump using isotopic measurements have been performed on rabbit urinary bladder[8] and turtle colon.[9] The basic protocol was to add millimolar concentrations of Na^+ to the mucosal chamber, and to measure the bidirectional Na^+ fluxes using radioisotopes. The difference in the flux then represents active Na^+ transport. Ouabain was used to demonstrate that this net flux was only due to the Na^+ pump. These data were then fitted by kinetic models, which have been previously used to describe Na^+ transport via this system, and yield values for the maximum Na^+ flow through the pump, the number of Na^+ ions that bind to activate the pump, as well as a binding rate constant. K^+ transport was studied in a similar manner for the turtle urinary bladder. The general conclusion was that the pump was electrogenic, transporting three Na^+ out for two K^+ in.

Lewis and Wills[10] took advantage of the electrogenic nature of the pump, i.e., the net current flow produced by the pump, and measured pump current as a function of cell Na^+ concentration or serosal K^+ concentration. Because the measurements are electrical in nature, i.e., measuring changes in membrane potential due to the current-generating pump, one must take precautions against membrane potential changes caused by depletion or buildup of ions (particularly K^+) in the intracellular compartment and lateral intercellular spaces. In addition, since the rate of pump turnover is also dependent on lateral intercellular space (LIS) K^+ concentration, LIS K^+ concentrations must remain near constant. To minimize changes in LIS K^+ concentration, Lewis and Wills[10] replaced the serosal NaCl–Ringer's solution with a K^+–Ringer's solution of identical composition to the mucosal solution (after mucosal nystatin addition). Next they added aliquots of NaCl, first to the serosal and then to the mucosal solutions. The hyperpolarization (V_T) produced by mucosal NaCl addition was then divided by the transepithelial resistance to calculate the pump current. The current can then be plotted against the mucosal Na^+ concentration to generate the relationship between Na^+ concentration and pump current. Such a plot is valid if the mucosal Na^+ concentration is equal to the cell Na^+ concentration (this is also one of the assumptions of the tracer experiments described above). Both experimental results and computer modeling demonstrated that this was not true.[10,11] Cell Na^+

[8] D. C. Eaton, A. M. Frace, and S. U. Silverthorn, *J. Membr. Biol.* **67**, 219 (1982).
[9] K. L. Kirk, D. R. Halm, and D. C. Dawson, *Nature (London)* **287**, 237 (1980).
[10] S. A. Lewis and N. K. Wills, *J. Physiol. (London)* **341**, 169 (1983).
[11] R. Latta, C. Clausen, and L. C. Moore, *J. Membr. Biol.* **82**, 67 (1984).

concentration was significantly lower than mucosal concentration; the discrepancy was most prominent at high (60 mM) mucosal Na$^+$ concentrations, where cell concentration was determined to be 60% lower than the mucosal concentration. Figure 5a demonstrates the relationship between mucosal and cell Na$^+$. Because the relationship is nonlinear, but one assumes that mucosal Na$^+$ equals cell Na$^+$, then both the rate constants and maximum current generated will be in error, as might be the case for the model used to describe the relationship between Na$^+$ concentration and pump current. If Na$^+$ is added to the mucosal solution with an anion (e.g., SO$_4^{2-}$ or gluconate) which cannot pass through the nystatin pathway, the discrepancy between mucosal and cell Na$^+$ will be greater, as also illustrated in Fig. 5a. A more significant effect is on cell volume (Fig. 5b). Addition of NaCl maintains the cell near isovolumic, while addition of sodium gluconate causes a progressive decrease in cell volume. Recent studies on toad urinary bladder[12] suggest that a decrease in cell volume results in a loss of basolateral pump activity (by either pump inactivation or internalization). If this also occurs in other preparations it will have profound effects on the apparent kinetic parameters of the pump estimated by either electrical or isotopic techniques.

Serosal K$^+$ activation of the pump was performed using a different protocol and different antibiotic. First, normal NaCl–Ringer's solution are added to both sides of the preparation. Next gramicidin D is added to the mucosal solutions. This increases apical membrane permeability to Na$^+$ but (importantly) not to Cl$^-$. The response of the pump to serosal K$^+$ concentrations is studied by removing all K$^+$ from the mucosal solution, allowing Na$^+$ to load into the cell (for 5 min), and then adding a small amount of K$^+$ (as KCl) to the serosal solution. Using microelectrodes one can monitor the basolateral membrane potential and resistance, and thus calculate a current induced by the pump. The serosal solution is returned to a zero K$^+$–Ringer's solution, Na$^+$ is allowed to load into the cell, and the next higher concentration of K$^+$ is added. Serosal K$^+$ concentration is then plotted against current.

The concentration dependence of ouabain inhibition can also be studied using the above gramicidin-treated preparation. In brief, one measures the hyperpolarization induced by adding K$^+$ (7 mM) to a K$^+$-free serosal solution. The serosal solution is replaced with a K$^+$-free Ringer's solution, an aliquot of ouabain added (10^{-8} M), and the hyperpolarization is measured when K$^+$ (7 mM) is added to the serosal solution. The K$^+$ is washed out and the next higher concentration of ouabain is added. After a saturat-

[12] S. A. Lewis, A. G. Butt, M. J. Bowler, J. P. Leader, and A. D. C. Macknight, *J. Membr. Biol.* **83**, 119 (1985).

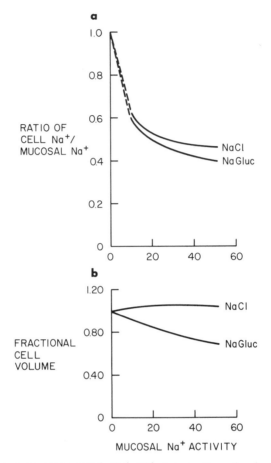

FIG. 5. (a) A computer-generated relationship between the ratio of cell Na^+ activity to mucosal Na^+ as a function of the activity of mucosal Na^+. The experimental condition is that the apical membrane has been permeabilized with nystatin when both mucosal and serosal solutions contain high K^+ and an impermeant anion. Na^+ is added to both mucosal and serosal solution with a permeant anion (Cl^-) or an impermeant anion [gluconate (Gluc)]. As the activity of mucosal Na^+ is increased, the ratio of activities decreases, meaning that less Na^+ is in the cell than in the mucosal bath. Even at the beginning, mucosal activity of 10 mM cellular Na^+ is ~6 mM. The decrease in this ratio is greater when Na^+ is added with an impermeant anion, although less than one might expect. (b) The effect on fractional cell volume (ratio of cell volume at any mucosal Na^+ to cell volume at $Na^+ = 0$) of symmetrical Na^+–anion addition. The experimental conditions are the same as in a. When Na^+ is added with a permeant anion, the maximum increase in cell volume is 6%. In contrast, when Na^+ is added with an impermeant anion (gluconate), cell volume progressively decreases by 32% at 52 mM sodium gluconate. This decrease in cell volume with sodium gluconate accounts for the greater than expected ratio of cell Na^+ to mucosal Na^+ activity shown in a.

ing dose of ouabain, the decrease in pump current is plotted against ouabain concentrations and curve is fit by the appropriate kinetic model. For more details, see Lewis and Wills.[10]

Electroneutral Processes

Electroneutral processes cannot be studied electrically, but require the use of isotopic flux measurements. To date only one report in the literature has appeared. Hamilton and Eaton[13] reported the existence of a NaCl (K) cotransporter which is furosemide sensitive. These researchers used the nystatin-treated preparation previously described.

Other Uses of Ionophores

The emphasis of this chapter was to describe the use of ionophores in studying the mechanism of ion transport by intact epithelia. Below we list some other uses of ionophores, particularly how they have been applied to epithelial membrane vesicle preparations.

As noted in Table I, although membrane vesicles (apical or basolateral membrane) are useful tools for studying both electrolyte and nonelectrolyte transport using radioactive isotopes, without the use of ionophores to dissipate ionic gradients or establish membrane potentials, the interpretation of electrolyte or nonelectrolyte fluxes would be equivocal.

A voltage can be established across a vesicle membrane by addition of the K^+-specific carrier, valinomycin, to a suspension of vesicles with an internal K^+ concentration which can be greater, equal to, or less than the bulk solution. Such ionophore-induced membrane voltages allow the discrimination between electroneutral and electrogenic transport processes. That valinomycin is able to establish membrane potentials has been measured using potential-sensitive fluorescent dye (see Wright[14]). Preexisting (nonionophore-induced) membrane potentials can be short-circuited by the uncoupler FCCP. Last, ionophores such as nystatin, amphotericin B, nigericin, or monensin can be used to load vesicles with small cations (and for the polyenes) anions, while A23187 can load vesicles (as well as intact epithelia) with Ca^{2+}.

The polyene antibiotics (amphotericin, nystatin, and filipin) bind to membrane sterols and produce membrane "bulges" which can be viewed using freeze-fracture techniques (see Tillack and Kinsky[15] and Verkleij *et*

[13] K. L. Hamilton and D. C. Eaton, *Fed. Proc., Fed. Am. Soc. Exp. Biol.* **42**, 1353 (1983).
[14] E. M. Wright, *Am. J. Physiol.* **246**, F363 (1984).
[15] T. W. Tillack and S. C. Kinsky, *Biochim. Biophys. Acta* **323**, 43 (1973).

al.[16]). Stetson and Wade[17] used this phenomenon to study the sterol distribution in both plasma (apical and basolateral membranes) and intracellular membranes of the toad urinary bladder in controls and after antidiuretic hormone stimulation.

Summary

In this chapter we have described the use of antibiotics for studying epithelial membrane resistances and permeabilities, as well as active and passive transport processes. We placed particular emphasis on sources of artifacts that could result in erroneous determination of any of the above parameters, as well as methods for assessing such artifacts (where possible). Although the number of possible artifacts seems large, it is not our intention to dissuade researchers from using this probe of epithelial function, but rather to point out how data might be incorrectly interpreted.

Acknowledgments

We thank Dr. C. Clausen for providing us with the computer-estimated values of intracellular Na^+ and cell volume. This work was supported by NIH Grants AM33243 (S.A.L.) and AM29962 (N.K.W.).

[16] A. J. Verkleij, B. de Kruijff, W. F. Gerritsen, R. A. Demel, L. L. M. Van Deenen, and P. H. J. Ververgaert, *Biochim. Biophys. Acta* **291**, 577 (1973).
[17] D. L. Stetson and J. B. Wade, *J. Membr. Biol.* **74**, 131 (1983).

[37] Cultures as Epithelial Models: Porous-Bottom Culture Dishes for Studying Transport and Differentiation

By JOSEPH S. HANDLER, NORDICA GREEN, and RODERICK E. STEELE

Interest in transport by epithelia in culture has increased enormously since the demonstration that cultured epithelia perform vectorial transport.[1,2] The critical innovation was based on the use of a permeable attachment surface so that the solutions on both sides of the cultured epithelium were separate and could be sampled, as they are in a standard Ussing

[1] D. S. Midfeldt, S. T. Hamamoto, and D. R. Pitelka, *Proc. Natl. Acad. Sci. U.S.A.* **7**, 1212 (1976).
[2] M. Cereijido, E. S. Robbins, W. J. Donlan, C. A. Rotunno, and D. D. Savatini, *J. Cell Biol.* **77**, 853 (1978).

chamber for flux and short-circuit current measurements. A second benefit of the permeable surface for epithelial cell culture is increased differentiation.[3] The stimulus to differentiation is probably the accessibility of nutrients, hormones, and growth signals to the basal surface of the epithelium. In culture, as *in situ,* the basal plasma membrane of an epithelium forms the attachment to the underlying surface, and the apical plasma membrane faces the free solution (e.g., *in situ,* urine, salivary secretion, intestinal contents; in culture, growth medium). When epithelia cultured on plastic tissue culture dishes grow out to confluence and form competent tight junctions, the junctions impair exchange between the medium, which is added to the apical surface of the epithelium and the basolateral plasma membrane, normally the membrane that has transporters for nutrients and receptors for hormones and growth factors. When cells are grown on a porous surface, the dish can be arranged so that medium is added below the porous surface and can reach the nutrient surface of the cells. Presumably that results in more differentiation than is routinely seen in cultures on plastic tissue culture dishes. Improved growth and differentiation has been demonstrated as well when epithelial cells are grown on a tissue-specific natural supporting stroma laid down on a plastic tissue culture dish.[4] However, that arrangement is not suitable for study of transepithelial transport. Some epithelial cultures demonstrate further differentiation when grown on floating collagen gels. Cells are seeded and grow to confluence when the gels are attached to a plastic tissue culture dish. When the culture is confluent, the gel is released mechanically and floats in the medium. Most epithelial cultures contract the gel shortly after it is released, and change from a flattened squamous epithelium to a columnar epithelium. In some cultures the shape change is associated with differentiation exceeding that seen if contraction of the floating gel is prevented.[5] Compared to porous-bottom culture dishes (vide infra), contracted floating collagen gels are not as easy to manipulate into an Ussing chamber for transport studies. Floating collagen gels and tissue-specific natural supporting stromas (reconstituted basement membrane rafts) have been covered in an earlier volume in this series.[4]

In this chapter, methods are described for fabricating porous-bottom culture dishes with porous bottoms composed of cellulose ester filters (Millipore), polycarbonate filters (Nucleopore), collagen membranes,[6] and

[3] J. S. Handler, A. S. Preston, and R. E. Steele, *Fed. Proc., Fed. Am. Soc. Exp. Biol.* **43,** 221 (1984).
[4] L. M. Reid and M. Rojkind, this series, Vol. 58, p. 263.
[5] J. M. Shannon and D. R. Pitelka, *In Vitro* **17,** 1016 (1981).
[6] R. E. Steele, A. S. Preston, J. P. Johnson, and J. S. Handler, *Am. J. Physiol.* **251,** C136 (1986).

a similar arrangement in which the porous growth surface is a placental amnion that has been denuded of its epithelial cells.[7] In all of these culture dishes, the medium above the cells is separate from the medium below the porous membrane, facilitating transport measurements. They can be manipulated easily under sterile conditions in a laminar flow hood so that short-circuit current, potential difference, and transepithelial resistance as well as other noninvasive measurements (e.g., cAMP release, oxygen consumption) can be performed and the culture returned to the incubator so that the same culture can be studied many times. Figure 1 illustrates filter-bottom, collagen-bottom, and amnion-bottom culture dishes.

Materials

Polycarbonate tubing, 1.0-in. inner diameter, 1.25-in. outer diameter (inner area for cell culture = 5 cm²). Wider diameter tubing has been used as well. Polycarbonate is preferred because it can be boiled without deformation, an aid in cleaning and sterilization.

Polycarbonate sheets, 0.01 and 0.03 in. thick.

Filters, cellulose esters, pore size = 0.45 μm, outer diameter = 47 mm (HAWP, Millipore Corp.); and polycarbonate, pore size = 0.8 μm, outer diameter = 47 mm (Nucleopore Corp.). Other pore sizes have been used successfully. These were chosen for convenience (HAWP) and for maximum total pore area (Nucleopore).

Type I collagen from calf skin (Worthington, Cooper Biomedical).

Silicone rubber cement (e.g., General Electric RTV 118).

Thimble-type polyethylene stoppers, such as those sold by A. H. Thomas, catalog no. 8758-F27. These stoppers have a slight inner projecting ridge at the top. This size, 17 × 11 mm (top and bottom outer diameters), fits conveniently in the wells of a six-well tissue culture dish. Smaller and larger areas have been used.[7]

Disposable sterile tray (e.g., Linbro trays from Flow Laboratories, catalog no. 76-622-05).

Fabrication of Dishes

With the exception of amnion-bottom dishes (vide infra), porous-bottom culture dishes are formed by attaching a flat porous membrane to the bottom of a polycarbonate ring so that the ring forms the side of the dish. The bottom of the dish is maintained above the floor of its container by attaching three short spacers or feet to the ring.

[7] N. Green, A. Algren, J. Hoyer, T. Triche, and M. Burg, *Am. J. Physiol.* **249,** C97 (1985).

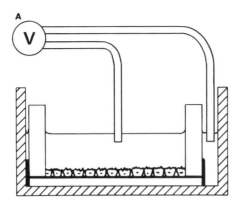

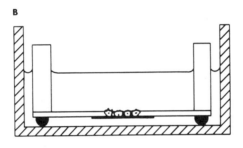

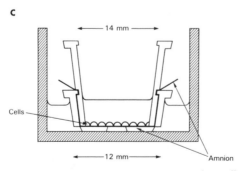

FIG. 1. (A) Confluent epithelium on a cellulose ester-coated or collagen-coated polycarbonate filter-bottom dish. Bridges are in place for measurement of transepithelial potential difference. The dish is moved to the side of the well so that a bridge can be placed in the solution outside the dish. (From Ref. 6.) (B) Confluent epithelium on a collagen membrane-bottom dish. (From Ref. 6.) (C) Amnion-bottom dish. (From Ref. 7.) For simplicity, the feet in A and B are drawn 180° apart, instead of 120°.

Cellulose Ester Filter-Bottom Culture Dishes (Fig. 1A). The polycarbonate tubing is cut into 0.5-in. rings (this will yield a dish with a side wall 0.5 in. in height). The rings are often dirty after cutting. They are cleaned by boiling in detergent (Alconox) solution and then are rinsed well. Dry cellulose ester filters can be attached to rings with silicone rubber cement. The filters swell when wet, causing the bottom of the dish to bulge. This can be avoided by wetting the filters before attachment. Wet size is achieved in a few minutes if the filters are soaked in 30% ethanol in water. Wet filters are attached to rings by softening the attachment surface of the ring and then pressing the softened polycarbonate ring into the wet filter. Softening is accomplished with a 5-sec dip of the bottom of the ring in chloroform. Excess chloroform is allowed to evaporate for 5 sec, and then the ring is pressed into the wet filter, which is placed on a hard, smooth surface. The dish is soaked in distilled water for 1–2 hr. Excess filter protruding outside the ring is removed by pulling the excess filter down. It tears easily at the outer edge of the ring. If there are unsealed areas they become evident at this point. Feet are formed by cutting 0.01-in.-thick polycarbonate sheets into thin strips 0.1 in. wide. Feet are attached by dipping the end of a thin strip of polycarbonate in chloroform and then holding the softened end of the strip against the outside of the ring near the attachment of the filter. After three feet are attached about 120° apart, the feet are trimmed so that they project about 1 mm below the filter. Dishes are sterilized by immersion in 2.5% glutaraldehyde for 30–60 min (70% ethanol has also been used successfully). The dishes are removed from glutaraldehyde under sterile conditions in a laminar flow hood and rinsed well in sterile distilled water. Dishes can be stored in phosphate-buffered saline solution (PBS) under sterile conditions. Alternatively, dry dishes can be sterilized with γ-irradiation and stored dry.

Polycarbonate Filter-Bottom Culture Dishes (Fig. 1A). Polycarbonate filters do not swell when wet, so dry filters can be cemented to rings using silicone rubber cement. Excess filter outside the ring is trimmed away with a single-edge razor blade. Most cells do not attach well to polycarbonate, so the filters must be coated with a material such as collagen to aid cell attachment. After the filter is attached to the ring and the cement has dried, 0.15 ml of a solution of 3 mg/ml of Type I collagen is spread over the surface of the filter inside the dish. The dish is exposed to NH_3 fumes for 5 min and dried for several hours. The attachment of feet and sterilization are performed as for cellulose ester filter-bottom cups.

Collagen Membrane-Bottom Culture Dishes (Fig. 1B). Collagen membranes are not stiff enough to cover the entire 5-cm² bottom of the standard ring without support. Therefore, the unsupported area is reduced by gluing to the bottom of the ring a polycarbonate disk with a conical hole in its center. The disk is prepared by cutting 1.25-in.-diameter circles from 0.03-in.-thick polycarbonate sheets with a punch. A conical hole with the smaller diameter 1 to 10 mm is drilled in the center of the disk. The smaller diameter determines the area for cell growth on the collagen membrane which will be glued across the smaller hole. The bottom of the disk (smaller hole) is sanded with waterproof abrasive paper to roughen the surface around the hole. The disk is oriented so that the larger diameter of the center hole is up toward the ring and is cemented to the ring using silicone rubber cement. Three drops of silicone rubber cement are placed near the outer edge of the bottom of the disk to form feet. The cement hardens overnight.

Collagen membranes are formed by spreading a solution of Type I collagen, 7 mg/ml, on a smooth surface at 0.1 ml/cm². A 3.25 × 4-in. piece of glass fits in a 15-cm tissue culture dish and forms a membrane of about 70 cm². The collagen solution is allowed to dry (this may take 2 days or more). The dry collagen membrane is reconstituted by covering it with 100 ml of a 3.5% solution of concentrated NH₄OH. After 30 min the membrane is carefully stripped off the glass and the glass removed. After 15 more min the NH₄OH solution is carefully aspirated and the membrane rinsed twice with 10 ml of H₂O. The last water rinse is quickly replaced with 70% ethanol. The transition from 3.5% NH₄OH to 70% ethanol should not take more than a minute. The collagen membrane can be stored in 70% ethanol for many months. Disks of this collagen membrane are cut to size in the 70% ethanol dish with a cork borer; 9-mm discs are used for 5-mm holes. The disks are transferred to a 10-cm tissue culture dish containing 70% ethanol to leave unwanted cuttings behind. The collagen disks are rinsed quickly in distilled water and placed in 2.5% glutaraldehyde for 5 min to stiffen them by cross-linking the collagen. The stiffened collagen disks are rinsed in distilled water and placed in PBS.

The area on the bottom of the disk that has been roughened with abrasive paper is coated near the hole with collagen (7 mg/ml) and allowed to dry overnight; 30 μl is used for a 5-mm hole. The collagen-coated surface is soaked in 3.5% NH₄OH for 30 min, and then rinsed in distilled water and allowed to dry. The area near the hole is recoated with collagen solution (a very thin layer this time). A precut stiffened collagen disk is rinsed in distilled water and lifted from the water using a microscope slide and forceps to get the membrane flat and as free of water as possible (water interferes with bonding). The disk is placed over the hole and on top of the collagen coating the plastic disk. The collagen attached to the disk is exposed to NH₃ gas for 5 min. Finally, the dish is sterilized by immersion in 2.5% glutaraldehyde for 30 min. The dish is rinsed well in PBS and stored in PBS.

Amnion-Bottom Culture Dishes (Fig. 1C). Membranes prepared from amnions are held in an embroidery-hoop-type arrangement. Each holder is fabricated from two hollow polyethylene stoppers of the same size. Using a lathe, a circle is scored on the outer surface of one (the top or inner) stopper. The location of the scoring is determined by nesting stoppers one inside the other. The score is made where the inner ridge of the outer stopper meets the outside of the inner stopper. For 19 × 11-mm stoppers (A. H. Thomas, cat. no. 8758-F27) this point is 8 mm from the wide top. The stopper is cut across 13 mm from the top. The bottom (outer) stopper is cut 9 mm from the top. The two pieces can be fit together at this point as the stoppers might be stacked for storage. The inner projecting ridge at the top of the bottom stopper should lock into the outer ridge that has been cut into the upper stopper. The amnion will be held firmly between them at this point and will lie across the cut bottom of the top stopper. Using a nibbler, a 2-mm-high notch is cut into the bottom (cut edge) of the bottom stopper so that in each quadrant there is about 50° of notch. The remaining stopper in each quadrant will act as feet. Holders are sterilized by gas sterilization.

Amnions are collected under sterile conditions at cesarian section deliveries and thereafter are prepared under sterile conditions. Amnions are immediately rinsed in sterile PBS supplemented with antibiotic–antimycotic mixture (Gibco). They are denuded of their epithelial cell layer by incubation for 45 min in PBS containing 0.5% Triton X-100 at room temperature with gentle shaking. Thereafter, all procedures are performed in a laminar flow hood. Manipulations are most easily performed wearing sterile surgical gloves. To mechanically remove adhering epithelial cells, the amnion is placed on a flat sterile tray such as a Limbro tissue culture tray and both surfaces are wiped several times with gauze wet with PBS. The denuded amnion is rinsed five times in PBS and stored at 37° in medium containing antibiotics.

Each amnion is inserted into several holders. Before insertion into holders, debris adhering to a stored amnion is removed by gently wiping both surfaces of the amnion with gauze wet with PBS. Before placing the amnion in holders, it is laid out as a flat sheet on a sterile tray. The orientation of the amnion is determined by phase-contrast microscopy of a sample cut from the flattened amnion. The amnion should be oriented so that the surface with the basement membrane is down. The top (inner) half of the holder is placed under the amnion so that the narrow end is up, against the amnion. A slight excess of amnion is placed over the opening by poking the amnion down slightly into the holder with a finger tip. This will prevent tearing when the other half of the holder is put in place and stretches the amnion. The lower (outer) part of the holder is then inverted, placed over the amnion, and gently pushed down until the two parts of the holder are snug and the amnion stretched between them. The amnion should now be flat across the cut end of the inner part of the holder. Other areas of the amnion are placed in other holders. Finally, the amnion between holders is cut with scissors so that each amnion and its holder are separate. Excess amnion is trimmed. Amnions in their holders can be stored under sterile conditions in PBS with antibiotic–antimycotic mixture.

Cell Culture

Each 1.25-in.-diameter porous-bottom culture dish is used in the well of a 6-well culture dish (Costar, Cambridge, Massachusetts, or Nunc, Denmark) [dishes from Falcon (Oxnard, California) wells are too shallow for 0.5-in.-high cups]. Before seeding, the porous bottom is soaked in growth medium for 2–24 hr to allow attachment factors in the medium to permeate the filter. Permeable-bottom culture dishes are usually seeded more densely (one-half to twice the cell density found on confluent plastic tissue culture dishes) than plastic tissue culture dishes used to propagate cells. We usually place 2 ml of medium on the apical surface and 2 ml on the basal surface of seeded dishes. This results in the same hydrostatic pressure on both sides of the porous bottom. The two solutions do not make contact except through the filter. Care should be taken that drops of medium that fall on the upper edge of the ring do not form an electrical bridge between the solutions inside and outside the dish. We have found that transepithelial resistance occasionally falls when hydrostatic pressure on the basal surface exceeds that on the apical surface. Therefore, we add solution first to the apical surface and withdraw solution first from the basal surface. For cultures on collagen membrane-bottom culture dishes the order of replacing solutions is reversed because the risk of the collagen membrane tearing away from the disk is greater than the risk of damaging the epithelium. Amnion bottom culture dishes are handled similarly. Dishes with 12 or 24 wells are more appropriate for smaller holders. The amount of medium on each side of the membrane should be adjusted so that there is not a large hydrostatic gradient across the membrane.

Comments

Filter-bottom cups are easiest to fabricate on a large scale and have been used for many studies of epithelial development,[8] transport,[9] and for

[8] M. A. Lang, J. Muller, A. S. Preston, and J. S. Handler, *Am. J. Physiol.* **250,** C138 (1986).
[9] J. P. Johnson, R. E. Steele, F. M. Perkins, J. B. Wade, A. S. Preston, S. W. Green, and J. S. Handler, *Am. J. Physiol.* **241,** F129 (1981).

cell fractionation to prepare vesicles for flux studies.[10] Cell outlines can be seen on amnions and occasionally on collagen-coated polycarbonate filters. Cellulose ester filter-bottom dishes can be coated with collagen using the technique described for coating polycarbonate filter-bottom dishes. Collagen membranes allow the best visualization of cells with phase-contrast microscopy. They have been used for electrophysiological studies using intracellular microelectrodes and patch clamps.[11] Amnion-bottom dishes and collagen-bottom dishes have been particularly valuable for growing primary cultures from very few cells.[7,12]

[10] S. Sariban-Sohraby, M. Burg, W. P. Wiesmann, and J. P. Johnson, *Science* **225**, 745 (1884).
[11] S. E. Guggino, W. B. Guggino, B. A. Suarez-Isla, N. Green, and B. Sacktor, *Fed. Proc., Fed. Am. Soc. Exp. Biol.* **44**, 443 (1985) (Abstr.).
[12] M. B. Burg, N. Green, S. Sohraby, R. Steele, and J. Handler, *Am. J. Physiol.* **242**, C229 (1982).

[38] Volume Regulation in Epithelia: Experimental Approaches

By ANTHONY D. C. MACKNIGHT and JOHN P. LEADER

Introduction

The study of volume regulation in epithelia requires the application of a range of techniques, many of which are discussed in detail in this volume. No chapter dealing with this topic can hope to cover in detail all of these techniques, nor would it be profitable to do so. Rather, we shall first provide a framework within which volume regulation in epithelia can be considered. We shall then focus on some of the techniques applied to determine cell volume and cell water and ions, and finish by discussing some of the preparations which have been used to study the regulation of cell volume in epithelia.

Basic Aspects of Cellular Volume Regulation

Whatever the cell type, certain fundamental questions relating to volume regulation must be answered. In many areas progress has been disappointingly slow and arguments raised in the early years of this century remain unresolved. Chief among these is the perennial problem of the role of the plasma membrane in determining cell composition and, thereby, water content. The early proponents of the membrane theory were happy to ascribe a dominant role to the membrane.[1] This remains the conventional view and one which will be adopted here. But the alternative approach, which attributes both the unequal distributions of diffusible ions and the water content to properties of the cellular cytoplasm, has always had support. For example, Moore et al.[2] wrote, in 1912, "The varying concentrations of sodium, potassium, chlorine and phosphatic ions within and without the cell are an expression of specific affinities of the definite colloids of each particular cell-type for these ions, and do not mean that there is a membrane acting as a closed gate to these ions" and "sodium ions can readily be washed out of cells because the cell proteins have no affinity for them, while potassium ions are retained to the last."

Today, perhaps the most widely promulgated interpretation of this type is that propounded by Ling over the past 20 or so years (see Ling[3] for a full discussion). In his association–induction hypothesis, the accumulation by cells of potassium rather than sodium is attributed to the preferential adsorption of potassium by fixed anionic groups on proteins. Cellular water content is determined by two factors, the multilayer polarization of cell water by proteins whose configurations are affected by ATP, and salt linkages formed between adjacent charged or fixed groups on the same or neighboring proteins.

Despite the fundamental differences between the two hypotheses, it has proved surprisingly difficult to obtain experimental evidence which allows unequivocal rejection of one of them. However, the continued success of the membrane-based model in explaining experimental results, such as alterations in the electrical potential and the movements of ions across membrane vesicles, and the isolation, purification, and reconstitution in liposomes and lipid bilayers of membrane-derived proteins which can generate net ion movements, justifies the emphasis to be placed in this chapter on the role of membrane-based processes in regulating cellular volume.

This is not to imply, however, that all cell ions are simply in free solution in an aqueous phase whose properties are identical to extracellular water. Certainly divalent ions may be immobilized or bound on charged groups on cellular constituents, or sequestered in subcellular compartments such as endoplasmic reticulum and mitochondria. Whether some of the predominant diffusible univalent ions, potassium, sodium, and chloride, are similarly bound or compartmentalized remains to be resolved.

[1] J. Bernstein, *Pfluegers Arch.* **92,** 521 (1902).
[2] B. Moore, H. E. Roaf, and A. Webster, *Biochem. J.* **6,** 110 (1912).
[3] G. N. Ling, "In Search of the Physical Basis of Life." Plenum, New York, 1984.

Direct comparisons of data obtained with analytical techniques which allow estimation of total cellular ions (chemical and isotopic analysis, electron microprobe analysis), with ion-sensitive microelectrodes which estimate ion activities directly, and with other techniques which reflect the extent to which the freedom of ion movements are restricted (such as NMR spectroscopy) should ultimately resolve this question. At present it is customary to regard the major univalent ions as being more or less in free solution with activity and osmotic coefficients similar to those of the same ions in the extracellular fluid.

The state of water within cells also remains controversial. While it is established that the activity of water in the vicinity of a surface is depressed and that this water can differ in its solvent properties from bulk water, the debate, discussed by Clegg,[4] focuses on the extent of this altered water within cells. Since, under a variety of experimental conditions, cell water appears to behave as predicted for bulk water, we shall adopt the pragmatic approach of discussing it in these terms while recognizing the somewhat arbitrary nature of this choice.

A second but related issue, which has been the subject of debate from the early years of cell physiology, concerns the nature of the cytoplasm. Though many accounts treat the cytoplasm as a watery soup in which float the cellular organelles, as knowledge of the complex and highly organized cytoskeleton grows, such a view becomes increasingly untenable. An alternative is to regard the cytoplasm as a more or less highly organized gel. Observations, such as the lack of Brownian motion in cytoplasm of living cells viewed by phase-contrast or Nomarski optics (and its prompt appearance in portions of the cell following uptake of water),[5] and the gel-like nature of cytoplasm extruded from cells such as the squid giant axon,[6] lend credence to this view.

Though debate continues about the nature of the cytoplasm, it must be emphasized that the common approach, which is to describe water movements between cells and surrounding fluids in terms of hydrostatic and osmotic forces and the ease with which water passes across the plasma membranes, is valid equally for both models. Thus

$$J_v = L_p (\Delta P - \sigma \Delta \pi) \tag{1}$$

where J_v is the volume flow per unit time, ΔP and $\Delta \pi$ are the differences in hydrostatic and osmotic pressures, respectively, between cell and extracellular fluids, L_p is the hydraulic conductivity of the membrane, and σ is the

---

[4] J. S. Clegg, *Am. J. Physiol.* **246**, R133 (1984).
[5] D. R. DiBona, *J. Membr. Biol. (Spec. Issue)*, 45 (1978).
[6] C. S. Spyropoulos, *J. Membr. Biol.* **32**, 19 (1977).

reflection coefficient which relates the hydraulic conductivity to the coefficient of osmotic flow, L_{pd} ($\sigma = -L_{pd}/L_p$).

Expansions of Eq. (1) contain two osmotic pressure terms. In the usual approach, which regards the cytoplasm as a solution, Σc^i is the total contribution of impermeant cellular solutes, and $\Sigma \Delta c^p$ is the difference in concentrations between extracellular and cellular permeant solutes [i.e., $\Sigma \sigma \Delta c^p = \Sigma \sigma (c_{ec}^p - c_c^p)$].

$$J_v = L_p \left[\Delta P - RT(\Sigma c^i - \Sigma \sigma \Delta c^p) \right] \tag{2}$$

where R and T have their usual meaning. Since the cellular impermeant solutes must have a reflection coefficient of 1 by definition, σ is omitted from this term in the equation. In conventional terms we can regard Σc^i as including the osmotic contributions of the counterions which balance any charges on the impermeant solutes (the Donnan excess).

When we apply this same formulation to consider exchanges of water between a gel-like cytoplasm and the extracellular fluid, the physical meaning attributed to Σc^i will be influenced by the model of the gel which is adopted. In a gel, water activity will be depressed by dilution with the gel molecules, by elastic deformation of the gel network, and by interactions between fixed charges on the network. These effects, together, are formally equivalent to the depression of water activity by cellular impermeant solute and can therefore be represented as Σc^i. Water activity within the get will also be depressed by the dissolved permeant solutes. The differences between this effect in the gel and in the medium are represented in the term $\Sigma \sigma \Delta c^p$.

In both models, volume flow will be zero if $\Delta P = RT(\Sigma c^i - \Sigma \sigma \Delta c^p)$. In such a situation a hydrostatic pressure gradient would result if the plasma membrane or surrounding structures restrained cellular swelling. Under such conditions with a gel, ΔP is called the swelling pressure of the gel. In the absence of a hydrostatic pressure, J_v will be zero when the osmolalities of the cell and extracellular fluids are identical, i.e., $RT(\Sigma c^i + \Sigma \sigma c_c^p) = RT \Sigma \sigma c_{ec}^p$. In a gel, in the absence of a restraint of any kind on swelling, the equilibrium volume of the gel has been reached under these conditions.

It is generally accepted that the plasma membranes of animal cells are unable to withstand hydrostatic pressure gradients.[7] It is not possible experimentally to prove that there is no difference in osmolality between cells and extracellular fluids, because even a difference of 1 mOsm/kg water, which would lie within the experimental error of the available methods, would require a hydrostatic pressure of 2.5 kPa (19.3 mm Hg). Neverthe-

[7] E. N. Harvey, *Protoplasmatologia* 2, E5 (1954).

less, the best estimates (for example, in Ref. 8) are in accord with the view that osmolalities of medium and cell are equal.

Given equality in hydrostatic pressure, how is the cellular volume maintained constant? The conventional view, discussed and analyzed at length elsewhere,[9] is that the osmotic effect of the impermeant cellular solute (sometimes referred to as colloid osmotic pressure because at least some of these solutes are of large molecular weight) is offset by the energy-dependent exclusion of sodium from the cells.[10,11] This has been referred to as the double-Donnan[12] or pump leak hypothesis.[13,14] The extrusion of sodium is commonly assumed to reflect the activity of the ouabain-inhibitable, Mg-dependent Na^+,K^+-ATPase or sodium pump, though other explanations have been put forward.

We are now in a position to provide a simple formulation of factors involved in determining cellular volume in terms of the membrane-based model. Such a formulation has been published by Jakobsson[15] and will simply be outlined here.

Two essential requirements in the steady state are for osmotic equilibrium between cells and extracellular fluid, and for cell electroneutrality. Thus the sum of cellular concentrations of diffusible and nondiffusible solutes must equal medium osmolarity, and the sum of the charges on cellular sodium and potassium must equal those on cellular chloride and on impermeant cellular solutes. Measurements of cellular univalent diffusible ions (Na^+, K^+, and Cl^-) and cellular water content, and of medium osmolarity, enable estimation of impermeant solute concentrations and of their average charge. Such procedures provide minimum estimates in that they assume similar activity and osmotic coefficients in cell and extracellular fluids.

The remaining relationships depend upon the assumptions made about the system. For example, we can assume that chloride is distributed passively at electrochemical equilibrium. Knowledge of medium chloride concentration and of either cell chloride concentration or plasma membrane potential is then required. In addition, simplifying but reasonable assumptions that the electric field within the membrane is linear and that

[8] R. H. Maffly and A. Leaf, *J. Gen. Physiol.* **42**, 1257 (1959).

[9] A. D. C. Macknight and A. Leaf, *Physiol. Rev.* **57**, 510 (1977).

[10] T. H. Wilson, *Science* **120**, 104 (1954).

[11] A. Leaf, *Biochem. J.* **62**, 241 (1956).

[12] A. Leaf, *Ann. N.Y. Acad. Sci.* **72**, 396 (1959).

[13] D. C. Tosteson and J. F. Hoffman, *J. Gen. Physiol.* **44**, 169 (1960).

[14] D. C. Tosteson, *in* "The Cellular Functions of Membrane Transport" (J. F. Hoffman, ed.), p. 3. Prentice-Hall, New York, 1964.

[15] E. Jakobsson, *Am. J. Physiol.* **238**, C196 (1980).

pump flux is directly proportional to cellular sodium concentration allow derivation of relationships between the passive and active fluxes of sodium and of potassium.

The five equations derived by Jakobsson[15] allow examination of the factors that determine the steady-state volume in cells in which sodium and potassium are actively transported by the Na^+,K^+-ATPase, chloride is in electrochemical equilibrium, and only conductive pathways allow passive ion movements across the plasma membrane. They were also extended by him to examine the behavior of such cells in conditions under which volume will alter (e.g., inhibition of the sodium pump, increased medium potassium concentration). However, inasmuch as cotransport and countertransport mechanisms also influence cell ion concentrations, steady-state volume, and the rates of change of volume, the equations presented by Jakobsson[15] require modification and extension.

Furthermore, these equations were not designed specifically to examine cellular volume in epithelia. They therefore ignore one of the major attributes of these cells—their polarity. Several models have now been developed in attempts to examine the available data from studies of a variety of epithelia.[16-19] These adopt as their framework the three-compartment model elaborated by Koefoed-Johnsen and Ussing.[20] Though they differ somewhat, the major contribution of this approach is to focus attention on the specific information required for the complete characterization of epithelial solute and water transport, including the maintenance and regulation of cellular volume.

In particular, we need to know for both apical and basolateral plasma membranes (1) the types and properties of the pathways through which solutes and water can move and (2) the membrane potentials. We also need to know the properties of the junctional pathway for solutes and water. Finally, we need to determine cellular solutes and water, specifying for each solute its osmotic contribution, and also, for each ion, its thermodynamic activity.

Detailed discussion of the techniques for assessing the properties of plasma membranes and junctional pathways are presented in this volume and have been reviewed elsewhere.[21,22] We discuss here techniques for the measurements of cellular volume and of cellular water and solutes, and

[16] M. M. Civan and R. J. Bookman, *J. Membr. Biol.* **65**, 63 (1982).
[17] H. G. Ferreira and K. I. G. Ferreira, *Proc. R. Soc. London B* **218**, 309 (1983).
[18] R. Latta, C. Clausen, and L. C. Moore, *J. Membr. Biol.* **82**, 67 (1984).
[19] V. L. Lew, H. G. Ferriera, and T. Moura, *Proc. R. Soc. London B* **206**, 53 (1979).
[20] V. Koefoed-Johnsen and H. H. Ussing, *Acta Physiol. Scand.* **42**, 298 (1958).
[21] A. D. C. Macknight and J. P. Leader, *Clin. Lab. Sci.* **18**, 339 (1983).
[22] J. Phillips and S. Lewis (eds.), *J. Exp. Biol.* **106**, 1 (1983).

comment also on aspects of measurement of membrane potentials in epithelial cells.

Techniques for Analysis of Cell Volume and Composition

Cell volume may be examined directly using a wide range of microscopy techniques applied to fixed tissues, or, more recently, to living preparations. Alternatively, cell water and solute contents can be determined by a variety of techniques. Most epithelial cells are composed predominantly of water which represents some 70 to 80% of the volume of the cells *in vivo* or when incubated in isosmotic extracellular-type media *in vitro.* In any epithelium studied over hours *in vitro,* it is presumed that cell solid matter remains relatively constant so that changes in cell volume and in cell water content can be regarded as equivalent.

The advantages of examining changes in volume directly in living preparations lie particularly in the ability to follow virtually instantaneously the changes in volume which result from changes in medium osmolality or from the applications of hormones or drugs to the epithelium. The disadvantages are that it is not possible to obtain direct evidence about the linkages between these changes in volume and any changes in cell solute contents which accompany (and may be responsible) for them. The best that can be done is to alter the ionic composition of the bathing media and observe the associated modifications of the volume responses of the cells. Otherwise, techniques for measuring cell ions directly must be employed in parallel experiments, as has been done, for example, by Fisher and Spring[23] in their studies of volume regulation in *Necturus* gallbladder.

An interesting alternative method for monitoring rapid changes in cellular volume has been developed by Reuss.[23a] Epithelial cells of the gallbladder of *Necturus* were loaded with tetramethylammonium by exposure of the apical surface to a high concentration of this substance in the presence of the polyene antibiotic nystatin. After a short time, the nystatin was removed and intracellular tetramethylammonium was found, using an ion-selective electrode, to be between 2 and 15 mM, and to remain unchanged for several hours. Changes in cellular volume could then be monitored by determining the changes in intracellular tetramethylammonium concentration following changes in osmolality of the mucosal bathing solution. Such measurements allowed not only determination of the hydraulic conductivity of the apical membrane, but a second, potassium-sensitive microelectrode placed in the bath allowed estimation of the thickness of the "unstirred layer" at the cell surface. This ingenious

[23] R. S. Fisher and K. R. Spring, *J. Membr. Biol.* **78,** 187 (1984).
[23a] L. Reuss, *Proc. Natl. Acad. Sci. U.S.A.* **82,** 6014 (1985).

method, which is facilitated in this preparation by the large cells, does not appear to have been adopted for use in other tissues.

Each approach, therefore, has its advantages and limitations. Some of these are discussed in more detail below. This discussion is of necessity selective not exhaustive, and we cite references to illustrate particular approaches.

Applications of Microscopy to the Study of Volume Regulation in Epithelia

Fixed Tissues. Conventional light and electron microscopy techniques have been applied to the study of various aspects of epithelial function as discussed in detail by DiBona *et al.*[24] These have ranged from simple qualitative estimates of the effect of various treatments on cellular volume, for example those of Bobrycki *et al.*[25] and of Lewis *et al.*,[26] to detailed analysis of morphological components of a variety of tight and leaky epithelia.

Accurate measurements of cellular shape and volume in fixed and sectioned material can be obtained by the technique of stereological analysis, an introductory account of which is given by James.[27] In brief, stereology provides a general means of obtaining three-dimensional information from two-dimensional objects. The process of analysis proceeds in several stages. First, representative sections of the tissue are prepared. Test lattices, consisting of a sheet bearing a line or point pattern, are superimposed upon the sections, or images of them, and analysis is performed by recording the interaction between the two. As an example, Welling and Welling[28] used this method to determine the shape and volume of epithelial cells from the rabbit renal proximal tubule. For these measurements, a series of transmission electron micrographs, printed at a magnification of between 5000 and 20,000 times, were divided by a series of fine parallel lines into a set of concentric parallel zones, each zone representing a constant percentage of cell height from basement membrane to apical cell surface, excluding the brush border. A grid of perpendicular and parallel lines was then superimposed on the photograph and the number of intersections between the intercellular channels and grid lines, or between cell membranes and grid lines, was counted manually for each zone. These data could then be used to make a three-dimensional model of cell shape for each zone. Since the

[24] D. R. DiBona, K. L. Kirk, and R. D. Johnson, *Fed. Proc., Fed. Am. Soc. Exp. Biol.* **44**, 2693 (1985).

[25] V. A. Bobrycki, J. W. Mills, A. D. C. Macknight, and D. R. DiBona, *J. Membr. Biol.* **60**, 21 (1981).

[26] S. A. Lewis, A. G. Butt, J. M. Bowler, J. P. Leader, and A. D. C. Macknight, *J. Membr. Biol.* **83**, 119 (1985).

[27] N. T. James, *Semin. Ser. — Soc. Exp. Biol.* **3**, 9 (1977).

[28] L. W. Welling and D. J. Welling, *Kidney Int.* **9**, 385 (1976).

technique is basically a statistical procedure for evaluating line length, the measurement error is proportional to the square root of the number of intersections counted. It follows that either a small number of intersections must be measured in a large number of photographs, or a large number of intersections recorded from fewer. Such a procedure is tedious, and Simone et al.[29] describe a computer-assisted method for achieving similar ends. The accuracy of such analyses depends heavily on two further limitations. First, the choice of samples for study must be random. If the sections chosen are atypical, then no increase in the number of test points can compensate. Second, and more generally, considerable precautions must be taken in all procedures which involve harsh chemical treatments such as fixation, embedding, and staining to assure that these do not materially distort or change the parameters to be measured. Even where only comparative measurements are involved, small changes in procedure may lead to large changes in final measurement. Even with modern methods using fixatives in isosmotic solutions, it seems best to regard this approach as allowing comparisons to be made for any one preparation between different conditions of incubation rather than providing absolute values for cell volumes.

As well as providing information about cellular volume, conventional techniques of light and electron microscopy have also been used to examine changes in size of cellular organelles under conditions in which cell volume is altered (see, for example, Refs. 30–32).

Living Preparations. In recent years, considerable attention has focused on the direct measurement of cellular volume in epithelia using light microscopy. MacRobbie and Ussing[33] used skin from the frog, *Rana temporaria,* which was tied with thread to a grooved plastic ring, the outer surface of the skin being uppermost. The skin, placed on a microscope stage with the inner (serosal) surface in contact with a glass-wool pad moistened with saline, which rested in a Lucite dish, was viewed through a water-immersion lens ($\times 50$). The thickness of the skin was determined by measuring, with the calibrated fine-focusing control of the microscope, the vertical distance between two reference points, one on its outer surface, the other on a pigment cell immediately below the epithelium. By careful selection of a measuring site where a superficial spot coincided with the sharp outline of a melanophore, it was possible to determine the thickness of the skin between the two levels to within 1 μm. Since the area of the skin

[29] J. N. Simone, L. W. Welling, and D. J. Welling, *Lab. Invest.* **41,** 334 (1979).
[30] C. L. Vôute, K. Møllgård, and H. H. Ussing, *J. Membr. Biol.* **21,** 273 (1975).
[31] M. A. Russo, G. D. V. Van Rossum, and T. Galeotti, *J. Membr. Biol.* **31,** 267 (1977).
[32] G. D. V. Van Rossum and M. A. Russo, *J. Membr. Biol.* **59,** 191 (1981).
[33] E. A. C. MacRobbie and H. H. Ussing, *Acta Physiol. Scand.* **53,** 348 (1961).

exposed was constant, the change in height could be interpreted as a change in volume. Whittembury[34] used a similar method and also reported that with 10 successive measurements of thickness an accuracy better than 1 μm could be obtained. The need for repeated measurements to get good accuracy means that this method is really only appropriate to relatively thick tissues and to measurements of new steady states following an experimental challenge, or to investigations where there is reason to suppose that volume changes will be slow (of the order of minutes).

In recent years three notable advances have stimulated the study of epithelia using the light microscope. These are (1) the development of differential interference contrast (DIC) optics,[35] (2) the development of rapid techniques for image storage and manipulation by computers, and (3) the construction of chambers permitting microscopic observation of epithelial sheets while separately and independently perfusing both sides of the preparation. As a consequence of these developments, interest has been rearoused in the optical study of cells.[36] To illustrate this approach for the study of volume regulation, recent work on the mammalian renal tubule and the gallbladder of *Necturus* will be mentioned briefly.

The technique of isolating and perfusing portions of the mammalian renal tubule has been available since the pioneering studies of Burg and associates.[37] Although ultrastructural studies of fixed material subjected to experimental treatment had been reported,[38,39] it is not until recently that this powerful technique has been combined with the analytical capabilities afforded by optical techniques. The advantages of DIC optics are twofold. First, although it does not convey as much information as phase contrast, DIC gives an artificially high contrast at edges, thus permitting accurate definition of the boundaries of cell and organelle. Second, it gives a very shallow depth of field at high magnifications (about 0.5 μm) and thus permits excellent resolution of "optical sections" of relatively thick objects. Grantham and associates[40,41] used DIC optics to visualize isolated segments of rabbit proximal renal tubules whose ends had been crimped into

[34] G. Whittembury, *J. Gen. Physiol.* **46**, 117 (1962).
[35] R. D. Allen, G. B. David, and G. Nomarski, *Z. Wiss. Mikrosk. Mikrosk. Tech.* **69**, 193 (1969).
[36] P. de Weer and B. M. Salzberg (ed.), *Soc. Gen. Physiol. Ser.* **40**, 1–480 (1986).
[37] M. B. Burg, J. Grantham, M. Abramow, and J. Orloff, *Am. J. Physiol.* **210**, 1293 (1966).
[38] C. E. Ganote, J. J. Grantham, H. L. Moses, M. B. Burg, and J. Orloff, *J. Cell Biol.* **36**, 355 (1968).
[39] J. J. Grantham, C. E. Ganote, M. B. Burg, and J. Orloff, *J. Cell Biol.* **41**, 562 (1969).
[40] J. Grantham, M. Linshaw, and L. Welling, *in* "Epithelial Ion and Water Transport" (A. D. C. Macknight and J. P. Leader, eds.), p. 339. Raven, New York, 1981.
[41] M. A. Linshaw and L. W. Welling, *Am. J. Physiol.* **244**, F172 (1983).

holding pipets so that the lumens were collapsed. Eyepiece micrometer measurements of the tubular diameter could then be used to determine tubular cell volume. The tubules were firmly held so that rapid changes of the bathing solution were possible. More recently[42] this technique has been improved by the rapid recording of a television image of the tubule, and subsequent reproduction and measurement of the captured image. Screen distortion was minimized by arranging the tubule in the vertical plane of the camera. Since the frame speed of the television system used was $\frac{1}{60}$ sec, relatively rapid changes in volume could be followed.

A somewhat similar approach has been adopted by DiBona and colleages[43,44] using isolated and perfused renal cortical collecting tubules of the rabbit. In this setup, tubules were visualized during perfusion using DIC optics. A novel feature of the experimental arrangement was the use of a water-immersion lens as the front lens of the condenser. Although this lens had a relatively low numerical aperture (0.75), the resolution was not significantly impaired, and the longer working distance which this afforded allowed room for the holding pipets. For determination of cellular volume, the microscope was adjusted so that the opposing luminal walls were sharply defined, and cell thickness was minimal. Photographic images were recorded and the cellular and luminal areas of a chosen region were calculated by placing the images on a digitizing pad linked to a computer, and tracing round the margins. The use of a computer with a large dynamic range of density adjustments greatly assisted the delineation of edges. Cellular volume could then be derived by treating the tubule as a right circular cylinder. To reduce the initially high water permeability of the collecting duct, experimental data were collected from tubules incubated at 25°. To minimize the effect of unstirred layers, the bath was flushed rapidly during solution changes. Even so, the time required to change the bathing solution was about 45 sec. However, this system could be simply modified to record and analyze relatively rapid changes in cellular volume.

Several important technical improvements to this system were made by Strange and Spring.[45] A new design for the luminal perfusion system allowed extremely rapid (halftimes, 70-150 msec) changes of luminal perfusate, and rapid flow through the bath reduced basolateral unstirred layers to a minimum. Careful attention to insulation allowed the experiments to be performed at 37°. Cell volume was determined by recording a series of "optical sections" at 1.2-μm intervals of cells located on the

[42] D. J. Welling, L. W. Welling, and T. J. Ochs, *Am. J. Physiol.* **245**, F123 (1983).
[43] K. L. Kirk, D. R. DiBona, and J. A. Schafer, *J. Membr. Biol.* **79**, 53 (1984).
[44] K. L. Kirk, J. A. Schafer, and D. R. DiBona, *J. Membr. Biol.* **79**, 65 (1984).
[45] K. Strange and K. R. Spring, *J. Membr. Biol.* **96**, 27 (1987).

bottom of the tubule. This series of recordings took too long (1-2 sec) to obtain information on rates of cell volume change. This was achieved by imaging a cross section of individual cells in the lateral wall of the tubule and recording images of this cross section at 15 frames per second, during solution changes.

Whittembury and associates are also applying optical techniques to study changes in volume in isolated renal tubules.[45a-47] Initially, improved image resolution was achieved using an optical image splitter.[45a,46] However, it was not easy to follow the time course of the changes in volume as precisely as one would have liked because of the need for manual adjustment of the image splitter. To overcome the problem of adequate resolution of the changes in volume with time, these workers have now developed an alternative approach.[47] Segments of single isolated rabbit renal tubules with collapsed lumens are suspended in a special chamber, crimped at each end with micropipets, and stained with either neutral red or methylene blue. This staining increases the contrast between the tubule, which appears dark, and the bright background. A television camera attached to a microscope feeds its signal to a special processor which allows the area of the dark segments in each frame to be analyzed. The resulting signal is recorded on a pen recorder. Using this approach, an optical resolution close to 0.03 μm could be achieved and changes in the dark image area of 1% were easily resolved. In practice it proved possible to detect accurately changes in volume resulting from a change in bath osmolality of as little as 10 mOsm/liter. Furthermore, these workers were able to change the medium composition in their chamber very rapidly (95% replacement in about 85 msec).

The possibilities afforded by the application of modern microscopy techniques to quantitative studies of sheets of epithelial is best illustrated by the work of Spring and associates (reviewed by Spring[48-48b]). The design and construction of a micro-Ussing-type chamber[49] has permitted the collection of continuous quantitative information over several hours.

In the Spring-Hope chamber, flat epithelial sheets are mounted between two concentric rings of "delrin" and are held in place by laterally intruded 27-gauge needles. For technical details of the construction and

[45a] E. González, P. Carpi-Medina, and G. Whittembury, *Am. J. Physiol.* **242**, F321 (1982).

[46] P. Carpi-Medina, E. González, and G. Whittembury, *Am. J. Physiol.* **244**, F554 (1983).

[47] P. Carpi-Medina, E. González, B. Lindemann, and G. Whittembury, *Pfluegers Arch.* **400**, 343 (1984).

[48] K. R. Spring, *Fed. Proc., Fed. Am. Soc. Exp. Biol.* **44**, 2526 (1985).

[48a] K. R. Spring, *Mol. Physiol.* **8**, 35 (1985).

[48b] K. R. Spring, *Pfluegers Arch. Suppl. 1* **405**, S23 (1985).

[49] K. R. Spring and A. Hope, *Science* **200**, 54 (1978).

use of the chamber, reference should be made to the original papers.[49,50] When mounted on the microscope stage and viewed with Nomarski optics, cell boundaries can be readily visualized. Because of the narrow vertical resolution provided by Nomarski optics, the boundaries of the cells appear sharp and give a clear image as the focal plane is shifted from the mucosal to serosal surface. Due to the small volume of the mucosal chamber, at normal flow rates the total volume of the chamber is washed out about three times per second. It is possible, therefore, to subject the preparation to rapid changes of solution under continuous observation. As Spring and Hope[49] point out, the full possibilities of such a system could not be realized without facilities for accumulating and handling large amounts of data. In the Spring–Hope system, this is achieved by mounting a high-resolution television camera on the microscope and recording successive images as the focusing adjustment of the microscope is moved automatically through a series of focal planes from mucosa to serosa. At the end of an experiment, the images are recalled and a simple program in a microcomputer permits tracing of the outline of a chosen cell at successive depths. The contained area is calculated and successive areas integrated to give a cellular volume, A computer-based analytical system for these calculations has been presented recently.[51] In spite of the difficulties anticipated in the use of this technique (for example, the difficulty in tracing around irregular boundaries, and of deciding the upper and lower limits of the cells), it has provided highly reproducible data in the *Necturus* gallbladder. Spring and Persson[52] report that sequential measurements of the same cell give values for cell volume reproducible to within about 2%.

For the study of cell volume regulation in flat translucent sheets of epithelial tissue, the technique developed by Spring is clearly a significant advance over other methods. The caveat must be included, however, as Spring points out, that inasmuch as it is cell volume which is measured, experiments must be devised which introduce measurable changes in that parameter.

In spite of the successful application of this technique to the gall bladder of *Necturus,* few other tissues have been investigated using the Spring–Hope chamber. The principal reasons for this are probably that, first, the chamber is difficult to set up and to use and, second, that there are few tissues which are as suitable for this system as is the *Necturus* gallblad-

[50] K. R. Spring and A. Hope, *J. Gen. Physiol.* **73**, 305 (1979).

[51] D. J. Marsh, P. K. Jensen, and K. R. Spring, *J. Microsc.* **137**, 281 (1985).

[52] K. R. Spring and B. E. Persson, *in* "Epithelial Ion and Water Transport" (A. D. C. Macknight and J. P. Leader, eds.), p. 15. Raven, New York, 1981.

der. Most interesting epithelia are either too irregular to be easily mounted in the chamber, or have cells which are too small to apply the same techniques for the study of cellular volume. For example, when epithelial sheets from the toad urinary bladder are mounted in the Spring–Hope chamber, contraction of the underlying smooth muscle frequently causes the epithelial cells to move out of the focal plane. This movement may be reduced by stretching the bladder tightly, but the cells are then so thin that accurate measurement of the cell thickness is impossible. Kachadorian, Sariban-Sohraby, and Spring[52a] overcame this problem by making a large number of loosely mounted preparations and choosing only those which remained stationary. Davis and Finn,[53,54] on the other hand, chose to use the frog urinary bladder, which is better suited to such optical studies.

One weakness of the DIC system which should be mentioned is that its operation is optimal when at least some fraction of the incident light passes round the preparation rather than through it, thus generating maximal interference. In consequence, observation of flat sheets of epithelia does not give the high contrast shown by dispersed cells. In addition, subserosal connective tissue adds to light scattering, which further reduces contrast information. In spite of this, very high resolution of cell outlines and their contents is possible.

Contrast can be enhanced and thus the information content of the optical image may be improved by a combination of DIC and fluorescence techniques. Spring and Smith[55] have reviewed the technical difficulties of deriving quantitative information from fluorescence microscopy. Studies may use autofluorescence[56] to identify specific cell types in a mixed epithelium, fluorescent dyes, which give a quantitative measure of the activity of specific intracellular ions, or intravital dyes, such as the dye, $3,3'$–dihexyloxacarbocyanine iodide, $diOC_6$-(3), to locate cellular components. For example, Foskett[57] has used the calcium-sensitive fluor, Fura-2, in combination with DIC optics to record simultaneously changes in cellular volume and intracellular calcium activity during stimulation of isolated pancreatic acinar cells. Development of this technique is likely to generate new understanding of the role of the cellular metabolic machinery in adjustment of cellular volume.

[52a] W. A. Kachadorian, S. Sariban-Sohraby, and K. R. Spring, *Am. J. Physiol.* **248**, F260 (1985).

[53] C. W. Davis and A. L. Finn, *Science* **216**, 525 (1982).

[54] C. W. Davis and A. L. Finn, *Fed. Proc., Fed. Am. Soc. Exp. Biol.* **44**, 2520 (1985).

[55] K. R. Spring and P. D. Smith, *J. Microsc.* **147**, 265 (1987).

[56] M. Köhler and E. Frömter, *Pfluegers Arch.* **403**, 47 (1985).

[57] J. K. Foskett, *IUBS 2nd Int. Congr. Comp. Physiol. Biochem. (Abstr.)* **416** (1988).

In spite of the ingenious technical developments (reviewed by Kam[57a]) which have in the past few years generated such a rapid advance in the study of epithelial cell volume regulation by optical microscopy, it is likely that the next few years will see even more rapid progress. In particular the development of the confocal scanning laser microscope[57b,58] will permit marked improvements in both temporal and spatial resolution.

Techniques for Analysis of Cell Composition

There are four principal methods by which cellular ions and water can be determined.

1. Chemical analysis of the epithelial cells—the cells must be separated from any underlying supporting tissues, and contamination by extracellular fluid corrected for by using a suitable extracellular marker.

2. Isotopic techniques—the steady-state distribution of isotope can provide important information about the exchangeability of ions or water and about their source from mucosal or interstitial fluid. Measurements of the rate of uptake or of washout of isotope may also provide additional information about ion and water fluxes across mucosal and basolateral plasma membranes. In practice, interpretation of the data obtained has often proved difficult.

3. X-Ray microanalysis—this technique provides an important new approach to the study of cellular ions.

4. Ion-selective microelectrodes—these can provide unique information about ion activities in cellular compartments. Unlike the other techniques, measurements are made in living cells, and serial observations in the same preparation are possible.

An additional technique, nuclear magnetic resonance (NMR) spectroscopy, has been applied in an effort to characterize the states of water, sodium, and potassium within cells. There is still controversy about the interpretation of the signals obtained from cells[59] and the technique will not be discussed here. However, the development of anionic paramagnetic shift reagents and their use to shift the NMR absorption of extracellular

[57a] Z. Kam, Q. Rev. Biophys. 20, 201 (1987).
[57b] I. J. Cox, J. Microsc. 133, 149 (1984).
[58] G. J. Brakenhoff, H. T. M. van der Voort, E. A. van Sprosen, and W. Nanninga, Scanning Microsc. 2, 33 (1988).
[59] M. Shporer and M. M. Civan, Curr. Top. Membr. Transp. 9, 1 (1977).

sodium away from the resonance of intracellular sodium is an exciting new development. A recent example of the application of this approach is provided by Wittenberg and Gupta.[60]

Before discussing each of these techniques in more detail, there are two points that should be noted:

1. By dividing the amount of an ion by the amount of water, it is possible to derive a concentration of cellular ions from chemical analysis. Such a derived value may have little meaning. Only the ion-selective microelectrodes can provide a direct estimate of cellular ion activity. Although results obtained for cellular ions in X-ray microanalysis have been expressed in relation to cell water, this technique measures an amount, not an activity.

2. These techniques are not applicable equally to all the preparations to be discussed. Chemical analysis can be performed on them all, though its validity is dependent on minimizing contamination of the epithelial cell sample with other cell types, and on the adequacy of the extracellular markers. Isotopic techniques are critically dependent upon adequate and uniform mixing of the isotope with all of the available extracellular ion or solute studied. The thickness of the preparation and any unstirred layers, as well as cellular heterogeneity, may compromise the conclusions that can be drawn from the experimental results. Microelectrode techniques may be applied, in principle, to the surface cells of slices, isolated tubules, intact epithelia, separated sheets of epithelial cells, and single cells. The limitations of this approach reflect problems of cellular size, accessibility, and mobility and motility of the preparation, as will be discussed. X-Ray microanalysis, likewise, can be applied to any of these preparations, but is limited to cells on or near the surface which freeze rapidly enough so that ice crystal damage is minimized.

Chemical Analysis. Standard techniques for chemical analysis have been applied widely to both epithelial and nonepithelial tissues. Immediately upon removal from the animal or following *in vitro* incubation in an appropriate medium, tissue samples are blotted when possible to remove adhering fluid. A hardened ashless filter paper (Whatman 541, 542, or equivalent) is suitable for this and no measurable contamination of the tissue results. Gentle pressure can safely be exerted during the blotting for it is very difficult to exert sufficient pressure with the finger to actually squeeze fluid out of most tissue samples. When dealing with cell or tubule suspensions, or with separated sheets of epithelial cells, it may not be possible to blot the samples. The best that can then be done is to use some

[60] B. A. Wittenberg and R. K. Gupta, *J. Biol. Chem.* **260,** 2031 (1985).

type of centrifuge tube, such as that described by Burg and Orloff[61] for handling suspensions of isolated renal tubules, and obtain a packed pellet, the extracellular contamination of which can be corrected for with a suitable marker (see below).

Tissue water contents are equated with the loss of weight when tissue is dried under conditions which do not cause sublimation of solutes or oxidation of organic constituents. Drying to a constant weight in a hot air oven at 105°, over phosphorus pentoxide *in vacuo*, or by freeze drying all prove satisfactory in most instances. In our experience, at least 3 mg or more of dried weight is required for accurate determination of water content using this technique. With such small samples it is essential that tubes be weighed several times at each stage (when empty, with wet tissue, and with dried tissue) and that several tubes without tissue be carried through the procedures to allow corrections to be made for changes in room temperature or humidity which may affect the weights from day to day. In addition, weighing tubes containing tissue must not be left to stand in the laboratory unsealed. Apart from the risk of possible contamination, water may be lost from the small sample. We have not, however, encountered problems of loss of weight by wet tissues during the weighing itself, when the tubes are unsealed. Any good stable balance capable of accurate weighing to 0.1 ± 0.05 mg should be adequate for these studies. We have, however, encountered a significant problem in using the newer electronic balances with both glass and plastic tubes. Rather than a stable reading, the apparent weight can vary by many milligrams over several minutes. This behavior appears to result from static electricity on the tubes and precautions must be taken to obviate this problem if accurate reproducible results are to be obtained.

An alternative to drying of tissues is to determine water content from the distribution of tritiated water. This technique is particularly useful when too little tissue is available for accurate weighing; it has been used, for example, to determine the water content of isolated renal tubules[37] and renal papillary collecting duct cells.[62] Correction for extracellular water is, of course, required. Values are usually referred to protein content of the sample when there is insufficient tissue for accurate weighing. Though the ratio will differ somewhat from epithelium to epithelium, about 70% of the tissue dry weight is contributed by protein as measured using conventional techniques.

Tissue contents of ions and other solutes are determined by extraction from the tissue. Though concentrated nitric acid is preferred by some, a

---

[61] M. B. Burg and J. Orloff, *Am. J. Physiol.* **203**, 327 (1962).

[62] D. Dworzack and J. J. Grantham, *Kidney Int.* **8**, 191 (1975).

careful study of extraction procedures has established that 0.1 M nitric acid is adequate.[63] Provided the dried sample can be spread relatively thinly over the surface of the tube, we have found that sodium, potassium, and chloride can be completely extracted overnight at room temperature and without the need to shake the tubes during extraction. (They must, of course, be shaken thoroughly to ensure complete mixing immediately prior to analyses.) Extracellular markers may prove more difficult to extract from dried tissue. We normally extract tissue for several days when we analyze for radioactive inulin, and we have found that some extracellular markers, notably polyethylene glycol 4000, are not readily extracted from dried tissue.[64] In this case extraction from undried tissue was required to fully recover the marker.

Sodium and potassium are measured readily by flame photometry, and chloride is measured by conductometric techniques.[65] These analyses can be carried out directly on the tissue extracts. To avoid the problem of neutralizing the nitric acid we prepare our standards in 0.1 M HNO_3. The volume of extract is chosen so that the samples contain between 0 and 0.5 mmol/liter of ion, and standards of 0.1, 0.2, 0.3, 0.4, and 0.5 mmol/liter of NaCl and KCl are used. Over this range the flame photometer readings are not directly proportional to concentration and calibration curves must be used to provide accurate data. One advantage of using the chloride salts for the flame photometer standards is that, since the chloride titration time is linearly related to chloride concentration, the accuracy of these standards is readily checked by chloride titration. It is important to appreciate the possibility that tissue chloride may be overestimated in some tissues by the conductometric technique, which also titrates sulfhydryl groups unless more elaborate methods for extraction and sample preparation are used.[66] In microsamples, helium glow photometry has been used to determine sodium and potassium,[37] but it has been difficult to obtain consistent and reproducible results. Divalent ions (calcium and magnesium) are measured by atomic absorption spectrometry. Details of techniques for microanalysis of a variety of solutes are given by Greger et al.[67]

To enable estimation of cellular contents of water and ions, tissue values must be corrected for the adherent extracellular medium. Much of this can be removed by gentle blotting as discussed above. Alternatively, extracellular ions can be washed from samples by immersion in ice-cold isosmotic solutions, free of the predominant extracellular ions, for rela-

[63] J. R. Little, *Anal. Biochem.* **7**, 87 (1964).
[64] D. J. L. McIver and A. D. C. Macknight, *J. Physiol. (London)* **239**, 31 (1974).
[65] E. Cotlove, H. V. Trantham, and R. L. Bowman, *J. Lab. Clin. Med.* **51**, 461 (1958).
[66] D. D. Macchia, P. I. Polimeni, and E. Page, *Am. J. Physiol.* **235**, C122 (1978).
[67] R. Greger, F. Lang, F. G. Knox, and C. Lechene, *Methods Pharmacol.* **4B**, 105 (1978).

tively short periods, which need to be determined for each preparation (usually about 30 sec proves satisfactory). However, correction for extracellular water is still required with washed tissues. Measurement of the extracellular space in isolated epithelia is the major problem with chemical analysis. A wide range of markers has been used, including nonmetabolized carbohydrates of various molecular weights (mannitol, sucrose, raffinose, inulin), charged molecules ([61]Co-labeled EDTA, sulfate), and the inert polyethylene glycols (900 and 4000). In general, inulin remains as satisfactory a marker as any in most preparations. However, it equilibrates relatively slowly (minutes to an hour) in tissues, so that, where brief exposure is required (seconds), mannitol may be preferred. The availability of all these compounds as radioisotopes simplifies the measurement of their concentrations in medium and tissue extracts. However, we have noted that samples containing tritiated mannitol, sucrose, or raffinose may lose considerable counts on drying at 105°,[68] so that, where possible, the ^{14}C-labeled solute is preferred. Even with this label, however, we have sometimes encountered problems with relatively small-molecular-weight solutes from losses of counts after drying, and recommend that medium samples should always be dried and then analyzed, as are tissue samples, to check on this possibility and establish the validity of using a particular isotopically labeled extracellular marker. As mentioned in the discussion on the use of radioactive tracers for determining membrane potential, it is possible, if necessary, to use potassium content of dried and undried tissues to calculate the water contents of undried tissue samples indirectly.

Theoretically, it is best to use a radioactive tracer molecule in the presence of an excess of unlabeled molecule to minimize problems which might arise from binding or sequestration of tracer in some small subcompartment. In practice, however, we have not found this to be necessary when using inulin as an extracellular marker.

Measurements of the volume of distribution of a marker which equilibrates throughout the whole extracellular compartment and remains exclusively within that compartment allow calculation of cellular water from measured tissue water content. Estimation of cellular ions, however, requires the additional assumptions that all ions in the extracellular space are in diffusional equilibrium with the incubation medium and that there is no appreciable binding of ions to the extracellular matrix. This remains the major problem in interpreting data from chemical analysis of epithelia. There is no difficulty with potassium, for more than 98% of this ion is cellular when tissues are incubated in a typical balanced sodium chloride medium. Furthermore, as illustrated in Fig. 1, the direct proportionality

[68] P. M. Hughes and A. D. C. Macknight, unpublished observations (1979).

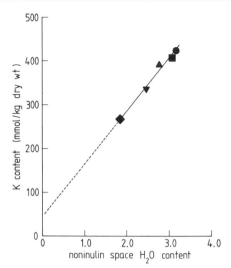

FIG. 1. The relationship between cell potassium content and noninulin space H_2O content in a variety of tissues. (●) Toad bladder, (■) rabbit colon, (▲) *Necturus* gallbladder, (◆) rat renal cortex, and (▼) mouse diaphragm. The symbols represent mean values obtained in experiments in our laboratory. (Reproduced by permission from Ref. 157.)

between noninulin space water and tissue potassium in several tissues, with an estimated cellular potassium concentration of some 135 to 150 mmol/ liter for each case, argues strongly that noninulin space water is close to, if not identical with, cell water. However, there remain problems with sodium and with chloride in matching the values obtained from chemical analysis with those measured in some preparations by electron microprobe analysis or estimated from ion-selective microelectrodes.

Isotopic Techniques. Only brief comments will be made in this section which relate to the determination of cellular ions. With the epithelial preparation in a steady state, the rates of equilibration of cellular with extracellular ions can be assessed. Either the tissue is exposed to the isotope and the rate of accumulation measured, or the tissue is first loaded with the isotope and the rate of loss to the isotope-free bathing medium (wash out) is determined. In such studies, techniques of compartmental analysis[69] enable one, in principle, to extract from the overall pattern of exchange, the kinetics of exchange and the size of individual compartments.[70,71] Normally, at least two compartments can be recognized—cellular and

[69] A. K. Solomon, *in* "Mineral Metabolism" (C. L. Comer and F. Bronner, eds.), p. 119. Academic Press, New York, 1960.
[70] A. L. Finn and M. L. Rockoff, *J. Gen. Physiol.* **57,** 326 (1971).
[71] B. Biagi, E. Gonzalez, and G. Giebisch, *Curr. Top. Membr. Transp.* **13,** 229 (1980).

extracellular. Sometimes further compartments are apparent. With appropriate preparations in which transepithelial transport, isotope content, and isotopic fluxes can be determined together with the rate of cellular uptake or loss, it is theoretically possible to derive the cellular ion contents as well as the unidirectional fluxes of the transported ion across both apical and basolateral membranes of the transporting cells. However, in practice, such studies have not always provided satisfactory data, the derived cellular contents being high,[72] possibly due to trapping of isotope within the extracellular compartments.

Isotopes may be used simply to determine the extent of exchangeability of ions in epithelial preparations. Here tissues are incubated with isotope until a steady-state distribution is achieved. Assuming the same specific activity in tissue and in medium, the content of isotope may be compared with the content from chemical analysis. Again, to obtain cellular values the major interpretative problem, as with chemical analysis, is to define the extracellular ion content. A second assumption is that both isotopic and nonisotopic species have equal access to the entry sites on the plasma membranes. In some epithelial (toad urinary bladder,[73] bullfrog choroid plexus,[74] rabbit ileum,[75] and descending colon[76]), much of the cellular potassium exchanges very slowly with medium potassium. This finding may reflect preferential recycling of cellular potassium diffusing across the basolateral plasma membrane, i.e., this unlabeled potassium may have preferential access to the entry sites on the outer surface of the plasma membrane.[77]

X-Ray Microanalysis. The principles of X-ray microanalysis and its application to biological specimens are reviewed in detail elsewhere.[78-81] This type of analysis can be carried out using suitably equipped scanning or transmission electron microscopes or dedicated electron microprobes. Briefly, a stream of high-energy electrons striking a specimen results in the emission of high-energy, short-wavelength photons (X-rays) of two types. One type, arising from electron transitions within the excited atoms, results

[72] A. D. C. Macknight, D. R. DiBona, and A. Leaf, *Physiol. Rev.* **60,** 615 (1980).
[73] B. A. Robinson and A. D. C. Macknight, *J. Membr. Biol.* **26,** 269 (1976).
[74] T. Zeuthen and E. M. Wright, *J. Membr. Biol.* **60,** 105 (1981).
[75] H. N. Nellans and S. G. Schultz, *J. Gen. Physiol.* **68,** 441 (1976).
[76] R. A. Frizzell and B. Jennings, *Fed. Proc., Fed. Am. Soc. Exp. Biol.* **36,** 360 (1977).
[77] M. M. Civan, *in* "Epithelial Ion and Water Transport" (A. D. C. Macknight and J. P. Leader, eds.), p. 107. Raven, New York, 1981.
[78] T. A. Hall and B. L. Gupta, *J. Microsc.* **136,** 193 (1984).
[79] P. L. Moore, J. A. V. Simson, H. L. Bank, and J. D. Balentine, *Methods Achiev. Exp. Pathol.* **11,** 138 (1984).
[80] R. B. Moreton, *Biol. Rev.* **56,** 409 (1981).
[81] R. Rick, A. Dörge, and K. Thurau, *J. Microsc.* **125,** 239 (1982).

in an energy emission characteristic of the particular element. The other type (Bremsstrahlung) is due to deceleration of the incident electrons within the specimen and not to electron transitions, and yields a continuous energy spectrum, or continuum, sometimes also referred to as uncharacteristic radiation. The thinner the specimen, the smaller is the Bremsstrahlung.

The emitted X-rays can be detected either by a wavelength-dispersive or an energy-dispersive system. Though the wavelength-dispersive system gives a high resolution and thus provides good separation of adjacent energy peaks, it has a low efficiency and can detect only a narrow portion of the total spectrum at any one time. Improvements in energy-dispersive detectors in recent years suggest that these will become the standard detectors for analysis of biological specimens. In these detectors the emitted X-rays impinge on a lithium-drifted silicon crystal coated on both surfaces by a thin layer of gold, across which a high voltage (1000 V) is placed. This crystal is usually isolated from the electron microscope vacuum by a thin beryllium window. (Though windowless detectors are available, and have the advantage of allowing measurement of elements of lower atomic weight than fluorine, they are much more susceptible to contamination and damage and are not generally recommended.) When X-rays strike the crystal, a current pulse arises which is proportional to the X-ray energy. After amplification of this current pulse, it is converted to a voltage pulse which is measured in a pulse-height analyzer and stored. A major advantage of this technique is that a complete energy spectrum over the range of energies of interest can be accumulated over the time course of the analysis. This spectrum contains both the characteristic elemental peaks and the continuum radiation.

To apply the technique to cell analysis, however, requires suitable specimens. The aim must be to retain the normal ionic distributions and water content within the cells during the preparation. As a first step, the tissue is rapidly frozen at low temperature. Depending on the nature of the specimen, rapid freezing may best be achieved either by plunging it into a liquid cryogen or by slamming it onto a precooled surface. A variety of cryogens have been employed.[82] Liquid nitrogen should be avoided since its vaporization forms a gaseous film over the specimen, which slows freezing appreciably. Probably the best cryogen is liquid propane prepared by cooling propane gas with liquid nitrogen. To this can be added a small volume of isopentane.[83] The isopentane–propane mixture remains liquid at the temperature of liquid nitrogen, rather than freezing as propane alone

[82] H. Plattner and L. Bachmann, *Int. Rev. Cytol.* **79**, 237 (1982).
[83] B. Jehl, R. Bauer, A. Dörge, and R. Rick, *J. Microsc.* **123**, 307 (1981).

does at $-180°$. Slamming tissue against a copper block precooled by liquid nitrogen (or liquid helium) can also provide very rapid freezing, and relatively simple devices have been described.[84] The major aim must be to avoid excessive ice-crystal formation. The quality of freezing is readily assessed by freeze substitution.[85] Even with the best systems, it is not possible to obtain a tissue depth of more than about 10 to 20 μm free of detectable ice-crystal damage by electron microscopy. However, if one is simply concerned with comparing average cellular and extracellular compositions rather than with determining the precise distributions of elements between cytosol and organelles, a greater thickness of tissue is available for analysis.

The frozen tissue must be sectioned at a low enough temperature to prevent translocation of elements during the cutting. This is a controversial topic. The thinner the section, the lower the sectioning temperature can be. From a consideration of the states of ice and given the possible losses of water by sublimation, however, it is generally claimed that one must prepare sections at temperatures lower than $-70°$ to $-80°$.[86] Fortunately, it is possible to prepare 1-μm-thick sections at these temperatures. Such sections are thick enough to be readily analyzed in a scanning electron microscope fitted with an X-ray-detecting system, yet thin enough to allow electrons in the impinging electron beam to be transmitted through the specimen. This simplifies considerably the analysis of the data.

Tissue sections may be analyzed in the microscope in either the hydrated or the dehydrated state. Analysis of hydrated sections has the advantage that elements contained within areas devoid of matrix (tubular lumens, lateral intercellular spaces) should retain their distribution. In addition, analysis of a hydrated section followed by analysis of the same section after dehydration *in situ* allows estimation of water content. Analysis of hydrated sections, however, requires that the section retains its hydration during its preparation and that the temperature remains low (below $-100°$) during the transfer of the section from the microtome to the microscope, and within the microscope during analysis. Special transfer systems and cold stages are now commercially available. A disadvantage with frozen hydrated sections is that identification of structure in scanning and transmission electron micrographs is very difficult. It thus is not easy to equate measured values for elements with structural components. The distribution of phosphorus has been used recently to localize nuclei, and the distribution of the potassium signal can be equated with

[84] A. Verna, *Biol. Cell.* **49**, 95 (1983).
[85] H. Plattner and H. P. Zingsheim, *Subcell. Biochem.* **9**, 1 (1983).
[86] T. A. Hall and B. L. Gupta, *Q. Rev. Biophys.* **16**, 279 (1983).

cells.[87] However, the disadvantage of a poor image from hydrated sections can be overcome by reanalyzing the same section after dehydration in the microscope.

Freeze-dried sections can be prepared by standard techniques. Exposure to 10^{-6} torr at $-80°$ for 12 hr is more than sufficient for 1- to 2-μm-thick sections. These sections can then be analyzed without the necessity for a cold stage in the microscope. However, inasmuch as mass loss caused by the electron beam may be minimized at low temperatures, there seems to be an advantage in also analyzing freeze-dried sections at $-150°$. During freeze drying, the losses of water from matrix-free structures (tubular lumens and other extracellular compartments) cause the elements which were in solution to accumulate on the nearest solid surfaces. Therefore, localized regions of high elemental concentration may be found adjacent to plasma membranes. This is a major disadvantage if one is attempting to critically examine the distribution of elements in these regions. However, there is no evidence that, at the level of resolution of the technique, migration of cellular elements occurs to any significant extent during freeze drying. Though not comparable in detail to the structural resolution obtainable in appropriately fixed tissues, the scanning and transmission electron micrographs obtainable from freeze-dried sections allow adequate resolution of cellular and extracellular compartments and enable identification of different cell types in epithelial preparations. At present, it appears that the best approach is to prepare frozen hydrated sections which can be analyzed, then dehydrated within the microscope, before reanalysis and obtaining electron micrographs.

In attempting to allocate the measured elemental energies to particular tissue localities, it is possible to locate the specimen within the microscope so that the electron beam is centered on the area of interest. The size of the beam can be adjusted, though a minimum diameter is required to carry sufficient current to excite adequate X-ray production. In practice, beams of diameter of the order of 100 nm are commonly used, and areas of 1 to 2 μm are scanned for 100 to 200 sec. Since all of the X-rays produced need not come from the actual area through which the electron beam passes, care is required to minimize scattering of the electron beam and other artifacts.[88]

Having obtained data, this must be analyzed and quantified. This requires correction of the measured elemental peaks for continuum radiation and the expression of the extracted elemental X-ray energies in terms of tissue contents. There are several approaches to correct for the contin-

[87] A. J. Saubermann, P. Echlin, P. D. Peters, and R. Beeuwkes, *J. Cell Biol.* **88,** 257 (1981).
[88] C. P. Lechene, *Fed. Proc., Fed. Am. Soc. Exp. Biol.* **39,** 2871 (1980).

uum. The one we favor is that developed by Fuchs and Fuchs,[89] which is based on the peak-to-background method introduced by Hall.[90] The background is removed by digital filtering in combination with a linear least-squares procedure for peak deconvolution. Suitable computer programs for this are available[89] or can be developed, and provide isolated peak intensities for the elements of interest.

Having identified the intensity of the specific elemental radiation, this must be expressed in terms of tissue composition. Two techniques have been applied in epithelial studies. The simpler approach is to place a thin layer (approximately 10 to 20 μm in thickness) of a solution of known composition and mass which approximates tissue mass onto the surface of the tissue just prior to freezing. In studies of frog skin and toad urinary bladder, solutions of 20% albumin (w/v)[91,92] or dextran[93] in the Ringer's solution which bathed the mucosal surface were used. Each cryosection then has its own control area of known composition and of the same thickness as the sectioned tissue. Comparison of the intensity of the elemental signals from the albumin layer and from the tissue allows direct quantification of the tissue concentrations of elements. Furthermore, in freeze-dried sections, the dry weight of the tissue (and therefore its water content per unit per dry weight) can be estimated by comparison of the Bremsstrahlung intensities of cell and standard. Quantification using this method requires that the sections analyzed are of uniform thickness, and, in freeze-dried sections, that no local changes in density occur during the drying.

The alternative technique for quantification of tissue composition was developed by Hall.[90] The Bremsstrahlung is proportional to the total electron dose and the thickness of the specimen. In thin sections, the value of the Bremsstrahlung is directly related to the weight of the material excited by the electron beam, so that by normalizing the characteristic X-ray line intensity to the intensity of the Bremsstrahlung, and by comparison of this ratio with that obtained separately from known standards, it is possible to estimate elemental concentrations, provided that mass loss during the irradiation and contamination of the specimen are avoided. Measurement

[89] H. Fuchs and W. Fuchs, "Scanning Electron Microscopy," Vol. 2, p. 377. SEM Inc., Chicago, Illinois, 1981.

[90] T. A. Hall, *Phys. Tech. Biol. Res.* **1A**, 17 (1971).

[91] A. Dörge, R. Rick, U. Katz, and K. Thurau, *in* "Epithelial Ion and Water Transport" (A. D. C. Macknight and J. P. Leader, eds.), p. 127. Raven, New York, 1981.

[92] R. Rick, A. Dörge, A. D. C. Macknight, A. Leaf, and K. Thurau, *J. Membr. Biol.* **39**, 247 (1978).

[93] M. M. Civan, T. A. Hall, and B. L. Gupta, *J. Membr. Biol.* **55**, 187 (1980).

of cell ions in both hydrated and dehydrated sections allows calculation of cellular water content:

$$\text{Cell } H_2O \text{ (kg/kg dry wt.)} = \frac{\text{cell ions dehydrated (mmol/kg dry wt.)}}{\text{cell ions hydrated (mmol/kg wet wt.)}} - 1$$

(3)

Though it is customary to express the results from X-ray microanalysis in terms of elemental concentrations, it must be emphasized that the quantity of an element in the area excited is detected, whether that element was free in solution or part of the cellular structure. Thus, like chemical analysis, it allows no distinction between "bound" and "free" ions within cells. The advantages of the technique (ability to measure the composition of different cell types within a complex epithelium, ability to measure composition in cytoplasm and larger cellular organelles differentially, and the independence of the measurements from extracellular contamination) are such that, as the techniques are refined and the sensitivities of detection (especially for sodium) are improved, much important new information about epithelial composition and cell volume regulation under different conditions will be obtained.

Ion-Selective Microelectrodes. Ion-selective intracellular microelectrodes have two advantages over other analytical methods: they measure the gradient of electrochemical activity directly, and they may be used nondestructively either to monitor intracellular ionic activity over long periods or to investigate rapid changes in cellular composition. Detailed discussion about the types of electrodes available and their construction, calibration, and use can be found elsewhere.[21,94-96]

Ion-selective microelectrodes can now be fabricated routinely using either special ion-selective glasses or by introducing liquid ion-exchange resins into the tips of conventional microelectrodes. These hydrophobic liquids effectively form a membrane separating a reference solution in the microelectrode from the external solution. Liquid ion exchangers are of two types. The first, which includes the well-known resins manufactured by Corning for potassium (477317) and chloride (477315) determination, consists of a hydrophobic electrolyte dissolved in an organic hydrophobic solvent. Armstrong[96] discusses qualitatively the operation of these systems,

[94] R. C. Thomas, "Ion-Sensitive Microelectrodes: How to Make and Use Them." Academic Press, New York, 1978.

[95] N. W. Carter and L. R. Pucacco, *Methods Pharmacol.* **4B**, 195 (1978).

[96] W. McD. Armstrong, *in* "Epithelial Ion and Water Transport" (A. D. C. Macknight and J. P. Leader, eds.), p. 85. Raven, New York, 1981.

which is treated in a more strictly theoretical manner by Eisenman and colleagues.[97] As Armstrong[96] points out, the discriminant properties of membranes depend either (in cases where the membrane solvent has a relatively high dielectric constant) on the solvent properties or (when the solvent dielectric constant is low) on the particular combination of organic electrolyte and solvent chosen. Although this approach has the advantage that a particular combination can be chosen for a specific purpose, it has the limitation that such combinations can only be found by a laborious process of trial and error. Furthermore, because discrimination is determined by relatively nonspecific factors, such systems tend to have relatively low selectivity.

The second type of membrane incorporates neutral ligands in which the selectivity depends on the formation of highly specific complexes between a particular species of ion and a carrier molecule. The nature of these complexes depends to a large extent upon the structure of the carrier molecule. This raises the possibility that such molecules can be specifically synthesized for a particular cation. Such an approach, utilized by Simon and associates,[98,99] has already lead to the synthesis of highly specific carriers for the sodium and calcium ions, and further developments in this area can be anticipated.

It is important to stress that when an ion-selective microelectrode is advanced from the bathing medium into a cell and a successful penetration is made, any change in potential that is recorded is the algebraic sum of two potentials in series: (1) the cellular membrane potential and (2) the change in electrode potential due to a change in ionic activity at the electrode tip. Thus, it is necessary to know the intracellular activities of any interfering ions to which the electrode is also sensitive. Determination of the effect of interfering ions may be achieved easily if a specific electrode is available to measure the intracellular activities of those ions. Where this is not possible, estimates must be made to relate measured concentrations to calculated activities. Because this is a significant source of error, the need for highly selective electrodes becomes obvious.

To compare the activity of an ion measured by an ion-selective microelectrode with estimates of the intracellular concentration determined from chemical analysis or X-ray microanalysis, it is necessary to relate, in some way, concentration and activity in the intracellular environment. In bulk solutions this relation is defined by $a = \gamma c$, where a is the activity, c is the concentration of the particular ion, and γ is the activity coefficient, a

[97] G. Eisenman (ed.), "Glass Electrodes for Hydrogen and Other Cations." Dekker, New York, 1977.
[98] M. Guggi, M. Oehme, E. Pretsch, and W. Simon, *Helv. Chim. Acta* **58**, 2417 (1976).
[99] W. Simon and W. E. Morf, *Membranes* **2**, 329 (1973).

function of the ionic strength and composition of the solution. Although the intracellular fluids are isosmotic with interstitial fluid, the presence of appreciable concentrations of polyvalent electrolytes within the cells might be expected to increase the intracellular ionic strength as compared with the external fluid, which is made up largely of simple electrolytes. Thus the activity coefficients for ions within the cells may differ appreciably from those in the bathing media. Furthermore, the measured activity is that in the microenvironment of the electrode tip and need not reflect the activity throughout the cell. As Table I shows, it is often difficult to match data from ion activity measurements with data obtained by other techniques, unless one uses activity coefficients lower than those predicted for simple electrolyte solutions. The problem is especially marked for sodium and chloride.

These results serve to illustrate some of the major interpretative difficulties which are faced in determining cellular ionic composition. What is the true intracellular ion content? What is the correct cellular activity coefficient for each of the ions under study? If different cellular activity coefficients are required for different ions to reconcile the data, does this indicate differential binding or compartmentalization of ions? Only the application of a variety of techniques can provide answers to these questions.

Another problem deserves brief comment, for it relates directly to the difficulties in studying cellular volume regulation. A measured change in the intracellular activity of an ion will result from either a change in the transmembrane movement of that ion or from an opposing movement of water. Thus measurement of change of ionic activity within a cell cannot

TABLE I

RELATIONSHIPS BETWEEN MEASURED CELL ION ACTIVITIES AND DERIVED
CELL ION CONCENTRATIONS

Tissue	Activities (mmol/liter)			Concentrations (mmol/liter)			Required activity coefficients		
	Na^+	K^+	Cl^-	Na^+	K^+	Cl^-	Na^+	K^+	Cl^-
Necturus gallbladder[a]	13	92	26	62	136	52	0.21	0.68	0.50
Rat skeletal muscle[b]	7	99	5	26	168	26	0.27	0.59	0.18

[a] From J. P. Leader, A. D. C. Macknight, D. R. Mason, and W. McD. Armstrong, *Proc. Univ. Otago Med. Sch.* **59,** 82 (1981).

[b] From J. P. Leader, J. J. Bray, A. D. C. Macknight, D. R. Mason, D. McCaig, and R. G. Mills, *J. Membr. Biol.* **81,** 19 (1984).

per se distinguish between changes in the intracellular content of that ion and changes in the amount of intracellular water. One must, in addition, have independent measures of both of these. Fisher and Spring[23] have studied changes in cell ion activities in response to changes in medium osmolality in *Necturus* gallbladder, and used previously measured changes in cell volume under the same conditions to calculate cell ion contents. The difficulties with these techniques are such that this approach can yield qualitative rather than truly quantitative data.

A particularly elegant example of the application of both optical and ion-selective electrode studies to the regulation of cellular volume is given by the work of Larsen, Spring, and colleagues.[99a,99b]

Techniques for Measurement of Plasma Membrane Potentials

Interpretation of the results of analysis of cellular composition in relation to the regulation of cellular volume is critically dependent on a knowledge of potentials across both the apical and basolateral membranes. Not only is such knowledge essential for the calculation of cell ion activities, as discussed above, but, since it contributes to the driving force for ion movements across the membranes, the membrane potential is an integral part of equations which relate cell solute concentrations to membrane fluxes of the ions. Since the measurement of membrane potentials is not otherwise dealt with in this volume, some discussion of the available techniques is provided.

Conventionally, plasma membrane potentials are measured using microelectrodes. Particularly in epithelia, it has often proved difficult to obtain satisfactory results with microelectrodes, so that, as well as refinements of this technique, alternative techniques are being applied to assess these potentials.

Microelectrode Techniques. Liquid junction potentials. In the conventional method for measurement of transmembrane potentials, the microelectrode is filled with a concentrated electrolyte solution (usually concentrated KCl) and electrical contact is established with solution through a chlorided silver wire. The circuit is completed through a salt bridge connecting the bathing solution with a second liquid/metal junction. If this second metal/liquid junction is also made of a chlorided silver wire in contact with concentrated KCl solution, and the salt bridge consists of a KCl solution of the same strength in agar, then, when the microelectrode and the salt bridge are both in contact with the bathing medium, the circuit should be symmetrical and record zero potential. The microelectrode can

[99a] E. H. Larsen, H. H. Ussing, and K. R. Spring, *J. Membr. Biol.* **99**, 25 (1987).
[99b] N. S. Willumsen and E. H. Larsen, *J. Membr. Biol.* **94**, 173 (1986).

then be advanced into the cell and the transmembrane potential determined as the difference in potential recorded before and after traversing the membrane. An important assumption, which is often ignored, underlies this convention. When both the electrode and the salt bridge are in contact with the bathing medium there are two liquid junction potentials in the circuit, one between the electrode tip and the medium and one at the point of contact of the salt bridge with the medium. At these points, diffusion of ions down their concentration gradients will set up diffusion potentials, whose magnitudes may be calculated by the use of the appropriate equations. In most physiological experiments, the liquid junctions are between a concentrated electrolyte solution and a dilute saline solution. Here the size of the junction potential is determined largely by the properties of the more concentrated electrolyte solution. Since the mobilities of K^+ and Cl^- are almost equal, this junction potential is small, generally of the order of 2–3 mV. In addition, since the circuit is symmetrical, the two diffusion potentials from electrode and from the salt bridge cancel so that no potential difference is recorded in the circuit. When the electrode is inserted into the cell, however, the two liquid junctions are not symmetrical. Within the cell much of the anionic contribution to electroneutrality is from large organic ions whose mobilities will be much lower than that of chloride, and thus the liquid junction potential will now differ from that between the electrode and the external medium. Thus, when the electrode traverses the cell boundary, the change in potential recorded is the algebraic sum of the membrane potential and any change in the liquid junction potential at the electrode tip. Indirect methods have suggested that the change in liquid junction potentials may lead to an underestimate of the true transmembrane potential by as much as 2 to 3 mV.[100] Where alternative solutions are used for filling the micropipet, the magnitude of this difference may be even larger.[101] In view of other sources of uncertainty, such as membrane damage or depolarization caused by ion leakage from the electrode tip, this problem is usually ignored. However, where the external bathing solution is changed to one having a markedly different composition, as, for example, when relative ion permeabilities are being investigated or the effects on cell volume of anisosmotic media are being studied, liquid junction potentials cannot be ignored.

Diffusional exchange of solute in the reference salt bridge with the bathing medium may give rise to large, spurious potentials, whose magnitude is incalculable. This difficulty has been discussed by Barry and Dia-

[100] U. V. Lassen and B. E. Rasmussen, in "Membrane Transport in Biology" (G. Giebisch, D. C. Tosteson, and H. H. Ussing, eds.), Vol. 1, p. 169. Springer-Verlag, Berlin and New York, 1978.

[101] T. Hironaka and S. Morimoto, J. Physiol. (London) 297, 1 (1979).

mond[102] and by Way and Diamond.[103] One solution to this problem is to maintain the reference liquid junction at a constant composition by maintaining a slow flow of solution through the bridge. To avoid accumulation of KCl in the bathing solution, this flowing liquid junction is generally made of dilute NaCl. The liquid junction potential can then be accurately calculated using the Henderson equation or variants thereof.[102] Unfortunately, data for the motilities of ions at 37° are incomplete, and, for many organic ions, are lacking altogether. Because of this difficulty, and the tediousness of calculating liquid junction potentials, such possible sources of error, though they may be considerable, are often neglected.

A further troublesome source of possible error in making transmembrane potential measurements in epithelia during changes in bathing solutions lies in the development of further diffusional potentials in regions of restricted diffusion. The effect of "unstirred layers" in the bulk bathing solutions may be reduced by adequate design of the incubation chamber and by sufficient stirring, but, where measurements of potential are made across basolateral membranes of epithelial cells, it may be necessary to take account of the fact that a variable amount of connective tissue intervenes. If the serosal bathing solution is changed, then during the period of equilibration in the new solution, the measured potential between a microelectrode in an epithelial cell and a reference electrode in the bath will be influenced, not only by any change in transmembrane potential, but also by the potential developed as ionic gradients in the serosal connective tissue are dissipated. For accurate and rapid measurements of changes in the membrane potential alone, it may be necessary to site the reference electrode much closer to the membrane. An example of this approach is provided by Fisher.[104]

Tip potentials. A more troublesome source of error lies in the presence of so-called "tip potentials." These may be defined as the potential, other than that due to a liquid junction potential, existing between a microelectrode with intact tip, and an equivalent microelectrode with a broken tip immersed in the same solution. Since the early work of Adrian[105] and Nastuk and Hodgkin,[106] it has become clear that tip potentials may arise from two sources, i.e., either from surface conduction between the inner surface of the glass and the contained electrolyte, or from current flow through the glass in the region of the electrode tip. Thus one might expect, and this seems to be true in practice, that as microelec-

[102] P. H. Barry and J. M. Diamond, *J. Membr. Biol.* **3**, 93 (1970).
[103] L. W. Way and J. M. Diamond, *Biochim. Biophys. Acta* **203**, 298 (1970).
[104] R. S. Fisher, *Am. J. Physiol.* **247**, C495 (1984).
[105] R. H. Adrian, *J. Physiol. (London)* **133**, 631 (1956).
[106] W. L. Nastuk and A. L. Hodgkin, *J. Cell. Comp. Physiol.* **43**, 39 (1950).

trodes are pulled to finer and finer tip diameters with correspondingly greater electrical resistances, tip potentials tend to increase. A disconcerting aspect of this phenomenon is that tip potentials may appear, or change, during the course of an experiment. Furthermore, since ion selectivity of the electrode glass itself may contribute to the tip potential, the stability of the tip potential before and after insertion into a cell is no guide to its contribution to the measured intracellular potential. To a large extent this problem may be overcome by using electrodes of a lower resistance; indeed, large initial tip potentials often fall after one or two cellular penetrations have been made, presumably due to breakage of extremely fine tips. As Thomas[94] points out, the assumption that very high-resistance electrodes are also very sharp is not necessarily true. Modern techniques for beveling the tips of electrodes can give electrodes which will penetrate small cells often as easily as much finer electrodes, and, because of their lower resistance, these give much lower electrical noise, with consequent ease of recording, and have correspondingly smaller and more stable tip potentials. A number of systems, most of them homemade, have been described in the literature. In our laboratory, a system similar to that of Brown and Flaming[107] involving abrasion of the tip of the electrode in an abrasive slurry while monitoring electrode resistance has been found to give excellent results for both single-barrelled and double-barrelled electrodes. With smaller cells, where it may not be possible to use beveled electrodes, the reported measurements of potential made with high-resistance electrodes should be treated with caution.

As an alternative to the use of concentrated electrolyte solutions for filling microelectrodes, Thomas and Cohen[108] have used the carrier potassium tetrakis(p-chlorophenyl)borate dissolved in n-octanol in the tips of silanized microelectrodes. They found that the selectivity of this ion exchanger was such that it did not discriminate between sodium and potassium ions. Since in many cellular systems the sum of the activities of sodium and potassium ions is approximately equal inside and outside the cells, it should be possible to use an electrode of this type as a potential-sensing probe without the problems of junctional and tip potentials. In addition, the hydrophobic silanized surface of such a microelectrode seems to penetrate cell membranes more easily and to make a better seal. In our laboratory we have made transmembrane potential measurements using a double-barrelled electrode in which one barrel was unsilanized and filled with KCl, and the other was silanized and filled with ionophore, to investigate the Thomas–Cohen neutral ionophore microelectrode as a potential probe. Though there was excellent agreement between the potential mea-

[107] K. T. Brown and D. G. Flaming, *Neuroscience* **2**, 813 (1977).
[108] R. C. Thomas and C. J. Cohen, *Pfluegers Arch.* **390**, 96 (1981).

surements from the two barrels in the extensor digitorum longus muscle of the mouse, in the epithelial cells of the *Necturus* gallbladder, the KCl barrel gave consistently higher values by about 5 mV.[109] The reasons for this difference are unclear but it indicates that results obtained using the neutral ionophore electrode should be interpreted with caution. A disadvantage of these electrodes is that they have a much higher impedance than conventional ones, so it is difficult to measure their resistance before and after penetration of a cell. Thus criteria for a valid penetration are more difficult to establish.

Impalement Artifacts. The technique for penetration of large, excitable cells by microelectrodes has changed little since the early experiments of Ling and Gerard.[110] The electrode is placed close to the tissue and then advanced by some micromanipulative arrangement in small steady steps. Penetration of the plasma membrane is marked by a sudden shift of potential which, thereafter, remains stable. In epithelial cells, penetration of the membrane is not simple. First, epithelia available as sheets are generally positioned in such a way that only one surface of their polarized cells is accessible. Usually the basolateral plasma membranes are obscured by the underlying connective tissue subjacent to the basement membrane, and, although workers have on occasion penetrated epithelial cells either by carefully dissecting away the connective tissue or by advancing a microelectrode through it,[111-114] the difficulties are considerable. Penetrations have, therefore, usually been made from the mucosal side in such epithelia. In contrast, where the epithelium is tubular, only the serosal surface is available for impalement. Since there is often relatively little connective tissue surrounding the tubule, puncture is somewhat easier.

Second, the cells are often small and the experimental arrangement generally requires that mucosal and serosal sides be perfused independently. Also the tissue underlying the epithelium frequently incorporates smooth muscles, which means that considerable technical ingenuity may be required to maintain the cells in a stable position. This may be achieved, in part, by supporting the tissue from the serosal side on a more or less rigid meshwork, and by maintaining a slight hydrostatic pressure of 1 to 2 cm of

[109] P. J. Donaldson, A. G. Butt, J. P. Leader, and A. D. C. Macknight, *Proc. Univ. Otago Med. Sch.* **63,** 57 (1985).
[110] G. Ling and R. W. Gerard, *J. Cell. Comp. Physiol.* **34,** 383 (1949).
[111] R. S. Fisher, D. Erlij, and S. I. Helman, *J. Gen. Physiol.* **76,** 447 (1980).
[112] J. T. Higgins, B. Gebler, and E. Frömter, *Pfluegers Arch.* **371,** 87 (1977).
[113] J. Navarte and A. L. Finn, *J. Gen. Physiol.* **75,** 323 (1980).
[114] S. G. Schultz, S. M. Thompson, and Y. Suzuki, *in* "Epithelial Ion and Water Transport" (A. D. C. Macknight and J. P. Leader, eds.), p. 285. Raven, New York, 1981.

water across the tissue (mucosa positive). This appears to be adequate to suppress spontaneous muscular movements in some tissues.[115]

A number of artifacts may arise during the impalement of epithelial cells. Some of the following artifacts are most frequently encountered. First, on approaching the cells, the electrode may record a small variable negative potential which increases as the microelectrode is advanced, but which remains stable when the electrode is still. Such recordings appear to be the result of pressure artifacts arising from contact either with an overlying layer of dead cells, or with mucous. When a cell is penetrated, it is frequently found that the potential rises rapidly to a peak and then almost immediately decays. It may then remain stable, decay further, or rise slowly to a steady value over the next several minutes. Such variability has led to the careful search by many workers for objective means of defining what should be regarded as a stable potential. Ideally, the microelectrode should rapidly penetrate the plasma membrane and reach a point of contact with intracellular liquid, with the formation of an insulating seal at the point of penetration. If the tip of the electrode becomes partially plugged with cytoplasm or the seal between the glass surface of the electrode and the membrane is imperfect, the recorded potential will change. Artifacts of the first kind may be detected by monitoring the microelectrode resistance before, during, and after impalement. If the resistance of the microelectrode rises significantly when in the cell (by more than a few megohms), then the tip is probably plugged with cellular contents and the measured potential should be disregarded. One symptom of this is that the potential recorded rises as the electrode is pushed further into the cell.

If a good insulating seal is not established between the electrode and the penetrated membrane, current leakage through the gap created will tend to depolarize the cells. Because many, if not all, epithelial cells are more or less tightly coupled electrically, the rate of decay of potential may be relatively slow. Conversely, a sudden leakage of cellular contents, followed by sealing, may be accompanied by a slow return to a stable resting potential. The adequacy of the membrane seal can be objectively assessed in many cases by determining the so-called voltage–divider ratio. If a pulse of current is passed across the tissue, then the voltage drop across the mucosal and serosal plasma membranes will be proportional to their relative resistances. Usually, a greater voltage drop occurs across the mucosal membrane reflecting its greater electrical resistance. While it is advanced up to the cell, the microelectrode will show no response to transepithelial current pulses. Once inside the cell it will record the voltage drop across the mucosal membrane in response to a current pulse superimposed upon the

[115] L. Reuss and A. L. Finn, *J. Gen. Physiol.* **64,** 1 (1974).

transmucosal potential. Inadequate sealing causing paraelectrode leaks will greatly attenuate the transmucosal voltage drop. Examples of these types of artifacts are illustrated by Armstrong and Garcia-Diaz.[116]

The objective assessment of epithelial plasma membrane potentials, therefore, requires first that the electrode resistance should be substantially unchanged before, during, and after cell impalement; that the steady potential recorded by the electrode in the bathing medium should not have changed by more than 1 or 2 mV before and after penetration; that the recorded intracellular potential should remain at a steady value for some minutes; and that the measured voltage–divided ratio should be consistent and within expectations. Ideally, measured potentials should be confirmed, where possible, by penetration from the alternative surface. In very few tissues has it been possible to satisfy all these criteria. Additionally, the rate of perfusion across the surface of the tissue during the measurements may influence the magnitude of the recorded potential.

In some epithelia it has proved to be extremely difficult to obtain reliable penetrations. In these cases it may be possible to use alternative methods, instead of microelectrodes.

Alternative Methods for Membrane Potential Measurement. In recent years a number of alternative approaches have been developed to determine plasma membrane potentials, and some of these have been applied to epithelia.[117]

In essence, there are two alternatives to the use of microelectrodes to measure membrane potential. One is to take advantage of the fact that any ion which can permeate membranes rapidly will reach an equilibrium between medium and membrane-bounded compartments which is dependent upon the transmembrane potential. Determination of the equilibrium ratio will then permit calculation of the potential. There are, however, several limitations to this apparently simple approach. First, the substance whose distribution is to be measured must have the following properties:

1. It must cross plasma membranes rapidly enough to reach an equilibrium distribution within an experimentally acceptable time.

2. It must be capable of detection in minute quantities. The concentration in bulk solution is determined relatively easily, and the limitation is the measurement of intracellular concentration. Since the cells are usually polarized such that the internal compartment is electrically negative to the exterior, passively distributed cations will be accumulated within cells and

---

[116] W. Armstrong and J. F. Garcia-Diaz, *in* "Epithelial Ion and Water Transport" (A. D. C. Macknight and J. P. Leader, eds.), p. 43. Raven, New York, 1981.

[117] J. P. Leader and A. D. C. Macknight, *Fed. Proc., Fed. Am. Soc. Exp. Biol.* **41,** 54 (1982).

anions excluded. For accurate measurements, therefore, it is preferable to use a cationic substance.

3. It must be nontoxic at high concentrations. Since cations will be accumulated inside cells, being concentrated approximately 10-fold for each 60 mV of potential difference, it follows that cells will be subject to relatively high concentrations of the chosen marker. If these affect metabolism, then the measured potential may not reflect the normal value.

4. It must not be bound to cellular molecules nor accumulated in subcellular compartments. For example, accumulation by mitochondria as a consequence of a potential difference, between cytosol and the interior of the mitochondrion, can yield erroneous results.[118,119]

These requirements may be met, at least in part, by a number of substances (chiefly organic lipophilic ions) which may be detected either by using radioactive tracers or by their fluorescence. However, this approach is only valid if the tissue is short-circuited, in which case the membrane potentials across both apical and basolateral plasma membranes are the same, or if the substance can diffuse across only one of the two membranes.

Radioactive Tracers. The lipophilic cations triphenylmethyl- and tetraphenylphosphonium have been prepared as tritiated compounds at high specific activity. Frequently it is necessary to carry out some preliminary experiments to confirm that the substances can penetrate cell membranes. In many cases penetration is assisted by the presence of the lipophilic anion tetraphenylboron. Several important technical difficulties hinder the ready use of these cations. First, they can be used accurately only on a homogeneous population of cells. Second, analysis can only be achieved by cellular destruction, and so these substances are unsuited to dynamic experiments. Third, because of the possibility of exchange of tritium between the labeled compound and the aqueous solvent, the experiments can be only of limited duration and, therefore, equilibration should be achieved rapidly. Fourth, heating samples containing triphenylmethylphosphonium, or removing water by evaporative freeze drying, causes loss of radioactivity from the samples to a variable extent. It is, therefore, necessary to determine concentration of intracellular triphenylmethylphosphonium indirectly. Leader and Macknight,[117] in their studies on the *Necturus* gallbladder and on the urinary bladder of the toad, related the potassium content of samples extracted without drying to that of samples whose water content was known. In the gallbladder of *Necturus*

---

[118] D. Lichtshtein, K. Dunlop, H. R. Kaback, and A. J. Blume, *Proc. Natl. Acad. Sci. U.S.A.* **76**, 2580 (1979).

[119] S. Schuldiner and H. R. Kaback, *Biochemistry* **14**, 5451 (1975).

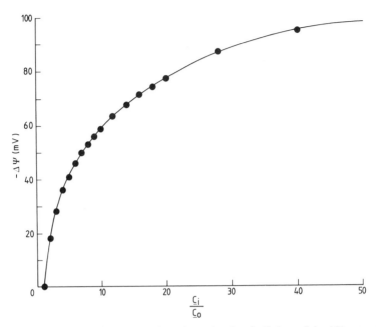

FIG. 2. The relationship between the estimated ratio of cellular activity (C_i) to medium activity (C_o) and the derived membrane potential ($-\Delta\psi$) for a cation distributed at equilibrium. Because the ratio is a logarithmic function of potential difference, quite large changes in the measured ratio make only a small difference to the estimated potential difference. (Reproduced by permission from Ref. 117.)

they found good agreement between the distribution of triphenylmethylphosphonium ions and measured membrane potentials. It must be noted that, with membrane potentials within the usual physiological range (some 50 to 80 mV or so), the measured distribution ratio of triphenylmethylphosphonium, between 10:1 and 20:1, is so large that the calculated transmembrane potential will not be affected markedly by a considerable degree of nonpotential-dependent distribution (Fig. 2). By the same token, however, it must be admitted that this limits the accuracy of the measurement.

As an alternative to the determination of intracellular TPMP, an extracellular ion-specific electrode has been used to monitor redistribution. A tetraphenylphosphonium-sensitive electrode was designed by Kamo, Muratsugau, Hongoh and Kobatake,[119a] and was used by them to determine the transmembrane potential of mitochondria, by monitoring the change in concentration of the ion in the external medium following addition of a

[119a] N. Kamo, M. Muratsugau, R. Hongoh, and Y. Kobatake, *J. Membr. Biol.* **49**, 105 (1979).

known volume of a mitochondrial concentration of the ion in the external medium following addition of a known volume of a mitochondrial suspension. Binding of the ion to the mitochondrial membrane was found to be negligible.

The problem raised by nonpotential-dependent distribution can be solved, at least in part, by determining the distribution ratio achieved when the transmembrane potential is brought to a known value, as, for example, by treating the plasma membranes with valinomycin. Under these conditions, the membrane is permeable predominantly to potassium ions, and, if the internal and external concentrations of potassium are known, then the membrane potential can be calculated and the difference between that and the measured ratio used to estimate the contribution of nonpotential-dependent distribution. Such experiments must be done with great care, however, since altering the potential will create driving forces for other ion movements, and if these ions are redistributed at similar rates to the marker substance, the calculated potential will be in error.

Fluorescence detection. As an alternative to the use of radioactive tracers, the ability to make quantitative measurements on trace amounts of fluorescent chemicals has provided another method of measurement of membrane potentials. Waggoner and co-workers have reported the synthesis of a number of dyes which, being lypophilic and bearing charge, will distribute themselves between cells and the medium.[120] Within cells their fluorescence is quenched and, therefore, the total fluorescence of the system (medium and cells) is altered by an amount which is, in part, a function of the transmembrane potential. If that potential is changed, then the redistribution of the dye will be signaled by a change in the fluorescence of the system. This has been exploited by Hoffman and Laris[121] to give a technique which provides a sensitive monitor of changes of potential in red blood cells. Two dyes in particular, 3,3″-dipropylthiacarbocyanine iodide [diSC$_3$-(5)] and diOC$_6$-(3), have been used, because they give by far the largest responses of those tested to date. The technique, however, is subject to several disadvantages. First, a large number of experiments are necessary as a preliminary to discover the optimum concentration of cells and dye. Second, it is necessary to calibrate the system so that a change in fluorescence can be equated with a change in potential. The low permeability of red blood cells to potassium ions facilitates the design of experiments in which, by the addition of valinomycin, the membrane potential can be driven to a known value over a large range. In epithelia such an approach has been unsuccessful.[117] Third, red blood cells have a limited number of

[120] A. G. Waggoner, *Annu. Rev. Biophys. Bioeng.* **8**, 47 (1979).
[121] J. F. Hoffman and P. C. Laris, *J. Physiol. (London)* **239**, 519 (1974).

intracellular structures for dye accumulation and can be obtained readily in suspension. This is not true of epithelia. When tested in a number of different ways with epithelial tissues, these dyes are rapidly adsorbed onto intracellular structures and onto extracellular material, including mucous and connective tissue, and even at high external concentrations (up to 1 mmol), all the dye is removed from solution.[117] Fourth, a number of studies have shown that the dyes may act as metabolic inhibitors or cause photodynamic damage.[120] Though attempts have been made,[122] it seems that this technique is not directly applicable to most epithelia. The dyes, however, offer other useful possibilities for epithelial investigation, for example, their use in assessing changes in cell volume, as mentioned earlier in this chapter.

Waggoner[120] has reported a second class of dyes which appear to offer more exciting possibilities for measurement of membrane potentials and which are beginning to be exploited in epithelia. These compounds are lipid-soluble dyes which dissolve in membranes and which show a change in optical signal when the electric field in which they are situated is changed. This change may be measured as a change in fluorescence at the appropriate wavelength, a change in absorbance or reflectance, or as a change in the rotation of polarized light. Cohen et al.[123] have tested many hundreds of these dyes, using as a model the action potential of squid axon.

There has been a steady improvement in the size of the signal as new dyes are tested. In a recent review, Grinvald et al.[124] point out that the styryl dyes recently synthesized in his laboratory give a signal about 120 times larger than those in the pioneering experiments of Salzberg et al.[125] They are also less phototoxic by a factor of between 200 to 300. Using these new dyes, Grinvald et al.[126] have been able to record signals from the finest of individual cell processes.

Although voltage-clamp experiments using the squid axon have shown that the size of the change of optical signal is linearly related to the magnitude of the change in electric field in which the dye is oriented,[120] the absolute magnitude of the fluorescence is dependent upon extraneous factors such as the partition coefficient of the dye between membrane lipids and the bathing medium. Thus an absolute calibration of fluorescence in terms of transmembrane voltage is not possible unless the voltage can be measured independently. Nevertheless, important information can

[122] C. N. Graves, G. Sachs, and W. S. Rehm, *Am. J. Physiol.* **238**, C21 (1980).
[123] L. B. Cohen, B. M. Salzberg, V. Davila, W. N. Ross, D. Landowne, A. C. Waggoner, and C.-H. Wang, *J. Membr. Biol.* **19**, 1 (1974).
[124] A. Grinvald, R. Hildesheim, I. C. Farber, and L. Anglister, *Biophys. J.* **39**, 301 (1982).
[125] B. M. Salzberg, H. V. Davila, and L. B. Cohen, *Nature (London)* **246**, 508 (1973).
[126] A. Grinvald, A. Fine, I. C. Farber, and R. Hildesheim, *Biophys. J.* **42**, 195 (1983).

be obtained. Recently, Crowe[127] has reported that it is possible to record optical signals from the apical membrane of the epithelial cells of the toad urinary bladder using the styryl dye RH160.[126] The sizes of the signals (when measured as a fraction of steady fluorescence) were linearly related to the magnitude of a transepithelial voltage pulse, were abolished by the application of nystatin (which inserts channels into the apical membrane and lowers its resistance), and were greatly enhanced by the addition of amiloride (which, by competitively binding with sodium entry sites, raises the apical membrane resistance and hence increases the fraction of the applied transepithelial voltage which appears across the apical membrane). Thus this method can provide data from which the relative and absolute resistances of the apical and basolateral membranes can be calculated, even without a knowledge of the absolute transmembrane potential differences.

Epithelial Preparations for the Study of Cell Volume Regulation

Aspects of cell volume regulation have been studied in a variety of epithelia. As well as reflecting the specific interests of different investigators, this is also a consequence of the difficulties in applying the available techniques to different epithelia. To illustrate the principles and the problems involved, we shall here discuss advantages and disadvantages of several commonly used preparations. We shall not deal with epithelial cells in culture since these are discussed by Handler *et al.* in this volume.[128] Nor will we provide step-by-step accounts of the details involved in the handling of each preparation. The references cited provide this.

Mammalian Renal Epithelia

Few epithelial tissues are amenable to simple treatments for the determination of their composition by chemical techniques, because they are small or are not easily isolated from supporting tissues. Exceptions include the kidney, from which tissue slices are readily prepared. The use of the slice technique for studying volume regulation in renal epithelia was pioneering by Robinson[129] and by Mudge.[130] More recently, suspensions of renal tubules[61,131] and of isolated cells[132] have been prepared, and individual tubules have been dissected from the kidney and incubated and perfused *in vitro*.[37]

---

127 W. E. Crowe, *Proc. Univ. Otago Med. Sch.* **63**, 12 (1985).
128 J. S. Handler, N. Green, and R. E. Steele, this volume [37].
129 J. R. Robinson, *Biochem. J.* **45**, 68 (1949).
130 G. H. Mudge, *Am. J. Physiol.* **165**, 113 (1951).
131 R. S. Balaban, S. P. Soltoff, J. M. Storey, and L. J. Mandel, *Am. J. Physiol.* **238**, F50 (1980).
132 A. Vandewalle, B. Kopfer-Hobelsberger, and H.-G. Heidrich, *J. Cell Biol.* **92**, 505 (1982).

Each preparation has its advantages and disadvantages. The renal cortical slice is easily obtained from kidneys removed from the animal, cut into segments, and kept in chilled sodium Ringer's solution prior to slicing. A description of a general procedure for preparing slices is provided by Cohen,[133] who reports in detail a modification of the earlier method of Deutsch.[134] We have found this technique, which involves free-hand slicing with a razor blade, to be perfectly satisfactory for the routine preparation of cortical slices down to 0.15 mm thick. Any contamination by medullary elements is clearly visible through the frosted glass slide, which is applied to the upper surface of the segment during the slicing. We have not found any advantage for the production of renal slices in using devices designed specially for the automatic production of tissue slices. In our experience slices cut by such devices, including the Stadie–Riggs system,[135] are often rather too thick. It is also less easy to detect any medullary contamination. However, for more friable tissues, in particular liver, with which we have experience, a slicer of the type described by McIlwain and Buddle[136] can, when properly set up, provide excellent reproducible relatively thin (about 0.3 mm) sections.

To obtain consistent and reproducible results, slices should first be incubated in a relatively large volume of well-stirred oxygenated medium for at least 15 min. This allows the cells to establish a steady-state composition and it removes red blood cells and damaged material on the surface of the slice.[137] Cells damaged by slicing lose their contents into the medium and equilibrate with extracellular fluid. Such damaged cells do not appear to affect the steady-state composition of the slices as judged by a direct comparison of the composition of outermost slices, cut only on one surface, with deeper slices which have both surfaces cut.[138]

Kidney slices have the advantages of being easy to handle with sufficient tissue, even in thin slices (0.2 mm or so), for chemical analysis of water and ions by the conventional methods discussed earlier. Microelectrode studies can also be carried out with them, though relatively few such studies have been reported.[139,140] In addition, X-ray microanalysis should

---

[133] P. P. Cohen, in "Manometric Techniques" (W. W. Umbreit, R. H. Burris, and J. F. Stauffer, eds.), p. 135. Burgess, Minneapolis, Minnesota, 1945.

[134] W. Deutsch, J. Physiol. (London) **87**, 56P (1936).

[135] W. C. Stadie and D. C. Riggs, J. Biol. Chem. **154**, 687 (1944).

[136] H. McIlwain and H. L. Buddle, Biochem J. **53**, 412 (1953).

[137] J. R. Robinson, J. Physiol. (London) **158**, 449 (1961).

[138] A. D. C. Macknight, Proc. Univ. Otago Med. Sch. **45**, 38 (1967).

[139] F. Proverbio and G. Whittembury, J. Physiol. (London) **250**, 559 (1975).

[140] A. Kleinzeller, J. Nedvídková, and A. Knotková, Biochim. Biophys. Acta **135**, 286 (1967).

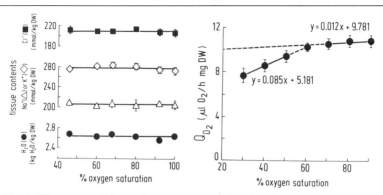

FIG. 3. Tissue composition and oxygen consumption (Q_{O_2}) in rat renal cortical slices (0.2 to 0.3 mm thick) incubated at 25°. Note that Q_{O_2} remains relatively constant at P_{O_2} greater than 60%, but declines slowly below that. Nevertheless, at P_{O_2} greater than 40%, tissue composition remains constant. Note also that the Q_{O_2} of rat renal tissue is about double that of rabbit. This study, therefore, provides a more critical test of the adequacy of tissue oxygenation at 25° than would a study of rabbit renal cortical slices. (Data redrawn from Ref. 144.)

be readily applied to the preparation, as it can be to other renal preparations.[141]

Kidney slices have the disadvantage that they cannot be incubated at 37° without developing some degree of hypoxia, as judged by loss of tissue potassium. However, thin slices (0.2 to 0.3 mm) can be incubated at 25°, maintaining a constant cell composition with a high cell potassium content and a stable oxygen consumption with medium oxygen saturations of greater than 50% (Fig. 3). Another disadvantage also reflects the slice thickness in that studies of the early (seconds to minutes) responses to medium perturbations, such as changes in medium osmolality or the addition of radioisotopes, are not possible. Diffusional delays within the slice interstitial fluid mean that cells at the surface of the slice will equilibrate much more rapidly with the medium than will those toward the center of the slice.

Cellular heterogeneity is also a complication, but, since some 80% or so of the cells are proximal tubular, this is not a major problem in most studies. Of more concern is the absence of a patent lumen in the proximal tubules. Within seconds of the interruption of the blood supply, filtrate disappears from the lumen and the slices are, therefore, prepared from kidneys without patent proximal tubular lumens. This also applies to many preparations of tubular suspensions. However, by perfusing the kidney with solutions containing collagenase *in situ* prior to removal, it is

[141] K. Thurau, F. Beck, J. Mason, A. Dörge, and R. Rick, in "Epithelial Ion and Water Transport" (A. D. C. Macknight and J. P. Leader, eds.), p. 137. Raven, New York, 1981.

possible to prepare isolated tubules with patent lumens.[131,142] This suggests that under *in vivo* conditions the lumens remain patent as a consequence of the tubular hydrostatic pressure holding the collagenase-sensitive basement membrane under tension. Removal of this pressure with interruption of glomerular filtration would, therefore, contribute to the collapse of the lumen. If this interpretation is correct, it seems likely that kidney slices with open lumens could also be prepared from kidneys which were first perfused with collagenase solutions *in situ*. We are presently investigating this possibility.

However, even though the lumens in slices are collapsed, there is indirect evidence that an appreciable transepithelial transport of sodium continues in this preparation. If, as seems true for other epithelia, basolateral membrane sodium permeability is low relative to apical membrane permeability, the extent of ouabain inhibition of oxygen consumption should allow a comparison of the sodium fluxes across the apical plasma membranes in different preparations. In suspensions of isolated rabbit tubules with open lumens incubated at 37°, Balaban *et al.*[131] reported a decrease of almost 70% with maximal doses of ouabain, whereas, in rabbit renal cortical slices incubated at 25°, both Whittam and Willis[143] and Cooke[144] found an inhibition of 40%. These data suggest, therefore, that the renal cortical slices are transporting sodium from lumen to serosal medium at about 50% of the rate of the isolated tubule suspension.

Suspensions of isolated tubules may be prepared, as originally described,[61] by removing kidneys, separating the appropriate part (cortex or medulla), and exposing slices or portions thereof to collagenase. For the cortex, this approach yields proximal cortical tubules, the majority of which have collapsed lumina. A better approach, modified from Balaban *et al.*,[131] which yields mostly proximal tubules, the majority of which (some 87%) have patent lumens, is described in detail by Soltoff and Mandel.[145] Briefly, in an anesthetized rabbit, the kidneys are perfused *in situ* with a slightly hyperosmotic solution (which contains collagenase) for some 10 min. This perfusion is preceded and followed by 5- to 10-min perfusions without the enzyme. Kidneys are then removed and the cortex is cut into pieces and stirred at 4° for 30 min in an isosmotic extracellular-type medium. This disperses the tubules. The suspension is then filtered through gauze to remove any unseparated fragments. Following several centrifugation and washing steps, a satisfactory preparation is obtained.

[142] P. Vinay, A. Gougoux, and G. Lemieux, *Am. J. Physiol.* **241,** F403 (1981).

[143] R. Whittam and J. S. Willis, *J. Physiol. (London)* **168,** 158 (1963).

[144] K. R. Cooke, *Q. J. Exp. Physiol.* **64,** 69 (1979).

[145] S. P. Soltoff and J. Mandel, *J. Gen. Physiol.* **84,** 601 (1984).

The advantages of tubular suspensions include the minimal diffusional delay in exchanges between cells and medium. This makes them ideal for examining isotope exchange. It is also possible to obtain adequate oxygenation at 37° and, in tubules with patent lumens, metabolic studies can be combined with studies of isotope exchange and the effects of inhibitors to examine relationships between metabolism, transepithelial transport, and factors such as sodium pump activity, which can affect cell composition.[146,147] Against these advantages must be set the disadvantages, the most important of which are the technical difficulties of separating tubules from medium and of determining cell water and ions on relatively small samples. There is also the problem of possible impairment of cell function as a consequence of exposure to collagenase. This may be of great importance when examining regulation of cell volume, for it has been suggested that the basement membrane plays an important role in volume regulation by restricting cell swelling.[148] As with renal slices, cellular heterogeneity can be minimized by careful preparation. Even so, the preparations examined by Soltoff and Mandel[145–147] contained about 15% of tubules which were not of proximal origin, a value similar to that for renal cortical slices.

Isolated individual tubules can be dissected from portions of kidney. Such dissected tubules seem to be most easily obtained from rabbit kidneys and the bulk of the reported studies have utilized tubules from this species. Full descriptions of this technique can be found elsewhere.[149] Briefly, slices from the appropriate part of the kidney are incubated in chilled media. Using a binocular microscope and fine forceps, individual tubules are dissected out and then transferred to appropriate containers for incubation with or without perfusion of the lumen, as required.

The advantages of isolated tubules include minimal diffusional delay between cells and medium, and incubation at 37°. The tubules have not been subjected to enzymatic treatment and it is possible to perfuse the lumen and collect the perfusate. For studies of cellular volume, however, the major advantage is the fact that it is possible to visualize directly changes in cell volume following manipulations of the bathing medium, as discussed earlier.

The major disadvantage of this preparation lies in the minute quantities of tissue, making accurate analysis of cell composition extremely difficult. In addition, the lumina are collapsed in the absence of perfusion and it is difficult to obtain satisfactory tubules from species other than the rabbit.

[146] S. P. Soltoff and J. Mandel, *J. Gen. Physiol.* **84**, 623 (1984).
[147] S. P. Soltoff and J. Mandel, *J. Gen. Physiol.* **84**, 643 (1984).
[148] M. A. Linshaw, *Am. J. Physiol.* **239**, F571 (1980).
[149] A. M. Chonko, J. M. Irish III, and D. J. Welling, *Methods Pharmacol.* **4B**, 221 (1978).

It is instructive to compare data obtained for cell water and ions from the different types of renal preparation (Table II). It is apparent that, if one is content to work at 25° rather than at 37°, renal slices remain excellent preparations as judged by the relatively high K^+/Na^+ ratio and the relatively low cell water, sodium, and chloride contents under steady-state conditions. In our experience the first evidence of impairment of cell function is loss of tissue potassium (which must be from the cells) in exchange for sodium, followed by uptakes of sodium, chloride, and water. Note that, when compared to the recent data from renal tubule suspensions incubated at 37°, cells incubated at 25° contain somewhat more potassium (and also sodium). This does not necessarily indicate that cells at 25° are more viable *in vitro*. It may reflect alterations in the balance between active and passive solute movements at these two temperatures.

In summary, when selecting a renal preparation with which to study cell volume regulation in this epithelium, the following points should be considered. First, thin renal slices are suitable for experiments which can be performed at 25°, and in which steady-state volumes rather than transient changes in volume or composition (including isotope exchanges) are

TABLE II

COMPARISON OF DERIVED CELL ION AND WATER CONTENTS FROM
CHEMICAL ANALYSIS IN PREPARATIONS FROM RABBIT RENAL CORTEX

Tissue	Temperature (°C)	H_2O (kg/kg dry wt.)	Concentration (mmol/kg dry wt.)		
			Na^+	K^+	Cl^-
Slices[a]	25	2.08	86	347	108
Suspensions[b]	25	2.14	147	285	121
Closed lumens[c]	28	2.36	194	253	153
Open lumens[d]	25[e]	2.07[f]	126	300	—
	37	1.73	89	238	—
Single tubules[g]	25	2.37	165	261	132

[a] From Cooke and Macknight, *J. Physiol. (London)* **349,** 135 (1984).

[b] From M. B. Burg, E. F. Grollman, and J. Orloff, *Am. J. Physiol.* **206,** 483 (1964).

[c] From M. Paillard, F. Leviel, and J. P. Gardin, *Am. J. Physiol.* **236,** F226 (1979).

[d] From Ref. 145.

[e] Recalculated using protein/dry weight ratio of 0.71 given in the reference.

[f] H_2O contents at 25° calculated on the assumption that gains of Na and K, compared with 37°, were accompanied by isosmotic water uptake.

[g] From Ref. 37.

to be examined. Their major advantages lie in their ease of handling and the simplicity of analysis. Much important information continues to be obtained with them. Second, suspensions of renal tubules are particularly suited to studies at 37° in which aspects of metabolism and transport are to be explored and in which isotope exchanges are examined. Third, individual isolated tubules are the preparations of choice for the direct visualization of changes in cell volume. Only in this preparation, with luminal perfusion, is it possible to measure transepithelial movements of ions and water directly *in vitro*.

Epithelial Sheets

Many epithelia can be obtained as sheets suitable for mounting in Ussing-type chambers. Among the most widely studied include epithelia from the gastrointestinal tract, from amphibian skin, and from urinary bladder. The principles applied to the study of transepithelial movements of solutes and water and the techniques for voltage clamping epithelia have been documented exhaustively (for a recent review with references, see Ref. 21) and will not be dealt with here. Instead, we shall highlight some of the techniques available and the problems which arise in attempting to assess cell volume in sheets of epithelia.

First, it is not possible to estimate epithelial cell composition from chemical analysis of the whole tissue, for the epithelial cells contribute but a small fraction to the total mass (for example, epithelial cells represent only about 10% of the total dry mass in the thin translucent toad urinary bladder).[150] Two approaches have been employed to deal with this problem. Cells can be removed from the underlying tissues and then incubated and collected for analysis. This usually involves treatment of the tissue with an enzyme, such as crude collagenase, which breaks down the basement membrane and allow sheets of epithelial cells to be removed by gentle scraping with a glass slide. An example of this approach, with a detailed description of the technique, is provided in Macknight *et al.*[151] Efforts have been made to obtain single, isolated epithelial cells by combining this approach with exposure to calcium-free media containing a chelating agent such as EGTA [ethylene glycol bis(β-aminoethyl ether)-N,N,N,N'-tetraacetic acid] and mechanical agitation. Phase-contrast microscopy of such cells reveals that they round up with the development of intracellular vesicles and blebbing of the plasma membranes. We feel that

[150] A. D. C. Macknight, M. M. Civan, and A. Leaf, *J. Membr. Biol.* **20**, 387 (1975).
[151] A. D. C. Macknight, D. R. DiBona, A. Leaf, and M. M. Civan, *J. Membr. Biol.* **6**, 108 (1971).

it is better to work with small sheets of isolated epithelial cells rather than to attempt to obtain individual cells from such sheets.

There are, however, major disadvantages in working with these isolated epithelial sheets. It is no longer possible to quantify transepithelial solute and water movements. In addition, both the apical and basolateral plasma membranes are exposed to any changes in the ionic or nutrient composition of the medium and to any isotopes, hormones, or drugs added to the medium.

To overcome some of these difficulties, techniques have been developed by which transepithelial transport across an intact epithelial cell layer can be monitored and the cells then obtained for analysis. In mammalian intestine it is possible simply to remove the epithelial cell layer, together with a small portion of the muscularis mucosae, as an intact sheet by scraping with a glass slide, and to mount this in a chamber.[152] This approach works well with colon and can also be achieved with ileum. These preparations, in fact, seem more viable than do samples of the whole thickness of the gut wall. This is presumed to reflect difficulties in providing adequate oxygenation at 37° and the presence of unstirred layers in subepithelial tissue. However, jejunum prepared in this way is very friable and it is difficult to obtain a sufficient area of undamaged epithelium to provide adequate material for subsequent chemical analysis.

In frog skin, pretreatment with collagenase followed by dissection or by exposure to a gradient of hydrostatic pressure allows separation of the epithelial cell layer as a sheet from the supporting layers.[153-155] This preparation is becoming increasingly popular, particularly for microelectrode studies and for experiments in which rapid effects from drugs and hormones are required. In the much thinner toad urinary bladder, incubation of the whole tissue in chambers can be followed by scraping of the mucosal surface with a glass slide, which removes virtually all the epithelial cells with minimal contamination by submucosal elements.[151] However, to obtain sufficient epithelial cells for analysis, large hemibladders are required. We routinely use hemibladders large enough to cover an area of 8 cm^2 in the Ussing chamber, mounting the tissues in the chamber as loosely as possible. This normally provides more than 3 mg (dry weight) of epithelial cell scrapings. In addition, we always perfuse the heart with isosmotic sodium–Ringer's solution prior to removal of the bladders, to wash the cells out of the bladder's blood vessels. Otherwise, the scrapings obviously can be contaminated by red blood cells.

[152] R. A. Frizzell, M. J. Koch, and S. G. Schultz, *J. Membr. Biol.* **27**, 297 (1976).
[153] J. Aceves and D. Erlij, *J. Physiol. (London)* **212**, 195 (1971).
[154] R. H. Parsons and T. Hoshiko, *J. Gen. Physiol.* **57**, 254 (1971).
[155] R. M. Rajerison, M. Montegut, S. Jard, and F. Morel, *Pfluegers Arch.* **332**, 302 (1972).

Conclusions

The application of the techniques described here to a variety of epithelia, together with the use of other techniques discussed in this volume, is beginning to enable a new and dynamic view of the maintenance and regulation of volume in epithelial cells (for recent reviews, see Refs. 48, 156, and 157).

The steady-state volume reflects the balance between the inflows and outflows of permeant solutes superimposed upon a metabolically dependent constancy of nondiffusible impermeant cell solutes. At any time, cell water content is dependent upon the total cell osmoles, as water moves rapidly across plasma membranes to maintain isosmolality between cells and interstitial fluid.

In isosmotic sodium–Ringer's solution the volume that cells establish may reflect the most energetically favorable state for the required rate of transepithelial solute transport. Some, if not all, epithelial cells maintain a remarkably constant volume despite large changes in rates of transcellular transport. Such constancy requires parallel changes in the rates of ion uptake across the apical plasma membrane and of ion extrusion across the basolateral membrane.[158] In a sodium-absorbing epithelium, the equality of sodium uptake and sodium extrusion could simply reflect the properties of the Na^+,K^+-ATPase, with the tendency for cell sodium to change, thereby altering the rate of activity of individual pump units. Alternatively, however, the numbers of pump units available in the basolateral membrane may be matched to the rate of sodium entry. In addition, basolateral membrane potassium permeability would have to change in parallel with pump activity for cell potassium to remain constant. Both the changes in the rates of sodium extrusion and potassium diffusion from the cells may, therefore, result from the activation of quiescent membrane units or from the insertion into or removal from the membrane of units stored elsewhere in the cell. It is also possible, though highly speculative, that the pump units in fact contain a pathway for potassium diffusion from the cells.[159]

In addition to these adjustments of basolateral membrane properties, apical plasma membrane permeability is also influenced by alterations in transepithelial sodium transport. Increases in cell sodium inhibit further sodium entry to the cells across the apical membrane. This may reflect an

[156] P. Kristensen and H. H. Ussing, *in* "The Kidney: Physiology and Pathophysiology" (D. W. Seldin and G. Giebisch, eds.), p. 173. Raven, New York, 1985.
[157] R. Gilles, A. Kleinzeller, and L. Bolis (eds.), *Curr. Top. Membr. Transp.* **30**, 3–271 (1988).
[158] S. G. Schultz, *Am. J. Physiol.* **241**, F579 (1981).
[159] B. M. Anner, *Biochem. Int.* **2**, 365 (1981).

increase in cytosolic calcium activity as a consequence of the decreased gradient for calcium–sodium countertransport across the basolateral membrane.[160] This capacity for apical membrane sodium permeability to respond to changes in cell sodium provides further protection of cellular volume.

As well as this matching of ion influxes and effluxes under isosmotic conditions, many epithelial cells appear also to be able to minimize changes in volume under anisosmotic conditions, an ability that they share with other cell types.[161,162] This may involve the same mechanisms as are involved in minimizing changes in volume under isosmotic conditions (see, for example, Ref. 26).

Further understanding of epithelial cell volume regulation will come not simply from the application of the techniques presented in this chapter. It will also require the use of a wide range of techniques, many of which are discussed in other chapters in this volume, which allow the properties of the apical and basolateral plasma membranes to be explored, and which will enable us to understand how these properties are regulated and controlled. At the level of whole animal physiology, homeostatic mechanisms are now well characterized. It appears that at the cellular level, also, negative feedback mechanisms may operate to control the volume of epithelial, and, perhaps, other cells.

Acknowledgments

Work in the authors' laboratory is supported by a program grant from the Medical Research Council of New Zealand.

[160] H. Chase, Jr., *Am. J. Physiol.* **247,** F869 (1984).
[161] E. K. Hoffman, *Fed. Proc., Fed. Am. Soc. Exp. Biol.* **44,** 2513 (1985).
[162] A. W. Siebens, *in* "The Kidney: Physiology and Pathophysiology" (D. W. Seldin and G. Giebisch, eds.), p. 91. Raven, New York, 1985.

[39] Scanning Electrode Localization of Transport Pathways in Epithelial Tissues

By J. Kevin Foskett and Carl Scheffey

Much of our understanding of the mechanisms for water and solute transport by epithelia has been as a result of studies of intact, isolated epithelial tissues. An important advantage of intact preparations is that the cells which comprise the tissues remain polarized, joined at their apical boundaries by tight junctions, thereby permitting study of vectorial transport of solute and water. Most epithelial tissues are composed of more than one functionally distinct cell type, so net vectorial transport is the integrated result of the absorptive and secretory activities of the various cell types which comprise them. As such, a proper understanding of the specific transport functions of each individual cell type is necessary. Although there are several approaches to this problem, most involve invasive (lectin binding, dye loading) or disruptive (cell culture, fixation for observations of cell structure, enzyme localization) procedures which preclude studying the various cell types in the native intact epithelium. However, some noninvasive methods have been recently employed. Pathways for water flow across the amphibian urinary bladder have been localized by mass spectrometer analysis of fluid collected at discrete sites immediately above the epithelium.[1] Pathways for ionic current flow across an epithelium have been noninvasively localized optically in transmitted light by monitoring the effects of transepithelial passage of current on individual cell volume[2,3] and electrically by measuring the distribution of extracellular current density immediately above the surface of the epithelium (references below). The present chapter confines itself to a discussion of this latter approach.

Electrophysiological Methods for Scanning Extracellular Current Density in Epithelia

"Scanning micropipet" methods measure voltage gradients in the medium above the epithelial surface. The voltage gradients are interpreted either implicitly or explicitly as current densities. The conversion factor from voltage gradient to current density is the conductivity of the medium, so that interpretation of the measurement as absolute current density is always subject to an assumption about the local conductivity of the medium. The rationale of the technique is that by measuring a voltage gradient over a short enough distance, close enough and perpendicular to the tissue surface, a good representation of the current density through the epithelial surface is obtained. The physics of electric fields implies that the representation becomes perfect as the distance approaches zero. Unfortunately, the signal-to-noise ratio concomitantly goes to zero. Thus, finite distances must be used, which places a limitation on the spatial resolution, i.e., transepithelial current away from the point of recording contributes to

[1] J. A. Jarrell, J. G. King, and J. W. Mills, *Science* **211**, 277 (1981).
[2] R. G. O'Neil and R. A. Hayhurst, *Am. J. Physiol.* **248**, F449 (1985).
[3] J. K. Foskett and H. H. Ussing, *J. Membr. Biol.* **91**, 251 (1986).

the measurements. The extracellular recording approach has been taken using essentially two different methodologies in which either electrolyte-filled glass microelectrodes or metal-filled electrodes have been employed.

Glass Microelectrodes

Frömter[4] used a glass microelectrode to attempt to localize current flow to the tight junctions between cells in the amphibian gallbladder. Under microscopic observation the recording electrode was scanned along and immediately above or just touching the surface of the epithelium and its voltage was monitored with respect to a fixed reference electrode in the bath. Similar experiments were done by Hudspeth.[5] In each case, the amount of current which had to be passed across the tissue to generate a sufficient voltage at the tip of the electrode was very large: 1.2 or 0.5 mA/cm^2, respectively. Such current densities may disrupt the normal physiological parameters of the tissue. Bindslev et al.[6] showed that currents of such magnitudes influence the properties of the paracellular pathway, rendering it more permeable, and they and Reuss and Finn[7] showed that the cell membrane resistances are also changed.

Glass Microelectrodes with Signal Averaging

In an attempt to overcome this problem, Cereijido and co-workers[8,9] employed digital signal averaging techniques. Using a similar electrode geometry, a set of 512 (or 1024) pulses of current was passed across the tissue and the voltage deflections at the tip of the exploring electrode were accumulated in a digital averager. The electrode was raised 8 μm and a second set of pulses was passed, and the electrode potentials were recorded, averaged, and subtracted from the first series. As a result of these signal averaging techniques, they were able to reduce the current density passed across the tissue to 20–50 μA/cm^2. The significant advantages of this approach are that physiologically tolerable levels of current density are employed and that the use of a < 1-μm tip of the electrode allows a fairly high degree of spatial resolution. A significant disadvantage is that the temporal resolution is poor. Since digitizing 500–1000 waveforms in each of two electrode positions requires over 15 sec, only steady-state conduct-

[4] R. Frömter, J. Membr. Biol. 8, 259 (1972).

[5] A. J. Hudspeth, Proc. Natl. Acad. Sci. U.S.A. 72, 2711 (1975).

[6] N. Bindslev, J. M. Tormey, R. J. Pietras, and E. M. Wright, Biochim. Biophys. Acta 332, 286 (1974).

[7] L. Reuss and A. L. Finn, J. Membr. Biol. 25, 115 (1975).

[8] M. Cereijido, E. Stefani, and A. M. Palomo, J. Membr. Biol. 53, 19 (1980).

[9] M. Cereijido, E. Stefani, and B. C. de Ramirez, J. Membr. Biol. 70, 15 (1982).

ances can be examined using this approach. Foskett and Ussing[3] have recently taken a similar approach in an attempt to localize Cl⁻ conductance to the mitochondria-rich cell in the isolated frog skin, but simply averaged 1024 voltage readings rapidly obtained at each electrode height during constant transepithelial voltage-clamp conditions. Sampling time was thus reduced to ~ 1 sec to provide a measurement with noise levels of ± 2 μV.

Vibrating Probe Technique

Noise levels can be reduced and response time is faster using the vibrating probe technique. As such, this technique is the preferred one in attempts to localize physiological currents in intact epithelia. The rest of this chapter will be concerned with a discussion of this method.

Unlike the techniques described above, the vibrating probe technique uses a metal electrode to measure the voltages. It has been successfully applied to a variety of epithelia, including the fish opercular membrane,[10-12] toad skin,[13,14] and cultured MDCK cells.[15] A piezoelectric element is used to vibrate the electrode at 100–6000 Hz along a line 5–20 μm long, near the surface of the epithelium, so that a plot of its position versus time is a sine wave. The voltage signal of the vibrating electrode then has a sine wave component proportional to the voltage gradient along the line of vibration. Computational electronics is used to deduce the voltage gradient from the recorded electrode signal.

Because the electrode's vibration stirs the medium, it is easy in most preparations to justify an assumption that the local resistivity of the medium near the electrode is the same as the bulk resistivity of the medium. (Indeed, the stirring might possibly dissipate a physiologically present salt accumulation or depletion.)

The presence of a physically small source of current can be detected on the epithelium by moving the probe over the surface and finding a localized peak of current density above the source. As a practical requirement in such experiments, the distance between the probe and the epithelial surface must be precisely controlled. (This is just as true for the other scanning microelectrode methods mentioned above.) The recorded current density

[10] J. K. Foskett and C. Scheffey, *Science* **215**, 164 (1982).
[11] C. Scheffey, J. K. Foskett, and T. E. Machen, *J. Membr. Biol.* **75**, 193 (1983).
[12] J. K. Foskett and T. E. Machen, *J. Membr. Biol.* **85**, 25 (1985).
[13] U. Katz, W. Van Driessche, and C. Scheffey, *Biol. Cell* **55**, 245 (1985).
[14] C. Scheffey and U. Katz, *in* "Ionic Currents in Development" (R. Nuccitelli, ed.), p. 213. Liss, New York, 1986.
[15] K. Sugahara, J. H. Caldwell, and R. J. Mason, *J. Cell Biol.* **99**, 1541 (1984).

is inversely proportional to the square of the distance from a localized source, so control of the distance is a demanding task. The epithelium must be as flat as possible, so that the probe may be moved around in the plane over it, keeping the distance between the two constant within 5 μm without necessitating frequent readjustments of the height of the probe above the tissue.

In order to estimate the size of the source, the current density it produces could be measured at a known distance above it, and its total current deduced as $2\pi r^2$ times the observed current density. To do this accurately requires very precise measurement of the distance between probe and tissue (because the estimate is proportional to the square of that distance).

Vibrating Probe Construction. Electrodes for the vibrating probe technique have a metal conductor surrounded up to the tip by an insulator of glass, epoxy, or organic polymer. The tip of the electrode is a ball of platinum black, providing a high-capacitance contact with the medium. Procedures for making such electrodes have been in the literature for decades because of their applicability to extracellular recording in the brain cortex. In our studies with the vibrating probe we have used metal-filled glass microelectrodes and insulated metal electrodes. The techniques for making metal-filled glass electrodes are based on procedures outlined by Svaetichin,[16] Gesteland *et al.,*[17] Frank and Becker,[18] and Zimmermann.[19] A borosilicate glass microelectrode is pulled on a standard microelectrode puller and cut 2 cm from the tip to leave a short glass funnel. It is necessary to break back the tip of the electrode to 1 – 3 μm in diameter, which is done according to Zimmermann.[19] Two glass slides are cemented to each other and the edge of this sandwich is brought into focus on the stage of a microscope at 40 $\times$ magnification. The glass funnel is attached with dental wax to a glass slide on the microscope's moving stage, with its tip directed toward the slide sandwich. The electrode is then driven slowly and deliberately against the edge of the slide sandwich. The tip breaks piece by piece as the stage is advanced, until the desired tip diameter is achieved. The electrode is next filled with metal. Wood's metal is used because of its low melting point (70°). A wire of Wood's metal is first prepared by melting the metal and sucking it up into a length of polyethylene (PE) tubing of inner diameter equal to the inner diameter of the shank of the electrode. A short

[16] G. Svaetichin, *Acta Physiol. Scand. Suppl.* **24,** 86 (1951).
[17] R. C. Gesteland, B. Howland, J. Y. Lettvin, and W. H. Pitts, *Proc. Inst. Radio Eng.* **47,** 1856 (1959).
[18] K. Frank and M. C. Becker, *Phys. Tech. Biol. Res.* **5,** 22 (1964).
[19] G. Zimmermann, *Vison Res.* **18,** 869 (1978).

($\sim 4-5$ mm) piece of the wire is inserted into the shank. The tapering portion of the electrode is held against a hot piece of metal. As recommended by Zimmermann,[19] we have used an aluminum foil-covered 5-W resistor connected to a variable transformer. Under low-power microscopic observation, a stiff wire that closely fits into the back end of the funnel is pushed against the Wood's metal until the front end of the Wood's metal wire melts and flows into the tip. At this time a check is made to ensure that bubbles or gaps are not present in the metal near the tip. If none is present, then electrical contact is made at the back end of the Wood's metal by melting it there and inserting a 125-μm gold wire. Alternatively, manufacture can start with commercially available insulated stainless-steel or platinum electrodes. Parylene insulation with thickness of at least 1 μm is recommended. Stainless steel cannot be readily soldered, but electrical connections to it can be made with silver epoxy.

To minimize electrode impedance, the tip of the electrode is next platinized. By whichever manner the electrode is insulated, the procedure for platinization of the tip is essentially the same. If a ball of Wood's metal formed at the tip of the electrode when it flowed into the tip, it should be removed by gently tapping the electrode. The tip is then dipped for 1 sec into concentrated HCl and rinsed with distilled water, and is then placed in the plating solution. This platinizing solution consists of ~ 10 mg/ml H_2PtCl_6 and ~ 100 μg/ml $Pb(CH_3COO)_2$ in distilled water. Under microscopic observation, current (~ 50 nA) is passed out of the tip from a Pt wire also in the platinizing solution. The actual amount and duration of the current depends on the desired final tip size. In our studies of the opercular membrane,[10-12] tips with diameters of $5-7$ μm were necessary. Best results were obtained by platinizing with very low currents for a long time ($2-4$ min) followed by a rapid deposition of platinum by higher currents for a brief period (~ 5 sec). The first stage of deposition is maintained until the tip has grown to $3-4$ μm. The second stage produces a spongy, thick layer with high capacitance. This two-stage deposition is similar to that described by Jaffe and Nuccitelli,[20] but results in smaller tips since the individual layers are smaller. As an optional step which may help to keep the electrode's capacitance more stable, the tip can be gold plated before the platinization. The tip is placed in 0.2% AuKCN and current is passed into it from a Pt electrode until a 2-μm-diameter ball has formed at the tip. It is extremely useful to be able to monitor the electrode impedance during the entire plating procedure. This can be accomplished by passing a 1-kHz, 1-nA square wave of current into the tip during the platinization proce-

---

[20] L. F. Jaffe and R. Nuccitelli, *J. Cell Biol.* **63,** 214 (1976).

dure. Since this current is small compared to the platinization current, the impedance measurement has no effect on the platinization process. The response of the tip to the impedance current is the sum of a square wave of amplitude proportional to the convergence resistance and a triangle wave inversely proportional to tip capacitance. In practice, the desired impedance of the electrode determines the extent of platinization necessary. The second stage of deposition results in a dramatic reduction in the tip impedance as the spongy layer is laid down. It is necessary to monitor both the impedance and the appearance of the tip to achieve the proper balance between desired size of the tip and the level of electrode noise (i.e., impedance) which can be tolerated. Impedance measurements cannot be made while the probe is recording currents. However, it is useful to check the impedance of the tip periodically during the course of an experiment to make sure that the tip has not fractured or accumulated excessive debris.

Probe Vibration. To measure extracellular currents, the electrode is vibrated along an excursion up to ~ 50 μm long by a piezoelectric bimorph element. We have used lead zirconium titanate bimorphs (PZT-5H) from Vernitron (Bedford, Ohio) which are 1 to 2 in. long $\times$ 0.25 in. wide. Bimorphs of the parallel type are preferred to provide extra protection against field leakage through the front of the electrode holder. The bimorph is mounted in a grounded holder which has a holding rod at one end to allow it to be maneuvered with micromanipulators. The bimorph is grounded on both sides and the inside face is connected to a lead to provide the driving voltage.

A voltage applied to the bimorph causes it to flex. The bimorph is mounted in the holder by securing it at one end, which results in movement at the other end when a voltage is applied. One method to secure the bimorph in the holder is to glue a small insulating block (phenolic) to the inside base of the holder at the back end to which the bimorph is secured with a support screw. A hole in the front face of the holder provides access to the free front end of the bimorph to allow a 2-cm-long glass rod to be attached with glue to the grounded side of the bimorph. To this glass rod the electrode is attached with dental wax (metal-filled glass electrode) or by a socket connector (commercial insulated electrode).

To cause the electrode to vibrate, a sine wave of up to 20 V peak-to-peak at one of the bimorph's resonance frequencies is applied to the ungrounded face of the bimorph. Since the resonance frequency is dependent upon the mass to be vibrated, it is necessary to adjust the vibration frequency for each electrode. The length of the probe excursion and the particular resonance frequency used depend on the experimental requirements. Smaller excursions provide higher spatial resolution, but noise levels increase. Vibration frequency determines (ultimately) the response time of the probe but higher frequencies may, since the vibration stirs the adjacent medium, damage or disturb cells.

Principle of the Vibrating Probe

Metal electrodes have more noise than fluid-filled electrodes at frequencies below 10 Hz, but much lower noise in the frequency band above 100 Hz. The vibrating probe takes advantage of that low noise at higher frequencies, by a trick that creates a higher frequency signal from a standing potential gradient. It moves the electrode up and down in the potential gradient, to generate a sine wave voltage signal at the frequency of vibration (above 100 Hz). For small voltage fields, a view of the electrode's signal on the oscilloscope might leave this sine wave buried in the noise. However, the amplitude of even such a small sine wave can be measured by the use of computational circuitry called a lock-in amplifier.

The least noisy way to measure the amplitude of a sine wave component in the vibrating electrode signal is given by Fourier analysis: multiply into the signal a sine wave whose phase and frequency match those of the electrode's tip movement, and take the average over time of the product. A lock-in amplifier does just that. It "locks in" to the sine wave powering the piezoelectric crystal. It then adds a phase delay to that sine wave that is adjusted by the experimenter to match the phase delay between the sine wave driving the crystal and the sine wave of movement for the electrode tip. (The phase delay is due to mechanical lags in the system.) An average over time is then taken of the product between the phase-adjusted sine wave and the amplified signal from the vibrating electrode. The output of the lock-in amplifier is then a voltage proportional to the voltage field along the line of vibration of the tip of the electrode.

That average over time is electronically implemented as a low-pass filter. The time constant for that low-pass filer sets the response time for the vibrating probe, and also sets the noise level. The stage of the low-pass filter that sets the response time is usually a simple RC filter (one pole), although it may be preceded by faster stages to remove components of the product at the frequency of vibration. When the output filter is one pole with a time constant τ, the attainable root-mean-square (rms) noise of the probe (in A/cm^2) is given by

$$\text{rms noise} = 8.5 \times 10^{-5}/[L(D\tau\rho)^{1/2}] \tag{1}$$

where L is the length of the line of vibration (in μm), ρ is the resistivity of the medium (in Ωcm), and D the diameter of the electrode tip (in μm).

The lock-in output time constant τ is seldom profitably set above 10 sec, because of drift in the output signal. For scanning over epithelia, a time constant of 2 sec is usually convenient; for examining the faster components of a response after a voltage-clamp step, the response time might be set to 100 msec or even less. With $\tau = 2$ sec, noise is typically less than 0.5 $\mu A/cm^2$.

The derivation of Eq. (1) assumes that the metal–solution interface at the electrode's surface does not contribute to the electrode's voltage noise at the frequency of vibration.[11,21] The following procedure helps to ensure the approximate realization of that assumption: (1) maximize the electrode's capacitance during its manufacture; (2) with the electrode not vibrating, examine the voltage noise at the lock-in output. The voltage noise at the lock-in output (preferably viewed with τ near 1 sec) will tend to decrease as the frequency of the oscillator is increased. A minimum noise should be reached at a frequency below 2 kHz; seek a resonant frequency of the bimorph with noise near that minimum.

Some further fine adjustments of the frequency may be necessary to minimize the lock-in output noise, in order that the lock-in output is minimally affected by sources of periodic interference. Interference (both electrostatic and inductive) at the harmonics of half the power line frequency is often present, not only from power lines but also from video sync signals. However, only interference with frequencies within $10/\tau$ Hz of the frequency of vibration is significant to the noise at the output of the lock in. Thus, although the interference should be minimized by other means as much as possible, its effect on the vibrating probe signal can often be eliminated by small changes in the frequency of vibration. After minimizing the noise with the vibration off, the lock-in output noise with vibration on is examined. If the lock-in noise is then more than twice the theoretical minimum, further adjustment of the oscillator frequency may help. Noise far in excess of the theoretical minimum can occur if there is a resonance at the frequency of vibration in the signal wire, in the electrode mount, in the bimorph mount, or in any of the tubes connected to the chamber.

When the vibration is turned on, it produces an offset at the lock-in output. This offset is never zero, even in the absence of electrical current at the electrode tip. It seems to be due in part to microphonics in the signal cable and in the electrode itself. It can be ignored as long as it remains constant during the experiment and as long as only differences in current density are measured. If it is desired to know absolute current density, then some conditions, including the position for the probe, must be found where the current density is known to be zero, so that the difference the lock-in output at that position and the test position can be measured. Such a "reference" condition must always be defined for a vibrating probe

---

[21] C. Scheffey, Ph.D. thesis. University of California, Berkeley, 1981.

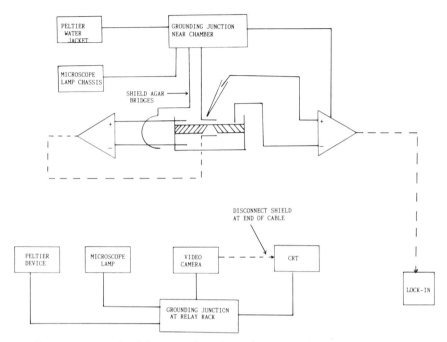

FIG. 1. An example of the grounding scheme for an experimental setup. A "stellate" arrangement avoids ground loops. Grounds that must be kept close together (in potential) are connected in small groups with short wires. Most critical of these is the group near the experimental dish, which must keep the circuit ground close to the grounded side of the bath (the upper side). The final grounding arrangement for any rig must ultimately be worked out by trial and error; connections shown in this figure do not necessarily represent hard and fast rules.

experiment, and all current densities are measured relative to the current density at that point.

System Integration and Grounding

The vibrating probe is sensitive to voltages below 1 μV. This means not only that it is sensitive to pickup from electrostatic fields, but that the grounding and interconnection of the electronic devices involved must be done carefully. A complete treatment on the principles of proper grounding is beyond the scope of this review; the reader is referred to Purves.[22] Instead, an example of a properly grounded experimental setup will be described, shown in Fig. 1. For simplicity, only connections to the circuit

[22] R. D. Purves, "Microelectrode Methods for Intracellular Recording and Iontophoresis." Academic Press, New York, 1982.

grounds (commons) are shown. If the chassis for a circuit is not connected to circuit ground, it should be grounded separately. If the circuit ground is connected to the chassis, then the chassis should be insulated from its surroundings, e.g., a relay rack, or a mounting post for a preamplifier that is grounded by another path.

The vibrating probe voltage is measured differentially with respect to a reference electrode by a low-noise preamplifier. We recommend a preamp with a capacitatively coupled input stage based on the dual FET U401, similar to the source-follower circuits shown by Evans[23] and by Scheffey.[24,25]

Both the preamplifier for the voltage clamp and the current-passing electrode on the top side of the epithelium are connected directly to the circuit ground of the probe preamplifier, using thick, short wires. This scheme requires that the voltage clamp monitors the current it passes by measuring the voltage across a resistor in series with the other current-passing electrode. If the clamp uses the other common method for monitoring the current, a virtual ground, then the virtual ground must be connected to the current-passing electrode on the top side of the bath, and the circuit ground for the virtual ground circuit should be directly connected to the probe preamplifier ground.

The entire experimental setup rests on a grounded plate, and pieces of metal near the chamber (e.g., the microscope stage) are grounded. All these grounds are connected together within a meter of the preparation. The transepithelial voltage-measuring calomel electrodes and their agar bridges should be surrounded in part with grounded shields.

Signals of vibration frequency must be minimized in the transepithelial voltage signal, since such signals would cause the clamp to modulate the current it passes at the frequency of vibration, giving the vibrating electrode a signal that will be detected by the lock-in amplifier in the absence of a voltage field near its tip. For the purpose of working on this problem, the transepithelial voltage signal from the clamp can be connected into the input of the lock-in amplifier, with the clamp "off" (tissue open circuit). The shift at the lock-in output in response to turning on the probe vibration is then a quantitative indicator of the amplitude of the vibration-coherent signal in the measured transepithelial voltage. Three possible mechanisms for the cross-coupling are (1) capacitative coupling from the

---

[23] A. D. Evans (ed.), "Designing with Field Effect Transistors," p. 91. McGraw-Hill, New York, 1981.
[24] C. Scheffey, in "Ionic Currents in Development" (R. Nuccitelli, ed.), p. 3. Liss, New York, 1986.
[25] C. Scheffey, in "Ionic Currents in Development" (R. Nuccitelli, ed.), p. xxv. Liss, New York, 1986.

bimorph to the chamber, or to any of the conductive pathways connected to it, e.g., the perfusion tubes or the agar bridges to the calomel electrodes; (2) inductive coupling between the bimorph power line and cables involved carrying the transepithelial voltage signal; and (3) acoustic coupling near the chamber. Of these problems, the first is the most likely problem. It should be solved by adding electrostatic shielding.

Additional comments on vibrating probe methodology are found in Refs. 24 and 25.

Application of Vibrating Probe Technique to Epithelia

Experimental Chamber and Optics. The localization of conductance pathways and sources of current in an intact epithelium in general requires that the epithelium be studied using the Ussing chamber technique. The ability to measure transepithelial voltage and to pass current across the tissue allows the current/voltage properties of specific pathways to be determined. The other general requirement is that the pathways of interest —specific cell types, cell junctions, glands, etc.— be able to be visualized so that the tip of the electrode can be accurately positioned closely above ($<20 \mu$m) the surface of the pathway of interest. The physical size of the pathway of interest and the optical characteristics of the tissue will determine the specific requirements for the design of the perfusion chamber to support the tissue as well as the microscopic optics required.

The chamber and optical setup described below satisfy many of the requirements to measure small currents immediately above the surface of an isolated epithelium. Figure 2 is a diagram of the system. Following isolation of the tissue and any procedures designed to eliminate some of the connective tissue (e.g., collagenase treatment), the tissue is stretched apical side down across a central aperture of a circular plate. A smaller annular disk is pressed into the aperture of the plate to thereby hold the tissue securely in position. The plate–tissue assembly, with apical side up, is transferred to the perfusion chamber, whose aperture (area, ~ 0.5 cm^2) is covered by a Nytex mesh to support the tissue. The Nytex mesh should provide as much open area as possible and the mesh openings should be small enough to support the tissue but large enough that diffraction patterns at the edges of the Nytex do not overlap to obscure the view of the entire window. We have used mesh with ~ 200-μm openings with $\sim 60\%$ total open area. The mesh is held in position by a thin layer of silicon grease and the tissue–plate assembly which sits on it. The tissue–plate assembly is held in position by a layer of silicon grease smeared between the plate and the chamber. The grease serves another critical function of electrically sealing the edge of the tissue. As an additional measure to

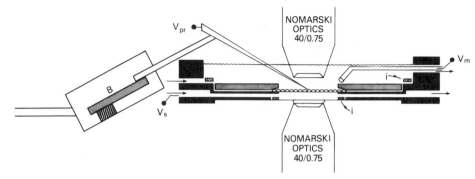

FIG. 2. Chamber and optical setup for vibrating probe (V_{pr}) measurements of extracellular current density immediately above the surface of an intact epithelium. The chamber is placed on the stage of a microscope. Transepithelial voltage is measured with a pair of agar bridges (V_m, V_s); current is passed by Ag–AgCl wires (i). Baths can be continuously perfused (arrows). The vibrating probe is vibrated by a piezoelectric bimorph (B) secured in a holder positioned to achieve nearly vertical vibration of the tip. A reference electrode for the probe is also in the mucosal bath (not shown).

secure the tissue–plate assembly to the chamber, horizontal pins can be used. When the tissue–plate assembly is secured in the chamber, the upper, apical surface of the epithelium is nearly flush with the upper surface of the floor of the chamber, greatly facilitating access of the microelectrode to the tissue from as horizontal an approach as possible.

The chamber contains ports to perfuse both sides of the tissue, to measure transepithelial voltage, and to pass current across the tissue. The bottom, serosal compartment contains an inlet port to solutions from an overhead reservoir, and an outlet port on the other side of the aperture serves to drain it. Perfusion is by gravity feed. Transepithelial hydrostatic pressure gradients can be eliminated by observing the tissue while the outflow resistance of the serosal bath is manipulated. The combined use of a support and controls over the hydrostatic pressure help to ensure that the tissue is as flat as possible, which, as discussed above, is essential in scanning experiments. Perfusion of the apical side is from another overhead reservoir via a port into the upper bath. The upper bath is drained by continuous aspiration. Since the mass of the electrode cannot be too great if it is to be vibrated properly, the upper bath must be shallow enough not to reach the back end of the electrode. However, the tissue should have 1–3 mm of medium over it, because vibrating probe measurements can be noisy if made closer than 0.5 mm away from the fluid surface.[20] The perfusion lines should have a resistive barrier within 50 cm of the chamber, within an area protected from electrostatic fields.

Transepithelial voltage is measured by a pair of calomel electrodes connected to the tissue by agar bridges. The bridge on the serosal side is inserted into the serosal inflow perfusion port. The bridge on the mucosal side ends near the tissue and is held in place by a port in the wall of the upper part of the chamber. Current is passed across the tissue by a pair of chlorided silver wires which encircle the aperture on either side of the tissue. The wire is flat, measuring 0.005 in. thick × 0.0625 in. wide, which minimizes the vertical aspect of the chamber to reduce the necessary working distance for the objective lens in an inverted microscope, or for the condenser lens in an upright microscope, while preserving a large electrode surface area. Preliminary experiments revealed that this geometry ensures that a uniform current density is passed across the entire aperture at the level of the tissue.

The distance from the surface of the tissue to the bottom of the chamber, which is a coverslip for maximum optical quality is ∼1.5 mm. This allows use of either an upright or inverted microscope with fairly high resolution and magnification by employing as both condensor and objective lenses a pair of Zeiss 40/0.75 water-immersion lenses which have working distances of 1.6 mm. These lenses provide a good combination of working distance, magnification, and resolution. The optical magnification which is necessary to be able to visualize specific pathways sets constraints on the scaling considerations of the overall chamber design. Higher magnification and higher resolution are usually associated with shorter working distances. In an upright microscope, use of a short-working-distance objective lens will require that the electrode approach the tissue from a very shallow angle. This explains why, in the chamber described above, the height of the walls of the chamber was kept low, the distance from the center of the aperture to the walls was made sufficiently great, and the height of the walls of the aperture opening itself was reduced (or eliminated) on the side of the tissue being approached with the vibrating probe. A longer working-distance lens, such as the Leitz 32× and Zeiss UD-40×, will allow a steeper approach and relaxation of some constraints on the chamber design. Localization of sites of current density is most precise when the probe vibration is perpendicular to the surface of the tissue. A steep electrode approach will require a compensatory modification of the angle of the holder of the piezoelectric bimorph, as shown in Fig. 2. In fact, for all geometries it is desirable for the electrode holder to be angled as shown in Fig. 2, since it results in the line of vibration being close to vertical (as a consequence of the fact that the tip vibrates within an arc defined by the location of the center of the bimorph).

The microscope is equipped with Wollaston prisms and a polarizer and analyzer to provide differential interference contrast (DIC). The combination of fairly high numerical aperture and DIC will often result in improved optics by enhancing contrast and providing a fairly shallow depth of field. However, the optical characteristics of the tissue will determine the microscopic requirements. For example, it may not be necessary to have high magnification. In our studies of chloride currents in the opercular epithelium,[10-12] a 10× objective combined with a standard long-working-distance condenser was sufficient because the cells of interest (chloride cells) are large and the tissue is nearly transparent. Similarly, tissue optical properties may be such that DIC is ineffective, so other contrast enhancement techniques, including oblique illumination, phase contrast, or video, could be used.

In addition to allowing the tissue to be clearly observed, good optics serve another critical function to permit the tip of the vibrating probe to be clearly resolved. For a typical measurement of current over a specific site on an epithelium, it is necessary to position the tip of the probe $7-15$ μm above the surface of the tissue. To quantitate the current, it is necessary to know this height as precisely as possible. This is routinely done by determining the difference in the microscope focus micrometer reading when the focus is at the center of the tip versus when the focus is at the surface of the tissue, correcting for the difference in refractive indices if an interface between two media is being viewed through. The accuracy of this measurement is determined by contrast and by the depth of field of the optics (the latter being determined by the combined numerical apertures of the objective and condensor lenses). The error in the determination can be checked by two sets of experiments. In the first, the distance between the probe and a fixed microelectrode, measured optically, can be compared with that measured electrically. In the second, the distance measured optically can be compared by manually lowering the probe and recording its displacement until it touches the epithelial surface (determined by noise on the probe output).

Finally, although Fig. 2 has been discussed as employing an upright microscope, it is obvious that the microscope could have been an inverted one. The inverted microscope offers the advantage that the objective lens is out of the way of the electrode and relatively high numerical aperture condenser lenses with long working distances are available. In addition, the objective does not have to look through an air–water interface to see the tissue. This optical advantage may be offset by the fact that the lens must look through any connective tissue to see the epithelium.

Calibration. Before mounting the tissue in the chamber, the probe is calibrated, to adjust the phase delay in the lock-in amplifier (between the sine wave powering the piezoelectric crystal and the sine wave to be multiplied into the signal) and to determine the constant of proportionality to convert the voltage at the output of the lock-in amplifier into a current density along the line of vibration.

The electrode tip is placed near the aperture where the tissue will be mounted and the vibration resonances are observed. A resonant frequency is chosen high enough to give the desired noise performance from the probe, and that allows vibration that appears to be fairly close to vertical. It is important that the depth of the medium be the same as the depth that will be used in the experiment, because the resonance properties of the probe may to an extent be dependent upon its depth. This is also the reason for doing the calibration at or near the tissue aperture.

Phase adjustment for the lock-in amplifier may be accomplished with a saline-filled glass electrode positioned within 30 μm of the probe. The lock-in output is set to a fast time constant and current is passed out of the glass microelectrode to get a large response at the output of the lock in. The phase of the lock in is adjusted to give minimal response. Then the phase is changed by 90°, to give the optimal size of response.

The saline-filled electrode should have a sharp tip, such as those used in intracellular recording. A tip with too low a resistance can give rise to artifacts, because the stray capacitance from the bimorph can cause injection of significant levels of current at the frequency of vibration, which is in turn detected by the lock-in amplifier.

The direction of vibration will be determined during the calibration procedure, but because it may be necessary to examine the direction of vibration for several different frequencies of vibration, faster ways of approximately determining the direction may be useful:

1. If the electrode's excursion amplitude (for a given frequency and bimorph power amplitude) is known from experience when vibrating the electrode in the horizontal direction (when that excursion is easy to see), then the direction of vibration may be estimated by viewing how much smaller the horizontal component of probe vibration is when the probe is moved in the nearly vertical direction that will be used in the experiment. This method becomes more accurate if the probe is illuminated with a strobe light.

2. If the depth of field for the optics is shallow enough, then both the horizontal and vertical components of the line of vibration may be estimated. Again, a strobe light improves the estimates.

The direction of vibration may be more accurately determined using a glass electrode as a current source. The response of the probe to current injected from 50 μm directly below it is compared with the response to current injected from 50 μm away in a horizontal direction, with the glass microelectrode at the same height as the probe. The ratio of the two responses can be used to determine the direction of vibration: response to horizontal current/response to vertical current = tan(angle away from vertical), and the current density at the probe for each of the current injections I is

$$\text{Current density} = I/4\pi(50 \ \mu\text{m})^2 \qquad (2)$$

allowing one to calculate the constants of proportionality between current density and probe response for the horizontal and vertical directions. These constants of proportionality are then used to routinely convert changes in the lock-in output into changes in current density.

Recording Currents from Cells. Following the calibration procedures, the tissue is mounted in the chamber and the perfusion is initiated. While the tissue electrical parameters are stabilizing, the regions of the tissue of interest are brought into focus and the optics optimized. The probe is brought into a position approximately 250 μm above the surface of the epithelium.

A typical experimental protocol is shown in Fig. 3. The probe is vibrated at the same excursion and frequency as during the calibration procedures and its output is recorded at 250 μm above the surface while the tissue is at the resting transepithelial potential, i.e., transepithelial current is zero. The signal under these conditions defines the zero-current

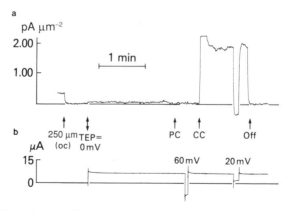

FIG. 3. Typical protocol used to measure short-circuit current and conductance of individual cells in intact epithelia. (a) Probe output (converted to a current density); (b) tissue (fish opercular membrane) short-circuit current; 250 μm (oc) shows the probe response when it is positioned 250 μm above the tissue under open-circuit (zero-current flow) conditions, defining the zero-current probe output. TEP = 0 mV indicates that the tissue is voltage clamped to 0 mV. PC, Probe positioned over pavement cell; 60 mV, tissue rapidly clamped to 60 mV, serosa positive; CC, probe positioned over chloride cell; 20 mV, tissue rapidly clamped to 20 mV, serosa positive; Off, vibration off. (From Ref. 12.)

probe output. This procedure should be carried out as often as once every several minutes during an experiment, depending on to what extent the probe output level is drifting and on the size of the current densities to be measured. To localize active sources of current, the tissue is then voltage clamped to 0 mV. The probe is lowered to immediately above the surface of the tissue. If there is a perfusion system it may be necessary to turn it off during probe measurements, because (1) the probe's output is usually dependent to some extent on the depth of the electrode in the medium, and changes in that depth as a result of perfusion may introduce errors into the probe output; and (2) the perfusion may cause the epithelium to move, interfering with attempts to scan along the surface at a constant height above the tissue.

In Fig. 3, the probe is initially positioned 11 μm above a surface, a so-called pavement cell in the fish opercular membrane. The probe output is not different from that recorded at 250 μm above the tissue under open-circuit conditions, indicating that the pavement cell does not generate a measurable short-circuit current (see below). The transepithelial voltage was then rapidly stepped from 0 to 60 mV, serosal side positive, to provide a driving force for passive current flow. The lack of response of the probe output indicates that the pavement cell has an extremely low ionic conductance. The probe was then positioned over a different cell type, a so-called chloride cell. The probe output reveals that this cell is generating an appreciable current. The response of the probe to a rapid step of the transepithelial voltage to 20 mV indicates that chloride cells are extremely conductive sites as well. To permit quantitation of these signals, the vibration is turned off and the height of the tip of the probe above the tissue is recorded as described earlier.

Interpretation of Data from Vibrating Probe Experiments

When the probe reports nonzero current over the epithelium, it must be interpreted in terms of the current density across the epithelium. A question that obviously appears from the above experimental procedure concerns the validity of the zero-current reference. That is, if the probe output is the same over the epithelium as it was during the back-off, out-of-clamp reference procedure, then does that mean that the electrogenic transport under that probe position is zero? These questions will be answered here for specific idealized classes of transport by epithelia. Many of the comments will clearly apply to other kinds of electric field scanning as well.

A Single Current Source on an Insulating Surface. Consider first the situation of a single cell actively pumping an ionic current across an epithelium that is a perfect insulator everywhere else. In the first approximation, the cell's current (in pA) is given by

$$I = 2\pi h^2 c \qquad (3)$$

where h is the height (in μm) of the probe's center of vibration above the cell, and c is the current density (in pA/μm^2) recorded by the probe. The approximation of Eq. (3) came from the assumptions that (1) the electric field is uniform along the line of probe excursion, which is not accurate when the length of the line of vibration is less than twice h, and (2) the cell is small enough at the surface to be considered a point source of current. This assumption begins to cause an error in the measurement as the cell's diameter as the surface exceeds $h/4$.

Departure from these assumptions can be dealt with in a straightforward numerical way. If the cell's diameter is large enough that its field cannot be simulated as a single point source, then the electric field near it can be calculated by simulating it with a population of current sources. Each of these current sources creates an electrical field with a potential that drops off as the reciprocal of the distance from the source. The potential at any point above the cell may then be calculated by summing the fields of the set of points making up the cell, and the relation between current of the cell and lock-in output may be calculated by simulation of the lock-in amplifier's function: the potential of each point along the path of vibration is calculated and multiplied by the sine of its phase in the vibration cycle. This calculation may sound laborious but it may be completed in a short time on a microcomputer. It will usually yield only a small correction to Eq. (3). In the case that h equals the length of the line of vibration, and the source of current has a diameter up to $h/2$, the total correction is less than 10%. Assumptions 1 and 2 mentioned above are for the sake of completeness, because they may be significant in some geometries.

The single most important source of error in estimating the size of an isolated current source is the error in estimating the height of the probe above the tissue surface. Referring to Eq. (3), judging the height as 7 μm when the true height is 10 μm results in a 50% error in the estimated size of the current source.

Single Active Source on a Surface of Passive Conducting Cells. If the current source described above were located on a real epithelium, one might be concerned that the surrounding cells were not good enough insulators to make the calculations presented above useful. They almost always are. Even a 5-μm-diameter cell pumping 10 nA into a saline with a resistivity of 100 Ω cm creates only 0.5 mV of potential at its boundary and only tens of microvolts 5 μm away. The currents flowing back across the epithelium under such small voltages are usually negligible; they may be calculated if the resistance of the pathways is known.

In the calculations presented above and herein, the back-off, out-of-clamp protocol clearly provides a correct level for the output of the lock-in equivalent to zero current.

Homogeneous Current Density on the Surface of an Epithelium. Suppose one cell type covers most of the epithelial surface, and that a current is passing through it (either active or passive). If the current from each cell is proportional to the cell's surface area, then the electric field above the surface is very simple. It is the same everywhere. As one goes up from the surface, the current density in the medium stays constant, until the height above the tissue becomes comparable with the depth of the medium.

The back-off and exit-clamp condition should give an accurate reference for the zero-current level of probe output, in this case. Open-circuiting the tissue should send its transepithelial voltage to the reversal potential for the homogeneous current, giving zero-current density everywhere.

Set of Active Current Sources Sparsely Distributed among Passively Insulating Cells. This is the situation analyzed by the vibrating probe for opercular membrane.[10-12] The estimation of the size of each current source is much the same problem as in the case of a single current source on an insulating surface mentioned above, except for one complication: the current density measured by the probe over a current source will be contaminated with current that spreads from other cells. Suppose we measure the current density in the vertical direction at a height h above a current source. Then the contribution to the probe signal of a current source of the same size at a distance d away is about h^3/d^3 times smaller. (If the direction of vibration has a horizontal component, then the contribution of the distant current source can be even more; see appendix C of Scheffey *et al.*[11])

If there are several current sources at a distance of only several times h, their contribution can be quite significant. Figure 4 shows calculated simulations to illustrate this point. Figure 4a shows the current densities measured by a probe as it moves past a point current source at heights of 10 and 20 μm. Figure 4b shows the current densities measured by the probe in the presence of a square array of sources, all the same size, spaced 40 μm apart. At the 20-μm height, the isolated point source can be located, but the recording in the array of sources is difficult to interpret. Even at the 10-μm height, the current density recorded between the cells appear to have a baseline value of one-fourth that recorded at the peaks. This potentially misleading situation has been seen in opercular membrane (cf. Refs. 10 and 11).

When making a probe measurement at a distance from the epithelium out of clamp, this situation becomes very much like the situation described above (homogeneous current density). Far enough from the tissue, the probe senses an average of the current over the surface of the epithelium.

a

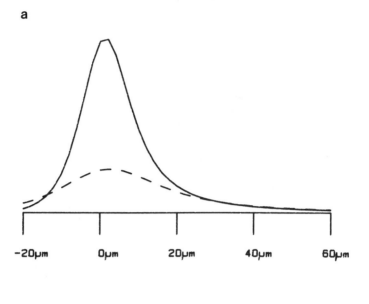

-20µm   0µm   20µm   40µm   60µm

b

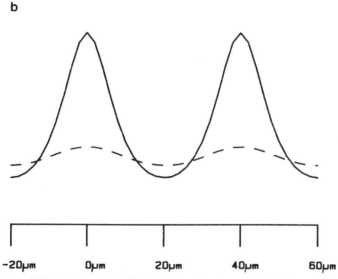

-20µm   0µm   20µm   40µm   60µm

FIG. 4. (a) Calculated responses of a probe vibrating at an angle 20° from vertical moving past a single source of current on an otherwise insulating surface. Abscissa shows displacement of the probe in the horizontal direction, ordinate shows response of the probe in arbitrary units. Solid line, probe at a height 20 µm above the surface; dashed line, height 10 µm. (b) Responses calculated as in previous figure for scans of probe at the same two heights (20 µm, solid line; 10 µm, dashed line), but this time for an array of current sources instead of a single one. Sources at 0 and 40 µm are part of a square-arranged array.

With the voltage clamp not passing current, that average is zero, and the probe gets a true zero-current reference.

The caveat must be made that some experimental situations might not allow the probe to go far enough away to see a true average of transepithelial current.

An Isolated Source of Current, of Small Geometry, in the Midst of a Uniform Current Density on the Epithelium. Here the best procedure to estimate the size of a source of current is to keep the probe at a fixed height above the tissue, and compare the probe output over the source with the probe output away from a source. The uniform current density due to the surrounding cells then contributes equally to the two probe positions; the surrounding discrete current sources contribute approximately equally. The difference between the two positions then represents mostly a contribution from the source under study, and its size may be estimated from Eq. (3).

Prospects for Future Vibrating Probe Studies of Epithelia

The vibrating probe technique is ideally suited for partitioning currents and conductances among the various cell types in heterogeneous epithelia, and we expect that it will find increasing utilization in such studies. Although it has been employed only on flat-sheet epithelia to data, there is no technical reason why it cannot also be used to examine heterogeneity of cell types in tubular epithelia. We also expect the vibrating probe technique to be useful in studies of homogeneous epithelia. For instance, the vibrating probe has been employed recently to localize current flow to domes in cultured epithelial cells.[15]

Section V

Modification of Cells

[40] Experimental Control of Intracellular Environment

By PETER F. BAKER[1] and DEREK E. KNIGHT

Introduction

The experimental analysis of cell function often requires preparations intermediate between intact cells and cell-free systems. Thus, although a range of experimental manipulations including microinjection,[1a] cell fusion[2-4] transient permeabilization,[5,6] and hypertonic endocytosis[7,8] permit substances to be introduced into intact cells, it is virtually impossible with these techniques to exert long-term control over specific variables. These aims are much more easily achieved in cell-free systems; but here there is always a very real possibility that important factors may have been lost during preparation. In addition, some intracellular events, such as exocytosis, have yet to be reconstituted satisfactorily in the test tube. To overcome these problems, various types of permeable cell preparations have been developed. Such preparations have much to offer and, under appropriate conditions, permit a moderate degree of control over intracellular processes that are otherwise inaccessible to experimental manipulations.

Two general approaches have been adopted: (1) techniques that permit extracellular and intracellular fluids to equilibrate and (2) techniques that permit separate control over intracellular and extracellular environments. In both instances, the size range of molecules that can be controlled depends on the precise technique used. More information on these different techniques and their relative advantages and disadvantages are given below, but before embarking on this discussion, it is important to consider

---

[1] Deceased.

[1a] R. W. Meech, *in* "Techniques in Cellular Physiology" (P. F. Baker, ed.), pp. 1–16. Elsevier, Amsterdam, 1981.

[2] D. Fisher and A. H. Goodall, *in* "Techniques in Cellular Physiology" (P. F. Baker, ed.), pp. 1–16. Elsevier, Amsterdam, 1981.

[3] S. J. Doxsey, J. Sambrook, A. Helenius, and J. White, *J. Cell Biol.* **101**, 19 (1985).

[4] R. A. Schlegel and M. C. Rechsteiner, *in* "Microinjection and Organelle Transfer Techniques: Methods and Applications" (J. E. Celiz, A. Graessman, and A. Layter, ed). Academic Press, Orlando, Florida, 1985.

[5] S. Cockcroft and B. D. Gomperts, *Nature (London)* **279**, 541 (1979).

[6] D. E. Knight, *in* "Techniques in Cellular Physiology" (P. F. Baker, ed.), pp. 1–20. Elsevier, Amsterdam, 1981.

[7] R. A. Jorgenson and R. C. Nordie, *J. Biol. Chem.* **255**, 5907 (1980).

[8] C. Y. Okada and M. Rechsteiner, *Cell* **29**, 33 (1982).

TABLE I
TABLE I
COMPOSITION OF MEDIUM IN WHICH ADRENAL MEDULLARY CELLS
HAVE BEEN RENDERED PERMEABLE[a]

Medium	Concentration	Comments
Potassium glutamate	140 mM	K can be replaced by Na, but the nature of the major anion is very important. Cl⁻ is inhibitory. Glutamate can be replaced by acetate and some workers have used PIPES
MgATP	5.0 mM	Essential
PIPES, pH 6.6	20.0 mM	pH can be varied from 6.0 to 8.0; other buffers can be used
EGTA	0.4 mM	Can be increased to 50 mM without altering the results. BAPTA is preferable to EGTA when a very low free Ca is required
Free Mg	2.0 mM	—
Ionized Ca	<100 nM	—

[a] This medium has proved to be entirely satisfactory for adrenal medullary cells, but under some conditions it may be necessary to add a high-molecular-weight substance such as dextran (molecular weight, 10,000) to offset the intracellular colloid osmotic pressure.

the choice of solution to which the intracellular environment is to be exposed.

Choice of Intracellular Solution

Once the plasma membrane is breached, the cytosolic components of low molecular mass will tend to exchange with those present in the suspension medium. This raises an absolutely fundamental problem. What artificial medium will preserve intracellular function? It seems likely that this may well depend on the cytosolic function under investigation, but we have been surprised at how difficult it is to find a fully satisfactory medium. Table I describes the medium that we have found to be best for preserving calcium-dependent membrane turnover in "leaky" bovine adrenal medullary cells. Even in this medium, however, calcium can release only 30% of the total catecholamine, and we suspect that our medium lacks factors essential for the mobilization of the rest of the catecholamine.

The inclusion of MgATP, the maintenance of an ionized Ca less than 10^{-7} M, and a low concentration of chloride are all important features of this medium. The same features have proved to be of importance for the preservation of cortical granule exocytosis in sea urchin eggs,[9] serotonin release from platelets,[10] and particle movement in crab axons.[11,12]

It may also be necessary to replace key macromolecules such as enzymes when the permeabilization procedure permits them to escape.

Techniques That Permit Equilibration of Intracellular and Extracellular Fluids

Included among these techniques are partial or complete removal of the plasma membrane either mechanically or by detergents; local, nonselective, permeabilization of the plasma membrane by exposure either to brief high-voltage pulses or pore-forming chemicals; and selective permeabilization by exposure to suitable ionophores. The first two of these techniques are discussed in more detail below.

Partial or Complete Removal of Plasma Membrane

Mechanical Methods. In those circumstances where measurement can be made on a single cell, mechanical disruption of the plasma membrane provides a simple way to access the cell interior. The ease with which this can be achieved depends on the cell size. In giant axons, protoplasm can be squeezed out in a relatively undisturbed state,[13,14] providing an excellent preparation for biochemical studies of relatively undisturbed protoplasm.[15-17] This preparation has recently achieved considerable attention in studies of the mechanism of fast axoplasmic transport.[18,19]

Extruded axoplasm is one of the few preparations from which a pure sample of protoplasm can be obtained without contamination with extracellular fluid or exposure to an artificial intracellular fluid. In other large cells the plasma membrane can be removed by dissection, but it is hard to avoid some contamination of the cell contents with extracellular fluid. It is preferable under such circumstances to bathe the cell in an artificial intra-

[9] P. F. Baker, D. E. Knight, and M. Whitaker, *Proc. R. Soc. London B* **207,** 149 (1980).

[10] D. E. Knight and M. C. Scrutton, *Thromb. Res.* **20,** 437 (1980).

[11] R. J. Adams, *Nature (London)* **297,** 327 (1982).

[12] R. J. Adams, P. F. Baker, and D. Bray, *J. Physiol. (London)* **326,** 7P (1982).

[13] P. F. Baker, A. L. Hodgkin, and T. I. Shaw, *J. Physiol. (London)* **164,** 330 (1962).

[14] R. S. Bear, F. O. Schmitt, and J. Z. Young, *Proc. R. Soc. London B* **123,** 520 (1937).

[15] P. F. Baker and W. W. Schlaepfer, *J. Physiol. (London)* **276,** 103 (1978).

[16] P. F. Baker and T. I. Shaw, *J. Physiol. (London)* **180,** 424 (1965).

[17] R. J. Lasek, *Curr. Top. Membr. Transp.* **22,** 39 (1984).

[18] R. D. Allen, J. Metuzals, I. Tasaki, S. T. Brady, and S. Gilbert, *Science* **218,** 1127 (1983).

[19] S. T. Brady, R. J. Lasek, and R. D. Allen, *Science* **218,** 1129 (1983).

cellular fluid before beginning to dissect off the plasma membrane. Wholesale loss of soluble components of protoplasm can be minimized by carrying out the whole permeabilization process under liquid paraffin. This approach has been applied with great success in investigations of the contractile machinery of muscle (see Refs. 20 and 21).

Detergents. In those cases where measurements cannot be made on a single cell, or where the cell is just too small for micromanipulation, the plasma membrane can be breached by exposure to detergents, especially those that complex with membrane-bound cholesterol. As detergents will tend also to affect other membrane systems inside the cell — although these, in general, contain less cholesterol — it is important to use concentrations of detergent as low as compatible with effecting permeabilization and to remove the detergent, for instance by binding to albumin, once permeabilization has been achieved. No one detergent has proved superior, and successful results have been obtained using filipin, Brij 58, digitonin, saponin, and even Triton.[7,22-25] The choice of detergent must be governed by the system under investigation, and this is particularly important if it is desired to retain "normal" function of intracellular membrane systems.

In screening different detergents, it is convenient to have a simple assay for permeabilization. This can be chosen to detect the degree of permeabilization achieved. Thus the release of ^{42}K or, more simply, ^{89}Rb (because of its longer half-life), previously loaded into the cell via the Na^+-K^+ exchange pump, is a good reporter of ionic permeability; leakage of the cytosolic enzyme lactate dehydrogenase is a good marker for protein permeability. Alternatively, a normally impermeant molecule can be placed in the extracellular fluid and its access to the cell interior monitored. One such marker that has found favor in some laboratories is ethidium bromide, which forms a fluorescent complex with DNA, but any labeled molecule can be used provided the cells can be separated rapidly from the bulk of the extracellular fluid, for instance by centrifugation through oil.

Detergents permit the generation of a wide spectrum of permeabilities, from small lesions to virtually complete removal of the plasma membrane. Provided the cellular process under investigation is unaffected by low concentrations of detergent, this method is a very simple and convenient

[20] A. Fabiato, *J. Gen. Physiol.* **85,** 189 (1985).
[21] R. Natori, *Jikeikai Med. J.* **1,** 119 (1954).
[22] W. Z. Cande, K. McDonald, and R. L. Meeusen, *J. Cell Biol.* **88,** 618 (1981).
[23] G. Fiskum, S. W. Craig, G. L. Decker, and A. L. Lehninger, *Proc. Natl. Acad. Sci. U.S.A.* **77,** 3430 (1980).
[24] H. S. Gankema, E. Laanen, A. K. Groen, and J. M. Tager, *Eur. J. Biochem.* **119,** 409 (1981).
[25] E. Murphy, K. Coll, T. L. Rich, and J. R. Williamson, *J. Biol. Chem.* **255,** 6600 (1980).

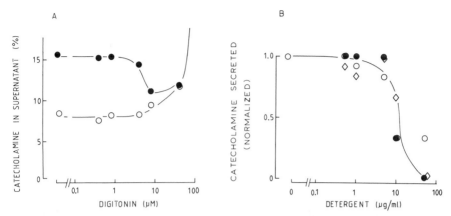

FIG. 1. Effect of detergents on Ca-dependent secretion from isolated bovine adrenal medullary cells. (A) Cells were rendered leaky[29] and incubated for 3 min with various concentrations of digitonin. The Ca^{2+} was then raised to 10 μM for a further 15 min and the catecholamine in the supernatant determined (●). Cells not challenged with Ca^{2+} (○). Temperature, 37°. (B) Cells were rendered leaky as in A and incubated with various concentrations of digitonin (●), Brij 58 (○), or saponin (◇) for 5 min before being challenged with 10 μM Ca^{2+} for 15 min. The catecholamine released as a result of the Ca^{2+} challenge is shown and is normalized to that seen in the absence of detergents. Temperature, 37°.

way to gain access to the cell interior; however, when membranes are involved, detergents must always be viewed with considerable caution. This is not to say that detergent permeabilization cannot be used for investigation of intracellular membrane systems, because detergent-permeabilized cells have been used for a variety of *in situ* studies of mitochondria and endoplasmic reticulum (see, for instance, Refs. 25–27), but there must always be a possibility of modified function. This is well illustrated in Fig. 1, which shows the effects of various detergents on Ca^{2+}-dependent exocytosis in chromaffin cells. These cells were first made permeable by exposure to brief high-voltage discharges (see below) and subsequently were exposed to different concentrations of detergent before challenge with calcium.[28,29] The Ca^{2+}-dependent membrane fusion event is inhibited rapidly and completely by relatively low concentrations of detergent. Figure 1 also shows that higher concentrations of detergent cause lysis of the secretory vesicles. Even despite the very limited useful life of such preparations,

[26] G. M. Burgess, I. S. McKinney, A. Fabiato, B. A. Leslie, and J. W. Putney, *J. Biol. Chem.* **258**, 15336 (1983).

[27] H. Wakasugi, T. Kimura, W. Haase, A. Kribben, R. Kauffmann, and I. Schulz, *J. Membr. Biol.* **65**, 205 (1982).

[28] P. F. Baker and D. E. Knight, *Philos. Trans. R. Soc. London B* **296**, 83 (1981).

[29] D. E. Knight and P. F. Baker, *J. Membr. Biol.* **68**, 107 (1982).

cell models generated by detergent have been used with some success in investigation of exocytosis.[30-32]

Local Nonselective Permeabilization by Exposure to Brief High-Voltage Pulses (Electropermeabilization)

Just as mechanical methods provide a very simple and clean way to remove the plasma membrane from a suitable cell, exposure to brief high-voltage pulses provides a similarly simple and clean way to create localized nonselective pores in the plasma membrane. The theoretical basis of the technique and its experimental application are described below.

Theoretical Considerations: Effect of an Electric Field on a Cell Suspension. If two flat parallel metal plates are positioned facing each other 1 cm apart and at a potential difference of, say, 1000 V, anything placed between them is said to experience an electric field of 1000 V/cm. When a cell, which may be considered as a hollow, thin-walled sphere, is exposed to such an electric field, a voltage will be imposed across its membrane. The magnitude of this potential depends on the intensity of the field, E (V/cm), and the size of the cell, r (cm), and will vary around the circumference of the cell. With reference to Fig. 2 the voltage at point P (V_p) across the membrane of a spherical cell is given by the simplified equation,

$$V_P = CrE \cos \Theta \quad \text{volts}$$

where the value of the constant C depends on the relative conductivities of the extracellular fluid, the cytosol, and the membrane. For most cells the conductivities of the extra- and intracellular fluids are much greater than that of the membrane and the dimensions of the cell are much greater than the membrane thickness. Under these conditions the constant term C approaches a value of 1.5 and the equation becomes

$$V_P = 1.5rE \cos \Theta \quad \text{volts} \tag{1}$$

Clearly the potential difference imposed across the membrane will be a maximum at points A and B in Fig. 2 corresponding to $\cos \Theta = \pm 1$ and will decrease around the cell to positions F and G where the electric field has no influence on the membrane ($\cos \Theta = 0$). If, therefore a cell 20 μm in diameter is placed in a field of, say, 2000 V/cm, a transmembrane potential of 3 V would occur at opposite ends of the cell in line with the electric field, i.e., at points A and B. It is this large potential that is thought to cause the membrane to break down and become permeable to solutes.

[30] J. C. Brooks and S. Treml, *J. Neurochem.* **40**, 468 (1983).
[31] L. A. Dunn and R. W. Holz, *J. Biol. Chem.* **258**, 4989 (1983).
[32] S. P. Wilson and N. Kirshner, *J. Biol. Chem.* **258**, 4994 (1983).

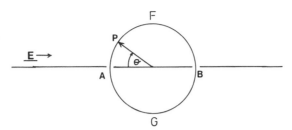

Fig. 2. Diagrammatic representation of a cell in an electric field E (see the text).

Studies on nerve, muscle, and artificial membranes suggest that a potential difference of 0.3 V is sufficient to bring about membrane breakdown insofar as the membrane exhibits a greatly reduced resistance. However, other studies using cell suspensions, artificial membranes, and plant cells[6,33-36] indicate strongly that a transmembrane voltage of about 1 V is necessary to cause a marked increase in membrane permeability. It is probable that this rather wide range of voltages reflects different criteria on which breakdown is assumed to have occurred, but the possibility of different mechanisms cannot be excluded.[37,38] Since changes in permeability to solutes of low molecular mass seem to be associated with membrane voltages of about 1 V, this figure can be used as a guide to calculate from Eq. (1) the field strength required to gain access to the cytosol.

Figure 3 describes the maximum voltage imposed across the membranes of various sizes of cells and organelles as a result of exposing them to electric fields. As expected from Eq. (1), the smaller the cell, the larger the field required to charge up the membrane to any given voltage. Figure 3 also includes some particularly instructive data relating to adrenal medullary cells. It compares the influence of the applied field on the membrane potential across the plasma membrane (cell diameter, 20 μm) with that across the membranes of two groups of intracellular organelles, the chromaffin granules (approximate diameter, 0.2 μm), which contain the secretory product catecholamine, and the mitochondria (approximate diameter, 1 μm). When a cell suspension is exposed to a field of 2000 V/cm, the plasma membrane should experience a transmembrane potential of 3 V, which ought to bring about breakdown, whereas the potential experienced by the membranes of the chromaffin granules and mitochondria should be

[33] P. F. Baker and D. E. Knight, *Nature (London)* **276,** 620 (1978).
[34] A. J. H. Sale and W. A. Hamilton, *Biochim. Biophys. Acta* **163,** 37 (1968).
[35] U. Zimmerman, G. Pilwat, and F. Reimann, *Biophys. J.* **14,** 881 (1974).
[36] U. Zimmerman, P. Scheurich, G. Pilwat, and R. Benz, *Angew. Chem.* **20,** 325 (1981).
[37] R. Benz, F. Beckers, and U. Zimmermann, *J. Membr. Biol.* **48,** 181 (1979).
[38] J. N. Crowley, *Biophys. J.* **13,** 711 (1973).

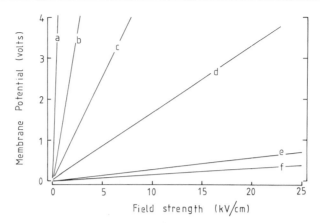

FIG. 3. The expected membrane potential [from Eq. (1)] across (a) sea urchin eggs 100 μm in diameter, (b) adrenal medullary cells 20 μm in diameter, (c) red blood cells 7 μm in diameter, (d) platelets 2 μm in diameter, (e) mitochondria 1 μm in diameter, and (f) chromaffin granules 0.2 μm in diameter as a result of exposure to various field strengths.

only 30 and 150 mV, respectively, which is insufficient to bring about breakdown.

The dependence of breakdown in an intense electric field on particle size is an exciting and potentially very useful aspect of the high-voltage technique, as it predicts that the cytosol can be accessed without disturbing the structure and function of the intracellular organelles. This has been demonstrated in such diverse cells as adrenal medullary cells,[29,33] sea urchin eggs,[9] and platelets,[10] wherein access has been achieved without altering the secretory mechanism or calcium-sequestering ability of the organelles.

Methodology. Two methods have been used routinely to subject isolated cells to electric fields: (1) passing cells through a Coulter counter[36] and (2) discharging a capacitor through a cell suspension. The main disadvantage of using the Coulter counter method is the length of time taken to harvest a worthwhile number of leaky cells, and for this reason we routinely use the second method, which is described below. Our experimental setup is illustrated in Fig. 4.

The capacitor is charged to a known voltage, isolated from the power supply, and discharged through a cell suspension between metal electrodes. The switch must discharge the capacitor in a single event. It can be simply two metal contacts pushed together manually. The electric field in the cell suspension will decay exponentially with a time constant determined by the product of the capacitance and the resistance of the cell suspension. A

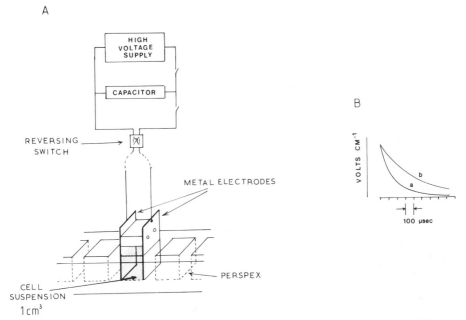

FIG. 4. (A) The experimental setup used to expose a suspension of cells to a brief electric field. (B) The time course of decay of the electric field applied to cell suspensions made up in the solution listed in Table I (a) and the same solution with the bulk of the potassium glutamate replaced by sucrose (b) when the applied voltage is 2 kV and the capacitor is 2 μF.

2-μF capacitor discharged across electrodes 1 cm apart and containing 1 ml of cell suspension having a resistance of about 100 Ω will produce a field through the suspension decaying with a time constant of 200 μsec. The duration of the field may be increased by increasing either the resistance of the suspending medium or the size of the capacitor. Partial replacement of the ions in the suspending medium by sucrose is an easy way to increase the resistance. When a series of discharges is given, the reversing switch (Fig. 3) should reduce the effects of electrode polarization. In practice, however, for short pulse durations and large voltages, the polarization effect is negligible.

Cells 20 μm in diameter can be permeabilized by exposure to 2 kV/cm. Smaller cells need proportionately higher voltages; this can be achieved by exposure to the same or a smaller field applied over a proportionately shorter distance. For example, the 20 kV/cm needed to access platelets is most easily obtained by applying 2 kV over 1 mm.

The following points must also be borne in mind (for further details, see Knight[6]).

1. The duration of exposure to the field must be long enough for the membrane to charge up. In practice this is a few microseconds at most. To some extent the choice of strength and duration of the applied field will depend on whether the experimental protocol requires the cells to reseal or remain leaky. In general, resealing becomes less likely as the strength and duration of the field are increased; resealing, however, is much more common in some cell types than in others, and it is not possible to generalize. Thus red cells, macrophages, and HeLa cells reseal more readily than adrenal medullary cells at 37° (although they remain permeable at 0°).

2. The discharge of the capacitor will cause the cell suspension to heat up. For 2000 V/cm, the heating is approximately 1° per discharge.

3. Increasing the number of discharges does not necessarily increase the number of effective pores in the plasma membrane, because once the plasma membrane has been breached it becomes progressively more difficult to charge up the remaining intact portions of membrane. In practice, the maximum number of pores that can be produced is probably not more than 20. The effective diameter of the pores can be calculated from the rates of uptake or loss of suitable markers.[29] In the specific case of bovine adrenal medullary cells exposed to 2000 V/cm and assuming two pores per discharge, the pores have an effective diameter of 4 nm.

4. The method is applicable to cells attached to culture dishes or to electrically conducting culture beads, to cells embedded in agar, or even to tissue slices, although immobilization of the cells may limit the number of lesions that can be made.

The two main disadvantages of the high-voltage technique are that accessibility is largely restricted to small molecules and that, even for these molecules, the rate of change in intracellular solute composition is dependent on the diameter of the cell and the number and sizes of the pores that have been created by exposure to the electric field. Diffusional equilibrium will be fastest in very small cells, such as platelets. In larger cells it may take a few minutes even for quite small molecules. The problem of diffusional equilibrium can be overcome to some extent by equilibrating the leaky cells at 0° and then initiating the process under investigation by imposing a sudden temperature jump, in which the preparation is warmed rapidly to 37°. An alternative approach, for which ATP is required, is to equilibrate with caged ATP and initiate the reaction by liberating the ATP by flash photolysis.

Local, Nonselective, Permeabilization by Exposure to Pore-Forming Chemicals

In addition to detergents, exposure of cells to a variety of chemicals can lead to increased permeability. These include (1) activation of endogenous receptors such as the ATP^{4-} receptor[39-40a]; (2) exposure to exogenous pore-forming agents such as certain viruses (see Ref. 41), complement,[42] and bacterial toxins[43,44]; and (3) general decalcification (see Refs. 43 and 45).

In the case of decalcification, the range of molecules to which the cell becomes permeable is very variable and the mechanism by which this is produced is not well defined (see Refs. 46 and 46a). Exposure of cells to low calcium concentration is known to increase membrane permeability, at least one component of which is associated with rather nonspecific ion fluxes through calcium channels (see Refs. 47–49), but this seems unlikely to account for all the reported effects of decalcification on membrane permeability. For these reasons, decalcification should perhaps be used with extreme caution and regular checks should be made to establish whether the altered permeability persists in media of different composition.

The other agents are all well-proved and effective ways to generate local, nonselective, pores rather similar in size to those generated by exposure to brief high-voltage pulses. The use of ATP^{4-} requires a suitable receptor and these are present in only a few cell types; the other agents are probably effective on a wide range of cells although they have, so far, been little used. In the case of viruses, a significant amount of foreign (viral) membrane is inserted into the cell surface and it may be necessary to establish what effect this has on cell function.

[39] S. Cockcroft and B. D. Gomperts, *Biochem. J.* **188,** 789 (1980).
[40] L. A. Heppel, G. A. Weisman, and I. Friedberg, *J. Membr. Biol.* **86,** 189 (1985).
[40a] J. F. Aiton and J. F. Lamb, *J. Physiol. (London)* **248,** 14P (1975).
[41] B. D. Gomperts, J. M. Baldwin, and K. J. Micklem, *Biochem. J.* **210,** 737 (1983).
[42] E. S. Schweitzer and M. P. Blaustein, *Exp. Brain Res.* **38,** 443 (1980).
[43] B. F. McEwan and W. J. Arion, *J. Cell Biol.* **100,** 1922 (1985).
[44] M. Thelestam and R. Möllby, *Biochim. Biophys. Acta* **557,** 156 (1979).
[45] B. Sakmann and E. Neher, "Single Channel Recording." Plenum, New York, 1983.
[46] D. J. Miller and G. L. Smith, *J. Physiol (London)* **346,** 74P (1984).
[46a] H. Streb, R. F. Irvine, M. J. Berridge, and I. Schulz, *Nature (London)* **306,** 67 (1983).
[47] W. Almers, E. W. McCleskey, and P. T. Palade, *J. Physiol. (London)* **353,** 565 (1984).
[48] W. E. Almers and E. W. McCleskey, *J. Physiol. (London)* **353,** 585 (1984).
[49] P. Hess and R. W. Tsien, *Nature (London)* **309,** 453 (1984).

Techniques That Permit Separate Control over Intracellular and Extracellular Environment

A major drawback to the techniques described in the previous section is that *both* sides of the cell membrane are exposed to the *same* solution. As physiological extracellular and intracellular fluids are very different, this means that the plasma membrane is subject to very unphysiological conditions. This can be avoided by the techniques of intracellular dialysis and perfusion (for a review, see Ref. 50). Initially developed for large cells such as squid axons[13,51–53] and barnacle muscle,[33] they have been progressively miniaturized for application to much smaller cells. The combination of the patch-clamp technique in the whole-cell recording mode with perfusion of the interior of the patch pipet has made possible the general application of perfusion/dialysis techniques to cells 10 μm or less in diameter (see Refs. 45, 54, and 55).

The critical factor with these techniques is whether the intracellular process under investigation can be assayed with adequate sensitivity in a single cell. They are ideally suited to measurement of electrical parameters such as conductance and capacitance, but they can also be combined with suitable optical methods for monitoring other aspects of cell function. A particularly interesting application is the measurement of membrane turnover by exocytosis and endocytosis, which can be followed at the single-cell level by monitoring fluctuations in membrane capacitance, small step increases, implying membrane addition by exocytosis, and step decreases, implying membrane retrieval by endocytosis.[56,57]

[50] P. G. Kostyuk and O. A. Krishtal, *Methods Neurosci.* **5** (1984).
[51] P. F. Baker, A. L. Hodgkin, and T. I. Shaw, *Nature (London)* **190,** 885 (1961).
[52] F. J. Brinley and L. J. Mullins, *J. Gen. Physiol.* **50,** 2303 (1967).
[53] T. Oikawa, C. S. Spyropoulos, I. Tasaki, and T. Teorell, *Acta Physiol. Scand.* **52,** 195 (1961).
[54] S. G. Cull-Candy, *Recent Adv. Physiol.* **10,** 1 (1984).
[55] O. P. Hamill, A. Marty, E. Neher, B. Sakmann, and F. J. Sigworth, *Pfluegers Arch.* **391,** 85 (1981).
[56] J. M. Fernandez, E. Neher, and B. D. Gomperts, *Nature (London)* **312,** 453 (1984).
[57] E. Neher and A. Marty, *Proc. Natl. Acad. Sci. U.S.A.* **79,** 6712 (1982).

[41] Implantation of Isolated Carriers and Receptors into Living Cells by Sendai Virus Envelope-Mediated Fusion

By ABRAHAM LOYTER, N. CHEJANOVSKY, and V. CITOVSKY

Introduction

Infection of animal cells by enveloped viruses is achieved in two ways. The large majority of the viruses, such as those belonging to the togavirus or myxovirus groups, are taken into intracellular compartments by endocytosis.[1] Fusion between the viral envelope and the lysosomal membrane is triggered by the intralysosomal low pH and leads to the introduction of the viral nucleocapsid into the cell cytoplasm.[1] Alternatively, penetration of animal cells by enveloped viruses belonging to the paramyxovirus group, such as New Castle Disease virus (NDV) or Sendai virus, occurs following fusion of the viral envelope with the cell plasma membrane.[2,3] Thus, viruses belonging to the latter group are characterized by the following unique properties which allow their use as biological carriers.[1] (1) Fusion occurs with external cell surfaces, namely cell plasma membranes. (2) Virus–membrane fusion does not require any special environment and takes place at pH 7.4. Binding of Sendai virus particles to cell surface sialic acid residues of membrane glycolipids and glycoproteins is mediated by the viral binding protein, hemagglutinin/neuraminidase (HN) glycoprotein, while for fusion the presence of the viral fusion (F) protein is required.[4,5] Recently, various experimental approaches have shown that the HN glycoprotein, besides serving as the viral binding protein, also plays an active role in the fusion process itself.[6-8] The two viral glycoproteins, i.e., the HN and the F polypeptides, are located within the viral envelope.[5] This implies that (1) fusion of the virus particles with living cells should result in the insertion of the viral glycoproteins into the cell plasma membranes,[5]

---

[1] P. W. Choppin and A. Scheid, *Rev. Infect. Dis.* **2**, 40 (1980).

[2] J. White, M. Kilejian, and A. Helenius, *Q. Rev. Biophys.* **16**, 151 (1983).

[3] R. G. Kulka and A. Loyter, *Curr. Top. Membr. Transp.* **12**, 365 (1979).

[4] G. Poste and C. A. Pasternak, *in* "Membrane Fusion" (G. Poste and G. L. Nicolson, eds.), p. 215. Elsevier/North-Holland, Amsterdam, 1978.

[5] A. Loyter and D. J. Volsky, *in* "Membrane Reconstitution" (G. Poste and G. L. Nicolson, eds.), p. 215. Elsevier/North-Holland, Amsterdam, 1982.

[6] N. Miura, T. Uchida, and Y. Okada, *Exp. Cell Res.* **141**, 409 (1982).

[7] T. D. Heath, F. J. Martin, and B. A. Macher, *Exp. Cell Res.* **149**, 163 (1983).

[8] O. Nussbaum, N. Zakai, and A. Loyter, *Virology* **138**, 185 (1984).

and (2) membrane vesicles which contain these two envelope glycoproteins should be fusogenic, as are intact virions; that is, they should fuse with recipient cells and transfer the viral glycoproteins into the cell plasma membranes.

Fusogenic, reconstituted Sendai virus envelopes (RSVE) can be obtained after solubilization of intact virus particles with nonionic detergents such as Nonidet P-40[9] or Triton X-100.[10,11] Detergent-insoluble material can be removed by centrifugation, leaving the viral envelope phospholipids and glycoproteins (HN and F polypeptides) in the supernatant. Removal of the detergent from the clear supernatant by either dialysis using Spectrapor-2 tubing[9,10] or by the direct addition of SM-2 Bio-Beads[11] leads to the formation of empty envelopes, which resemble the envelopes of intact Sendai virions but are devoid of the viral nucleocapsid.[5]

During the last few years, we have shown that RSVE can serve as an efficient carrier for the introduction of either soluble macromolecules[12] or membrane components[5,13] into eukaryotic cells. It has been shown that when soluble macromolecules such as protein or DNA molecules are present in the detergent-solubilized fraction, they are trapped within the vesicle formed after removal of the detergent.[12,14,15] Such loaded fusogenic vesicles have been used for the introduction of proteins and DNA molecules into living cultured cells.[5,12,15] If a membrane component is added to the detergent-solubilized mixture of the viral glycoproteins and phospholipids, it is inserted into the viral envelope formed after removal of the detergent, resulting in the formation of "hybrid" fusogenic viral envelopes.[5,13,16-19] Fusion of such "hybrid" vesicles with living cells in

[9] Y. Hosaka and K. Shimizu, *Virology* **49**, 627 (1972).
[10] D. J. Volsky and A. Loyter, *FEBS Lett.* **92**, 190 (1978).
[11] A. Vainstein, M. Hershkovitz, S. Israel, S. Rabin, and A. Loyter, *Biochim. Biophys. Acta* **773**, 181 (1984).
[12] A. Vainstein, A. Razin, A. Graessmann, and A. Loyter, this series, Vol. 101, p. 492.
[13] A. Loyter, D. J. Volsky, M. Beigel, H. Ginsburg, and Z. I. Cabantchik, *in* "Transfer of Cell Constituents into Eukaryotic Cells" (J. E. Celis, A. Graessmann, and A. Loyter, eds.), p. 143. Plenum, New York, 1980.
[14] T. Uchida, M. Yamaizumi, E. Mekada, and Y. Okada, *Biochem. Biophys. Res. Commun.* **87**, 371 (1979).
[15] A. Loyter, A. Vainstein, M. Graessmann, and A. Graessmann, *Exp. Cell Res.* **143**, 415 (1983).
[16] D. J. Volsky, Z. I. Cabantchik, M. Beigel, and A. Loyter, *Proc. Natl. Acad. Sci. U.S.A.* **76**, 5440 (1979).
[17] A. Projansky-Jakobovitz, D. J. Volsky, A. Loyter, and N. Sharon, *Proc. Natl. Acad. Sci. U.S.A.* **77**, 7247 (1981).
[18] M. Beigel and A. Loyter, *Exp. Cell Res.* **148**, 95 (1983).
[19] N. Mazurek, P. Bashkin, A. Loyter, and I. Pecht, *Proc. Natl. Acad. Sci. U.S.A.* **80**, 6014 (1983).

culture results in transplantation of the inserted components into the plasma membrane of the recipient cells.[16-19] For example, the purified and isolated anion channel of the human erythrocyte (band 3)[16] or the cromolyn-binding protein (CBP) from rat basophilic leukemia cells (RBS-2H3)[19] has been inserted into RSVE. Incubation of RSVE–band 3 hybrid vesicles with Friend erythroleukemia cells (FELC) resulted in transplantation of the band 3 molecules into the plasma membrane of these cells.[13,16] Similarly, incubation of hybrid viral vesicles containing the CBP with a variant of basophilic cells lacking this protein resulted in fusion-mediated transplantation of the CBP into plasma membranes of the variant cells. Membranes rich in lectin receptor as well as in H-2 antigens were coreconstituted with Sendai virus envelopes, which were then fused with living cells whose membranes lacked these components.[17]

This method has also permitted the transfer of viral receptor from receptor-positive to receptor-negative cells.[5,15,20,21] Simian virus 40 (SV40) receptors from African monkey kidney cells were first inserted into Sendai virus envelopes, and the hybrid vesicles formed were then fused with SV40-unsusceptible cells.[15] In the same manner, membranes rich in receptors for Epstein–Barr virus (EBV) from human lymphoma Raji cells were coreconstituted with RSVE. Fusion-mediated implantation of the EBV receptors into membranes of receptor-negative cells converted the latter susceptible to infection by EBV.[20,21] Thus, it appears that RSVE may serve as a convenient and efficient vehicle for the introduction of ion channels as well as virus, lectin, or hormone receptors, or any other membrane components of well-defined structure or function into plasma membranes of living animal cells.

Procedures and Results

A detailed description of the methods and techniques, as well as of the buffers and reagents used to obtain purified preparations of Sendai virions and for the efficient reconstitution of their envelopes, has recently been published in *Methods of Enzymology*.[12] Therefore, in the present chapter only a short description of these methods will be given, with the exception of new methods and techniques, which will be described in full detail.

Propagation and Purification of Intact Sendai Virus Particles

Sendai virus particles are estimated by hemagglutination titer (Salk's pattern method), namely, by their ability to agglutinate chicken erythro-

[20] I. M. Shapiro, G. Klein, and D. J. Volsky, *Proc. Herpes Virus Workshop, 5th* (1980).
[21] D. J. Volsky, I. M. Shapiro, and G. Klein, *Proc. Natl. Acad. Sci. U.S.A.* **77,** 5453 (1980).

cytes at room temperature.[22] Generally, about 10^4 hemagglutinating units (HAU) are present per milligram of viral protein.

High amounts of virions can be obtained by propagation in the allantoic cavity of 10-day-old chicken fertilized eggs, following injection of 2 HAU (in about 0.2 ml) per each fertilized egg.[12] The allantoic fluid is collected after 48–72 hr of incubation and then centrifuged at low (1400 g, 10 min) and high speed (100,000 g, 30 min) in the cold.[12] The pellet is suspended in Solution Na (160 mM NaCl, Tricine–NaOH, pH 7.4), homogenized, and then loaded on 15% (w/v) sucrose cushions, following centrifugation at 100,000 g for 1 hr. The pellet which contains purified virus particles is suspended in Solution Na to give about 1×10^4 HAU/ml. About 1×10^5 HAU of Sendai virions can be obtained from each 100 ml of allantoic fluid.[12] The virus is stored at $-70°$.

Quantitative Estimation of the Fusogenic Activity of Intact Sendai Virions

It is advisable that before experiments with RSVE or "hybrid" envelopes are performed, the fusogenic activity of the intact virus particles be tested. This is especially important in cases in which huge amounts of stored virions will be used. The ability of the virus to agglutinate chicken erythrocytes does not necessarily mean that it possesses fusogenic activity. The traditional and conventional method for studying the fusogenic activity of Sendai virions has been to estimate its hemolytic activity at pH 7.4[3,5] It has been well established that the ability of Sendai virions to hemolyze red blood cells reflects its fusogenic activity, i.e., its ability to fuse with cell plasma membranes.[23] However, the viral hemolytic activity is not directly correlated to the viral fusogenic activity, and depends substantially on the source and age of the red blood cells used as well as on the age and history of the virions.[23] Early-harvested virions (obtained from the allantoic fluid collected after 12–24 hr of incubation) although fusogenic, possess very little if any hemolytic activity. Recently, we[24,25] as well as others[26] have shown that fluorescence dequenching methods can be used for a quantitative estimation of the fusogenic activity of RSVE or of intact Sendai virions. When fluorescent molecules such as octadecylrhodamine B chloride (R_{18}) or N-4-nitrobenz-1,2-oxa-1,3-diazolphosphatidylethanolamine (N-NBD-PE) are inserted at certain surface densities into envelopes of

[22] Z. Toister and A. Loyter, J. Biol. Chem. **248,** 422 (1973).

[23] Y. Shimizu, K. Shimizu, N. Ishida, and M. Homma, Virology **71,** 48 (1976).

[24] N. Chejanovsky and A. Loyter, J. Biol. Chem. **260,** 7911 (1985).

[25] V. Citovsky and A. Loyter, J. Biol. Chem. **260,** 12072 (1985).

[26] M. C. Harmsen, Y. Wilschut, G. Scherphof, C. Hulstaert, and D. Hoekstra, Eur. J. Biochem. **149,** 591 (1985).

intact Sendai virions or RSVE, respectively, their fluorescence is reduced because of self-quenching.[25,26] Dilution of the fluorescent molecules due to fusion processes results in fluorescence dequenching.[27]

Insertion of R_{18} into Envelopes of Intact Sendai Virions. The fluorescent molecule R_{18} was inserted into envelopes of intact virus essentially as described before.[26] Briefly, 7 μl of 6 mg/ml of ethanolic solution of R_{18} (Molecular Probes, Inc., Junction City, Oregon) is rapidly injected into 700 μl of Solution Na containing 1.7 mg of purified Sendai virus. The suspension is then incubated for 15 min at room temperature in the dark. Under such conditions, the viral membrane contains about 3 mol% of R_{18} per total viral lipids. At the end of the incubation period, the virions are washed with 20 volumes of Solution Na (100,000 g, 30 min at 4°) and resuspended in the same buffer to give 0.2 mg protein/ml.

Fusion between Sendai Virions and Human Erythrocyte Ghost Membranes or Cultured Living Cells. Incubation of fluorescent intact virions with human erythrocyte ghosts (HEG) at 37° resulted in a relatively high degree of fluorescence dequenching (Table I). Fluorescence dequenching was obtained also following incubation of fluorescent virions with living cultured cells such as hepatoma tissue culture cells (HTC). As the degree of fluorescence dequenching is directly related to the extent of virus–cell fusion,[24] it can be inferred from the results of Table I that about 40% of the virus particles in the preparation fused with either HEG or HTC. Very little fluorescence dequenching was observed when trypsinized, phenyl-methylsulfonyl fluoride (PMSF)- or dithiothreitol (DTT)-treated fluorescent virions were incubated with HEG or HTC. It has been shown that all of the above treatments greatly inhibit the fusogenic activity of the virus. Trypsin, PMSF, or DTT virions are neither able to lyse, infect, nor fuse animal cells.[5] These results further emphasize the view that the fluorescence dequenching observed indeed reflects a process of virus–membrane fusion.

Fusion between Intact Sendai Virions and Phospholipid Vesicles (Liposomes). As mentioned above, the first step in the virus–cell membrane fusion process is binding of virus particles (or RSVE) to cell surface receptors, i.e., membrane sialoglycolipids or sialoglycoproteins.[28] Evidently, the degree of virus cell binding will significantly affect the extent of virus–membrane fusion. In cases where a direct estimation of the viral fusogenic activity is required without the involvement of its binding ability (in experiments in which cells with low amounts of virus receptors are used), it

[27] A. J. Schroit and R. E. Pagano, *Cell* **23**, 105 (1981).
[28] R. Rott and H. O. Klenk, *in* "Virus Infection and Cell Surface" (G. Poste and G. L. Nicolson, eds.), p. 47. North-Holland, Amsterdam, 1977.

TABLE I

QUANTITATIVE ESTIMATION OF THE FUSOGENIC ACTIVITY OF INTACT SENDAI VIRIONS:
USE OF FLUORESCENTLY LABELED SENDAI VIRUS ENVELOPES[a]

Virion	Hemolysis (%)	R_{18} dequenching (%) after incubation with			
		HEG	HTC	PC/cholesterol	PS
Sendai	78	46	38	43	36
PMSF–Sendai	7	4	3	7	38
DTT–Sendai	8	6	4	9	32
Trypsinized Sendai	4	2	3	7	8

[a] Sendai virions were labeled with R_{18} as described in the text and in Ref. 26. To inhibit their hemolytic and fusogenic activities, Sendai virions were incubated for 30 min at 37° with either trypsin (60 μg/mg viral protein),[11] 3 mM DTT, or 7 mM PMSF.[24] After the incubation, the viral preparations were washed with 20 volumes of Solution Na and resuspended with the same buffer. Human erythrocyte ghosts (HEG) were prepared from freshly drawn human erythrocytes, type O.[24] Hepatoma tissue culture cells (HTC), clone CM22, were grown in suspension in modified Swim's 77 medium, supplemented with 10% newborn calf serum. Large unilamellar liposomes, i.e., phosphatidylcholine/cholesterol (PC:cholesterol 1:0.5 molar ratio) or phosphatidylserine (PS), were prepared by removal of the detergent from the octylglucoside solution of the appropriate lipid (10:1 detergent:lipid molar ratio) as described before.[25] Untreated and treated fluorescently labeled Sendai virions (1.0 μg) were incubated with either HEG (200 μg), HTC (10⁶ cells), or human erythrocytes (150 μl, 2–3%, v/v) for 30 min at 37°. The same fluorescent (R_{18}-labeled) Sendai virions (1.0 μg) were incubated with the liposomes (400 μM phospholipids) of the indicated composition for 45 min at 37°. At the end of the incubation period, the degree of hemolysis was estimated as described earlier,[22] and the extent of fluorescence dequenching (R_{18}-DQ), as before.[26] The degree of fluorescence was estimated (excitation at 560 nm and emission at 590 nm) before and after the addition of 0.1% Triton X-100. The degree of fluorescence in the presence of the detergent was considered to represent an infinite dilution of the R_{18} molecules, i.e., 100% dequenching. All fluorescence measurements were carried out with a Perkin–Elmer MPF-4 spectrofluorometer with a 520-nm high-pass filter and narrow excitation slits to reduce light scattering.

is recommended to use liposomes composed of phosphatidylcholine (PC) and cholesterol as a recipient membrane. It has been well established that Sendai virions (and RSVE) are able to fuse efficiently with liposomes composed of PC and cholesterol but lacking any sialic acid-containing components such as sialoglycoproteins or sialoglycolipids.[25,29]

The results in Table I clearly show that incubation of fluorescently labeled, intact Sendai virions with liposomes composed of PC and cholesterol resulted in a high degree of fluorescence dequenching. Moreover, the results in this table show that the fluorescence dequenching obtained

[29] M.-C. Hsu, A. Scheid, and P. W. Choppin, *Virology* **126,** 361 (1983).

indeed reflects the viral fusogenic activity needed for its infectivity as well as for fusion with human erythrocyte membranes or with living cultured cells. Inactive (trypsin-, PMSF-, or DTT-treated) virus particles, which are unable to hemolyze human erythrocytes or to fuse with either their membranes or those of HTC, also failed to fuse with PC/cholesterol liposomes. On the other hand (Table I), the PMSF- and DTT-treated virions fused efficiently with liposomes composed of negatively charged phospholipids, as can be inferred from the high degree of fluorescence dequenching obtained upon incubation of these treated Sendai virions and liposomes composed of phosphatidylserine (PS). Only treatment with trypsin blocked the ability of Sendai virions to fuse with the negatively charged liposomes, indicating that, for such fusion to occur, the presence of intact viral glycoproteins is required.

Thus, for studying viral fusogenic activity, we do not recommend the use of negatively charged phospholipid vesicles, which are frequently used for studies of liposome–liposome fusion.[30] It seems that even noninfective, inactive virus particles such as PMSF- or DTT-treated virions are able to fuse with such negatively charged phospholipids, probably due to an unspecific interaction between the Sendai viral glycoproteins and the phospholipid negatively charged groups. From our results, it should be inferred that only fusion between the virus and PC/cholesterol liposomes reflects the fusogenic activity needed for infection and fusion with cell membranes.

Reconstitution of Sendai Virus Envelopes

Reconstitution of Fusogenic Viral Envelopes. Fusogenic RSVE can be obtained essentially as described before[10-12] and are summarized schematically in Fig. 1. Briefly, a pellet of Sendai virus particles (3–20 mg viral protein) is dissolved in 0.2–1.3 ml of Triton X-100 in 100 mM NaCl, 50 mM Tris–HCl, pH 7.4, and 0.1 mM PMSF (Triton: viral protein, 2:1, w/w, with a final concentration of Triton X-100 of 2–4%). After shaking for 1 hr at 20°, the detergent-insoluble material is removed by centrifugation (100,000 g for 60 min).

Removal of Triton X-100 can be achieved by dialysis in Spectrapor-2 tubing for 48–72 hr in the cold against a buffer containing 10 mM Tris–HCl, pH 7.4, 2 mM MgCl$_2$, 2 mM CaCl$_2$, 1 mM NaN$_3$, and 1.2 g/liter of SM-2 Bio-Beads or 1 mg/ml of bovine serum albumin.[12] Under these conditions, about 0.015–0.03% of the Triton X-100 remains after 60 hr of dialysis in the cold.[12] The turbid suspension obtained—which contains resealed RSVE—is centrifuged (100,000 g, 30 min) and the resulting pellet is suspended in Solution Na to give 1 mg/ml and is stored at −70°.

---

[30] J. Wischut, N. Duzgunes, R. Fraley, and D. Papahadjopoulos, *Biochemistry* **19**, 6011 (1980).

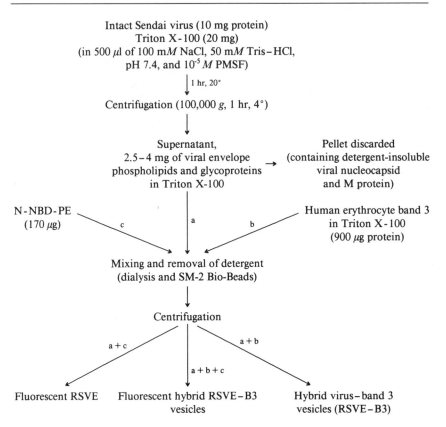

Intact Sendai virus (10 mg protein)
Triton X-100 (20 mg)
(in 500 μl of 100 mM NaCl, 50 mM Tris–HCl,
pH 7.4, and 10^{-5} M PMSF)

1 hr, 20°

Centrifugation (100,000 g, 1 hr, 4°)

Supernatant,   Pellet discarded
2.5–4 mg of viral envelope   →   (containing detergent-insoluble
phospholipids and glycoproteins   viral nucleocapsid
in Triton X-100   and M protein)

N-NBD-PE   Human erythrocyte band 3
(170 μg)   c   a   b   in Triton X-100
                                                (900 μg protein)

Mixing and removal of detergent
(dialysis and SM-2 Bio-Beads)

Centrifugation

a + c   a + b

a + b + c

Fluorescent RSVE   Fluorescent hybrid RSVE–B3   Hybrid virus–band 3
                         vesicles   vesicles (RSVE–B3)

FIG. 1. A scheme summarizing the coreconstitution method for obtaining RSVE–B3, fluorescent RSVE–B3, or fluorescent RSVE.

An alternative method to remove the Triton X-100 is by the direct addition of SM-2 Bio-Beads to the detergent-solubilized mixture of the viral phospholipids and glycoproteins,[11] thus avoiding the long dialysis step, as follows. For each 10 mg of Triton X-100, 70 mg of SM-2 Bio-Beads is added. After 2–3 hr of incubation at 20° with vigorous shaking, a second 70 mg SM-2 Bio-Beads is added. Following an additional incubation period of 12–14 hr at room temperature (20°) with vigorous shaking, the soluble phase is removed using a 1-ml tuberculin syringe into which the SM-2 Bio-Beads cannot penetrate. The SM-2 Bio-Beads are washed twice with about 0.5 ml of Solution Na. The turbid suspension obtained (containing RSVE) is centrifuged (100,000 g, 30 min) and the resulting pellet is suspended in Solution Na to give about 1 mg/ml. Using ^{3}H-labeled Triton X-100, it was demonstrated that only about 0.005% of the detergent remained in the RSVE suspension.[11] From each milligram of intact virus particles, about 300 μg of detergent-solubilized viral glycoproteins and 80–100 μg of reconstituted envelopes were obtained.

Coreconstitution: Formation of "Hybrid" Reconstituted Fusogenic Viral Envelopes (RSVE–B3). The main features of the method for preparation of fusogenic "hybrid" vesicles, i.e., viral envelopes bearing nonviral polypeptides, are summarized in Fig. 1. The principal stages of the method are described using a human erythrocyte anion-exchange system, namely, band 3 (B3), as a model system. The human erythrocyte band 3 was the first membrane protein which was used to develop and demonstrate the various aspects of the fusion-mediated implantation technique,[16] due to the following reasons[16,31]: (1) it is a transmembrane polypeptide of 100,000 MW which fulfills a well-defined transmembrane function; (2) band 3 can be isolated in a pure form and is functionally intact in detergent solution; and (3) it can be radiolabeled, thus allowing a quantitative estimation of its fate during the various stages of the implantation method.

Human erythrocyte ghosts are extracted with Triton X-100 to yield a detergent solution of the B3 protein (purification procedures as described in Ref. 31). Intact Sendai virus is solubilized with Triton X-100, and detergent-insoluble material is removed as described above, leaving in the clear supernatant mainly the phospholipids and the HN and F glycoproteins (see also Fig. 1). For the formation of RSVE-B3 hybrid vesicles, 2500 μg of the detergent-solubilized viral envelope glycoproteins is mixed with 900 μg of the detergent-solubilized human erythrocyte band 3 (Fig. 1, pathway a + b).

It is recommended that the viral protein/nonviral protein weight ratio in the coreconstitution system should be kept at about 2.0 and above. It has been noticed that this ratio cannot be further decreased without affecting the fusogenic activity of the coreconstituted vesicles.

The detergent can be removed from the mixture either by dialysis in Spectrapor-2 tubing or by the direct addition of SM-2 Bio-Beads, as described above. When Triton X-100 is removed by the direct addition of SM-2 Bio-Beads, it is important to check whether the added nonviral polypeptides are also removed. In a typical experiment, about 50% of the Triton X-100-solubilized glycoproteins are reconstituted in the envelopes during removal of the detergent.[10] The rest (50%) are adsorbed by the Bio-Beads. When the amount of polypeptides removed by the SM-2 Bio-Beads exceeds 60–70%, it is recommended to employ the dialysis method (dialysis with Spectrapor-2 tubing) for removal of the detergent. The turbid suspension obtained after removal of the detergent, containing the "hybrid" vesicles (Fig. 1, pathway a + b), is centrifuged (100,000 g, 30 min)

[31] Z. I. Cabantchik, P. A. Knauf, and A. Rothstein, *Biochim. Biophys. Acta* **515**, 239 (1978).

and the pellet obtained is suspended in Solution Na to give about 1 mg/ml; this is stored at $-70°$ until use.

Band 3 can be stoichiometrically labeled with the inhibitor of its anion-exchange activity, namely, tritiated 4,4′-diisothiocyanostilbene-2,2′-disulfonic acid ([³H]H₂DIDS).[32] Hence, the distribution of band 3 can be studied by following the fate of the radiolabeled polypeptide.[16]

The formation of hybrid vesicles consisting of human erythrocyte band 3 and the viral envelope glycoproteins can also be demonstrated by the use of radiolabeled band 3.[5,16] As can be seen in step A of Table II, anti-Sendai virus antiserum can precipitate up to about 80% of the total [³H]H₂DIDS-labeled band 3 protein present in the RSVE–B3 preparation.

For coreconstitution, it is possible to use, in addition to purified membrane proteins such as human erythrocyte band 3[16] or the CBP of rat basophilic leukemia cells,[19] extracts of whole plasma membranes.[17,18,20]

TABLE II
DETECTION OF HUMAN ERYTHROCYTE BAND 3 AND FLUORESCENT PHOSPHOLIPID MOLECULES IN SENDAI VIRUS ENVELOPES FORMED BY CORECONSTITUTION SYSTEM[a]

Anti-Sendai virus antibodies incubated with	Precipitated % of total	
	[³H]H₂DIDS–Band 3	Fluorescence units
Step A[a] RSVE + [³H]H₂DIDS–Band 3 (mixture)	1.4	—
RSVE–[³H]H₂DIDS–Band 3 hybrid vesicles	82	—
Step B[b] Fluorescent PC vesicles	—	10
Fluorescent PC vesicles + RSVE (mixture)	—	12
Fluorescent RSVE	—	80

[a] Hybrid vesicles (RSVE–[³H]H₂DIDS–Band 3) were obtained as described in the text, Fig. 1, and previously.[16] For immunoprecipitation, the two systems were incubated with 40 μg of anti-Sendai virus antibodies for 30 min at 37°, after which the virus was precipitated for 2 min at 10,000 g in an Eppendorf microfuge. The amount of radioactivity in the pellet was determined as before.[5,16]

[b] Fluorescent RSVE (bearing N-NBD-PE, 94:6 viral lipids:N-NBD-PE) were prepared as described in the text, Fig. 1, and earlier.[24,25] Fluorescent PC vesicles were prepared by mixing 250 μg of PC (from egg yolk, type V-E, Sigma) with a chloroform solution of 10 μg N-NBD-PE. The mixture was evaporated to dryness and the thin layer obtained was suspended in 250 μl of PBS. Small unilamellar vesicles were formed by sonication of the turbid suspension (2 min at 4°) in Heat-System Ultrasonic 350 sonicator. External, nonincorporated N-NBD-PE was removed by sieve chromatography, using Sephadex G-25. For immunoprecipitation, a mixture of 5 μg RSVE and 5 μg fluorescent PC or 5 μg of fluorescent RSVE was incubated with 40 μg of anti-Sendai virus antibody and precipitated as described in footnote a above. The fluorescence units associated with the pellet were determined. The NBD fluorescence was measured at 469 nm excitation and 545 nm emission. It should be noted that, as opposed to liposomes composed of PC/cholesterol, liposomes composed of only PC do not fuse with Sendai virions.[25]

[32] Z. I. Cabantchik and A. Rothstein, J. Membr. Biol. 15, 207 (1974).

Evidently, more is known about a particular membranal function than about the molecular entity representing that function. The procedure usually includes homogenization of cells swollen in hypotonic buffer (10 mM Tris–HCl, pH 7.4), followed by centrifugation through a steep sucrose gradient.[17] The pure plasma membrane fractions thus obtained are up to 70-fold enriched in membrane markers such as 5'-nucleotidase and alkaline phosphatase, as compared to the homogenate.[17] Generally, 0.5–1.0 mg membrane protein could be recovered from 10^9 tumor cells[20,21] and about 10 times less was recovered from the same amount of lymphocytes.[17]

The use of membrane extracts presumably decreases the coreconstitution efficiency of a particular protein of interest, since it is diluted by other membrane proteins. To compensate for this dilution, the virus protein/nonvirus protein weight ratio should be kept, in the reconstitution system, close to 2.0–2.5 or above,[17,20] so as not to affect the fusogenic activity of the hybrid vesicles formed after removal of the detergent. This imposes a further limitation to the loading of the coreconstitution system when whole extracts of cell plasma membranes are used. To increase the efficiency of the system, it is recommended to employ a procedure that will allow enrichment of a particular protein.

Quantitative Estimation of RSVE or of Hybrid Vesicles' Fusogenic Activity

The experimental manipulations employed for obtaining reconstituted viral envelopes, and especially hybrid vesicles, may significantly affect the fusogenic activity of the vesicles formed. Various detergents are employed for extracting or purifying the nonviral membrane polypeptides, which are used for the coreconstitution experiments. These detergents, as well as the presence of the foreign polypeptides within the viral envelopes, may significantly modify or even block the ability of the virus envelopes to interact and to fuse with cell plasma membranes. Therefore, before incubation with cell culture, the fusogenic activity of these vesicles should be studied by a quantitative, simple, quick, and reliable method. This can be achieved by observing the interaction between fluorescently labeled RSVE (or hybrid vesicles) and liposomes composed of PC and cholesterol.[25] Increase in the fluorescence degree (fluorescence dequenching) is due to fusion between the viral envelopes and the liposome membranes. We highly recommend this assay system because (1) it is quick and simple, requiring no special membrane preparations other than phospholipid vesicles, and (2) RSVE (and hybrid vesicles) may fuse efficiently with liposomes composed of only PC and cholesterol but lacking virus receptors. Thus, this system offers a direct means to study the virus fusogenic activity, independently of its binding activity, which may be modified during the coreconstitution process.

Preparation of Fluorescent RSVE or Hybrid Vesicles. Sendai virus particles (10 mg protein) are dissolved with 4% (w/w) Triton X-100 as described above. After centrifugation, the clear supernatant—containing 2.5–3 mg of viral glycoproteins and phospholipids—is added to a thin film of 170 μg of N-NBD-PE (Avanti Biochemicals, Birmingham, Alabama), as described elsewhere[24] (94:6 viral lipids:N-NBD-PE). The thin film of fluorescent molecules is obtained after a chloroform solution of N-NBD-PE is dried under nitrogen. Following the addition of the detergent solution containing the viral glycoproteins, the mixture obtained is vigorously shaken for 5 min. At this stage, to obtain fluorescent RSVE, the detergent is removed (Fig. 1, pathway a + c) or, for the formation of fluorescent hybrid vesicles, a nonviral membrane component is added before the removal of the detergent (Fig. 1, pathway a + b + c). The detergent is removed as described above, either by dialysis in Spectrapor-2 tubing or by direct addition of SM-2 Bio-Beads. The fluorescent RSVE or hybrid vesicles obtained are collected by centrifugation (100,000 g, 30 min) and the pellet obtained is suspended either in Solution Na or in phosphate-buffered saline (PBS) to give 1–2 mg/ml. Formation of RSVE bearing N-NBD-PE is evident from the results in Table II, step B, showing that antiviral antibodies were able to precipitate about 80% of the fluorescent molecules added to the viral reconstitution system. Centrifugation of a mixture containing antiviral antibodies and fluorescent PC vesicles, or antiviral antibodies, fluorescent PC, and RSVE, resulted in the precipitation of only 10–12% of the fluorescent molecules (Table II, step B).

Fusion of Fluorescent RSVE (or Hybrid Vesicles) with Phospholipid Vesicles (Liposomes). Fluorescent RSVE, similar to intact virions, promote a high degree of lysis when incubated with human erythrocytes (Table III). Although it has been well established that virus-induced lysis reflects a process of virus–membrane fusion,[23] in some cases it may be misleading, especially as reconstituted viral envelopes contain residual amounts of detergents.[26]

The results in Table III show that incubation of fluorescent RSVE with liposomes composed of PC and cholesterol resulted in a high degree of fluorescence dequenching. The view that the fluorescence dequenching observed reflects, in a quantitative measure, the fusogenic activity of the virus envelope that is required for its infectivity and for fusion with biological membranes is inferred from the following observations (Table III). (1) No fluorescence dequenching was observed when trypsinized, PMSF- or DTT-treated viral envelopes were incubated with PC/cholesterol liposomes. All these treatments are known to block the virus' ability to fuse with cell plasma membranes or to hemolyze red blood cells.[4,5] (2) Incubation of fluorescent vesicles containing only the viral F or HN glycoprotein

TABLE III
FUSION OF FLUORESCENT RSVE WITH LIPOSOMES COMPOSED OF PC/CHOLESTEROL:
QUANTITATIVE ESTIMATION OF RSVE FUSOGENIC ACTIVITY

Preparation	Hemolysis (%)	PC/cholesterol (N-NBD-PE dequenching, % of total)
Fluorescent RSVE[a] treated with		
None	86	42
Trypsin	6	2
PMSF	9	6
DTT	4	4
Vesicles[b] bearing only the viral		
F glycoprotein	6	4
HN glycoprotein	8	4

[a] Fluorescent RSVE (RSVE bearing N-NBD-PE, 94:6 viral lipids:N-NBD-PE) was prepared as described in the text, Fig. 1, and before.[24,25] Liposomes composed of PC/cholesterol were obtained as described in Table I and earlier.[25] The RSVE were treated with trypsin, PMSF, and DTT as described in Table I for intact Sendai virions. Untreated and treated fluorescent RSVE (0.5 μg) were incubated for 30 min at 37° with PC/cholesterol liposomes (250 μM in PC) in a final volume of 550 μl of Solution Na. At the end of the incubation period, the degree of fluorescence was estimated (N-NBD excitation at 469 nm and emission at 543–546 nm) before and after addition of 0.1% Ammonyx-LO (Onyx Corp.). The degree of fluorescence in the presence of detergent was considered to represent an infinite dilution of the N-NBD-PE; that is, 100% dequenching. The degree of hemolysis was estimated as described in Table I and elsewhere,[22] using freshly drawn human erythrocytes.

[b] For preparation of vesicles bearing only the viral F or HN glycoprotein, these polypeptides were isolated from the detergent solution of the viral particles as previously described.[8] The F- and HN-bearing vesicles were obtained after removal of the detergent by the direct addition of SM-2 Bio-Beads. Fluorescent F and HN vesicles were obtained exactly as described for the preparation of fluorescent RSVE.[25] Fluorescent F and HN vesicles (0.5 μg) were incubated with PC/cholesterol liposomes (250 μM in PC) as described for fluorescent RSVE in footnote a. Hemolysis was as determined in footnote a and Table I.

with the PC/cholesterol liposomes did not result in fluorescence dequenching. It has been well established that both glycoproteins are required to allow fusion of the viral envelopes with cell plasma membranes.[2,4,5] The observation (Table III; see also Ref. 24) that the viral HN glycoprotein is required for fusion with PC/cholesterol liposomes lacking viral receptors further emphasizes the view[7,8] that this glycoprotein, besides being the virus binding protein, also plays an active role in the fusion process.

In view of the above, incubation with PC/cholesterol liposomes offers a simple method to prove that the RSVE or hybrid vesicles are fusogenic and contain the two viral envelope glycoproteins in the appropriate proportions. Inhibition by the various inhibitors proves that such fluorescence dequenching is not due to lipid–lipid exchange but to virus–liposome fusion.

Selection of an Appropriate Detergent for Preparation of Fusogenic RSVE or Hybrid Vesicles

Sendai virions, like other enveloped viruses, can be solubilized by various detergents.[2,5,10] Incubation of Sendai virus particles with either ionic or nonionic detergents results in the complete solubilization of the viral envelopes.[5] Sendai virions can be dissolved by octylglucoside (see Table IV), deoxycholate, cholate, and lysolecithin of different compositions, as well as by Nonidet P-40 and Triton X-100.[5] In all cases, removal of the detergent either by sieve chromatography (using Sephadex columns), by dialysis, or by direct addition of SM-2 Bio-Beads resulted in the formation of vesicular membrane structures which resembled envelopes of intact virions.[5] Nevertheless, in our hands (see results in Table IV), and as can be inferred from studies by other groups,[14,33] only solubilization by Nonidet P-40 or Triton X-100 failed to result in an irreversible inactivation of the viral fusogenic activity. However, since it is almost impossible to completely remove Trtion X-100 from the viral glycoprotein preparations, some of the RSVE properties, such as its lytic activity,[1,26] were attributed to a residual amount of Triton left in the viral vesicles.

Furthermore, for implantation experiments, a nonviral membrane component is coreconstituted with the viral glycoprotein to give hybrid vesicles (Fig. 1, pathway a + b). Isolation and purification of a nonviral membrane polypeptide also necessitates the use of various detergents.[5] Residual amounts of such detergents may be carried with the nonviral polypeptides to the coreconstitution system and cause inactivation of the viral fusogenic activity. Hence, to establish an efficient fusion-mediated transplantation system, it is crucial to study the fusogenic activity of the RSVE or hybrid vesicles obtained, as mentioned earlier.

The results in Table IV show that the RSVE obtained by solubilization of intact virions by Triton X-100 were able to bind to (cause agglutination) and hemolyze human erythrocytes. In addition, Table IV shows (as can also be inferred from the results in Table III) that incubation of the fluorescent RSVE with PC/cholesterol liposomes resulted in a high degree of fluorescence dequenching. Almost the same degree of fluorescence dequenching was obtained when the fluorescent RSVE were incubated with negatively charged liposomes (liposomes composed of PS). On the other

[33] M.-C. Hsu, A. Scheid, and P. W. Choppin, *Virology* **95**, 476 (1979).

TABLE IV

FUSOGENIC ACTIVITY OF RSVE OBTAINED FROM TRITON X-100 OR OCTYLGLUCOSIDE-SOLUBILIZED
VIRAL PARTICLES: SELECTION OF APPROPRIATE DETERGENT

Preparation	Fluorescent RSVE treated with[c]	Hemagglutination	Hemolysis (%)	PC/cholesterol PS (N-NBD-PE dequenching, % of total)	
RSVE obtained from	None	+	78	42	46
Triton X-100-solu-	Trypsin	+	6	8	8
bilized Sendai	PMSF	+	8	7	45
virions[a]	DTT	−	6	7	48
RSVE obtained from	None	+	4	7	38
octylglucoside-solu-	Trypsin	+	3	7	8
bilized Sendai	PMSF	+	4	6	35
virions[b]	DTT	−	4	8	35

[a] RSVE were obtained from Triton X-100 solubilized Sendai virions and labeled with N-NBD-PE as described in Table III, Fig. 1, and elsewhere.[24,25]

[b] Sendai virions were dissolved with octylglucoside (Sigma) exactly as described for solubilization with Triton X-100, following removal of the detergent by slow dialysis, as described before.[26]

[c] The two preparations of RSVE were treated with trypsin, PMSF, and DTT as described in Table I for intact Sendai virions. The degree of hemolysis and hemagglutination were estimated exactly as previously described,[22] except that human erythrocytes were used instead of chicken erythrocytes. For fluorescence dequenching measurements, the N-NBD-PE-labeled RSVE preparations were incubated with PC/cholesterol liposomes or PS liposomes (250 μM in phospholipid each) exactly as described in Table III.

hand, the fluorescent RSVE that were obtained after solubilization of intact virions with octylglucoside[26] could agglutinate (bind to) human erythrocytes but failed to hemolyze them (Table IV). Moreover, incubation of these RSVE with liposomes composed of PC and cholesterol did not result in any significant fluorescence dequenching. Fluorescence dequenching was obtained only when these RSVE were incubated with liposomes composed of PS. However, this fluorescence dequenching was not subject to inhibition by either PMSF or DTT (Table IV). These results support the view that incubation of fluorescent vesicles, containing the Sendai virus glycoproteins, with PC/cholesterol liposomes, but not with negatively charged liposomes,[34] offers the most reliable system to study which detergent should be used in order to reconstitute viral envelopes that will retain their ability to fuse with biological membranes.

[34] V. Citovsky, N. Zakai, and A. Loyter, *Exp. Cell Res.* **166**, 27a (1986).

Fusion of Hybrid Reconstituted Vesicles with Target Cells: Quantitation of Hybrid Vesicle–Cell Fusion

Fusion of the hybrid reconstituted vesicles bearing a nonviral exogenous protein with target cultured cells resulted in fusion-mediated implantation of the vesicle proteins into the plasma membrane of target cells.[5,16-19] The implanted proteins can be detected by methods which analyze either their structure or function.[5] Using ^{3}H-labeled band 3, it has been observed that an average of 5×10^5 band 3 polypeptides can be integrated into membranes of each FELC.[18] The ^{3}H-labeled band 3 proteins are detectable up to about 72 hr after implantation.[18]

The ^{125}I-labeled EBV receptor-rich membranes can be detected on target cultured cells as long as 36 hr after implantation.[20] However, the possibility that these numbers represent vesicles and radiolabeled material that has been absorbed to the cell surface or taken into the cell by endocytic-like processes cannot be excluded.[18] Thus, in order to avoid unnecessary experimental work and misleading results, it is suggested that before any implantation experiments are performed with new cell lines or with new hybrid vesicles, the ability of the cell to interact first with intact Sendai virus (Table I) or with the hybrid vesicles (Table V) should be analyzed. Evidently, in cases where the intact virions fail to fuse with a certain cell line, there is little chance that such cells will interact with hybrid vesicles and will become susceptible to fusion-mediated transplantation experiments. A direct method for following the fusion activity of intact Sendai virus is described in Table I.

Similar to intact virions, fluorescent RSVE or hybrid fluorescent vesicles can also be incubated with cell cultures at 37°, after which the degree of fluorescence can be estimated.[35] The results in Table V show that incubation of fluorescent RSVE (RSVE bearing N-NBD-PE) with FELC results in fluorescence dequenching. Employing this system, all the experimental conditions required for implantation of membrane proteins can be analyzed with the use of FELC as recipient cells.

Maximal fluorescence dequenching (45% increase in the fluorescence degree, maximal RSVE–cell fusion) was obtained when 0.5 μg of RSVE were incubated with 10^6 FELC at 37° for 25–30 min (Table V). Significantly less (3–6%) fluorescence dequenching was observed upon incubation of FELC with unfusogenic RSVE, i.e., trypsinized-, PMSF-, or DTT-treated viral envelopes (Table V).

Based on our results,[24,25] it can be assumed that the degree of fluorescence dequenching can be considered as a direct measure of the extent of

---

[35] N. Chejanovsky, Y. I. Henis, and A. Loyter, *Exp. Cell Res.* **164,** 353 (1986).

TABLE V
FUSION OF RSVE OR HYBRID (RSVE–B3)
VESICLES WITH LIVING CULTURED CELLS AS
MONITORED BY FLUORESCENCE
DEQUENCHING METHODS[a]

Treatment	N-NBD-PE dequenching (% of total) after incubation with	
	Aedes albopticus (insect cell line)	FELC
Fluorescent RSVE		
None	7	45
Trypsin	ND	7
PMSF	ND	6
DTT	ND	3
Fluorescent RSVE–B3 hybrid vesicles		
None	6	48
Trypsin	ND	8
PMSF	ND	5
DTT	ND	6

[a] Fluorescent RSVE and fluorescent RSVE–B3 hybrid vesicles were prepared as described in Fig. 1 and in Table III, and treated with trypsin, PMSF, and DTT as described in Table I for intact Sendai virions. Friend erythroleukemia cells (FELC) were grown in Dulbecco's modified Eagle's medium, supplemented with 10% fetal calf serum, up to 5×10^5 cells/ml, then washed twice in PBS and resuspended in PBS to give 1.5×10^7 cells/ml. Monolayers of *A. albopticus* cells (kindly donated by Prof. R. Rott, Institute of Virology, Giessen, Federal Republic of Germany) were grown in Dulbecco's modified Eagle's medium with 10% fetal calf serum and 0.3% yeast extract. Untreated and treated fluorescent RSVE and RSVE–B3 hybrid vesicles (0.5 μg) were incubated with 1.5×10^6 FELC or *A. albopticus* cells for 40 min at 37° in a final volume of 1.6 ml of PBS. The fluorescence in the different systems was recorded before and after addition of 0.1% Ammonyx-LO, and the degree of fluorescence dequenching (N-NBD-PE dequenching) was calculated by taking the degree of fluorescence in the presence of Ammonyx-LO as 100% dequenching (see Table III). ND, Not determined.

RSVE–cell fusion; therefore, in the experiments described in Table V, about 45% of the RSVE in the population fused with FELC. The results in Table V also show that the fluorescence degree (extent of fluorescence dequenching) obtained by incubation of hybrid RSVE–B3 vesicles with FELC is very close to that obtained with RSVE, thus showing that the insertion of the polypeptide band 3 into the viral envelopes did not significantly modify their fusogenic activity. The results in Table V also show that no change in the fluorescence intensity was observed when either fluorescent RSVE or fluorescent hybrid vesicles were incubated with the insect cell line *Aedes allopticus*. Indeed, these cells are noted to be unsusceptible to infection by enveloped viruses belonging to the myxovirus or paramyxovirus groups[36] and are therefore resistant to fusion with RSVE.[24]

Fusion-Mediated Implantation of Exogenous Protein into Plasma Membranes of Target Cells: Estimation of a New Function

It is clear that in most cases the protein transferred into the recipient membranes, by use of hybrid vesicles as carriers, retains its ability to function in the new environment.[5] This has been demonstrated in several systems. Insertion of synergenic H-2^b membrane components from insulin responder mice into allogeneic spleen cells from H-2^k nonresponder mice endows the latter with the capacity to present beef insulin to H-2^b responder T lymphocytes.[37] Receptor for EBV and for SV40 has been inserted into receptor-negative cells with the use of hybrid vesicles and with the subsequent application of EBV or SV40 to the receptor-implanted cells.[15,20] CBP was implanted into membranes of variant basophilic cells which were defective in it. Fusion-mediated implantation led to restoration of Ca^{2+} uptake and degranulation capacity of the variant after IgE-mediated stimulation.[19]

The results in Table VI show that the insertion of band 3, the erythrocyte anion exchange system, into FELC led to a marked and specific stimulation of anion exchange in the target cells.[5,16] This can be demonstrated by estimation of ^{36}Cl efflux (at 1°) from ^{36}Cl-loaded FELC. As can be seen in Table VI, efflux of ^{36}Cl from FELC containing the transferred erythrocyte band 3 was significantly higher than that observed with control Friend cells. The results of Table VI also show that ^{36}Cl efflux from the ^{36}Cl-loaded FELC can be abolished by the addition of the specific inhibitor

---

[36] V. Stollar, B. D. Stollar, R. Koo, K. H. Harrap, and R. W. Schlesinger, *Virology* **69**, 104 (1976).

[37] A. Prujanski-Jakobovitz, A. Frankel, N. Sharon, and I. R. Cohen, *Nature (London)* **292**, 666 (1981).

TABLE VI

FUSION-MEDIATED IMPLANTATION OF HUMAN ERYTHROCYTE BAND 3 INTO PLASMA
MEMBRANES OF FELC: STIMULATION OF ^{36}Cl EFFLUX[a]

	^{36}Cl efflux from ^{36}Cl-loaded FELC[b] (% of total)			
Time of incubation (hr)	Control, unfused cells	Incubated with RSVE–B3		[^{3}H]H$_2$DIDS associated with FELC[c] (% of total)
		−DIDS	+DIDS	
4	4	43	4	96
72	3	24	4	52

[a] Hybrid vesicles, namely RSVE–B3 or RSVE–[^{3}H]H$_2$DIDS-labeled B3, were obtained as described in Fig. 1 and before.[5,16,18] Hybrid vesicles (100 μg) were incubated with $5 \times 10^7 - 1 \times 10^8$ FELC in 1 ml of Solution Na, containing 10 mM CaCl$_2$, first for 15 min at 4° and then for 20 min at 37°. At the end of the incubation period, the cells were washed with a warm Solution Na and then resuspended in 1 ml of Dulbecco's modified Eagle's medium, containing 10% fetal calf serum, and allowed to grow in a CO$_2$ incubator.[16,18]

^{36}Cl efflux determination. After 4 and 72 hr of incubation, samples of 1 ml were removed, centrifuged, washed, suspended in 0.3 ml of Solution Na, and loaded with Na^{36}Cl for 10 min at 37°, as described before.[16] At the end of the incubation, ^{36}Cl efflux was estimated in the presence or absence of 5 mM DIDS as described earlier.[16]

Determination of radioactivity ([^{3}H]H$_2$DIDS–B3) associated with FELC. At the end of the incubation period, the cells were washed twice with Solution Na, and the pellet obtained was dissolved in Pico-Fluor (Packard) and its radioactivity was determined by a scintillation counter.[18] One hundred percent radioactivity is considered as the amount of counts found to be associated with FELC at the end of the fusion process (30 min of incubation at 37°).

[b] For ^{36}Cl efflux, FELC were incubated with RSVE–B3 hybrid vesicles.

[c] For estimation of [^{3}H]H$_2$DIDS–band 3 associated with FELC, the cells were incubated with RSVE–[^{3}H]H$_2$DIDS–B3.[16]

4,4′-dinitro-2,2′-stilbenedisulfonic acid (DNDS). These results perhaps provide the most direct proof for the complete and transmembrane insertion of the band 3 polypeptide into the cell membrane matrix. The inhibitor of anion exchange in the erythrocyte, DIDS, which binds only to the cell surface part of the band 3 molecule,[31] inhibited approximately 80–90% of the ^{36}Cl exchange activity in the band 3-implanted FELC (Table VI; see also Refs. 13 and 16). These results indicate that most of the band 3 molecules are inserted in their original orientation. Similar to intact virus–cell fusion, hybrid–cell fusion is also accompanied by cell lysis, the degree of which depends on the cell type and the amount of vesicles applied to cells. Extensive cell lysis reduces the target cell's viability and thereby the overall implantation efficiency. Therefore, each new system requires preliminary testing to determine the maximal amount of hybrid vesicles that can be applied to a given number of cells without significantly affecting the target cell's viability.

The results in Table VI also show that a correlation exists between the amount of band 3 found implanted in FELC and the extent of ^{36}Cl efflux from these cells. A high degree of ^{36}Cl efflux is observed 4 hr after the implantation, when a relatively high amount of band 3 is present in FELC cells. On the other hand, a reduction of ~50% in the ^{36}Cl efflux was observed after 72 hr of cell growth, during which about 50% of the band 3 protein was degraded, as determined by the use of $[^3H]H_2DIDS$-labeled band 3 (Table VI).

Conclusion

Figure 2 summarizes the main steps suggested for the successful use of hybrid vesicles containing the Sendai viral envelope glycoproteins as a carrier for the implantation of exogenous proteins into cell plasma membranes.

1. Before initiating any reconstitution experiments and formation of hybrid vesicles, the fusogenic activity of the purified, intact Sendai virus should be tested. A direct method for studying the viral fusogenic activity, independently of its virus–cell-binding ability, is to follow the fusion of fluorescent, intact virions with PC/cholesterol liposomes. If positive, i.e., the virus shows a high degree of fusion (fluorescence dequenching), the same preparation of the fluorescent-intact virions should be used to study its fusion ability with the target cells. Sendai virions fuse with a wide range of cell types. The only requirement for fusion with Sendai virions as well as with hybrid vesicles is the presence of sialic acid residues on cell surface glycoproteins and glycolipids. In cases when fusion between the virions and certain cultured cell lines is very low due to poor virus–membrane interaction, it is suggested to try to increase the extent of virus binding by the addition of specific anticell antibodies or polypeptide hormones to the virus envelopes. It has recently been shown that RSVE, bearing specific anticell antibodies or insulin molecules, were able to interact and fuse with neuraminidase-treated cells, namely, virus receptor-depleted membranes.[38,39]

[38] A. G. Gitman and A. Loyter, *J. Biol. Chem.* **259**, 9813 (1984).
[39] A. G. Gitman, A. Graessmann, and A. Loyter, *Proc. Natl. Acad. Sci. U.S.A.* **82**, 309 (1985).

(a) Intact Sendai virions → (b) Fluorescent hybrid vesicles → (c) Hybrid vesicles

A. Preparation of fluorescent (bearing R_{18}) intact virions	A. Preparation of fluorescent hybrid fusogenic vesicles by coreconstitution system	A. Preparation of hybrid vesicles

B. Quantitative estimation of the viral fusogenic activity using fluorescence dequenching methods and
  1. PC/cholesterol liposomes
  2. Cultured cells of interest
	B. Quantitative estimation of hybrid vesicles' fusogenic activity with
	  1. PC/cholesterol liposomes
	  2. Cultured cells of interest
		B. Fusion of hybrid vesicles with cultured cells of interest for transplantation of nonviral proteins into cell plasma membranes

FIG. 2. Schematic representation of the sequence of experiments suggested for the construction of workable fusogenic hybrid vesicles.

2. After reconstitution or coreconstitution of viral envelopes and hybrid vesicles, respectively, it is again desirable to test the RSVE (or hybrid vesicle) fusogenic activity by first assessing interactions with liposomes composed of PC and cholesterol and then with target cells. This, again, can be estimated by employing fluorescent RSVE and monitoring the increase in the fluorescence degree following interaction with the liposomes and the cultured cells.

It is anticipated that in the near future many aspects of this method will be modified, such as developing new methods which will allow fusion of viral envelopes with cells that are not susceptible to fusion with Sendai virions. One way to achieve this goal, as mentioned above, is by reconstitution of "targeted" viral envelopes. These are reconstituted viral envelopes bearing specific antibodies or other ligands, which under certain conditions may substitute the viral binding protein, i.e., the HN glycoprotein.[38,39] We expect that the fusion-mediated transplantation method will be used to elucidate many unresolved questions:

1. Questions related to target cell recognition by immunological effectors: The qualitative and quantitative requirements for target cell recognition by cellular immunological components such as CTL or natural killer (NK) cells, antibodies, and complement are still poorly understood. Using the coreconstitution approach, it is possible to transfer various components from target cells into other unrelated cell types and to study their structure as well as their function.

2. Questions related to elucidation of the structure and function of cell surface receptors for many of the animal enveloped viruses.

3. Questions related to establishing biological assays for many as-yet unknown cell surface receptors of the polypeptide hormones.

Acknowledgment

This work was supported by grants (to A.L.) from the National Council for Research and Development, Israel; from the GSF, Munich, Federal Republic of Germany; and from International Genetic Sciences Partnership, Israel.

[42] Resonance Energy Transfer Microscopy: Visual Colocalization of Fluorescent Lipid Probes in Liposomes

By Paul S. Uster and Richard E. Pagano

Introduction

Resonance energy transfer (RET) is characterized by the transfer of photon energy from a donor fluorochrome to an acceptor molecule that is in close physical proximity. The donor fluorescence is quenched, and, if the acceptor is an appropriate fluorescent molecule, it will then fluoresce as if excited directly.[1] Because RET decreases in proportion to the inverse sixth power of the distance between the two probes, this phenomenon is effective only when the donor and acceptor molecules are within 100 Å of each other.[2] RET therefore provides a convenient method for probing inter- and intramolecular distances in proteins[3] and in membranes.[4-9] RET experiments are typically conducted in solution using a spectrofluorimeter to monitor fluorescence intensity changes at a given wavelength,[10-12]

[1] R. A. Badley, *in* "Modern Fluorescence Spectroscopy" (E. L. Wehry, ed.), Vol. 2, pp. 91–168. Plenum, New York, 1976.

[2] L. Stryer and R. P. Haugland, *Proc. Natl. Acad. Sci. U.S.A.* **58,** 719 (1967).

[3] L. Stryer, *Annu. Rev. Biochem.* **47,** 819 (1978).

[4] H. Ellens, J. Bentz, and F. C. Szoka, *Biochemistry* **24,** 3099 (1985).

[5] G. A. Gibson and L. M. Loew, *Biochem. Biophys. Res. Commun.* **88,** 141 (1979).

[6] D. K. Struck, D. Hoekstra, and R. E. Pagano, *Biochemistry* **20,** 4093 (1981).

[7] P. S. Uster and D. W. Deamer, *Arch. Biochem. Biophys.* **209,** 385 (1981).

[8] P. Vanderwerf and E. F. Ullman, *Biochim. Biophys. Acta* **596,** 302 (1980).

[9] P. M. Keller, S. Person, and W. Snipes, *J. Cell Sci.* **28,** 167 (1977).

[10] J. S. Becker, J. M. Oliver, and R. D. Berlin, *Nature (London)* **254,** 152 (1975).

[11] J. D. Pardee, P. A. Simpson, L. Stryer, and J. A. Spudich, *J. Cell Biol.* **94,** 316 (1982).

[12] D. L. Taylor, J. Reidler, J. A. Spudich, and L. Stryer, *J. Cell Biol.* **89,** 362 (1981).

but until now this technique has not been used as a visual microscopic tool to observe directly the sites of donor quenching and sensitized acceptor fluorescence within a specimen and to record this distribution on photographic film. In this chapter we outline the necessary modifications to a standard epifluorescence microscope for RET microscopy of one particular donor/acceptor pair, and demonstrate the feasibility of this technique using fluorescent lipid probes in liposomes.

Materials

Lipids

Dioleoylphosphatidylcholine (DOPC), N-4-nitrobenz-2-oxa-1,3-diazolylphosphatidylethanolamine (N-NBD-PE), and Lissamine Rhodamine B-labeled phosphatidylethanolamine (N-Rh-PE) are available from commercial sources (Avanti Polar Lipids, Birmingham, Alabama; Molecular Probes, Eugene, Oregon). N-Sulforhodamine dioleoylphosphatidylethanolamine (N-SRh-PE) is synthesized and purified using dioleoylphosphatidylethanolamine and sulforhodamine sulfonyl chloride (Texas Red).[13]

Preparation of Unilamellar Liposomes

For spectrophotofluorimetric measurements, small unilamellar liposomes containing the appropriate fluorescent lipids are made by the ethanol injection procedure.[14] The dried lipids are dissolved in ethanol (5 μmol/ml) and are rapidly injected into 10 mM 4-(2-hydroxyethyl-1-piperazineethanesulfonic acid) (HEPES)-buffered calcium- and magnesium-free Puck's saline at pH 7.4. The liposomes are dialyzed overnight at 4° against 18 mM HEPES-buffered Eagle's minimal essential medium without indicator, pH 7.4, and diluted to the required concentration with buffer.

Fluorescence spectra and intensities are recorded using a standard spectrophotofluorimeter such as an Aminco-Bowman (SLM-Aminco Inc., Urbana, Illinois).

Preparation of Macroscopic Liposomes for RET Microscopy

A mixture of two populations of macroscopic liposomes mounted on a single microscope slide is prepared for RET microscopy as follows. Chloroform stock solutions containing 2 μmol/ml of either DOPC/N-NBD-PE

[13] P. S. Uster and R. E. Pagano, J. Cell Biol. 103, 1221 (1986).
[14] J. M. H. Kremer, M. W. J. v. d. Esker, C. Pathmamanoharan, and P. H. Wiersema, Biochemistry 17, 3932 (1977).

(99/1, mol/mol) or DOPC/N-NBD-PE/N-SRh-PE (98/1/1, mol/mol/mol) are first made. One solution is added dropwise (5 × 1 μl) to a microscope slide warmed to 40° on a thermal block, while the other is added similarly to a #1 glass coverslip, also at 40°. After drying, the coverslip is inverted onto the slide and 10 μl of 10 mM HEPES-buffered, calcium- and magnesium-free Puck's saline (pH 7.4) containing 0.4% (w/v) low-melting agarose is added to the side of the coverslip. The resulting liposomes are allowed to hydrate for 30 min at 40°, and the slide is then cooled to room temperature to solidify the agarose matrix. The macroscopic, oligolamellar liposomes are transparent to slightly gray by phase-contrast microscopy.

Modifying the Fluorescence Microscope for RET Microscopy

Any conventional fluorescence microscope can be modified with appropriate excitation and emission filters to define the three optical channels required for RET microscopy.[15] Table I describes the optimal filter combinations for observing RET on Zeiss equipment when the donor fluorophore is NBD (or fluorescein) and the acceptor fluorophore is SRh or Rh. In the donor or NBD channel, the specimen is excited with blue light (428–444 nm) and the resulting green NBD fluorescence is observed through a narrow window (515–565 nm) which passes only green light. In the acceptor or SRh channel, the specimen is excited with green light (540–552 nm) and the consequent red fluorescence is observed through filters which only pass red (>610 nm) light. In the transfer channel, the specimen is illuminated with the blue light appropriate for excitation of NBD, but observation of the emitted light is restricted to red wavelengths that are characteristic of SRh fluorescence. Thus, any fluorescence seen in the transfer channel is due to energy transfer between donor and acceptor probe molecules.

Resonance Energy Transfer of Lipid Probes in Liposomes

Spectroscopic Considerations

In order to use RET microscopy as a sensitive, unequivocal assay of probe colocalization, the donor fluorescence emission should overlap the absorption spectrum of the energy acceptor significantly. In addition, the

---

[15] In our studies, a Zeiss IM-35 epifluorescence microscope, equipped with Planapo 40/0.9 NA and 63/1.4 NA objectives and an HBO 100 mercury lamp, is used. Electronic shutters control specimen illumination, and the intensity of the exciting light is attenuated as required, with neutral density filters. Black and white photomicrographs are recorded on Kodak Tri-X panatomic film and push-processed to ASA 1600 using Diafine developer.

TABLE I

OPTICAL FILTERS FOR RET MICROSCOPY USING NBD AS DONOR AND
SULFORHODAMINE AS ACCEPTOR FLUOROCHROMES[a]

| | Zeiss filter pack components | | | Spectral window (nm) | |
| | | | | | |
Channel	Exciter filter	Dichroic beam splitter	Barrier filter	Excitation wavelengths	Emission wavelengths
Donor	BP436/17	FT510	BP515–565	428–444	515–565
Acceptor	BP546/12	FT580	LP620	540–552	≥610
Transfer	BP436/17	FT510	LP620	428–444	≥610

[a] Reproduced from *The Journal of Cell Biology*,[13] 1986, Vol. 103, pp. 1221–1234 by copyright permission of The Rockefeller University Press.

fluorescent donor probe should be excited at wavelengths where the molar absorption of the acceptor probe is relatively small. This minimizes direct excitation of the acceptor and increases the fraction of acceptor probe available for energy transfer. These criteria are met by the donor/acceptor pairs (NBD/SRh or NBD/Rh) used in the present study.

It has been established that the efficiency of energy transfer is proportional to the inverse sixth power of the distance of separation between donor fluorochrome and acceptor.[2] For the NBD (or fluorescein) donor, RET to an energy acceptor is not appreciable beyond a 100 Å separation.[16] Thus, RET between fluorescent lipids will not be detected unless both the donor and acceptor probes are in the same bilayer.[6–8] This is demonstrated in Fig. 1. The emission spectrum of mixed, equimolar amounts of DOPC/ N-NBD-PE (99/1, mol/mol) and DOPC/N-SRh-PE (99/1) liposomes illuminated at 470 nm exhibits the characteristic strong NBD emission at 515–565 nm, and essentially no SRh fluorescence at 600–630 nm (Fig. 1A). However, when a third population of DOPC/N-NBD-PE/N-SRh-PE (98/1/1) liposomes is illuminated, the intense green emission of NBD is quenched strongly and the fluorescence of SRh is increased substantially (Fig. 1B). Thus, we see that the transferred photon energy is reemitted at the acceptor's own characteristic emission peak as if it had been excited directly. This sensitized emission (which is indicated by the peak appearing at 600–630 nm) is concomitant with donor quenching, and is a characteristic of RET, distinguishing it from other mechanisms of fluorescence quenching.[1]

In a bilayer, RET between fluorescent probes is increased by increasing the surface density (mol%) of these molecules. Our laboratory[13] and other

[16] D. D. Thomas, W. F. Carlsen, and L. Stryer, *Proc. Natl. Acad. Sci. U.S.A.* **75**, 5746 (1978).

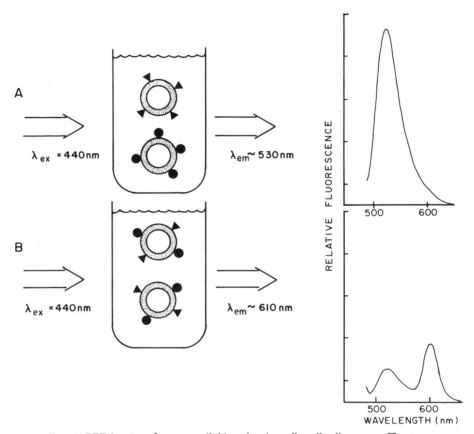

FIG. 1. RET between fluorescent lipid probes in unilamellar liposomes. Fluorescence emission spectra were determined for (A) singly labeled [DOPC/N-NBD-PE (99/1, mol/mol)] and [DOPC/N-SRh-PE (99/1, mol/mol)] liposomes mixed in equimolar amounts or (B) doubly labeled [DOPC/N-NBD-PE/N-SRh-PE (98/1/1, mol/mol/mol)] liposomes ($\lambda_{ex} =$ 470 nm). Total nanomoles of N-NBD-PE and N-SRh-PE were the same in both spectra. (▲) N-NBD-PE; (●) N-SRh-PE.

investigators[17] have confirmed that the transfer efficiency is dependent on acceptor probe surface density and is independent of the donor probe surface density. The efficiency of energy transfer (E) is calculated from donor quenching using the equation

$$E = 1 - F/F_0 \tag{1}$$

where the steady-state relative fluorescence intensity is measured in the presence (F) and absence (F_0) of the acceptor.[17] The transfer efficiency is

[17] B. K.-K. Fung and L. Stryer, *Biochemistry* **17**, 5241 (1978).

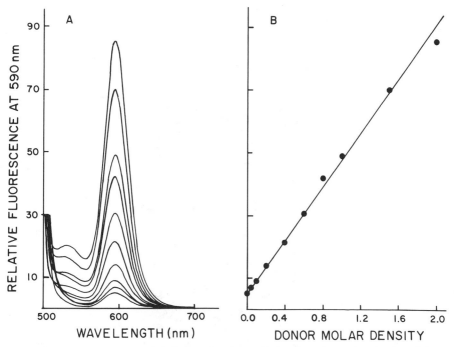

FIG. 2. Sensitized acceptor fluorescence is proportional to the molar density of the fluorescent donor membrane probe. DOPC liposomes (50 μM total lipid) were prepared containing 1 mol% of the fluorescent acceptor N-Rh-PE and 0.0, 0.05, 0.1, 0.2, 0.4, 0.6, 0.8, 1.0, 1.5, and 2.0 mol% of N-NBD-PE. Each sample was excited at 435 nm, and the fluorescence emission spectra were recorded (A). From these spectra, the plot of fluorescence intensity at 590 nm versus donor molar density (B) was obtained.

constant over a wide range of donor surface densities, even when the surface density of the energy donor is in excess of that of the acceptor. In contrast, the sensitized acceptor fluorescence is directly proportional to the molar density of the donor, as shown in Fig. 2, for the donor/acceptor pair, NBD/Lissamine Rhodamine B.[18] From these findings we see that to maximize RET a labeled bilayer should have an appreciable amount of acceptor (0.5–2.0 mol%) and donor (>0.1 mol%) probes.

Observations of Energy Transfer through the Microscope

RET microscopy can be readily demonstrated using the modified microscope described in Table I and a mixture of singly (donor alone; NBD–) and doubly (donor + acceptor; NBD + SRh) labeled liposomes.

---

[18] Similar findings were reported in Ref. 13, using NBD/SRh as the donor–acceptor pairs.

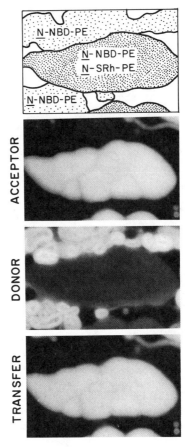

FIG. 3. Energy transfer microscopy of fluorescent lipids in liposomes. (Top) Schematic representation of fluorescently labeled model membranes containing DOPC/N-NBD-PE (99/1, mol/mol), or DOPC/N-NBD-PE/N-SRh-PE (98/1/1, mol/mol/mol). (Acceptor) Direct excitation of SRh in the acceptor channel ($\lambda_{ex} = 546$ nm, $\lambda_{em} \geq 610$ nm) verified the location of N-SRh-PE. (Donor) Direct excitation of N-NBD-PE in the donor channel ($\lambda_{ex} = 436$ nm, $\lambda_{em} = 515$–565 nm). Note quenching of NBD fluorescence in the doubly labeled membranes. (Transfer) Sensitized N-SRh-PE fluorescence was visualized in the transfer channel ($\lambda_{ex} = 436$ nm, $\lambda_{em} \geq 610$ nm).

Only the doubly labeled liposomes should exhibit strong donor quenching and enhanced emission of the fluorescent acceptor. This is indeed the case, as seen in Fig. 3. The top panel of this figure schematically represents the field of liposomes photographed in the different microscope channels. The location of the acceptor is verified by direct excitation ($\lambda_{ex} = 546$ nm) of N-SRh-PE in the acceptor channel (acceptor panel). When NBD is observed in the donor channel, its fluorescence is seen to be highly quenched only in the doubly labeled liposomes (see donor panel). Finally, when the specimen is observed in the transfer channel, the doubly labeled liposomes show strong red fluorescence (see transfer channel). A slight background from the singly (NBD) labeled liposomes in the transfer channel is observed under some illumination conditions because NBD emits a small but detectable amount of photon energy in the red region of the visible spectrum. In control experiments, no fluorescence is seen when liposomes labeled only with *N*-SRh-PE are observed in the transfer channel.

We conclude that both donor quenching and sensitized acceptor fluorescence in model membranes can be seen directly in the fluorescence microscope with the appropriate choice of donor and acceptor fluorochromes, and microscope filters. In principle, RET microscopy can be used to enhance our ability to study the spatial distribution of fluorescent probes, since the resolution of conventional light microscopy is a few tenths of a micrometer while the working scale for RET is ≤ 100 Å. The filter packs described in Table I are suitable for a variety of fluorescent donor (dansyl, fluorescein, and NBD) and acceptor (rhodamine, sulforhodamine, and some of the carbocyanine dyes) probes. Other donor–acceptor probe combinations can be developed with appropriate changes in the microscope optics. Finally, although we have restricted this chapter to a discussion of RET between membrane-bound fluorescent lipids in model membranes, elsewhere we have shown that this method can be extended to visualize RET between membrane-bound fluorescent lipids (or lipids and proteins) in living cells.[13]

Acknowledgment

Supported by U.S.P.H.S. Grant GM-22942.

[43] Permeabilizing Mammalian Cells to Macromolecules

By Gary A. Weisman, Kevin D. Lustig, Ilan Friedberg, and Leon A. Heppel

This chapter is primarily concerned with procedures that increase the plasma membrane permeability of mammalian cells to normally impermeant compounds. Permeabilization methods described include treatment of cells with adenosine 5'-triphosphate (ATP),[1] ionophores, Sendai virus, detergents, and *Staphylococcus aureus* α-toxin. Physical methods for altering the normal permeability barrier of cultured cells, including scrape-loading, treatment with hypertonic medium, and transient exposure to an electric field, are also briefly discussed. Several procedures for monitoring the plasma membrane permeability of cells in monolayer culture are described.

Measurement of Alterations in Plasma Membrane Permeability to Normally Impermeant Compounds

Alterations in the plasma membrane permeability of cultured cells can be detected by assaying for the release of a prelabeled pool of normally impermeant metabolites or by measuring the uptake of normally impermeant probes into the cells.

Method 1. Intracellular pools of uridine or adenine nucleotides are labeled by incubating cells with [³H]uridine or [³H]adenosine, respectively. Briefly, cells are labeled at 37° for 2–4 hr in fresh growth medium containing 1.0 μM ³H-labeled nucleoside (0.4 Ci/mmol). Labeled cells are washed three times with an isotonic, buffered saline solution, to remove extracellular label, and are incubated under conditions that promote permeabilization (see below). Physical parameters of the assay medium, including temperature, pH, osmolarity, and ionic strength, can be varied to optimize the extent of permeabilization. The efflux of the labeled intracellular pool is monitored by withdrawing samples from the assay medium at various intervals after the addition of the permeabilizing agent. Untreated cultures serve as controls. The medium remaining at the end of the experiment is aspirated and the residual intracellular pool is extracted by incubating the cells for 20 min with ice-cold 5% (w/v) trichloroacetic acid (TCA). The radioactivity in all samples and the TCA-soluble pool is determined using liquid scintillation spectrometry.

The radioactivity released from the cells is calculated using Eq. (1).

$$R_n^T = R_n\{[(T/S) + 1] - n\} + \sum_{i=0}^{i=n-1} R_n \tag{1}$$

where n is the sample number (e.g., 1–8), R_n^T is the total radioactivity

---

[1] Abbreviations: ATP, adenosine 5'-triphosphate; TCA, trichloroacetic acid; FITC, fluorescein isothiocyanate; HeLa, cell line derived from human cervical carcinoma; CHO, cell line derived from Chinese hamster ovaries; HEPES, 4-(2-hydroxyethyl)-1-piperazineethanesulfonic acid; Tris, 2-amino-2-(hydroxymethyl)-1,3-propandiol; araCTP, 1-β-D-arabinofuranosylcytosine 5'-triphosphate; CCCP, carbonyl cyanide-*m*-chlorophenylhydrazone; EAT, cell line derived from Ehrlich ascites tumors; BHK, cell line derived from baby hamster kidney; MEM, modified Eagle's medium.

released at the time the nth sample is withdrawn, R_n is the measured radioactivity of the nth sample, T is the initial volume of assay medium, and S is the sample volume.

The percentage of the labeled intracellular pool released from the cells at the time the nth sample is withdrawn is determined by Eq. (2).

$$(\% \text{ of pool released})_n = \frac{R_n^T \times 100}{R_f^T + R_{TCA}} \tag{2}$$

where R_{TCA} is the radioactivity in the residual TCA-soluble pool, R_f^T is the total radioactivity released at the time the last sample is withdrawn, and R_n^T is defined in Eq. (1). A plot of percentage of the pool released versus time gives an indication of the rate of efflux of the normally impermeant, labeled compounds from the intracellular pool.

It is necessary to verify that labeled compounds released by the cells are actually impermeant nucleotides. Alternatively, the label released into the assay medium may be due to carrier-mediated efflux of nucleosides formed by dephosphorylation of intracellular nucleotides. The radioactivity appearing in the medium from an intracellular ^{3}H-labeled pool can be separated into nucleotides and nucleosides by chromatography on polyethyleneimine-coated cellulose columns.[2]

Method 2. Alterations in plasma membrane permeability can also be monitored by following entry of normally impermeant probes into cells. Fluorescent, radioactive, and colorimetric probes of various sizes can be used to examine the kinetics of channel formation and to approximate the size of membrane channels. Fluorescein isothiocyanate (FITC)-labeled dextrans are particularly suitable for measuring permeability changes in mammalian cells since they are commercially available in a variety of molecular weights (4–70 kDa) and are stable under physiological conditions. Radioactively labeled inulin (5.2 kDa), trypan blue (961 Da), lucifer yellow (457 Da), ethidium bromide (394 Da), 6-carboxyfluorescein (376 Da), *p*-nitrophenyl phosphate (263 Da), and other normally impermeant probes (see below) are also commercially available.

Cells permeabilized in the presence of normally impermeant probes should be resealed and washed free of extracellular probe prior to extraction of the intracellular soluble pool. Cells permeabilized by a variety of procedures can be resealed by incubation in an isotonic medium, pH 7.4, containing 1–2 mM calcium or magnesium. Intracellular levels of the probe can be determined by spectrofluorometry, liquid scintillation spectrometry, spectroscopy, or microscopy, depending upon the nature of the probe. For example, FITC–dextran levels can be measured spectrofluorometrically after extraction of cells with 0.1% (v/v) Triton X-100. FITC–

[2] R. P. Magnusson, A. R. Portis, Jr., and R. E. McCarty, *Anal. Biochem.* **72,** 653 (1976).

dextran fluorescence can be measured at an excitation wavelength of 490 nm (bandpass = 5.4 nm) and an emission wavelength of 520 nm (bandpass = 5.4 nm), and levels of FITC–dextran are then quantitated by comparison to the fluorescence of FITC–dextran standard solutions. It can be verified that FITC–dextrans are incorporated intact by chromatographing cell extracts on Sephadex G-75 columns against FITC–dextran and FITC standards. Uptake of [^{3}H]inulin or other radiolabeled probes can be quantitated by liquid scintillation spectrometry of extracts obtained by incubating cells for 20 min with ice-cold 5% (w/v) TCA. Uptake of probes should be expressed relative to cell number or total protein, since many permeabilization procedures induce some loss of cells from the support matrix.

When using certain probes, it is not necessary to seal the cells to verify that the probe entered the cells intact. For example, uptake of the normally impermeant phosphate ester, p-nitrophenyl phosphate, into permeabilized cells leads to generation of p-nitrophenol in the assay medium due to the action of intracellular phosphatases.[3] Levels of p-nitrophenol in the medium can be detected spectrophotometrically at 410 nm.

Permeabilizing Agents

ATP

Mouse fibroblasts transformed spontaneously or virally,[4] human cervical carcinoma (HeLa) cells,[5] mouse melanoma cells,[6] and transformed Chinese hamster ovary (CHO) cells[7] became freely permeable to prelabeled pools of nucleotides and sugar phosphates when incubated in an alkaline (pH 7.4–8.2), isotonic buffered (HEPES, Tris) medium containing low levels of divalent cations (25–75 μM calcium and magnesium) and 150–250 μM ATP. In transformed fibroblasts, the membrane channels induced by ATP were resealed when the ATP-containing medium was removed and replaced with an isotonic, buffered medium, pH 6.8, containing 20 mM glucose, 1 mM calcium, and 1 mM magnesium.[8] Sealing was also promoted by the addition of cofactors and substrates (for glycolysis and oxidative phosphorylation) that were lost during the permeabilization pro-

---

[3] E. Rozengurt and L. A. Heppel, *Biochem. Biophys. Res. Commun.* **67**, 1581 (1975).
[4] E. Rozengurt and L. A. Heppel, *J. Biol. Chem.* **254**, 708 (1979).
[5] N. R. Makan, *Exp. Cell Res.* **114**, 417 (1978).
[6] T. Kitagawa and Y. Akamatsu, *Jpn. J. Med. Sci. Biol.* **35**, 213 (1982).
[7] T. Kitagawa and Y. Akamatsu, *Biochim. Biophys. Acta* **649**, 76 (1981).
[8] P. Dicker, L. A. Heppel, and E. Rozengurt, *Proc. Natl. Acad. Sci. U.S.A.* **77**, 2103 (1980).

cedure.[9] The finding that several cycles of ATP-induced permeabilization and resealing did not affect subsequent cell viability[10] has led to experiments in which small charged molecules of biochemical interest were sealed into ATP-permeabilized cells. Impermeant inhibitors of DNA, RNA, and protein synthesis such as 1-β-D-arabinofuranosylcytosine 5'-triphosphate (araCTP) (440 Da), guanylyl-β,γ-methylene diphosphonate (521 Da), and sparsomycin (266 Da), respectively, were sealed inside transformed mouse fibroblasts, resulting in complete inhibition of cell growth.[11] Treatment of transformed mouse fibroblasts with ATP for periods longer than 30 min induced an increase in the plasma membrane permeability to FITC–dextrans (4 and 9 kDa) and [^{3}H]inulin (5.2 kDa).[12] Cells remained viable and grew normally even when sealed 60 min after ATP addition.

ATP has also been used to increase the plasma membrane permeability of a variety of untransformed cells to normally impermeant compounds. Rat mast cells incubated with greater than 3 μM ATP^{4-} released prelabeled intracellular pools of nucleotides and sugar phosphates[13] and incorporated 6-carboxyfluorescein (376 Da), [^{57}Co]HEDTA (401 Da), but not [^{3}H]inulin (5.2 kDa).[14] Addition of magnesium promoted sealing of ATP-induced channels in mast cells.[15] Mouse macrophages were permeabilized to normally impermeant, fluorescent compounds such as 6-carboxyfluorescein (376 Da), lucifer yellow (457 Da), and fura-2 (831 Da), but not trypan blue (961 Da), after treatment with millimolar levels of ATP.[16] ATP-permeabilized macrophages were sealed by removing the ATP-containing medium or by the addition of magnesium to the assay medium. Untransformed 3T3 mouse fibroblasts and CHO cells, resistant to ATP in buffers of physiological ionic strength ($\mu = 0.14$), released a prelabeled pool of intracellular nucleotides when treated with ATP ($> 500 \mu M$) in assay medium of reduced ionic strength ($\mu = 0.07$).[17] In addition, human blood platelets incorporated aequorin (20 kDa), a calcium-sensitive photoprotein, when incubated with 10 mM EGTA and 5 mM ATP.[18]

[9] N. R. Makan, *J. Cell. Physiol.* **101**, 481 (1979).
[10] E. Rozengurt, L. A. Heppel, and I. Friedberg, *J. Biol. Chem.* **252**, 4584 (1977).
[11] T. Kitagawa, *Biochem. Biophys. Res. Commun.* **94**, 167 (1980).
[12] A. S. Saribas, G. A. Weisman, and K. D. Lustig, unpublished results (1988).
[13] S. Cockcroft and B. D. Gomperts, *Nature (London)* **279**, 541 (1979).
[14] J. P. Bennett, S. Cockcroft, and B. D. Gomperts, *J. Physiol. (London)* **317**, 335 (1981).
[15] B. D. Gomperts, *Nature (London)* **306**, 64 (1983).
[16] T. H. Steinberg, A. S. Newman, J. A. Swanson, and S. C. Silverstein, *J. Biol. Chem.* **262**, 8884 (1987).
[17] G. A. Weisman, K. D. Lustig, and L. A. Heppel, unpublished results (1987).
[18] P. C. Johnson, J. A. Ware, P. B. Cliveden, M. Smith, A. M. Dvorak, and E. W. Salzman, *J. Biol. Chem.* **260**, 2069 (1985).

In transformed mouse fibroblasts[19] and untransformed mouse macrophages,[20] extracellular ATP induced an increase in Na^+ and K^+ permeability and a significant reduction in the plasma membrane potential prior to the formation of membrane channels to normally impermeant compounds. A reduction in intracellular ATP levels induced by treatment with a variety of uncouplers, energy transfer inhibitors, or inhibitors of oxidative phosphorylation led to a reduction in the minimum concentration of extracellular ATP required to permeabilize transformed mouse fibroblasts.[4] Reduction of the intracellular ATP concentration most likely acted synergistically with extracellular ATP by decreasing the availability of substrate to active transporters that serve to maintain an electrochemical gradient across the plasma membrane. Treatment of transformed mouse fibroblasts with 1 mM ouabain, a specific inhibitor of the plasma membrane Na^+,K^+-ATPase, also reduced the concentration of extracellular ATP required to induce permeabilization.[19] Ouabain alone served to decrease cellular sodium and potassium gradients, underscoring the critical role of these ions in the maintenance of a plasma membrane barrier to impermeant molecules. Treatment of transformed mouse fibroblasts with antagonists of calmodulin, a calcium-binding protein involved in the regulation of the plasma membrane Ca^{2+}-ATPase, among other enzymes,[21] also decreased the effective permeabilizing concentration of extracellular ATP.[22]

ATP-induced permeabilization in transformed mouse fibroblasts[3,10] and macrophages[16] was inhibited below pH 7.4 and 25°. The permeabilizing effects of ATP on transformed mouse fibroblasts were antagonized by addition of 20 mM glucose, a treatment which served to maintain intracellular ATP levels in the presence of uncouplers, energy transfer inhibitors, or inhibitors of respiration.[4] Permeabilization of fibroblasts by ATP was also inhibited by addition of calcium and magnesium or analogs of calcium, including lanthanum and terbium.[22,23] Calcium and magnesium apparently prevented permeabilization by chelating free ATP (ATP^{4-}), the active substrate for permeabilization in fibroblasts,[24] mast cells,[25] and mac-

[19] G. A. Weisman, B. K. De, I. Friedberg, R. S. Pritchard, and L. Heppel, *J. Cell. Physiol.* **119**, 211 (1984).

[20] S.-S. J. Sung, J. D.-E. Young, A. M. Origlio, J. M. Heiple, H. R. Kaback, and S. C. Silverstein, *J. Biol. Chem.* **260**, 13442 (1985).

[21] W. Y. Cheung, *Science* **207**, 19 (1980).

[22] B. K. De and G. A. Weisman, *Biochim. Biophys. Acta* **775**, 381 (1984).

[23] N. R. Makan, *J. Cell. Physiol.* **106**, 49 (1981).

[24] L. A. Heppel, G. A. Weisman, and I. Friedberg, *J. Membr. Biol.* **83**, 251 (1985).

[25] S. Cockcroft and B. D. Gomperts, *Biochem. J.* **188**, 789 (1980).

rophages.[16] However, the inhibitory effect of lanthanides in fibroblasts cannot be due entirely to a reduction in the ATP^{4-} concentration, since inhibition was observed at lanthanide concentrations well below the ATP concentration.[22] ATP-induced permeabilization in transformed mouse fibroblasts was also inhibited by incubation of cells in chloride-free, thiocyanate-containing medium,[26] a treatment which also inhibited a plasma membrane Na^+,K^+,Cl^--cotransporter that regulates cell volume.[27] Diuretics, such as furosemide and bumetanide, also inhibit the cotransporter and prevented permeabilization by ATP at diuretic concentrations greater than 1 mM and after preincubation periods of 30–60 min. Dithiothreitol (> 3 mM) inhibited ATP-induced permeabilization after 60-min preincubation periods.[24]

Ionophores

Ionophores are compounds which enhance the transfer of ions across membranes either by forming ion channels in the membrane or by acting as ion carriers.[28] In addition to ion transfer, several ionophores induced an increase in plasma membrane permeability to normally impermeant compounds.[29] This permeability change was elicited by electrogenic (but not nonelectrogenic) ionophores, and was dependent upon a reduction in the plasma membrane potential.[30] A sharp increase in membrane permeability to normally impermeant compounds induced by electrogenic ionophores occurred when the plasma membrane potential dropped below a threshold value. Membrane depolarization appears to underlie the effects of a variety of permeabilizing agents.[31]

Electrogenic, channel-forming ionophores that induce permeabilization can be classified according to their mode of interaction with the plasma membrane. The "quasi" ionophore gramicidin, which is directly incorporated into membranes,[28] caused efflux of a prelabeled pool of nucleotides from transformed mouse fibroblasts at concentrations greater than 5 μM.[30] Another class of channel-forming ionophores, the polyene antibiotics, interacts with membrane components such as sterols.[29] Exam-

[26] G. A. Weisman, B. K. De, and R. S. Pritchard, *J. Cell. Physiol.* **138,** 375 (1989).

[27] J. A. McRoberts, S. Erlinger, M. J. Rindler, and M. H. Saier, Jr., *J. Biol. Chem.* **257,** 2260 (1982).

[28] B. C. Pressman, *Annu. Rev. Biochem.* **45,** 501 (1976).

[29] A. W. Norman, A. M. Spielvogel, and R. G. Wong, *Adv. Lipid Res.* **14,** 127 (1976).

[30] I. Friedberg, G. A. Weisman, and B. K. De, *J. Membr. Biol.* **83,** 251 (1985).

[31] C. L. Bashford, G. M. Alder, G. Menestrina, K. J. Micklem, J. J. Murphy, and C. A. Pasternak, *J. Biol. Chem.* **261,** 9300 (1986).

ples of polyene antibiotics include amphotericin B, nystatin, and filipin, all of which increased the membrane permeability of yeast cells and erythrocytes to cytoplasmic components.[32,33] Studies in which amphotericin B caused both cell death and permeabilization to low-molecular-weight compounds such as nucleotides indicated that these events are separable, since permeabilization could be maintained without causing cytotoxicity.[34] Filipin ($50 \mu M$) has been used to permeabilize rat hepatocytes to inulin (5.2 kDa), trypan blue (961 Da), sucrose (342 Da), and glycerol 3-phosphate (170 Da).[35]

Carrier-type ionophores that caused an increase in membrane permeability to normally impermeant compounds included valinomycin, which induces the rapid electrogenic transfer of potassium across membranes,[36] and the uncoupler carbonyl cyanide-m-chlorophenylhydrazone (CCCP), which transfers protons electrogenically.[37] Treatment of HeLa cells with $1 \mu M$ valinomycin was used to introduce hygromycin B (550 Da), a normally impermeant inhibitor of protein synthesis.[38] CCCP, at concentrations greater than $3 \mu M$, induced efflux of an intracellular pool of labeled nucleotides in transformed mouse fibroblasts.[30] A23187, an ionophore which electrogenically transports calcium and magnesium more effectively than sodium and potassium,[39] also caused efflux of intracellular nucleotide pools in mouse fibroblasts at concentrations greater than $1 \mu M$.[40] Carboxylic, carrier-type ionophores, such as monensin and nigericin, mediate the electrically neutral exchange of protons for sodium and protons for potassium, respectively.[41] These nonelectrogenic ionophores did not affect plasma membrane permeability to normally impermeant metabolites in mouse fibroblasts,[30] but did cause entry of hygromycin B (550 Da) and α-sarcin (16.8 kDa) in HeLa cells.[42] When HeLa cells were treated with these nonelectrogenic ionophores, an increase in entry of the protein syn-

[32] W. C. Chen, D. L. Chou, and D. S. Feingold, *Antimicrob. Agents Chemother.* **13**, 914 (1978).

[33] J. Bratjburg, G. Medoff, G. S. Kobayashi, S. Elberg, and C. Finegold, *Antimicrob. Agents Chemother.* **18**, 586 (1980).

[34] R. Mosckovitz, S. Chaitchik, and I. Friedberg, *Regular Meet. Int. Union Physiol. Sci.,* p. 287 (1984).

[35] H. S. Gankema, E. Laanen, A. K. Groen, and J. M. Tager, *Eur. J. Biochem.* **119**, 409 (1981).

[36] P. W. Reed, this series, Vol. 55, p. 435.

[37] P. G. Heytler, this series, Vol. 55, p. 462.

[38] M. Alonso and L. Carrasco, *Eur. J. Biochem.* **109**, 535 (1980).

[39] P. W. Reed and H. A. Lardy, *J. Biol. Chem.* **247**, 6970 (1972).

[40] I. Friedberg, G. A. Weisman, and L. A. Heppel, unpublished results (1986).

[41] B. C. Pressman and M. Fahim, *Annu. Rev. Pharmacol. Toxicol.* **22**, 465 (1982).

[42] M. Alonso and L. Carrasco, *Eur. J. Biochem.* **127**, 567 (1982).

thesis inhibitors occurred under conditions of membrane hyperpolarization. An increase in membrane potential correlated with entry, but not efflux, of normally impermeant compounds.

Sendai Virus

Addition of Sendai virus to a variety of mammalian cell lines (2–500 $HAU/10^6$ cells) caused an increase in the plasma membrane permeability of normally impermeant metabolites such as phosphoryl choline, sugar phosphates, and nucleotides. Permeabilization by Sendai virus has been demonstrated in normal and transformed mouse fibroblasts, HeLa cells, lung and nasal cultures, freshly isolated brain cells, peripheral human lymphocytes, and other cell lines.[43] Cells permeabilized by Sendai virus could be resealed by addition of calcium. The concentration of intracellular calcium required to induce secretion of histamine from rat mast cells was determined by introducing defined amounts of calcium-containing buffers into virally permeabilized cells.[44] Treatment with high concentrations of virus (>500 $HAU/10^6$ cells) was used as a means to introduce compounds with molecular weights above 1000 into cultured cells. High concentrations of UV-inactivated virus enabled entry of T_4 endonuclease (20 kDa) into *Xeroderma pigmentosum* cells[45] and of fragment A of diphtheria toxin (22 kDa) into human FL and mouse L cells.[46] Fusion of viral envelopes containing fragment A with the plasma membrane of mouse L cells led to entry of the toxin into the cells.[47]

Efflux of normally impermeant compounds from cells treated with Sendai virus occurred after a lag period of several minutes.[48] During this lag period, an increase in the membrane permeability of Na^+ and K^+ occurred, resulting in plasma membrane depolarization.[49] The permeabilizing effects of Sendai virus were antagonized by addition of calcium[50] or Con A.[51] Virus-induced permeabilization was temperature dependent (only occurring above 15°[43]) and pH dependent (occurring at pH 7.4 but

[43] C. A. Pasternak, *in* "Membrane Processes" (G. Benga, H. Baum, and F. A. Kummerow, eds.), p. 140. Springer-Verlag, New York, 1983.

[44] B. D. Gomperts, J. M. Baldwin, and K. J. Micklem, *Biochem. J.* **210**, 737 (1983).

[45] K. Tanaka, M. Sekiguchi, and Y. Okada, *Proc. Natl. Acad. Sci. U.S.A.* **72**, 4071 (1975).

[46] M. Yamaizumi, T. Uchida, and Y. Ikada, *Virology* **95**, 218 (1979).

[47] T. Uchida, Y. Miyake, M. Yamaizumi, E. Mekada, and Y. Okada, *Biochem. Biophys. Res. Commun.* **87**, 371 (1979).

[48] C. L. Bashford, K. J. Micklem, and C. A. Pasternak, *J. Physiol. (London)* **343**, 100P (1983).

[49] C. C. Impraim, K. A. Foster, K. J. Micklem, and C. A. Pasternak, *Biochem. J.* **186**, 847 (1980).

[50] C. A. Pasternak and K. J. Micklem, *Biochem. J.* **140**, 405 (1974).

[51] K. J. Micklem and C. A. Pasternak, *Biochem. J.* **162**, 405 (1977).

not at pH 5.3).[52] The permeability change induced by Sendai virus was dependent upon the presence of the viral envelope, which consisted of a phospholipid bilayer containing multiple copies of two integral glycoproteins (F and HN) and an intrinisic protein (M),[53] and required fusion of the envelope with the plasma membrane.[54] Permeabilization was not dependent upon entry of viral proteins into the cells and was induced by infective or noninfective, UV-irradiated virus.[43] Permeabilization by Sendai virus has many features in common with the ATP-induced permeabilization of transformed mouse fibroblasts, although there were significant differences.[55] Only ATP-induced permeabilization was influenced by changes in the ionic strength of the medium, by inhibition of a Na^+,K^+,Cl^--cotransporter and by preincubation of cells with dithiothreitol.

Detergents

Lysolecithin. Lysolecithin (30–50 μg/ml) permeabilized baby hamster kidney (BHK) cells, CHO cells, Chinese hamster lung fibroblasts, Chinese hamster embryo fibroblasts, and rat macrophages to normally impermeant compounds such as trypan blue (961 Da).[56] The concentration of lysolecithin required was dependent on the cell line tested and on cell density. The permeability change occurred within 2 min after addition of lysolecithin to monolayer or suspension cultures and permeabilized cells could be resealed by incubation in growth medium containing 10% (v/v) calf serum.[56] Normally impermeant compounds, including trypan blue, deoxynucleotides, and folyl and methotrexate polyglutamates, were sealed inside hepatoma cells after permeabilization with 1 mg/ml lysolecithin.[57] Inhibitors of DNA polymerase, such as araCTP (440 Da) and *N*-ethylmaleimide (125 Da) were introduced into BHK cells permeabilized by lysolecithin, enabling the evaluation of the role of the polymerase in DNA replication and repair.[58]

Tween 80 and Tween 60. Ehrlich–Lettre ascites cells were permeabilized to nucleotides and amino acids after a 30-min incubation in ascites

[52] K. Patel and C. A. Pasternak, *Biosci. Rep.* **3**, 749 (1983).

[53] D. Fisher and A. H. Goodall, *in* "Techniques in Cellular Physiology" (P. F. Baker, ed.), p. 115/1. Elsevier/North-Holland, County Clare, Ireland, 1981.

[54] A. M. Wyke, C. C. Impraim, S. Knutton, and C. A. Pasternak, *Biochem. J.* **190**, 625 (1980).

[55] Z. Sternberg and L. A. Heppel, *Biochem. Biophys. Res. Commun.* **148**, 560 (1987).

[56] M. R. Miller, J. J. Castellot, and A. B. Pardee, *Exp. Cell Res.* **120**, 421 (1979).

[57] M. Balinska, W. A. Samsonoff, and J. Galivan, *Biochim. Biophys. Acta* **721**, 253 (1982).

[58] J. J. Castellot, M. R. Miller, D. M. Lehtomaki, and A. B. Pardee, *J. Biol. Chem.* **254**, 6904 (1979).

fluid containing 250 mM sucrose and 1% (v/v) Tween 80.[59] Permeability changes induced by Tween 80 were reversible and cells grew normally and remained viable as transplants following detergent treatment. DNA replication[60] and ribonucleotide reductase activity[61] were studied in CHO cells rendered permeable to deoxyribonucleoside triphosphates by incubation in an isotonic buffer containing 1% (v/v) Tween 80. RNA synthesis in Friend erythroleukemia cells permeabilized by exposure to hypotonic buffer containing 1% (v/v) Tween 80 was measured by monitoring the incorporation of radiolabeled uridine 5'-triphosphate into RNA.[62] Ehrlich ascites tumor (EAT) and hepatoma cells became permeable to proteins as large as serum albumin when incubated for 30–60 min in a phosphate buffer, pH 7.3, containing 1% (v/v) Tween 60.[63] The cells could be resealed at 37° by incubation in a physiological medium (Medium 199) containing 20% (v/v) bovine serum.

Dextran Sulfate 500. Dextran sulfate 500 (2–20 μg/mg cells) was used to increase the plasma membrane permeability of EAT cells to rubidium, ATP, erythrosin B (880 Da), and sorbitol (182 Da) and to decrease the active transport of α-aminoisobutyric acid.[64] The cells were permeabilized with dextran sulfate 500 in a buffered, isotonic solution and were resealed by incubation with ascites fluid from tumor-bearing mice. Addition of calcium and glucose partially restored the normal permeability barrier in permeabilized cells.[64] Dextran sulfate 500 has been used to introduce orotic acid (156 Da) into EAT cells, enabling the evaluation of the role of cytoplasmic 5-phosphoribosyl-1-pyrophosphate levels in *de novo* UMP biosynthesis.[65]

Staphylococcus aureus α-Toxin

The 34-kDa protein, α-toxin, secreted by pathogenic staphylococci, induced an increase in plasma membrane permeability in a number of cell types.[66] An increase in the membrane permeability of rabbit ileum smooth muscle to calcium, strontium, and ATP induced by *Staphylococcus aureus* α-toxin was used to study the effect of these metabolites on muscle tension

[59] E. R. M. Kay, *Cancer Res.* **25,** 764 (1965).
[60] D. Billen and A. C. Olson, *J. Cell Biol.* **69,** 732 (1976).
[61] R. G. Hards and J. A. Wright, *Arch. Biochem. Biophys.* **220,** 576 (1983).
[62] T. E. Gilroy, A. L. Beaudet, and J. Yu, *Anal. Biochem.* **143,** 350 (1984).
[63] A. G. Malenkov, S. A. Bogatyreva, V. P. Bozhkova, E. A. Modjanova, and J. M. Vasiliev, *Exp. Cell Res.* **48,** 307 (1967).
[64] M. Kasahara, *Arch. Biochem. Biophys.* **184,** 400 (1977).
[65] J. J. Chen and M. E. Jones, *J. Biol. Chem.* **254,** 2697 (1979).
[66] A. W. Bernheimer, *Biochim. Biophys. Acta* **344,** 27 (1974).

and relaxation.[67] Rat hepatocytes incubated with α-toxin (4–35 human hemolytic units/13–21 mg dry weight of cells) for 5–10 min at 37° and pH 7.0 were rendered 93–100% permeable to trypan blue while exhibiting a loss of 22% of the cytoplasmic lactic dehydrogenase activity.[68]

Physical Methods for Permeabilizing Cells

Scrape Loading

Monolayer cultures of untransformed mouse fibroblasts were loaded with fluorescently labeled ovalbumin, immunoglobulin G, and dextrans with molecular weights as high as 2,000,000 by scraping the cells from their substratum in the presence of the compounds.[69] Approximately 50–60% of the scraped fibroblasts remained viable after the procedure and 40% of the surviving cells contained the normally impermeant compounds. After scrape loading, the cells required 12 hr to regain their normal morphology. Human neutrophils, a macrophage-like cell line, and HeLa cells also incorporated fluorescently labeled dextran (70 kDa) after scrape loading, although subsequent cell viability was not assessed.

Hypertonic Treatment

Monolayer cultures of BHK cells became permeable to normally impermeant compounds when incubated at 37° for 40–50 min in modified Eagle's medium (MEM) supplemented with 4.2% (w/v) NaCl.[70] The effects of DNA precursors and hydroxyurea on DNA synthesis were investigated in the permeabilized cells. Cells permeabilized by hypertonic treatment were resealed by incubation in MEM containing 10% (v/v) calf serum and grew normally after a short lag period. A normally impermeant inhibitor of DNA synthesis, araCTP (440 Da), was sealed into BHK cells after hypertonic treatment, resulting in the inhibition of DNA synthesis.

Electric Field

Transient exposure of a number of cell lines to an electric field has been shown to increase plasma membrane permeability to normally imper-

---

[67] P. Cassidy, P. E. Hoar, and W. G. L. Kerrick, *Biophys. J.* **21,** 44a (1978).

[68] B. F. McEwen and W. J. Arion, *J. Cell Biol.* **100,** 1922 (1985).

[69] P. L. McNeil, R. F. Murphy, F. Lanni, and D. L. Taylor, *J. Cell Biol.* **98,** 1556 (1984).

[70] J. J. Castellot, M. R. Miller, and A. B. Pardee, *Proc. Natl. Acad. Sci. U.S.A.* **75,** 351 (1978).

meant compounds.[71] The calcium dependence of catecholamine release from bovine adrenal medullary cells was studied by introducing defined amounts of calcium-containing buffers into cells permeabilized by exposure to a transient electric field.[72] Suspensions of mouse L cells deficient in the thymidine kinase gene incorporated linear or circular plasmid DNA containing the gene after exposure of the cells to a series of electrical impulses at 20°.[73] Approximately 95 transformants per 10^6 cells were obtained using this procedure.

[71] D. E. Knight, in "Techniques in Cellular Physiology" (P. F. Baker, ed.), p. 113/1. Elsevier/ North-Holland, County Clare, Ireland, 1981.

[72] P. F. Baker and D. E. Knight, J. Physiol. (London) **316,** 7P (1981).

[73] E. Neumann, M. Shaefer-Ridder, Y. Wang, and P. H. Hofschneider, EMBO J. **1,** 841 (1982).

Author Index

Numbers in parentheses are footnote reference numbers and indicate that an author's work is referred to although the name is not cited in the text.

Subject Index

A

inhibition of glutamine and histidine
uptake into, 143
isolated using free-flow electrophoresis,
530–531
spontaneously arising, 602–603
Elastase, in production of isolated suspensions of cells, 485
Electrodiffusion, simple, 69–72
Electron paramagnetic resonance, probes,
structures of, 359
Electron spin resonance probes, in
determination of surface potential,
373–374
Electroosmosis, 346–347
Electroosmotic fluid velocity, 346–347
Electropermeabilization, 868–869
methodology, 824–826
theoretical considerations, 822–824
Electrophoresis, 346
Electrophoretic mobility, principles,
346–349
Electrostatic cell sorting, 557–559
droplet deflection and sample collection,
558–559
sort decision, 557
sort timing and droplet charging, 558
sort timing and droplet formation, 557
Electrostatic energy, of ion movement
through integral membrane protein,
64–65
Electrostatic potentials
adjacent to membranes
formation, 342
measuring, 342–364
measurement, experimental considerations, 349–351
Elutriation, 462
advantages of, 482–483
behavior of cells during, effect of choice
of enzyme, 486
cell collection, 487
cell load, 487
cell separation by, 482–497
clumping of cells, 486–487
composition of media, 486–487
definition of, 482
dispersion media, 485–486
factors affecting, 482–485
of intestinal cells, 494–495
of liver cells, 495–496

of lung cells, 495
monitoring size distribution of cells being
separated, 489–490
practical considerations of, 486–488
preparation of cells for, 485
profile, for cell types, 490–491
rotor preparation, 486
separation protocols, 488–490
supplements, 485–486
Elutriator chamber, design of, 483–484
Elutriator rotor, 484
development of, 482–483
source, 483
Emulphogen, 20
Endocytosis, 10
of enveloped viruses, 829
measurement of membrane turnover by,
828
Energy transduction. See also Oxidative
phosphorylation
degree of coupling, 405
localized versus delocalized chemios-
mosis, 439–444
phenomenological stoichiometry,
405–406
thermodynamic parameters, 403–406
Enniatin A
as ion carrier, 274
structure of, 275
Enniatin B
as ion carrier, 274
structure of, 275
Enzymatic reaction, rate equation for,
164–165
Enzymes, 63
catalytic activity of, and intrinsic binding
energy in transition state, 154–155
cellular
flow cytometric analysis, 559–560
histochemical detection, 560–561
immunofluorescent detection methods,
559–560
membrane-bound, effect of surface
potential on, 388
for obtaining single-cell suspension,
513–514
that catalyze coupled vectorial processes
binding specificities, 147
reversible reactions at active sites of,
147–148

Rhodopsin, 5

RNA
flow cytometric fluorescence analysis,
561–563
as marker for nonparietal cells, 453

Rose and Bryant ratio, 303

RSVE. *See* Sendai virus, RSVE

Rubidium-8, partially relaxed Fourier
transform (PRFT) spectra, as RbCl in
presence of gramicidin channels,
295–297

Rubidium-87
nuclear magnetic resonance
background data, 322
excess longitudinal relaxation rate plots
for channel interactions, 325–328
rate constant data for channel
interactions, 331
nuclear magnetic resonance properties of,
317
relaxation properties of, quantities
relevant to, 318–320

Rubidium chloride, nuclear magnetic
resonance
carbonyl carbon chemical shifts, 315–316
site (carbon-13) chemical shift, 316–317
site chemical shift data, 315

Rubidium ion
interactions with gramicidin transmembrane channel, binding and rate
constants for, 325–328
single-channel currents and conductances,
339–341

S

Saponin, 820

Sarcoplasmic reticulum, calcium pump, 4

SBPL. *See* Soybean phospholipids

Secretagogues, enzyme secretion caused by,
590–591

Sedimentation, of cells, 462
theory of, 463–464

Sendai virus
as biological carrier, 829
envelopes, 866
reconstitution, 835–839
fusion of RSVF or hybrid vesicles with
living cultured cells, monitored by

fluorescence dequenching, 844–846
fusogenic viral envelopes
hybrid, 830–831
reconstituted, formation of, 837–839
reconstituted, 830
reconstitution of, 835–837
hemolytic activity, fusogenic activity of,
quantitative estimation of, 832
hybrid vesicles
fluorescent
fusion with phospholipid vesicles,
840–842
preparation of, 840
fusogenic activity, quantitative
estimation, 839–842
preparation of, selection of detergent
for, 842–843
intact virions, fusogenic activity of,
quantitative estimation of, 832
particles, propagation and purification of,
831–832
permeabilization of cells with, 858,
865–866
RSVE (fusogenic envelopes, reconstituted)
fusogenic activity, quantitative
estimation, 839–842
preparation of, selection of detergent
for, 842
RSVF, fluorescent
fusion with phospholipid vesicles,
840–842
preparation of, 840
testing RSVG, fusogenic activity, 849

Sendai virus envelope-mediated fusion,
829–850

SH reagents, as inhibitors for transport
systems, 17

Simian virus 40 receptor
fusion-mediated implantation into plasma
membranes of target cells, 846
transfer from receptor-positive to
receptor-negative cells, 831

Single-site gated-pore mechanism, 16

Single-site reorientation model, 16

Site-directed mutagenesis, 14

Small unilamellar liposomes, preparation of,
ethanol injection procedure, 851

Small unilamellar vesicles
freeze–thaw cycling of, 196
fusion-mediated size increase of, 196